ANNUAL REVIEW OF ENTOMOLOGY

ANNUAL REVIEW OF ENTOMOLOGY

THOMAS E. MITTLER, *Editor*
University of California, Berkeley

CARROLL N. SMITH, *Editor*
U.S. Department of Agriculture, Ret.

VINCENT H. RESH, *Associate Editor*
University of California, Berkeley

VOLUME 22

1977

ANNUAL REVIEWS INC. 4139 EL CAMINO WAY PALO ALTO, CALIFORNIA 94306

ANNUAL REVIEWS INC.
Palo Alto, California, USA

International Standard Book Number: 0-8243-0122-6
Library of Congress Catalog Card Number: A56-5750

REPRINTS

The conspicuous number aligned in the margin with the title of each article in this volume is a key for use in ordering reprints. Available reprints are priced at the uniform rate of $1 each postpaid. The minimum acceptable reprint order is 10 reprints and/or $10.00, prepaid. A quantity discount is available.

PRINTED AND BOUND IN THE UNITED STATES OF AMERICA

ANNUAL REVIEWS INC. is a nonprofit corporation established to promote the advancement of the sciences. Beginning in 1932 with the *Annual Review of Biochemistry,* the Company has pursued as its principal function the publication of high quality, reasonably priced Annual Review volumes. The volumes are organized by Editors and Editorial Committees who invite qualified authors to contribute critical articles reviewing significant developments within each major discipline.

Annual Reviews Inc. is administered by a Board of Directors whose members serve without compensation.

Annual Reviews are published in the following sciences: Anthropology, Astronomy and Astrophysics, Biochemistry, Biophysics and Bioengineering, Earth and Planetary Sciences, Ecology and Systematics, Energy, Entomology, Fluid Mechanics, Genetics, Materials Science, Medicine, Microbiology, Nuclear Science, Pharmacology and Toxicology, Physical Chemistry, Physiology, Phytopathology, Plant Physiology, Psychology, and Sociology. The *Annual Review of Neuroscience* will begin publication in 1978. In addition, two special volumes have been published by Annual Reviews Inc.: *History of Entomology* (1973) and *The Excitement and Fascination of Science* (1965).

PREFACE

With almost any enterprise, an occasional backward look is in order. In a field as broad, complex, and dynamic as entomology, recourse to hindsight as a means of appraising performance in the art of reviewing seems particularly appropriate. So as we approach the quarter-century mark in the publication of the *Annual Review of Entomology*—the Editorial Committee's plans for Volume 24 were made in December 1976—a scrutiny of what has gone before may be useful.

It is generally understood that a broad and balanced coverage of developments significant to entomology is a primary objective of *ARE.* But it is also apparent that the range of topics appropriate for review each year cannot be covered adequately in a single volume; thus, the desired breadth and balance must be achieved through a sequence of volumes. To meet this particular planning problem, selection of topics for *ARE* volumes is customarily made on the basis of some 20 general categories. The categories currently used as planning guidelines are indicated in the cumulative index of chapter titles, which is found on the last pages of the volume. Not all categories are represented in each volume; some topics appear only in alternate years and a few may be scheduled at greater time intervals, depending upon developments in a particular specialty. The system is flexible and the scope of the categories, as well as the regularity of scheduling topics pertaining to them, may be modified from time to time to reflect changes in research emphasis or in the presumed importance of different lines of inquiry. However, it does give reasonable assurance of breadth and balance in subject-matter coverage, even though Editorial Committee myopia may occasionally cause topics ripe for review to be overlooked.

Achieving breadth and balance in review articles involves more than thorough subject-matter coverage in the planning stage. It is not always possible to find authors willing to attempt a review of complex and dynamic topics. And all too frequently, particularly of late, scientists who initially accepted invitations to prepare reviews have defaulted, for one reason or another, or have asked to postpone publication because of the inability to meet editorial deadlines. Some of these problems have seriously affected the contents of certain volumes.

Then there is the matter of the scientific points of view of those responsible for preparing reviews, and these may be as numerous as the potential reviewers. Entomology, like other branches of science, is not unaffected by national or regional influences, and it is to be expected that the relative significance attached to different entomological developments will vary, depending upon the regional or national background of the reviewer. And from the point of view of national or geographic representation in the preparation of review articles, the *ARE* record might be improved by a more cosmopolitan orientation. Volumes 1 through 21 contain 406 articles prepared by 519 authors. A tabulation of these contributions reveals a striking imbalance in the national affiliations of the writers. Except for the United Kingdom, relatively few reviews came from Western European or Asiatic countries. Germany, The Netherlands, and Switzerland combined account for 28 reviews, six reviews came from France, and three each came from Czechoslovakia, Italy, and Sweden. Japanese scientists have prepared five reviews, and four came from scientists in the USSR. On a percentage basis, scientists in the United States have been responsible for nearly 58% of the reviews, and the combined contributions from English-speaking countries (Australia, Canada, the United Kingdom, and the United States) amount to a whopping 83% of the reviews. Certainly these data do not reflect geographic distribution of entomological expertise. Presumably the limited number of contributions from non-English countries reflects communication inconveniences associated with language differences. We may hope that problems of this sort can be overcome. Toward that end, the Editorial Committee earnestly solicits suggestions for future issues of *ARE* regarding potential reviewers of significant developments in entomology of a worldwide basis.

Comparison of the title pages of successive volumes of the *Review* shows changes in the composition of its Editorial Committee. These result from the fact that each member may serve for only a five-year term and that one member retires each year. However, few changes have occurred among the Editors who serve ex officio on the Committee. Their possible reappointment after five-year terms has permitted the necessary continuity in the Committee and in other matters relating to the *Review.* After 21 years of unceasing effort on behalf of the *Review,* Ray F. Smith has relinquished his leadership role as senior Editor. The entomological community throughout the world is deeply indebted to Ray for his devotion to *Annual Review of Entomology* since its first publication in 1956.

Louise Libby, who has served us so well for the past two years as Assistant Editor, has been succeeded by Roberta Little. We thank them and the compositors and printers for once again attaining the high standards established by the *Annual Review of Entomology.*

THE EDITORIAL COMMITTEE

CONTENTS

INDEXES

Ann. Rev. Entomol. 1977. 22:1–22

BIOLOGICAL CONTROL OF FOREST INSECTS

❖6117

Hubert Pschorn-Walcher
Commonwealth Institute of Biological Control, European Station,
CH-2800 Delémont, Switzerland

INTRODUCTION

Biological control of forest insects in the classical sense (i.e. by importation of natural enemies against introduced pest species) has mainly been applied in the temperate regions of the world; to date very few projects have been undertaken in tropical forests. The great majority of projects concerned North America (Canada and the United States), which probably has the highest proportion of introduced forest insects, chiefly of European origin. Biological control programs undertaken in these two countries have been reviewed in detail (15, 17, 25, 63), as have been the relatively few projects undertaken in Europe (43), Australia (119), Africa (42), and the Southeast Asian and Pacific Regions (88).

Biological control of forest insects in general, or of some pest species in particular (diprionid sawflies, gypsy moth), has been dealt with in this review series (4, 11, 18, 55) and in a number of other publications (e.g. 5, 24, 38, 62, 73, 77, 104). The general principles of classical biological control of terrestrial insects have been the subject of a number of review articles (7, 16, 37, 101, 106, 108), recent books (20, 23, 36, 40, 48, 53, 69, 70, 98, 107), and authoritative papers (49, 61, 94, 97, 110, 116, 124).

The scope of this review is restricted to biological control of forest insect pests introduced from one continent into another by means of importation of their natural enemies from their home countries, i.e. to what is often called classical biological control. But rather than reviewing individual major projects completed or underway, I want to deal mainly with the strategy and principles of classical biological control as they apply to forest insects because, in my opinion at least, these principles are not necessarily identical with those applying to pests of intensively managed agricultural crops, which have formed the bulk of the biological control projects of the world. The aim of this paper is thus to review step by step the procedures and their underlying philosophy of biological control in the forest ecosystem, from the early planning stages of a project through the exploration phase to the final stages of introduction and colonization, as outlined in the flow diagram of Figure 1. Because of this restriction to the classical approach, two other areas of biological control by parasites and predators remain undiscussed: (*a*) the use of exotic natural enemies from allied host species against introduced or native pests, and (*b*) the utilization of native biological control agents against native pests.

The first approach has been discussed in detail by Pimentel (72), who found that more than one third of pest species were controlled by an introduced natural enemy of an allied pest species or genus. He suggests that parasites that have been long associated with their natural host lose their ability to severely limit host numbers, because of the tendency for parasites and hosts to evolve a degree of homeostasis, i.e. a stable and balanced economy between the interacting species. Although this theory of a coevolution between host and enemies appears to be attractive on theoretical grounds, this method has had little or no success against forest pests, especially when native host species were involved (8, 17). However, only a few and often superficial trials of this kind have been made so far, and thus it would be premature to abandon Pimentel's theory because of the preponderance of failures. It is conceivable that importing exotic natural enemy species against a native pest may be more difficult in forest ecosystems than in agroecosystems, because of the presumably more stable and diversified parasite complexes of native forest insects. At any rate, careful studies on both ends of the projects are required (e.g. see 122) to elucidate empty parasitological niches in the natural enemy complex of the target pest and to find species capable of fitting into these or filling a niche more effectively than a native enemy species. A special case is the method called *adaptation-importation,* utilized with the European pine shoot moth. Two species of nearctic parasites, *Itoplectis conquisitor* and *Elachertus thymus,* which adapted to the introduced shoot moth in North America, have recently been imported to Europe, the native home of the pest, in the hope that they would complement the limited efficacy of the native enemy complex of the moth; to date, however, no success has been achieved (39, 43).

The second approach has a long history in forest protection, especially in Europe, but with few exceptions, most experiments to control forest pests through inundative releases of native species of natural enemies have been fragmentary and of too short a duration to be conclusive (43). Exceptions are the use and manipulation of red wood ants of the genus *Formica* in Europe (and recently in Canada) (32, 41, 43, 71); bird protection measures, which are standard practice in many European forest stands (10, 36, 43, 54); and perhaps the use of egg parasites of the genus *Trichogramma,* chiefly in East-European countries, the Soviet Union, and mainland China (91; R.M. Prentice, personal communication).

FOREST ECOSYSTEMS AND BIOLOGICAL CONTROL

Forest communities differ in several important aspects from intensively managed agroecosystems, and the implications involved with regard to biological control are briefly outlined in the following:

1. Forests are usually long-lived ecosystems with a high stability in space and time. Because of their long evolutionary history and their stratification, they exhibit a high degree of diversity in both plant and animal communities. For the exploration phase of a biological control project, this diversity has a particular advantage:

usually a rich complex of natural enemies can be expected to exist. For the importation phase, however, forest ecosystems may have the disadvantage that introduced species have fewer chances to find empty niches and to escape competition from their native relatives.

2. Forest communities are often extended rather uniformly over large areas, with gradual boundaries between the different forest types. The advantage is that parasite and predator complexes of forest pests frequently exhibit only minor regional differences. On the other hand, the vast surfaces covered, e.g. by the boreal coniferous forests and some of their insect pests, make any release program of biological control agents a long-term adventure.

3. Until recently, many forests and their species composition have been relatively little disturbed or altered by human activities. This has encouraged the evolution and preservation of highly structured, well-balanced parasite-host complexes, whereas with many agricultural pests continued human interference since ancient times has often interrupted the evolution of stable parasite-host associations. Many crops and their pest species have been moved around the world, the latter losing in the process many of their original parasites and acquiring some new ones, but overall their complexes of natural enemies appear often to be rather immature. Crop alternation, especially in annual crops, varietal differences, and differences in cultivation and plant protection practices often lead to pronounced regional, or even local, differences in the parasite-predator complexes of agricultural pest insects, compared to those of their forest counterparts.

These differences between natural enemy associations in forest vs intensively managed agricultural ecosystems would seem to have considerable bearing on the biological control strategies to be applied. With forest insects, detailed preintroduction studies into the structure of their parasite-predator complexes appear to be both warranted and profitable. Highly evolved, phylogenetically old host-parasite systems are frequently characterized by a high degree of redundancy, and the goal of an explorer in biological control should be to unravel the intricate network of interrelationships and competitive interactions between the different members of the complex in order to be able to select a series of promising control agents with minimal interference but optimal integration with one another. Natural enemy complexes of forest insects (and this applies also to pest insects from seminatural agroecosystems) have, in my opinion, a higher degree of predictability upon which to make a more sophisticated choice of a species combination for introduction against an introduced pest. With many agricultural pests, especially those under intensive care, predictability is lower because of the more erratic structure of their parasite-host associations, and an educated hit-or-miss type of approach with regard to parasite-predator introductions may often provide a solution more quickly than preintroduction studies in greater depth.

An explorer in a biological control project of certain types of agricultural pests might therefore continue his research in the home country of the pest straight from box IV to box VII, as indicated by the broken arrow in the flow-diagram shown in Figure 1, bypassing partially or entirely the route through boxes V and VI, whereas when working on a forestry project the detour via these two boxes may be the more profitable, safer, though slower, way.

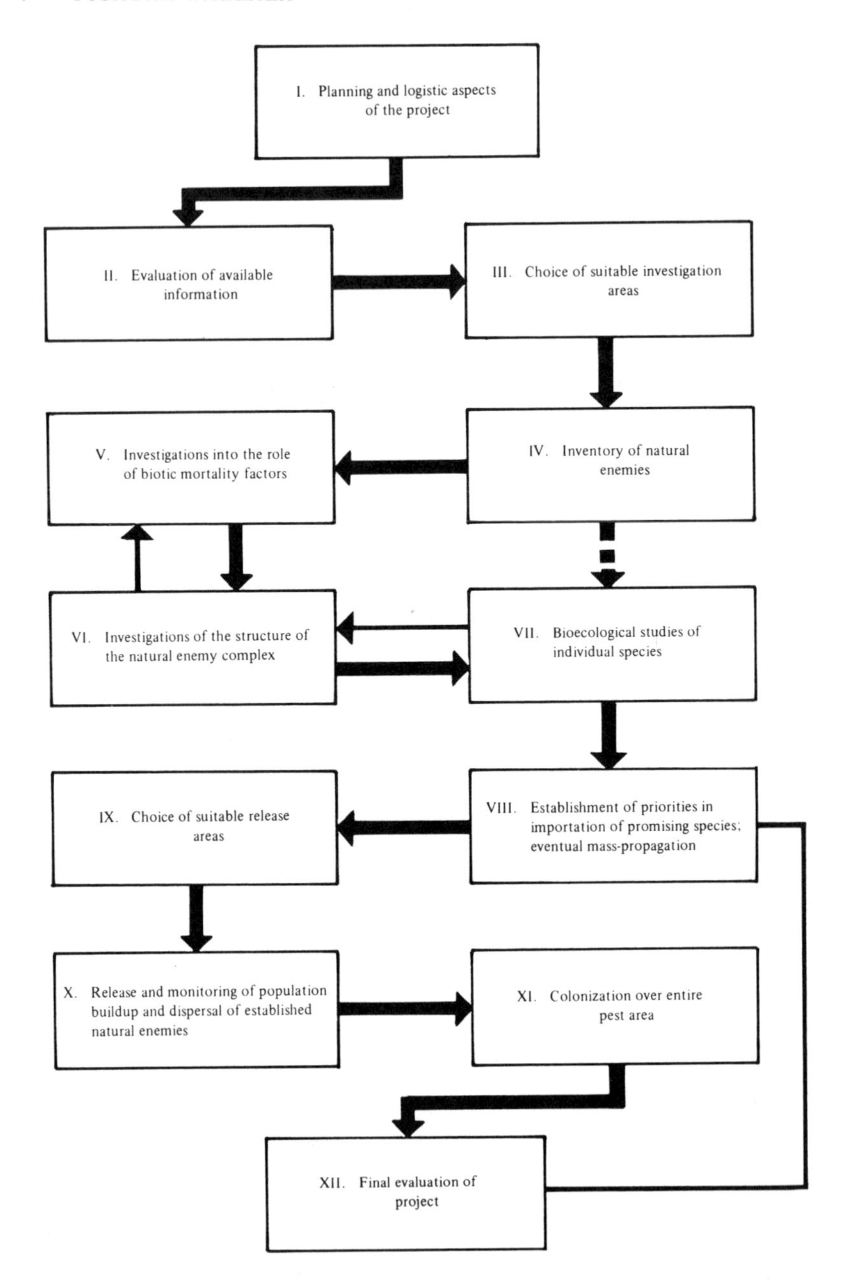

Figure 1 Flow diagram of a biological control project against a forest insect pest.

PLANNING AND LOGISTIC ASPECTS OF A PROJECT

One of the first steps in any planning of a biological control program is to establish whether the target pest species is introduced or native. Especially with species found today both in North America and Eurasia, it is often difficult to say whether their Holarctic distribution is of an ancient or a recent nature, i.e. caused by recent immigration or introduction from one continent into the other. Lindroth (57) has given five criteria to identify introduced insect species, and a sixth criterion has been added since (78).

1. Historical criterion. This applies when the history of introduction is definitely known, as with the gypsy moth in North America.
2. Geographical criterion. Introduced insect species often exhibit a patchy, immature distribution pattern because they might not have had time enough to colonize the entire distribution areas of their host plants. Relevant cases are the balsam woolly aphid, *Adelges piceae,* and the alder sawfly, *Eriocampa ovata,* in North America. Both are restricted to widely disjunct distribution areas, e.g. *A. piceae* to the Maritime Provinces of Canada and the northeastern United States, the Appalachian Mountains, and the Pacific Northwest.
3. Ecological criterion. Insect species associated with the human environment, such as cockroaches, are frequently introduced.
4. Biological criterion. Highly specific phytophagous insects occurring mainly or exclusively on exotic tree species are usually introduced. A case in point is the mountain ash sawfly, *Pristiphora geniculata,* in North America; it infests principally European mountain ash. Another typical example is the larch insects of Great Britain. As the European (and Japanese) larch was originally not native to the British Isles, but introduced during the last centuries, monophagous larch insects such as the five sawfly species now associated with this tree in Britain must either be introduced or recent invaders from the continent.
5. Taxonomic criterion. This may apply to species holding an isolated taxonomic position, i.e. having no close relatives among the native fauna, as, for example, the European spruce sawfly, *Gilpinia hercyniae,* and the introduced pine sawfly, *Diprion similis,* which belong to a group of diprionid genera originally entirely restricted to Eurasia.
6. Parasitological criterion. The lack of a typical parasite complex often indicates that the host species has not evolved in the area where characteristic parasites are absent. With the European birch leafminer *Fenusa pusilla,* for example, only very polyphagous chalcidoid species are found in North America, but typical sawfly parasites of the ichneumonid subfamilies Tryphoninae and Scolobatinae (almost entirely restricted to sawfly hosts) are missing (28).

The greater the number of criteria that apply, the higher the probability that the pest species in question has been introduced.

The next step is then to establish the geographical origin of the target species. This is usually easier with forest insects than with agricultural pests, whose host plants have been moved around by humans since ancient times. With monophagous species, the original home country of the host plant is necessarily also the native home of the pest. With oligophagous species, the question may be more complicated, but taxonomic and evolutionary (i.e. the presence of closely related taxa) and/or parasitological evidence (i.e. the occurrence of a rich, saturated parasite complex) may provide valuable clues.

A detailed taxonomic study of intraspecific variation in a widespread species might help to narrow the range of the area from which a species has been introduced. A good example is provided by the recent analysis of the strains of the larch sawfly, *Pristiphora erichsonii,* occurring in North America (121). Whether this species is native, a natural immigrant, or introduced into the New World has long been in debate (19, 68, 78). Two of the four strains found in North America appear to be old Eurasian immigrants across the Bering Land bridge, but the two others were accidentally introduced from Britain in 1913 when sawfly cocoons only partially parasitized by *Mesoleius tenthredinis* were released. The two accidentally established strains, one resistant to *M. tenthredinis,* have since spread over most of boreal North America and have been predominant in the heavy outbreaks recorded after World War II.

Of importance in the later planning stages are the logistic aspects of a project (51). These encompass administrative, technical, and other considerations, such as the accessibility (physically and politically) of potential survey areas in the native home of the pest, the availability of local aid and research facilities, customs and quarantine regulations, and efficient postal services, which might influence the choice of suitable investigation areas and certain aspects of research.

EVALUATION OF AVAILABLE INFORMATION

Ordinarily, this entails an intensive search through the existing literature both for the target host and its natural enemies, but it has to be remembered that many of the older parasite-host records are prone to error (e.g. 80, 92).

Further valuable information can be obtained from local forest entomologists, taxonomic specialists, and museum collections. In many countries amateur entomologists (who frequently publish less than they know, in contrast to some professionals) are often an excellent source of information on the distribution and abundance of the target species and also have good advice on collecting and rearing methods.

CHOICE OF SUITABLE INVESTIGATION AREAS

It is usually assumed that potential biological control candidates should be obtained from a region within the native home of the pest that is climatically similar to the land of introduction. The best guide for a climatic comparison of the two areas in question is the well-known Klimadiagramm Weltatlas of Walter & Lieth (115), which contains pictorial climatograms from about 8000 stations of the world.

Apart from climatic data, a phytosociological comparison of the two regions is often useful, especially when the forest biota are floristically similar to each other, as is the case with Eurasia and North America, where many Holarctic plant species serve as useful indicators of floristic and ecological similarity. A comparison along these lines has been made for the balsam woolly aphid project (76). In many countries, detailed maps of vegetation types are now available, as are maps on nature reserves rich in plant and animal species providing valuable clues for promising habitats.

Another important consideration for the selection of suitable study and collection areas is to distinguish between primary (autochthonous) and secondary (allochthonous) distribution areas of the host plant and target insect. A good example is the European larch, *Larix decidua,* and its insect fauna, the original distribution of which was confined to the central Alps of Europe, with some outposts in the Carpathian Mountains and in Poland. During the last few centuries larch was planted all over Europe, and many of its monophagous pest species have since invaded the new distribution area. Their natural enemies have lagged behind, and thus today we often find only an impoverished parasite complex in these secondary distribution areas. For instance, the larch casebearer, *Coleophora laricella,* has some 20 parasite species in the Alps, but only about 10 in North Germany and 9 in Sweden (52). Of the 11 parasites of the larch sawfly, 3 are lacking in central Bavaria, just 100 km north of the autochthonous larch forests of the Tyrol, and further north their number decreases rapidly to 3 in Britain and one in Sweden (83).

With such pest species it is essential to survey their original distribution area because it is only in their native home that a mature, well-balanced complex of natural enemy species can be expected to exist. However, surveys in the secondary distribution areas may be a valuable source for additional collections of some species. First, because the host may be more common in these man-made plantations (e.g. the larch sawfly is exceedingly scarce in the Alps but may develop to pest proportions in regions where larch was originally not native); and, second, because genetically different parasite populations may be involved in the different areas, e.g. the Bavarian strain of *Mesoleius tenthredinis* is resistant to encapsulation by its host, *Pristiphora erichsonii,* whereas the native alpine strain is not (86, 105). Furthermore, secondary distribution areas of the pest species may be more suitable from a climatical point of view, and occasionally new parasite species may have adapted to the new host.

INVENTORY OF NATURAL ENEMIES

Once a choice of suitable investigation areas has been made, the next step is to establish an inventory, as complete as possible, of the parasites, predators, and diseases associated with the target pest. In this endeavor, two requirements are essential, namely, to take into account that regional differences in the natural enemy complex might exist, and that there might also occur qualitative and/or quantitative differences in the composition of the parasite complex under high and low host densities, respectively.

1. With pest species that have been widely distributed since ancient times, e.g. over Europe or the entire northern Palaearctic area, regional differences in their parasite complexes are sometimes surprisingly indistinct; for example, with ermine moths (*Yponomeuta* spp.), the major parasites are virtually the same in Europe and Japan (79). Usually, however, the quantitative importance of the individual species

of a complex may vary considerably between different regions (and with agricultural insects often from plantation to plantation). For example, ectoparasites of the subfamily Ephialtinae attacking the European pine shoot moth, *Rhyacionia buoliana,* are quite prominent in eastern, continental Europe but are of little importance in the maritime areas of northwestern Europe (92), whereas the parasite complex of the birch casebearer, *Coleophora serratella,* is dominated by the braconid *Orgilus punctulator* in the warmer parts of central Europe, but by *Apanteles* spp. in the cooler parts, yet in northern Europe *Campoplex* spp. and *Habrocytus semotus* are the dominant species (29). Qualitative changes (absence of major parasites from certain regions within the natural home of the pest) appear to be less common, at least with forest insects, being often restricted to minor, less-specific species of the complex.

2. The composition of the parasite complex frequently changes with the abundance of the host species, that is, there are parasite species that are better adapted to low host densities and others that predominate during an outbreak of the host. A classical case is the successful biological control of the winter moth, *Operophtera brumata,* in Canada where a combination of a high (*Cyzenis albicans*) and a low host density parasite (*Agrypon flaveolatum*), both introduced from Europe, has been responsible for the termination of the mass outbreak and the subsequent maintenance of low levels of the pest population (30, 31). With the colonial European pine sawfly, *Neodiprion sertifer,* field experiments with prearranged patterns of host density and long-term population studies have shown that the two ichneumonids, *Lamachus eques* and *Synomelix scutulatus,* possess the highest host-finding ability and are thus an important element of the parasite complex under low host densities. However, they usually fail to build up when the host population increases (i.e. they lack a numerical response) and are then largely replaced by two other species of larval parasites, *Lophyroplectus luteator* and *Exenterus abruptorius,* better adapted to high sawfly densities (84).

Generally speaking, many parasite complexes appear to be composed of a sequence of r- and K-strategists (34, 35, 74) working in succession both within a given host generation (i.e. early larval parasites are mainly r-strategists, whereas late larval or cocoon parasites are often K-strategists) and between different generations (i.e. low-host density parasites appear to be mainly r-selected, while high host density species are more often K-selected). Any inventory of biological control agents must therefore make sure that low host density parasites are not missed by restricting the survey to outbreak situations of the pest because these are easier and less time-consuming to work upon.

INVESTIGATIONS INTO THE STRUCTURE OF THE ENEMY COMPLEX

A complex of natural enemies (e.g. parasites) is usually composed of a set of species acting in a sequence, i.e. attacking different stages of the host. The members of a group of parasite species occupying the same parasitological niche (123, 125) form a parasite guild (74, 90) (e.g. the guilds of egg, larval, and pupal parasites or the guild of hyperparasites). As species belonging to the same parasite guild overlap significantly in their niche requirements, competition between them is common. Not infrequently, however, different guilds may also overlap and interfere with each other; for example, the egg-larval parasites may interfere with the true larval parasites, or, more often, larval parasites (emerging often only from the host pupa or cocoon) may interfere with the true pupal (cocoon) parasites. This network of interrelationships between the different species is conveniently termed the *structure* of the parasite complex (123).

The higher the number of species in the complex and in each of its parasite guilds, the greater are usually the interactions between the different species in the form of super-, multi-, and hyperparasitism. Studies into the structure of a parasite complex in the home country of a pest should aim at elucidating these different interactions as far as possible in order to establish a kind of pecking order between members of the same parasite guild (i.e. between ecological homologues) and between different guilds. As a result of such investigations, it is often found that the most highly adapted and best-synchronized parasite species are intrinsically inferior in competition with less-specialized members of the guild, but that they may be extrinsically superior in such attributes as host-finding ability, reproductive capacity, and numerical response to host density, which largely compensate for the losses sustained in direct, intrinsic competition. This phenomenon of balanced competition (125) has been recorded for the larval parasites of *Neodiprion sertifer* (81, 84), *Pristiphora erichsonii* (83, 85), the European fir budworm (*Choristoneura murinana*) (121), and the European pine shoot moth (*Rhyacionia buoliana*) (93), to mention just a few examples from our own work, and is probably typical for all parasite complexes in which r-selected and K-selected species coexist by virtue of differences in their competitive abilities at the intrinsic and extrinsic levels, respectively.

On the basis of such studies into the structure of a complex of natural enemies, it has been suggested that intrinsically inferior species should be introduced first in order to give them a chance to demonstrate their full control potential in the absence of competing species (36, 87). As Zwölfer (125) states, such a procedure would not only provide more insight into the interactions of a host-parasite system but would also help to avoid unnecessary parasite introductions and prevent irreparable mistakes, such as might have occurred, for example, with the introduction of the largely cleptoparasitic ichneumonids, *Temelucha interruptor* and *Eulimneria rufifemur,* into Canada, to the detriment of the highly adapted braconid, *Orgilus obscurator,* a very effective parasite of the European pine shoot moth and an efficient searcher at low host densities (1, 85, 93, 99). An experiment to measure the efficacy of *Orgilus* by releasing it on an island where competing internal parasites were still absent remained inconclusive because other factors caused the host density to decrease rapidly (99).

INVESTIGATIONS INTO THE POPULATION-DYNAMICAL ROLE OF BIOTIC MORTALITY FACTORS

Any biological control program rests on the assumption that there exist natural enemies capable of controlling the target pest. Studies in the home country of a pest should therefore aim at taking a census of both the host and parasite-predator populations in different localities and under different situations, e.g. with regard to host density and other variable parameters, with the view to assess the population-dynamical impact of the biotic mortality factors involved.

As Varley (109) has pointed out, simple routine census assessments can provide a valuable picture of pest and parasite fluctuations over the years; this is merely descriptive but a useful first step. More detailed figures are needed for a life table of the pest species. The cause of population change can then be revealed by a key factor analysis. Further studies of density-dependent relationship in the pest's life table can identify stabilizing factors and help to explain population levels (66, 67, 113). From a study of the natural enemies present their life tables can be derived, and populations are modeled so as to make useful assumptions on the population-dynamical role of the parasite-predator complex associated with a given target pest.

The disadvantage of such a life table and key factor analysis, even if primarily restricted to the importance of biotic mortality factors, is that it takes considerable time and manpower to obtain realistic results. The first can be alleviated by taking subsamples from widely differing host densities in any one generation, but this procedure is not applicable when a density-dependent relationship is the result of behavioral (functional) responses of natural enemies within a single host generation (46). In practice, one often has to be content with establishing whether a given complex, or species, of natural enemies exhibits a positive or negative correlation with the host density. Demonstration of an inverse relationship to host density, e.g. as with the predators of the balsam woolly aphid in Europe (26), is a discouraging sign, and indeed the establishment of several European predators against this aphid in various parts of North America (14) has so far had little benefit.

This is not the place to go into the detailed pros and cons of population-dynamical studies and models, especially because this subject has been repeatedly reviewed in this review series (45, 112) and elsewhere. Although the increasing use of these modern approaches to biological control will undoubtedly be scientifically rewarding and helpful in putting both foreign exploration and the release program on a sounder basis, it must be admitted that any predictions derived from a multivariate analysis or population model are no guarantee that the biological control agents selected will be effective in the area of introduction; for instance, a long-term key factor analysis of the winter moth in England (111, 113) did not reveal the potential of two common European parasite species as controlling factors of the pest, yet, after their establishment in eastern Canada, *Cyzenis albicans* and *Agrypon flaveolatum* became key factors in successfully reducing winter moth populations to extremely low levels (30, 31).

BIOECOLOGICAL STUDIES OF INDIVIDUAL SPECIES

Bioecological studies include an array of investigations into such questions as host searching and host specificity, synchronization with the host, physiological tolerance to different climatic factors, and genetic aspects of natural enemy populations. For the study of host finding, the experimental method of exposing, in the field, artificial populations of the pest species to the attack of natural enemies has much to be recommended. This has been done with the colonial European pine sawfly, *Neodiprion sertifer*, and with gypsy moth colonies (84; H. Pschorn-Walcher, unpublished data).

Host specificity might either be phylogenetically or ecologically determined. Phylogenetic host specificity occurs primarily with host-parasite groups that have obviously been closely associated during their long history of coevolution, e.g. certain sawfly families and their parasites of the subfamilies Scolobatinae and Tryphoninae (82) or aphids and the subfamily Aphidiinae (59). Ecological host specificity, on the other hand, is common with host groups that live in concealed environments, such as gallmakers (2), leafminers, except in the Agromyzidae where coevolution with the Alysiinae seems to have taken place (44), and woodborers. In these cases, host specificity may be determined by such secondary characters as the host plant or the shape and seasonal appearance of a gall; hence ecologically similar but taxonomically unrelated host species may have similar or virtually identical parasite complexes.

Synchronization with the host and the determination of the host stage(s) susceptible to parasite attack are important elements to study, especially in biological control programs that involve transfer of natural enemies to a region with alternate seasons (e.g. from the Northern to the Southern Hemisphere) or with a different daylength regime (100). Investigations into the diapause behavior of hosts and enemies are also a prerequisite to their successful manipulation in a mass-breeding and release program.

Studies into the temperature and moisture requirements of natural enemy species, including determination of cold-hardiness, might help to single out unsuitable control agents, to make a better choice of suitable release areas, and to make crude predictions about the potential spread of an established enemy in, for example, a harsh, boreal environment.

Genetic studies of natural enemy populations have been inexcusably neglected in biological control in general and in that of forest insects in particular, although relatively simple modern methods such as the isoenzyme technique (114) are now available. Ideally, such studies should be made on any population of natural enemies prior to as well as after successful introduction (118) in order to obtain more information on such important basic questions as the suitability of different genetic strains or the role that genetic processes (the so-called founder principle) play in the successful adaptation of a species to the new environment. In some cases a genetic polymorphism is already obvious morphologically, e.g. in the females of the lady beetle, *Aphidecta obliterata,* introduced from Europe into North America for the biological control of the balsam woolly aphid (27) (and in both sexes of many other coccinellids). In other species, polymorphism is only physiologically expressed, e.g. of the two strains of *Mesoleius tenthredinis* in Europe, the Bavarian strain (recently introduced into Canada) is resistant against encapsulation by its host, the larch sawfly, whereas the endemic, alpine strain is heavily encapsulated (86, 105). Genetic aspects of natural enemy introductions have been discussed by a number of authors (12, 13, 33, 56, 58, 60, 65, 89, 95, 120), and genetic problems in the mass production of biological control agents have recently been reviewed by Mackauer (60).

ESTABLISHMENT OF PRIORITIES IN IMPORTATION OF PROMISING SPECIES—EVENTUAL MASS PROPAGATION

The establishment of priorities in an importation program rests on the assumption that there exist valid criteria by which the most promising species of a given complex of natural enemies can be identified and that these criteria can be evaluated for the majority of species by means of ecological studies preceding importation, as described in the foregoing chapters. Although a matter of considerable debate, the characteristics of a potentially successful natural enemy most frequently mentioned in the literature (e.g. 20, 22, 109, 110) include: (*a*) a high searching ability enabling the species, when at low population density, to kill enough pests to prevent pest increase when the pest is also uncommon (110); (*b*) a high reproductive rate, although a high egg potential may mainly save time in establishment and is advantageous only when the parasites themselves suffer heavy mortality; (*c*) a high degree of host specificity, although this may imply an ecologically marginal status (58); (*d*) a good synchronization with the host; and (*e*) a high degree of adaptability to a wide range of ecoclimatic conditions.

These characteristics are frequently combined in a group of parasites known as r-strategists, and consequently the suggestion has been made that r-selected species of natural enemies, which are good colonizers and able to penetrate into and to adapt to a new environment, should be given high priority in any introduction program (3, 34, 64, 74). This suggestion has, in my opinion, much to recommend it, especially because r-strategists usually have low competitive abilities at the intrinsic level, i.e. in multiple parasitism (84, 87, 124), so that if they are imported first, they have a chance to demonstrate their full potential. In the home country of the pest, this potential is often impaired through interference by competitively superior, K-selected members of the parasite complex. Species under r-selection are also supposed to be adapted to some degree of inbreeding (12, 13) and should thus be better able to go through the bottleneck of initial colonization in small numbers, including prerelease laboratory breeding.

The preintroduction studies outlined in the previous chapters should aid in recognizing r-selected species, especially in the undisturbed, well-balanced parasite complexes of forest pests. Natural enemies of the r-selected type should show up particularly in areas or years of low host populations. Their development is usually intimately geared to that of the host, being often in long contact with the latter, e.g. by attacking the early larval stage of the host but emerging only from the mature larvae or pupae. This is frequently coupled with a high reproductive rate expressed, for example, in the ichneumonids in the high number of ovarioles present in their females (75). This is an adaptation to the sustained losses which early-attacking parasites suffer in competition with later-attacking ones and/or from hyperparasitism. Host specificity is usually well developed (in the form of monophagy or oligophagy) and often of the phylogenetic type described before.

In the biological control of forest insect pests r-selected types of natural enemies have repeatedly proven superior to more polyphagous, less-well-adapted K-strategists. To mention three examples: (*a*) In the European spruce sawfly, *Gilpinia hercyniae,* two typically r-selected introduced parasites, *Exenterus vellicatus* and *Drino bohemica,* have largely replaced a K-selected, more polyphagous species, *Dahlbominus fuscipennis,* in Canada. Although released in over 880 million individuals, this latter species has been important only in the early phase of the colonization program, as long as sawfly populations were still high (9). (*b*) In the biological control program of the larch sawfly in Canada, only the two parasite species most typically r-selected (*Olesicampe benefactor, Mesoleius tenthredinis*) have become successful colonizers, whereas four other species, representing more K-selected types, that were released have failed entirely (83, 105). (*c*) With the winter moth in Nova Scotia, the situation has been similar to *b* (30, 31). In the last two examples the introduced parasites have reached considerably higher rates of parasitism in the area of introduction than in the home country of the pest, where they suffer heavily from competition by later attacking parasite species and/or hyperparasites (85, 86).

In highly diversified parasite complexes, e.g. of Lepidoptera or sawflies, several parasite species of the r-selected type may coexist within a parasite guild. In Europe, for example, the early larval parasites representing ecological homologues, the three species of *Apanteles,* and the two species of *Campoplex* all attack the young mining stages of their host, *Coleophora serratella,* though with different regional frequencies (29). Here a choice is more difficult to make (except perhaps on the basis of their intrinsic competitive abilities), and it seems best to introduce the whole set of ecological homologues, though preferably first into different but ecologically similar areas to allow for a comparison of the colonization success of the individual (or groups of) species involved. This practice is currently followed in Newfoundland where *Apanteles* spp. and *Campoplex* spp. have been released in ecologically similar but geographically isolated areas (A. G. Raske, personal communication).

The question of single vs multiple-species introductions has been and still is a matter of debate. Some authors (e.g. 87, 102, 103, 117, 123) have been advocates of the single-species (or a preselected combination of species) philosophy, while others (e.g. 47, 50, 113) speak in favor of multiple-species introductions; the most outspoken supporter of this idea is probably DeBach (21, 22), who contends that ecological preintroduction studies need not precede importations of natural enemies and that all species of any promise should be imported and tried, since it is impossible to predict which one(s) would be effective. I would agree with DeBach as far as natural enemies of agricultural pests of highly artificial and periodically disturbed environments (including pests of doubtful origin) are concerned, but I would argue that with most forest pests (and agricultural insects of less disturbed habitats) characterized by more stable, highly evolved, and diversified parasite complexes, preintroduction studies to establish priorities in importation of promising species are not only useful but necessary.

A compromise between the two extreme schools of thought might finally prove to be the superior strategy, both from a practical and theoretical standpoint. Since on the average only about 2–4 species out of 10 released have become established in biological control programs (15, 20, 63, 119), a single species introduction is obviously taking a considerable risk of total failure. Importations of most or all prospects without proper prescreening studies risks irrevocable mistakes (as with the cleptoparasite *Temelucha interruptor* already mentioned), as well as time and effort being wasted on such ineffective species as the European predators of the balsam woolly aphid. As Eichhorn (26) has shown, these live only on the interest but do not attack the capital of their host populations.

Theoretically competitive interactions between different parasite species and guilds may stabilize the parasite complex per se, but may conceivably lower the ability of highly adapted, but competitively often inferior, species of the complex (often the r-strategists) to exert their full control potential (87, 117, 123, 125). Hence an optimal combination of natural enemies with little interference between its members but a high degree of coaction might be the ideal, whenever the situation allows it.

The continued debate about the pros and cons of different biological control strategies suffers because experimentation in biological control has been inexcusably neglected in most release programs. The examples cited in favor of one or the other strategy are largely the same in most publications dealing with this subject (and this paper is no exception), but a better-founded theory will only be arrived at when many more carefully planned and documented examples of releases of single species or different combinations of species in geographically distinct but ecologically comparable situations have become available. The crux is that pest problems are not very suitable for an experimental approach because there is usually considerable pressure to obtain a quick solution of the problem (and hence the temptation to introduce many species of control agents in the hope that one or the other will be a quick success). I would therefore suggest that we should perhaps try biological control of insect species (introduced as well as native ones) of little or no economic concern, which would be more amenable to experimental testing of conflicting strategies of biological control, such as the single- vs multiple-introduction approach, the use of monophagous vs polyphagous parasites, the influence of hyperparasites, the suitability of natural enemies from closely related hosts, and other controversies of this kind.

CHOICE OF SUITABLE RELEASE AREAS

Little is said here about choosing suitable release areas because the factors are largely self-evident: areas climatically and ecologically similar to the home country of the introduced control agents are obviously the best choice. Logistic considerations, however, may influence or restrict the choice made. If life-table plots are available, as they have been with some of the releases of larch sawfly parasites in Canada (105), releases into these plots greatly facilitate monitoring and interpreting population buildup and effectiveness of the release.

Beirne (7), in an analysis of about 75 biological control attempts in Canada, concluded that semi-isolated release sites are preferable and that repeated releases in different stands have a greater chance to be successful, providing the release sites do not lack factors needed for the introduced agents to survive.

RELEASES AND MONITORING OF POPULATION BUILDUP AND DISPERSAL OF ESTABLISHED NATURAL ENEMIES

In biological control programs against forest insects, direct inoculative releases, usually of freshly imported natural enemies, have dominated the scene. Mass-breeding and release programs [e.g. with *Dahlbominus fuscipennis* in Canada (63), with gypsy moth parasites in the United States (55), and with the *Sirex*-infesting nematode, *Deladenus siricidicola,* in Australia (6)] have been relatively uncommon, considering that forest insects are often more difficult to rear and that the vast surfaces involved can be a formidable obstacle to inundative releases.

For inoculative releases in the forest environment of Canada, Beirne (7) has shown that about 60% of species that averaged over 800 individuals per release became colonized, but only about 15% of those that averaged less than 800 per release became colonized, implying that the greater the number released (and the greater the number of releases per species), the greater the likelihood of establishment. Beirne contends that a number of failures of the past could be successes in the future if the species is re-introduced in quantities sufficiently large to make colonization a probability.

Releases of exotic natural enemies limited in size (less than about 500 individuals) may not provide enough genetic diversity for the colony to be successful and may impair mating and host finding (60). In order to improve the genetic variability, laboratory crossings with different sources of the larch casebearer parasite, *Chrysocharis laricinellae,* are currently being conducted in Oregon. This approach has been little used in the biological control of forest pests (R. B. Ryan, personal communications). Space limitation could mean that a species fails to establish at one site but would be successful at another, and time limitation may be detrimental when the release coincides with a particularly unsuitable period or when temporary establishment is thwarted by subsequent inclemencies, such as a cold winter before the colonized population has had time to adapt to the new environment. Perseverance appears therefore to be a necessary requisite of biological control projects, especially in forest habitats.

Experimental setup of release programs and follow-up studies on population buildup and dispersal have been surprisingly poor in many biological control projects, and much useful information on the reasons of successes or failures of individual species has thus been lost forever. Only recently have life-table and other population studies been employed (e.g. 30, 31, 105) in assessing the colonization success obtained.

Modeling of data on population buildup and dispersal obtained after successful colonization is a further useful tool for interpreting the effectiveness of established biocontrol agents. A model developed recently for the larch sawfly in central Canada has shown that the introduced larval parasite, *Olesicampe benefactor,* was the key factor in the termination of the pest's outbreak, in spite of heavy hyperparasitism by the Holarctic *Mesochorus dimidiatus.* When the two parasites were not included and the model ran for extended periods, outbreak levels of the sawfly were commonly reached (N. G. H. Ives, personal communication).

COLONIZATION OVER ENTIRE PEST AREA

The extended range of many forest pests makes colonization over the entire pest area essential to many biological control programs in order to speed up dispersal of the natural enemies established. Usually, and rightly so, material of the colonized population has been used for redistribution, because this population may have undergone profound changes in its gene pool as an adaptation of the species to the new environment and may thus be better suited for relocation than freshly imported stock. A case in point is the colonization of the European ichneumonid, *Olesicampe benefactor,* in central Canada. When originally released in 1961 on the eastern side of Lake Winnipeg, the univoltine species took 4 yr for parasitism to become significant, but in 1966 an explosive population buildup produced a parasitism of over 90%, which remained at this high level for the following 5 yr. Dispersal, too, was slow at the beginning (less than 3 km in the first 5 yr), but later it increased rapidly to an estimated 80–130 km per year between 1970 and 1974 (105; J. A. Muldrew, personal communication). Relocations of *Olesicampe* from Manitoba to the Maritimes were successful more quickly, indicating that a strain better adapted to North American conditions had developed in central Canada. As the introduced populations were derived from many small, scattered populations of Austria and Switzerland and from a larger source in pre-Alpine Bavaria, a high degree of genetic diversity should have been present in the imported stock, facilitating adaptive re-structuring of the gene pool during the precarious early phase of colonization (105).

FINAL EVALUATION OF THE PROJECT

Two important points have been sadly neglected in the past. The first is a critical appraisal of the success or failure of a project. The degree of success has usually been stated in such vague terms as complete or partial control, but demonstration that the introduced species of natural enemies have been the causative agents of control achieved has often remained a tacit assumption, at least in many older biological control programs. This situation has improved since more powerful monitoring techniques such as the life-table approach have become available. Among others, the winter moth and larch sawfly projects in Canada (30, 31, 105) can be mentioned as examples of follow-up studies of a particularly high standard.

In some cases, monitoring has been discontinued too early, especially with projects with no immediate or only a partial success. Valuable data on further dispersal and on the resurgence of both pest and parasite-predator populations have been lost, as in the case of the evolution and/or dispersal of the resistant strains of the larch sawfly in Canada (105, 121), which appeared after the temporarily successful importation of *Mesoleius tenthredinis* from the British Isles before World War I.

Other forest pests have subsided after successful biological control, but it is often not clear whether the natural enemies or some other factor, such as an abatement of the pest's aggressiveness, have kept them at a low level ever since. This course of events may satisfy the forester, but the biological control worker should be interested in probing more deeply into this question. There are almost innumerable examples of old establishments of natural enemies, both in forestry and agriculture, for which no recent assessment of their present status has been made.

The axiom that in scientific work one can learn as much from failures as from successes has received little attention in biological control. Rarely have the causes of failures been analyzed but there is little doubt that failures of many releases in the past have been caused by faulty procedures such as inadequate numbers released (8), lack of proper timing of the release, or release of a highly specific species such as *Lophyroplectus luteator* against the wrong host, *Gilpinia hercyniae,* in Canada [recent releases of this ichneumonid against its proper host, *Neodiprion sertifer,* have been immediately successful (17)]. I would concur with Messenger & van den Bosch (65) that in these and many more cases the possibility exists that a new sample of the natural enemy, collected from a more appropriate locality or host population, or under different circumstances than the original collection, may finally give better results, as may the introduction of other species of biological control agents not yet tried or not adequately tested in the past.

A second, even more important point that has been and still is neglected by biological control workers is the economics of biological control. Balance sheets of costs and results are virtually nonexistent, except in the cases of agricultural pests (e.g. 20, 70, 96). In some of these, the economic benefits derived from a rather small outlay of funds have often been staggering, and there can be little doubt that the returns from the investment of funds in biological control projects of forest pests must also have been very satisfactory. As biological control is largely dependent on the support from federal, state, and other public agencies, it is of paramount importance that a proper assessment of the cost-benefit ratio of successful projects be made in the course of their final evaluation in order to better advertise the results achieved and to clearly demonstrate both to administrators and the public the economic benefits obtained from biological control in the past and those still obtainable in the future.

ACKNOWLEDGMENTS

Thanks are expressed to Dr. F. J. Simmonds, Director, the Commonwealth Institute of Biological Control, and to my former colleague Dr. H. Zwölfer (now at the University of Bayreuth, West Germany) for their helpful assistance in the preparation of the flow diagram (Figure 1), and to the editors for the correction and improvement of the English style of the manuscript.

Literature Cited

1. Arthur, A. P., Steiner, J. E. R., Turnbull, A. L. 1964. The interaction between *Orgilus obscurator* (Nees) and *Temelucha interruptor* (Grav.), parasites of the pine shoot moth *Rhyacionia buoliana* (Schiff.). *Can. Entomol.* 96: 1030–34
2. Askew, R. R. 1961. On the biology of the inhabitants of oak galls of Cynipidae in Britain. *Trans. Soc. B. Entomol.* 14:237–68
3. Askew, R. R. 1975. The organization of chalcid-dominated parasitoid communities centered upon endophytic hosts. In *Evolutionary Strategies of Parasitic Insects,* ed. P. W. Price, pp. 130–53. London: Plenum. 224 pp.
4. Balch, R. E. 1958. Control of forest insects. *Ann. Rev. Entomol.* 3:449–68
5. Balch, R. E. 1960. The approach to biological control in forest entomology. *Can. Entomol.* 92:297–310
6. Bedding, R. A., Akhurst, R. J. 1974. Use of the nematode *Deladenus siricidicola* in the biological control of *Sirex noctilio* in Australia. *J. Aust. Entomol. Soc.* 13:129–35
7. Beirne, B. P. 1962. Trends in applied biological control of insects. *Ann. Rev. Entomol.* 7:387–400
8. Beirne, B. P. 1975. Biological control attempts by introductions against pest insects in the field in Canada. *Can. Entomol.* 107:225–36
9. Bird, F. T., Elgee, D. E. 1957. A virus disease and introduced parasites as factors controlling the European spruce sawfly. *Can. Entomol.* 89:371–78
10. Bösenberg, K. 1970. Zur Bedeutung der Vogelwelt im Rahmen der biologischen Schädlingsbekämpfung, besonders im Wald. See Ref. 24, pp. 71–82
11. Buckner, C. H. 1966. The role of vertebrate predators in the biological control of forest insects. *Ann. Rev. Entomol.* 11:449–70
12. Carson, H. L. 1971. Speciation and the founder principle. *Stadler Genet. Symp.* 3:51–70
13. Carson, H. L. 1973. Reorganization of the gene pool during speciation. In *Genetic Structure of Populations. Popul. Genet. Monogr,* ed. N. E. Morton, 3:274–80. Honolulu: Univ. Hawaii Press. 205 pp.
14. Clark, R. C., Greenbank, D. O., Bryant, D. G., Harris, J. W. E. 1971. *Adelges piceae* (Ratz.), balsam woolly aphid. See Ref. 17, pp. 113–27
15. Clausen, C. P. 1956. Biological control of insect pests in the continental United States. *US Dept. Agric. Tech. Bull.,* No. 1139. 151 pp.
16. Clausen, C. P. 1958. Biological control of insect pests. *Ann. Rev. Entomol.* 3:291–310
17. Commonwealth Agricultural Bureaux 1971. Biological control programmes against insects and weeds in Canada 1959–1968. *Tech. Commun. Commonw. Inst. Biol. Control.* No. 4. 266 pp.
18. Coppel, H. C., Benjamin, D. M. 1965. Bionomics of the nearctic pinefeeding diprionids. *Ann. Rev. Entomol.* 10: 69–96
19. Coppel, H. C., Leius, K. 1955. History of the larch sawfly, with notes on origin and biology. *Can. Entomol.* 87:103–11
20. DeBach, P., ed. 1964. *Biological Control of Insect Pests and Weeds.* London: Chapman & Hall. 844 pp.
21. DeBach, P. 1971. The theoretical basis of importation of natural enemies. *Proc. Int. Congr. Entomol., 13th* 2:140–42
22. DeBach, P. 1972. The use of imported natural enemies in insect pest management ecology. *Proc. Tall Timbers Conf. Ecol. Anim. Control Habitat Manage.* 3:211–33
23. DeBach, P. 1974. *Biological Control by Natural Enemies.* Cambridge, Engl: Cambridge Univ. Press. 323 pp.
24. Deutsches Entomologisches Institut 1970. Biologische Bekämpfungsmethoden von Forstschädlingen. *Tagungsber. Dtsch. Akad. Landwirtschaftwiss. Berlin,* No. 110. 202 pp.
25. Dowden, P. B. 1962. Parasites and predators of forest insects liberated in the United States through 1960. *U.S. Dept. Agric. Handb.* No. 226. 70 pp.
26. Eichhorn, O. 1968. Problems of the population dynamics of silver fir woolly aphids, genus *Adelges* (=*Dreyfusia*). *Z. Angew. Entomol.* 61:157–214
27. Eichhorn, O., Graf, P. 1971. Sex-linked colour polymorphism in *Aphidecta obliterata* L. *Z. Angew. Entomol.* 67:225–31
28. Eichhorn, O., Pschorn-Walcher, H. 1973. The parasites of the birch leaf-mining sawfly (*Fenusa pusilla* Lep.) in central Europe. *Tech. Bull. Commonw. Inst. Biol. Control,* No. 16, pp. 79–104
29. Eichhorn, O., Pschorn-Walcher, H., Schröder, D. 1971. Gegenwärtige Projekte der biologischen Bekämpfung

verschleppter Forstschädlinge. *Anz. Schädlingskd.* 44:145–52

30. Embree, D. G. 1966. The role of introduced parasites in the control of winter moth in Nova Scotia. *Can. Entomol.* 98:1159–68
31. Embree, D. G. 1971. *Operophtera brumata* (L.), winter moth. See Ref. 17, pp. 167–75
32. Finnegan, R. J. 1975. Introduction of a predacious red wood ant, *Formica lugubris,* from Italy to Eastern Canada. *Can. Entomol.* 107:1271–74
33. Force, D. C. 1967. Genetics in the colonization of natural enemies for biological control. *Ann. Entomol. Soc. Am.* 60:727–29
34. Force, D. C. 1972. r- and K-strategies in endemic host-parasitoid communities. *Bull. Entomol. Soc. Am.* 18:135–37
35. Force, D. C. 1974. Ecology of insect-host parasitoid communities. *Science* 184:624–32
36. Franz, J. M. 1961. Biologische Schädlingsbekämpfung. In *Handbuch der Pflanzenkrankheiten,* ed. P. Sorauer, 6:1–302. Berlin: Parey. 627 pp. 2nd. ed.
37. Franz, J. M. 1961. Biological control of pest insects in Europe. *Ann. Rev. Entomol.* 6:183–200
38. Franz, J. M. 1965. Forest insect control by biological measures. *FAO/IUFRO Symp. Intern. Danger For. Dis. Insects,* Vol. 2. 21 pp.
39. Franz, J. M. 1970, 1971. Biological and integrated control of pest organisms in forestry. *Unasylva* 24:37–46; 25:45–56
40. Franz, J. M., Krieg, A. 1972. *Biologische Schädlingsbekämpfung.* Berlin, and Hamburg: Parey. 208 pp.
41. Gösswald, K. 1951. *Die rote Waldameise im Dienste der Waldhygiene: Forstwirtschaftliche Bedeutung, Nutzung, Lebensweise, Zucht, Vermehrung und Schutz.* Lüneburg: Metta Kinau. 160 pp.
42. Greathead, D. J. 1971. A review of biological control in the Ethiopian region. *Tech. Commun. Commonw. Inst. Biol. Control,* No. 5. 162 pp.
43. Greathead, D. J., ed. 1976. A review of biological control in western and southern Europe. *Tech. Commun. Commonw. Inst. Biol. Control,* No. 7. In press
44. Griffiths, G. C. D. 1964. The Alysiinae parasites of the Agromyzidae. *Beitr. Entomol.* 14:823–914
45. Harcourt, D. G. 1969. The development and use of life-tables in the study of natural insect populations. *Ann. Rev. Entomol.* 14:175–96
46. Hassell, M. P. 1966. Evaluation of parasite and predator responses. *J. Anim. Ecol.* 35:65–75
47. Hassell, M. P., Varley, G. C. 1969. New inductive population model for insect parasites and its bearing on biological control. *Nature* 223:1133–37
48. Huffaker, C. B., ed. 1971. *Biological Control.* New York and London: Plenum. 511 pp.
49. Huffaker, C. B., Kennett, C. E. 1969. Some aspects of assessing the efficiency of natural enemies. *Can. Entomol.* 101: 425–47
50. Huffaker, C. B., Messenger, P. S., DeBach, P. 1971. The natural enemy component in natural control and the theory of biological control. See Ref. 48, pp. 16–67
51. Inman, R. E. 1970. Problems in searching for and collecting control organisms. In *Proc. Int. Symp. Biol. Control Weeds 1st. Misc. Publ. Commonw. Inst. Biol. Control* 1:105–8. 110 pp.
52. Jagsch, A. 1972. Populationsdynamik und Parasiten-Komplex der Lärchenminiermotte, *Coleophora laricella* Hbn. im natürlichen Verbreitungsgebiet der Europäischen Lärche, *Larix decidua* Mill. *Z. Angew. Entomol.* 73:1–42
53. Kilgore, W. W., Doutt, R. L., eds. 1967. *Pest Control: Biological, Physical and Selected Chemical Methods.* New York: Academic. 477 pp.
54. Kluyver, N. N. 1954. Die biologische Schädlingsbekämpfung durch Vögel. *Festschr. Vogelschutzwarte Essen-Altenhunden,* pp. 27–31
55. Leonard, D. E. 1974. Recent developments in ecology and control of the gypsy moth. *Ann. Rev. Entomol.* 19: 197–230
56. Levins, R. 1969. Some demographic and genetic consequences of environmental heterogeneity for biological control. *Bull. Entomol. Soc. Am.* 15:237–40
57. Lindroth, C. H. 1957. *The Faunal Connections between Europe and North America.* Stockholm: Almqvist & Wiksell. 344 pp.
58. Lucas, A. M. 1969. The effect of population structure on the success of insect introductions. *Heredity* 24:151–54
59. Mackauer, M. 1961. Zur Frage der Wirtsbindung der Blattlaus-Schlupfwespen. *Z. Parasitenkd.* 20:576–91
60. Mackauer, M. 1976. Genetic problems in the production of biological control agents. *Ann. Rev. Entomol.* 21:369–85

61. Callan, M. E. 1969. Ecology and insect colonization for biological control. *Proc. Ecol. Soc. Aust.* 4:17–31
62. McGugan, B. M. 1962. Biological control of forest insects: the role of parasites and predators. *Proc. World For. Congr., 5th* 2:935–40
63. McLeod, J. H., McGugan, B. M., Coppel, H. C. 1962. A review of the biological control attempts against insects and weeds in Canada. *Tech. Commun. Commonw. Inst. Biol. Control,* No. 2. 216 pp.
64. Messenger, P. S. 1964. Use of life tables in a bioclimatic study of an experimental aphid-braconid wasp host-parasite system. *Ecology* 5:119–31
65. Messenger, P. S. van den Bosch, R. 1971. The adaptability of introduced biological control agents. See Ref. 48, pp. 69–92
66. Morris, R. F. 1959. Single factor analysis in population dynamics. *Ecology* 40:580–88
67. Morris, R. F. 1963. Predictive population equations based on key-factors. *Mem. Entomol. Soc. Can.* 32:116–29
68. Nairn, L. D., Reeks, W. A., Webb, F. E., Hildahl, V. 1962. History of larch sawfly outbreaks and their effect on tamarack stands in Manitoba and Saskatchewan. *Can. Entomol.* 94: 242–55
69. National Academy of Sciences 1969. *Principles of Plant and Animal Pest Management and Control.* Washington: Nat. Acad. Sci. Publ., 1965. 508 pp.
70. Ordish, G. 1967. *Biological Methods in Crop Pest Control.* London: Constable. 242 pp.
71. Pavan, M. 1959. Attività italiana per la lotta biologica con formiche del gruppo *Formica rufa* contro gli insetti dannosi alle foreste. *Collana Verde* 4. 78 pp.
72. Pimentel, D. 1963. Introducing parasites and predators to control native pests. *Can. Entomol.* 95:785–92
73. Prebble, M. L. 1960. Biological control in forest entomology. *Bull. Entomol. Soc. Am.* 6:6–8
74. Price, P. W. 1973. Parasitoid strategies and community organization. *Environ. Entomol.* 13:415–26
75. Price, P. W. 1974. Strategies for egg production. *Evolution* 28:76–84
76. Pschorn-Walcher, H. 1958. Climatic and biocoenotic aspects for the collections of predators of *Dreyfusia piceae* in Europe. *Proc. Int. Congr. Entomol., 10th.* 4:801–5
77. Pschorn-Walcher, H. 1961. Biological control of forest insects: Recent work and future aspects. *Unasylva* 15:3–7
78. Pschorn-Walcher, H. 1963. Historisch-biogeographische Rückschlüsse aus Wirt-Parasiten-Assoziationen bei Insekten. *Z. Angew. Entomol.* 51:208–14
79. Pschorn-Walcher, H. 1964. On the parasites of some injurious Lepidoptera from northern Japan. *Tech. Bull. Commonw. Inst. Biol. Control.* 4:24–37
80. Pschorn-Walcher, H. 1965. The ecology of *Neodiprion sertifer* (Geoff.) and a review of its parasite complex in Europe. *Tech. Bull. Commonw. Inst. Biol. Control* 5:33–97
81. Pschorn-Walcher, H. 1967: Biology of the ichneumonid parasites of *Neodiprion sertifer* (Geoff.) in Europe. *Tech. Bull. Commonw. Inst. Biol. Control* 8:7–52
82. Pschorn-Walcher, H. 1969. Die Wirtsspezifität der parasitischen Hymenopteren in ökologisch-phylogenetischer Betrachtung (unter besonderer Berücksichtigung der Blattwespenparasiten). *Ber. 10. Wandervers. Dtsch. Entomol.* 80:55–63
83. Pschorn-Walcher, H. 1973. Die grosse Lärchenblattwespe (*Pristiphora erichsonii* Htg) in Europa und ihre biologische Bekämpfung in Nordamerika. *100-Jahre Hochsch. Bodenkultur, Wien, Fachveranst.* 4:451–71
84. Pschorn-Walcher, H. 1973. Die Parasiten der gesellig lebenden Kiefern-Buschhornblattwespen als Beispiel für Koexistenz und Konkurrenz in multiplen Parasit-Wirt-Komplexen. *Verh. Dtsch. Zool. Ges.* 66:136–45
85. Pschorn-Walcher, H., Schröder, D., Eichhorn, O. 1969. Recent attempts at biological control of some Canadian forest insect pests. *Tech. Bull. Commonw. Inst. Biol. Control* 11:1–18
86. Pschorn-Walcher, H., Zinnert, K. D. 1971. Investigations on the larch sawfly (*Pristiphora erichsonii* Htg.) in central Europe. Part II. Natural enemies: their biology and ecology and their role as mortality factors in *P. erichsonii. Tech. Bull. Commonw. Inst. Biol. Control* 14:1–50
87. Pschorn-Walcher, H., Zwölfer, H. 1968. Konkurrenzerscheinungen in Parasitenkomplexen als Problem der biologischen Schädlingsbekämpfung. *Anz. Schädlingsk* 41:71–76
88. Rao, V. P., Ghani, M. A., Sankaran, T., Mathur, K. C. 1971. A review of biological control of insects and other pests in

South-East Asia and the Pacific Region. *Tech. Commun. Commonw. Inst. Biol. Control* No. 6. 149 pp.

89. Remington, C. L. 1968. The population genetics of insect introduction. *Ann. Rev. Entomol.* 13:415–26
90. Root, R. B. 1967. The niche exploitation pattern of the blue-gray gnatcatcher. *Ecol. Monogr.* 37:317–50
91. Schieferdecker, H. 1970. Zum Stand der *Trichogramma*-Forschung in Europa und deren weitere Aufgaben. See Ref. 24, pp. 137–75
92. Schröder, D. 1966. Zur Kenntnis der Systematik und Oekologie der "*Evetria*"-Arten. *Z. Angew. Entomol.* 57:333–429; 58:279–308
93. Schröder, D. 1974. A study of the interactions between the internal larval parasites of *Rhyacionia buoliana*. *Entomophaga* 19:145–71
94. Simmonds, F. J. 1959. Biological Control—Past, present and future. *J. Econ. Entomol.* 52:1099–1102
95. Simmonds, F. J. 1963. Genetics and biological control. *Can. Entomol.* 95: 561–67
96. Simmonds, F. J. 1967. The economics of biological control. *J. R. Soc. Arts* 115:880–98
97. Simmonds, F. J. 1972. Approaches to biological control problems. *Entomophaga* 17:251–64
98. Stehr, F. W. 1975. Parasitoids and predators in pest management. In *Introduction to Insect Pest Management,* ed. R. L. Metcalf, W. Luckmann, pp. 147–88. New York: Wiley. 587 pp.
99. Syme, P. D. 1971. *Rhyacionia buoliana* (Schiff.), European pine shoot moth. See Ref. 17, pp. 194–205
100. Tauber, M. J., Tauber, C. A. 1976. Insect seasonality: diapause maintenance, termination, and postdiapause development. *Ann. Rev. Entomol.* 21:81–107
101. Thompson, W. R. 1956. The fundamental theory of natural and biological control. *Ann. Rev. Entomol.* 1:379–402
102. Turnbull, A. L. 1967. Population dynamics of exotic insects. *Bull. Entomol. Soc. Am.* 13:333–37
103. Turnbull, A. L., Chant, D. A. 1961. The practice and theory of biological control of insects in Canada. *Can. J. Zool.* 39:697–753
104. Turnock, W. J., Muldrew, J. A. 1971. Parasites. In *Toward Integrated Control,* 59–87. *Proc. Ann. Northeast. For. Insects Work Conf., 3d.* 129 pp.
105. Turnock, W. J., Muldrew, J. A. 1971. *Pristiphora erichsonii* (Hartig), larch sawfly. See Ref. 17, pp. 175–94
106. van den Bosch, R. 1971. Biological control of insects. *Ann. Rev. Ecol. Syst.* 2:45–66
107. van den Bosch, R., Messenger, P. S. 1973. *Biological Control.* New York: Intex Educational Publ. 180 pp.
108. van den Bosch, R., Stern, V. M. 1962. The integration of chemical and biological control of arthropod pests. *Ann. Rev. Entomol.* 7:367–86
109. Varley, G. C. 1970. The need for life tables for parasites and predators. In *Concepts of Pest Management,* ed. R. L. Rabb, F. E. Guthrie, pp. 59–68. Raleigh, NC: N. Carolina State Univ. Press
110. Varley, G. C. 1974. Population dynamics and pest control. In *Symp. Br. Ecol. Soc., 13th,* ed. D. Price Jones, M. E. Solomon, pp. 15–26. Oxford: Blackwell
111. Varley, G. C., Gradwell, G. R. 1968. Population models for the winter moth. In *Insect Abundance,* ed. T. R. E. Southwood, pp. 132–42. *Symp. R. Soc. Entomol. London* 4. 160 pp.
112. Varley, G. C., Gradwell, G. R. 1970. Recent advances in insect population dynamics. *Ann. Rev. Entomol.* 15:1–24
113. Varley, G. C., Gradwell, G. R. 1971. The use of models and life-tables in assessing the role of natural enemies. See Ref. 48, pp. 93–112
114. Wagner, R. P., Selander, R. K. 1974. Isoenzymes in insects and their significance. *Ann. Rev. Entomol.* 19:117–38
115. Walter, H., Lieth, H. 1964, 1967. *Klimadiagramm-Weltatlas,* Vols. 1, 2. Jena: Fischer
116. Waterhouse, D. F., Wilson, F. 1968. Biological control of pests and weeds. *Sci. J.* 4:31–37
117. Watt, K. E. F. 1965. Community stability and the strategy of biological control. *Can. Entomol.* 97:887–95
118. Whitten, M. J. 1970. Genetics of pests in their management. See Ref. 109, pp. 119–34
119. Wilson, F. 1960. A review of the biological control of insects and weeds in Australia and Australian New Guinea. *Tech. Commun. Commonw. Inst. Biol. Control,* No. 1. 102 pp.
120. Wilson, F. 1965. Biological control and the genetics of colonizing species. In *Genetics of Colonizing Species,* ed. H. G. Baker, G. L. Stebbins, pp. 287–306. New York and London: Academic. 588 pp.

121. Wong, H. R. 1974. The identification and origin of the strains of the larch sawfly, *Pristiphora erichsonii,* in North America. *Can. Entomol.* 106:1121–31
122. Zwölfer, H. 1961. A comparative analysis of the parasite complexes of the European fir budworm, *Choristoneura murinana* (Hb.) and the North American spruce budworm, *C. fumiferana* (Clem.). *Tech. Bull. Commonw. Inst. Biol. Control* 1:1–162
123. Zwölfer, H. 1963. The structure of the parasite complexes of some lepidoptera. *Z. Angew. Entomol.* 51:346–57
124. Zwölfer, H. 1967. Insect introduction and biological control. *Proc. Pap. IUCN Tech. Meet., 10th: IUCN Publ. New Ser.* 9:141–50
125. Zwölfer, H. 1971. The structure and effect of parasite complexes attacking phytophagous host insects. In *Dynamics of Populations,* ed. P. J. den Boer, G. R. Gradwell, pp. 405–18. Wageningen: Ctr. Agric. Publ. Docum.

Ann. Rev. Entomol. 1977. 22:23–32

CONTEMPORARY VIEWS ON THE INTERRELATIONSHIPS BETWEEN FLEAS AND THE PATHOGENS OF HUMAN AND ANIMAL DISEASES

❖6118

V. A. Bibikova
Martsinovsky Institute of Medical Parasitology and Tropical Medicine,
Ministry of Public Health, Moscow, USSR 119830

The fleas, like any other group of parasitic, bloodsucking arthropods, have been exposed to contact with various pathogenic agents throughout the protracted period of their existence. As a result of this association, the present interrelationships between the fleas and the pathogens of zoonoses and anthroponoses have evolved.

Various indices are used to appraise the relationship between a vector and a causative agent (pathogen): the capacity of the causative agent to reproduce in the vector, the duration of their joint sojourn, the pathogenic effect, etc. However, these relationships most vividly manifest themselves in the mechanism of transmission that has evolved historically for the given disease.

It is in the formation and realization of the mechanism of transmission that the maximal available adaptive possibilities between the pathogen and the vector are most vividly expressed, since the parasitic nature of the pathogen presupposes, as a condition of the existence of the species, the obligatory and multiple transmission from one host to another.

Such a methodological approach to the assessment of the relationships between the vector and the causative agent of the transmissive disease stems from the recognition of the ecological principle of existence of natural foci (40) and the theory of the mechanism of transmission (28). For the time being the extent of our knowledge of the peculiarities of transmission permits us to consider relationships of only six species of pathogens with the order Siphonaptera (Table 1).

Table 1 Peculiarities of transmission of certain pathogens by fleas

Disease and pathogen	Localization of pathogen in the flea	Reproduction of pathogen in flea	Duration of existance in vector	Main method of transmission	Pathogenic effect on the flea
Myxomatosis, *Fibromavirus myxomatosis*	digestive tract	–	up to 100 days	mechanical inoculation	+
Tularemia, *Francisella tularensis*	digestive tract	–	several days	mechanical inoculation	–
Murine typhus, *Rickettsia mooseri*	digestive tract	+	throughout lifetime	specific contamination	–
Murine trypanosomiasis, *Trypanosoma lewisi*	digestive tract	+	throughout lifetime	specific contamination	–
Salmonellosis, *Salmonella enteritidis*	digestive tract	+	up to 40 days	specific inoculation	+
Plague, *Yersinia (Pasteurella) pestis*	digestive tract	+	from several months to more than a year	specific inoculation	+

MYXOMATOSIS

The *Fibromavirus myxomatosis* virus affects hares and rabbits in Australia, Europe, and South America and is transmitted by the fleas of these animals, *Spilopsyllus cuniculi,* or mosquitoes. The virus does not reproduce in fleas and is gradually excreted from the organisms. The long survival, up to 100 days, of the pathogen in the intestine and the mouth organs of the flea is noted only in cold temperatures (4, 22, 43). The virus is passed on through the contaminated mouth organs of the flea. Rather interesting conclusions about the reasons for inactive transmission of myxomatosis virus by the flea are given in the above-cited investigations. A remarkable peculiarity of the rabbit flea is its almost immobile position on the ears of the rabbit, where, because of peculiarity of the blood circulation, they can rarely acquire the pathogen. When infected fleas have a blood meal on the ears of rabbits, the incubation period of the disease lasts 20 days instead of 2–3 days. Moreover, the ovaries of the fleas that had a blood meal from the sick rabbits develop abnormally. The evolution of the rabbit flea, according to Rothschild & Ford (43), was along the line of strictly local parasitizing and the reduction of contacts with the virus, since the mass destruction of the hosts is not favorable to the parasite.

TULAREMIA

This disease is widespread in the Old and New Worlds. Fleas are capable of ingesting a large quantity of *Francisella tularensis* microbes (up to 100,000); however, these do not reproduce and soon perish because they do not leave the digestive tract. There are no microbes in infected fleas after 2 months at a temperature of 4–7°C, and at 18–25°C they cannot be detected after 20 days (23). Although other investigators note longer periods of time (25, 45–47), they all agree about the absense of the reproduction of these microbes in the flea; thus the transmission of the tularemia microbe is purely mechanical. The efficiency of the transmission of tularemia by fleas is very low. For instance, out of 116 experiments (23), only six were positive.

According to some authors (23, 39) transmission is obtained in the first 5 days. Mechanical contamination through the excrement of the infected fleas is of some importance. The significance of fleas in the epizootiology and epidemiology of tularemia is limited; the role of the main vector of tularemia belongs to other groups of Arthropoda.

ENDEMIC OR MURINE EXANTHEMATOUS TYPHUS

This disease is spread throughout the Old and New Worlds, especially in port cities in the subtropical and tropical zones. The causative agent of the disease is *Rickettsia mooseri*. Rats and mice contract the disease; sometimes humans do also. The fleas of rats, mice, and human dwellings are easily infected with rickettsiae, but the persistence of the causative agent in the flea is limited to the intestinal tract only. The flea is infected for life, and the pathogen does not harm the flea. The rickettsiae reproducing in the body of the flea, especially if it feeds regularly, are excreted with the feces over a long period of time (15, 16, 21, 37). Investigators note the high concentration and pathogenicity of rickettsiae in the feces of the vector. The efficiency of infection of the warm-blooded host is the highest when the rickettsiae are implanted on the mucous membranes.

In murine typhus there is no active mechanism of transmission of the pathogen directly to the blood of the animal, but this is compensated for by the stability of the rickettsiae despite unfavorable effects of the environment and their very high virulence. The infectious dose for a man is one fifth of a single excrement of the flea. The method of transmission of the pathogen is the specific contamination, deprived of deep mutual adaptation of the pathogen and the vector.

MURINE TRYPANOSOMIASIS

Trypanosoma lewisi is the causative agent of trypanosomiasis, which occurs in rats in almost all the corners of the world. The vectors of this disease are the fleas of rats, as well as fleas of human dwellings and of cats, dogs, and mice. The circulation of the causative agent is only within the intestine of the flea. The most detailed study of the life cycle of trypanosomes in fleas was made by Minchin & Thompson (36). Within the first 6 hr after entering the stomach of the flea with the blood, trypanosomes apparently undergo some sort of physiological change. The confirmation of this is that, if the trypanosomes are administered to the rats during this period, they cannot induce the disease. After 6 hr they enter the epithelial cells of the stomach; each trypanosome acquires a pearlike form in the cell vacuole and reproduces through repeated divisions. The epithelial cells in which the trypanosomes accumulate finally break, making it possible for the parasites to get into the intestinal lumen. The intracellular phase lasts from 18 hr to 5 days and can be in any part of the stomach epithelium. The trypanosomes that emerge from the epithelial cells can again penetrate them, repeating the described process of reproduction. After this, they move to the posterior sections of the intestinal tract (into the anal and rectal intestine), where they change their structure and take on a crithidial form. The investigators call this the rectal phase; it is characterized by polymorphicity—the formation of herpetomonad, crithidial, and trypanosome forms. Finally all these forms are excreted from the intestine to the environment. The majority of investigators consider that only small trypanosome forms can infect animals.

Unlike the pathogens discussed before, trypanosomes not only reproduce in fleas, but also go through a certain cycle of morphophysiological changes. But during this too, the fleas do not transmit the causative agent through biting and only passively contaminate the environment or the skin of the animal. Susceptible animals contract trypanosomiasis while licking the flea excrements. The method of trypanosome transmission is typical specific contamination.

SALMONELLOSIS

Out of a very extensive group of salmonelloses we deal here only with the disease in rats cause by *Salmonella enteritidis*. The fate of this pathogen in the digestive tract of the murine fleas has been studied in sufficient detail (26). The salmonellae reproduce in the stomach of the flea and during defecation contaminate the environment to a great extent. It has been established that the exit portals for the pathogen are both the anus and the mouth. The pathogen does not exit immediately, but only after reproduction and a significant accumulation in the intestine of the flea. The salmonellae are accumulated in the proventriculus of the flea, which is situated before the stomach. This round, muscular organ is lined with chitin bristles, which in the normal state do not obstruct the passage of the blood to the stomach; during contraction they act as a barrier to the reverse flow of blood.

The massive reproduction of the salmonellae apparently takes place in the anterior of the stomach; from here the microorganisms penetrate into the proventriculus and obstruct its normal function. They fill the gaps among the bristles, mechanically hindering the process of pulsation of the proventriculus, which leads to the free outflow of blood to the anterior sections of the intestine. Thus fleas transmit salmonellae not only by the method of specific contamination but also by specific inoculation, with the direct, active introduction of the pathogen into the blood of the animal, as is the case during the bloodsucking of a sand fly infected by *Leishmania*. However, besides the direct inoculation of the causative agent, the fleas themseleves suffer from this method of transmission. The same investigators note that disruption of the function of the proventriculus suppresses the vitality of the flea; the insects noticeably lose their tonus and elasticity of the intestinal canal, the esophagus expands abnormally and is displaced, and the fleas perish much earlier than they normally do.

PLAGUE

The plague is widespread in the desert and steppe zones of the world and its only specific vectors are among the Siphonaptera (29, 41, 42, 46, 47). The essence of the main, most effective method of transmission of the causative agent of plague by fleas lies in specific inoculation from an insect with a blocked intestinal track. This phenomenon of blocking the intestinal tract is so important and interesting that it needs special consideration and detailed analysis. The block was first described by Bacot & Martin (5), and then by other investigators (10, 17, 25, 29, 34, 35, 38, 41, 46). The reproducing bacilli, *Yersinia pestis,* occupy the anterior position in the stomach and form a viscous accumulation, which wholly closes the lumen of the flea proventriculus. Usually the fleas, which take numerous blood meals, fail in their next attempt to feed because the blood that is ingested into the esophagus comes up against the bacillary block in the proventriculus and cannot go further. After many futile muscular efforts, the blocked flea, during the blood meal, regurgitates the content of the frontal sections of the esophagus together with the plague bacilli. Such fleas remain hungry and constantly attempt to bite and feed, which raises the possibility of numerous transmissions of the pathogen. The poor state of the fleas is apparently aggravated by poisoning with the products of the vital activity of the microbes—endotoxins, exotoxins, and metabolic products. Soon (an average of 5–10 days after the blocking) the exhausted and starved insect perishes. During this period, the blocked flea can infect dozens of animals, which in turn infect new fleas, and thus the usual cycle of maintenance of the pathogen in the natural plague focus continues to exist.

In addition to other methods of transmission of the plague microbe by fleas, there is specific contamination. This occurs when the body of the warm-blooded host or the environment is contaminated by the flea excrement and also when the flea is crushed. But because of the biological features of the flea and the pathogen and the level of reactivity of rodents, this method of infection is obviously of secondary importance.

Lately there has been a lively discussion of the method of mechanical inoculation of the causative agent of plague through flea mouth organs that are contaminated by bacilli (during the first days after the insect is infected). The American investigators (17, 31) consider this method of transmission to be especially important at the time of acute epizootics. Until recently, it was unclear how an assessment could be made of the quantitative and qualitative preservation of the microbes outside and inside the mouth organs. However, it is obvious that, without the proof of contamination and of at least relatively short survival of the bacilli on the proboscis of the flea, we cannot uphold the possibility of the mechanical transmission of the causative agent, since one intake during the first day after an infective feeding does not exclude other ways of transmissions. Experiments with direct contamination of the proboscises and investigations of the vitality of the bacilli contained in periodic washings from them conducted with a special apparatus for dosed feeding (11) showed that under moderate temperatures the bacilli remain viable for only 3 hr. Since the overwhelming majority of flea vectors do not feed again within this period of time, they consequently cannot carry out mechanical inoculation of the bacilli. The role of mechanical inoculation was obviously greatly overemphasized.

The significant element in plague transmission is still the specific inoculation of the pathogen through the bite of a blocked flea. This method of transmission is ensured by a whole complex of adaptations of the pathogen to the organism of the flea: the change of the optimum temperature for the plague bacilli in relation to the optimum temperature of the corresponding species of vectors, the possibility of selection of bacilli varying in virulence during the reproduction in the flea, and the occurrence in the plague bacilli of biological properties that ensure the effective development of the block in the digestive tract of the flea (3, 9, 13, 17, 24, 29, 31, 34, 35, 38, 42, 44).

The frequency of blocking depends not only on the species of fleas and the environmental conditions but also on the properties of the plague bacillus strain. Formerly, however, the influence of biological properties of the pathogen strain on the efficiency of blocking was not taken into consideration. The assessment of this influence was greatly facilitated after discovering the determinants of virulence in *Y. pestis* strains and developing the methods used in their discovery (19, 30).

One determinant of virulence is the ability of bacilli to grow in the form of pigmented colonies in a culture medium with gemine (determinant P). Jackson & Burrows (30) pointed out the dense consistency of pigmented colonies, which did not give homogeneous suspension in phosphate buffer (the majority of bacilli remained glued together in the form of very dense flakes). At the same time, the colonies of nonpigmented variants were distinguished by soft consistency, and in phosphate buffer they formed homogeneous suspension. This peculiarity of P^+ and P^- colonies indicates a tendency of P^+ variants to give compact growth in the presence of gemine due to forces of adhesion among bacilli. In the intestinal tract of the flea, the protein component of hemoglobin is disrupted under the influence of proteolitic ferments, releasing the blood pigment with subsequent formation of gemine or closely related ferrous compounds.

It seems that the plague bacillus strain, capable of compact growth in the form of pigmented colonies in the medium with gemine—determinant of virulence of P^+, according to Burrows (19)—can block the proventriculus of the flea most effectively. This hypothesis proved to be correct. Out of 214 fleas (*Xenopsylla cheopis*) infected by the P^+ strain, 33 became blocked; at the same time, none of 186 fleas infected by the P^- strain formed the block (9, 33). Moreover, the results of other investigators (32) who showed a rather effective blocking of *X. cheopis,* irrelevant to the property of pigment formation, confirmed our hypothesis.

We see the essence of correlation of signs of pigment formation with the capacity to form a block of proventriculus in the following way. Bacilli of P^+ variants, capable of reacting with gemine, form dense lumps, becoming fixed among the bristles of the proventriculus. The formation of pigment itself from this point of view is only a morphological expression of the capacity of P^+ variants to react with gemine, which leads to the changes of physicochemical properties of the cell wall and ensures firm adhesion of the bacilli required for their stable consolidation in the proventriculus of the flea. The formation of the above-mentioned property of the pathogen may be considered an adaptive sign, emerging in the course of mutual evolution of the components of the natural focus; as a result, the properties, encouraging the realization of the most effective mechanism of transmission and preservation of the species, were fixed in the plague bacillus by natural selection (8, 9, 14).

It is worth mentioning that the term vector capacity is obviously insufficient for expressing the function of fleas in the plague foci, where their role goes beyond the mere transmission of the pathogen, and extends to their long retention of the pathogenic organisms and their interaction with the biological properties of the plague bacilli; in fact, their relationship is that embodied in the concept of a true host.

Finally, I should like to direct attention to the existence of phenotypic changes in the *Y. pestis* population in view of its alternating parasitism in the rodent and the flea. The conditions of existence of plague bacilli in vertebrates and arthropods are quite different. That is why at each of the two ecological stages different adaptive mechanisms function in the bacilli and facilitate the sequential change of the host. In the organisms of warm-blooded animals, those properties of the bacillus that determine its high virulence and resistence to phagocytosis are of the foremost importance (20). On the other hand, in the organism of the flea the greatest significance is in the ability of the bacilli to cause blocking of fleas, determined by the existence in R form and by the determinant P. The obvious differences in properties of the microbe population from the point of view of their significance as adaptations to different living conditions found its expression in the evolutionary-fixed phenomenon of phenotypic variability of plague pathogens (1, 2, 6, 9, 12–14, 33). The major factor out of a number studied, which induces the phenotypic variability of the bacilli when hosts are changed, is the temperature. In the vertebrate organisms, plague bacilli live as a rule at 37°C, and in arthropods almost always at a lower temperature. The ability of the bacilli to synthesize FI and VW antigens phenotypically is realized, as a rule, at 37°C (18, 19).

Genetically conditioned regulation of the process of antigen FI and VW biosynthesis is, as seems to be the case for plague bacilli, a means of realization in specific circumstances of the most economic of the complex of metabolic reactions.

The presence of phenotypic variability of populations of plague bacilli in connection with natural alternation of generations in rodents and fleas provides an explanation of the process by which populations of the plague pathogen developed as a parasitic species (12, 13).

CONCLUSION

In studying the mechanism of transmission of different pathogens in the order Siphonaptera, our attention is attracted by certain general laws. The pathogens in the flea organism live and reproduce only within the limits of the digestive tract. Not a single one penetrates into the body cavity, the salivary glands, or the gonads of the insect. That is why there is no transovarial transmission of the pathogens in fleas, just as there is no transphasal transmission during metamorphosis.

The most widespread method of transmission of infective material among fleas is mechanical or specific contamination, i.e. the fleas, according to Beklemishev's terminology (7), serve as mere disseminators of the pathogens. Such transmission is not perfect and requires other compensatory factors to ensure the circulation of a pathogen: the development of persistence of the pathogen in the environment, an increase in its pathogenicity, etc. Such a form of transmission is basic in murine typhus and trypanosomiasis and is secondary in plague and tularemia.

Another form of transmission in fleas is mechanical inoculation, which takes place in tularemia and myxomatosis. And although the inoculation in itself contains elements of active introduction of the pathogen into the organism of the susceptible animal, it shows the simple and primitive relations existing between the vector and the pathogen. In the majority of cases, the mechanical transmission of the pathogen is facultative and moreover not the only mode of infection.

Another method is specific inoculation of the pathogen, which occurs in plague and apparently in salmonellosis. This mechanism of transmission shows a greater degree of adaptation of the pathogen to the vector; it promotes the greater epizootiological efficiency of the vector and a greater manifestation of the parasitic features of the pathogen in regards to the host. It is important to note that, within the limits of the order Siphonaptera, even this most specialized specific mechanism of transmission is far from being perfect, i.e. the vector does not have a sufficient level of adaptation that would ensure a normal life span.

All the forms of the described methods of transmission of the pathogens and the relations between them and the fleas give evidence that the relationships are not as highly evolved as those that came into existence between pathogens (spirochetes, viruses, protozoa) and ticks and mosquitoes (i.e. the most ancient groups of vectors).

The primitiveness and imperfection of the relations between the pathogens in question and the fleas makes it possible to postulate the comparatively recent establishment of the fleas as vectors. What little we know from the literature about the geological age of the fleas indicates their fairly recent origin. This group appeared later than ticks (Gamasoidea, Argasidae, Ixodidae) and later than mosquitoes. The appearance of the order Siphonaptera goes back to the end of the Mesozoic era or the beginning of the Tertiary period (7); and this means that their adaptation as the vectors of pathogens of certain diseases formed even later.

Literature Cited

1. Akiev, A. K. 1969. On physiological variability of the plague microbe. *Probl. Osobo Opasnykh Infekt.*, pp. 22–25
2. Akiev, A. K. 1969. Contemporary situation and recent problems of studying of mechanism of preservation of the plague pathogen in non-epizootic years. *Itogi Rab. Protivochumn. Uchrezd.* 1964–1968 *Perspectivi Dalneishei Deyatelnosti* (theses), pp. 88–92
3. Alutin, I. M., Sorokina, L. Y. 1970. Changes in quantity of the plague microbe in the organism of rat fleas depending on the cycle of feeding. *Sb. Perenoschiki Osobo Opasnykh Infekt. Borba,* pp. 15–23
4. Andrewes, C. H. 1954. Myxomatosis in Britain. *Nature* 174 (4429):529–30
5. Bacot, A., Martin, C. 1914. Observations on the mechanism of transmission of plague by fleas. *J. Hyg.* (*Plague Suppl.*) 13(3):423–39
6. Bayer, A. P., Akiev, A. K., Suvorova, A. E. 1974. Morphology of the virulent plague microbe in the organism of the flea *Xenopsylla cheopis. Probl. Osobo Opasnich Infekt.* 6(40):60–63
7. Beklemishev, B. N. 1943. On the relationship between systematic position of the pathogen and the vectors of transmissive diseases of the vertebrates and man. *Med. Parazitol Parasit. Bolezni* 17(5):385–99

8. Bibikova, V. A. 1965. Conditions for the existence of the plague microbe in fleas. *Cesko. Parazitol.* 12:41–46
9. Bibikova, V. A. 1974. Transmission of plague by fleas. *Izd. Med. Moscow,* p. 187
10. Bibikova, V. A., Alexeev, A. N. 1969. Infectivity and block-formation in dependence on the amount of plague microbes taken up by the flea. *Parasitologia* 3 (3):196–203
11. Bibikova, V. A., Alexeev, A. N., Khrustselevskaya, N. M. 1967. On the absence of mechanical inoculation of the plague pathogen by fleas. *Mater. Nauch. Konf. Protivochumn. uchrezhd. Srednei Asii Kasakhstana, Posv. 50 Velikoy Okt. Soc. Rev., Alma-Ata,* pp. 268–70
12. Bibikova, V. A., Klassovsky, L. N. 1968. On phenotypic variability of the plague pathogen in connection with the change of hosts. *Parasitologia* 2/3: 209–14
13. Bibikova, V. A., Klassovsky, L. N. 1972. The development of plague microbe population in fleas. *Parasitologia* 6(3):229–36
14. Bibikova, V. A., Khrustselevskaya, N. M. 1967. The relationship between the pathogen and the vector in the model of the plague. *Zh. Microbiol. Epidemiol. Immunol.* 2:111–16
15. Blanc, G., Baltazard, M. 1937. Nontransmission a l'homme du typhus murium par piqures des puces infectees (*Xenopsylla cheopis, Pulex irritans*). *Bull. Acad. Med.* 117:434–46
16. Blanc, G., Baltazard, M. 1940. Comportument du virus du typhus epidemique ches les puces *Xenopsylla cheopis* et *Pulex irritans. Bull. Acad. Med.* 123:7–8, 44
17. Burroughs, A. 1947. Sylvatic plague studies. The vector efficiency of nine species of fleas compared with *Xenopsylla cheopis. J. Hyg.* 45:371–96
18. Burrows, T. W. 1962. Genetics of virulence in bacteria. *Br. Med. Bull.* 18(1):69–73
19. Burrows, T. W. 1963. Virulence of *Pasteurella pestis* and immunity to plague. *Ergeb. Mikrobiol. Immunitas* 37:59–106
20. Cavanaugh, D. C., Randall, R. 1959. The role of multiplications of *Pasteurella pestis* in mononuclear phagocytes in the pathogenesis of flea-borne plague. *J. Immunol.* 83(4):348–63
21. Ceder, E., Dyer, R., Rumreich, A., Badger, L. 1931. Typhus fever in feces of infected fleas (*Xenopsylla cheopis*) and duration of infectivity of fleas. *Public Health Rep.* 46:3103–6
22. Chapple, P., Lewis, N. 1965. Myxomatosis and rabbit flea. *Nature* 207(4995):388–89
23. Dudolkina, L. A. 1954. Experimental data on the ability of the rodent fleas of human dwellings to transmit and store tularemia infection. *Izv. Irkutsk. Nauchno-Issled. Protivochumn. Inst. Sib. Dal'nogo Vostoka* 12:53–57
24. Eskey, C. R. 1938. Recent developments in the our knowledge of plague transmission. *Public Health Rep.* 53(1): 49–57
25. Eskey, C. R., Haas, V. H. 1940. Plague in the Western part of the United States. *Public Health Bull.* 254:1–83
26. Eskey, C. R., Prince, F., Fuller, F. B. 1949. Transmission of *Salmonella enteritidis* by the rat fleas *Xenopsylla cheopis* and *Nosopsyllus fasciatus. Public Health Rep.* 64(30):933–41
27. Golov, D. A., Tiflov, V. E. 1934. On the problem of the role of water rat (*Arvicola amphibius*) fleas *Ceratophyllus walkeri* in the epidemiology of tularemia. *Tesisy Dokladov Vseross. Konf. Microbiol. Epidemiol.,* pp. 33–38
28. Gromashevsky, L. V., ed. 1958. *Mechanisms of Transmission of the Infection.* Kiev: Medizin Izdat USSR. 332 pp.
29. Ioff, I. G. 1941. *Problems of the Ecology of Fleas in Connection With Their Epidemiological Significance.* Piatigorsk: Ordjonikidz Kraev. Izdat. 113 pp.
30. Jackson, S., Burrows, T. W. 1956. The pigmentation of *Pasteurella pestis* on a defined medium containing haemin. *Br. J. Exp. Pathol.* 37:570–76
31. Kartman, L., Prince, F. M., Quan, S., Stark, H. E. 1958. New knowledge on the ecology of sylvatic plague. *Ann. NJ Acad. Sci.* 70(3):668–711
32. Kartman, L., Quan, S. 1964. Notes on the fate of avirulent *Pasteurella pestis* in fleas. *Trans. R. Soc. Trop. Med. Hyg.* 58:363–65
33. Klassovsky, L. N., Bibikova, V. A. 1968. Problems of ecology of plague and pseudotuberculosis microbes. Communication 4. On the property of the plague causative agent, ensuring the possibility of transmissive transmission. *Zh. Microbiol. Epidemiol. Immunol.* 4:118–122
34. Kondrashkina, K. I. 1969. Fleas: do they suffer from plague? *Probl. Osobo Opasnykh. Infect.* 5:212–22

35. Kondrashkina, K. I., Kuraev, I. I., Zakharova, T. A. 1968. Some problems of mutual adaptation of the plague microbe with the organism of fleas. *Parasitologia* 2(6):543–48
36. Minchin, E., Thompson, J. 1915. The rat-trypanosoma, *Trypanosoma lewisi,* in its relation to the rat-flea, *Ceratophyllus fasciatus. Quart. J. Microsc. Sci.* 60:463–92
37. Mooser, H., Castaneda, R. 1932. The multiplication of the virus of Mexican typhus fever in fleas. *J. Exp. Med.* 55:307–23
38. Novokreshchenova, N. S., Kochetov, A. H., Starozhitskaya, G. S. 1969. The dependence of the formation of the plague block in fleas on the intensity of their infection with the plague pathogen and feeding conditions. *Parasitologia* 3:203–13
39. Olsuf'ev, N. G., Tolstukhina, E. N. 1941. The role of the flea *Ctenophthalmus assimilis* Tasch. in the transmission and storage of tularemia infection. *Archiv. Biol. Nauk* 63:1–2, 81–88
40. Pavlovsky, E. N. 1948. Guide to human parasitology. *Izd. AN SSSR* 2:581–1022
41. Pollitzer, R. 1954. *Plague.* Geneva: WHO. 695 pp.
42. Reports on plague investigations in India. The mechanism by means of which the flea cleans itself of plague bacilli. 1908. *J. Hyg.* 8:1–17
43. Rothschild, M., Ford, B. 1965. Observations on gravid rabbit fleas *Spilopsyllus cuniculi* (Dale) parasitising the hare (*Lepus europaeus* Pallas), together with further speculations concerning the course of myxomatosis at Ashton, Northants. *Proc. R. Entomol. Soc. London Ser. A* 40(7–9):109–17
44. Steinhaus, E. A., ed. 1963. *Insect Pathology: An Advanced Treatise,* Vols. 1 and 2. New York: Academic. 661 pp. and 689 pp.
45. Tiflov, V. E. 1959. The role of fleas in epizootology of tularemia. *Tr. Nauchno-Issled. Protivochumn. Inst. Kavk. Zakavk.* 2:363–90
46. Tiflov, V. E. 1960. The significance of fleas in distribution of diseases. *Tr. Nauchno-Issled. Protivochumn. Inst. Kavk. Zakavk.* 4:18–35
47. Tiflov, V. E. 1964. The fate the bacterial cultures in the organism of the flea. *Ectoparasites* 4:181–99

Ann. Rev. Entomol. 1977. 22:33–51

BIOLOGY AND BIONOMICS OF BLOODSUCKING CERATOPOGONIDS

❖6119

D. S. Kettle
Department of Entomology, University of Queensland,
St. Lucia, Queensland 4067, Australia

In recent years there has been increasing attention paid to three genera [*Culicoides, Leptoconops, Forcipomyia* (*Lasiohelea*)] of ceratopogonids, whose females feed on warm-blooded vertebrates. This habit makes them pests of man and livestock. They are a nuisance to humans living on the coast and in poorly drained inland areas, but they are most important as vectors of virus diseases that are transmitted to domestic animals. *Culicoides* are proven vectors of bluetongue to sheep and cattle (93) and have been incriminated in the transmission of horsesickness to equines and ephemeral fever to cattle (26). The involvement of bloodsucking ceratopogonids in the transmission of human and animal diseases was reviewed in 1965 (60) and their bionomics and control in 1962 (59) and 1969 (61). The present review refers only to work covered in the latter two articles where necessary for the appreciation of more recent research. Current interest in ceratopogonids has been fostered by a biannual news letter, *Ceratopogonid Information Exchange,* edited by J. Boorman, Animal Virus Research Institute, Pirbright, Woking, Surrey, U.K.

TAXONOMIC PREREQUISITES

Ecological studies are dependent on a sound taxonomy, and in the case of the Ceratopogonidae this is developing satisfactorily. Adequate keys are available for the recognition of subfamilies, tribes, and genera of Ceratopogonidae (123), and in most zoogeographical areas there have been attempts to produce comprehensive faunistic works. Approximately one quarter (924 of 3870) of the ceratopogonid species described are in the genus *Culicoides* (1123), and efficient subgenera are required to handle this wealth of material. Present subgenera do not meet this need, and consequently nearly two thirds (72 of 113) of the species of *Culicoides* recorded in the USSR fall into one (*Oecacta*) of the 19 subgenera described (46).

In areas where the species list is largely complete, attention has been given to resolving troublesome taxa. Biological differences between populations of *Leptoconops torrens* led to the reestablishment of the synonym *L. carteri* (5, 121) through the use of quantitative taxonomic methods involving nonmetric, multidimensional scaling. The effect of this was to supplement the initial discriminating character—microscopic eye pubescence—with six morphometric characters. Comparable differences have been described for *L. becquaerti* (79, 80) where differences in wing length correlated with autogeny and anautogeny. Unlike *L. torrens* and *L. carteri,* which occupy different breeding sites, the two forms of *L. becquaerti* occupy the same site. It would be valuable to have a similar quantitative analysis carried out on the two forms of *L. becquaerti* and resolve their relationship.

The *Culicoides* (*Selfia*) group of species constitute a major portion of the *Culicoides* fauna of western North America. These species are readily identified in the male but are virtually inseparable in the female. Atchley (4) has applied quantitative methods to the separation of females of the three more common and widely distributed species (*C. denningi, C. hieroglyphicus, C. jamesi*). Forty-three characters (25 adult and 18 pupal) were used to discriminate between them. Separation was very good (97.7%) using all characters, but pupal characters played a major role. In their absence, the 25 adult characters achieved only 86.6% separation (4). Better separation was achieved (96.3%) using three adult and four pupal characters, but its value to ecological studies is weakened by including characters from different stages of development.

Although the three *Selfia* species show similar patterns of morphological variation in pupae, males, and females throughout their geographical range, the dependence of pattern on climate was quite different. In *C. denningi,* variation was highly dependent, with pupae and adults showing the same phenetic clustering. In *C. hieroglyphicus,* clustering was more apparent in adults than in pupae, but there was still considerable dependence on climate. Throughout the geographic range of *C. jamesi,* adults conformed to a more-or-less uniform phenotype, which was almost independent of climate, while pupae showed clustering by localities (2, 3).

The situation in *C. austropalpalis* is quite different. Two forms of larvae occurring in the same breeding sites gave rise to two forms of adults (65). These are size variants differing in larval head length and adult wing length. Univariate statistical analysis of a further 15 characters in the female and 14 in the male provided no clear-cut difference except for a tendency for the larger form to have more distinct patterning in the apical third of the wing. The characters used were those in common taxonomic practice and included nine ratios (six female and three male), which Atchley (4) deliberately omitted. It would be interesting to subject the *C. austropalpalis* data to an analysis similar to that used on the *Selfia* species. There is one very real difference, however: in *C. austropalpalis,* as in *L. becquaerti,* the two forms coexist in the same breeding site, whereas in *Selfia* the three species have minimal overlap in breeding sites (1).

The need to differentiate between sibling species has prompted the search for additional taxonomic characters, and use is now being made of female cibarial armature (14) and number and distribution of ascoides on the male antenna (24).

Many dipterous larvae possess polytene chromosomes which have proved invaluable in the separation of sibling species in related nematocera, e.g. Simuliidae (31) and Culicidae (23), but at present it is not even known if ceratopogonids have such chromosomes. This could be a rewarding field for research.

Detailed knowledge of the larval biology is essential for a full understanding of bionomics. Without this information it is impossible to construct life tables or to identify key factors limiting pest populations. Such studies are dependent on sound taxonomy, and although the characters separating the immature stages at subfamily level have been known for nearly 50 years, the taxonomy of larvae and pupae, with the exception of the Forcipomyinae, is very poorly understood. Individual descriptions are scattered throughout the literature, but there is a dearth of substantial papers. V. M. Glukhova (in preparation) describes larvae of 14 genera and 100 species of ceratopogonine midges. The addition of *Paradasyhelea* (66) means that larvae are known for only 15 out of 59 genera in the subfamily (123).[1]

With respect to descriptions of specific immature stages the situation is even worse. In *Culicoides,* the most widespread and abundant of the bloodsucking ceratopogonids, 88 larval descriptions are available for a known world fauna of 924 species, i.e. less than 10%. Most of these are Palaearctic (40 spp) (39, 40) or Nearctic (25 spp) (54) species, and a start has been made on the Ethiopian (8 spp) (101) and Australian (12 spp) (67) faunas, but elsewhere virtually no work has been done. Our knowledge of pupae is at a comparable level.

The development of simple techniques for the rearing of ceratopogonid larvae (73, 116) should encourage the taxonomic study of the immature stages. Larvae and pupae may be kept under continuous observation and exuviae recovered for detailed study. These methods have proved suitable for rearing several species from egg to adult. They offer the possibility of rearing the progeny of wild-caught, gravid females and obtaining information not only on the morphology of the larvae and pupae but also on variation in the adult progeny of individual females.

Since 1971 there has been increased interest in the genus *Leptoconops,* resulting in descriptions of three larvae and five pupae (21, 22, 43, 78). This almost doubles the number of associations available, which now total eight and ten, respectively. This represents about 10% of the 81 species described (123).

BIOLOGY AND BIONOMICS OF ADULTS

Reproduction

Our knowledge of bloodsucking ceratopogonids has reached the stage where emphasis in research has shifted from general biological faunistic surveys to in-depth studies on particular aspects of the bionomics of individual species. Successive components of the reproductive biology from blood feeding and mating to oviposition and hatching are examined.

[1]The summary states that there are "78 genera and subgenera" in the Ceratopogonidae (123). In fact, the paper classifies 63 genera and subdivides 8 of them into 57 subgenera.

BLOOD FEEDING Although autogeny is well established in *Culicoides* and *Leptoconops,* most species require a blood meal to mature the first batch of eggs (41), and all require one for the second and subsequent egg batches. The source of the blood meal is of paramount importance and determines which species are pests or disease vectors. Precipitin tests have been used to determine host preferences (102), but blood-fed specimens form only a small proportion of any collection (109, 110). *C. variipennis,* a very large species, takes a blood meal of 0.56 mg (119), which is about 20% that of a mosquito. This complicates identification, but Murray (98) has developed a technique permitting the testing of one meal against 30–40 antisera. Few, if any, species are host-specific. Hair & Turner (48) collected 15 *Culicoides* spp. from humans and four other mammals, and from five avian species. Only one of the 11 species taken in any number was attracted to only a single host, namely *C. hollensis*. Although *Culicoides* species are opportunist feeders there are some hosts that elicit no response. Thus, while thousands of *C. brevitarsis* were collected off a single cow, the collector and other nearby humans were rarely bitten (19, 115). Choice of host was influenced by the height at which a species searched; for example *C. arboricola,* fed on caged rabbits and turkeys elevated to 7 and 15 m, but not when they were on the ground, whereas *C. sanguisuga* fed on both hosts at all three levels (118). Hunger raised the intensity of illumination at which feeding occurred (52) and might be expected also to increase host range.

Species differ in the site on the host at which they feed. *C. subfascipennis* and *C. obsoletus* fed almost exclusively on the belly of cattle, *C. punctatus* fed both there (67%) and on the back (31%), whereas *C. chiopterus* was almost completely restricted to the legs (103). On standing cattle, *C. brevitarsis* landed most abundantly on the ridgeline near the tail. Numbers decline steadily towards the head and more abruptly down the flanks with none occurring on the head or belly (19). This observation plus the fact that largest numbers occurred on the flanks of lying animals implies that *C. brevitarsis* approached its host from above, whereas *C. obsoletus* and *C. chiopterus* came from below. Differential site selction must be taken into account if collections are to represent accurately the relative abundance of attacking species.

Host-seeking females also discriminated between individuals of the same species. More *C. sanguisuga* were collected from large than from small goats, but this difference became nonsignificant when allowance was made for surface area (53). *C. sanguisuga* did not distinguish between hosts of various colors, nor did *C. furens* or *C. barbosai* (71, 72), but this is to be expected in members of a genus that is predominantly nocturnal or crepuscular.

Of greater interest was the observation that *C. furens, C. barbosai,* and *L. becquaerti* were differentially attracted to four human collectors (C, K, L, and S) simultaneously. All three species were weakly attracted to S but differed in their response to the others. *C. barbosai* landed equally on all three ($K \simeq L \simeq C$), *L. becquaerti* preferred K and L to C ($K \simeq L > C$), and *C. furens* discriminated between all three ($C > K > L$) (69, 71, 72). Additional evidence was produced to show that these responses could be modified in certain circumstances (62, 71). These observations showed that each species responded differently to the same individual host and therefore that the mechanism of discriminating between hosts must be different, i.e.

that the various cues provided by the host produced quantitatively different responses according to the species.

SWARMING AND MATING Mating in flight is widespread throughout the Diptera (29) and is often accompanied by swarming of the males. Glukhova & Dubrovskaya (42) observed swarming in 14 species and recognized two major groups. In 11 species, swarming was unconnected to the presence of a host, and this was considered to be the primitive condition. In three other species, swarming occurred adjacent to hosts on which females were feeding. These species also swarmed in the absence of hosts. A third group of species, including *C. melleus* (84), mate without swarming. In *C. nubeculosus* and *C. puncticollis,* swarming is facultative, a fact which may reduce its value as a discriminating character between closely related species such as *L. carteri* and *L. torrens* (121).

The behavior of *C. brevitarsis* does not fit easily into the above classification. It formed swarms throughout pasture fields in the absence of hosts, but its swarming pattern was modified in the presence of its host (19). Near cattle swarms were larger, more numerous (about six times), closer together, nearer the ground, and more or less spherical. This description puts *C. brevitarsis* in group 2, but swarming and blood feeding were not simultaneous. Swarming occurred in the hour before sunset when little blood feeding was taking place. Most feeding occurred immediately after sunset (19), which could indicate a very closely coordinated system, with swarming and presumably mating occurring just before sunset and blood feeding shortly afterwards. In support of this suggestion, nearly all (97%) nulliparous *C. brevitarsis* on cattle were inseminated (19).

Downes (29) described a typical swarm as having individuals facing into the wind and oscillating up- and downwind across a marker. With increasing wind speeds (>3.6 m sec^{-1}) swarms flew lower and finally settled. Only in very low wind speeds (<0.2 m sec^{-1}) did flight follow the outlines of the marker. Swarms of *C. brevitarsis* did not have this pattern. They were not aligned upwind but over the middle of the sharply defined shadow/sunlight boundary of the marker. Most swarms were formed at wind speeds below 1.2 m sec^{-1} and swarming ceased at 1.9 m sec^{-1}. In the swarms individuals were not orientated upwind but rather towards the sun. When the marker was modified, swarms of *C. brevitarsis* immediately dispersed but reformed almost immediately after it was restored. This requires no comment, but the formation of a swarm within 10 sec of its apparent removal by netting was unexpected (19). It implies that swarms represent dynamic phenomena, with individuals entering and leaving them like humans in rush hour public transit. Downes (29) has emphasized the need to relate mating assemblies with breeding sites. *C. brevitarsis* breeds in bovine dung in pastures where pats are likely to be numerous but scattered over a large area. Dispersal from swarm to swarm would ensure outbreeding and involve little risk of becoming lost, while speed would be essential to offset declining male potency, which in *C. melleus* begins when the male is 8 hr old (50).

Linley and his co-workers Hinds and Adams have provided a detailed picture of the mating behavior of *C. melleus,* which was sexually active on emergence and did not swarm but instead mated on contact. The overwhelming impression was of speed. Male potency reached a peak at 4–8 hr, and by 96 hr the number of spermatozoa ejaculated at first mating was inadequate (<600) to achieve maximum fertility in the female (50, 88). Newly emerged males achieved the fastest mating times after contacting the female (8 min 23 sec) in spite of taking the longest time to achieve coupling (31 sec) (84). Young males (1–2 hr) completed a second mating in 30 min, but older males (24–36 hr) required an hour, there being a delay in forming the spermatophore. Mated females resisted a second mating by running away, kicking, or tipping the abdomen; such activities were incipient in young virgin females and became more obvious with age (85). Female resistance prolonged mating and reduced the number of sperm transferred (89). As time between successive matings of the female increased, the number of sperm transferred in the second mating rose until after 96 hr it was virtually the same as in the first (90). Complete mixing of sperm occurred in the female, and the contribution of each male to the progeny of a twice-mated female was in direct proportion to the sperm transferred (82).

In the male the first mating was the most productive, achieving maximum sperm transfer and fertility. Thereafter male potency declined markedly. Using the data of Linley & Hinds (88), the regression of sperm number (Y) on the second to the seventh successive matings (X) was $\hat{Y} = 1063 - 142X$, i.e. 142 fewer sperm were transferred with each successive mating. The agreement with the data is excellent ($r^2 = 0.994$). On extrapolation 921 sperm should have been transferred in the first mating, but 1156 actually were, 25% more than expected.

AUTOGENY Autogeny in bloodsucking Ceratopogonids is now well documented. Glukhova & Dubrovskaya (41) have recorded autogeny in about a quarter of the species examined in the USSR (5 of 21). However, there was uncertainty about *C. circumscriptus,* which in the Donets region and Kirgizstan was anautogenous, but Glukhova (38) had previously found it to be autogenous in Karelia, as had Becker (8) in the United Kingdom. A taxonomic problem is involved because the closely related *C. salinarius* is autogenous and could easily have been confused with *C. circumscriptus.* Work is proceeding to resolve this point. In *C. riethi* the expression of autogeny was unaffected by starvation in the fourth instar, all emerging females being autogenous,but the number of eggs laid was reduced to 28–54% of that laid by females fed in the fourth instar (41).

Some species, *C. puncticollis* (41) and *C. brevitarsis* (17), are anautogenous, and the evolutionary value of autogeny and anautogeny needs to be considered. Autogeny decreases mortality in the first gonotrophic cycle, maintains the species in the absence of hosts, and enables a population to maintain its hold on a breeding site by maturing eggs without dispersal. It is particularly valuable where breeding sites have a patchy, irregular distribution or where host density is low.

Autogeny requires the accumulation of food reserves, particularly protein, by the larva for egg maturation. This could be expected to prolong the larval stage. Anautogeny would be favored by abundance of hosts and ephemeral breeding sites on which a population could maintain itself only temporarily. Food reserves would be used for searching for oviposition sites, not for maturing eggs. *C. brevitarsis* breeds in bovine dung (20), an ephemeral habitat. Each generation has to find new breeding sites, which are only available near cattle. Cattle provide both a blood meal and oviposition sites. There is no advantage in *C. brevitarsis* being autogenous. Indeed, anautogeny might shorten larval life by eliminating the need to acquire substantial food reserves. Emergence of *C. brevitarsis* from dung in the field was complete in 24 days in the summer (18).

FECUNDITY AND POTENCY Fecundity varies both within and between species. Intraspecific variation has been shown to be in part size-dependent (87), and there is evidence that the same applies to interspecific variation. Data on fecundity are rarely accompanied by estimates of size, e.g. wing length measurements. However, there is a significant correlation ($r = 0.679$, $P<0.05$) between Service's data on fecundity for ten species (109) and their wing length as taken from Campbell & Pelham-Clinton (16). In order to build up a useful body of data, it would be helpful if workers would cite wing length and time of year when giving fecundities.

In autogenous *C. furens* fecundity was a function of time of year (87). This in turn was partially a function of size, with lower seasonal temperatures (March, 16°C; September, 26°C) producing larger adults which laid more eggs (80 in March, 43 in September). When standardized for size by using wing length/fecundity regression coefficients, there was still a twofold range in monthly fecundity. It was minimal in February (26.2) and maximal in March (54.8). From March to February there was a steady fall in standardized fecundity (Y.) I have fitted the following linear regression to these data, numbering the months (X) 1 through 12, beginning with March. The regression was $\hat{Y} = 52.9 - 2.22X$; $r^2 = 0.912$. The standardized fecundity declined by 2.2 ova per month from March to February and then increased sharply. This annual boost in fecundity is presumably a peculiarity of species breeding all year round, although it was not obvious in the unstandardized fecundity of *C. melleus* (91). It should be noted that the seasonal decline was a property of the population and not of the individual, and it would be of interest to know whether any comparable seasonal change in fecundity occurs in species with restricted breeding periods. In *C. melleus* it has been established that there are seasonal changes in male potency, similar but not identical to those in female fecundity (91).

OVIPOSITION AND FERTILITY No observations have been made on oviposition in the field. This is precluded by the very small size of the female and the relatively enormous extent of the breeding sites. Even when the breeding site was so limited that it could be observed all the time (e.g. dung pats), Campbell never saw oviposition by *C. brevitarsis* (18). This was understandable after it was realized that ovipositions were of the order of three to four per day and that oviposition occurred over the whole 24 hr, i.e. one female every six to eight hours. Information on oviposition had to be obtained indirectly by observing the resulting adult emergence. Oviposition occurred throughout the diel, reaching a maximum between 1400 to 1600 hr, continued high until midnight, and was minimal from then until 0700 hr. The oviposition rate into dung pats was constant over the first seven days after being dropped. Oviposition then ceased; there was no increase in number of adults emerging from pats exposed more than seven days. This abrupt cessation could have been caused by the crust becoming too thick for oviposition.

Eggs of *Culicoides* are susceptible to dessication and are laid on moist or wet surfaces (99). Those of *L. spinosifrons* have been found down to 90 mm in damp sand, with the greatest concentration between 30 and 60 mm (33). Eggs of *Culicoides* have a layer of tall ansulae on the concave surface and a sparser layer of shorter ones on the convex surface; these were considered to act as suckers attaching eggs to the substrate (9). Examination of the eggs of *C. brevitarsis* has indicated that ansulae on the concave surface almost certainly function as a plastron facilitating respiration under wet conditions (17). A similar structure had previously been described on the eggs of *Musca domestica* (51).

The greatest density of eggs of *C. brevitarsis* occurred at the center of the dung pat and decreased towards the periphery (18). Notwithstanding this gradient, the greater area of the peripheral zone more than compensated for the lower density of oviposition. Shape of the dung pat was quite unimportant to oviposition *C. brevitarsis,* and adults were reared from pats of all shapes and sizes, with two exceptions: few adults emerged from pats that were surrounded by a metal sleeve the same height as the pat or from pats sunk flush with the ground. These observations indicated the importance of visual stimuli to gravid *C. brevitarsis,* which probably approached dung pats from above, as postulated earlier in its landing on cattle for feeding. However, oviposition also occurred at night when visual stimuli would be at a disadvantage.

Fertility in *C. circumscriptus* averaged 89% and was usually in excess of 90%, but occasionally it fell as low as 13% (9). Reduced fertility occurred among eggs deposited after decapitation, but it was not clear whether the 13% observation came into that group. The females were wild-caught and could have been inadequately inseminated. The same cannot be said for *C. melleus* females which had mean maximum fertility of only 66% after mating with males at maximum potency (88). It seems an extraordinary waste of resources for a species to have evolved finely adjusted mating behavior only to nullify the advantage by having one third of its eggs sterile. This result may not reflect field performances because the authors state that "*C. melleus* females do not oviposit readily in the laboratory" (88). In *C. obsoletus* and *C. impunctatus,* mean and maximum fertilities were 44.9% and 82.2%, and 50.5%, and 89.6%, respectively (108). Are such low fertilities normal or laboratory induced?

Bionomics of Adults

Quantitative ecological studies depend upon an efficient method of sampling, which requires knowledge of the insects' behavior in the field. Little is known about the resting habits of *Culicoides,* but several species of *Leptoconops* burrow into sand near their breeding sites (32, 78). Collecting from a number of habitats, Muradov (97) found resting *C. similis* and *C. circumscriptus* only on trees and shrubs, *C. firuzae* only in wells, and eight other species mainly in animal shelters. Active adults may be captured when landing on host or in traps with or without attractants. Bias is unavoidable in such methods. Light traps may be made more efficient and selective for *Culicoides* by the use of black light (10, 77), but they are less effective at dawn and dusk. During the night moonlight may introduce a pseudolunar cycle (114). The addition of radiating strips of netting improved suction trap catches of mosquitoes but had virtually no effect on catches of *Culicoides* (111). Service (109) compared three sampling methods and obtained widely differing results. The numbers of species collected and the percentages of *C. impunctatus* were sticky trap, 7 (75%), human bait, 9 (42%), and light trap, 12 (19%). Catches were also influenced by trap height. Service (110) obtained linear regressions for log female density against log height for four *Culicoides* spp. Collections of *L. spinosifrons* on sticky traps dropped sharply with height; most were collected within 0.6 m of the ground and none above 2.25 m (78). In addition to these controllable variables, there remains the powerful modifying influence of meteorological conditions.

24-HOUR CYCLE Species of *Lasiohelea* and *Leptoconops* are mainly diurnal, whereas *Culicoides* spp. have a large nocturnal component in their activity (32, 36, 45, 62, 78). Gornostaeva (44) correlated spiracle size with time of activity. The diurnal *Lasiohelea sibirica* has smaller spiracles than the crepuscular and nocturnal *C. obsoletus.* The diurnal cycles of *Lasiohelea sibirica, L.* nr *nipponica, Leptoconops becquaerti,* and *L. spinosifrons* have morning and late afternoon peaks, separated by reduced activity around noon (13, 32, 45, 70). In *L. spinosifrons* the peaks were influenced by sunlight, and hence morning peaks predominated on east-facing beaches and afternoon peaks on west-facing ones (33). Empty nulliparous and parous females have basically similar 24-hr cycles, although nulliparous *L. spinosifrons* were more active in the middle of the day (32) and nulliparous *C. marksi* flew earlier in the evening (36). *Culicoides* spp. commonly have peaks at sunset and sunrise and variable nocturnal activity. Comparable results for *C. brevitarsis* were obtained by truck trap (36) and bait animal (19); activity was concentrated at sunset and then declined to give little activity after midnight or at dawn. *C. marksi, C. austropalpalis* (36), *C. furens,* and *C. barbosai* (62) showed continuous activity throughout the night with sporadic outbursts. In *C. furens* these occurred at 2100 and 2400. Using suction traps Service (110) obtained only single peaks of activity at dusk (2100) in four *Culicoides* spp. In the vicinity of poultry *C. arakawae* maintained a more or less uniform level of activity throughout the night (74).

Skierska (113) has defined the conditions under which *Culicoides* attacks occurred in Poland: temperature of 8.4° to 22.2°C; relative humidity (RH) $> 50\%$; illumination 84–4230 lux; wind speed < 3 m sec^{-1}. Such limits are functions of the number of readings and the size of the *Culicoides* populations observed. Therefore, some workers have sought a relationship in which unit change in a factor produced a proportional change in *Culicoides* activity. Kettle (63) showed that following the dawn peak changes in the biting rates of *C. furens* and *C. barbosai* could be largely accounted for (86% and 90%, respectively) by changes in wind speed and temperature. Increases of wind speed of 0.66 and 0.92 m sec^{-1} halved the respective biting rates of *C. furens* and *C. barbosai.* Increases of temperature of 1.5°C within the range 24°C to 30°C had the same effect. There was evidence that the effect of temperature was dependent upon its magnitude. Thus below 21°C, temperature and biting rate were positively correlated, between 21° and 24°C they were independent, and above 24°C they were negatively correlated. *L. becquaerti* continued to bite in winds of 4.46 to 5.35 m sec^{-1} because the inhibiting effect of wind was offset by positive responses to associated increases in temperature and illumination (70).

SEASONAL CYCLE In the tropics adults may be found year round (64), whereas in cooler climates adults are seasonal in their appearance (13) and most species overwinter as larvae (56, 112). *C. denningi* hibernates under ice in the sandy bottom of the Saskatchewan River (37). In the spring its larvae move to the shoreline for pupation in great abundance (3000 m^{-2}). Temperate region species usually have one or two generations per year (13, 109,111), but virtually nothing is known about the processes by which these cycles are determined in ceratopogonids. In Florida, where *C. furens* breeds year round, Linley, Evans & Evans (86) found that temperature in the breeding sites influenced adult production. In tidal ditches emergence reached a maximum early in spring, while in impoundments maximum emergence occurred in the summer and continued until late in the fall. In warmer areas rainfall was often the climatic factor that influenced adult abundance. This applied to *C. schultzei* in India (106), and to five East African species (120), but not to *Culicoides* in Israel (12), where breeding sites were man-made rather than climatic.

Coastal species often, but not always, have cycles related to the tides. Thus the biting rates of *C. furens* and *C. barbosai* varied with the lunar (tidal) cycle, whereas that of *L. becquaerti* showed no such correlation because its numbers were more dependent on rainfall (64, 70). The emergence cycle of *L. spinosifrons* was determined both by the tide cycle and slope of the beach (33). When the difference between the two high tides of each lunar month was small and the beach nearly flat, peaks of emergence occurred every 14 days. When the tidal difference was larger and the slope steeper, emergences occurred every 28 days (32). *C. marmoratus* breeds in areas covered at high water springs (HWS). Adult emergence occurred year round, and Kay (58) has made the interesting suggestion that if oviposion is to occur at HWS, then adult emergence must take place sufficiently earlier to permit ovarian development; hence the relationship of emergence to tide cycle should vary seasonally.

DISPERSAL AND SURVIVAL *L. spinosifrons* dispersed only a modest distance inland (400 to 600 m) (33, 78) and even less so along the beach. Using Muradov's (96) data on the biting rates of *C. puncticollis* and *C. schultzei* at different distances from their breeding sites, I have calculated regressions of log catch (Y) on distance (X) in units of 100 m. They gave an excellent fit to the data, accounting for 96.8% and 95.6% of the variation in Y. The regression coefficients and their standard errors were *C. puncticollis* -0.1112 ± 0.0068 and *C. schultzei* -0.1860 ± 0.0133. They are significantly different and indicate that the respective biting rates of *C. puncticollis* and *C. schultzei* halved for every 270 m and 162 m.

Females may be readily classified on external examination as empty, blood-fed, half-gravid, or gravid (110). Dyce (35) has shown that parous *Culicoides* females are distinguished by the possession of burgundy-red pigment in the abdomen. This character enables empty and blood-fed grades to be further subdivided, giving six age groups in all. Further grading requires careful dissection and examination of the ovarioles to determine the number of previous ovarian cycles. Skierska (113) has reported four cycles in a number of *Culicoides* spp. and Mirzaeva (95) three in *C. sinanoensis.* Duval (32) found that changes in fat-body and malpighian tubules in *L. spinosifrons* were too variable for use as aging criteria, and instead he depended on relict bodies.

Kay (58) determined the probability of survival of *C. marmoratus* in summer and winter, obtaining daily rates of 0.48 and 0.76, respectively. The probability of an individual completing a second ovarian cycle was very low (<0.003). Mirzaeva (95) obtained comparable survival of *C. sinanoensis* over two ovarian cycles (first to third), with 4–5% survival from one cycle to the next. Uniparous females (59%) outnumbered nullipars (37%), suggesting that *C. sinanoensis* was relatively inactive in the first cycle and might be autogenous.

BIONOMICS OF IMMATURES

The immature stages of bloodsucking midges have limited mobility, and their habitats result from selection by the gravid female. Eggs of only one species, *L. spinosifrons,* have been found in nature (33). In most species the egg stage is of short duration, but in *L. spinosifrons* it lasts a minimum of 12 days at 30°C and may extend for 6 months at 95% RH, during which time hatching may be induced by submersion in water or exposure to a saturated atmosphere. Most pupae of *L. spinosifrons* were found at depths of 30–60 mm (33), and those of *C. brevitarsis* were most numerous in the wet spongelike zone of dung (19). In both species pupae and larvae showed similar distributions, but *C. melleus* larvae moved higher up the beach before pupation and consequently pupae had a different distribution (83). Before eclosion pupae moved to the surface, which in *C. brevitarsis* occurred mainly between 1200 and 1600.

Few observations have been made on the natural food of ceratopogonid larvae. Laurence & Mathias (78) have assessed the food of *L. spinosifrons* larvae by direct examination of gut contents and the availability of organisms in the habitat. This species feeds on microorganisms and decaying organic material. The numerous long needlelike teeth on its epipharynges may be an adaptation to a small-particle diet. Larvae of *C. subimmaculatus* have been found containing a whole polychaete worm or whole desmids. To date, 12 species of *Culicoides* have been observed readily feeding on small nematodes in the laboratory (73). Larvae of six other genera[2] of vermiform ceratopogonids feed similarly. This habit may be correlated with the possession of lightly built pharynges bearing small blunt teeth on both halves of the dorsal epipharyngeal comb. Nevill (99) has described cannibalism in *C. milnei* and *C. nivosus,* but over several years of working with ceratopogonid larvae I have never observed this phenomenon; cannibalism could either be peculiar to these species or to the conditions under which they were maintained.

[2]*Alluaudomyia, Bezzia, Palpomyia, Paradasyhelea, Sphaeromias,* and *Stilobezzia.*

Culicoides larvae are apneustic, relying on cutaneous respiration. They therefore avoid anaerobic conditions and occur mainly at the surface in muddy habitats: for example, few *C. circumscriptus* occurred below a depth of 20 mm (117). In sandy soils with larger air spaces, larvae can descend deeper, and their distribution appears to be related to moisture and temperature. Larvae of *L. spinosifrons* occurred down to 150 mm; most occurred between 30 to 60 mm, in sandy habitats above high water neaps (HWN) (33). Larvae of *C. melleus,* another sand dwelling species, were concentrated in the top 12 mm, but this species occupied an intertidal habitat largely below HWN, and soil moisture presented no problem because the habitat was inundated twice in every 24–25 hr (83). The habitats of *L. kerteszi* and *L. becquaerti* had their surfaces kept damp by capillary action from subsoil water; in sand this means that the water table must be within 500 mm of the surface (80, 107). Larvae of *L. mediterraneus* were normally in the top 10–30 mm but occurred down to 70–80 mm, depending upon the level of the adjoining river (43). Too much water was damaging to some species, and flooding greatly reduced larval populations of *L. borealis* in the valley of the North Don river (30).

Larvae of *C. brevitarsis* were not limited in bovine dung by moisture content, which ranged from 174% to 537% of the dry weight. Consequently they were found throughout the pat, and their distribution was influenced more by temperature. The time for 50% mortality of *C. brevitarsis* larvae was 1 hr at 42°C and 2 hr at 41°C. These temperatures are often exceeded in the upper levels of dung pats that are exposed to full insolation. When surface temperatures reached 43°C, larvae descended deeper into the pat (19). A smaller response was shown by larvae of *C. melleus,* which descended slightly deeper in the morning and afternoon in response to temperature and illumination (83).

Morphological Adaptations

There are two structural features that appear to be associated with particular larval habitats. Larvae that live in almost pure sand have reduced sclerotization of the head capsule. In *Leptoconops* the capsule is unsclerotized and support is given by paired ventral rods (78). In *C. melleus* (55), *C. molestus,* and *C. subimmaculatus* (67), sclerotization of the head capsule is restricted to anterior (subgenal band) and posterior (postoccipital ridge) encircling hoops, which are connected by three to five longitudinal struts. It is suggested that these arrangements provide flexibility and elasticity, enabling the head capsule to deform when squeezing between sand grains and then to recoil from deformation. *C. phlebotomus* has adopted a different strategy. Its larval head capsule is particularly heavily chitinized (104) and presumably forcibly displaces sand grains or restricts itself to the larger spaces between grains.

Some ceratopogonid larvae, e.g. *Isohelea,* have long perianal setae (68). In *Culicoides* these have been described only in larvae found in tree holes, e.g. *C. angularis* (67), *C. guttipennis, C. flukei, C. villosipennis* (54), but not all tree-hole-breeding species (e.g. *C. borinqueni*) (92) have them. It is suggested that long perianal setae amplify the oscillatory movements of the larval body, and thereby permit faster swimming, which could be advantageous to predators or to escaping prey in a confined habitat.

Breeding Sites

Information on the breeding sites of bloodsucking ceratopogonids has mostly been obtained by trapping newly emerged adults. Potential breeding sites were either sampled by placing emergence traps in the field or by removing samples to the laboratory and collecting any adults produced. Although both methods have given useful qualitative information—Skierska (112) has reared 32 species; Hair, Turner & Messersmith (49) 21 species; Kremer, Hirtzel & Molet (76) 16 species; and Davies (28) 10 species—they are quantitatively unsound. Emergence traps modify the microhabitat of the covered area and may even enhance population estimates (27), whereas removal of material from the habitat changes many conditions by creating an artificially enclosed system. Little attempt has been made to recover and identify the immatures. Extraction techniques have been used mainly when the habitat was considered to be monospecific, which limited its application. A quantitative technique, based on numbers of pupae collected in an hour (57), would be heavily biased unless all the species involved have identical seasonal cycles, which is unlikely.

There is no accepted uniform method of classifying breeding sites. Descriptions vary according to the individual's perspective, and neat categories are upset by the existence of intermediate habitats. Certain faunas appear to be well defined and have minimal or no overlap with others. These occur in habitats such as dung, rotting vegetation, tree holes, river beds, and fungi. Only species of the *Avaritia* subgenus have been bred from unadulterated dung: *C. chiopterus, C. dewulfi* (68), *C. brevitarsis* (20), and *C. pallidipennis* (100). The *copiosus* group of species are associated with rotting cacti (122), and *C. loughnani* accompanied the successful *Cactoblastis cactorum* when it was introduced into Australia to control prickly pear (34). The majority of *Culicoides* spp. breed in the ecotone between aquatic and terrestrial habitats. These may be grouped on salinity, pollution, degree of exposure to sun, soil composition, and vegetation, but the categories grade into one another. Vegetation is likely to prove the best criterion, because it is a visible expression of not only present but also past conditions.

The faunas of acid, oligotrophic moorland soils differ from those of alkaline, eutrophic lowland habitats; *C. albicans* and *C. impunctatus* are characteristic of the former (75). Saline coastal habitats in the tropics and subtropics are often major sources of anthropophilic species, including *C. furens, C. barbosai,* and *C. insignis* in the Neotropical region; *C. ornatus**, *C. molestus**, and *C. subimmaculatus** in the Australian region; and *C. belkini** and *C. pelilouensis* in the Pacific region (* E. J. Reye, personal communication). Coastal breeding sites are either muddy or sandy in character. There is a fairly sharp division between mud- and sand-dwelling species, e.g. *C. furens* and *C. barbosai* in mud and *C. phlebotomus* and *C. molestus* in sand. Some saltwater species, such as *C. riethi* can tolerate salinities of 66 to 230 g l^{-1} (25). Generaly the subgenera of *Leptoconops, Holoconops* and *Styloconops* breed in almost pure sand and *Leptoconops* breeds in clay-silt soils (32, 94). The critical importance of tidal range with respect to larval distribution has been neatly demonstrated by Duval, Rajaonarivelo & Rabenirainy (33), who correlated the changing distribution of *L. spinosifrons* larvae with changes in beach topography. Such associations are common in tidal species, for instance, *C. molestus* is mainly concentrated in a narrow vertical belt around mean HWN (E. J. Reye, P. B. Edwards, personal communication).

Certain species are adaptable with respect to their breeding sites. *C. stellifer* occurred in three of the four main categories of breeding sites recognized by Hair, Turner & Messersmith (49): (*a*) tree holes; (*b*) sunlit, dung-polluted mud; and (*c*) heavily shaded mud covered with decaying leaves. *C. odibilis* and *C. cubitalis* have been categorized as "ubiquitous" species, occurring with 26 and 22 other species, respectively (75).

Attempts to identify key factors on which to classify breeding sites have given diverse results. Kremer & Callot (75) emphasized the importance of soil composition, especially sand content, which they consider to be a more important factor than moisture or organic matter. Callot et al (15) found *C. cubitalis* to favor sandy sites with variable or constant moisture content, while five other species were more or less restricted to silty, temporary pools. Battle & Turner (7) used a nutrient index which took into account the amount of organic material and four elements (Ca, Mg, K, P), and Kardatzke & Rowley (57) adopted organic matter and sodium ion concentration as their criteria. Progress in this area will depend upon workers obtaining comparable data, which in turn requires agreement on the main physical and chemical variables to be measured.

CONCLUSION—POPULATION MODELLING

Although quantitative data on bloodsucking ceratopogonids have greatly increased recently, they are nevertheless inadequate for model building. Some information is available on fecundity and its moderation by temperature; fewer data exist on egg and adult mortalities and virtually none on larval mortality, notwithstanding that this stage has the longest duration. Records of parasites in ceratopogonids have been reviewed by Bacon (6), but there is no information on predators. With the establishment of laboratory colonies of nine species—*Leptoconops kerteszi* (105), *Lasiohelea taiwana, C. arakawae, C. schultzei* (116), *C. nubeculosus, C. variipennis, C. riethi* (11), *C. furens* (81), and *C. guttipennis* (47)—there is the opportunity to investigate the effect of environmental factors such as temperature, food, and parasites on population growth. Many factors may be evaluated realistically only in the field, but laboratory studies should both complement field observations and direct attention to the critical areas requiring field study.

Acknowledgments

I wish to thank past and present colleagues who have assisted me, especially M. M. Elson, E. J. Reye, and my secretary, Mrs. R. Crombie.

Literature Cited

1. Atchley, W. R. 1970. A biosystematic study of the subgenus *Selfia* of *Culicoides*. *Univ. Kans. Sci. Bull.* 49:181–336
2. Atchley, W. R. 1971. A statistical analysis of geographic variation in the pupae of three species of *Culicoides*. *Evolution* 25:51–74
3. Atchley, W. R. 1971. A comparative study of the cause and significance of morphological variation in adults and pupae of *Culicoides*: a factor analysis and multiple regression study. *Evolution* 25:563–83
4. Atchley, W. R. 1973. A quantitative separation of the females of *Culicoides* (*Selfia*) *denningi*, *C.* (*S*). *hieroglyphicus* and *C.* (*S*). *jamesi*. *J. Med. Entomol.* 10:629–32
5. Atchley, W. R. 1974. A quantitative taxonomic analysis of *Leptoconops torrens* and *L. carteri*. *J. Med. Entomol.* 11:467–70
6. Bacon, P. R. 1970. Natural enemies of the Ceratopogonidae—a review. *Tech. Bull. Commonw. Inst. Biol. Control* 13:71–82
7. Battle, F. V., Turner, E. C. 1972. Some nutritional and chemical properties of the larval habitats of certain species of *Culicoides*. *J. Med. Entomol.* 9:32–35
8. Becker, P. 1960. Observations on the feeding and mating of *Culicoides circumscriptus* Kieffer. *Proc. R. Entomol. Soc. London Ser. A* 35:6–11
9. Becker, P. 1961. Observations on the life cycle and immature stages of *Culicoides circumscriptus* (Kieff.). *Proc. R. Soc. Edinburgh Ser. B* 67:363–86
10. Belton, P., Pucat, A. 1967. A comparison of different lights in traps for *Culicoides*. *Can. Entomol.* 99:267–72
11. Boorman, J. 1974. The maintenance of laboratory colonies of *Culicoides variipennis* (Coq.), *C. nubeculosus* (Mg) and *C. riethi* Kieff. *Bull. Entomol. Res.* 64:371–77
12. Braverman, Y., Galun, R. 1973. The occurrence of *Culicoides* in Israel with reference to the incidence of bluetongue. *Refu. Vet.* 30:121–27
13. Buyanova, O. F. 1960. On the biology of the blood-sucking midges of the genus *Lasiohelea* in the Krasnoyarsk region. *Med. Parazitol. Parazit. Bolezni* 29:702–5
14. Callot, J., Kremer, M., Geiss, J.-L. 1972. Iconographie de l'armature cibariale de 22 espèces de *Culicoides* liste des espèces qui en sont depourvu. *Ann. Parasitol. Hum. Comp.* 47:759–62
15. Callot, J., Kremer, M., Geiss, J.-L., Hirtzel, C. 1974. Étude écologique de quelques *Culicoides* des Vosges. *Ann. Parasitol. Hum. Comp.* 49:649–51
16. Campbell, J. A., Pelham-Clinton, E. C. 1960. A taxonomic review of the British species of *Culicoides* Latreille. *Proc. R. Soc. Edinburgh Ser. B* 67:181–302
17. Campbell, M. M., Kettle, D. S. 1975. Oogenesis in *Culicoides brevitarsis* Kieffer and the development of a plastron-like layer on the egg. *Aust. J. Zool.* 23:203–18
18. Campbell, M. M., Kettle, D. S. 1976. Some factors involved in *Culicoides brevitarsis* Kieffer finding and ovipositing in dung. *Aust. J. Zool.* 24:75–85
19. Campbell, M. M. 1974. *The biology and behaviour of Culicoides brevitarsis Kieffer with particular reference to those features essential to its laboratory colonisation.* PhD thesis. Queensland Univ., Brisbane, Australia. 255 pp.
20. Cannon, L. R. G., Reye, E. J. 1966. A larval habitat of the biting midge *Culicoides brevitarsis* Kieffer. *J. Entomol. Soc. Queensland* 5:7–9
21. Clastrier, J. 1971. Isolment et description de la larve de *Leptoconops* (*Leptoconops*) *irritans* Noé, 1905. *Ann. Parasitol. Hum. Comp.* 46:737–48
22. Clastrier, J. 1973. Le genre *Leptoconops*, sous-genre *Holoconops* dans le Midi de la France. *Ann. Soc. Entomol. Fr.* 9:895–920
23. Coluzzi, M., Sabatina, A. 1967. Cytogenetic observations on species A and B of the *Anopheles gambiae* complex. *Parassitologia* (*Rome*) 9:73–88
24. Cornet, M. 1974. Caractères morphologiques utilisés pour l'identification des *Culicoides*. *Cah. ORSTOM Ser. Entomol. Med. Parasitol.* 12:221–29
25. Damian-Georgescu, A., Spataru, P. 1971. Metamorfoza la *Culicoides riethi* Kieff., 1914. *Stud. Cercet. Biol. Ser. Zool.* 23:531–36
26. Davies, F. G., Walker, A. R. 1974. The isolation of ephemeral fever virus from cattle and *Culicoides* midges in Kenya. *Vet. Rec.* 95:63–64
27. Davies, J. B. 1966. An evaluation of the emergence or box trap for estimating sand fly (*Culicoides* spp) populations. *Mosq. News* 26:69–72
28. Davies, J. B. 1973. Sand flies breeding near Las Cuevas and Maracas beaches.

J. Trinidad Field Nat. Club, pp. 53–67
29. Downes, J. A. 1969. The swarming and mating flight of Diptera. *Ann. Rev. Entomol.* 14:271–98
30. Dubrovskaya, V. V. 1972. On the classification of foci of distribution of bloodsucking midges according to the hydrological conditions of the Donets region. *Parazitologiya* 6:103–6
31. Dunbar, R. W. 1969. Nine cytological segregates in the *Simulium damnosum* complex. *Bull. WHO* 40:974–79
32. Duval, J. 1971. Etude écologique du cératopogonide halophile *Styloconops spinosifrons* (Carter 1921) des plages de Nossi-Bé en vue d'une lutte rationnelle au moyen d'insecticide. *Cah. ORSTOM Ser. Entomol. Med. Parasitol.* 9:203–20
33. Duval, J., Rajaonarivelo, E., Rabenirainy, L. 1974. Écologie de *Styloconops spinosifrons* sur les plages de la côte Est de Madegascar. *Cah. ORSTOM Ser. Entomol. Med. Parasitol.* 12: 245–58
34. Dyce, A. L. 1969. Biting midges reared from rotting cactus in Australia. *Mosq. News* 29:644–49
35. Dyce, A. L. 1969. The recognition of nulliparous and parous *Culicoides* without dissection. *J. Aust. Entomol. Soc.* 8:11–15
36. Dyce, A. L., Standfast, H. A. 1972. A comparison of the flight patterns of *Culicoides* and other blood-sucking Nematocera. *Abstr. Int. Congr. Entomol. Canberra, 14th*, pp. 279–80
37. Fredeen, F. J. H. 1969. *Culicoides* (*Selfia*) *denningi*, a unique river-breeding species. *Can. Entomol.* 101:539–44
38. Glukhova, V. M. 1958. On the gonotrophic cycle in midges of the genus *Culicoides* in the Karelian ASSR. *Parazitol. Sb.* 18:239–54
39. Glukhova, V. M. 1968. Systematic review of larvae of the genus *Culicoides*. *Parazitologiya* 2:559–67
40. Glukhova, V. M. 1969. Description of larvae of midges of the genus *Culicoides*. *Parazitologiya* 3:461–67
41. Glukhova, V. M., Dubrovskaya, V. V. 1972. On autogenic maturation of eggs of blood-sucking midges. *Parazitologiya* 6:309–19
42. Glukhova, V. M., Dubrovskaya, V. V. 1974. On the swarming flight and mating in blood-sucking midges. *Parazitologiya* 8:432–37
43. Glukhova, V. M., Kravez, G. A., Smatov, Zh. 1975. Morphology and biology of the early stages of biting midges *Leptoconops* (*H*) *mediterraneus* Kieff. *Parazitologiya* 9:190–96
44. Gornostaeva, R. M. 1964. On the adaptation of biting midges to different conditions of air humidity. *Zool. Zh.* 43:145–47
45. Gornostaeva, R. M. 1967. The diurnal activity of attacks of *Lasiohelea sibirica* Bugan midges in the area of Krasnoyarsk hydro power station construction. *Med. Parazitol. Parazit. Bolezni* 36:11–17
46. Gutsevich, A. V. 1973. *The blood-sucking midges (Ceratopogonidae) Fauna USSR, Insecta, Diptera,* Vol. 3, part 5. Leningrad: Science Press. 270 pp.
47. Hair, J. A., Turner, E. C. 1966. Laboratory colonization and mass procedures for *Culicoides guttipennis*. *Mosq. News* 26:429–33
48. Hair, J. A., Turner, E. C. 1968. Preliminary host preference studies on Virginia *Culicoides*. *Mosq. News* 28:103–7
49. Hair, J. A., Turner, E. C., Messersmith, D. H. 1966. Larval habitats of some Virginia *Culicoides*. *Mosq. News* 26:195–204
50. Hinds, M. J., Linley, J. R. 1974. Changes in male potency with age after emergence in the fly, *Culicoides melleus*. *J. Insect Physiol.* 20:1037–40
51. Hinton, H. E. 1967. The respiratory system of the egg-shell of the common housefly. *J. Insect Physiol.* 13:647–51
52. Humphreys, J. G., Turner, E. C. 1971. The effect of light intensity on feeding activity of laboratory reared *Culicoides guttipennis* (Coquillett). *Mosq. News* 31:215–17
53. Humphreys, J. G., Turner, E. C. 1973. Blood feeding activity of female *Culicoides*. *J. Med. Entomol.* 10:79–83
54. Jamnback, H. 1965. The *Culicoides* of New York State. *NY State Mus. Sci. Serv. Bull.* 399:1–154
55. Jamnback, H., Wall, W. J., Collins, D. L. 1958. Control of *Culicoides melleus* (Coquillett) in small plots with brief descriptions of the larvae and pupae of two coast *Culicoides*. *Mosq. News* 18:64–70
56. Jones, R. H. 1967. An overwintering population of *Culicoides* in Colorado. *J. Med. Entomol.* 4:461–63
57. Kardatzke, J. T., Rowley, W. A. 1971. Comparisons of *Culicoides* larval habitats and populations in Central Iowa. *Ann. Entomol. Soc. Am.* 64:215–18
58. Kay, B. H. 1973. Seasonal studies of a population of *Culicoides marmoratus*

(Skuse) at Deception Bay, Queensland. *J. Aust. Entomol. Soc.* 12:42–58
59. Kettle, D. S. 1962. The bionomics and control of *Culicoides* and *Leptoconops*. *Ann. Rev. Entomol.* 7:401–18
60. Kettle, D. S. 1965. Biting ceratopogonids as vectors of human and animal diseases. *Acta Trop.* 22:356–62
61. Kettle, D. S. 1969. The ecology and control of blood-sucking ceratopogonids. *Acta Trop.* 26:235–47
62. Kettle, D. S. 1969. The biting habits of *Culicoides furens* (Poey) and *C. barbosai* Wirth and Blanton. I. The 24-h cycle, with a note on differences between collectors. *Bull. Entomol. Res.* 59:21–31
63. Kettle, D. S. 1969. The biting habits of *Culicoides furens* (Poey) and *C. barbosai* Wirth and Blanton. II. Effect of meteorological conditions. *Bull. Entomol. Res.* 59:241–58
64. Kettle, D. S. 1972. The biting habits of *Culicoides furens* (Poey) and *C. barbosai* Wirth and Blanton. III. Seasonal cycle, with a note on the relative importance of ten factors that might influence the biting rate. *Bull. Entomol. Res.* 61:565–76
65. Kettle, D. S., Elson, M. M. 1975. Variation in larvae and adults of *Culicoides austropalpalis* Lee & Reye in S. E. Queensland. *J. Nat. Hist.* 9:321–36
66. Kettle, D. S., Elson, M. M. 1975. The immature stages of *Paradasyhelea minuta* Wirth and Lee with a note on adult antennal sensilla and a discussion on the relationships of the genus *Paradasyhelea* Macfie. *J. Aust. Entomol. Soc.* 14:255–61
67. Kettle, D. S., Elson, M. M. 1976. The immature stages of some Australian *Culicoides* Latreille. *J. Aust. Entomol. Soc.* 15:In press
68. Kettle, D. S., Lawson, J. W. H. 1952. The early stages of British biting midges *Culicoides* Latreille and allied genera. *Bull. Entomol. Res.* 43:421–67
69. Kettle, D. S., Linley, J. R. 1967. The biting habits of *Leptoconops becquaerti*. I. Methods; standardisation of technique, preferences for individuals, limbs and positions. *J. Appl. Ecol.* 4:379–95
70. Kettle, D. S., Linley, J. R. 1967. The biting habits of *Leptoconops becquaerti*. II. Effect of meteorological conditions on biting activity: 24h and seasonal cycles. *J. Appl. Ecol.* 4:397–420
71. Kettle, D. S., Linley, J. R. 1969. The biting habits of some Jamaican *Culicoides*. I. *C. barbosai* Wirth & Blanton. *Bull. Entomol. Res.* 58:729–53
72. Kettle, D. S., Linley, J. R. 1969. The biting habits of some Jamaican *Culicoides*. II. *C. furens* (Poey). *Bull. Entomol. Res.* 59:1–20
73. Kettle, D. S., Wild, C. H., Elson, M. M. 1975. A new technique for rearing individual *Culicoides* larvae. *J. Med. Entomol.* 12:263–64
74. Kitaoka, S., Morii, T. 1964. Hourly flying and biting activities of *Culicoides arakawae* and *Culicoides odibilis* attracted to light traps in chicken houses. *Bull. Nat. Inst. Anim. Health* 49:16–21
75. Kremer, M., Callot, J. 1963. Les cératopogonidés des terrains tourbeaux des Hautes Vosges. In *Le Hohneck*, pp. 315–21. Strasbourg: Assoc. Philomathique d'Alsace et de Lorraine
76. Kremer, M., Hirtzel, C., Molet, B. 1974. Résultats préliminaires d'une étude écologique des *Culicoides* des Sansouires de Camargue. *Ann. Parasitol. Hum. Comp.* 49:653–56
77. Kwan, W. H., Morrison, F. O. 1973. Light trap catches of *Culicoides* at Lac Serpent, Quebec. *Phytoprotection* 54: 72–77
78. Laurence, B. R., Mathias, P. L. 1972. The biology of *Leptoconops* (*Stylોconops*) *spinosifrons* (Carter) in the Seychelles Islands, with descriptions of the immature stages. *J. Med. Entomol.* 9:51–59
79. Linley, J. R. 1968. Autogeny and polymorphism for wing length in *Leptoconops becquaerti* (Kieff.) *J. Med. Entomol.* 5:53–66
80. Linley, J. R. 1969. Seasonal changes in larval populations of *Leptoconops becquaerti* (Kieff.) in Jamaica with observations on the ecology. *Bull. Entomol. Res.* 59:47–64
81. Linley, J. R. 1968. Colonisation of *Culicoides furens* (Poey). *Ann. Entomol. Soc. Am.* 61:1486–90
82. Linley, J. R. 1975. Sperm supply and its utilisation in doubly inseminated flies, *Culicoides melleus*. *J. Insect Physiol.* 21:1785–88
83. Linley, J. R., Adams, G. M. 1972. Ecology and behaviour of immature *Culicoides melleus* (Coq.). *Bull. Entomol. Res.* 62:113–27
84. Linley, J. R., Adams, G. M. 1972. A study of the mating behaviour of *Culicoides melleus* (Coquillett). *Trans. R. Entomol. Soc. London* 124:81–121
85. Linley, J. R., Adams, G. M. 1974. Sexual receptivity in *Culicoides melleus*.

Trans. R. Entomol. Soc. London 126: 279–303
86. Linley, J. R., Evans, F. D. S., Evans, H. T. 1970. Seasonal emergence of *Culicoides furens* at Vero Beach, Florida. *Ann. Entomol. Soc. Am.* 63:1332–39
87. Linley, J. R., Evans, H. T., Evans, F. D. S. 1970. A quantitative study of autogeny in a naturally occurring population of *Culicoides furens* (Poey). *J. Anim. Ecol.* 39:169–83
88. Linley, J. R., Hinds, M. J. 1974. Male potency in *Culicoides melleus* (Coq.). *Bull. Entomol. Res.* 64:123–28
89. Linley, J. R., Hinds, M. J. 1975. Quantity of the male ejaculate influenced by female unreceptivity in the fly, *Culicoides melleus. J. Insect Physiol.* 21: 281–85
90. Linley, J. R., Hinds, M. J. 1975. Sperm loss at copulation in *Culicoides melleus. J. Entomol. Ser. A* 50:37–41
91. Linley, J. R., Hinds, M. J. 1976. Seasonal changes in size, female fecundity and male potency in *Culicoides melleus. J. Med. Entomol.* 13:In press
92. Linley, J. R., Kettle, D. S. 1964. A description of the larvae and pupae of *Culicoides furens* (Poey) and *Culicoides hoffmani* Fox. *Ann. Mag. Nat. Hist.* 7:129–49
93. Luedke, A. J., Jones, R. H., Jochim, M. M. 1967. Transmission of bluetongue between sheep and cattle by *C. variipennis. Am. J. Vet. Res.* 28:457–60
94. Majori, G., Bernardini, F., Bettini, S., Finizio, E., Pierdominici, G. 1971. Ricerche sur Ceratopogonidi nel Grossetano: Nota V-Caratteristiche geologiche dei focolai di *Leptoconops* spp. *Riv. Parassitol.* 32:277–91
95. Mirzaeva, A. G. 1974. The age composition of females of *Culicoides sinanoensis* Tok. from coniferous broad leaved forests of south maritime territory. *Parazitologiya* 8:524–30
96. Muradov, Sh. M. 1972. The flight range of midges from their breeding places. *Parazitologiya* 6:189–90
97. Muradov, Sh. M. 1973. Daytime resting places of midges. *Veterinaryia (Moscow)* 7:33–34
98. Murray, M. D. 1970. The identification of blood meals in biting midges (*Culicoides*). *Ann. Trop. Med. Parasitol.* 64:115–22
99. Nevill, E. M. 1967. *Biological studies on some South African Culicoides species and the morphology of their immature stages.* MSc. thesis. Pretoria Univ., Pretoria, South Africa. 73 pp.
100. Nevill, E. M. 1968. A significant new breeding site of *Culicoides pallidipennis* Carter, Ingram and Macfie. *J. S. Afr. Vet. Med. Assoc.* 39:61
101. Nevill, E. M. 1969. The morphology of the immature stages of some South African *Culicoides* species. *Onderstepoort J. Vet. Res.* 36:265–84
102. Nevill, E. M., Anderson, D. 1972. Host preferences of *Culicoides* midges in South Africa as determined by precipitin tests and light trap catches. *Onderstepoort J. Vet. Res.* 39:147–52
103. Nielsen, B. O. 1971. Some observations on biting midges attacking grazing cattle in Denmark. *Entomol. Scand.* 2:95–98
104. Painter, R. H. 1926. Biology, immature stages and control of the sand flies (biting Ceratopogoninae) at Puerto Castella, Honduras. *United Fruit Co. Med. Dept. Ann. Rep.* 15:245–62
105. Rees, D. M., Lawyer, P. G., Winget, R. N. 1971. Colonization of *Leptoconops kerteszi* Kieffer by anautogenous and autogenous reproduction. *J. Med. Entomol.* 8:266–71
106. Reuben, R. 1965. Note on the seasonal prevalence of *Culicoides schultzei* (Enderlein): Synonym *Culicoides oxystoma* Kieffer. *J. Bombay Nat. Hist. Soc.* 62:308–9
107. Rioux, J. A., Descous, S. 1965. Détection du biotype larvaire de *Leptoconops (Holoconops) kerteszi* Kieffer, 1908, dans le "Midi" méditerranéan. *Ann. Parasit. Hum. Comp.* 40:219–29
108. Service, M. W. 1968. Blood digestion and oviposition in *Culicoides impunctatus* Goetghebuer and *C. obsoletus* (Meigen). *Ann. Trop. Med. Parasitol.* 62:325–30
109. Service, M. W. 1969. Light trap catches of *Culicoides* spp. from southern England. *Bull. Entomol. Res.* 59:317–22
110. Service, M. W. 1971. Adult flight activities of some British *Culicoides* species. *J. Med. Entomol.* 8:605–9
111. Service, M. W. 1974. Further results of catches of *Culicoides* and mosquitoes from suction traps. *J. Med. Entomol.* 11:471–79
112. Skierska, B. 1973. Faunistic-ecological investigations on blood-sucking midges of the Polish coastal area. *Bull. Inst. Mar. Med. Gdansk* 24:113–33
113. Skierska, B. 1974. Cératopogidés de la côte de la Pologne. *Ann. Parasitol. Hum. Comp.* 49:641–43
114. Stam, A. V. 1965. Influence de la périodicité lunaire sur l'activité de quelques

nématoceres hématophages. *Proc. Int. Congr. Entomol., 12th, London,* pp. 300–1

115. Standfast, H. A., Dyce, A. L. 1968. Attacks on cattle by mosquitoes and biting midges. *Aust. Vet. J.* 44:585–86
116. Sun, W. K. C. 1974. Laboratory colonisation of biting midges. *J. Med. Entomol.* 11:71–73
117. Tabaru, Y., Iked, Y., Hasegawa, M., Hattori, K. 1973. Studies on the control of the biting midges in Hokkaido. III. On the chemical control and vertical distribution of the larvae of *Culicoides circumscriptus* Kieffer in Saroma, shore of Lake Saroma, Hokkaido. *Jpn. J. Sanit. Zool.* 23:215–18
118. Tanner, G. D., Turner, E. C. 1973. Vertical activities and host preferences of several *Culicoides* species in a south western Virginia forest. *Mosq. News* 34:66–70
119. Tempelis, C. H., Nelson, R. L. 1971. Blood feeding pattern of midges of the *Culicoides variipennis* complex in Kern County, California. *J. Med. Entomol.* 8:532–34
120. Walker, A. R., Davies, F. G. 1971. A preliminary survey of the epidemiology of bluetongue in Kenya. *J. Hyg.* 69:47–60
121. Wirth, W. W., Atchley, W. R. 1973. *A review of the North American Leptoconops. Texas Tech. Press. Grad. Studies* No. 5, Lubbock, Tex. 57 pp.
122. Wirth, W. W., Hubert, A. A. 1960. Ceratopogonidae reared from cacti with a review of the *copiosus* group of *Culicoides. Ann. Entomol. Soc. Am.* 53: 639–58
123. Wirth, W. W., Ratanaworabhan, N. C., Blanton, F. S. 1974. Synopsis of the genera of Ceratopogonidae. *Ann. Parasitol. Hum. Comp.* 49:595–613

Ann. Rev. Entomol. 1977. 22:53–78

DYNAMIC ASPECTS OF INSECT-INSECTICIDE INTERACTIONS[1]

❖6120

W. Welling
Laboratory for Research on Insecticides, Wageningen, The Netherlands

INTRODUCTION

The interaction between insecticide and living organism is quite a complicated process. Several steps can be distinguished: uptake, translocation, activation, inactivation, and interaction with the target. Some of these individual steps are consecutive, for example, some organophosphorus compounds have to be activated before they can react with the target, but others take place simultaneously. The eventual effect exerted by a certain quantity of a toxic compound depends on its intrinsic reactivity with the target(s) and on its concentration as a function of time at the target site. The actual form of this concentration-time profile is determined not only by the efficiency of the single steps in the intoxication process but also by their interactions. For example, the rate at which a compound reaches a certain detoxication site can affect the proportion degraded.

To further our understanding of the toxicological process we need an integrated quantitative description treating the single steps as parts of a dynamic system. It is the purpose of this review to discuss the more recent literature on insecticide-target interactions from this point of view, in the hope of collecting material necessary to build such a system. Hence more emphasis is placed on mode of entry and factors determining bioavailability than on an extensive description of the precise way in which an insecticide can be degraded. Finally, some attention is paid to attempts to integrate the existing knowledge on isolated processes by using models.

[1]The following abbreviations are used in this review: CNS, central nervous system; KT_{50}, time to cause 50% knockdown; OP, organophosphorus; AChE, acetylcholinesterase; and K_m, Michaelis-Menten constant.

MODE OF ENTRY OF CONTACT INSECTICIDES

The way along which a contact insecticide reaches its target is still a matter of controversy. Two opposed views are held; the most generally accepted one is penetration through the cuticle into the hemolymph, which rapidly transports it throughout the insect body. From the hemolymph an insecticide spreads to all kinds of tissues, including that containing the target, and during this process it may be sorbed onto proteins, dissolved and stored in lipid-rich phases, activated (if necessary), or detoxified. The other view is brought forward and defended mainly by Gerolt (33–36). He suggests that a topically applied insecticide moves laterally in the integument and enters the target (the central nervous system) via the tracheal system. Figure 1 depicts both models of entry schematically.

Arguments in favor of the latter theory can be divided into two groups: (*a*) arguments based on observations that seem to contradict the hemolymph-carrier concept, and (*b*) those based on data that support directly lateral movement and entry via the tracheae. The following summary of Gerolt's experiments, starting with the first group, tries to evaluate his conclusions, taking into account the results obtained by other investigators.

1. If an insecticide is transported by hemolymph, the rate of distribution throughout the insect body should be independent of the site of application, and one would not expect the site of application to affect the speed of action to any large extent.

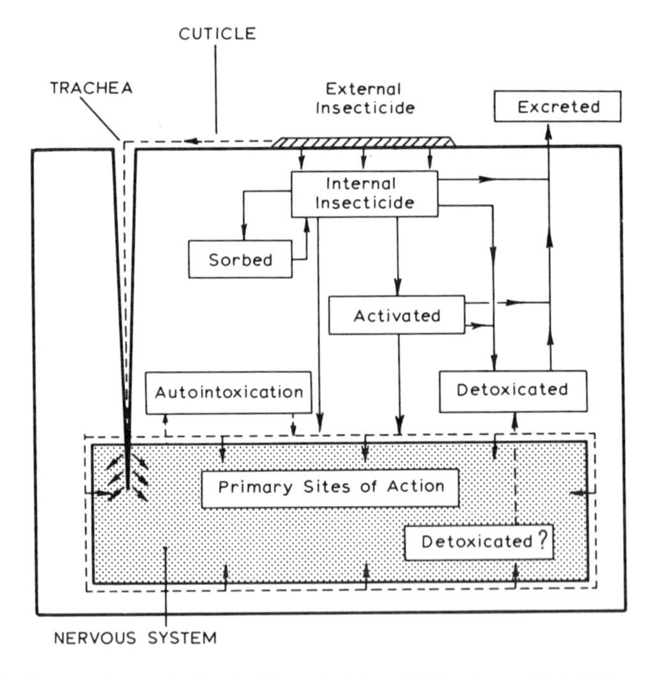

Figure 1 Entry and spread of topically applied insecticides. From Burt (18).

It is, however, a general observation that the reverse is true [(1,2); for a review of the older literature see (33)]. Knockdown times of house flies for both dieldrin and dichlorvos are smallest when applied nearest to the CNS (ventral prothorax) and the less-effective places are those farthest away from it (33). No connection was found between thickness of the integument and speed of action; application on the abdomen of the locust *Schistocerca gregaria,* with a thickness of 3–4 mg/cm^2, gave six- to sevenfold longer knockdown times than did application on the thorax, with a thickness of about 11 mg/cm^2. Ahmed & Gardiner (1–3) made detailed investigations of the relationship between site of application of malathion and toxicity to desert locusts. They found that several factors play a role, e.g. penetration rate into and accumulation in the underlying tissues and spreading over the cuticle. These factors may interact with detoxication. Since the rates of these processes differ from site to site, a dependence of toxicity on site of application is clearly not incompatible with the supposed transport function of hemolymph.

2. A second point put forward by Gerolt is that one must expect injection of olive oil into the hematocoel to absorb insecticide from the hemolymph and thus increase KT_{50}. This was tested with four insecticides, which differed considerably in polarity, but the results did not live up to the expectation. Increase of KT_{50} was negligible, which disproves, according to Gerolt, that insecticides are normally transported by blood. Similar experiments were carried out by Collins et al (25), but although their experimental data agree remarkably well with Gerolt's, Collins et al reached the opposite conclusion.

This type of experiment has been criticized by Olson (56). He showed that on partitioning of dieldrin between olive oil and buffer containing protein, the insecticide shows a strong preference for the aqueous phase as a result of binding to protein. Injection of olive oil should therefore affect dieldrin concentration in hemolymph, and thus affect KT_{50}, only slightly.

3. If insecticides are transported to their site of action by hemolymph, then blockade of hemolymph circulation must delay speed of knockdown on application of insecticides on the abdomen. Gerolt tested this by applying a ligature (waxed cotton thread) around the waist of house flies. Reduction of hemolymph flow was assessed by injection of tritiated water into the thorax, and with ligatured flies leakage of radiolabel from thorax to abdomen was extremely small. The effect on KT_{50} for several types of insecticides was either small or absent however—at most an increase of two- to threefold. It could only be observed when the insects were rapidly knocked down, i.e. at KT_{50} less than an hour in nonligatured flies. These experiments are difficult to evaluate. If there is transport by hemolymph, it is certainly not the rate-limiting factor in translocation of externally applied insecticide. One ought to know, therefore, to what extent flow of hemolymph must be reduced in order to evoke such effects on KT_{50}. In this connection it is worth mentioning that Yeager & Munson (77) showed that heart cauterization inhibits appearance of DDT symptoms in cockroaches.

4. Insecticides usually are more toxic upon injection than upon topical application. This is often used as an argument in favor of hemolymph being the carrier. Gerolt argues that many of such experiments may have been incorrectly interpreted.

He showed that if dieldrin dissolved in acetone is injected, the amount of acetone per se affects the toxic activity; the smaller the quantity of organic solvent, the lower the speed of action. With the oxime carbamate methomyl, the LD_{50} for house flies is 0.13 μg per fly when injected in 0.1 μl of acetone and 0.8 μg per fly when dissolved in water. For a fair comparison of injection vs topical application, one should therefore avoid the use of organic solvents as much as possible. Gerolt found that introduction of crystalline dieldrin or OP insecticides that were adsorbed on filter paper into the abdominal cavity of house flies increased KT_{50} in comparison with external exposure four to ten times, depending on the compound used; it was shown that this could not be ascribed to a low rate of dissolution.

To understand how this boosting effect of organic solvents on injected toxicants is brought about, perhaps one could think of an increased bioavailability due to less adsorption on blood proteins in the presence of organic solvent [e.g. see Krasner (40) for effects of organic solvents on drug adsorption by proteins]. Anyhow, these important observations made by Gerolt should be borne in mind when evaluating experiments described in literature, since injection of insecticides dissolved in organic solvents is common practice. It is clear, therefore, that one should be very careful in using differences in toxicities and speed of action as an argument in favor of transport by hemolymph.

In another type of experiment, the effect of way of entry on toxicity of an oxime carbamate for house flies was assessed. Equitoxic doses of the radioactive labeled compound were either applied topically (0.25 μg) or injected (0.13 μg), and blood concentrations were determined for up to 2 hr after treatment. According to Gerolt, it is to be expected that equitoxic doses, whether injected or applied topically, must give the same hemolymph concentration if hemolymph is the carrier. From the time profiles of radioactivity, it is obvious that this expectation did not come true: when injected, methomyl leveled off to a threefold-higher concentration in the hemolymph than when applied topically. However, metabolic degradation of this compound is extremely fast, and data are lacking to show that radioactivity in the hemolymph is an adequate measure for unchanged insecticide.

5. From the absence of passage of dieldrin through isolated integument, Gerolt concluded that this route is not used in living insects either. It remains to be seen whether his experimental setup, i.e. integument in contact with saline, is a good model for such measurements. Olson (56) has shown that the solubility of lipophilic compounds like dieldrin is 200–400 times larger in hemolymph than in saline. This explains why the transfer rate of dieldrin, taken up by the cuticle, to saline is negligible. Polles & Vinson (60) studied penetration of endrin through isolated integuments of the tobacco budworm (*Heliothis virescens*) in contact with hemolymph and found penetration rates comparable with those observed in living larvae. The alternative explanation, brought forward by Olson for dieldrin, probably does not apply to a hydrophilic compound such as methomyl that passes through isolated house fly integument no more readily than dieldrin (35).

We come now to the second group of Gerolt's arguments, those directly supporting his theory of lateral movement through the integument and entry via the tracheae.

6. If an insecticide, applied on a house fly abdomen, moves laterally through the cuticle and reaches the CNS via the tracheae of the thorax, then an absorbent in intimate contact with the cuticle should affect the speed of action. This assumption was tested by providing house flies with a ring of beeswax around the waist before application of dieldrin. Such wax-protected flies were indeed found to be insensitive to dieldrin applied on the abdomen, whereas with the same amount of insecticide applied on the thorax, knockdown was only slightly delayed. Similar results were obtained with cockroaches (*Blatella germanica*), provided that with this insect the abdominal spiracles were first sealed. Wax rings also protected against OP compounds and carbamates. In another experiment it could be shown that part of the dieldrin, applied on the abdomen, could be recovered from the thorax of unprotected flies, but not from the thorax of wax-protected flies, by washing with methanol.

7. When methomyl is topically applied on the house fly abdomen, it is partially recoverable from the external thorax cuticle with methanol washings. Appropriate controls proved that this carbamate did not reach the thorax by mechanical transfer. Injected methomyl could not be recovered from the thorax. Similar results were obtained with several methomyl analogues. The evidence, summarized in this and the preceding paragraph, is indeed difficult to reconcile with transport by hemolymph, and it is Gerolt's strongest argument.

8. House fly larvae, exposed to a dry deposit of dieldrin, accumulate this compound in the tracheae, whereas several other tissues such as gut, fat body, and Malpighian tubes, as well as hemolymph, do not. It must be noticed, however, that many other investigators could not detect such an accumulation in the tracheae after topical application of various insecticides to adult house flies (5, 61, 64). Burt et al (24) have tried to verify Gerolt's hypothesis directly by introducing pyrethrin I into the tracheae, close to where these organs enter the nervous system, and monitoring for abnormal electrophysiological symptoms. Their results do not confirm a tracheal entry. Finally, it has been observed that upon intoxication of house flies with OP insecticides, inhibition of AChE in the thoracic ganglion proceeds from the periphery to the center of this organ and is not (initially at least) restricted to those areas where the tracheoles enter the ganglion (see below for more detailed description of these observations).

In summary, it is my opinion that the data collected by Gerolt are not sufficient to disprove the transport role of hemolymph. Some of his experiments are open to alternative interpretations or could not be confirmed by other investigators. Since according to most investigators translocation of insecticides to the CNS via the tracheae is open to question (and this is an essential part of Gerolt's theory), the alternative view seems still preferable. Strong evidence in favor of the latter has not been invalidated, i.e. the concentration of diazoxon and dimethoxon found in hemolymph after topical application of an LD_{50} dose, is sufficient to block conduction through synapses of metathoracic ganglia (23, 27). Furthermore, very recent results obtained by Gerolt (37) suggest that the movement of insecticides in insects is mediated by a process that requires the expenditure of metabolic energy, and this seems to favor the idea of hemolymph transport more than the idea of lateral movement through the cuticle. However, the mode of entry of contact insecticides may be much more complex than a simple hemolymph transport model suggests. Consequently, since knowledge of the mode of entry is essential to the construction of any toxicodynamic model, attempts directed at a reconciliation of the opposing views would be highly recommendable.

PENETRATION

Measurement Techniques

The outcome of measurements of rate of penetration into insect integuments is strongly dependent on the experimental method used. The usual technique is application of a small volume of solution of an insecticide in a volatile solvent, followed by washing off the remaining insecticide after an appropriate interval with a large volume of solvent. Both the nature of the solvent (57, 62) and the amount of it (3, 43) used for application affect penetration rate. In this way the disappearance of insecticide from the outer integument layer is measured, so it is small wonder that the physicochemical properties of the wash-fluid determine to a certain extent how effectively the applied material can be recovered. Elliott et al (30) observed that the pyrethroid 2,3-dimethyl benzyl chrysanthemate applied on the mustard beetle could be recovered much less effectively with methanol than with hexane. They assumed that hexane is a better solvent for lipids than methanol or acetone and therefore could destroy waxy areas and improve extraction. Some researchers distinguish between material that is extractable with water and with organic solvent (57). According to Olson (55), an insecticide present in the so-called A-layer of the epicuticle (mobile grease) can be extracted with water, whereas with a subsequent wash with an organic solvent, material present in the deeper integument layers (B-layer) is recovered. The rate of disappearance of DDT from A- and B-layer of the American cockroach is distinctly different.

The techniques mentioned so far only give information about the amount of insecticide no longer present on or in the outer integument layer. This quantity need not be equal to that transferred to hemolymph, since part of it is probably still present in the cuticle but not extractable by washing an insect externally. In order to study the amount of insecticide that has passed the cuticle completely, O'Brien (53) advocates the use of the so-called disk technique: some time after application of the toxicant a disk or patch of integument, large enough to contain any solute that has spread laterally from the point of application, is removed and analyzed for its content of insecticide.

Matsumura (48) discriminates between malathion, which can be washed off with acetone from the cuticle of *Periplaneta americana,* and malathion adsorbed on cuticular proteins, which can be recovered by extraction of the whole insect body with warm water.

Pretreatment of insect cuticle is another factor influencing penetration rate. Removal of epicuticular lipids by 5 min contact with silicic acid preparations retards malathion penetration in susceptible house flies, and to a smaller extent in resistant ones (6), thereby diminishing the difference in penetration rate between flies of the two strains. Abrasion of cockroach (*P. americana*) cuticle with silicic acid had no effect on the total amount of paraoxon penetrating in the first hour after application, although the initial uptake by the cuticle seems to be accelerated (56).

Finally, it should be mentioned that the difference between amount applied and amount washed off is not necessarily equivalent to amount penetrated, since a considerable proportion of the dose can get lost by evaporation (28).

Penetration Kinetics

If the logarithm of the percentage of insecticide not yet penetrated is plotted vs time, a straight line is sometimes obtained. Penetration is said then to be first-order with respect to the insecticide, because it is mathematically easy to prove that such a linear relation will hold true, if the net transport from the outer to the inner integument layer is proportional to the external concentration. The first-order differential equation describing this process is then:

$$dc/dt = -kc, \quad 1.$$

from which follows, after integration,

$$\ln(c/c_0) = -kt \quad 2.$$

in which c is the concentration of the penetrating compound, t is time, c_0 concentration at zero time, and k is a constant. From Equation 2 follows that the half-life of penetration is 0.69 k^{-1}.

Usually, net transport does not depend on the external concentration alone, but rather on the concentration difference on both sides of the barrier:

$$dc_1/dt = -k_1c_1 + k_2c_2 \quad 3.$$

in which c_1 and c_2 are the external and internal concentrations, respectively, and k_1 and k_2 are constants. Formally, the process quantified by Equation 3 is also first-order, but a plot of $\ln(c_1/c_0)$ vs time is curvilinear. Consequently, deviation from straight lines in plotting logarithm c versus time does not imply that penetration is not a first-order process.

Buerger (15, 16) has developed a theoretical model in order to explain linear log c-time plots on penetration of nonelectrolytes through integuments. This model describes the integument as an arbitrary number of serial and parallel compartments. The rate of transport is proportional to the concentration in the various compartments. Furthermore, it is assumed that the concentrations in all compartments, except those in the first and last one, are in steady state, and that the concentration in the last compartment (i.e. inside the insect) is negligible with respect to those in the others.

One wonders how realistic and consistent these assumptions are. First, back-diffusion is neglected, but this is not always allowed as was shown later on by Olson (55): DDT injected into American cockroaches could be recovered from the epicuticular lipids. Furthermore, Buerger's model implies, that steady state is very rapidly attained and that the amount of penetrating compound needed to saturate the integumental tissue (this is the corollary of his steady state assumption) is negligible with respect to the amount applied externally. However, Matsumura (48) showed that proteins from insect integuments can bind large amounts of malathion, and this also holds true for many other apolar insecticides. Finally, a major theoretical objection to this model is that the steady state assumption implies that the rate of disappearance from the outermost cuticle layer equals the rate of translocation to the innermost compartment. Since this rate is constant (proportional to a constant concentration), the penetration rate must be constant. Hence, the model does not predict linear plots of logarithm of concentration vs time.

Since in the majority of literature reports log c-time plots are curved rather than straight lines, it seems reasonable to suppose that linear plots are to be considered as a borderline case of the more general situation, characterized by an initial large penetration constant that gradually diminishes. Comparison of Equations 1 and 3 shows that when back-diffusion is negligible (i.e. when penetration is slow with respect to metabolism and/or excretion), linear plots are to be expected. The results of Schmidt & Dahm (63) and of Lindquist & Dahm (45) illustrate this. When their experimental data are plotted as logarithm of concentration vs time, straight lines are obtained for the penetration of piperonyl butoxide and DDT in the Madeira roach, with very long half-lives: 1.1 and 7.5 days, respectively. In a report of Uchida et al (73) on penetration of dimethoate in four insect species (house fly, milkweed bug, Colorado potato beetle, and American cockroach), all penetration curves deviate distinctly from linearity, but the largest deviation is found with house flies, which have the highest penetration rate and the smallest breakdown rate.

In deriving Equation 2 it has been tacitly assumed that the penetration constant k is independent of the concentration. However, many authors have reported that this constant decreased with increasing amounts applied. Results with DDT, dimethoate, and malathion have been summarized by Olson & O'Brien (57). Farnham et al (32) found a fivefold reduction in k in the case of diazinon penetration in SKA house flies by increasing the dose 100 times. This strain is very resistant to organophosphorus insecticides, and part of the resistance is caused by reduced penetration. In a susceptible strain these authors found that the proportion of diazinon penetrating per time unit was constant over a dose range from 0.01 to about 0.3 μg per fly; however, when the dose was increased further, the penetration constant dropped sharply. The inheritable penetration factor (mentioned above) causing resistance has been studied by Sawicki & Lord (62). The efficiency of this factor in reducing penetration is strongly dependent on dose, since the difference between strains with and without this factor disappeared at higher dosages. The delaying effect also seems to be correlated with the polarity of the compound: the difference is larger with diazinon than with diazoxon, and for the very polar molecule dimethoate the two strains do not differ at all (R. M. Sawicki, personal communication).

The slope of the penetration curve can also be affected by adsorption of the diffusing substances on cuticular proteins. This is a rapid process, reaching equilibrium very soon after the insecticide has been applied. The penetration curve then shows a very steep initial part (i.e. a very large penetration constant), changing—sometimes abruptly—into a part with much smaller slope. Matsumura's data are a good example for this (48). Penetration curves with very steep initial parts have also been reported by other investigators (30, 41, 43) for pyrethroids and benzene hexachloride isomers, although the mechanism was not clarified. Literature data show that both pyrethroids (24) and benzene hexachloride isomers (42, 43) are strongly adsorbed by proteins, which suggests that a similar mechanism, as elucidated by Matsumura, may be involved. Buerger & O'Brien (17) investigated penetration of DDT, famphur, and dimethoate through the cuticle of three insect species with the disk technique mentioned earlier and concluded that penetration rate is proportional to concentration, after an initial short period of very rapid uptake. Since this technique measures the amount of the insecticide lost by the whole integument and not only disappearance into the deeper integument layers, adsorption by proteins seems less probable as an explanation for this fast initial decrease of integumental insecticide concentration. Its actual magnitude, however, is highly variable, and taking into account the low zero-time recovery, for which seemingly no correction was applied, it may be absent altogether.

Relation between Penetration and Physicochemical Parameters

Penetration rate of insecticides through insect integuments is positively correlated with their polarity. For four insecticides and one noninsecticide, Olson & O'Brien (57) have determined both penetration velocity and partition coefficient for a system of olive oil and water. The insecticides used were dimethoate, paraoxon, dieldrin, and DDT, and the partition coefficients ranged from 0.34 to 316, in this order. They found that an increasing partition coefficient ran parallel to an increasing half-life of penetration. Their results have been used by Penniston et al (59) to quantify this relationship. They could show that this relationship is parabolic:

$$\log T_{1/2} = a(\log P)^2 + b \log P + c \qquad 4.$$

in which $T_{1/2}$ is the half-life of penetration, P is the olive oil/ H_2O partition coefficient, and a, b and c are constants.

Consequently, there is a certain value for log P for which log $T_{1/2}$ is minimal or penetration rate is maximal (a is positive); this value of log P is $-$ 1.46. It seems that this low value only has theoretical significance, since all known insecticides have log P values larger than $-$ 1.46 for olive oil/H_2O. The correlation coefficient of Equation 4 (0.998) is surprisingly high considering the limited number of compounds used and their widely differing structures.

The physiological interpretation of such a parabolic dependence of penetration on polarity has been sought in the behavior of molecules passing biological membranes. With decreasing polarity, molecules penetrate more easily into lipid-rich membranes, but on the other hand the higher the polarity, the faster they can leave such barriers. Hence, there must be an optimum in polarity at which passing through membranes is maximal. Olson & O'Brien explain increasing penetration rate with increasing polarity in a very similar way. They point out that the inner layers of insect integument are more polar than the lipid-rich outer layers. Partitioning of an insecticide from the epicuticle into the procuticle is therefore favored by increasing polarity. If, however, an insecticide dissolved in a very polar solvent (e.g. water) is applied, it is to be expected that partitioning into the epicuticle is rate-limiting, and this is favored by low polarity.

Besides partitioning, adsorption onto proteins may affect penetration rate. A combination of the octanol/water partition coefficient and a binding constant for adsorption of benzene hexachloride isomers on bovine serum albumin or nylon powder gave a better description of penetration and translocation of these insecticides in *P. americana* than partition coefficient alone (42, 43).

Insecticides when ingested enter the insect body by absorption from the alimentary tract. Speed of penetration through the midgut wall has been measured by Shah et al (66) with dimethoate and some dialkoxy analogues. Tobacco hornworm larvae (*Manduca sexta*) had a considerably higher penetration rate than the cockroach (*Blaberus cranifer*), but the most important conclusion from their work is that on decreasing polarity of the compounds studied, the penetration rate constant also diminished. Metabolism of some compounds in the gut wall can be an interfering factor in such investigations (65).

In 1967 O'Brien (52) wrote "one would have imagined that our knowledge of the factors which determine . . . integument penetration into insects would be substantial . . . but in fact an astonishing deficiency exists in our knowledge of this area, offering rich dividends to future investigators." Although some progress has been made since then, it seems that these dividends are still far from being paid in full. In particular a systematic comparison of the various methods for measuring penetration and a more extensive exploration of the relationship between penetration and physicochemical properties might be highly informative.

BIOAVAILABILITY

The term bioavailability is used in the pharmacological literature in several meanings (76). Here it is defined as the rate and extent to which an insecticide is absorbed and becomes available for the site of action.

Absorption on Proteins and Other Tissue Components

During penetration of the cuticle, transport by hemolymph, and distribution throughout the insect body, an insecticide can be absorbed by tissue components, such as proteins, and this affects the concentration at the target site. Lord (46) has estimated the proportion in which several insecticides, mainly OP compounds, are absorbed on tissue solids of house flies, by homogenizing whole insects and adding insecticide to suspensions of sedimentable material in buffer. The proportion of diazinon absorbed turned out to be independent of the concentration (0–50 μM) and absorption was completely reversible. Diazoxon, however, was much less absorbed and reached a saturation level at about 30 μM. The absorptive properties of the tissue solids can be partly ascribed to their lipid content, but on extraction with acidified chloroform-methanol mixtures, part of the absorptive capacity is retained. Increasing lipophilicity of the insecticide favors absorption. Lord's results, extrapolated to the in vivo situation, mean that the ratio of bound to free in solution varies from 3 for malaoxon to $\sim$1000 for dieldrin at equilibrium. These values underestimate the true in vivo situation because no correction was applied for binding to soluble tissue constituents.

A similar technique was used by Burt & Lord (23) and Burt et al (24) to determine absorption of diazoxon and pyrethrin I, respectively, on tissue solids of a cockroach (*P. americana*). The ratio of diazoxon bound to 1 g of solids (dry weight) to that in solution in 1 ml of buffer in equilibrium with it is 3.7. In other words, in vivo 40% of the total amount of diazoxon present is bound and 60% is free in solution in tissue fluids. For pyrethrin I, however, the absorptive capacity of tissue components is much higher and the ratio of bound to free in solution is 30,000 for whole tissue solids and 20,000 for nerve cord.

Sorption of four vinyl phosphates on solids of house fly heads has been measured by Boyer (9), using equilibrium dialysis. His results could be summarized by a Freundlich absorption isotherm:

$$\log (x/m) = \log K + (1/n) \log C \qquad 5.$$

in which x/m is the amount of OP insecticide bound per milliliter of absorbing suspension, C is the equilibrium concentration, and K and n are constants. If n does not differ too much from unity, then it follows from Equation 5 that the amount sorbed is proportional to concentration. The higher the solubility in water, the lower is the propensity to be absorbed. For tetrachlorvinphos, Boyer's data can be used to calculate that the ratio of bound to free in solution in a homogenate of 25 fly heads per ml varies from 2100 to 2900 at a concentration range of 10–0.1 μM. Extrapolated to in vivo this would mean that about 2–3% of the total amount of insecticide is free in solution.

Since hemolymph plays an important role in insecticide transport, binding to hemolymph proteins influences the speed at which the toxicant accumulates at the target site. Binding to hemolymph has been studied by several investigators (10, 56, 67, 75). For technical reasons the experimental animal usually is the American cockroach.

According to Olson (56), the solubility of dieldrin in water is 0.13–0.4 μM, but in hemolymph it is 57 μM, 200–400 times higher. High and low molecular weight proteins are involved in binding. Water solubility of tetrachlorvinphos is 30 μM, in hemolymph 220 μM (10). With equilibrium dialysis it was shown that the relationship between quantity absorbed and concentration follows a Freundlich isotherm (see Equation 5, above). There are at least two binding sites, namely, one with a dissociation constant of 0.36 μM and a maximal binding capacity of 0.16 μmole ml^{-1} hemolymph, and another with 5.0 μM and 0.52 μmole ml^{-1}, respectively. According to Winter et al (75), there is only one electrophoretically distinguishable protein that binds DDT, whereas Skalsky & Guthrie (67) find that more than one protein takes part in binding and moreover that DDT and dieldrin are absorbed by different protein molecules.

Tissue and Hemolymph Concentrations and Physiological Effects

In this section we restrict ourselves to those publications that enable us to relate accumulation data with effects caused by action on the target. This choice implies that a number of studies, in particular on chlorinated insecticides, dealing with accumulation and distribution only are not reviewed here. Recent reviews by Brooks

(14, 14a) cover this literature adequately. Studies with AChE inhibitors are first discussed and subsequently some results obtained with pyrethroids are summarized.

Burt & Lord (23) found that one hour after topical application of an LD_{50} dose of diazoxon to *P. americana,* the total, internal concentration reaches a maximum and then decreases rapidly, probably as a result of degradation (23). Hemolymph has a similar concentration-time profile. There is a close agreement between intoxication symptoms and electrophysiological disturbance in the metathoracic ganglion, whereas the sixth abdominal ganglion remains completely unaffected (23).

On continuous irrigation of exposed metathoracic ganglia with buffer solutions of diazoxon, a linear relationship between the logarithm of diazoxon concentration and the logarithm of blocking time (i.e. irrigation time needed to attain complete blockade of axonic conduction through the ganglion for the first time) was established. Comparison of two strains showed that a higher blocking time was found in insects with a lower LD_{50}, which implies that susceptibility at the site of action does not by itself determine susceptibility of the insect as a whole (21, 23).

From combined histochemical and electrophysiological work (21, 22), it appeared that AChE inhibition starts before conduction block can be observed, but that in this period inhibition is restricted to the peripheral part of the metathoracic ganglion. At the beginning of conduction block AChE inhibition has reached the neuropile; roughly 30% of the enzyme has then been inactivated.

Since the maximal diazoxon concentration in hemolymph is, as far as symptoms are concerned, equivalent to about the same concentration in irrigation experiments, the authors conclude that their results indicate that diazoxon in the CNS is in approximate equilibrium with that in hemolymph. It is interesting to compare these results with those obtained with dimethoate and its oxon analogue, dimethoxon, in similar experiments (27). Diazoxon and dimethoxon are about equally toxic to *P. americana* (LD_{90} of 2.6 and 3.5 μg per insect, respectively), but the former inhibits cockroach AChE 270 times faster than the latter. The speed of development of toxic symptoms also differs considerably. On application of an LD_{90} dose of diazoxon, the insects are already badly affected after 2 hr, but with dimethoxon this takes 10–20 hr more.

Just as with diazoxon the toxic symptoms caused by dimethoxon are closely correlated with conduction block in the metathoracic ganglion, but not with disruption of electrical activity in the sixth abdominal ganglion. In irrigation experiments with dimethoxon, a 200-fold higher concentration than with diazoxon is needed to block conduction through the metathoracic ganglion at the same time (20 min). However, if the same concentration of both compounds (e.g. 20 μM) is used, the ratio of blocking times is only 36. The authors do not elaborate on this point, but it seems plausible to explain it by differences in penetration rate. Perhaps dimethoxon, which is more polar than diazoxon, penetrates the ganglion faster than diazoxon, but this relative difference is diminished at dimethoxon concentrations much higher than that of diazoxon (it is admitted that the assumed relationship between polarity and penetration rate into the ganglion is largely hypothetical).

The discrepancy between equal toxicity and large difference in inhibition rate of AChE for diazoxon and dimethoxon is explained by a remarkable distinction in metabolic stability between the two compounds. Whereas the tissue concentration of diazoxon passes through a maximum after 1 hr, dimethoxon increases steadily over a period of 16 hr.

Miller et al (49) have used disruption in the coordination between the activation of motor units in the flight muscle of house flies (uncoupling) as a measure of disturbance of CNS activity. They found that in general carbamates were faster-acting than OP insecticides, whether administered topically on the flies or perfused on exposed thoracic ganglia, but speed of action was in no way correlated with toxicity. Dichlorvos was an exception in that it exerted its toxic effect very fast. As soon as the first uncoupling symptoms became visible, the amount of insecticide present in the thoracic ganglion was determined. Small amounts of the carbamate carbofuran were found (approximately 30 ng per ganglion), but much larger amounts of the OP compounds leptophos and monitor (approximately 1000 ng) were found. This distinction between carbofuran and both OP insecticides could be explained neither by difference in the rate of accumulation in the ganglion nor by difference in polarity. The authors suggest that the fast-acting compounds may have a different site of action in the ganglion, i.e. carbamates cause uncoupling before reaching the neuropile, whereas OP compounds might cause inhibition in the neuropile itself before uncoupling or before other symptoms of poisoning are produced. The probability of this hypothesis would be much higher if differences in degradation rates and reaction with AChE could be ruled out.

There are several papers dealing with the relationship between intoxication symptoms, and site and degree of AChE inhibition. According to Molloy (50), in house flies it is the thoracic ganglion that shows the first signs of AChE inhibition after topical treatment with a lethal dose of diazinon. The inhibition starts in the peripheral part of the ganglion (perineurium) and proceeds from there into the neuropile in later stages of intoxication. Even with flies that are badly affected or prostrate, inhibition is seldom complete. In the initial stage of the intoxication process (inhibition restricted to the peripheral part of the ganglion), about one third of the AChE activity is lost.

Similar results have been obtained by Farnham et al (31) using diazoxon. There is a close correlation between inhibition of AChE activity in the thoracic ganglion and external symptoms of poisoning. In distinction to Molloy, they find that in completely paralyzed flies almost no enzyme activity is left. With young flies (0-1 day old) of a diazoxon-resistant strain it was observed that at a dose that killed eventually 50% of the flies there was a considerable percentage recovery from paralysis. These flies had a normal AChE activity 24 hr after treatment.

With a few other OP insecticides Booth & Metcalf (7) came to essentially the same conclusions. They also included two carbamates in their experiments (aldicarb and *m*-isopropylphenyl methylcarbamate). The difference with the former type of AChE inhibitors is that with carbamates inactivation of the target enzyme in the central part of the thoracic ganglion was never observed, even in dead flies.

Biochemical and histochemical studies with house flies by Brady & Sternburg (12) and Brady (11) have shown that the level of AChE inhibition at knockdown is strongly dependent on the OP compound used. Earlier observations by Van Asperen

(4) point in the same direction. The very reactive inhibitor tepp (tetra ethyl pyrophosphate) knocks the flies down in 26 min, but with an equally toxic dose of dicrotophos it takes ten times as long to reach the same stage of intoxication. Although tepp is about 5000 times more reactive towards AChE in vitro than is dicrotophos, the extent of thoracic AChE inactivation with tepp was only half of that with dicrotophos, and, still more surprisingly, flies surviving treatment with dicrotophos had much more AChE inactivated than flies at knockdown with tepp. At knockdown with tepp, only AChE in the perineurium of the thoracic ganglion was strongly diminished, with dicrotophos also that in the neuropile. According to these investigators, these results present a paradox, which compels one to question the generally held view that inhibition of AChE at cholinergic synapses in the neuropile is solely or even primarily responsible for the toxicity of OP compounds to the house fly; components in the cell body region other than synapses or axons may be critically sensitive. Results obtained in studies with red flour beetle (*Tribolium castaneum*) indicate that the situation is similar to that in house flies (78).

Brain AChE of house flies seems to be better protected from OP inhibitors than thoracic AChE, but this protection is only temporary. On administration of tepp the brain enzyme is only half as inhibited as that of the thorax at knockdown (26 min), but with dicrotophos there is no difference (knockdown time 265 min). With a series of other OP insecticides (all *O,O*-dimethyl S-arylphosphorothiolates) the distinction between brain and thorax AChE susceptibility is even more pronounced: no inhibition at all of brain AChE at knockdown (8). The same compounds used on house crickets (*Acheta domestica*), however, gave a somewhat different picture: partial inactivation of peripheral thoracic AChE (increasing from metathoracic to prothoracic ganglion) and of brain enzyme in the hyperactive state (44), and almost complete inhibition at the same site at knockdown (8, 44).

Part of the AChE activity from head and thoracic ganglion of house flies does not sediment on high-speed centrifugation of homogenates. This soluble enzyme can be resolved in three to four isozymes on electrophoresis. Tripathi & O'Brien (72) studied these isozymes after in vivo treatment with an LD_{50} of four OP insecticides. Between 20 and 1280 min after dosing, the inhibition of the thoracic enzymes is always higher than of the head enzymes. The maximal inhibition is found between 20 and 80 min. One of the thoracic isozymes remains inhibited, even in surviving flies, whereas the others recover partially. One should realize in interpreting these data that in particular from 80 min onwards they present an average of dead and surviving flies. Zettler & Brady (79) found that in flies knocked down with tepp, the soluble thoracic enzymes are inhibited to the same extent as the whole homogenate, but with dicrotophos they are inhibited to a much smaller extent.

The significance of these results for our understanding of the intoxication process is difficult to evaluate. Only a small percentage of total AChE activity in the thoracic ganglion is soluble (78), and, moreover, the localization and physiological role of these isozymes is still obscure. Since with the head enzymes the order of sensitivity in vivo differs from the order of reactivity with these OP compound in vitro, it was concluded that they differ in their location within the brain (72). No data are available for thoracic isozymes.

The picture emerging from the above-summarized results is far from complete and is full of contradictions. As far as house flies are concerned, the toxic symptoms correlate reasonably well with AChE inhibition in the thoracic ganglion, but not with that in the brain. The results with house crickets indicate that this conclusion may not be true for all insects. There is no one fixed level of AChE inactivation corresponding with knockdown: the more reactive the OP compound is in vitro, the lower that level is. The relationship of inactivation of synaptic AChE with knockdown and death is puzzling. Whereas in some instances flies are knocked down or die with unaffected synaptic AChE, in other cases flies survive OP treatment with a considerable portion of their synaptic AChE blocked. It is obvious that future research should concentrate on elucidation of these inconsistencies. It might be useful, in addition to studying the site and extent of AChE inhibition in relation to development of toxic symptoms, to take into consideration the rate of penetration of the active compound from the hemolymph into the ganglion and its accumulation and distribution there.

Elliot et al (30) have described the penetration and internal accumulation of 24 pyrethroids in the mustard beetle (*Phaedon cochleariae*). Penetration is characterized by three phases. A very rapid initial one, in which about 40% of the topically applied dose is no longer extractable within a few seconds, is followed by a second phase in which the penetration rate constant gradually diminishes; finally there is a third phase in which the rate of penetration is proportional to the percentage of the amount remaining externally. Internally the compounds accumulate rapidly during phase 1 and 2, attain a plateau in phase 3, and then disappear gradually from the tissues. Insects under nitrogen (diminished oxidative degradation) show a lower penetration rate in phase 3.

In order to explain their results, the authors assume that during phase 3 penetration balances detoxication; in other words, the internal concentration is then in steady state and the slope of that part of the penetration vs time plot is therefore indicative of the detoxication rate. However, this assumption seems to be inconsistent with the experimental data: if the internal concentration is constant, then according to their model the elimination rate would be constant and thus also the penetration rate. In contrast to this consequence of what was assumed, however, a penetration rate proportional to external concentration is found.

Burt & Goodchild and Burt et al (19, 24) have studied penetration and accumulation of pyrethrin I in *Periplaneta americana* and related internal concentrations with disturbances of electrophysiological behavior of the CNS. When the proportion remaining on the outside (after topical administration) is plotted semilogarithmically against time, at least three consecutive phases can be distinguished, each with a half-life of penetration longer than the preceding one. The internal concentration reaches a plateau level after about 4 hr, and it is then 18% of the amount penetrated (72%), which points to a substantial degradation.

After application of an LD_{95}, the toxic symptoms correlate well with increased spontaneous activity in the sixth abdominal ganglion, but not with disturbance of axonal conduction. Four hours after application, when all insects are prostrate, this is still normal.

Attempts to determine the pyrethrin I content of nerve cords and hemolymph failed because the concentrations present were below the detection limit. However, hemolymph collected from cockroaches, treated 1–2 hr previously, contained sufficient toxicant to evoke a tenfold-increased spontaneous activity in the sixth abdominal ganglion and to reduce conduction of nerve impulses in giant fibers of untreated insects. On irrigation of exposed ganglia with saline solutions of pyrethrin I, it was found that the pyrethrin concentration in the saline, which was needed for inducing similar symptoms (as with hemolymph from treated cockroaches), had to be at least tenfold higher than what could maximally be present in hemolymph. It is not clear how to explain this discrepancy.

INTERACTIONS

The various processes that contribute, whether positively or negatively, to the accumulation of a toxic product at the site of the target can interact, and the degree of interaction affects the eventual toxicity. Two types of interaction may be distinguished. The first one is exemplified by O'Brien's well-known opportunity factor (54). If a detoxication enzyme is saturated with substrate, then it will degrade a larger proportion if this substrate is conveyed to the detoxication site at lower rate. A decrease of penetration rate, for example, may provide the degrading enzyme with an opportunity to increase its efficiency, which results then in an increased LD_{50}. For the same reasoning to apply, the detoxication system need not be saturated; the only condition is that the reaction rate increases less than proportionally to the substrate concentration ($S \ll K_m$ for Michaelis-Menten kinetics). In general we might call this type of interaction kinetic interaction. It is not restricted to penetration and detoxication but may also be expected between an activation and a subsequent detoxication reaction.

The relationship between rate of entry and toxicity has been studied more systematically by Sun & Johnson (70, 71). They used a congeneric series of monocrotophos analogues [*O,O*-dimethyl *O*-(3-hydroxy N-alkyl-*cis*-crotonamide) phosphates] and determined LD_{50} values for house flies after application, infusion, and injection, with and without the detoxication inhibitor sesamex. In particular for the higher members they found that LD_{50} (topical application) $>$ LD_{50} (infusion) $>$ LD_{50} (injection) and that synergism factors decreased in the same order. This is indicative of strong interaction between rate of entry and rate of detoxication: the faster the compounds enter, the smaller is the fraction detoxified and therefore the lower is the efficiency of the synergist. The ratio of the LD_{50} on infusion over that on injection is thus a useful measure for detoxication, and this is reflected in a very good correlation between this ratio and the degree of synergism on testing each compound in oil spray. No correlation was observed between toxicity and bimolecular rate constant for AChE inhibition, even when the compounds were injected and synergized with sesamex.

Another interesting example of interaction between penetration and detoxication has been reported with several carbamates. The LD_{50} for house flies was much higher on topical application than on spraying in oil; the oil seemed to synergize these compounds. This phenomenon is explained by assuming that kerosene enhances penetration and thus diminishes interaction. For this sort of synergism, the term quasi-synergism was coined to distinguish it from synergism caused by inhibition of detoxication.

In studies on the relationship between dose and speed of knockdown, Burt & Goodchild (20) observed that the same amount of pyrethrin I or of bioresmethrin, whether injected or applied topically, was needed to kill 50% of the insects 24 hr after treatment. This suggests, according to these authors, that by this time similar proportions of the dose had been eliminated from the insects, irrespective of the mode of administration. Consequently, there is no interaction between speed of entry and detoxication, and one might guess that degradation rate is proportional to the substrate concentration, or in other words that $[S] << K_m$. This seems a reasonable assumption in view of the very low concentration of pyrethrin I in solution, as mentioned earlier.

Although the relationship between knockdown and kill by pyrethroids is far from completely understood, there is no reason to suppose that different mechanisms are responsible. Pyrethroids that are good knockdown agents are usually more polar than good killing agents (13). One could speculate that higher polarity means faster accumulation at the target site and therefore rapid knockdown, but for a better understanding one also needs information on the relationship between polarity and detoxication rate and affinity to the target.

A rather complicated, but very intriguing, example of interaction is the shorter time to knockdown in resistant house flies, as compared with susceptible ones, on injection with parathion (29). The flies are resistant to OP insecticides as a result of increased oxidative detoxication, but in vitro they also have an increased capacity to activate parathion to paraoxon. This capacity does not seem to show up on topical application, but on injection the tissues are flooded with parathion and the resistant flies produce paraoxon at a higher rate than the susceptible.

The second type of interaction is the biochemical and biophysical one. The former occurs when an insecticide itself, its activation product, or its degradation product inhibit some step in the chain of events between contact and lethal reaction. An example is the inhibition of oxidative degradation of an organophosphate by its thionophosphate precursor (58, 74). If intoxication is enhanced, we might speak of positive feedback interaction; negative feedback interactions have not yet been reported, to our knowledge.

An example of biophysical interaction has been discussed by Boyer (9). Binding of the AChE inhibitor tetrachlorvinphos to indifferent protein lowers its concentration and thus its reaction rate with the target enzyme. Whether this results in protection and increased LD_{50} depends on the system degrading tetrachlorvinphos (if any) being saturated or not. Decrease of inhibitor concentration delays AChE in reaching its critical, lethal level, but if the degradation rate is proportional to substrate concentration, then this delay does not result in an increased total amount of toxic compound degraded, i.e. there is no change in LD_{50}. With carbamates, where the decarbamylation rate of AChE usually is appreciable, or with insecticides that react reversibly with the target, the above argument does not necessarily hold, and the interaction may be much less easy to predict.

MODELS

Since there are so many processes that contribute to the final lethal action of an insecticide, qualitative descriptions soon fall short of explanatory power and quantitative approaches must be substituted. A prerequisite, however, to any calculation is a simplification of the complex nature of a living organism; only those elements that are essential to describe adequately the insect-insecticide interaction are retained. This is called modeling. A serious obstacle is that we do not know beforehand which elements are essential and which are not.

Besides furthering our understanding of how insecticides perform their action, there is still another objective that might be served by modeling. If appropriate computation facilities are available, models can be used for simulating insecticide behavior (provided they are qualitatively correct), and with this technique we may learn to what extent variations in rates of the processes included in our model contribute to the final outcome. Distinguishing between sensitive and insensitive steps may give us a clue to how to manipulate chemical structures in order to obtain more selective insecticides. In view of the lasting need for such compounds, it seems worthwhile to try this rational approach, although success is by no means guaranteed.

We therefore conclude this review with a concise survey of insecticide models that have been described so far. In contrast to the very rich and extensive literature on the use of kinetic models in pharmacology (e.g. see 76), the number of publications on models of insecticide action is still very limited.

The fundamentals of such models have been outlined by Courshee (26) and by Sun (69). The former uses one- and two-compartment models, and the processes that determine accumulation are all first-order. Theoretical results are used to explain mortality of locusts caused by dieldrin. The mathematical approach of Sun is much less strict and is in fact a procedure to derive rates from data published by other investigators and to integrate these rates graphically. A rather confusing point in this publication is that no sharp distinction is made between rate and amount, and this has led to an incorrect presentation of penetration rate at zero time (zero instead of maximal) and consequently of the amount of active produced in Sun's Figures 1–6.

Hollingworth (39) described a multicompartment model to illustrate the difference in accumulation rate of fenitroxon (activation product of fenitrothion) in a susceptible and resistant strain of house fly. The resistant flies had a slightly lower penetration rate and a much higher degradation rate of fenitroxon. The model is depicted in Figure 2. There are five compartments (A–E) of which only the first three are essential to insecticidal action. A is the concentration of external fenitrothion, B of internal inactive precursor, C of internal fenitroxon, D and E of detoxication products; k_1 through k_4 are first-order rate constants. The dynamics are summarized by the following rate equations:

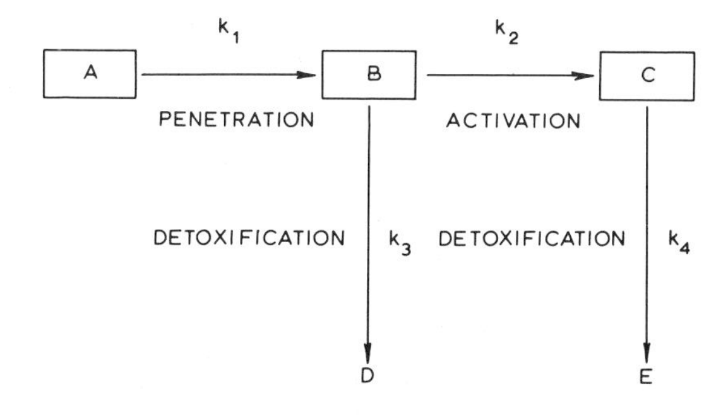

Figure 2 Kinetic model of penetration and metabolism of fenitrothion in house flies. From Hollingworth (39).

$$dA/dt = -k_1A \quad 6a.$$
$$dB/dt = k_1A - (k_2 + k_3)B \quad 6b.$$
$$dC/dt = k_2B - k_4C \quad 6c.$$
$$dD/dt = k_3B \quad 6d.$$
$$dE/dt = k_4C \quad 6e.$$

In this simple model an analytical solution for Equations 6a–e is not yet very difficult, but with increasing complexity of the model, analytical solutions are often no longer possible, and one must change over to numerical computation techniques. Figure 3 is a graphical representation of the solution of the five differential equations; it shows that in the resistant fly the concentration-time product of fenitroxon is much smaller than in the susceptible.

An essential element in the model of Figure 2 has been intentionally omitted, i.e. interaction of fenitroxon with AChE in the CNS. This step is of prime importance in the model designed by Green (38), in order to quantify the toxic effect of cholinesterase inhibitors in mammals and the therapeutic effect of some antidotes. Here the conception of minimum essential AChE was introduced: as soon as the percentage of remaining AChE activity drops below this level, an animal dies. This approach has been used by Smissaert et al (68) to calculate the minimum AChE fraction compatible with life in susceptible and resistant spider mites (*Tetranychus urticae*). The two strains differ in only one aspect, i.e. an altered AChE in the resistant strain with much smaller bimolecular rate constant k_i of inhibition by OP compounds and carbamates. A simple equation was derived to relate minimum AChE level P, k_i and LD_{50} of OP insecticides for both strains:

$$\ln P/\ln P' = (k_i \times LD_{50})/(k_i' \times LD_{50}') \quad 7.$$

in which the prime denotes the symbol values in one of the two strains. Assumptions implicit in equation 7 are (*a*) the amount of inhibitor detoxified is much larger than the amount used for irreversible AChE inactivation (shown to be a reliable assumption), and (*b*) all processes determining rate of accumulation of the inhibitor are first-order in both strains. For five organophosphates, differing widely in both

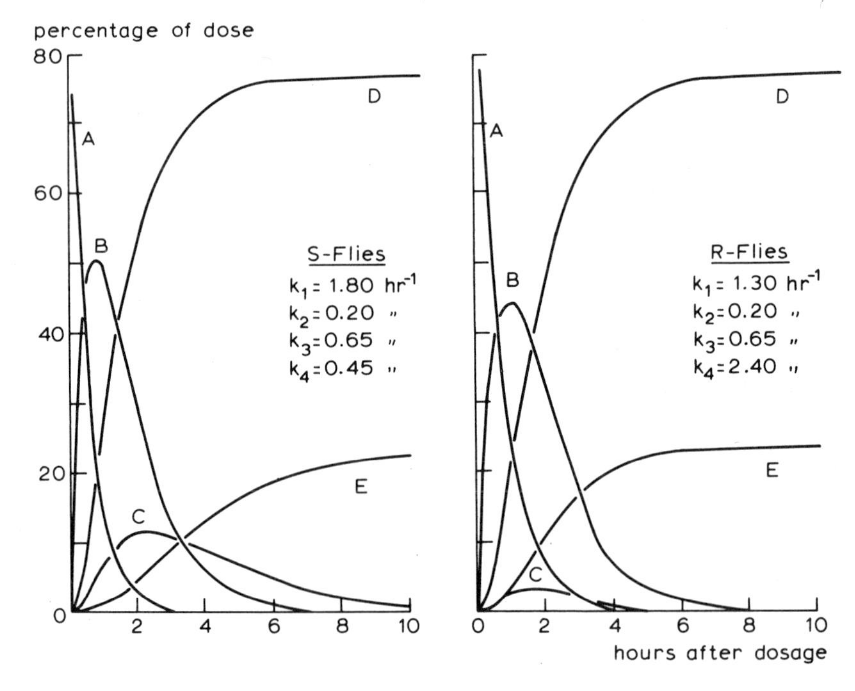

Figure 3 Time course of fenitrothion and its metabolites in susceptible (S) and resistant (R) house flies, calculated with the model depicted in Figure 2. From Hollingworth (39).

LD_{50} as well as k_i values the righthand part of equation 7 was remarkably constant. This seems to corroborate the assumptions made in deriving the equation.

By comparing LD_{50} values for the resistant strain and its F_1 hybrid with the susceptible one, the value of P for the resistant mites could be calculated. The F_1 has two types of AChE: one-half is the resistant type and the other half is the susceptible type. Because for a number of OP insecticides the ratio of the LD_{50} of the resistant strain over that of the hybrid was almost nonvariant, it was concluded that the inhibition of the susceptible-type AChE in the hybrid did not effect the LD_{50} value. It was calculated that the minimum AChE activity in the resistant strain was 16% of its normal value and in the susceptible strain 2.3%. Since the resistant strain has a lower AChE activity than the susceptible, it was to be expected that the minimum AChE activity, expressed as a proportion of the normal uninhibited activity, should be larger in the resistant strain.

The most complete model so far described is that developed by McFarlane (51) for action of oxime carbamates on house flies (see Figure 4). There are four compartments: cuticle, hemolymph, nerve tissue, and target, with volumes V_c, V_h, V_n and V_t. The denotations hemolymph and nerve tissue are used in a somewhat loose sense. They also include bound tissue water and fatty tissue (in which the compound may be stored), respectively. An amount Q_0 is administered topically at zero time

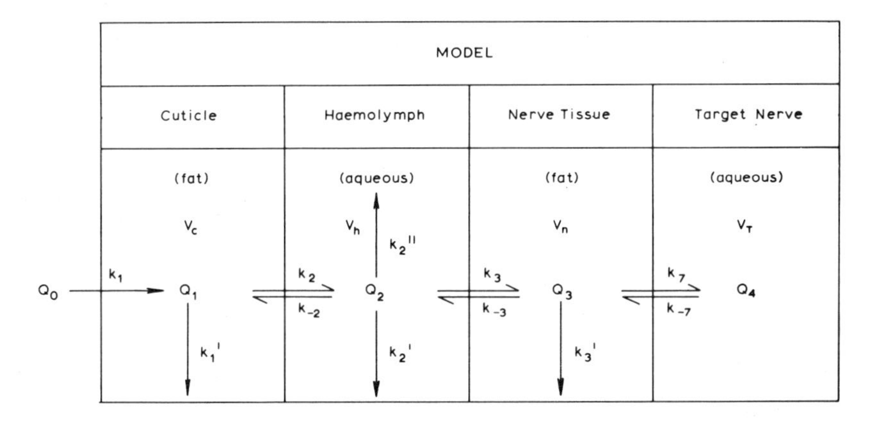

Figure 4 Kinetic model of penetration, distribution, metabolism, and reaction with the target of oxime carbamates in house flies. From McFarlane et al (51).

and penetrates with rate constant k_1 into compartment 1, where degradation (rate constant k_1') takes place. Transport to compartments 2, 3, and 4 is determined by rate constants $k_{\pm 2}$, $k_{\pm 3}$, and $k_{\pm 7}$, and degradation by k_2' and k_3'. Interaction with AChE at the target site (assumed to be the synaptic clefts) is given by the following scheme:

$$Q_4 + E \underset{k_{-4}}{\overset{k_4}{\rightleftharpoons}} EQ_4 \overset{k_5}{\rightarrow} EQ_4' \overset{k_6}{\rightarrow} E + Q_4'$$

in which k_4 and k_{-4} are the second- and first-order rate constants respectively, for formation and disassociation of the enzyme inhibitor complex EQ_4, k_5 is the carbamylation rate constant, and k_6 is the decarbamylation rate constant.

This model has been used to calculate the proportion of AChE inactivated as a function of both the amount of oxime carbamate applied and of time; the results are shown in Figure 5 in three dimensions. This could be compared with data on the percentage of flies knocked down, also as functions of dose and time. Since not all parameters are known with the same extent of certainty, and some even had to be guessed at, the correlation between computed and experimental results was considered satisfactory, at least at the qualitative level, and strongly suggested that the model reflected carbamate action on house flies.

All models, mentioned above, are based on the assumption of first-order kinetics (except for the reaction with AChE), and one wonders how safe this assumption is. A biochemical reaction is only first-order if the substrate concentration is much smaller than the Michaelis constant. As has been outlined earlier, presence of an opportunity factor implies that some process, be it degradative or activative, approaches saturation. The order of such a process is generally lower than one. On the other hand, first-order also implies substrate concentration much larger than

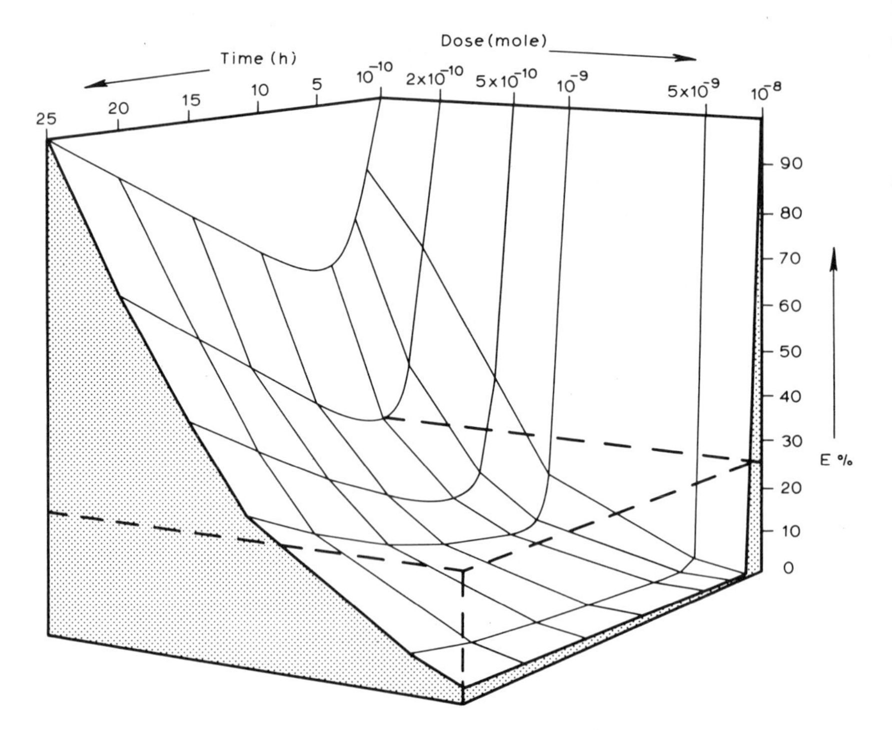

Figure 5 Three-dimensional relationship between time, dose, and percentage of AChE inactivated, calculated with the model depicted in Figure 4. From McFarlane et al (51).

enzyme concentration, and it remains to be seen whether this condition is always complied with.

It is obvious that the art of modeling of insecticide action is still in an early larval stage. Much has still to be learned about the relevance of the various processes reviewed here. Since at present none of the models discussed above has been put to the test, it is premature to discuss amply the parameters that have been used so far or that still have to be determined, and how this should be done. Models predicting the time course of insecticide concentrations in hemolymph and target tissue could be used to begin with. It is obvious that the subsequent step, i.e. introduction of the interaction with the target into such models, and explanation of toxic effects will be much more difficult, in particular if the target is not well defined. Work with AChE inhibitors seems therefore preferable to insecticides with another mode of action. The results obtained by Smissaert et al (68), however, show that starting the other way around, i.e. with effects on the target, may also be a useful approach, if appropriate insect strains are available. The link between observed biochemical or biophysical disturbances and mortality will be another hard nut to crack. This will at least necessitate the introduction of a statistical element in our deterministic models. We may hope that by taking advantage of the large amount of experience gained by pharmacologists in handling pharmacodynamic models, a major breakthrough can also be achieved in the field of toxicodynamic models for insecticidal action in the near future.

ACKNOWLEDGMENTS

I am very grateful to my colleagues Drs. F. J. Oppenoorth and H. R. Smissaert for reading my manuscript critically and for suggesting many useful improvements.

I thank Dr. N. R. McFarlane for sending me a preprint of his paper and permitting me to cite from it.

Literature Cited

1. Ahmed, H., Gardiner, B. G. 1968. Difference in susceptibility to malathion exhibited by various regions of the body of the desert locust. *Soc. Chem. Ind. Monogr.* 29:169–80
2. Ahmed, H., Gardiner, B. G. 1968. Variation in toxicity of malathion when applied to certain body regions of *Schistocerca gregaria* (Forsk.) *Bull. Entomol. Res.* 57:651–59
3. Ahmed, H., Gardiner, B. G. 1970. Penetration of malathion into locusts. *Pestic. Sci.* 1:217–19
4. van Asperen, K. 1960. Toxic action of organophosphorus compounds and esterase inhibition in houseflies. *Biochem. Pharmacol.* 3:136–46
5. Benezet, H. J., Forgash, A. J. 1972. Penetration and distribution of topically applied malathion in the house fly. *J. Econ. Entomol.* 65:53–57
6. Benezet, H. J., Forgash, A. J. 1972. Reduction of malathion penetration in house flies pretreated with silicic acid. *J. Econ. Entomol.* 65:895–96
7. Booth, G. M., Metcalf, R. L. 1970. Histochemical evidence for localized inhibition of cholinesterase in the house fly. *Ann. Entomol. Soc. Am.* 63:197–204
8. Booth, G. M., Lee, A.-H. 1971. Distribution of cholinesterases in insects. *Bull. WHO* 44:91–98
9. Boyer, A. C. 1967. Vinyl phosphate insecticide sorption to proteins and its effect on cholinesterase I_{50}-values. *J. Agric. Food Chem.* 15:282–86
10. Boyer, A. C. 1975. Sorption of tetrachlorvinphos insecticide (Gardona) to the haemolymph of *Periplaneta americana. Pestic. Biochem. Physiol.* 5:135–41
11. Brady, U. E. 1970. Localization of cholinesterase activity in housefly thoraces: inhibition of cholinesterase with organophosphate compounds. *Entomol. Exp. Appl.* 13:423–32
12. Brady, U. E., Sternburg, J. 1967. Studies on in vivo cholinesterase inhibition and poisoning symptoms in houseflies. *J. Insect Physiol.* 13:369–79
13. Briggs, G. G., Elliott, M., Farnham, A. W., Janes, N. F. 1974. Structural aspects of the knockdown of pyrethroids. *Pestic. Sci.* 5:643–49
14. Brooks, G. T. 1974. *Chlorinated Insecticides. II. Biological and Environmental Aspects,* pp. 197 Cleveland: CRC Press

14a. Brooks, G. T. 1976. In *Insecticide Biochemistry and Physiology,* ed. C. F. Wilkinson. New York: Plenum. In press

15. Buerger, A. A. 1966. A model for the penetration of integument by nonelectrolytes. *J. Theor. Biol.* 11:131–39
16. Buerger, A. A. 1967. A theory of integumental penetration. *J. Theor. Biol.* 14:66–73
17. Buerger, A. A., O'Brien, R. D. 1965. Penetration of non-electrolytes through animal integuments. *J. Cell. Comp. Physiol.* 66:227–34
18. Burt, P. E. 1972. Factors controlling the process of poisoning by topically-applied neurotoxic insecticides. *Proc. Int. Congr. Entomol. 14th, Canberra*
19. Burt, P. E., Goodchild, R. E. 1971. The site of action of pyrethrin I in the nervous system of the cockroach *Periplaneta americana. Entomol. Exp. Appl.* 14:179–89
20. Burt, P. E., Goodchild, R. E. 1974. Knockdown by pyrethroids: its role in the intoxication process. *Pestic. Sci.* 5:625–33
21. Burt, P. E., Gregory, G. E., Molloy, F. M. 1966. A histochemical and electrophysiological study of the action of diazoxon on cholinesterase activity and nerve conduction in ganglia of the cock-

roach *Periplaneta americana* L. *Ann. Appl. Biol.* 58:341–54

22. Burt, P. E., Gregory, G. E., Molloy, F. M. 1967. The activation of diazinon by ganglia of the cockroach *Periplaneta americana* L. and its action on nerve conduction and cholinesterase activity. *Ann. Appl. Biol.* 59:1–11
23. Burt, P. E., Lord, K. A. 1968. The influence of penetration, distribution, sorption and decomposition on the poisoning of the cockroach *Periplaneta americana* treated topically with diazoxon. *Entomol. Exp. Appl.* 11:55–67
24. Burt, P. E., Lord, K. A., Forrest, J. M., Goodchild, R. E. 1971. The spread of topically-applied pyrethrin I from the cuticle to the central nervous system of the cockroach *Periplaneta americana. Entomol. Exp. Appl.* 14:255–69
25. Collins, W. J., Hornung, S. B., Federle, P. F. 1974. The effect of topical and injected olive oil on knockdown by DDT, dieldrin, methyl parathion, and dimethoate in *Musca domestica. Pestic. Biochem. Physiol.* 4:153–59
26. Courshee, R. J. 1968. Kinetics of lethal effects of pesticides. *Soc. Chem. Ind. Monogr.* 29:220–40
27. Devonshire, A. L., Burt, P. E., Goodchild, R. E. 1975. Unusual features of the toxic action of dimethoate on *Periplaneta americana* and their causes. *Pestic. Biochem. Physiol.* 5:101–8
28. Devonshire, A. L., Needham, P. H. 1974. The fate of some organophosphorus compounds applied topically to peach-potato aphids [*Myzus persicae* (Sulz.)] resistant and susceptible to insecticides. *Pestic. Sci.* 5:161–69
29. El Bashir, S., Oppenoorth, F. J. 1969. Microsomal oxidations of organophosphate insecticides in some resistant strains of house-flies. *Nature* 223:210–11
30. Elliott, M., Ford, M. G., Janes, N. F. 1970. Insecticidal activity of the pyrethrins and related compounds III—Penetration of pyrethroid insecticides into mustard beetles (*Phaedon cochleariae*). *Pestic. Sci.* 1:220–23
31. Farnham, A. W., Gregory, G. E., Sawicki, R. M. 1966. Bioassay and histochemical studies of the poisoning and recovery of house-flies (*Musca domestica* L.) treated with diazinon and diazoxon. *Bull. Entomol. Res.* 57:107–16
32. Farnham, A. W., Lord, K. A., Sawicki, R. M. 1965. Study of some of the mechanisms connected with resistance to diazinon and diazoxon in a diazinon-resistant strain of houseflies. *J. Insect Physiol.* 11:1475–88
33. Gerolt, P. 1969. Mode of entry of contact insecticides. *J. Insect Physiol.* 15:563–80
34. Gerolt, P. 1970. The mode of entry of contact insecticides. *Pestic. Sci.* 1:209–12
35. Gerolt, P. 1972. Mode of entry of oxime carbamates into insects. *Pestic. Sci.* 3:43–55
36. Gerolt, P. 1975. Role of insect haemolymph in translocation of insecticides. *Pestic. Sci.* 6:223–38
37. Gerolt, P. 1975. Mechanism of transfer of insecticides in *Musca domestica. Pestic. Sci.* 6:561–69
38. Green, A. L. 1958. The kinetic basis of organophosphate poisoning and its treatment. *Biochem. Pharmacol.* 1:115–28
39. Hollingworth, R. M. 1971. Comparative metabolism and selectivity of organophosphate and carbamate insecticides. *Bull. WHO* 44:155–70
40. Krasner, J., Giacoia, G. P., Yaffe, S. J. 1973. Drug-protein binding in the newborn infant. *Ann. NY Acad. Sci.* 226:101–14
41. Kurihara, N., Nakajima, E., Shindo, H. 1970. In *Biochemical Toxicology of Insecticides,* ed. R. D. O'Brien, I. Yamamoto, pp. 41–50. New York: Academic
42. Kurihara, N., Uchida, M., Fujita, T., Nakajima, M. 1973. Studies on BHC isomers and related compounds V. Some physicochemical properties of BHC isomers (1). *Pestic. Biochem. Physiol.* 2:383–90
43. Kurihara, N., Uchida, M., Fujita, T., Nakajima, M. 1974. Studies on BHC isomers and related compounds VI. Penetration and translocation of BHC isomers in the cockroach and their correlation with physicochemical properties. *Pestic. Biochem. Physiol.* 4:12–18
44. Lee, A.-H., Metcalf, R. L., Booth, G. M. 1973. House cricket acetylcholinesterase: histochemical localization and in situ inhibition by *O, O*-dimethyl *S*-aryl phosphorothioates. *Ann. Entomol. Soc. Am.* 66:333–43
45. Lindquist, D. A., Dahm, P. A. 1956. Metabolism of radioactive DDT by the Madeira roach and European corn borer. *J. Econ. Entomol.* 49:579–84
46. Lord, K. A. 1968. Studies on the penetration and sorption of insecticides in insects. *Soc. Chem. Ind. Monogr.* 29:35–46

47. Lord, K. A., Molloy, F. M., Potter, C. 1963. Penetration of diazoxon and acetyl choline into the thoracic ganglia in susceptible and resistant houseflies and the effect of fixatives. *Bull. Entomol. Res.* 54:189–98
48. Matsumura, F. 1963. The permeability of the cuticle of *Periplaneta americana* (L) to malathion. *J. Insect Physiol.* 9:207–21
49. Miller, T., Kennedy, J. M., Collins, C., Fukuto, T. R. 1973. An examination of temporal differences in the action of carbamate and organophosphorus insecticides on houseflies. *Pestic. Biochem. Physiol.* 3:447–55
50. Molloy, F. M. 1961. The histochemistry of the cholinesterases in the central nervous system of susceptible and resistant strains of the house-fly, *Musca domestica* L., in relation to diazinon poisoning. *Bull. Entomol. Res.* 52:667–81
51. McFarlane, N. R., Paterson, G. D., Dunderdale, M. 1977. Modelling as a management tool in pesticide research. In *Crop Protection Agents—Their Biological Evaluation,* ed. N. R. McFarlane. London: Academic. In press
52. O'Brien, R. D. 1967. *Insecticides. Action and Metabolism,* p. 255. New York: Academic
53. Ibid, p. 258
54. Ibid, p. 267
55. Olson, W. P. 1970. Penetration of ^{14}C-DDT into and through the cockroach integument. *Comp. Biochem. Physiol.* 35:273–82
56. Olson, W. P. 1973. Dieldrin transport in the insect: an examination of Gerolt's hypothesis. *Pestic. Biochem. Physiol.* 3:384–92
57. Olson, W. P., O'Brien, R. D. 1963. The relation between physical properties and penetration of solutes into the cockroach cuticle. *J. Insect Physiol.* 9: 777–86
58. Oppenoorth, F. J., Voerman, S., Welling, W., Houx, N. W. H., van den Oudenweyer, J. W. 1971. Synergism of insecticidal action by inhibition of microsomal oxidation with phosphorothionates. *Nature* 223:187–88
59. Penniston, J. T., Becket, L., Bentley, D. L., Hansch, C. 1969. Passive permeation of organic compounds through biological tissue: a non-steady-state theory. *Mol. Pharmacol.* 5:333–41
60. Polles, S. G., Vinson, S. B. 1972. Penetration, distribution, and metabolism of ^{14}C-endrin in resistant and susceptible tobacco budworm larvae. *J. Agric. Food Chem.* 20:38–41
61. Quraishi, M. S., Poonawalla, Z. T. 1969. Radioautographic study of the diffusion of topically applied DDT-^{14}C into the house fly and its distribution in internal organs. *J. Econ. Entomol.* 62:988–93
62. Sawicki, R. M., Lord, K. A. 1970. Some properties of a mechanism delaying penetration of insecticides into houseflies. *Pestic. Sci.* 1:213–17
63. Schmidt, C. H., Dahm, P. A. 1956. The synthesis of C^{14}-labeled piperonyl butoxide and its fate in the Madeira roach. *J. Econ. Entomol.* 43:729–35
64. Sellers, L. G., Guthrie, F. E. 1971. Localization of dieldrin in house fly thoracic ganglion by electron microscopic autoradiography. *J. Econ. Entomol.* 64:352–54
65. Shah, A. H., Guthrie, F. E. 1971. In vitro metabolism of insecticides during midgut penetration. *Pestic. Biochem. Physiol.* 1:1–10
66. Shah, P. V., Dauterman, W. C., Guthrie, F. E. 1972. Penetration of a series of dialkoxy analogs of dimethoate through the isolated gut of insects and mammals. *Pestic. Biochem. Physiol.* 2: 324–30
67. Skalsky, H. L., Guthrie, F. E. 1975. Binding of insecticides to macromolecules in the blood of the rat and American cockroach. *Pestic. Biochem. Physiol.* 5:27–34
68. Smissaert, H. R., Abd El Hamid, F. M., Overmeer, W. P. J. 1975. The minimum acetylcholinesterase (AChE) fraction compatible with life derived by aid of a simple model explaining the degree of dominance of resistance to inhibitors in AChE "mutants." *Biochem. Pharmacol.* 24:1043–47
69. Sun, Y.-P. 1968. Dynamics of insect toxicology. A mathematical and graphical evaluation of the relationship between insect toxicity and rates of penetration and detoxication of insecticides. *J. Econ. Entomol.* 61:949–55
70. Sun, Y.-P., Johnson, E. R. 1969. Relationship between structure of several Azodrin insecticide homologues and their toxicities to house flies, tested by injection, infusion, topical application, and spray methods with and without synergist. *J. Econ. Entomol.* 62: 1130–35
71. Sun, Y.-P., Johnson, E. R. 1972. Quasi-synergism and penetration of insecticides. *J. Econ. Entomol.* 65:349–53

72. Tripathi, R. K., O'Brien, R. D. 1973. Effect of organophosphates *in vivo* upon acetylcholinesterase isozymes from housefly head and thorax. *Pestic. Biochem. Physiol.* 2:418–24
73. Uchida, T., Rahmati, H. S., O'Brien, R. D. 1965. The penetration and metabolism of ^{3}H-dimethoate in insects. *J. Econ. Entomol.* 58:831–35
74. Welling, W., DeVries, A. W., Voerman, S. 1974. Oxidative cleavage of a carboxylester bond as a mechanism of resistance to malaoxon in houseflies. *Pestic. Biochem. Physiol.* 4:31–43
75. Winter, C. E., Giannotti, O., Holzhacker, E. L. 1975. DDT-lipoprotein complex in the American cockroach haemolymph: a possible way of insecticide transport. *Pestic. Biochem. Physiol.* 5:155–62
76. Wagner, J. G. 1975. *Fundamentals of Clinical Pharmacokinetics.* Hamilton, Ill.: Drug Intelligence Publ. 461 pp.
77. Yeager, F. J., Munson, S. C. 1945. Physiological evidence of a site of action of DDT in an insect. *Science* 102:305
78. Zettler, J. L., Brady, U. E. 1970. In vivo cholinesterase inhibition in *Tribolium castaneum* with organophosphate insecticides. *J. Econ. Entomol.* 63: 1628–31
79. Zettler, J. L., Brady, U. E. 1975. Acetylcholinesterase isozymes of the housefly thorax: in vivo inhibition by organophosphorus insecticides. *Pestic. Biochem. Physiol.* 5:471–76

Ann. Rev. Entomol. 1977. 22:79–100

DYNAMICS OF LARCH BUD MOTH POPULATIONS[1,2]

❖6121

W. Baltensweiler, G. Benz, P. Bovey, and V. Delucchi
Department of Entomology, Swiss Federal Institute of Technology, Zürich, Switzerland

Ideally studies on the population dynamics of organisms should be holistic, general, precise, and realistic (64). Much too often, however, support for research on forest pests approximates the curve of the gradations in time and space, resulting in periodic gathering of bits and pieces of information that differ little from that gathered earlier or elsewhere (68). Therefore it is not surprising that lack of comprehensive, quantitative information on many generations of a field population has handicapped the progress of theoretical thinking on the dynamics of populations, which badly needs "more light and less heat" (86). It has been proposed that research efforts should be coordinated internationally on a few model insects in order to cope with complexity and diversity of population dynamics at a high level of efficiency (25). We consider that the larch bud moth, *Zeiraphera diniana,* would meet the criteria for being such a model insect (18).

The coincidence of large-scale insect outbreaks, the availability of effective insecticides, and economic expansion of industry and tourism after World War II promoted several extensive control actions against such forest pests as the spruce budworm, *Choristoneura fumiferana,* in eastern Canada and the larch bud moth, *Z. diniana,* in Switzerland. Public demand to protect the subalpine larch forests in the recreation area of the Engadine (Switzerland) against the recurring defoliation by the larch bud moth triggered the present research phase. This article is the first attempt to publish the results of the research period (1949–1975) comprehensively. In addition, information concerning previous outbreaks of *Z. diniana* is reviewed on a worldwide base.

[1]This work was supported mainly by the Swiss National Science Foundation and the Swiss Federal Institute of Technology (Zürich) during the period from 1949 to 1975. In addition, funds were received from the schweizerischer Fonds zur Förderung der Wald- und Holzforschung and the Swiss cantonal government of Graubünden and Wallis. The great help provided by the forest services of Austria, France, Italy, and Switzerland is thankfully acknowledged.

[2]Contribution No. 87 of the working group for the study of the population dynamics of the larch bud moth.

TAXONOMY

The larch bud moth was first described by the French entomologist Guenée (1845) as *Sphaleroptera diniana,* based on a specimen collected near Digne in the Department Basses Alpes. In 1846 the German Zeller and the English Douglas described two tortricids as *Grapholita pinicolana* and *Poecilochroma occultana;* these have since been recognized as conspecific with *S. diniana.* Hübner's (1796–1799) description of *Tortrix griseana* was based on a doubtful pictorial presentation of the moth and has been invalidated. Subsequently the species was included in several other tortricid genera, such as *Steganoptycha* Steph., *Enarmonia* Hb., *Semasia* (Steph.) Kenn., and *Eucosma* Hb., and was finally placed in the Holarctic genus *Zeriaphera* (37).

Z. diniana is Palaearctic in distribution. In Europe it is found from the Pyrenees, the Alps, and the Tatra Mountains, northward to England and Scandinavia. The subspecies *Z. diniana* var. *desertana* occurs throughout the boreal forests of Asia from the Ural Mountains to eastern Siberia (52). According to T. Oku (personal communication), *Z. diniana* found in Japan (65) resembles *Z. diniana* var. *desertana.* A larch bud moth feeding on several North American conifers was also associated with *Z. diniana* by various authors but was subsequently identified by Mutuura & Freeman (77) as *Z. improbana.* Unfortunately, they do not specify the characteristics used to distinguish the two species.

HOST PLANTS AND EPIDEMIC STATUS

Z. diniana is oligophagous. Feeding is restricted to conifer genera such as *Larix, Pinus,* and *Picea.* At defoliation densities, mature larvae occasionally feed on undergrowth *Abies.* Outbreaks occur periodically on *Larix decidua* in the Alps (16), on *L. russica* (= *L. sibirica*), and *L. gmelini* (= *L. dahurica*) in Siberia (83) at eight- to ten-year intervals. Defoliation usually does not last more than two years. F. Abgrall (personal communication) indicates for the first time the defoliation of an European larch plantation in 1974 and 1975 from the Pyrenees. *L. decidua* and *L. leptoleptis* were defoliated repeatedly but at irregular intervals in England (42) and Japan (65; K. Kamijo and T. Oku, personal communication). No defoliation but continuous presence of *Z. diniana* has been observed in Poland (102) on *L. decidua* var. *sudetica* and in Switzerland at altitudes below 800 m (56).

Outbreaks on pines and spruces appeared throughout Europe (Table 1) during the period from 1921 to 1934 and again from 1955 to 1972. They lasted regionally from five to ten years.

BIOLOGY

The larch bud moth is univoltine through its entire Palaearctic range. Unless specified, all results described in the following paragraph apply to *Z. diniana* that occur on larch in Switzerland. Moth flight occurs at altitudes below 1000 m in June and July (56); at higher altitudes it extends from July to October (69). The eggs are

Table 1 Outbreaks of *Z. diniana* on *Pinus* and *Picea* spp. in Europe

Host plant	Altitude (m)	Area (ha)	Period	Country	Reference
Pinus mugo	100	20	1902	Denmark	(54)
P. cembra	1900	1,000	periodic	Switzerland	(38)
P. silvestris	500–900	—	1921	Scotland	(42)
	500–900	100	1961; 1968–1969	Norway, Sweden	(41, 48)
	600	10	1964, 1967	England	(42)
P. contorta	750	200	1964, 1968, 1970	England, Scotland	(42)
	—	—	1921	Scotland	(42)
Picea abies	800	100,000	1924–1934	DDR/CSSR	(81, 82)
	200	10	1958	Finland	Kangas[a]
	1300	6,500	1956–1959	CSSR	(94)
	800	70,000	1965–1969	DDR/CSSR	(57, 93, 95)
	550	400	1968–1969	CSSR	(93)
P. sitchensis	—	—	1921	Scotland	(42)
	1500	100	1955–1957	England	(42)
	600	25	1965–1968	Scotland/England	(42)
	200	10	1968	Norway	(48)

[a] E. Kangas, personal communication.

deposited by means of a protractile ovipositor underneath bark scales and in cones (21) in the colline region, but at higher altitudes eggs are laid underneath the talli of the lichen *Parmelia exasperatula,* which cover three-year-old and older branches of larch. The eggs overwinter as embryos in which segmentation and gastrulation are just accomplished (27). Under optimum diapause conditions, which require a chilling temperature of 2°C for 120 to 210 days, egg mortality is only 4–7% and postdiapause development at 18°C is completed within six days. These temperature conditions produce similar effects as those found under field conditions in the subalpine region, where egg mortality is equally low. The cold-hardiness of *Z. diniana* eggs is remarkable. Bakke (13) determined the supercooling points of a Norwegian population on pine collected in November 1968 to be –46.5°C and in February 1969 to be –51.3°C. Eggs from Swiss larch-form populations tested in January 1970 exhibited somewhat higher supercooling points. Eggs from the lowland at 500 m supercooled at –38.4°C and from various subalpine provinces at –43.8°C (A. Bakke, personal communication). Temperatures of 18°C, 14°C, and 10°C applied for 30 days, each combined with subsequent storage temperatures at 2°C, were used to simulate field conditions at lower altitudes. Egg mortality after 180 days was 14–19%. But the egg stage could be prolonged up to 270 days before mortality rose above 36% (56).

At low altitudes, emergence of first-instar larvae occurs in April and extends over two to three weeks, whereas at the timberline emergence may last more than four weeks in May or the first part of June (17). Synchronization of emergence with the flushing of the larch is essential for successful installation of first-instar larvae; a mean needle length of 6–8 mm is optimal. Second- and third-instar larvae feed between the needles, and only the older third- and fourth-instars spin the needles together to form the characteristic fascicles within which the larvae rest during daytime. Feeding occurs mainly during night and early in the morning, with maximum activity around midnight (34). The mature fifth-instar larvae live along the branch axis in webbings, which often contain frass and dried needles. The larval period lasts 40–60 days and 4–6 needle clusters are inhabited but only partially consumed.

For pupation, the larvae descend to the ground at times of greatest light intensity, either by means of a silken thread or by dropping directly. They immediately enter the forest litter and construct a cocoon from humus and mineral particles. Metamorphosis lasts 25–36 days. Pupal weight varies considerably, even when the larvae are reared at optimum conditions. Female pupae exhibit a range from 33.7 to 38.0 mg, and male pupae vary from 28.7 to 31.5 mg (32, 100). The Siberian subspecies *Z. diniana* var. *desertana* reaches even higher weights. Naumenko (78) presents weights for female and male pupae of an outbreak population as 41.4 and 37.7 mg, respectively.

The moths emerge early in the morning, but flight activity does not begin before dusk (74) and lasts until midnight, if the temperature remains above 8°C (97). The female produces a male-attractive pheromone (31), which has been identified as *trans*-11-tetra-decenyl-acetate by the electroantennogram method (87), whereas the *cis*-isomer inhibits male attraction (31, 35). Fecundity is correlated with pupal weight (32) and varies considerably between generations. When reared at optimum conditions in the field, mean fecundity is as high as 174 eggs (maximum is 354 eggs) (100), but it decreases subsequent to defoliation by 75–85% (20, 32). Since *Z. diniana* has postmetabolic gametogenesis (30), the realization of the maximum egg potential is influenced by longevity (which ranges from 20 to 40 days in the subalpine region) (100), the status of nutrition during the moth stage (56, 69), the presence of stimuli for oviposition by insemination (30), the presence of the host plant (4, 30, 97), and weather factors (17, 100).

ECOTYPES AND LARVAL POLYMORPHISM

Bovey & Maksymov (38) described two sympatric colormorphs of *Z. diniana* that are distinguishable only during the final larval instar but are obviously related to their host plants: a dark morph, found mainly on *Larix decidua* and therefore called larch form, and an orange-yellow morph, which occurs on *Pinus cembra* and is called the cembran pine form. These authors also established the close synchronization of first-instar emergence with the bud break of their respective host plants, with the larch form hatching significantly earlier than the cembran pine form. It was concluded that selective survival on the host plants was of major importance for the existence of these two ecotypes (38).

Crossing experiments between the two extreme color forms yielded a full range of intermediate color phases. With the help of a classification scheme for colortypes, a great variability in colortype structure between years and sites was found for field populations of both the larch and the cembran pine forms (22). Heterozygotes show a pattern of dark and light areas on their head capsules, in which the light areas on a predominantly dark head capsule correspond with the dark areas of a predominantly light heterozygote. Relevant information from historic outbreaks on larch, pine, and spruce throughout Europe (see references cited Table 1), recent samples from England by Day (42), and examination of larval samples collected on spruce during the 1965 to 1969 outbreak in the Erzgebirge has revealed that the morphotypic structure of bud moth populations on spruce is intermediate between populations on larch and pine (19). Experimental results indicate that color polymorphism is induced by a complex environmental-genetic interaction. Coloration of the fifth-instar larvae follows Gloger's rule: cool temperatures raise the proportion of dark colortypes and warm temperatures induce light coloration. However, inbreeding and assortative mating for the dark analplate result in 100% black progeny, whereas sibmating for the light analplate or the black head capsule induces variable proportions of the pure traits. Developmental rates of morphotypically intermediate F_1 progeny from crossings between fast-developing larch and slow developing pine or spruce forms are also intermediate, and occasionally a sixth-larval instar may occur (W. Baltensweiler, in preparation).

Day & Baltensweiler (43) confirmed that when optimum conditions for food and temperature are provided the relative survival for offspring of dark parents is slightly higher than of intermediate parents. Nutritional stress, on the other hand, leads to an increase in frequency of intermediate morphotypes in subsequent generations. Thus, colormorphism is indicative of population fitness.

These results reveal that polymorphism in the bud moth expresses itself along two different time scales: (*a*) within the generation, where polymorphic plasticity allows individual acclimatization to the different temperature conditions of the late larval instar feeding site on larch, pine, or spruce; (*b*) between generations since the genetically controlled change in population fitness, triggered by nutritional stress, influences genotypic structure of subsequent generations.

POPULATION DYNAMICS IN THE ALPS

Methods of Study

The dynamics of abundance in successive generations, which is defined as the fluctuation, is described verbally by means of Schwerdtfeger's terminology (91) and quantitatively by various methods which complement each other in time and space.

Cartography of larch defoliation was used as an index of maximum density of larval populations in the alpine distribution area. In the first year of visible defoliation, this method delimits areas of fastest population growth. This task was carried out with the assistance of the forest services of France, Italy, Switzerland, and Austria over the last two gradation cycles (W. Baltensweiler, in preparation). Information on defoliation patterns prior to 1960 was recovered from the archives of the forest services and from other literature as far back as 1811.

The population census initiated by Auer in 1949 (67) provides a basic population estimate (76) for the larval stage of each generation on a weight basis (number of larvae per kg of larch twigs with foliage); this basic population density may be converted to an absolute estimate (11), or it may serve for quantification of the defoliation index (5). The census was intended to give information on the spatial pattern of population density in successive generations covering 62 km^2 of forests in the Engadine Valley. This valley is orographically a closed unit of 120 km^2 in the center of the alpine arch. Phytosociologically most of its forests belong to the subalpine *Larici pinetum cembrae* (49). The samples are taken in relation to topographic features (altitude, exposure, etc) and are weighted according to host tree density per unit area (8). The individual larch tree is considered as the sample unit and is selected at random each year. The proper sample, 1 kg of twigs with foliage, is extracted from three branches cut at three levels within the crown of the tree. In 1959 this large-scale population census was extended to four additional subalpine outbreak areas distributed along the alpine arch: the Briançonnais (France), the Goms (Switzerland), the Valle Aurina (Italy), and the Lungau (Austria) [see map in (44)]. Furthermore, 20 sample plots (12; W. Baltensweiler, in preparation) of restricted size were selected systematically along the altitudinal profile from the area of the cyclic fluctuation type within the subalpine larch-cembran pine forest to the area of the latent fluctuation within the broad leaf tree forest, the *Aro-Fagetum* (49) of the colline region of Switzerland. In these forests, the larch is grown in small patches from 1 to 100 ha in size. According to the size of the stands, 4 to 20 larch trees per stand are sampled, and up to 3 replicates within an area of 1000 km^2 were assumed to provide the desired representativeness.

Life table studies including a census of the egg stage, the larval stage, and the mature larva leaving the larch tree for pupation were conducted at three sites in the Engadine area and at one near Lenzburg (30 km west of Zurich) in the latent fluctuation area at 500 m. Moth dispersal was studied from 1971 to 1975 by a large-scale network of light and pheromone traps covering 30,000 km^2 in Switzerland (26).

Defoliation Pattern

Since 1850, information on 15 defoliation periods is available for the area of the canton Graubünden in southeastern Switzerland. Defoliation periods occurred in the Engadine Valley at intervals of 8.64 $\pm$ 0.29 years and the mean duration of a defoliation period amounts to 3.0 $\pm$ 0.22 years. Defoliation appears first at the warmest sites within the Engadine, i.e. in SW-SE exposure, and often only within a zone between 1700 to 2000 m altitude. The size of the area affected in the first year of an outbreak varies considerably among outbreaks; the warmer the weather has been prior to the particular year, the larger is the defoliated area. The regular temporal spatial sequence of defoliation patterns allows to classify the gradations in relation to weather (16).

Defoliation pattern within the subalpine forest zone reveals that the Engadine Valley in the central part of the alpine arch exhibits the most regular periodicity. There is some evidence that during the last century outbreaks appeared first in the Austrian Alps and spread in the following four years to the west, but in this century this trend was reversed. The gradation 1901 to 1903 serves as the pivotal point; this period was characterized by excessive precipitation throughout the Alps (17). The penultimate gradation was typical when damage occured from 1960 to 1962 in the

French Alps, from 1962 to 1964 in the Swiss Alps, and in 1965 only in the Austrian Alps. In contrast, the last outbreak was unusual because defoliation appeared in 1970 simultaneously in the French Alps and in the Southern Tyrolian Alps in the central part of the alpine arch (W. Baltensweiler, in preparation).

At a lower altitude, between 1200 to 1600 m defoliation occurred at irregular intervals only, the frequency being 0.4 to 0.6 as compared to the Engadine. Also, damage appeared one to four years later than in the subalpine zone (15).

Quantitative Characteristics of the Fluctuation Types

The larval population census provides unique information, since it covers three full generation cycles for the Engadine area (5, 12; C. Auer, in preparation), two complete cycles for four additional areas in the subalpine optimum region (12) and simultaneous densities for several local populations representing the irregular and latent fluctuation type in the montane and colline region of Switzerland. The fluctuations are described by four criteria (see Table 2): the mean density per cycle, the mean for the upper, and lower density limits, and their respective ranges. Based on the assumption that the logarithm of population size follows the normal distribution, about 90% of the observations fall within 1.5 standard deviations (sd) of the mean, i.e. within the range of 3 sd. The antilog$_{10}$ of 3 sd represents the multiplicative factor from the expected minimum to the expected maximum of the respective mean (101).

The first and the second cycle show remarkably similar values for all four criteria. The mean densities vary between 2 and 8 larvae/kg of foliage, which appears rather astonishing with respect to the 30,000-fold range between minimum and maximum density of a cycle. Maximum densities are even less variable (1.3- to 1.6-fold). In contrast, the 3rd cycle is rather aberrant; the Valle Aurina population remained at 8 larvae, well below the defoliation threshold of approximately 100 larvae. The Goms and Lungau populations, with 78 and 117 larvae, respectively, caused defoliation only locally, whereas the forests of the Briançonnais and Engadine were defoliated in the usual extensive manner. These differences in maximum density are even more remarkable as high and low peak densities alternate in a spatial sequence from west to east along the alpine arch (see Table 2—regions are arranged from west to east for the third cycle). Finally, minimum densities are quite similar throughout the subalpine area, with one exception: when the Engadine population dropped in 1967 to the extreme low level of 0.002 larvae per kg foliage, the very large range of 170,000 for the third cycle was induced.

The variability of the cyclic and latent fluctuation types is exemplified by several local populations (Table 2). In the Engadine, the Sils site in SE exposure had been defoliated ten times between 1878 and 1958, the Celerina site in NW exposure eight times. In the Brinzauls site 50 km to the northwest of the Engadine, defoliation occured only four times in the same period. Population dynamics of the Sils site reflects the proper cyclic type, whereas the latter two populations exemplify the aperiodic fluctuation type exhibited at suboptimum sites. The populations in Trimmis (near Chur) and Lenzburg (near Zurich), at a distance of 100 and 250 km, respectively, from the Engadine, are characteristic of the latent fluctuation type.

Table 2 Variability of the density of *Zeiraphera diniana*

Fluctuation type	Census period (year)	Region/locality	Census area (ha)	Altitude (m)	Mean density[a]			Density limits[a] Minimum			Density limits[a] Maximum		
					n	y	*	n	y	*	n	y	*
cyclic	1949–1976	Engadine	6200	1700–2100	28	2.40	35,270	3	0.015	196.29	3	274.2	1.64
1st cycle	1949–1958	Engadine	6200	1700–2100	10	2.82	28,807	1	0.018	—	1	331.8	—
2nd cycle	1959–1966	Engadine CH	6200	1700–2100	8	2.72	32,643	1	0.080	—	1	248.8	—
	1958–1965	Briançonnais F	8000	1500–2100	8	7.96	3,470						
	1958–1965	Goms CH	2600	1500–2100	8	2.68	13,614	4	0.067	13.71	4	247.2	1.33
	1960–1966	Valle Aurina I	3000	1500–2100	7	2.35	17,579						
	1961–1967	Lungau A	2500	1500–2100	7	5.41	10,876						
3rd cycle	1966–1975	Briançonnais			10	1.50	15,961	1	0.013	—	1	227.7	
	1966–1975	Goms			10	0.82	23,641	1	0.011	—	1	77.6	
	1967–1975	Engadine			9	2.81	171,279	1	0.002	—	1	249.6	
	1967–1975	Valle Aurina			9	0.81	1,097	1	0.021	—	1	7.6	
	1968–1975	Lungau			8	3.83	4,893	1	0.041	—	1	117.4	
Local populations													
cyclic													
optimum	1951–1975	Sils Engadine	10	1800	25	7.91	10,337	3	0.028	34.09	3	614.4	2.09
aperiodic	1951–1975	Celerina Engadine	10	1800	25	1.67	219,758	3	0.006	103.04	3	529.0	9.31
suboptimum	1951–1975	Brinzauls	10	1300	25	3.16	31,078	3	0.022	7.91	3	470.2	1.55
latent	1951–1975	Trimmis	10	700	25	0.95	2,966	3	0.018	33.23	3	26.4	5.58
	1966–1975	Lenzburg/Weiach	10 + 10	500	19	1.07	12	4	0.353	9.10	5	2.1	2.21

[a]Symbols: n, number of generations; y, number of fifth-instar larvae per 1 kg larch foliage; and *, range of y = antilog (3 × standard deviation).

Their mean densities of 1.1 larvae appear to be relatively high, but their maximum densities remain well below the defoliation threshold. The minimum density at Lenzburg is highest of all regional and local populations.

INFLUENCE OF CLIMATE AND WEATHER

It must be stressed that the development of the larch bud moth remains essentially unaltered, irrespective of the climatic conditions. The egg stage lasts ten months in the oak-beech forest at 500 to 800 m and 9 months in the larch-cembran pine forest at 1800 to 2200 m. The most obvious difference is the advancement of the active stages by about one or two months at low altitudes (17). As a consequence the egg stage is exposed to considerably higher temperatures, which prolongs diapause (56) at the cost of increased egg mortality (24). Above 2000 m, moth flight is delayed until September, but then temperatures below 8°C hinder flight and oviposition activity and limit population increase. Thus total developmental time of the bud moth fits best with the seasonal and diurnal course of temperature at altitudes of 1700–1900 m.

Radiation at these altitudes induces increased heterogeneity of the temperature regime and consequently of developmental processes and age structure of populations. Therefore impact of weather variability between years is effectively buffered and contributes to the climate and weather generally favoring population increase in the subalpine region (22). However, rainy and cold spring seasons, topographic features, and snow cover function to influence synchronization of first-instar emergence and bud break of the larch negatively and contribute to the consistent spatial pattern of differential population growth. The differentiation of optimum and suboptimum sites within the area of cyclic fluctuation is reflected in the development of mean densities in the course of two population cycles in relation to exposure and altitude (5, 8).

The sequential effect of temperature on bud moth development along the altitudinal profile in the Alps was used to formulate a weather pattern for maximum population increase. This approach was successful in explaining the population increase of *Z. diniana* on Norway spruce observed in central Europe (19) and on pine and larch in England (42).

DYNAMICS OF DISPERSAL

Local Dispersal

Olfactory stimulation from larch foliage is probably a necessary prerequisite for the female moth to release its sex pheromone and/or the male to respond to it. Males disperse by positive anemotaxis in a consistent nightly windfield (79, 96) which is induced by topography. It is also conceivable that the females aggregate by the same mechanism at localities such as gullies or mountain shoulders which are optimally suited for dispersion of the pheromone. This phenomenon contributes to the fact that the fastest population increases regularly occur at the same places.

Vagility of moths emerging in defoliated stands is very much increased, and since oviposition on larch trees without green foliage is significantly reduced, a redistribution of population density in inverse relation to defoliation intensity is induced (97). Such emigration and immigration on the local scale prolongs the outbreak phase. The great variability of the mean maximum density (Table 2) at the suboptimal Celerina site exemplifies this aspect.

Long-Range Dispersal

Conspicuous mass flights of *Z. diniana* on mountaintops or glaciers (50, 69), on the ice of the Arctic (53), or at the lights of cities (40) have been frequently reported. Invariably these observations were related to the presence of defoliated larch stands separated by distances up to 700 km. Each year during the last defoliation period in the Alps (from 1972 to 1974), W. Baltensweiler (in preparation) recorded three to four exodus flight periods that resulted in transit flights of more than 100 km. These flight periods are associated with indifferent barometric pressure over central Europe, light winds, and above-normal temperatures at 2000 m. On one occasion (26), migrating moths were grounded by a cold front and scattered all over the landscape and cities. During the following days, the moths gradually aggregated by appetitive flight behavior (66a) on the artificially grown larch trees within the broad leaf tree forest. These migrants not only increased the population density of the autochthonous population, but also altered the morphotypic population structure (W. Baltensweiler, in preparation).

FOOD PLANT RELATIONSHIPS

As in many other insect/plant relationships, the larch bud moth may deplete the larch to such an extent that the larval population loses its food resources and starves (75). Insect activity may also alter the plant physiologically and induce it to produce food which cannot fully satisfy the needs of the larvae. Tree mortality occurs after defoliation (5), and losses are particularly severe under certain weather conditions and/or after defoliation in two consecutive years (W. Baltensweiler and H. Rappo, in preparation). However, the fact that larch forests in the Alps have survived more than a dozen defoliation periods within the last 125 years (16) indicates that destruction of the basic resource is not an important factor in the population dynamics of *Z. diniana.*

Competition for Food and Space

Each larch tree furnishes a certain amount of food which allows the growth of a defined number of bud moth larvae which feed on the needle clusters. Their number depends on the dimensions of the crown or the length of the branches and twigs which bear a proportional number of clusters (11). Since larvae need a certain minimal amount of food to pupate and an even larger amount to reach normal pupal weights (32, 59), intraspecific competition for food becomes an important factor when the resources are depleted by high larval populations. Starvation then leads to heavy larval and pupal mortality, reduced weight of pupae and adults, and reduced fecundity (32, 36, 60, 97), probably as a direct consequence of the reduced fat body of the pupa. Fecundity may drop from an average number of 130 eggs per normally fed female to an average of 20 eggs per female in a defoliated forest (32). Besides furnishing food, the host plant may also act as an arena for intraspecific competition which involves territorial behavior like that observed for *Rhagoletis completa* (39) and *Sparganothis pilleriana* (88), or there may be competition for a more generalized requirement for space free of other insects of the same species. Similar to the pine looper, *Bupalus piniarius,* in which even infrequent contacts with other larvae decrease the size, fecundity, and fertility of the resulting females, irrespective of the superabundance of available food (63), it has been found that pupal weight of *Z. diniana* diminishes with increasing larval density (32, 60), and that 80% or more of the larvae may leave overpopulated trees as young larvae, when food is not yet a limiting factor (32). These observations have recently been confirmed by F. Omlin (unpublished data), who found that, in the culmination year 1973, an average of 66% of the larval populations left the trees before reaching the last larval instar and that food shortage led to additional mortality, resulting in an average total loss of 97.7% (92–100%) of the populations. This value, however, is exceptional. Despite the high mortality rate in a completely defoliated larch forest, the number of surviving adults is as a rule very high. Even though a substantial number of the early emerging moths migrate to green forests (97), the remaining moths with their reduced reproduction potential may still produce such a large number of eggs that, if only one third of them developed to fifth-instar larvae, defoliation would be repeated in the following year (F. Omlin, unpublished data). Fortunately for the subalpine larch, cumulative defoliation is the exception rather than the rule.

Changes in Tree Physiology Induced by Z. diniana

Defoliation of larch trees reduces the photosynthetic activity to such an extent that annual shoot growth is fully or partly inhibited (32). The same is true for cambial growth in the year(s) of defoliation and for one or two subsequent years (55). The latter effect may be regarded as part of the complex syndrome of delayed effects which have been described by Benz (32) and which may last for two to four years depending on the biotope and weather conditions. One of the most impressive delayed effects is the reduction of the food resources for *Z. diniana,* e.g. the length of needles may be reduced to less than 50% of normal (16; G. Benz, unpublished). These short needles appear to have a tougher texture and are usually darker than normal needles, though chlorotic signs may be found on some trees. Depending on the region and the year, the buds or even the fully grown needles may be covered with a layer of sticky oleoresin. The needles of some trees may grow to almost normal length, but 40–50% of the spurs do not produce needles at all. It seems that in these cases the reserves of the dead spurs are used for the growth of needles on the living spurs. Another regular aftereffect is a slowdown in the growth of larch needles.

Chemical analyses of needles from normal and defoliated larch trees showed that as a rule the latter contain a higher proportion of dry material, especially raw fiber (acid/detergent fiber) and less nitrogen (raw protein) (32; G. Benz, in preparation).

Induced Antibiosis

Feeding tests with larvae of *Z. diniana* indicated that the needles of defoliated trees are not as readily consumed as normal needles. This is partly because of the greater toughness of the former and partly because they contain more deterrents and/or less phagostimulants. Their nutritive value is reduced, not only because of the higher contents of indigestible fiber, but also because the nutrients are not as well digested, absorbed or assimilated (32). It is not known whether or not wound-induced proteinase inhibitors are involved, similar to those reported in solanaceous plants (61) or tannins that may reduce digestibility of proteins (51) while acting as deterrents.

Benz (32) found negative correlations between the fiber contents of the needles on one side and larval survival and pupal weight on the other side, whereas these parameters were positively correlated with the protein contents of the needles. Oleoresin on the needles may hinder the neonate larvae and may kill up to 100% of them (G. Benz, unpublished). Thus most of the above-mentioned aftereffects in defoliated larch trees have negative effects on *Z. diniana.* Since they reduce the rate of survival as well as the weight of larvae and pupae (and thus the fertility and fecundity of the adults), they may be regarded as induced transient antibiosis factors which temporarily increase the resistance of larch trees against *Z. diniana.*

As mentioned above, weather conditions and the biotope may influence the extent of the delayed reactions of defoliated trees. Some observations suggest that similar reactions of the larch may also be induced by certain weather conditions alone or in combination with stimuli, which the trees receive from the activity of bud moth larvae still far below defoliating density.

NATURAL ENEMIES

Parasitoids and Predators

The natural enemy complex associated with *Z. diniana* in western Europe comprises 94 species of parasitoids and several species of predators. Eight species (five *Gelis* and three *Mesochorus*) always act as hyperparasitoids and five others (four *Itoplectis* and *Mesopolobus subfamatus*) may be primary as well as secondary parasitoids.

The diversity of the natural enemy complex varies according to the altitude, and its greatest diversification occurs in the subalpine optimum zone of the host. In fact, 76 species of parasitoids have been obtained in the Engadine during the last three cycles of *Z. diniana* (14, 37, 44, 58, 85), whereas only 11 species have been observed in the lowland area near Lenzburg (56). Nine of the 11 species found in the lowland also occur at high altitude. As many as 29 parasitoids attacking *Z. diniana* in the Engadine parasitize other Lepidoptera associated with the larch tree; however, the most abundant parasitoid species are nearly monophagous, at least above 1600 m (44). Nearly 87% of the parasitoid species belong to the Ichneumonidae (the majority), Braconidae, and Chalcidae (Hymenoptera), and the rest to the Tachinidae

(Diptera). At the family level, the natural enemy complex of *Z. diniana* in Siberia appears to be the same. Florov (52) lists 33 species of the same four families and the Ichneumonidae are by far the most numerous of the complex. Day (42) presents a list of 29 parasitoids of *Z. diniana* on various host plants in England.

The complex of predaceous arthropods of *Z. diniana* has received much less attention than that of the parasitoids. It appears, however, that the species involved vary considerably by area and that only a few species are predators of *Z. diniana* in a given area. In the Engadine, some representatives of Dermaptera and Acarina were detected, whereas in the lowland areas near Zurich only one mirid and one neuropterous species were observed preying on *Z. diniana* (45).

The egg stage of *Z. diniana* has a duration of nine to ten months, with eggs exposed to the action of natural enemies, during the summer and possibly the following spring. Predaceous arthropods destroy a great number of eggs. In lowland areas of Switzerland, they eliminate in experimental egg populations up to 80% (56), but their impact on a sparse, widely dispersed autochthonous population is not known. The most important predator here is the mirid *Deraecoris annulipes.* In similar experiments at high altitudes where field populations reach comparable densities, it appears that predators play an important role in modeling the phase of the host cycle (45). The most important predaceous species in these localities are the mites *Balaustium murorum* and *Bdella vulgaris* (45). The impact of parasitoids, which exclusively belong to the genus *Trichogramma,* is insignificant. The degree of parasitism varies from 0.8 to 2.0% in lowland areas and from 0.6 to 6.8% at high altitudes.

Nearly all primary parasitoids of *Z. diniana* known today attack the larvae at different stages of development. Some of them have been observed parasitizing *Z. diniana* from the lowland to the upper limit of the forests, for instance, *Coccygomimus turionellae, Triclistus podagricus, Eubadizon extensor, Elachertus argyssa,* and *Dicladocerus westwoodii.* Despite their wide distribution, however, most have been observed as parasitoids of *Z. diniana* only in certain areas and to attack this host only occasionally. Seventy-five percent of the larval parasitoids from the Engadine belong to this category. The most common species attacking *Z. diniana* at high altitude are the ichneumonids *Phytodietus griseanae, Diadegma patens, Triclistus pygmaeus,* and *T. podagricus,* and the eulophids *E. argyssa, D. westwoodii,* and *Sympiesis punctifrons* (1, 14, 84). Several species of parasitoids attack the young host larvae, for instance, *T. podagricus, Apanteles* spp., and the eulophids. Most of the parasitoids are, however, associated with the last two free-living larval instars of *Z. diniana.* With the exception of the eulophids, the parasitoids are univoltine at high altitudes and overwinter in the litter as larvae (*P. griseana*), pupae (*T. pygmaeus*), or as free adults (*D. patens*)(14).

The degree of larval parasitism in lowland areas does not exceed 4.5% (56). At high altitudes, the parasitism as a rule remains below 10% during the phase of host density increase and augments progressively during the phase of host density decline, reaching a maximum of 70–80% two to three years after maximum host density. In general, the parasitoid complex tends to be dominated by the Ichneumonidae (in particular by *P. griseanae*) during the phase of highest host density, whereas the three species of eulophids become most important during the phase of population decrease and minimum density. However, the presence of parasitoids during the latter phase (one host larva per larch tree or less) is very difficult to ascertain. All other parasitoids are, at least in the Engadine, much less abundant (V. Delucchi and A. Renfer, in preparation). The loss of importance of eulophids in the course of host density increase is attributed by Aeschlimann (1, 2) to the distortion of sex ratio, increased larval mortality, and the insufficient paralysis of the host, and by Baltensweiler (14) to the behavior of the host.

Because of the difficulty to find pupae of *Z. diniana* in the soil, pupal parasitoids have been investigated only during a short phase of maximum host density in the Engadine as well as in the Simplon region (canton Wallis). Apart from the Gelinae, which are hyperparasitic in habit, pupae of *Z. diniana* seem to be attacked in the soil by a few species mainly represented by *Phaeogenes osculator* and some unidentified Ephialtinae (Ichneumonidae) (14). The degree of pupal parasitism may reach 20% locally, but in general it is less than 1%. The impact of predators on pupae has never been evaluated.

Entomopathogens

Mortality caused by microorganisms rarely plays an important role in the population dynamics of *Z. diniana* (6). A highly virulent *Baculovirus* sp. of the granulosis type (28, 29, 70, 71) reached epizootic levels during the culmination phase of the 1949 to 1958 gradation in the Engadine and reduced bud moth populations decisively (71, 72). Since then, the same virus has been found in populations of *Z. diniana* in France, Italy, and Austria, but it never again reached epizootic levels. During the 1959 to 1967 gradation, disease incidence in the Engadine was spotty and varied on individual trees from 0 to 25%, and the average mortality was 10% (G. Benz, in preparation). Granulosis incidence was negligible during the 1967 to 1977 gradation (89, 90). Results suggest the occurrence of viruses that were not apparent in a part of the bud moth populations of the Engadine (28; G. Benz, in preparation). These viruses may sometimes be activated by stress (90).

G. Benz (unpublished data) found an *Entomopoxvirus* sp. in bud moth larvae in the Engadine, but it never reached more than local significance. However, it may have been involved in the breakdown of Austrian *Z. diniana* populations in 1947 reported by Jahn (66), who thought the cause to be a polyhedrosis virus.

Besides viruses, Martignoni (57) described a sporozoan belonging to the microsporidian genus *Telohania,* from *Z. diniana.* G. Benz (unpublished data) found at least four species of microsporida: two *Octosporea* species, one *Telohania* species, and one *Nosema* species, which all together rarely killed more than 1% of the larval *Z. diniana* populations.

SYNTHESIS

Weather and climate in the subalpine region favor population growth of the bud moth for the most part (20). However, since no cyclic weather conditions with a periodicity of eight to ten years are known, it appears that weather may modify rather than release the cyclic gradations of *Z. diniana* (16). Other processes must therefore regulate its temporal abundance. The available data show that the subalpine larch forest and *Z. diniana* form an autoregulating life system in which the insect multiplies for four to five generations under predominantly favorable climatic and trophic conditions and thus, by transgressing the carrying capacity of the trees, changes the nutritional base in such a way that environmental resistance to the insect becomes more severe (32). During the phase of population increase, the dark colormorph proliferates, whereas subsequent to defoliation the slow-developing intermediate ecotype is selected for four to five generations (22). Thus a population cycle reflects directional selection for two different ecotypes. This highly dynamic negative-feedback system successfully protects larch stands from destruction. Since larvae at high density may destroy young cembran pine trees in the undergrowth of larch stands, the system involves a positive-feedback element which favors the larch and thus efficiently retards the natural succession from pure larch stands towards the larch-cembran pine climax forest (23).

Two incidences are known from the subalpine optimum area where bud moth populations declined or stagnated without having caused defoliation:

1. Populations of the light cembran pine form fluctuate in the Engadine Valley synchronously with the dark larch form, although the cembran pine is never defoliated, because of the selective feeding on the current year's shoots and needles only. The decline in population numbers of the pine form may therefore not be fully attributable to a density-induced deterioration in food quality, though relevant experiments have not been made. Instead, a significant shift from light to intermediate morphotypic population structure was observed; this is interpreted to be the result of large-scale panmixia between the adults of the two ecotypes.
2. Populations in the Valle Aurina, which is representative of the cyclic fluctuation type in the eastern part of Southern Tyrol, stagnated from 1971 to 1975 at a medium density of 5 larvae/kg (C. Auer, in preparation). These years were unusually cool and rainy during the spring season, and consequently the larch needles grew very slowly and exhibited a large raw fiber content (80). In this period, the frequency of intermediate morphotypes increased from 3% to 24%.

The example of the Valle Aurina population confirms that the trophic stress may select for the intermediate morphotype regardless of whether it is induced by defoliation or by weather. The fact that prolonged selection for the intermediate morphotype or panmixia leading to it eventually results in population decline or stagnation is equally important. Continuously high proportions of intermediate morphotypes in the area of latency indicate, in contrast to the subalpine optimum area, the prevalence of a permanent stress situation (22). A statistically significant increase of dark morphotypes at Lenzburg in 1973 reflects the impact of immigrants from the subalpine outbreaks areas (W. Baltensweiler, in preparation).

The problem of colormorphs in aperiodically fluctuating bud moth populations in England appears to be more complicated. Day (42) found that, in general, English bud moth populations hatch much too late for optimum survival on the introduced larch trees. However, it is postulated that the remarkable plasticity of postdiapause development and/or the adaptive characteristics of the ecotypes serve as the basic prerequisite for disruptive selection (W. Baltensweiler, in preparation). Selective survival on early-flushing larch or late-sprouting picea and/or pine would provide the mechanism for the generation of local host-specific populations. This interpretation needs further verification, since the older records on bud moth outbreaks on spruce and pine (42) do not provide conclusive information on color phenotype composition. The fact, however, that even a small increase in density of a larch population was associated with a significant shift from intermediate to dark morphotype composition (42) reveals the existence of a basically similar mechanism as in the cyclic fluctuation area. Interestingly enough, Naumenko (78) also mentions the preponderance of dark phenotypes in increasing populations in Siberia.

Thus the change in frequencies of colormorphs within the various fluctuation types discloses a consistent strategy to cope with the variability of the environment. This strategy, expressed in a simplified manner, relies on two ecotypes, which differ in susceptibility against stress conditions, whereby the ecotype homozygous for dark coloration is the more susceptible. Because ecologically relevant changes in genotype frequencies within four generations are also found in field and laboratory populations of *Drosophila* (47, 92), it may be concluded that the strategy to survive employed by *Z. diniana* is not unique.

The role played by natural enemies is rather uncertain. In the optimum developmental zone of *Z. diniana,* the degree of parasitism augments slowly during the phase of host population increase and becomes rather high at the beginning of the phase of population decline. On predators, data are available only from a phase of high host density, and it appears that they play a major role in the decimation of the eggs. It is, however, too early to speculate about the significance of the natural enemy complex within the context of a cyclic change of host densities. The interactions between natural enemies, climate, and the intraspecific competition of the host for space and food remain to be defined. For the time being, it can be assumed that the natural enemies do not directly intervene in the feedback mechanism of *Z. diniana.*

MODELS OF POPULATION DYNAMICS

Baltensweiler (19) formulated a verbal model of weather factors favoring population increase outside the subalpine optimum zone, in which cool summer temperatures and cold winters were of prime importance for high survival rates in the egg stage. Retrospective analysis of population increase of *Z. diniana* in central Europe (19), the Pyrenees, and England (42) confirmed this hypothesis. Multiple regression analysis based on similar reasoning led to the acceptance of the hypothesis that certain elements of temperature may act as key factors in determining egg populations along the altitudinal profile (24). Auer applied Morris's key-factor analysis to population data of the Engadine, using parasitism, disease, defoliation, and temperature as independent variables (6, 7). With regard to the high accountability of the model (coefficient of determination : 0.9762), Auer concluded that larval parasitism and defoliation dominated as regulating factors, whereas disease and temperature were of minor importance only. Varley & Gradwell (99) criticized these conclusions.

According to their view, Auer had failed to include the residual mortality which, as they postulated, contained the decisive mortality. However, in their analysis, they omitted the effects of defoliation which Auer had found to be important; thus the argument turns round and round (9).

The application of static models to dynamic processes with various feedbacks and time lags remains disputable. In order to avoid this principal disadvantage of key-factor analysis, Van den Bos & Rabbinge (98) attempted to construct a dynamic model of the larch bud moth in the Engadine. They applied the simulation technique proposed by de Wit & Goudriaan (46) by using the simulation language CSMP III. The model contains three subsystems: the larch, the larch bud moth, and the complex of parasitoids represented as a single hypothetical species. The authors succeeded in simulating the observed population cycle of the Engadine very closely and arrived at the same amplitude and a cycle phase of 8 years. However, the omission of a single subsystem or any parts of it drastically altered the simulated fluctuation of the bud moth. A sensitivity analysis to evaluate the relative weight of the applied parameters did not reveal any particular key factor. Therefore the authors conclude that the larch bud moth cycle is only to be explained by the interaction of many different factors.

We fully agree with the authors (46) that the available information is in many ways imperfect and incomplete, but we disagree with their conclusion that this model tests the validity of the various hypotheses proposed for the explanation of the bud moth cycle. In our opinion, too many assumptions were required in arriving at their model. Nevertheless, by improving the biological information and by applying system identification and state estimations, this method of analysis may prove valuable in future investigations.

MANIPULATION

The periodicity of the larch bud moth fluctuation provides unique possibilities to manipulate the population at various density levels. Small-scale experiments with the specific granulosis virus of *Z. diniana* during the phase of culmination in 1963 on 9 ha (G. Benz, unpublished data) and during progression in 1970 on 17 ha (89) led to population reductions of not more than 33%, indicating that the virus is too quickly inactivated and thus not a useful control agent. On the other hand, experiments on 21 ha in 1963 and 1964 with *Bacillus thuringiensis* caused up to 95% reduction in population density and thus proved to be a possible control agent (33). The inundative release of natural enemies has been attempted on a small scale of 10 ha (3, 84). The experiment was facilitated by the delayed appearance of peak densities of the bud moth and parasitoids within the Alps. This allowed the transfer of mass-collected adults of parasitoids to the Engadine area. When considering the poor results of parasitism obtained, the extent of the area being treated, and the problem of maintaining industrial production of parasitoids over years of minimum host density, the financial implications are evident.

Three treatments with DDT, Phosphamidon, and *B. thuringiensis,* respectively, each of them covering ∼1000 ha subalpine larch forest in Switzerland and France, were conducted at peak density. In two instances, the practical goal, i.e. the elimination of defoliation, was achieved (10, 73), but the reasons for the simultaneous collapse of the treated and control populations from the near-defoliation threshold are still controversial. There are indications that a change in the genetic structure, either due to selection by the treatment and/or a subtle change in food quality or immigration of post peak moths from neighboring areas, were responsible for the regression phase. Another *B. thuringiensis* treatment in the French Alps, on 2700 ha at minimum density of 0.044 larvae/kg, reduced the density by 62% (62; C. Auer, unpublished data), but the population increased in the three subsequent generations at the same rate as the control.

Thus the manipulation efforts did not succeed in altering the cycle fundamentally; at best a one-year impact was achieved. At high densities, long-range dispersal seems effective in homogenizing populations in a positive or negative sense. This phenomenon not only renders attempts to control outbreaks somewhat questionable, but it also complicates proper evaluation of control efforts. The strategy to manipulate optimum developmental areas at low density appears to be more promising; however, two conditions are imperative: (*a*) the area administered must be related to the entire area of the cyclic fluctuation type, and (*b*) it is estimated that the density level at the local optimum sites must be reduced by a factor of 1000 below the average minimum density (Table 2). Investigations by means of the pheromone-confusion technique are currently under way.

ACKNOWLEDGMENT

Dr. C. Auer was responsible for developing the sampling techniques of the larval census, the assessment of the larval population densities during the period 1949 to 1975, and the realization of the various large-scale treatments against the larch bud moth. His efforts and collaboration are gratefully acknowledged.

Literature Cited

1. Aeschlimann, J. P. 1969. Contribution à l'étude de trois espèces d'Eulophides parasites de la Tordeuse grise du Mélèze, *Zeiraphera diniana* Guénée (Lepidoptera: Tortricidae) en Haute-Engadine. *Entomophaga* 14:261–319
2. Aeschlimann, J. P. 1973. Efficacité des parasites Eulophides de *Zeiraphera diniana* Guenée en fonction de l'état de l'hôte. *Entomophaga* 18:95–102
3. Aeschlimann, J. P. 1975. Biologie, comportement et lâcher expérimental de *Triclistus pygmaeus* Cresson. *Mitt. Schweiz. Entomol. Ges.* 48:166–71
4. Altwegg, P. 1971. Ein semisynthetisches Nährmedium und Ersatzsubstrate für die Oviposition zur von der Jahreszeit unabhängigen Zucht des grauen Lärchenwicklers, *Z. Angew. Entomol.* 69:135–70
5. Auer, C. 1961. Ergebnisse zwölfjähriger quantitativer Untersuchungen der Populationsbewegung des Grauen Lärchenwicklers *Zeiraphera griseana* Hübner (=*diniana* Guenée) im Oberengadin (1949/60). *Mitt. Eidg. Anst. Forstl. Vers'wes.* 37:174–263
6. Auer, C. 1968. Erste Ergebnisse einfacher stochastischer Modelluntersuchungen über die Ursachen der Populationsbewegung des grauen Lärchenwicklers *Zeiraphera diniana* Gn. (=*Z. griseana* Hb.) im Oberengadin, 1949/66. *Z. Angew. Entomol.* 62:202–35
7. Auer, C. 1971. A simple mathematical model for "key-factor" analysis and comparison in population research work. In *Statistical Ecology,* ed. G. P. Patil, E. C. Pielou, W. E. Waters, 2:33–48. University Park, Penn: Penn. State Univ. Press. 420 pp.
8. Auer, C. 1971. Some analyses of the quantitative structure in populations

and dynamics of larch bud moth 1949–1968. See Ref. 7, pp. 151–73
9. Auer, C. 1972. Zur Diskussion über mathematisch-statistische Modelle. *Z. Angew. Entomol.* 70:1–7
10. Auer, C. 1974. Ein Feldversuch zur gezielten Veränderung zyklischer Insektenpopulationsbewegungen. *Schweiz. Z. Forstwes.* 125:333–58
11. Auer, C. 1975. Dendrometrische Grundlagen und Verfahren zur Schätzung absoluter Insektenpopulationen. *Mitt. Eidg. Anst. Forstl. Vers'wes.* 50:89–131
12. Auer, C. 1975. Jährliche und langfristige Dichteveränderungen bei Lärchenwicklerpopulationen (*Zeiraphera diniana* Gn.) ausserhalb des Optimumgebietes. *Mitt. Schweiz. Entomol. Ges.* 48:47–58
13. Bakke, A. 1969. Extremely low supercooling point in eggs of *Zeiraphera diniana* (Guenée). *Norsk. Entomol. Tidsskr.* 3:81–83
14. Baltensweiler, W. 1958. Zur Kenntnis der Parasiten des Grauen Lärchenwicklers (*Zeiraphera griseana* Hübner) im Oberengadin. *Mitt. Eidg. Anst. Forstl. Vers'wes.* 34:399–477
15. Baltensweiler, W. 1962. Die zyklischen Massenvermehrungen des Grauen Lärchenwicklers (*Zeiraphera griseana* Hb.) in den Alpen. *Proc. Int. Congr. Entomol. Vienna, 11th, 1960* 2:185–89
16. Baltensweiler, W. 1964. *Zeiraphera griseana* Hübner in the European Alps. A contribution to the problem of cycles. *Can. Entomol.* 96:792–800
17. Baltensweiler, W. 1966. The influence of climate and weather on population age distribution and its consequences. *Proc. FAO Symp. Integrated Pest Control, Rome, 1965* 2:15–24
18. Baltensweiler, W. 1968. Ein Modellobjekt tierökologischer Forschung: der Graue Lärchenwickler, *Zeiraphera griseana* (=*Semasia diniana*). *Biol. Rundsch.* 6:160–67
19. Baltensweiler, W. 1966. Zur Erklärung der Massenvermehrung des Grauen Lärchenwicklers (*Zeiraphera griseana* Hb. = *diniana* Gn.). *Schweiz. Z. Forstwes.* 117:466–91
20. Baltensweiler, W. 1968. The cyclic population dynamics of the grey larch tortrix, *Zeiraphera griseana* Hübner (=*Semasia diniana* Guenée). In *Insect Abundance. Symp. R. Entomol. Soc. London,* ed. T. R. E. Southwood, 4:88–97
21. Baltensweiler, W. 1969. Zur Verteilung der Lepidopterenfauna auf der Lärche des Schweizerischen Mittellandes. *Mitt. Schweiz. Entomol. Ges.* 42:221–29
22. Baltensweiler, W. 1971. The relevance of changes in the composition of larch bud moth populations for the dynamics of its numbers. *Proc. Adv. Study Inst. Dynamics of Numbers in Populations, Oosterbeek, 1970,* pp. 208–19
23. Baltensweiler, W. 1975. Zur Bedeutung des Grauen Lärchenwicklers (*Zeiraphera diniana* Gn.) für die Lebensgemeinschaft des Lärchen-Arvenwaldes. *Mitt. Schweiz. Entomol. Ges.* 48:5–12
24. Baltensweiler, W., Giese, R. L., Auer, C. 1971. The grey larch bud moth: its population fluctuation in optimum and suboptimum areas. See Ref. 7, pp. 401–20
25. Baltensweiler, W., Hochmut, R. 1976. Predicting changes in incidence and damage of insect pests. *Proc. FAO/IUFRO Symp. , 2nd, New Delhi, 1975.* In press
26. Baltensweiler, W., von Salis, G. 1975. Zur Dispersionsdynamik der Falter des Grauen Lärchenwicklers (*Zeiraphera diniana* Gn.). *Z. Angew. Entomol.* 77:251–57
27. Bassand, D. 1965. Contribution à l'étude de la Diapause embryonnaire et de l'Embryogenèse de *Zeiraphera griseana* Hübner (=*Z. diniana* Guenée). *Rev. Suisse Zool.* 72:429–542
28. Benz, G. 1962. Untersuchungen über die Pathogenität eines Granulosis-Virus des Grauen Lärchenwicklers *Zeiraphera diniana* (Guenée). *Agron. Glas.* 1962:566–74
29. Benz, G. 1964. Aspects of virus multiplication and average reduplication time for a granulosis virus of *Zeiraphera diniana* (Guenée). *Entomophaga Mém. Hors Serie No. 2* pp. 417–21
30. Benz, G. 1969. Influence of mating, insemination, and other factors on Oögenesis and oviposition in the moth *Zeiraphera diniana. J. Insect Physiol.* 15:55–71
31. Benz, G. 1973. Role of sex pheromone, and its insignificance for heterosexual and homosexual behaviour of larch bud moth. *Experientia* 29:553–54
32. Benz, G. 1974. Negative Rückkoppelung durch Raum-und Nahrungskonkurrenz sowie zyklische Veränderung der Nahrungsgrundlage als Regelprinzip in der Populations-dynamik des Grauen Lärchenwicklers, *Zeiraphera*

diniana (Guenée). *Z. Angew. Entomol.* 76:196–228

33. Benz, G. 1975. Action of *Bacillus thuringiensis* preparation against larch bud moth, *Zeiraphera diniana* (Gn.), enchanced by β-Exotoxin and DDT. *Experientia* 31:1288–90
34. Benz, G. 1976. Tagesrhythmische Fressaktivität der Larven des Lärchenwicklers, *Zeiraphera diniana* (Gn.). *Experientia* 32:In press
35. Benz, G., von Salis, G. 1973. Use of synthetic sex attractant of larch bud moth *Zeiraphera diniana* (Gn.) in monitoring traps under different conditions, and antagonistic action of *cis*-Isomere. *Experientia* 29:729–30
36. Bovey, P. 1966. Le problème de la Tordeuse grise du Mélèze (*Zeiraphera diniana* Gn.) dans les forêts alpines. *Bull. Murithienne* 83:1–33
37. Bovey, P. 1977. Gattungen Spilonota Steph. und *Zeiraphera* Treitschke. In *Die Forstschädlinge Europas. Lepidopteren,* ed: W. Schwenke, 3:In press. Hamburg:Parey
38. Bovey, P., Maksymov, J. K. 1959. Le problème des races biologiques chez la Tordeuse grise du Mélèze, *Zeiraphera* griseana (Hb.). *Vierteljahrsschr. Naturforsch. Ges. Zürich* 104:264–74
39. Boyce, A. M. 1934. Bionomics of the walnut husk fly *Rhagoletis completa. Hilgardia* 8:363–579
40. Burmann, K. 1965. Beobachtungen über Massenflüge des Grauen Lärchenwicklers (*Z. diniana* Gn.). *Anz. Schädlingsk.* 38:4–7
41. Christiansen, E. 1970. Insect pests in forests of the Nordic Countries 1961–1966. *Norsk. Entomol. Tidsskr* 17: 153–58
42. Day, K. R. 1976. *An ecological study of Zeiraphera diniana Gn. (= griseana Hb.) on conifers in Britain.* PhD thesis. Univ. London, London, England
43. Day, K. R., Baltensweiler, W. 1972. Change in proportion of larval colourtypes of the larchform *Zeiraphera diniana* when reared on two media. *Entomol. Exp. Appl.* 15:287–98
44. Delucchi, V., Renfer, A., Aeschlimann, J. P. 1974. Contribution à la connaissance des Lépidoptères associés au mélèze en haute altitude et de leurs parasitoides. *Rech. Agron. Suisse* 13: 435–51
45. Delucchi, V., Aeschlimann, J. P., Graf, E. 1975. The regulating action of egg predators on the populations of *Zeiraphera diniana* Guenée. *Mitt. Schweiz. Entomol.* 48:37–45
46. DeWitt, C. T., Goudriaan, J. 1974. *Simulation of Ecological Processes.* Wageningen: Ctr. Agric. Publ. Document. 159 pp.
47. Dobzhansky, T. 1948. Genetics of natural populations. XVI: Altitudinal and seasonal changes produced by natural selection in certain populations of *Drosophila pseudoobscura* and *D. persimilis. Genetics* 33:158–76
48. Ehnström, B., Bejer-Petersen, B., Löytyniemi, K., Tvermyr, S. 1974. Insect pests in forests of the Nordic countries 1967–1971. *Ann. Entomol. Fenn.* 40: 37–47
49. Ellenberg, H., Klötzli, F. 1972. Waldgesellschaften und Waldstandorte der Schweiz. *Mitt. Eidg. Anst. Forstl. Vers'wes.* 48:591–930
50. Escherich, K., Baer, W. 1909. Einiges über den Grauen Lärchenwickler, *Steganoptycha diniana* Gn. (*pinicolana Z.*). In *Tharandter zoologische Miszellen. Naturwiss. Z. Forst. Landwirtsch.* 7:188–94
51. Feeney, P. P. 1969. Inhibitory effects of oak leaf tannins on the hydrolysis of protein by trypsin. *Phytochemistry* 8:2119–26
52. Florov, D. N. 1942. *Steganoptycha diniana* Gn. *desertana* Caraja in Eastern Siberia. *Izvestiya Bull. Inst. Biol. Geogr.* Irkutsk 9:169–207 (In Russian)
53. Florov, D. N. 1952. Der Lärchenwickler (Tortricidae, *Semasia diniana* (Gn.). *Izvestiya Russ. Geogr. Soc.* 84:622–27 (In Russian)
54. Fritz, N. 1903. Two tortricids. *Tidskr. Skovvaesen* 15:38–48 (In Danish)
55. Geer, G. A. 1975. *Der Einfluss des Grauen Lärchenwicklers, Zeiraphera diniana (Gn.) auf den Zuwachs der Lärche, Larix decidua, (Mill.) im Oberengadin.* Dissert. ETH Zürich, Nr. 5499
56. Graf, E. 1974. Zur Biologie und Gradologie des Grauen Lärchenwicklers, *Zeiraphera diniana* Gn. im schweizerischen Mittelland. *Z. Angew. Entomol.* 6:233–51; 347–79
57. Geiler, H., Theile, J. 1966. Zur Problematik des erneuten Massenauftretens vom Grauen Lärchenwickler (*Zeiraphera diniana* Guenée, 1845) an Fichte im oberen Erzgebirge. *Arch. Forstwes.* 16:831–35
58. Gerig, L. 1960. Zur Morphologie der Larvenstadien einiger parasitischer Hymenopteren des Grauen Lärchen-

wicklers (*Zeiraphera griseana* Hübner). *Z. Angew. Entomol.* 46:121–77

59. Gerig, L. 1966. Ergebnisse über Fütterungsversuche am Grauen Lärchenwickler (*Zeiraphera griseana* Hb. = *diniana Gn.*). *Z. Angew. Entomol.* 58:139–43
60. Gerig, L. 1967. Physiologische Untersuchungen am Grauen Lärchenwickler *Zeiraphera diniana* Gn. (= *Z. griseana* Hb.) während einer Periode der Massenvermehrung. *Z. Angew. Entomol.* 59:187–211
61. Green, T. R., Ryan, C. A. 1971. Wound-induced proteinase inhibitor in plant leaves: A possible defense mechanism against insects. *Science* 175: 776–77
62. Grison, P., Martouret, D., Auer, C. 1971. La lutte microbiologique contre la tordeuse du mélèze. *Ann. Zool. Ecol. Animale,* No. hors-sér., pp. 91–121
63. Gruys, P. 1970. Growth in *Bupalus piniarius* (Lepidoptera: Geometridae) in relation to larval population density. *Verh. Rijksinst. Naturbeheer* 1:127
64. Holling, C. S. 1968. The tactics of a predator. In *Insect Abundance.Symp. R. Entomol. Soc. London,* ed. T. R. E. Southwood, 4:47–58
65. Issiki, S., Mutuura, A. 1961. Microlepidoptera injurious to coniferous trees in Japan. *Jpn. Assoc. For. Technol., Tokyo.* 47 pp. (In Japanese)
66. Jahn, E. 1949. Die Polyederkrankheit des grauen Lärchenwicklers *Grapholita* (*Semasia*) *diniana. Mikroskopie* 4: 346–54

66a. Johnson, C. G. 1969. *Migration and Dispersal of Insects by Flight.* London: Methuen 763 pp.

67. Kaelin, A., Auer, C. 1954. Statistische Methoden zur Untersuchung von Insektenpopulationen. *Z. Angew. Entomol.* 36:241–82; 423–61
68. Leonard, D. E. 1974. Recent developments in ecology and control of the gypsy moth. *Ann. Rev. Entomol.* 19: 197–229
69. Maksymov, J. K. 1959. Beitrag zur Biologie und Oekologie des Grauen Lärchenwicklers *Zeiraphera griseana* (Hb.) im Engadin. *Mitt. Eidg. Anst. Forstl. Vers'wes.* 35:279–315
70. Martignoni, M. E. 1954. Ueber zwei Viruskrankheiten von Forstinsekten im Engadin. *Mitt. Schweiz. Entomol. Ges.* 27:147–52
71. Martignoni, M. E. 1957. Contributo alla conoscenza di una granulosi de *Eucosma griseana* (Hübner) quale fattore limitante il pullulamento dell'insetto nella Engadina alta. *Mitt. Eidg. Anst. Forstl. Vers'wes.* 32:371–418
72. Martignoni, M. E., Auer, C. 1957. Bekämpfungsversuch gegen *Eucosma griseana* (Hübner) mit einem Granulosis-Virus. *Mitt. Eidg. Anst. Forstl. Vers'wes.* 33:73–93
73. Martouret, D., Auer, C. 1976. Effets de *Bacillus thuringiensis* Berliner chez une population de Tordeuse grise du mélèze, *Zeiraphera diniana* Guenée en culmination gradologique. *Ann. For.* In press
74. Meyer, D. 1969. Der Einfluss von Licht und Temperaturschwankungen auf Verhalten und Fekundität des Lärchenwicklers *Zeiraphera diniana* (Gn.). *Rev.Suisse Zool.* 76:93–141
75. Milne, A. 1962. On a theory of natural control of insect population. *J. Theoret. Biol. 3:19–50*
76. Morris, R. F. 1955. The development of sampling techniques for forest insect defoliators, with particular reference to the spruce budworm. *Can. J. Zool.* 33:225–94
77. Mutuura, A., Freeman, T. N. 1966. The North American species of the genus *Zeiraphera. J. Res. Lepid.* 5:153–76
78. Naumenko, A. T. 1968. The mass-outbreak of the larch bud moth (*Zeiraphera diniana* Gn.) in the Irkutsk-Region. *Sci. Res. No. 115. Forest Protection* 1:106–10 (In Russian)
79. Nägeli, W. 1971. Der Wind als Standortsfaktor bei Aufforstungen in der subalpinen Stufe (Stillbergalp im Dischmatal, Kanton Graubünden). *Mitt. Eidg. Anst. Forstl. Vers'wes.* 47:39–147
80. Omlin, F., Herren, H. R. 1976. Zur Populationsdynamik des Grauen Lärchenwicklers, *Zeiraphera diniana* Gn. im Valle Aurina: Lebenstafeluntersuchungen und Nahrungsverhältnisse während der Vegetationsperiode 1975. *Mitt. Schweiz. Entomol. Ges.* 49: In press
81. Pfeffer, A. 1930. The larch bud moth—Enarmonia (*Epinotia, Steganoptycha*) *diniana* Gn. (*pinicolana* Z.). *Lesn. Pr.* 9:1–24 (In Czech)
82. Prell, H. 1930. Der Graue Lärchenwickler (*Enarmonia diniana* Z.). *Tharandter Forstl. Jahrb.* 81:49–92
83. Rajgorodskaya, I. A. 1970. Lepidoptera. In *Pest of Siberian Larch,* ed. A. S. Rozkov, Academy of Sciences of the USSR, pp. 226–74. USSR Acad. Sci.
84. Renfer, A. 1974. *Caractéristiques biologiques et efficacité de Phytodietus*

griseanae Kerrich parasitoide de Zeiraphera diniana Guenée en haute montagne. Dissert. ETH Zürich, Nr. 5278

85. Renfer, A. 1975. Caractéristiques biologiques de *Phytodietus griseanae* parasitoide de la Tordeuse grise du mélèze *Zeiraphera diniana* en haute montagne. *Ann. Soc. ent. Fr.* (N.S.) 11:425–55
86. Richards, O. W. 1961. The theoretical and practical study of natural insect populations. *Ann. Rev. Entomol.* 6: 147–62
87. Roelofs, W. L., Cardé, R., Benz, G., von Salis, G. 1971. Sex attractant of the larch bud moth found by electroantennogram method. *Experientia* 27:1438–39
88. Russ, K. 1969. Beiträge zum Territorialverhalten der Raupen des Springwurmwicklers, *Sparganothis pilleriana* Schiff. *Pflanzenschutzberichte* 40:1–9
89. Schmid, A. 1973. *Beitrag zur mikrobiologischen Bekämpfung des Grauen Lärchenwicklers, Zeiraphera diniana,* (Gn.). Dissert. ETH Zürich, Nr. 5045
90. Schmid, A. 1974. Untersuchungen zur Trans-Ovum-Uebertragung des Granulosisvirus des Grauen Lärchenwicklers, *Zeiraphera diniana* und Auslösung der akuten Virose durch Stressfaktoren. *Entomophaga* 19:279–92
91. Schwerdtfeger, F. 1968. *Oekologie der Tiere. Demökologie.* Hamburg: Parey. 448 pp.
92. Shorrocks, B. 1970. Population fluctuation in the fruit fly (*Drosophila melanogaster*) maintained in the laboratory. *J. Anim. Ecol.* 39:229–53
93. Šrot, M. 1974. The outbreak of the larch bud moth (*Zeiraphera diniana* Gn.) in the CSSR from 1964–1969. *Vedecky Casopis* 20:241–58 (In Czech)
94. Stolina, M., Novakova, E. 1959. La protection des forêtes et la typologie forestière. Ceske Vysoke Uceni Technicke v Praze, *Sb. ved. Pr. Lek. Fakulty Lesn.* 2:189–206
95. Theile, J. 1967. Zur Massenvermehrung des Grauen Lärchenwicklers, *Zeiraphera diniana* Guenée, in Fichtenbeständen des Erzgebirges (Situation 1966). *Arch. Forstwes.* 16:831–35
96. Urfer-Henneberger, C. 1970. Neuere Beobachtungen über die Entwicklung des Schönwetterwindsystems in einem V-förmigen Alpental (Dischmatal bei Davos). *Arch. Met Meteorol. Geophys. Bioklimatol Ser. B* 18:21–42
97. Vaclena, K. 1975. *Untersuchungen zur Dispersionsdynamik des Grauen Lärchenwicklers, Zeiraphera diniana Gn.* Dissert. ETH Zürich, Nr. 5603
98. Van den Bos, J., Rabbinge, R. 1976. *Simulation of the Population Fluctuation of the Grey Larch Bud Moth. Simulation Monograph Series.* Wageningen: PUDOC. 87 pp.
99. Varley, G. C., Gradwell, G. R. 1970. Recent advances in insect population dynamics. *Ann. Rev. Entomol.* 15:1–24
100. Von Salis, G. 1974. *Beitrag zur Oekologie des Puppen-und Falterstadiums des Grauen Lärchenwicklers, Zeiraphera diniana (Gn.).* Dissert. ETH Zürich, Nr. 5265
101. Williamson, M. 1972. *The Analysis of Biological Populations.* London: Arnold. 180 pp.
102. Zukowski, R. 1957. Beobachtungen der Entstehung und Entwicklung einer Gradation mancher Gattungen von Schmetterlingen in der Zeit von 1949–1959 auf dem Gebiet des Pieniny-Nationalparkes. *Sylwan* 4:25–33 (In Polish)

Ann. Rev. Entomol. 1977. 22:101–19

TECHNIQUES FOR THE EVALUATION OF INSECT REPELLENTS: A CRITICAL REVIEW[1]

❖6122

C. E. Schreck
Insects Affecting Man Research Laboratory, Agricultural Research Service,
US Department of Agriculture, Gainesville, Florida 32604

INTRODUCTION

In technologically oriented societies, the issue of environmental contamination coupled with problems of insecticide resistance and the economic restraints imposed by inflation have led to an intense reexamination of new and old methods in pest control. One field of research experiencing renewed interest is that of insect repellents as a method of protecting humans, plants, and animals. The protection of civilian and military personnel from biting pests and vectors of disease in recreational and wilderness areas, where other control methods are not economical, remains a serious problem.

Although repellent research has been relatively meager when compared with other entomological studies, such as the use of insecticides, many diverse techniques have been proposed as methods of evaluating insect repellents. This paper attempts to present a critical appraisal of the current status of repellent testing techniques. The discussion is confined to repellent testing techniques used in the search for materials that protect humans from arthropod bites. Since several authors—Christophers (13), Dethier (16), and Garson and Winnike (21)—have provided good coverage of the history of repellent research, it is not dealt with in this paper, except to comment that a great deal of credit is due P. Granett and B. V. Travis for their pioneering work in repellent testing techniques. The concept "time until the first bite" (protection time) proposed by Granett (29) is still widely used around the world as an index of repellency. Travis (76) realized that protection time is not the

[1]This paper reflects the results of research only. Mention of a pesticide or a commercial or proprietary product in this paper does not constitute a recommendation or an endorsement of this product by the US Department of Agriculture.

sole criterion in recognizing a promising repellent. He reasoned that the number of bites received through time should also be considered. These scientists and others (31, 32, 76, 79) concluded that differences in test subjects and differences in insect species were important considerations. The significance of these considerations, recognized by these early investigators as being extremely important in determining the efficacy of repellents on different people in different situations, is still unresolved and leads to confusion in interpreting the results of repellent research done by different groups.

Although the major thrust in repellent research has historically been in time of war and mainly in the United States, Canada, and Western Europe, contemporary peacetime efforts have been given greater attention on a worldwide scale, perhaps because in many instances personal protection is much more economical than area control measures. The United States is no longer the only nation with a significant program of insect repellent research. In the past 20 years, the USSR has built a similar program which is widely recognized for its significant contributions; even as early as 1948 investigations were underway to develop better ways to evaluate repellents (50). In addition, Australia, The Peoples' Republic of China, Czechoslovakia, Kenya, India, Israel, and Poland have become more involved in repellent research, as evidenced by the numbers of reports originating from these countries in recent years. Several of these reports are included in the bibliography; this widely diversified input can be of great benefit to further our knowledge of the mode of action of insect repellents and ways in which to improve their efficacy. It is to be hoped that this expanding study will draw together, through the exchange of ideas, those scientists engaged in insect repellent research for the mutual benefit of all. It is timely then to discuss the techniques currently employed. The author's appraisals are candidly expressed, with the hope that constructive and creative new approaches might emerge.

Interest in repellent research concerned with mode of action and how repellents function was probably brought about with the development of *N,N*-diethyl-*m*-toluamide (deet), a repellent significantly better than any previously known compound or mixture of compounds (22–24). Until that time, the most effective repellents [with the exception of ethyl hexanediol (Rutgers 612)] were mixtures of compounds, each of which was repellent to specific groups of insects (69). It was perhaps justified then, after many years of research, to believe the universal repellent had finally been found. This attitude would certainly tend to reduce the search for newer and better repellents and influence the type of research that would answer questions as to the mode of action of a repellent (35), its limitations, and the causes of loss of repellency (70). Since deet was broadly effective against a wide range of biting pests, it was conceivable that a study of the relationship between the many species of arthropods that were repelled by deet might show why this chemical was effective. The results of this work can largely be found in the period from 1957 to the present. Contributions include vapor action studies by Bar-Zeev & Smith (8), Elliott (19), Khan & Maibach (41), Potapov & Vladimirova (56), and Rezaev, Davydova & Malolkin (58); temperature and humidity studies by Khan, Maibach & Skidmore (42), Vladimirova (81), and Wood (83); chemical absorption studies by

Gleiberman & Voronkina (26), Gleiberman et al (27), and Smith (64); physiology and behavior studies by Khan (40) and Vladimirova (82); studies on emission of CO_2 and human attractiveness by Gilbert, Gouck & Smith (25), Gouck & Bowman (28), Strauss, Maibach & Khan (73), and Wright (84). Although a great deal of information has been gathered in the last 15 years or so, to date the mode of action of insect repellents is still unresolved.

A Survey of Testing Procedures

A great many procedures for determining repellency have been described over the years, but few have withstood the test of time. Two excellent surveys of repellent testing techniques are by Shambaugh, Brown & Pratt (62) and by Smith (65). The former is a general review of tests conducted with a wide range of medically important arthropods, and the latter is a specialized review of mosquito repellent tests. To avoid repetition, the following discussion is mainly on techniques in general use at the present, with some commentary on those that have been proposed in the past.

In 1919, Bacot & Talbot (1) published what is probably the first well-planned laboratory evaluation of mosquito repellents. The forearm was treated from wrist to elbow with 1 g of the experimental material and then exposed in a cage of *Aedes aegypti* mosquitoes. The protection period was determined by testing at 2-, 3-, and 5-hr intervals after treatment. Except for some minor changes, this method is today's most commonly practiced test procedure worldwide and *A. aegypti* is still the most commonly used species. Two conclusions may be drawn: that no progress has been made over the years or that this type of testing procedure is a very sound one. There is some truth in both conclusions. Progress has been relatively slow except for the war years of the 1940s and 1950s. It was during these periods that high priorities were set to develop better, more broadly effective insect repellents. As a result, the most effective compounds currently in use were developed.

Consequently, a program of empirical research was followed, with little time left for basic studies. A great deal of this research was initiated at the Agricultural Research Service (ARS) Insects Affecting Man and Animals Research Laboratory, then located at Orlando, Florida, and now in Gainesville, Florida, as the Insects Affecting Man Research Laboratory. The initial outcome of this research was the development of the skin repellent mixture used by the U.S. Armed Forces, called M-250 (78), containing 60% dimethyl phthalate, 20% ethyl hexanediol, and 20% Indalone® (butyl-3,4-dihydro-2,2-dimethyl-4-oxo-2H-pyran-6-carboxylate). The mixture was later replaced by an improved mixture, M-2020 (68), containing 4 parts dimethyl phthalate, 3 parts ethyl hexanediol, and 3 parts dimethyl carbate. Subsequently, the clothing mixture, M-1960, containing 3 parts 2-butyl-2-ethyl-1,3-propanediol, 3 parts *N*-butylacetanilide, 3 parts benzyl benzoate and 1 part emulsifier (67), was developed for treatment of military uniforms for use against mosquitoes, ticks, chigger mites, and biting flies. Though the mixtures for skin application have been mostly replaced by deet in the United States and many other parts of the world, the clothing repellent formulation is still in use by American military personnel. In the time since the development of these repellents, large numbers of compounds have been screened, but none has been effective enough to justify replacing them.

As mentioned previously, the development of deet called attention to the dearth of knowledge about the nature and mode of action of repellents. Since a multipurpose repellent had been found, there was good reason for scientists to put greater emphasis on basic research and our subject matter concerning methods to bioassay candidate repellents.

Laboratory Testing

The following discussion is intended to acquaint the reader with the various test methods proposed in recent years and their value in the ultimate development of personal-use repellents.

As has been pointed out, the principal species used in tests for mosquito repellents is *A. aegypti.* This species gained popularity early (1) when it was recognized to be easily reared, an avid blood feeder, and a good biter in the laboratory. *Aedes aegypti* is so popular as a test insect that very little has been published regarding other mosquito species used in repellent testing in the laboratory. Since *A. aegypti* has been used for repellent testing almost exclusively, data produced by different laboratories have some common ground for comparison. Unfortunately, investigators have come to rely heavily on this single species for information on repellents; consequently, many compounds may have been overlooked because they were not effective against *A. aegypti.* Recent tests at this laboratory with three well known repellents and three mosquito species provide an example of the differential effectiveness of the compounds on the various species (Table 1.)

Obviously there can be great differences between mosquito species and their response to various chemicals. Furthermore, if only one species is used, compounds ineffective against that species might be effective against another. Though it is not always economically feasible to use several species in a screening program, it is important to keep in mind that the species used in the bioassay is only providing a very limited glimpse at the potential of any chemical being tested. When *Anopheles*

Table 1 Effectiveness of three repellents as skin applications against three species of mosquitoes

	Protection time in minutes[a]		
Repellent	*Aedes aegypti*	*Anopheles quadrimaculatus*	*Anopheles albimanus*
N,N-diethyl-*m*-toluamide (deet)	426	96	87
Dimethyl phthalate	53	415	42
Ethyl hexanediol (Rutgers 612)	130	380	158

[a]Average of six tests with six subjects and 250 mg of the indicated repellent per forearm.

quadrimaculatus mosquitoes were returned to this laboratory's plan of screening new chemicals a few years ago, the number of compounds for secondary testing increased by one third above those screened against *A. aegypti* alone. The result is more economical use of the chemicals screened and new groups of compounds to study. At the Gainesville laboratory, it has been a practice to test all compounds that were favorable in the initial screen against all arthropods routinely available in the laboratory and field, no matter what record they may have had with individual species; this is a safeguard against overlooking potentially good repellents. Another method involves using a borderline standard (dimethyl phthalate) in the screening test. The principle is to test against mosquitoes considered uniformly avid blood feeders but not avid enough to overcome the effectiveness of the borderline standard in the primary test. This method allows compounds of moderate effectiveness to be further examined to assure that they are not discarded prematurely. Recently, a repellent considered not very effective against *A. aegypti* was found to be highly specific for deer flies (*Chrysops* spp.) (61).

In addition to the obvious requirements that test insects be reared under standardized conditions, the density of the laboratory population of test insects is also important. When large numbers of mosquitoes are used (1500–3000) in a 35 X 45 cm laboratory test cage, the biting rate is found to be more predictable (all other things being equal) than when low numbers (6–500) are used. Our experience has been that the smaller the sample, the greater the variability.

In 1950 Travis published a paper (77) discussing the variations observed in insect repellent testing. He noticed variations with different broods of mosquitoes between wet and dry skin and reduction in protection time as the biting rate increased. Among other factors mentioned, he noted that, "When the data from different subjects were being studied, there seemed to be no significant difference in results due to the size of the arm or leg, in spite of the fact that large limbs received less repellent per unit area than did small limbs." Several investigators have chosen to use small areas of the forearm treated on a unit area basis in making repellent evaluations, e.g. Khan, Maibach & Skidmore (42), Shimmin et al (63), and Spencer (71), to assure that the treatment rate is always the same from subject to subject. In spite of this attempt at homogeneity, many other factors are operating to cause variability: skin characteristics, such as temperature, texture, color, absorption, evaporation, moisture, dryness, hairiness (48, 78), and the creep rate of a chemical across the skin surface, to name a few. The factors vary with each test subject in nearly infinite combinations, so instead of trying to control all of them, we chose to select larger numbers of test subjects and to make more test replications in order to gain greater validity in our test results; this procedure may apply to both laboratory and field testing. Shimmin et al (63) have reported results with separate treatments on several areas of the forearms in order to obtain direct comparisons of repellents tested concurrently. Two questions arise as to the validity of this method: first the possible interaction of two adjacent chemicals in either the liquid or vapor phases, and second, the creepage of the chemicals on the skin. Creep can perhaps be crudely defined here as the lateral movement over time of a compound when applied to the skin from an area of higher concentration to an area of lower concentration. This factor could vary with different compounds and might be a source of differences observed among test methods. For a more detailed discussion of creep, see Dethier (15).

The following discussion concerns various proposed methods of evaluating repellents as characterized by the type of attractant or host used. The headings are adopted from those used by Smith (65) in a similar discussion.

NO ATTRACTANT USED The nonattractant test has been used to evaluate new compounds as repellents of crawling arthropods such as ticks, fleas, and mites. Primary screening for repellency to fleas has been described by Fedder (20), Kashafudinov, Abramov & Il'na (37), Smith & Burnett (66), and Zolotarev (88). The tests were quite similar in that a strip of fabric impregnated with a chemical and an untreated control strip were lowered into a container into which fleas had been added. After a predetermined time, the strips were lifted and the fleas remaining on the cloth strips were counted and the percentage of repellency was calculated as follows:

$$\frac{\text{Number on control} - \text{number on treated}}{\text{number on control}} \times 100$$

An olfactometer for comparative testing of new repellents against fleas is described by Potapov (52). The mechanism operates on the above principle and can run nine new compounds replicated three times in each test.

The crawling behavior of ticks is put to good use in tests described by Smith & Burnett (66), in which treated cloth patches are placed on a paddle and touched to the bottom of a pen infested with nymphal *Amblyomma americanum.* The number of ticks crawling from an untreated strip of paper to a point midway on the treated patch provided data for calculating the percentage of repellency based on a comparison with an untreated control. Similar tests by Dremova et al (17, 18), Maslov (47), Zhukova (91) and Zolotarev & Elizarov (89) suggest a method of determining repellency of chemicals against *Ixodes persulcatus* by applying the candidate material in horizontal stripes of progressively increasing concentrations. Ticks then climb the vertically positioned fabric until they reach a concentration they cannot tolerate. This type of test might be valuable in determining minimum effective dosages (MED) for promising tick repellents to be used in field trials. However, there is some uncertainty as to whether the toxic properties of some compounds may cause behavior that might be interpreted as avoidance when actually it is a toxic reaction. Dremova et al (17) may have observed such an effect when two species of ticks, *I. persulcatus* and *Dermacentor pictus,* were exposed repeatedly to repellents.

Published techniques for screening materials used in clothing treatments against mites are rare. A method described by Cross & Snyder (14), still in use in the United States, is one of the few known references to a laboratory screening technique. A treated patch is placed on a 4 X 4 inch glass slide. A cell is then formed by placing a mason jar rubber ring on the cloth. Chigger mites are transferred to the cloth inside the ring and covered with another 4 X 4 inch glass slide. Effectiveness is determined by the time in which all the mites become moribund, within a 15-min observation period. This is not a test for repellency but rather a toxicant assessment, and it was devised out of necessity because the larval biting stage of the mite used in testing is barely visible without a lens and could easily escape or be blown from the test surface.

The discussion thus far has been concerned with crawling arthropods and some of the techniques devised to take advantage of their particular structure and habits. These tests are designed to measure contact repellency or toxicity, rather than vapor repellency. Though vapor repellency might occur, it would not easily be recognized in the tests described. However, since these species become attached to the host mainly by crawling, vapor repellency is not thought to play as important a role as it might against flying insects.

An apparatus described by Bovingdon (11) designed to screen repellents against flies and mosquitoes consisted of a small six-sided glass cell with half of its interior surface treated with a candidate repellent. Insects were released and observed as to which surface (treated or untreated) they preferred to rest on. From the data obtained, a so-called repellence quotient was calculated (11). The method does not take into account the chemical vapors present in the small space provided or their effects on behavior, as was noted by Bruce & Decker (12) and by Potapov & Bogdanova (53), nor does the method explain how to measure vapor repellency as distinguished from contact repellency.

Another method proposed by Bar-Zeev (2) consisted of a petri dish containing two treated filter papers, allowing mosquitoes to choose the least repellent resting site. Although the author claims this method features rapidity of execution, it is estimated to take ten times longer to run than the widely preferred method of Linduska & Morton (45), which was adopted after they had run tests for repellency measured by the landing rate of mosquitoes in a cage containing suspended cloth patches (treated and untreated). The tests were extremely erratic, and Linduska & Morton concluded that, "a valid measurement of repellency could be obtained only when the treated fabric was in contact with the body surface." Probably the greatest criticism of the above methods which use no attractants to evaluate repellents against flying insects is that no satisfactory data have been published to show a close correlation between this type of laboratory procedure and field tests with human subjects.

INANIMATE ATTRACTANTS In their repellent studies in 1947 Granett & Sacktor (33) demonstrated that light could be used to attract nymphal and adult *A. americanum* ticks. Potapov (51) described an olfactometer in the shape of a drum having 20 tube ports equally spaced on the sides of the drum and fitted with glass cylinders containing treated or untreated filter papers, through which the test insects had to fly or crawl to escape the darkened interior. The ports having the fewest insects were assumed to contain the chemical treatments with greatest repellency. Yeoman & Warren (87) described a feeding technique in which the stable fly, *Stomoxys calcitrans,* must penetrate a surface of paper tissue treated with repellent; the surface covered a molasses substrate. Wright (85) has proposed that the intrinsic (initial) repellency of a compound can be determined with a machine that produces warm moist air as an attractant to *A. aegypti* and then measures the amount of chemical vapor that must be added to the air stream to nullify attraction [Kellogg, Burton & Wright (39)]. He compared his method of determining repellency with that of Bar-Zeev & Ben-Tamar (3) who, with repellent tests on rabbits, found a number of compounds with longer protection times than deet. According to Wright (85), his results showed no necessary relation between the two tests. Table 2 is a comparison of the data of these test methods, with data from the Gainesville laboratory supplied by the author. The latter are results of the primary laboratory screening by Linduska & Morton (45) on cloth. The intent is to show the variability in results using three different test methods. To literally interpret Wright's (85) statement, "... although a long-lasting repellent must also have good intrinsic repellency if it is to be effective," with the data in Table 2, it is found that items 2 and 3 are similar in protection time on rabbits and in intrinsic repellency but are quite dissimilar in the cloth screening test. Comparing items 5 and 10, a similarity in the cloth screening data and a substantial difference between protection times on rabbits and between data on intrinsic repellency are noted. Items 6 and 7 are both shown as long-lasting repellents on cloth, with similar data on protection time but with widely dissimilar intrinsic repellency. Item 6 is a widely recognized and effective repellent (67). Accordingly, Wright's statement does not necessarily apply when data originating from different test methods are compared. One might conclude either that this statement is faulty or that the method or methods of collecting the data are not always reliable.

Methods using inanimate attractants obviously do not use a blood meal from a host to attract the test insects. It would seem that insects actively seeking a blood meal would be much more difficult to repel than those in the act of responding to light, heat, moisture, sugar, or a source of stimulation other than from a natural host.

Table 2 Comparison of data from three methods of primary screening of repellents with *Aedes aegypti* mosquitoes

		Screening method and attractant used		
Item no.	Repellent	Human cloth test[a] protection time in days	Rabbit skin test[b] protection time ratio to deet standard	Warm moist air intrinsic repellency[c] ratio to deet
1	*N,N*-Diethyl-*m*-toluamide	56	1.0	1.0
2	*N,N*-Diethyl-2-ethoxybenzamide	37	1.86	0.5
3	*N,N*-Dipropyl-2-(benzyloxy) acetamide	0	1.33	.5
4	1-Butyl-4-methylcarbostyril	501	1.3	2.0
5	*N,N*-Dipropyl-2-ethoxybenzamide	77	1.26	.3
6	2-Butyl-2-ethyl-1,3-propanediol	145	1.1	1.7
7	1,3-Bis(butoxymethyl)-2-imidazolidone	265	1.08	.6
8	*N,N*-Diethyl-2-chlorobenzamide	28	.82	1.2
9	Hexachlorophenol	17+	.72	.2
10	1,3-Propanediol, monobenzoate	88	.53	7.5
11	Diisobutyl malate	15	.39	2.5

[a]USDA-ARS unpublished data.
[b]Bar-Zeev & Ben-Tamar (3).
[c]Wright (86).

ANIMAL ATTRACTANTS Numerous methods have been proposed using animals as attractants. Tristan et al (80) report successful tests by covering the legs of camels with impregnated cloth to evaluate the effectiveness of repellents against fleas *Xenopsylla cheopis* and *Ceratophyllus laeviceps.* Probably the most practical use of animals in the laboratory is in the evaluation of repellents against crawling insects. This is principally because the bites of fleas, ticks, and mites can cause severe sensitization reactions on human subjects. Animals are also frequently used in secondary tests and when the toxicity of the chemicals is not known. The use of guinea pigs in tests with fleas, *X. cheopis,* and ticks, *Ornithodoros tholozani,* has been described by Bar-Zeev & Gothilf (5, 6). The animal was immobilized in a stanchion with its shaved abdomen exposed. Two measured areas were treated with a candidate repellent and a standard in paired tests. Another method described by Hadani (34) used the natural host, the gerbil, *Meriones shawi tristriami,* as an animal attractant in tests with three species of nymphal and larval ixodid ticks, *Hyalomma excavatum, Rhipicephalus secundus,* and *R. sanguineus.* The caged and restrained gerbil was dipped or sprayed with a repellent and placed on the periphery of a tick-infested area. Later, engorged and attached ticks were counted to determine protection afforded by the repellent.

Tests by Yeoman & Warren (87) used laboratory mice as hosts of *S. calcitrans.* They report that tactile repellency rather than vapor repellency is most important in seeking a satisfactory stable fly repellent. Roth, Mote & Lindquist (59) report using mice in a primary evaluation of repellents for a deer fly, *Chrysops discalis,* because only small quantities of the test chemicals were available. Lal, Ginocchio & Hawrylewicz (44) used mice to determine the amount of blood ingested by mosquitoes feeding on chemically treated skin when the toxicity of the chemicals was not known. The major problems in using mice in primary screening are the time required in handling the animals and the complexity of the test procedures the authors have described. They also admit the difficulty in correlating field and laboratory data and speculate that differing conditions in the field very often override what has been observed in the laboratory. Starnes & Granett (72) were aware of the above disadvantages and developed their own test method using rabbits. Cloth impregnations could be tested on the shaved abdomens of the rabbits at the rate of 40 a day against *S. calcitrans,* but once again little correlation was noted between laboratory and field evaluations. Further tests using rabbits (skin treatments) in laboratory screening of compounds against *Culex pipiens molestus* are reported by Bar-Zeev & Ben-Tamar (3). The tests showed several compounds were superior to the deet standard in protection time, but in subsequent field studies with the same mosquito species using human subjects, the same compounds were either not significantly different or were inferior to the deet standard (4). Two compounds gave similar results in both laboratory and field tests. The use of animals as hosts in laboratory repellent studies could result in the elimination of chemicals that might be effective on humans. As has been pointed out, there is a recognized need for laboratory animals in repellent research when the bites of the arthropods might cause undue pain or injury to the human host or when toxicology of the chemical is unknown. Dethier (16), however, feels, "there is no reason to judge that these animals are inferior to man as test animals; and with respect to ease of manipulation, etc, they are superior." Laboratory and field data that showed differences in tests with animals and humans as host subjects have already been given. Schmidt, Acree & Bowman (60) showed marked variation in the amounts of deet absorbed by individual guinea pigs, and Gleiberman et al (27) showed marked variation in the amounts of deet absorbed by rabbits. Variability in absorbtion rates of chemicals is an important factor to consider in choosing test subjects. The validity of using animals in laboratory repellent tests to reflect the action of repellents on humans in the field is questionable and requires much additional study.

HUMAN ATTRACTANTS In the search for human-use repellents, the ultimate beneficiary is, of course, the human. However, Dethier (16), Christophers (13), Kasmin, Roadhouse & Wright (38), Wright & Burton (86), and others have objected to the use of human subjects in laboratory tests because of the variability often noted in the test data. It has been argued by Dethier (16) that "there is no a priori reason why a greater basic variability should exist in repellent tests than in toxicity tests or any other type of biological experiment." This is debatable, since one test deals with mortality (a condition easily determined), while the other deals with behavior (a complex of interrelated factors), and it is difficult to believe the two are comparable. Insect behavior includes such a vast number of complexities that it cannot easily be manipulated nor can the particular variables in question be easily removed since the desired behavior may depend on several of them.

In 1969 Wright & Burton (86) implied the obsolescence of the human arm test developed by Granett (30) to evaluate repellents by replacement with a machine [Kellogg, Burton, & Wright (39)]. Today the use of human subjects in repellent assays around the world is as widely accepted as ever. This alone does not necessarily validate the use of human subjects, but it shows that other methods proposed over the years have not been able to replace the human as the primary bait in repellent evaluation. The promise of data without variability and eventual discovery of the ultimate repellent has been suggested time and again in the literature, but unfortunately the promise has not yet been fulfilled. Although an abundance of techniques has been proposed (literature already cited), there has been a dearth of highly effective new compounds since 1954; over 22 years have passed without the emergence of a promising new compound that is significantly better in all phases of activity than deet. This does not mean we cannot develop a better bioassay. It does mean that to date there is little chance of substituting nonhuman or artificial attractants for the human test subject since it has not been unequivocally demonstrated that the results of tests with these attractants can be correlated to human subjects. It must be remembered that no matter what types of tests are developed to satisfy one's sense of contribution and innovation, the final test is on the human subject.

There are also very practical aspects of repellent research to be considered. It is unreasonable to expect all laboratories to be able to afford elaborate and costly repellent testing techniques especially adapted for their particular insect species.

The current primary screen for insect repellents at this laboratory (45, 49) is conducted with a human subject wearing a treated cotton stocking over his arm.

Similar tests (9, 54, 87) have been described, and most are in current use (A. A. Potapov, of the Martsinovski Institute, Moscow, personal communication). Treated cloth is tested over untreated-cloth-covered skin so chemicals are not in contact with the skin surface and can be tested without prior skin irritation studies.

It is believed that the USSR and the United States are the only two countries in the world with extensive insect repellent synthesis and screening programs. Both countries have a primary laboratory screen followed by secondary laboratory tests on human skin with compounds having favorable toxicology, as well as primary and secondary field studies on skin. The use of human subjects from initial screening to final development is evident in these programs and is derived from a practical approach which reasons that a product for human use should, whenever possible in the course of development, be tested on humans. Standardization of such test procedures would be beneficial on an international level since information could be exchanged with the assurance that the data were derived in the same manner.

Field Testing

Most scientists engaged in research on repellents for human use will agree that field tests should be conducted with human subjects. Specialized testing such as described by Bar-Zeev & Gothilf (7) and Potapov & Vladimirova (55) understandably require animal test subjects or traps, but usually the field test is closer to the actual use test and should duplicate as nearly as possible the intended use of the repellent.

In order for field tests to be successful, one must be familiar with species behavior in the study area and develop methods to determine avidity. Biting rate counts on untreated skin can be taken with flying insects to judge avidity. Tick drags made of white flannel cloth can be pulled over the ground and low vegetation to find heavy infestations, while mites are often located by using black plates laid on the ground. There are numerous methods that have proven successful in attracting various species in the field. For instance, in tests with the deer fly (*Chrysops* spp.), intermittent walking and standing with exposed arms raised over the head or at right angles to the body has been shown to be very attractive to these flies (61). Slow movement and short rests in a squatting position has worked well with some daytime biting mosquitoes, whereas others biting at night prefer to attack the subject at rest. It is important that test methods be designed to accommodate the test species rather than the investigator. Exposed parts of the body such as the hands, face, legs, and back should be adequately protected with gloves, head net, and clothing in order for biting pressure to be concentrated on the exposed treated skin. Very often a repellent such as deet can be applied to these surfaces in very heavy populations of mosquitoes without apparent reduction in biting pressure.

As stated earlier, field evaluations on skin are conducted in the USSR after favorable toxicology has been established on test chemicals. The forearms of subjects are used in paired tests to determine protection time which is calculated to the first bite, confirmed by the second and third bites within 15 min. These tests have been described by Rakhmanova (57), Zolotarev & Kalakutskaya (90), and Ivanova & Stavrovskaya (36). When toxicology is not known, cloth strips are impregnated with chemicals and are tested over the skin of test subjects in the field [Bataev, Ivanova

& Vorobieva (9)]. To determine the numbers and species of flying insects attacking test subjects, an interesting hood apparatus was developed by Monchadsky & Radzivilovskaya (50) and later modified by Berezantsev (10). After a given time of exposure, the hood is dropped over the subject, and traps the attacking insects so that they may be counted and identified. The method is used when clothing repellents are tested and when very heavy populations of horse flies and black flies make it difficult to count and identify species.

Though there has been some doubt as to the reliability of observations made by human test subjects [Lal, Ginocchio & Hawrylewicz (44)] and the promotion of technological methodology by others, the author's experience has shown that a conscientious, well-trained team of test subjects can contribute much more than animals or mechanical devices. Their observations on behavior are often many times more valuable than the data recorded. Poor results with the human observer in repellent research usually results from lack of experience and communication with the test coordinator. Building a team of expert subjects requires time and is usually not within the scope of a short-term grant to study insect repellents. This factor has perhaps been a major obstacle in the development of standardized repellent testing techniques. Tests described by Traub & Elisberg (74, 75) are examples of well-planned and well-directed field tests and demonstrate the effort required to accomplish such tasks.

Intrinsic Repellency

The literature search for this discussion did not reveal when the term intrinsic repellency was first used to describe a particular property of a compound; however, Dethier used it in 1947 (15). The term has periodically been used in the literature by a number of researchers, but few have set about to propose a working definition. Garson & Winnike (21) define intrinsic repellency: " . . . if a known amount or concentration of the material demonstrates some degree of repellency independent of time. Such repellency is measured by employing an olfactometer or by testing the repellency of surfaces to which candidate agents have been applied, immediately after application." This definition, then, is proposing that immediate or initial repellency is intrinsic repellency and that there can be a degree of intrinsic repellency according to the quantity of chemicals used when tested against a given species of insect. To pursue this reasoning, if the chosen amount of a chemical used in a test as described above is at a level that is effective, it would then be said to have intrinsic repellency. If the same amount of a second chemical was tested and it was not repellent, it would be said to have no intrinsic repellency at this amount. However, a larger quantity of the compound may have repellency and would then be said to have a degree of intrinsic repellency less than that of the first compound. To go a step further, it would be safe to assume that untold millions of intrinsic repellents exist and that they may be determined simply by raising the concentration to a level that is effective against a given species of insect.

The widely known insect repellent, deet, is effective on cloth against *A. aegypti* mosquitoes from .004 mg/cm^2 to at least 1 mg/cm^2 in certain tests, while another compound shows repellency at only 1 mg/cm^2 or greater. By definition, both of these compounds are said to have intrinsic repellency. It would seem paradoxical to assign such a specialized term as intrinsic to characterize a large and wide variety of compounds requiring so many qualifiers to justify their classification; indeed, Garson & Winnike's definition misses some very important points. Not only is the amount of material in the test important, but so are the species to be tested, the strain of species, the number of test insects, their sex, whether they are mated, their age, the host, attractant, or standard used, time of day, temperature, humidity, and so on. The species and other particulars above are extremely important, for while one species may be repelled, another may not or may even be attracted. Kost et al (43) reported deet to be an attractant at lower concentrations. Dethier (16) found *iso*-valeraldehyde to be an attractant or repellent according to the concentration tested. Smith (65) refers to published accounts showing lactic acid to be a repellent for *A. aegypti* under some conditions and an attractant under others. Lutta et al (46) reported observing differences in their results and those of other investigators as to nonrepellency or repellency of diethylamide of caproic acid.

Without a universally standardized test procedure familiar to all insect repellent investigators with which to determine repellency, the term intrinsic repellency is unfortunately of little worth. By definition, it infers the fundamental nature of a substance, its inherent, essential, innate, real, and fixed nature. Therefore, in literal translation, it should be used only to characterize compounds repellent at all levels of concentration, even when approaching zero. Thus far, no such compound has been found.

It would seem, since the term intrinsic repellency has been used for so long, that it is widely considered viable. In the author's opinion, however, it cannot be used realistically in contemporary repellent research.

DISCUSSION

In reviewing the literature for this paper, it would seem that repellent research has undergone a series of stresses which have pulled in several directions at once. As a result, academic research and empirical research adherents have been working in discord for some time. Lutta et al (46), in fact, view it as "extreme empiricism" needing rectification. Of the academic group, the stress has been placed on detailed, sometimes trivial points, which contributed little to the discovery of new and better insect repellents, whereas the empirical group has stressed research based on experience rather than firmly established scientific principles. The empirical group, though dogmatic and opinionated at times, is probably closer to a realistic course of study because of emphasizing the practical aspects of the development of insect repellents and deemphasizing basic research; the latter, though essential as the foundation of any scientific study, has not been very rewarding in the past search for new repellents. Ideally, combining these approaches in proper proportions would greatly enhance a repellent research program, as long as the major objectives of such a study remain in the forefront. A productive program would include

(*a*) Broad spectrum screening of new compounds of unknown repellent activity with several species of insects including *A. aegypti* and at least one other species of mosquito of a dissimilar genus.
(*b*) Prudent toxicological evaluation of promising compounds.
(*c*) Chemical synthesis and study of repellents and related compounds identified in the screening program.
(*d*) Secondary testing on cloth and skin in laboratory tests, with as many test subjects and species as is economically practical and with compounds considered toxicologically safe.
(*e*) Field evaluation of above compounds against as many subjects and species as can be tested within the economic limitations of the program.
(*f*) Basic research on the physical characteristics, toxicology, and behavioral and physiological effects (including insect chemoreceptor studies) of promising repellents on both insects and humans.
(*g*) Cooperative worldwide sharing of information, ideas, and testing of species exotic or inaccessible to many researchers.

The screening program should be quick and unsophisticated and give a simple yes or no answer while allowing enough flexibility that compounds of borderline activity are not overlooked. The test should not require elaborate analysis. A simple grading system that classifies compounds is all that is necessary at this stage of development. Also included should be the human subject as an attractant source on which to judge repellency. The reasoning is quite obvious when looking for compounds to repel vectors of human disease and pest insects that attack humans, the test subject should be a human. The use of the human subject in the initial screen and subsequent secondary tests leaves little adjustment to be made when field tests are run. Criticism of the variability incurred in tests using human subjects may be unfounded. The nature of the variability may be an asset rather than a liability. To restrict the conditions of a test to such a point as to disregard the inherent diversity in the human makeup would minimize the practical value and the very nature of the test. The variability should indicate to the investigator the limitations of a compound when it is to be used under diverse conditions with many different people. Lest this statement be misunderstood, the author notes that he is not advocating careless testing, but rather carefully planned and executed tests as usual. This type of test can indeed differentiate promising compounds from those with limited value at an early stage in development, resulting in less wasted motion. Compounds of apparent limited value are not removed from further consideration as yet, however, because information can be collected that may be used to aid in the synthesis of related compounds of increased activity.

Since the development of deet, no new compounds have emerged that are so broadly effective as this compound. In fact, several approach its effectiveness, but none as yet have been shown to be superior as topical multipurpose repellents. The question is whether deet is at the upper limits of effectiveness of any compound that can be applied to human skin. Slight improvements in effectiveness level above that of deet are insufficient to warrant large expenditures of research funds, and frankly there is no indication of a significant breakthrough in the near future. It is encouraging, however, to find compounds that are outstanding as specific repellents for specific species or genera (61), for herein is an alternate research path to follow in which compounds of specialized value could be formulated to be effective repellents against several species confined to different areas of the world. Compounds that were heretofore discarded because they lacked multipurpose repellency might be put to good use.

CONCLUSIONS AND RECOMMENDATIONS

In summary, it is imperative to the future success of repellent research that two points be made. First, because of cost and uncertainties in development, the chemical synthesis of new compounds by industrial sources has been reduced dramatically in the past few years. A synthesis program is vital as a source of new compounds with repellent activity. As suggested earlier, a strong program to screen chemicals for their possible repellency is a prerequisite to any long-term repellent research program. Without a supply of new material providing new knowledge, investigations can stagnate. Objectivity in screening new chemicals is necessary because one cannot solely depend on compounds related to known repellents for new leads. It is also important that the sources of materials be innovative and diverse in order to obtain new information from which researchers can develop new ideas.

Second, it has been suggested from time to time (62; A. A. Potapov, personal communication) that there is a need for standardized techniques for evaluating repellency. Since repellents are not equally effective against all species of biting arthropods, it is important that as many species as possible be tested to determine how broadly effective a repellent may be. In order for testing to be comparable and for more meaningful data to be gathered, it is vigorously proposed that standardization of a few select repellent testing techniques be considered. Probably standardization has never materialized because of different opinions as to which method is best. Various institutions, including the World Health Organization, have overcome such differences and today use standardized insecticide evaluation as well as insecticide resistance determination.

The following few guidelines should be considered in order to obtain comparable results in a standardized test against mosquitoes:

(*a*) The tests should be made as practical as possible to be universally acceptable, as well as easily understood and easily performed.

(*b*) Preliminary screening should be done by human subjects primarily for a direct reading of effectiveness on humans, and secondarily for direct comparisons on the forearms between a standard repellent and an experimental compound. Relative effectiveness is thereby determined no matter what the test conditions or species are.

(*c*) Test materials should be applied on each arm in standardized amounts with uniform coverage. The suggested dosage is 250–500 mg of the chemical dissolved in ethanol per application.

(*d*) Species in test must be identified. Biting rates on untreated skin should be recorded. A predetermined minimum biting rate is necessary to assure adequate biting pressure on the test chemicals.
(*e*) Three to six replications, preferably on as many different subjects, are necessary to determine variability and the mean protection time given by the compounds tested.
(*f*) Candidate repellents in the laboratory and field can be evaluated in terms of protection time and overall effectiveness. This might be accomplished by allowing a total of ten bites to occur before a compound is withdrawn from the test. Whether tested at 30-min intervals in the laboratory or at continuous exposure in the field, one should determine the overall protective value of the compound. For example, one compound might have protection time of 300 min against a species of mosquito on the basis of the first confirmed bite (one bite followed by a second within 30 min) with the tenth bite occurring 60 min later (360 min from application); a second compound could have the same 300-min protection time against the same species but require an additional 180 min until the tenth bite. Then the effectiveness of the latter compound could be said to be greater than the former, although protection time was the same for both.

After having rather thoroughly reviewed the literature concerning insect repellents, one wonders whether the subject of repellent testing techniques should be treated as a science or as an art. The author admits to endorsing a little of both here, for the subject of evaluating insect repellents is not nearly so well understood as some would have us think. However, one should not ignore the possibility that basic research might reveal an entirely new concept to describe the mode of action of repellents.

Acknowledgments

The author acknowledges with special gratitude the kind assistance of A. A. Potapov of the Martsinovsky Institute of Medical Parasitology and Tropical Medicine, Moscow, in preparing a review of laboratory and field test methods practiced in the USSR, and of Nelson Smith and D. E. Weidhaas of this laboratory for suggestions in preparation of the manuscript.

Literature Cited

1. Bacot, A., Talbot, G. 1919. The comparative effectiveness of certain culicifuges under laboratory conditions. *Parasitology* 11:221–36
2. Bar-Zeev, M. 1962. A rapid method for screening and evaluating mosquito repellents. *Bull. Entomol. Res.* 53(3): 521–28
3. Bar-Zeev, M., Ben-Tamar, D. 1971. Evaluation of mosquito repellents. *Mosq. News* 31(1):56–61
4. Bar-Zeev, M., Ben-Tamar, D., Gothilf, S. 1974. Field evaluation of repellents against mosquitoes in Israel. *Mosq. News* 34(2):199–203
5. Bar-Zeev, M., Gothilf, S. 1972. Laboratory evaluation of flea repellents. *J. Med. Entomol.* 9(3):215–18
6. Bar-Zeev, M., Gothilf, S. 1973. Laboratory evaluation of tick repellents. *J. Med. Entomol.* 10(1):71–74
7. Bar-Zeev, M., Gothilf, S. 1974. Field evaluation of repellents against the tick *Ornithodorus tholozani* Labou. & Mégn. in Israel. *J. Med. Entomol.* 11(4):389–92

8. Bar-Zeev, M., Smith, C. N. 1959. Action of repellents feeding through treated membranes or on treated blood. *J. Econ. Entomol.* 52:263–67
9. Bataev, P. S., Ivanova, L. V., Vorobieva, Z. G. 1963. New repellents. *Med. Parazitol.* 2:209–16
10. Berezantsev, Yu. A., 1959. A dark hood for counting bloodsucking Diptera. *Med. Parazitol.* 1:98–99
11. Bovingdon, H. H. S. 1958. An apparatus for screening compounds for repellency to flies and mosquitoes. *Ann. Appl. Biol.* 46(1):47–54
12. Bruce, W. N., Decker, G. C. 1957. Experiments with several repellent formulations applied to cattle for the control of stable flies. *J. Econ. Entomol.* 50(6): 709–13
13. Christophers, S. R. 1947. Mosquito repellents being a report of work of the Mosquito Repellent Inquiry. Cambridge 1943–45. *J. Hyg.* 45(2):176–231
14. Cross, H. F., Snyder, F. M. 1949. Chigger control washing and aging tests of a selected group of compounds effective as clothing treatments against chiggers. *Soap Sanit. Chem.*, Feb. 1949
15. Dethier, V. G. 1947. *Chemical Insect Attractants and Repellents.* Philadelphia: Blakiston. 289 pp.
16. Dethier, V. G. 1956. Repellents. *Ann. Rev. Entomol.* 1:181–202
17. Dremova, V. P., Belan, A. A., Smirnova, S. N., Yantsen, M. M. 1968. Effect of some repellents on *Ixodes persulcatus* and *Dermacentor pictus. Parazitologiga* 2(5):430–32
18. Dremova, V. P., Tsetlin, V. M., Zhuk, E. B.,Yankovskis, E., Gorbatkova, V. V. 1969. Activity of repellents employed in aerosol form. *J. Hyg. Epidemiol. Microbiol. Immunol.* 13(1):64–72
19. Elliott, R. 1964. Kinetic response of mosquitoes to chemicals. *Bull. WHO* 31(5):657–68
20. Fedder, M. L. 1960. Laboratory study of insectrepelling properties of new repellents. *Proc. Centr. Inst. Disinfection and Sterilization* 13:278–86
21. Garson, L. R., Winnike, M. E. 1968. Relationships between insect repellency and chemical and physical parameters —a review. *J. Med. Entomol.* 5(3): 339–52
22. Gilbert, I. H., Gouck, H. K., Smith, C. N. 1955. New mosquito repellents. *J. Econ. Entomol.* 48(6):741–43
23. Gilbert, I. H., Gouck, H. K., Smith, C. N. 1957. New insect repellent. Part I. *Soap Chem. Spec.* 33(5):115–17; 129–33
24. Ibid, Part II. 33(6):95–99;109
25. Gilbert, I. H., Gouck, H. K., Smith, C. N. 1966. Attractiveness of men and women to *Aedes aegypti* and relative protection time obtained with deet. *Fla. Entomol.* 49(1):53–66
26. Gleiberman, S. E., Voronkina, T. M. 1972. Absorption of the repellent diethyltoluamide. *Med. Parazitol. Parazit. Bolezn.* 41(2):189–97
27. Gleiberman, S. E., Voronkina, T. M., Latyshev, V. I., Tsetlin, V. M. 1974. Comparative study of resorptive-excretory properties of some repellents. *Med. Parazitol. Parazit. Bolezn.* 43(4):446–55
28. Gouck, H. K., Bowman, M. C. 1959. Effect of repellents on the evolution of carbon dioxide and moisture from human arms. *J. Econ. Entomol.* 52(6): 1157–59
29. Granett, P. 1938. Comparison of mosquito repellency test under laboratory and field conditions. *Proc. Ann. Meet. NJ Mosq. Assoc., 25th*, pp. 51–57
30. Granett, P. 1940. Studies of mosquito repellents. I. Test procedure and method of evaluating test data. *J. Econ. Entomol.* 33(3):563–65
31. Granett, P. 1944. "Paired product testing" for the evaluation of mosquito repellents. *Proc. Ann. Meet. NJ Mosq. Exterm. Assoc., 31st*, pp. 173–79
32. Granett, P., Haynes, H. L. 1945. Insect repellent properties of 2-ethylhexanediol-1, 3. *J. Econ. Entomol.* 38(6):671–75
33. Granett, P., Sacktor, B. 1947. Testing tick repellents and observations of phototropic effects. *J. Econ. Entomol.* 40(2):259–63
34. Hadani, A., Ziv, M. 1971. Laboratory study of tick repellents and acaricides. *Final Report of Research Conducted under Grant Authorized by U.S. Public Law 480.* 66 pp.
35. Hocking, B., Khan, A. A. 1966. The mode of action of repellent chemicals against blood-sucking flies. *Can. Entomol.* 98:821–31
36. Ivanova, L. V., Stavrovskaya, V. I. 1966. Preliminary data on field testing of the new repellent piperidilamid-*m*-tolual acid. *Med. Parazitol.* 3:320–23
37. Kashafutdinov, G. A., Abramov, V. S., Il'ina, N. A. 1969. Chemical structure and insect repellency. *Med. Parazitol.* 38(1):48–51
38. Kasmin, S., Roadhouse, L. A. O., Wright, G. F. 1953. Studies in testing insect repellents. *Mosq. News* 13:116–23

39. Kellogg, F. E., Burton, D. J., Wright, R. H. 1968. Measuring mosquito repellency. *Can. Entomol.* 100:763–68
40. Khan, A. A. 1965. Effects of repellents on mosquito behavior. *Quaest. Entomol.* 1:1–35
41. Khan, A. A., Maibach, H. I. 1972. Insect repellents. 1. Effect on the flight and approach by *Aedes aegypti. J. Econ. Entomol.* 65(5):1318–21
42. Khan, A. A., Maibach, H. I., Skidmore, D. L. 1973. A study of insect repellents. 2. Effect of temperature on protection time. *J. Econ. Entomol.* 66(2):437–38
43. Kost, A. N., Teren'tev, P. B., Elizarov, Yu. A., Tsyba, I. F. 1971. Perception of organic compounds by *Aedes aegypti* mosquitoes. *Khemoresteptsiya Nasekomykh. Mater. Vses. Simp., 1st,* pp. 89–94
44. Lal, H., Ginocchio, S., Hawrylewicz, E. J. 1963. Procedure for bioassaying mosquito repellents in laboratory animals. *Proc. Soc. Exp. Biol. Med.* V113:770–72
45. Linduska, J. P., Morton, F. A. 1947. Determining the repellency of solid chemicals to mosquitoes. *J. Econ. Entomol.* 40(4):562–64
46. Lutta, A. S., Knyazeva, N. I., Lebedeva, G. A., Lyakhtinen, T. A. 1966. Testing of new repellents in Karelia. *Entomol. Obozr.* 45(2):317–25
47. Maslov, A. V. 1961. On methods of laboratory testing of acarorepellent effectiveness. *Med. Parazitol.* 3:312–15
48. McCulloch, R. N., Waterhouse, D. F. 1947. Laboratory and field tests of mosquito repellents. *Commonw. Aust. Counc. Sci. Indust. Res. Bull. No. 213.* 28 pp.
49. McGovern, T. P., Schreck, C. E., Jackson, J., Beroza, M. 1975. *n*-Acylamides and *n*-alkylsulfonamides from heterocyclic amines as repellents for yellow fever mosquitoes. *Mosq. News.* 35(2): 204–10
50. Monchadsky, A. S., Radzivilovskaya, Z. A. 1948. A new method of quantitative evaluation of the activity of bloodsucking insect attacks. *Parazitol. Sb.* M-L 9:147–51
51. Potapov, A. A. 1966. Olfactometer for comparative tests of repellents. *Med. Parazitol.* (1):69–73
52. Potapov, A. A. 1968. Olfactometer for repellent testing on fleas. *Med. Parazitol.* (1):97–9
53. Potapov, A. A., Bogdanova, E. N. 1974. Insecticide properties of repellents. *Med. Parazitol. Parazit. Bolezn.* 43(5): 573–78
54. Potapov, A. A., Koshkina, I. V. 1970. Primary laboratory and field tests of repellents against gnats on fabric. *Med. Parazitol.* (1):45–9
55. Potapov, A. A., Vladimirova, V. V. 1965. Comparative tests of repellents against horse flies and black flies using olfactometer and traps. *Biol. Med. Seriya,* (2):99–104
56. Potapov, A. A., Vladimirova, V. V. 1970. Mechanism of the action of repellent vapors on mosquitoes and other blood-sucking insects. *Parazitol. Parazit. Bolezn. 39(6):718–22*
57. Rakhmanova, D. I. 1959. On the methods of individual tests of repelling substances against bloodsucking Diptera. *Med. Parazitol. Ny.* 484–87
58. Rezaev, N. I., Davydova, Zh. V., Makolkin, I. A. 1969. Physiochemical properties of biologically active repellents. *Biol. Nauki* 8:49–67
59. Roth, A. R., Mote, D. C., Lindquist, D. A. 1954. Tests of repellents against tabanids. *USDA-ARS-33-2.* 10 pp.
60. Schmidt, C. H., Acree, F. Jr., Bowman, M. C. 1959. Fate of C^{14}-diethyltoluamide applied to guinea pigs. *J. Econ. Entomol.* 52(5):928–30
61. Schreck, C. E., Smith, N., Gouck, H. K. 1976. Repellency of *N,N*-diethyl-*m*-toluamide (deet) and 2-hydroxyethyl cyclohexanecarboxylate against the deer flies *Chrysops atlanticus* Pechuman and *Chrysops flavidus* Wiedemann. *J. Med. Entomol.* 13:115–18
62. Shambaugh, G. F., Brown, R. F., Pratt, J. J. Jr. 1957. Repellents for biting arthropods. *Adv. Pest Control Res.* 1:277–303
63. Shimmin, R., Bayles, S., Spencer, T., Akers, W., Grothaus, R. 1975. Foursite method for mosquito repellent field trials. *US Naval Med. Field Res. Lab., Camp Lejuene, NC, Bureau of Med. and Surgery, Navy Dept.,* Vol. 25(1):121–23
64. Smith, C. N. 1957. Insect repellents. *Soap Chem. Spec.* 34(2):105–12, 203; 34(3):126–33
65. Smith, C. N. 1970. Repellents for anopheline mosquitoes. *Misc. Publ. Entomol. Soc. Am.* 7(1):99–117
66. Smith, C. N., Burnett, D. Jr. 1948. Laboratory evaluation of repellents and toxicants as clothing treatments for personal protection from fleas and ticks. *Am. J. Trop. Med. Hyg.* 28(4):599–607
67. Smith, C. N., Cole, M. M. 1951. Mosquito repellents for application to clothing. *J. Natl. Malaria Soc.* 10(3):206–12

68. Smith, C. N., Cole, M. M., Lloyd, G. W., Selhime, A. 1952. Mosquito repellent mixtures. *J. Econ. Entomol.* 45(5):805–9
69. Smith, C. N., Gilbert, I. H. 1953. Effectiveness against mosquitoes of general purpose mixtures for application to clothing. *J. Econ. Entomol.* 46:671–74
70. Smith, C. N., Gilbert, I. H., Gouck, H. K., Bowman, M. C. 1963. Factors affecting the protection period of mosquito repellents. *USDA, ARS Tech. Bull. No. 1285.* 36 pp.
71. Spencer, T. S., Shimmin, R. K., Schoeppner, R. F. 1975. Field test of repellents against the valley black gnat, *Leptoconops carteri* Hoffman. *Calif. Vector Views* 22(1):5–7
72. Starnes, E. B., Granett, P. 1953. A laboratory method for testing repellents against biting flies. *J. Econ. Entomol.* 46(3):420–23
73. Strauss, W. G., Maibach, H. I., Khan, A. A. 1968. Drugs and disease as mosquito repellents in man. *Am. J. Trop. Med. Hyg.* 17(3):461–64
74. Traub, R., Elisberg, B. L. 1962. Field tests on diethyltoluamide (deet), a highly effective repellent against mosquitoes in the nipah palm-mangrove swamp in Malaya. *Pac. Insects* 4(2): 303–13
75. Traub, R., Elisberg, B. L. 1962. Comparative efficacy of diethyltoluamide skin-application repellent (deet) and M-1960 clothing impregnant against mosquitoes in the nipah palm-mangrove swamps in Malaya. *Pac. Insects* 4(2): 314–18
76. Travis, B. V. 1947. Relative effectiveness of various repellents against *Anopheles farauti* Laveran. *J. Nat. Malaria Soc.* 6(3):180–83
77. Travis, B. V. 1950. Known factors causing variations in results of insect repellent tests. *Mosq. News* 10(3):126–32
78. Travis, B. V., Morton, F. A. 1946. Treatment of clothing for protection against mosquitoes. *Proc. Ann. Meet. NJ Mosq. Exterm. Assoc., 33rd,* pp. 65–69
79. Travis, B. V., Morton, F. A., Cochran, J. H. 1946. Insect repellents used as skin treatments by the armed forces. *J. Econ. Entomol.* 39(5):627–30
80. Tristan, D. F., Rudenchik, Yu. V., Chekalin, V. B., Vishnyakova, L.K., Dzhumabekov, K. D. 1970. Use of flea and tick repellents. *Probl. Osobo Opasnykh Infek.* (5):124–27
81. Vladimirova, V. V. 1969. Sensitivity of *Aedes aegypti* mosquitoes to repellents depending upon environment. *Med. Parazitol. Parazit. Bolezn.* 38(2):214–18
82. Vladimirova, V. V. 1970. Sensitivity of mosquitoes to repellents in relation to their physiological state. *Med. Parazitol. Parazit. Bolezn.* 39(1):49–53
83. Wood, P. Wm. 1968. The effect of ambient humidity on the repellency of ethylhexanediol ('6–12') to *Aedes aegypti. Can. Entomol.* 100:1331–35
84. Wright, R. H. 1962. The attraction and repulsion of mosquitoes. *World Rev. Pest Control.* 1(4):1–12
85. Wright, R. H. 1975. Why mosquito repellents repel. *Sci. Am.* 233(1):104–11
86. Wright, R. H., Burton, D. J. 1969. Pyrethrum as a repellent. *Pyrethrum Post* 10(2):14–21
87. Yeoman, G. H., Warren, B. C. 1970. Repellents for *Stomoxys calcitrans* (L.), the stable fly: techniques and a comparative laboratory assessment of butyl methylcinchoninate. *Bull. Entomol. Res.* 59:563–77
88. Zolotarev, E. Kh., Elizarov, Yu. A. primary laboratory test of repellents on fleas. *Med. Parazitol.* 6:738–39
89. Zolotarev, E. Kh., Elizarnov, Yu. A., 1962. On laboratory testing of repellents against *Ixodes persulcatus* ticks. *Med. Parazitol.* 4:434–35
90. Zolotarev, E. Kh., Kalakutskaya, T. V. 1960. Repellent study. IX. Diethyltoluamides. Comparative evaluation of *ortho-*, *meta-* and *para*-isomer repellency against ticks and mosquitoes *Vestn. Mosk. Univ.* 3:18–21
91. Zhukova, L. I. 1962. Improved laboratory testing of repellents against *Ixodes persulcatus* ticks. *Med. Parazitol.* 4:436–38

Ann. Rev. Entomol. 1977. 22:121–38

HORMONAL REGULATION OF LARVAL DIAPAUSE

❖6123

G. Michael Chippendale
Department of Entomology, University of Missouri, Columbia, Missouri 65201

INTRODUCTION

This review examines recent findings about the endocrine basis of larval diapause. Evidence has accumulated to support the view that larval diapause is regulated by distinctive endocrine interactions. While the lack of the molting hormone, ecdysone, causes the arrest of morphogenesis in both diapausing larvae and pupae, the mechanism leading to the inactivity of the ecdysial glands appears to differ between these life stages. The recent discovery of a juvenile hormone (JH) involvement in the regulation of the larval diapause of two species (24, 144) indicates that at least some species of diapausing larvae retain activity within their endocrine system. This finding has underscored the need to further examine the extent of endocrine activity during larval diapause. For background information, several reviews are available on insect endocrinology (36, 44, 139) and on the physiology and ecology of diapause (35, 87, 121).

Hormonal failure and JH regulation are the two principal theories which have been developed to explain larval diapause. The hormonal failure theory was developed originally to explain the endocrine basis of pupal diapause (136–138), and proposes that diapause is instituted when the cerebral neurosecretory (NS) system becomes inactive and the corpora cardiaca stop releasing the ecdysiotropin that is required to stimulate the ecdysial glands to secrete ecdysone. Morphogenesis resumes after the corpora cardiaca renew their release of ecdysiotropin. The diapause state, therefore, is thought to be distinguished by complete inactivity in the endocrine system, including the corpora allata. This theory has been extended to explain the endocrine basis of larval diapause (56, 71, 135). Conversely, the JH regulation theory proposes that diapausing larvae, unlike diapausing pupae, retain actively secreting corpora allata and that JH initiates and maintains diapause by regulating the secretion of ecdysiotropin (24, 144). The corpora allata are thought to be controlled by ordinary and NS neurones from the brain. An intermediate titer of JH associated with diapause is believed to arrest morphogenesis by inhibiting ecdysiotropin synthesis, transport, or release. Diapause is presumed to terminate when a change in the secretory rate of the corpora allata and in the titer of circulating JH frees the cerebral NS system from inhibition. Although both theories attribute larval diapause to a lack of ecdysiotropin, the hormonal failure theory proposes that the lack of ecdysiotropin is caused by an inactivation of the cerebral NS system imposed by extero- and proprioceptive signals received by the central nervous system, whereas the JH regulation theory proposes that an interaction between the brain and the corpora allata controls the activity of the ecdysiotropin-producing system. Since different regulatory mechanisms may have evolved among larvae of the diverse insect species, these two theories may eventually prove not to be mutually exclusive.

A special endocrine situation which does not fit either the hormonal failure or the JH regulation theory exists in some parasitic Hymenoptera and Diptera. In these species, larval diapause is determined largely in the maternal generation by such factors as photoperiod, temperature, age, diet, and whether females arose from diapausing or nondiapausing larvae. For example, after exposure to short days and low temperatures, female adults of the chalcidoid wasp, *Nasonia vitripennis,* produce progeny which enter a larval diapause. Perhaps these females produce a diapause hormone which enters the egg cytoplasm, persists during the immature larval instars, and causes the developmental arrest in the final larval instar (103, 104, 109). Although the larval endocrine system may be the target of such a diapause hormone, its origin, nature, and mode of action remain unknown. A similar hormonal mechanism may also account for the maternally regulated larval diapause of the hymenopterous parasites *Coeloides brunneri* (99) and *Ooencyrtus entomophagus* (7) and the flies *Lucilia sericata* (34) and *Lucilia caesar* (98).

This review summarizes the available information about the hormonal control of larval diapause. Although relatively few species have been studied to date, the extent of our current knowledge in the following areas is covered: (*a*) environmental control over the endocrine system; (*b*) activity of the cerebral NS system, ecdysial glands, and corpora allata; (*c*) effect of JH mimic and ecdysone treatments; (*d*) probable mode of action of JH and interactions between the cerebral NS system and corpora allata; and (*e*) hormonal control of the metabolism of diapausing larvae.

PHENOLOGY OF LARVAL DIAPAUSE

Diapause is a state of developmental arrest that enables an insect to survive under adverse environmental conditions and synchronizes the active stages of its life cycle with available food sources (10). It is instituted following the exposure of sensitive prediapausing stages to inductive environmental conditions, primarily short-day photoperiods and low temperatures in temperate-zone species that enter a winter diapause. Diapause, therefore, is a genetically controlled life phase for which preparatory biochemical adjustments, including the accumulation of nutrient reserves, and behavioral adaptations, often including the location of a protective site, occur in advance. In the case of larval diapause, evidence is accumulating that both its induction and duration are under polygenic control (58, 68).

Table 1 Examples of insects in which diapause occurs during the various larval instars

Family and order	Genus and species	Diapausing instar	Reference
Dermestidae (Col.)	*Anthrenus verbasci*	Mid, last (mature)	17
Meloidae (Col.)	*Epicauta segmenta*	Coarctate	112
Calliphoridae (Dipt.)	*Calliphora vicina*	Last (3rd), mature	130
Cecidomyiidae (Dipt.)	*Contarinia sorghicola*	Last, mature	141
Chironomidae (Dipt.)	*Chironomus tentans*	Last	38
Culicidae (Dipt.)	*Aedes triseriatus*	Last (4th)	30
Culicidae (Dipt.)	*Chaoborus americanus*	Last (4th)	19
Culicidae (Dipt.)	*Toxorhynchites rutilus*	Last (4th)	20
Culicidae (Dipt.)	*Wyeomyia smithii*	3rd and 4th	72
Delphacidae (Hom.)	*Delphacodes striatella*	4th	69
Braconidae (Hymen.)	*Apanteles melanoscelus*	Last, mature	132
Cephidae (Hymen.)	*Cephus cinctus*	Last, mature	28
Chalcididae (Hymen.)	*Nasonia vitripennis*	Last (4th), mature	109
Diprionidae (Hymen.)	*Neodiprion rugifrons*	Last (eonymph)	70
Ichneumonidae (Hymen.)	*Pleolophus basizonus*	Last (eonymph)	49
Ichneumonidae (Hymen.)	*Triclistus pygmaeus*	Last (eonymph)	3
Mymaridae (Hymen.)	*Caraphractus cinctus*	Last (4th), mature	64
Pteromalidae (Hymen.)	*Hypopteromalus tabacum*	Last, mature	81
Tenthredinidae (Hymen.)	*Athelia rosae*	Last (eonymph)	101
Arctiidae (Lep.)	*Spilarctia imparilis*	7th	117
Gelechiidae (Lep.)	*Pectinophora gossypiella*	Last (4th), mature	2
Heterogeneidae (Lep.)	*Monema flavescens*	Last, mature	119
Lasiocampidae (Lep.)	*Dendrolimus pini*	2nd–5th	43
Lasiocampidae (Lep.)	*Malacosoma testacea*	1st, pharate	125
Lymantriidae (Lep.)	*Porthetria dispar*	1st, pharate	74
Noctuidae (Lep.)	*Agrotis c-nigrum*	4th	67
Noctuidae (Lep.)	*Busseola fusca*	Last, mature	126
Nymphalidae (Lep.)	*Limenitis archippus*	3rd	58
Olethreutidae (Lep.)	*Laspeyresia pomonella*	Last (5th), mature	50
Pyralidae (Lep.)	*Chilo suppressalis*	Last (5th), mature	41
Pyralidae (Lep.)	*Diatraea grandiosella*	Last (6th), mature	23
Pyralidae (Lep.)	*Ephestia calidella*	Last, mature	33
Pyralidae (Lep.)	*Ostrinia nubilalis*	Last (5th), mature	11
Pyralidae (Lep.)	*Plodia interpunctella*	Last, mature	16
Schoenobiidae (Lep.)	*Rupela albinella*	Last (6th), mature	128
Tortricidae (Lep.)	*Adoxophyes orana*	3rd	59
Tortricidae (Lep.)	*Archippus breviplicanus*	3rd or 4th	93
Tortricidae (Lep.)	*Grapholitha funebrana*	Last (4th), mature	102
Chrysopidae (Neur.)	*Chrysopa nigricornis*	Last (3rd), mature	120
Chrysopidae (Neur.)	*Chrysopa oculata*	Last (3rd), mature	97
Aeshnidae (Odon.)	*Aeshna cyanea*	Mid or last	105
Aeshnidae (Odon.)	*Anax imperator*	Last	32
Coenagrionidae (Odon.)	*Enallagma hageni*	Mid	62
Gryllidae (Orth.)	*Gryllus campestris*	Penultimate	63
Gryllidae (Orth.)	*Pteronemobius nitidus*	Late	75

Table 1 shows examples of species from eight insect orders which enter a larval diapause. Examples have been restricted to species that enter a true seasonally regulated diapause, most commonly under photoperiodic control. To date, the physiology of larval diapause among Diptera, Hymenoptera, and Lepidoptera has been studied the most extensively, but hormonal studies have been restricted to a few species within these orders. Table 1 illustrates that larval diapause can intervene at any point from the pharate first instar to the mature nonfeeding phase of the final instar, i.e. the so-called prepupal stage.

Some insects appear to gain an adaptive advantage from a genetic polymorphism of larval diapause. Although larval diapause typically lasts for less than one year, a prolonged diapause extending over two or more years may occur in individuals of some diapausing populations (96). The endocrine mechanism of a prolonged diapause, which is especially common among sawflies, has not yet been investigated (95). Another example of genetic polymorphism is found among populations of coniferous budworms, *Choristoneura* spp. (52). Although most individuals are univoltine and enter diapause as second-stage larvae, some are semivoltine and enter a second diapause in the fourth instar. The induction of the second diapause appears to be solely under genetic control. Larvae of the pitcher plant mosquito, *Wyeomyia smithii,* on the other hand, overwinter in the third instar, but may enter a second diapause in the fourth instar in the spring to prevent the premature completion of metamorphosis (72).

ENVIRONMENTAL CONTROL OVER THE ENDOCRINE SYSTEM

Since diapause is determined before the onset of the resting stage, larvae must accumulate and store environmental information which they receive during their sensitive prediapausing stages (47). For example, a high incidence of the mature larval diapause of the pink bollworm, *Pectinophora gossypiella,* was induced when eggs and first-stage larvae were exposed to only 7 short-day photoperiods and then transferred to continuous illumination. Exposure to a few inductive cycles apparently entrains some as-yet-unknown component of the endocrine system, leading to the onset of diapause in mature larvae (2). Similarly, when larvae of the cabbage moth, *Barathra brassicae,* were exposed to only 14 short-day photoperiods, their endocrine system was programmed for a pupal diapause (46). Other results have been presented to show that food consumption acts synergistically with long-day photoperiods to accelerate the termination of the larval diapause of the culicid mosquito, *Chaoborus americanus* (19). Food intake by diapausing larvae appears to initiate some unspecified endocrine reflex, leading to the resumption of morphogenesis.

Evidence has accumulated to show that cephalic photoperiodic receptors receive the signals which regulate the seasonal development of several species. The receptor pigments appear to be located in the brain itself rather than in the compound eyes or ocelli. Recent findings indicate that the visual pigment rhodopsin is not the photoperiodic receptor pigment of *Drosophila* spp. (140). Although the larvae of few species have been examined, it is clear that even seemingly opaque head capsules transmit sufficient light to stimulate underlying extraocular receptors (18). For example, light penetrating the larval head capsule of the large white butterfly, *Pieris brassicae,* has been shown to act directly on the brain to regulate the secretion of ecdysiotropin and the induction of pupal diapause (29).

Other results suggest that extracephalic receptors, perhaps working in conjunction with those in the brain, also receive environmental cues and regulate larval diapause. For example, it has been suggested that the larval diapause of the European corn borer, *Ostrinia nubilalis,* is controlled by a developmental hormone, proctodone, which is secreted rhythmically by the ileal epithelium. The target of the hormone was thought to be the ecdysiotropin-producing cells of the cerebral NS system (14). However, this hypothesis has not been substantiated (13, 25). Instead, the most recent evidence indicates that the observed secretory activity of the larval ileum results from the presence of autolysosomes that remodel the epithelium at the beginning of metamorphosis (25, 80). An interaction between cephalic and extracephalic oscillators has also been postulated to regulate the induction of the larval diapause of the pine moth, *Dendrolimus pini* (123).

A proprioceptive involvement in the regulation of larval diapause has not yet been demonstrated convincingly. However, proprioceptive signals have been shown to regulate the onset of the pupal molting cycle in some species. For example, nerve section, body constraint, and paralysis have been shown to delay or prevent pupation of larvae of the greater wax moth, *Galleria mellonella.* Proprioceptive input about body form and posture from abdominal stretch receptors appears to be required before ecdysiotropin or ecdysone is released at the end of the final instar (12, 37, 110). In addition, last-stage larvae of the tobacco hornworm, *Manduca sexta,* have been shown to release ecdysiotropin only after they attain a weight of about five grams, implying that the brain must receive specific proprioceptive signals about body size, form, and posture to initiate the pupal molting cycle (89). Although we can reasonably conclude that extero- and proprioceptive signals control the onset and termination of larval diapause through their action on the endocrine system, we are only now beginning to uncover the complex pathways which are involved.

ACTIVITY OF THE CEREBRAL NEUROSECRETORY SYSTEM

The brain controls the activity of the ecdysial glands and corpora allata and is the principal control center of diapausing larvae (36). Although the NS system plays a pivotal role in this cerebral control, the precise functioning of the system during growth, metamorphosis, and diapause is not yet fully understood. For example, we are still unable to distinguish between the site of synthesis, nature, and transportation of the various neurosecretions, including ecdysiotropin and allatotropin (106). Although light microscopic studies frequently have suggested that at least some NS cells remain active during larval diapause, the function of these cells remains unknown at present (28, 39, 67, 82, 131). For example, a comparative study of the cerebral NS system of five species which exhibit a larval diapause showed that NS material accumulated in the perkaryia of type A cells during diapause, leading to the suggestion that diapause is initiated and maintained because ecdysiotropin is not transported from these cells. However, other NS cells with unspecified functions appeared to retain their synthetic and transport activities during larval diapause (67). In contrast, ligatures and surgical operations performed on diapausing eonymphs of diprionid and tenthredinid sawflies suggest that the entire endocrine system, including the cerebral NS system, remains inactive during diapause (115). The hormonal basis of diapause in sawflies needs further study because of the presence of an uncommon three-step metamorphosis, i.e. prepupa (eonymph-pronymph)-pupa-adult (70, 101).

Since histological procedures did not uncover any obvious differences between the activity of the cerebral NS system of diapausing and nondiapausing larvae of *O. nubilalis,* larval brains were transplanted to obtain additional information (11, 31, 79). Brains of diapausing larvae of *O. nubilalis* implanted into diapausing recipients were shown to cause a premature termination of diapause. Reciprocal transplants of larval brains between *O. nubilalis* and the nondiapausing *G. mellonella* showed that decerebrated wax moth larvae pupated after receiving brains from diapausing corn borers, whereas decerebrated diapausing corn borers pupated after receiving wax moth brains only if they were held under long days. These experiments led to the suggestion that the ecdysiotropin-producing system of the diapausing corn borers is potentially active and that a blood-brain barrier prevents the transport of ecdysiotropin in diapausing larvae. Tissue injury during surgery was thought to disrupt the barrier between the NS cells and the circulatory system (11, 31). A recent ultrastructural study suggests that the type-1 perineurial cells of the brains of diapausing and nondiapausing larvae of *O. nubilalis* have different permeability properties that may be associated with the proposed blood-brain barrier (60).

Ultrastructural studies of the larval NS system have yet to provide information that might uncover the site of synthesis and specific nature of the various neurosecretions. This difficulty is caused partly by the problems associated with correlating the ultrastructural features of the cells with their light microscopic staining properties (42). An examination of the ultrastructure of the cerebral NS system of the slug moth, *Monema flavescens,* showed that marked changes occur in one cell type during the termination phase of its larval diapause (119). Perhaps these changes are caused by the transport of ecdysiotropin from the perikarya of the NS cells. Ultrastructural evidence suggested that the synthesis of granules in the cerebral NS system of fifth-stage larvae of the hemipteran, *Rhodnius prolixus,* continued during a starvation-induced quiescence (85). During the quiescence an equilibrium appeared to exist between synthesis and lysosomal hydrolysis of the NS granules. Resumed feeding led to the transport and release of the granules and a termination of the quiescence. The lack of similar examples serves to illustrate that much remains to be learned about diapause-associated changes in the ultrastructure of the cerebral NS system.

The immediate cause of changes that have been observed in the cerebral NS system during larval diapause is still not clear. Studies using saturniid moths initially led to the suggestion that brains of diapausing larvae and pupae retain ecdysiotropin within their NS perikarya because they lack cholinesterase activity and are electrically inexcitable (127). Later studies have shown, however, that changes in the cholinergic system are related to the metamorphosis of the central nervous system and are not involved causally in diapause (113, 124). For example, a comparative study of diapausing and nondiapausing larvae of *O. nubilalis* showed that the cholinergic system does not regulate activity within the cerebral NS system or control the induction and termination of diapause (73). Recent evidence suggests that the hemolymph JH titer is one factor regulating activity within the cerebral NS system (77, 122).

ACTIVITY OF THE ECDYSIAL GLANDS

It has generally been believed that the ecdysial glands of diapausing larvae remain continuously inactive, thus resulting in the complete absence of molting activity. However, recent evidence has shown that the ecdysial glands of some diapausing larvae must be activated periodically because these larvae undergo one or more stationary molts during diapause. To date, such stationary molts have been found in diapausing larvae of the neotropical corn borer, *Diatraea lineolata* (66), the Asiatic rice borer, *Chilo suppressalis* (144), the stem borers, *Coniesta ignefusalis* (51) and *Busseola fusca* (126), and the southwestern corn borer, *Diatraea grandiosella* (24). All these larvae exhibit typical diapause characteristics and must contain a JH titer that is sufficiently high to permit the retention of the larval form when molting occurs. These stationary molts do not lead to the termination of diapause. Though this molting activity provides clues about the evolution of the diapause phenomenon in these species, it does not appear to provide the insect with any adaptive value. Selection pressures presumably favor those larvae that do not utilize essential reserves in diapause-associated molts.

Changes have been observed in the ecdysial glands of diapausing larvae of several species. Comparisons have been made of the inactive ecdysial glands of diapausing larvae and the active ones of late diapausing or nondiapausing larvae. For example, the inactive ecdysial gland cells of diapausing larvae of the wheat stem sawfly, *Cephus cinctus,* have been found to contain small lobular nuclei and many large vacuoles, but to lack granular inclusions in the cytoplasm (28). In contrast, the large cells of active ecdysial glands of late diapausing larvae contain large spherical nuclei and many granular inclusions in the cytoplasm. A comparison of the function and fine structure of the ecdysial glands of diapausing and nondiapausing larvae of *D. grandiosella* showed that these glands are required to initiate apolyses and that each gland consists of about 30 large secretory cells (147). The glands receive efferent neurones, some of which are neurosecretory, from the central nervous system. Each active cell is bounded by a thin tunica propria and consists of a large nucleus and a dense cytoplasm containing evenly dispersed mitochondria, ribosomes, and glycogen particles. In contrast, each inactive cell is bounded by a thick tunica propria and contains a shrunken convoluted nucleus and an inner cytoplasmic zone of glycogen particles. The reactivation of the ecdysial glands of late diapausing larvae of *M. flavescens* has also been examined (118). Ultrastructural evidence suggests that the synthesis and transport of ribonucleic acid to direct ecdysone synthesis does not occur until ecdysiotropin has been released from the corpora cardiaca. Extensive folding of the nuclear membrane, which probably facilitates ribonucleic acid transport into the cytoplasm, was observed. Ecdysiotropin may activate adenyl cyclase, thereby leading to the synthesis of cyclic adenosine monophosphate, which in turn controls α-ecdysone synthesis in the ecdysial gland cell (129).

RESPONSE OF DIAPAUSING LARVAE TO EXOGENOUS ECDYSONE

Several investigators have observed the response of diapausing larvae to injections of ecdysone in an attempt to determine their hormonal status during diapause. A full evaluation of this status involves the monitoring of early-, mid-, and late-diapausing larvae after ecdysone injection. However, this experimental condition has been met only rarely. Although an ecdysone injection was found to cause diapausing larvae of the solitary bees, *Nomia melanderi* and *Megachile rotundata,* and the dragonfly, *Aeshna cyanea,* to resume morphogenesis, no systematic study of its effect was undertaken (61, 105). Other experiments have shown that newly diapaused larvae of *O. nubilalis* responded to exogenous ecdysone only when high dossages were injected (15). These treatments usually caused an aborted molting cycle in which moribund larval-pupal intermediates were formed. The treated insects usually showed more larval than pupal characteristics, suggesting that JH is present during diapause. Early diapausing larvae of *C. suppressalis* treated with β-ecdysone underwent normal larval molts whereas late diapausing larvae molted into larval-pupal intermediates (144). Similarly, a β-ecdysone injection into early- and mid-diapausing larvae of *D. grandiosella* induced only stationary larval molts. However, an ecdysone injection into the thoracoabdominal portion of a previously ligatured larva caused a premature termination of diapause, suggesting that the presence of a cephalic factor (probably JH) sustained diapause (145). The results of a β-ecdysone injection into ligatured and nonligatured larvae of the maize stem borer, *Chilo partellus,* also led to the conclusion that ecdysone terminates diapause only after endogenous JH has dissipated (107).

ACTIVITY OF THE CORPORA ALLATA

Convincing evidence showing that allatectomy causes premature termination of larval diapause is still lacking because of surgical difficulties. However, several indirect lines of evidence suggest that diapausing larvae of *C. suppressalis* and *D. grandiosella* retain active corpora allata. The results of surgical transplants of active ecdysial glands and brains into decapitated diapausing larvae and histological observations suggest that the corpora allata of *C. suppressalis* remain active during early diapause and mid-diapause, but gradually lose activity during the termination phase (40, 41, 82). Furthermore, a correlation was observed between the loss of activity of the corpora allata and the renewed activity of the cerebral type B cells of *C. suppressalis,* suggesting that the JH titer regulated activity within the cerebral NS system during diapause. Early diapausing larvae of *D. grandiosella* have also been shown to contain active corpora allata because a β-ecdysone injection made only 17 days after a head ligature induced an almost normal pupal molt (26).

Hemolymph JH Titer

Evidence is accumulating to show that larval diapause in some species is characterized by the presence of an intermediate titer of JH. Such hemolymph analyses provide information about the equilibrium which exists between rates of JH secretion from the corpora allata, tissue uptake, and enzymatic deactivation during diapause. It has been demonstrated that the hemolymph JH titer of last stage nondiapausing larvae of *C. suppressalis* rapidly declines from 2400 to 100 *Galleria* Units (GU) ml^{-1}, whereas that of diapausing larvae declines slowly to 300 GU ml^{-1} after 60 days at 25°C 12L:12D (144). Similarly, early diapausing larvae of *C. partellus* have been shown to contain an intermediate JH titer of 250–580 GU ml^{-1} hemolymph (108). A comparison of the JH titers in early- and late-diapausing larvae of *D. grandiosella* showed that the JH titer declines from 1500 GU ml^{-1} to 250 GU ml^{-1} after 140 days at 23°C (148). A chemical assay has confirmed the presence of an intermediate JH titer in the hemolymph of newly diapaused larvae of *D. grandiosella.* A new electron capture-gas chromatographic procedure for JH identification showed the following JH titers (ng ml^{-1}): JH I, 0.8; JH II, 2.6; and JH III, 1.0. Newly diapaused larvae therefore contain intermediate titers of all three known JHs. The principal component (JH II) makes up about 59% of the total. These analytical values are the equivalent of about 1700 GU ml^{-1} hemolymph and correspond reasonably well with the *Galleria* bioassay data (16a).

RESPONSE OF LARVAE TO EXOGENOUS JH

A JH involvement in the regulation of the larval diapause of *D. grandiosella* and *C. suppressalis* has been demonstrated convincingly from JH treatments of nondiapausing and diapausing larvae. The larval developmental program of *D. grandiosella* can be switched from nondiapause to diapause by treating last-stage larvae with a JH mimic (27, 145). Three JH mimics applied to early-last-stage nondiapausing larvae induced a high incidence of formation of the immaculate diapausing morph, which is a nonpigmented variant of the spotted nondiapausing morph (23). The mimics did not activate the recipient's corpora allata, but elevated the functional JH concentration, each having a different intrinsic hormonal activity. The pupation rate of the immaculate morphs appeared to depend upon the rate of deactivation of the mimics. A comparison of the pupation rates, metabolic reserves, oxygen consumption, and spermatogenesis revealed that JH-induced and normal diapausing larvae were in similar dormant states (146). Besides initiating diapause, periodic applications of a JH mimic to diapausing larvae of *D. grandiosella* were also shown to prolong diapause and increase the number of stationary larval ecdyses. These JH treatments caused immaculate morphs to molt into spotted ones which reverted to immaculate ones when the treatment was discontinued, suggesting that the JH titer is lower in the immaculate than spotted morphs (27, 145). Similar results have been reported for exogenous JH treatments of *C. suppressalis* larvae (144). In this case, larvae fed an artificial diet containing JH I (650 ppm) and held under long-day conditions entered diapause. Results from these experiments suggest that a substantial endogenous JH titer is necessary for the initiation of larval diapause in these species.

Although the role of JH in the larval diapause of *O. nubilalis* has yet to be examined systematically, preliminary results have shown that JH treatments of nondiapausing larvae prolong the final instar and retard spermatogenesis (143). In addition, JH treatments have been shown to induce a larval dormancy in species which do not enter a larval diapause. For example, JH or JH mimic treatment of last-stage larvae of the silkworm, *Bombyx mori,* the notodontid moth, *Cerura vinula,* and the hawk moth, *Sphinx ligustri,* delayed pupation, causing the formation of dauer larvae (5, 57). In addition, the larval quiescence of the khapra beetle, *Trogoderma granarium,* was induced by feeding wheat flour containing a JH mimic (88). Although these artificial dormancies result from an elevation of the functional JH titer of last-stage larvae, it is unknown at present whether the mode of action of JH is related to that in normal diapausing larvae (76).

REGULATION OF JH TITER

Where JH is involved in controlling the initiation and maintenance of larval diapause, efficient mechanisms must have evolved to regulate its titer. Recently some progress has been made in identifying these mechanisms, but their precise involvement in larval diapause has not yet been worked out. The larval diapause of *D. grandiosella* may intervene when the hemolymph JH titer falls to an intermediate level because of neural and NS controls operating on the corpora allata. During the termination phase of diapause, a change in these controlling signals may lead to the inactivation of the corpora allata. Concurrently, JH esterases and epoxide hydrolase may be activated, resulting in a decline of the JH titer to the threshold for a pupal molt (26).

The most convincing evidence showing a brain-centered control of the corpora allata has come from studies using reproductive females. Although notable exceptions exist, evidence suggests that the corpora allata of adult females of several species are stimulated by a cerebral neurosecretion (allatotropin) and inhibited by cerebral-originating neural impulses (36, 55). The allatotropin may be released from the corpora cardiaca and act via the hemolymph rather than be transported through NS axons which originate in the brain and terminate in the corpora allata (86). Cerebral regulation of larval corpora allata through neural and/or NS pathways has been demonstrated in the migratory locust, *Locusta migratoria* (45), *R. prolixus* (8), *B. mori* (84, 94), and *G. mellonella* (48, 111). In the case of *G. mellonella* the activity of the corpora allata declines during the last larval instar, presumably because allatotropin production ceases and the glands are inhibited by neural impulses from the brain.

The regulation of the titer of JH available to target cells also results from enzymatic deactivation. Recent findings have shown that JH is deactivated through ester hydrolysis, epoxide hydration, and sulfate or glucoside conjugation (4, 133). Esterases and an epoxide hydrolase capable of deactivating JH have been detected in hemolymph, fat body, epidermis, and intestinal epithelium (114). JH circulates in the hemolymph bound to a specific carrier protein, which appears to be synthesized in the fat body (91). The carrier protein maintains JH in solution and protects it from degradation in the early larval instars. However, JH bound to the carrier protein in *M. sexta* is susceptible to hydrolysis by a new esterase, which appears in the hemolymph at the end of the last larval instar (100).

ENDOCRINE INTERACTIONS OF DIAPAUSING LARVAE

Although much further exploration of the interactions is needed, current evidence suggests that the intermediate titer of JH present in diapausing larvae of *D. grandiosella* maintains diapause by inhibiting the ecdysiotropin-producing system. Since a head ligature prevented diapausing larvae from molting in response to topical applications of a JH mimic or JH I, tissues in the head, rather than the ecdysial glands, form the target for JH (26). JH applications to nonligatured diapausing larvae appear to induce larval molts because the elevation of the JH titer activates the ecdysiotropin-producing cells of the cerebral NS system. This observation leads to the suggestion that the intermediate JH titer typical of this larval diapause functions by inhibiting the ecdysiotropin-producing system. The intermediate JH titer may inhibit the synthesis or transport of ecdysiotropin or its release from the corpora cardiaca. The stationary larval ecdyses observed during the normal progress of larval diapause may occur because the JH titer of some larvae is sufficiently high to activate the ecdysiotropin-producing system. Although JH feedback to the cerebral NS system has important implications for integration within the endocrine system, it has received little attention. Among reproductive females, evidence exists that the JH titer controls activity within the cerebral NS system of the blowfly, *Calliphora erythrocephala,* and *L. migratoria* (77, 78, 122). Among larvae, evidence exists that the JH titer controls the activity of the ecdysiotropin-producing cells of last-stage larvae of *M. sexta* (90). After hornworm larvae reach a critical mass in the middle of the last instar, the corpora allata stop secreting JH, the circulating JH is deactivated, ecdysiotropin is released, and pupal apolysis is initiated.

Current evidence suggests that the larval diapause of *D. grandiosella* is regulated by an interaction between the cerebral NS system and the corpora allata (26). It is likely that the brain regulates the activity of the corpora allata, and a diapause-associated JH titer inhibits the release of ecdysiotropin. The ecdysiotropin-producing cells may be activated in the presence of a high or low titer of JH, but inhibited in the presence of an intermediate JH titer, typical of larval diapause. There are probably other reasons for the presence of active corpora allata during the larval diapause of some species, including a regulatory role for JH in nucleic acid, oxidative, or synthetic metabolism.

HORMONAL REGULATION OF THE METABOLISM OF DIAPAUSING LARVAE

The three phases of the metabolism of diapausing larvae (prediapause, diapause maintenance, and postdiapause) are probably controlled by a variation of hormonal titers. Most of our current information about the hormonal control of metabolism during post-embryonic diapause has been obtained from studies on diapausing pupae and has therefore been limited to an examination of the effects of ecdysone (142). However, a specific titer of JH, coupled with a lack of ecdysone in the maintenance phase of diapause, may be involved in the metabolic regulation of larval diapause.

Some of the metabolic processes associated with larval diapause which may be under hormonal regulation include the massive accumulation of nutrient reserves (21), and the low rate of spermatogenesis in prediapausing larvae (6, 83). Once diapause intervenes, hormonal regulation may be involved in the decreased rate of oxygen consumption (53), and in the reduced activity of fat body mitochondria (50). At the molecular level, hormonal regulation of macromolecular synthesis (54), glycolysis and protein hydrolysis (116), transamination (1), and ester hydrolysis (92, 116) may occur during larval diapause. Specific hormonal titers associated with the termination phase of diapause presumably restore the metabolic processes to a high rate (22). Generalizations about metabolic effects of JH must be made with caution, however, because such effects are known to vary with age, stage, and sex (65). Furthermore, the precise mode of action of JH is as yet unknown. It may directly control enzymatic reactions or have an indirect effect by controlling the permeability of nuclear, mitochondrial, or plasma membranes (9, 134). Uncovering the hormonal control of the metabolism of diapausing larvae still remains a major challenge.

CONCLUSIONS

Current evidence indicates that the corpora allata are involved in controlling the larval diapause of some species. The concept of hormonal failure and an inactive endocrine system during all cases of larval diapause no longer remains valid. In two species with a facultative mature larval diapause, the induction and maintenance of diapause has now been shown to be dependent upon the retention of active corpora allata. Larval diapause occurs in these species because the presence of active corpora allata and an intermediate JH titer causes an extreme prolongation of the instar. Precise data about the extent of the retention of active corpora allata among diapausing larvae will only be forthcoming when the endocrine mechanisms of several unrelated species have been examined. Nevertheless, the JH regulation theory now provides an acceptable explanation of the endocrine relationships in the diapausing larvae of several species. During the diapause of these larvae the brain apparently regulates the activity of the corpora allata, and an intermediate JH titer inhibits the ecdysiotropin-ecdysone system, probably by controlling activity within the cerebral NS system. Yet, despite the advances made over the past decade, much remains to be learned about the hormonal regulation of larval diapause before many generalizations can be formulated. Additional comparative studies among immature and mature diapausing larvae need to be undertaken. Aspects which warrant further attention include mechanisms of transduction of environmental signals, neural and NS regulation of the corpora allata, interactions within the endocrine system, and the hormonal control of metabolism during diapause. Beyond these topics, the probable hormonal control of the distinctive behavioral patterns of prediapausing larvae is a related, but neglected, research area.

ACKNOWLEDGMENTS

I thank Drs. J. E. Carrel and K. J. Judy for their helpful reviews of the manuscript. Investigations in my laboratory were supported in part by NSF grant BMS74-18155. This review is a contribution from the Missouri Agricultural Experiment Station, paper no. 7489.

Literature Cited

1. Abd El Fattah, M. M., Algauhauri, A. E. I., Rostom, Z. M. F. 1972. Some enzymes in the haemolymph of *Pectinophora gossypiella* during induction and termination of diapause. *J. Comp. Physiol.* 78:20–25
2. Adkisson, P. L. 1966. Internal clocks and insect diapause. *Science* 154: 234–41
3. Aeschlimann, J.-P. 1974. Hibernation chez trois espèces de métopiines: Hymenoptera, Ichneumonidae. *Entomol. Exp. Appl.* 17:487–92
4. Ajami, A. M., Riddiford, L. M. 1973. Comparative metabolism of the cecropia juvenile hormone. *J. Insect Physiol.* 19:635–45
5. Akai, H., Kiguchi, K., Mori, K. 1973. The influence of juvenile hormone on the growth and metamorphosis of *Bombyx* larvae. *Bull. Sericult. Exp. Stn. Tokyo* 25:287–305
6. Alexander, B. R., Chippendale, G. M. 1973. Spermatogenesis of the southwestern corn borer, *Diatraea grandiosella.* 1. Comparison of rates in prediapause and nondiapause larvae. *Ann. Entomol. Soc. Am.* 66:747–52
7. Anderson, J. F., Kaya, H. K. 1974. Diapause induction by photoperiod and temperature in the elm spanworm egg parasitoid *Ooencyrtus* sp. *Ann. Entomol. Soc. Am.* 67:845–49
8. Baehr, J. C. 1976. Étude du contrôle neuro-endocrine du fonctionnement du corpus allatum chez les larves du quatrième stade de *Rhodnius prolixus. J. Insect Physiol.* 22:73–82
9. Baumann, G. 1969. Juvenile hormone effect on bimolecular lipid membranes. *Nature* 223:316–17
10. Beck, S. D. 1968. *Insect Photoperiodism,* pp. 135–84. New York: Academic. 288 pp.
11. Beck, S. D. 1968. Environmental photoperiod and the programming of insect development. In *Evolution and Environment,* ed. E. T. Drake, pp. 279–96. New Haven, Conn: Yale Univ. 470 pp.
12. Beck, S. D. 1970. Neural and hormonal control of pupation in *Galleria mellonella. Ann. Entomol. Soc. Am.* 63: 144–49
13. Beck, S. D. 1974. Photoperiodic determination of insect development and diapause. 1. Oscillators, hourglasses, and a determination model. *J. Comp. Physiol.* 90:275–95
14. Beck, S. D., Alexander, N. J. 1964. Proctodone, an insect developmental hormone. *Biol. Bull.* 126:185–98
15. Beck, S. D., Shane, J. L. 1969. Effects of ecdysones on diapause in the European corn borer, *Ostrinia nubilalis. J. Insect Physiol.* 15:721–30
16. Bell, C. H., Walker, D. J. 1973. Diapause induction in *Ephestia elutella* (Hübner) and *Plodia interpunctella* (Hübner) with a dawn-dusk lighting system. *J. Stored Prod. Res.* 9:149–58

16a. Bergot, B. J., Schooley, D. A., Chippendale, G. M., Yin, C.-M. 1976. Juvenile hormone titer determinations in the southwestern corn borer, *Diatraea grandiosella,* by electron capture—gas chromatography. *Life Sci.* 18:811–20

17. Blake, G. M. 1963. Shortening of a diapause-controlled life-cycle by means of increasing photoperiod. *Nature* 198: 462–63
18. Bounhiol, J.-J., Moulinier, C. 1965. L'opacité crânienne et ses modifications naturelles et expérimentales chez le ver à soie. *C. R. Acad. Sci. Ser. D* 261: 2739–41
19. Bradshaw, W. E. 1970. Interactions of food and photoperiod in the termination of larval diapause in *Chaoborus americanus. Biol. Bull.* 139:476–84
20. Bradshaw, W. E., Holzapfel, C. M. 1975. Biology of tree-hole mosquitoes:

photoperiodic control of development in northern *Toxorhynchites rutilus* (Coq.). *Can. J. Zool.* 53:889–93
21. Chippendale, G. M. 1973. Diapause of the southwestern corn borer, *Diatraea grandiosella:* utilization of fat body and haemolymph reserves. *Entomol. Exp. Appl.* 16:395–406
22. Chippendale, G. M., Beck, S. D. 1967. Fat body proteins of *Ostrinia nubilalis* during diapause and prepupal differentiation. *J. Insect Physiol.* 13:995–1006
23. Chippendale, G. M., Reddy, A. S. 1972. Diapause of the southwestern corn borer, *Diatraea grandiosella:* transition from spotted to immaculate mature larvae. *Ann. Entomol. Soc. Am.* 65:882–87
24. Chippendale, G. M., Yin, C.-M. 1973. Endocrine activity retained in diapause insect larvae. *Nature* 246:511–13
25. Chippendale, G. M., Yin, C.-M. 1975. Reappraisal of proctodone involvement in the hormonal regulation of larval diapause. *Biol. Bull.* 149:151–64
26. Chippendale, G. M., Yin, C.-M. 1976. Endocrine interactions controlling the larval diapause of the southwestern corn borer, *Diatraea grandiosella. J. Insect Physiol.* 22:989–95
27. Chippendale, G. M., Yin, C.-M. 1976. Diapause of the southwestern corn borer, *Diatraea grandiosella:* effects of a juvenile hormone mimic. *Bull Entomol. Res.* 66:75–79
28. Church, N. S. 1955. Hormones and the termination and reinduction of diapause in *Cephus cinctus* Nort. *Can. J. Zool.* 33:339–69
29. Claret, J. 1966. Mise en évidence du rôle photorécepteur du cerveau dans l'induction de la diapause, chez *Pieris brassicae. Ann. Endocrinol.* 27:311–20
30. Clay, M. E., Venard, C. E. 1972. Larval diapause in the mosquito *Aedes triseriatus:* effects of diet and temperature on photoperiodic induction. *J. Insect Physiol.* 18:1441–46
31. Cloutier, E. J., Beck, S. D., McLeod, D. G. R., Silhacek, D. L. 1962. Neural transplants and insect diapause. *Nature* 195:1222–24
32. Corbet, P. S. 1956. Environmental factors influencing the induction and termination of diapause in the emperor dragonfly, *Anax imperator* Leach. *J. Exp. Biol.* 33:1–14
33. Cox, P. D. 1975. The influence of photoperiod on the life-cycles of *Ephestia calidella* (Guenée) and *Ephestia figulilella* Gregson. *J. Stored Prod. Res.* 11:75–85
34. Cragg, J. B., Cole, P. 1952. Diapause in *Lucilia sericata* (Mg.) Diptera. *J. Exp. Biol.* 29:600–4
35. de Wilde, J. 1970. Hormones and the environment. *Mem. Soc. Endocrinol.* 18:487–514
36. Doane, W. W. 1973. Role of hormones in insect development. In *Developmental Systems: Insects,* ed. S. J. Counce, C. H. Waddington, 2:291–497. New York: Academic. 615 pp.
37. Edwards, J. S. 1966. Neural control of metamorphosis in *Galleria mellonella. J. Insect Physiol.* 12:1423–33
38. Englemann, W., Shappirio, D. G. 1965. Photoperiodic control of the maintenance and termination of larval diapause in *Chironomus tentans. Nature* 207:548–49
39. Fraser, A. 1960. Humoral control of metamorphosis and diapause in the larvae of certain Calliphoridae. *Proc. R. Soc. Edinburgh Ser. B* 67:127–40
40. Fukaya, M., Kobayashi, M. 1966. Some inhibitory actions of corpora allata in diapausing larvae of the rice stem borer, *Chilo suppressalis* Walker. *Appl. Entomol. Zool.* 1:125–29
41. Fukaya, M., Mitsuhashi, J. 1961. Larval diapause in the rice stem borer with special reference to its hormonal mechanism. *Bull. Natl. Inst. Agric. Sci. Jpn.* 13C:1–32
42. Geldiay, S., Edwards, J. S. 1973. The protocerebral neurosecretory system and associated cerebral neurohemal area of *Acheta domesticus. Z. Zellforsch. Mikrosk. Anat.* 145:1–22
43. Geyspits, K. F. 1965. Photoperiodic and temperature reactions affecting the seasonal development of the pine moths, *Dendrolimus pini* L. and *D. sibiricus* Tschetw. *Entomol. Rev.* 44: 316–25
44. Gilbert, L. I., King, D. S. 1973. Physiology of growth and development: endocrine aspects. In *The Physiology of Insecta,* ed. M. Rockstein, 3:249–370. New York: Academic. 512 pp. 2nd ed.
45. Girardie, A. 1974. Recherches sur le rôle physiologique des cellules neurosécrétrices latérales du protocerébron de *Locusta migratoria migratorioides. Zool. Jahrb. Physiol.* 78:310–26
46. Goryshin, N. I., Tyshchenko, G. F. 1973. Accumulation of photoperiodic information during diapause induction in the cabbage moth, *Barathra brassicae* L. *Entomol. Rev.* 52:173–76
47. Goryshin, N. I., Tyshchenko, V. P. 1974. Memory link and its position in

the mechanism of photoperiodic reaction in insects. *Zh. Obshch. Biol.* 35:518–30
48. Granger, N. A., Sehnal, F. 1974. Regulation of larval corpora allata in *Galleria mellonella. Nature* 251:415–17
49. Griffiths, K. J. 1969. Development and diapause in *Pleolophus basizonus. Can. Entomol.* 101:907–14
50. Hansen, L. D., Harwood, R. F. 1968. Comparisons of diapause and nondiapause larvae of the coddling moth, *Carpocapsa pomonella. Ann. Entomol. Soc. Am.* 61:1611–17
51. Harris, K. M. 1962. Lepidopterous stem borers of cereals in Nigeria. *Bull. Entomol. Res.* 53:139–71
52. Harvey, G. T. 1967. On coniferous species of *Choristoneura* in North America. 5. Second diapause as a species character. *Can. Entomol.* 99:486–503
53. Hayes, D. K., Horton, J., Schechter, M. S., Halberg, F. 1971. Rhythm of oxygen uptake in diapausing larvae of the codling moth at several temperatures. *Ann. Entomol. Soc. Am.* 65:93–96
54. Hayes, D. K., Reynolds, P. S., McGuire, J. U., Schechter, M. S. 1971. Synthesis of macromolecules in diapausing and nondiapausing larvae of the European corn borer. *J. Econ. Entomol.* 65:676–79
55. Highnam, K. C. 1964. Endocrine relationships in insect reproduction. *Symp. Entomol. Soc. London* 2:26–42
56. Highnam, K. C., Hill, L. 1969. *The Comparative Endocrinology of the Invertebrates.* London: Arnold. 270 pp.
57. Hintze-Podufal, C. 1975. Über die morphogenetische Wirkungsweise von Juvenilhormon und analogen Substanzen auf die Adultentwicklung von Schwärmern und Spinnern. *Z. Angew. Entomol.* 77:286–91
58. Hong, J. W., Platt, A. P. 1975. Critical photoperiod and daylength threshold differences between northern and southern populations of the butterfly, *Limenites archippus. J. Insect Physiol.* 21: 1159–65
59. Honma, K. 1966. Photoperiodic responses in two local populations of the smaller tea tortrix, *Adoxophyes orana* Fischer von Röslerstamm. *Appl. Entomol. Zool.* 1:32–36
60. Houk, E. J., Beck, S. D. 1975. Comparative ultrastructure and blood-brain barrier in diapause and nondiapause larvae of the European corn borer, *Ostrinia nubilalis* (Hübner). *Cell Tissue Res.* 162:499–510
61. Hsiao, C., Hsiao, T. H. 1969. Insect hormones: their effect on diapause and development of Hymenoptera. *Life Sci.* 8:767–74
62. Ingram, B. R. 1975. Diapause termination in two species of damselflies. *J. Insect Physiol.* 21:1909–16
63. Ismail, S., Fuzeau-Braesch, S. 1976. Programmation de la diapause chez *Gryllus campestris. J. Insect Physiol.* 22:133–39
64. Jackson, D. R. 1963. Diapause in *Caraphractus cinctus* Walker, a parasitoid of the eggs of Dytiscidae. *Parasitology* 53:225–51
65. Janda, J. Jr., Sehnal, F. 1971. The influence of juvenile hormone on glycogen, fat and nitrogen metabolism in *Galleria mellonella* L. *Endocrinol. Exp.* 5:53–56
66. Kevan, D. K. M. 1944. The bionomics of the neotropical cornstalk borer, *Diatraea lineolata* Wlk. in Trinidad, B.W.I. *Bull. Entomol. Res.* 35:23–30
67. Kind, T. V. 1968. Functional morphology of insect neurosecretory system during active development and under various types of diapause. In *Photoperiodic Adaptations in Insects and Acari,* ed. A. S. Danilevskii, pp. 153–91. New Delhi: INSDOC (TT70-57120)
68. King, A. B. S. 1974. Photoperiodic induction and inheritance of diapause in *Pionea forficalis. Entomol. Exp. Appl.* 17:397–409
69. Kisimoto, R. 1958. Studies on the diapause of the planthoppers. 1. Effect of photoperiod on the induction and completion of diapause in the fourth larval stage of the small brown planthopper, *Delphacodes striatella* (Fallen). *Jpn. J. Appl. Entomol. Zool.* 2:128–34
70. Knerer, G., Marchant, R. 1973. Diapause induction in the sawfly, *Neodiprion rugifrons* Middleton. *Can. J. Zool.* 51:105–8
71. Lees, A. D. 1956. The physiology and biochemistry of diapause. *Ann. Rev. Entomol.* 1:1–16
72. Lounibos, L. P., Bradshaw, W. E. 1975. A second diapause in *Wyeomyia smithii:* seasonal incidence and maintenance by photoperiod. *Can. J. Zool.* 53:215–21
73. Mansingh, A., Smallman, B. N. 1967. Neurophysiological events during larval diapause and metamorphosis of the European corn borer, *Ostrinia nubilalis* Hübner. *J. Insect Physiol.* 13:861–67
74. Masaki, S. 1956. The effect of temperature on the termination of diapause in the egg of *Lymantria dispar* Linné. *Jpn. J. Appl. Entomol. Zool.* 21:148–57

75. Masaki, S., Oyama, N. 1963. Photoperiodic control of growth and wing-form in *Nemobius yezoensis* Shiraki. *Kontyû* 31:16–26
76. Masner, P., Hangartner, W., Suchy, M. 1975. Reduced titers of ecdysone following juvenile hormone treatment in the German cockroach, *Blattella germanica. J. Insect Physiol.* 21:1755–62
77. McCaffery, A. R., Highnam, K. C. 1975. Effects of corpora allata on the activity of the cerebral neurosecretory system of *Locusta migratoria migratorioides* R. & F. *Gen. Comp. Endocrinol.* 25:358–72
78. McCaffery, A. R., Highnam, K. C. 1975. Effects of corpus allatum hormone and its mimics on the activity of the cerebral neurosecretory system of *Locusta migratoria migratorioides* R. & F. *Gen. Comp. Endocrinol.* 25:373–86
79. McLeod, D. G. R., Beck, S. D. 1963. The anatomy of the neuroendocrine complex of the European corn borer, *Ostrinia nubilalis,* and its relation to diapause. *Ann. Entomol. Soc. Am.* 56:723–27
80. McLeod, D. G. R., Graham, W. G., Hannay, C. L. 1969. An observation on the ileum of the European corn borer, *Ostrinia nubilalis. J. Insect Physiol.* 15:927–29
81. McNeil, J. N., Rabb, R. L. 1973. Physical and physiological factors in diapause initiation of two hyperparasites of the tobacco hornworm, *Manduca sexta. J. Insect Physiol.* 19:2107–18
82. Mitsuhashi, J. 1963. Histological studies on the neurosecretory cells of the brain and on the corpus allatum during diapause in some lepidopterous insects. *Bull. Natl. Inst. Agric. Sci. Jpn.* 16C: 67–121
83. Mochida, O., Yoshimeki, M. 1962. Relations with development of the gonads, dimensional changes of the corpora allata, and duration of post-diapause period in hibernating larvae of the rice stem borer. *Jpn. J. Appl. Entomol. Zool.* 6:114–23
84. Morohoshi, S. 1975. The control of growth and development in *Bombyx mori.* 26. Growth and development controlled by the brain-corpora allata system. *Proc. Jpn. Acad.* 51:130–35
85. Morris, G. P., Steel, C. G. H. 1975. Ultrastructure of neurosecretory cells in the pars intercerebralis of *Rhodnius prolixus. Tissue Cell* 7:73–90
86. Moulins, M., Girardie, A., Girardie, J. 1974. Manipulation of sexual physiology by brain stimulation in insects. *Nature* 250:339–40
87. Müller, H. J. 1970. Formen der Dormanz bei Insekten. *Nova Acta Leopold.* 35:7–27
88. Nair, K. S. S. 1974. Studies on the diapause of *Trogoderma granarium:* effects of juvenile hormone analogues on growth and metamorphosis. *J. Insect Physiol.* 20:231–44
89. Nijhout, H. F., Williams, C. M. 1974. Control of moulting and metamorphosis in the tobacco hornworm, *Manduca sexta* (L.): growth of the last-instar larva and the decision to pupate. *J. Exp. Biol.* 61:481–91
90. Nijhout, H. F., Williams, C. M. 1974. Control of moulting and metamorphosis in the tobacco hornworm, *Manduca sexta* (L.): cessation of juvenile hormone secretion as a trigger for pupation. *J. Exp. Biol.* 61:493–501
91. Nowock, J., Goodman, W., Bollenbacher, W. E., Gilbert, L. I. 1975. Synthesis of juvenile hormone binding proteins by the fat body of *Manduca sexta. Gen. Comp. Endocrinol.* 27:230–39
92. Odesser, D. B., Hayes, D. K., Schechter, M. S. 1972. Phosphodiesterase activity in pupae and diapausing and non-diapausing larvae of the European corn borer, *Ostrinia nubilalis. J. Insect Physiol.* 18:1097–1105
93. Oku, T. 1970. Studies on life histories of apple leaf-rollers belonging to the tribe Archipsini. *Bull. Hokkaido Prefect. Agric. Exp. Stn.* 19:1–52
94. Oshiki, T., Morohoshi, S. 1973. The control of growth and development in *Bombyx mori.* 20. Neurosecretion of the brain-corpora allata system in the trimolters derived from tetramolting silkworms by temperature and moisture shocks. *Proc. Jpn. Acad.* 49:353–57
95. Philogène, B. J. R. 1971. Revue des travaux sur les formes de diapause chez les Tenthrédinoidés les plus communs. *Ann. Entomol. Soc. Quebec* 16:112–19
96. Powell, J. A. 1974. Occurrence of prolonged diapause in ethmiid moths. *Pan-Pac. Entomol.* 50:220–25
97. Propp, G. D., Tauber, M. J., Tauber, C. A. 1969. Diapause in the neuropteran *Chrysopa oculata. J. Insect Physiol.* 15:1749–57
98. Ring, R. A. 1967. Maternal induction of diapause in the larva of *Lucilia caesar* L. *J. Exp. Biol.* 46:123–36
99. Ryan, R. B. 1965. Maternal influence on diapause in a parasitic insect, *Coel-*

oides brunneri Vier. *J. Insect Physiol.* 11:1331–36

100. Sanburg, L. L., Kramer, K. J., Kezdy, F. J., Law, J. H., Oberlander, H. 1975. Role of juvenile hormone esterases and carrier proteins in insect development. *Nature* 253:266–67
101. Saringer, G. 1967. Investigations on the light sensitive larval instar determining the diapause of *Athalia rosae* L. (= colibri Christ) *Acta Phytopathol.* 2: 119–25
102. Saringer, G. 1971. The role of temperature, photoperiod and food quality in the diapause of *Grapholitha funebrana* Tr. *Acta Phytopathol.* 6:181–84
103. Saunders, D. S. 1965. Larval diapause of maternal origin: induction of diapause in *Nasonia vitripennis* (Walk.). *J. Exp. Biol.* 42:495–508
104. Saunders, D. S. 1975. Spectral sensitivity and intensity thresholds in *Nasonia* photoperiodic clock. *Nature* 253:732–33
105. Schaller, F., Andries, J. C., Mouze, M., Defossez, A. 1974. Nouveaux aspects du contrôle hormonal du cycle biologique des Odonates: recherches sur la larve d'*Aeshna cyanea* (Müller). *Odonatologica* 3:49–62
106. Scharrer, B., Weitzmann, M. 1970. Current problems in invertebrate neurosecretion. In *Neurosecretion in Invertebrates,* ed. W. Bargmann, B. Scharrer, pp. 1–23. New York: Springer-Verlag. 380 pp.
107. Scheltes, P. 1974. Aestivation diapause in some equatorial insect pests. *Ann. Rep. Intern. Cent. Insect Physiol. Ecol.* 2:91–96
108. Scheltes, P. 1976. Aestivation diapause in *Chilo partellus* (Swinhoe) and *Chilo orichalcociliella* (Strand). *Ann. Rep. Intern. Cent. Insect Physiol. Ecol.* 3:In press
109. Schneiderman, H. A., Horwitz, J. 1958. Induction and termination of facultative diapause in the chalcid wasps *Mormoniella vitripennis* (Walker) and *Tritneptis klugii* (Ratzeburg). *J. Exp. Biol.* 35:520–51
110. Sehnal, F., Edwards, J. S. 1969. Body constraint and developmental arrest in *Galleria mellonella* L.: further studies. *Biol. Bull.* 137:352–58
111. Sehnal, F., Granger, N. A. 1975. Control of corpora allata function in larvae of *Galleria mellonella. Biol. Bull.* 148:106–16
112. Selander, R. B., Weddle, R. C. 1972. The ontogeny of blister beetles. 3. Diapause termination of coarctate larvae of *Epicauta segmenta. Ann. Entomol. Soc. Am.* 65:1–17
113. Shappirio, D. G., Eichenbaum, D. M., Locke, B. R. 1967. Cholinesterase in the brain of the cecropia silkmoth during metamorphosis and pupal diapause. *Biol. Bull.* 132:108–25
114. Slade, M., Wilkinson, C. F. 1974. Degradation and conjugation of cecropia juvenile hormone by the southern armyworm (*Prodenia eridania*). *Comp. Biochem. Physiol.* 49B:99–103
115. Slama, K. 1964. Physiology of sawfly metamorphosis. 2. Hormonal activity during diapause and development. *Acta Soc. Entomol. Czech.* 61:210–19
116. Sluss, T. P., Sluss, E. S., Crowder, L. A., Watson, T. F. 1975. Isozymes of diapause and non-diapause pink bollworm larvae, *Pectinophora gossypiella. Insect Biochem.* 5:183–93
117. Sugiki, T., Masaki, S. 1972. Photoperiodic control of larval and pupal development in *Spilarctia imparilis* Butler. *Kontyû* 40:269–78
118. Takeda, N. 1971. Nuclear activation and synthesis of secretory substances in prothoracic glands. *Nucleus Cytoplasm* 15:1–9 (In Japanese; English abstr.)
119. Takeda, N. 1972. Activation of neurosecretory cells in *Monema flavescens* during diapause break. *Gen. Comp. Endocrinol.* 18:417–27
120. Tauber, M. J., Tauber, C. A. 1972. Larval diapause in *Chrysopa nigricornis:* sensitive stages, critical photoperiod, and termination. *Entomol. Exp. Appl.* 15:105–11
121. Tauber, M. J., Tauber, C. A. 1976. Insect seasonality: diapause maintenance, termination, and postdiapause development. *Ann. Rev. Entomol.* 21:81–107
122. Thomsen, E., Lea, A. O. 1969. Control of the medial neurosecretory cells by the corpus allatum in *Calliphora erythrocephala. Gen. Comp. Endocrinol.* 12:51–57
123. Tyshchenko, V. P. 1967. Photoperiodic induction of the diapause in caterpillars of the pine moth. *Entomol. Rev.* 46: 331–40
124. Tyshchenko, V. P., Mandelstam, J. E. 1965. A study of spontaneous electrical activity and localization of cholinesterase in the nerve ganglia of *Antheraea pernyi* Guer. at different stages of metamorphosis and in pupal diapause. *J. Insect Physiol.* 11:1233–39
125. Umeya, Y. 1950. Studies on embryonic

hibernation and diapause in insects. *Proc. Imp. Acad. Jpn.* 26:1–9

126. Usua, E. J. 1970. Diapause in the maize stemborer. *J. Econ. Entomol.* 63: 1605–10
127. Van der Kloot, W. G. 1955. The control of neurosecretion and diapause by physiological changes in the brain of the cecropia silkworm. *Biol. Bull.* 109: 276–94
128. Van Dinther, J. B. M. 1961. The effect of precipitation on the break of diapause in the white rice borer, *Rupela albinella* (Cr.) in Surinam (South America). *Entomol. Exp. Appl.* 4:35–40
129. Vedeckis, W. V., Bollenbacher, W. E., Gilbert, L. I. 1974. Cyclic AMP as a possible mediator of prothoracic gland activation. *Zool. Jahrb. Physiol.* 78: 440–48
130. Vinogradova, E. B. 1974. The pattern of reactivation of diapausing larvae in the blowfly, *Calliphora vicina. J. Insect Physiol.* 20:2487–96
131. Waku, Y. 1960. Studies on the hibernation and diapause in insects. 4. Histological observations of the endocrine organs in the diapause and non-diapause larvae of the Indian meal-moth, *Plodia interpunctella* Hübner. *Sci. Rep. Tôhoku Univ. Ser. 4* (*Biol.*) 26:327–40
132. Weseloh, R. M. 1973. Termination and induction of diapause in the gypsy moth larval parasitoid, *Apanteles melanoscelus. J. Insect Physiol.* 19:2025–33
133. Whitmore, D. Jr., Gilbert, L. I., Ittycheriah, P. I. 1974. The origin of hemolymph carboxylesterases 'induced' by the insect juvenile hormone. *Mol. Cell. Endocrinol.* 1:37–54
134. Wigglesworth, V. B. 1969. Chemical structure and juvenile hormone activity: comparative tests on *Rhodnius prolixus. J. Insect Physiol.* 15:73–94
135. Wigglesworth, V. B. 1970. *Insect Hormones.* San Francisco: Freeman. 159 pp.
136. Williams, C. M. 1946. Physiology of insect diapause: the role of the brain in the production and termination of pupal dormancy in the giant silkworm, *Platysamia cecropia. Biol. Bull.* 90:234–43
137. Williams, C. M. 1952. Physiology of insect diapause. 4. The brain and prothoracic glands as an endocrine system in the cecropia silkworm. *Biol. Bull.* 103:120–38
138. Williams, C. M. 1969. Photoperiodism and the endocrine aspects of insect diapause. *Symp. Soc. Exp. Biol.* 23:285–300
139. Willis, J. H. 1974. Morphogenetic actions of insect hormones. *Ann. Rev. Entomol.* 19:97–115.
140. Winfree, A. T. 1975. Unclocklike behavior of biological clocks. *Nature* 253:315–19
141. Wiseman, B. R., McMillian, W. W. 1973. Diapause of the sorghum midge, and location within the sorghum spikelet. *J. Econ. Entomol.* 66:647–49
142. Wyatt, G. R. 1972. Insect hormones. In *Biochemical Action of Hormones,* ed. G. Litwack, pp. 385–490. New York: Academic. 542 pp.
143. Yagi, S. 1975. Endocrinological studies on diapause in some lepidopterous insects. *Mem. Fac. Agric. Tokyo Univ. Educ.* 21:1–49 (In Japanese, English summary)
144. Yagi, S., Fukaya, M. 1974. Juvenile hormone as a key factor regulating larval diapause of the rice stem borer, *Chilo suppressalis. Appl. Entomol. Zool.* 9:247–55
145. Yin, C.-M., Chippendale, G. M. 1973. Juvenile hormone regulation of the larval diapause of the southwestern corn borer, *Diatraea grandiosella. J. Insect Physiol.* 19:2403–20
146. Yin, C.-M., Chippendale, G. M. 1974. Juvenile hormone and the induction of larval polymorphism and diapause of the southwestern corn borer, *Diatraea grandiosella. J. Insect Physiol.* 20: 1833–47
147. Yin, C.-M., Chippendale, G. M. 1975. Insect prothoracic glands: structure and function in diapause and non-diapause larvae of *Diatraea grandiosella. Can. J. Zool.* 53:124–31
148. Yin, C.-M., Chippendale, G. M. 1976. Hormonal control of larval diapause and metamorphosis of the southwestern corn borer, *Diatraea grandiosella. J. Exp. Biol.* 64:303–10

Ann. Rev. Entomol. 1977. 22:139–55

EXTENSION ENTOMOLOGY: A CRITIQUE[1]

❖6124

Charles Lincoln[2]
Department of Entomology, University of Arkansas, Fayetteville, Arkansas 72701

B. D. Blair[3]
Extension Entomologist, Ohio State University, Columbus, Ohio 43210

Extension entomologists are one category of specialists in the Cooperative or Agricultural Extension Services of the Land Grant Universities of the United States. Their primary function has been to develop and give leadership to extension education programs in entomology. In doing so they work through county extension personnel in their educational work with farmers, homemakers, and youth (through 4-H clubs). An increasing number of diverse demands have recently been placed upon extension entomologists by non-producer oriented groups.

The Extension Service serves one of the three basic functions in agriculture within a Land Grant Institution; the others are research and resident teaching. Administrative names and systems vary from state to state, but commonly research is in an agricultural experiment station and resident teaching is in a college of agriculture and home economics. Entomologists may work in all three areas, in any one area, or with joint appointments between two or three areas.

This paper is an attempt to give the origins, history, and current status of extension entomology in relation to agriculture, homemaking, and many recent demands and problems. The future of extension entomology is explored, and its role as one phase of our profession is evaluated.

[1]Published with approval of the Director, Arkansas Agricultural Experiment Station.

[2]Dr. Charles Lincoln served as extension entomologist in Arkansas from 1942 to 1950. Since then he has been engaged in experiment station research with resident teaching and regulatory responsibilities. He also served as Department Head for 18 years.

[3]Dr. B. D. Blair entered extension work in Ohio at the county level in 1954. He has served as extension entomologist at the state level since 1960 and has devoted 60% of his efforts to extension entomology and 40% to experiment station research since 1964.

We were greatly assisted in the preparation of this review by 33 respondents (32 extension entomologists and a volunteer from the only state without one) to a lengthy and involved questionnaire. Their answers serve as a basis for much of this paper, and our heartfelt thanks go to them. No attempt was made to review the literature because the authors saw no way to make a relevant review. We take full responsibility for interpreting the results of the questionnaires and for all other ideas and opinions expressed, recognizing that many of them may be controversial or reflect personal bias.

ORIGINS AND GROWTH

The Land Grant system dates from the Morrill Act of 1862. It provided grants of public land for the establishment and support of colleges for teaching of agriculture and mechanical arts. It soon became evident that there was not much to teach that qualified as scientific agriculture.

The Hatch Act of 1887 encouraged the development of experiment stations to conduct agricultural research. The results of such research furnished a body of information for use in resident teaching. It also provided information that could be put into practice by farmers. Farmers in general did not rush to the experiment stations seeking the latest scientific information. Farmer Institutes and similar devices were used in attempts to convey information. An understanding of the effectiveness of the demonstration method of teaching farmers gradually developed. This was formalized in the Smith-Lever Act of 1914. It provided that extension work in agriculture and home economics should be carried on by the Land Grant colleges in cooperation with the U.S. Department of Agriculture (USDA) and the local people. Extension work was to "aid in diffusing among the people of the United States useful and practical information on subjects relating to Agriculture and Home Economics and to encourage the application of the same."

The devastation caused by the cotton boll weevil and the inability of farmers to cope with it was an important catalyst in bringing the Agricultural Extension Service into being. By logical deduction, extension entomology should have been at the forefront of the early work in agricultural extension, but such was not the case. Of 33 responses to the questionnaire, only 5 reported an extension entomologist prior to 1920. Only one, North Carolina, was a cotton-producing state, and 8 of the 11 "boll weevil states" responded to the questionnaire. Eight states initiated extension entomology in the 1920s, five in the 1930s, eight in the 1940s, two in the 1950s, four in the 1960s, and one in the 1970s.

Growth in numbers of extension entomologists has varied greatly from state to state. According to a roster of January 1976 furnished by Paul W. Bergman, Federal Extension Entomologist, one state had 26 extension entomologists, amounting to 25.26 man-year equivalents. At the other extreme, seven states listed one extension entomologist, amounting to between 0.80 and 1.00 man-year equivalents, and one state listed no extension entomologist. In addition, the USDA had one fulltime extension entomologist. There were 286 extension entomologists, of which 173 gave all their time to extension work, 17 gave 75–90% of their time, 27 gave 50–70%, 31 gave 25–40%, and 38 gave 5–20%.

Extension entomologists work with all insects of economic importance, including the preservation of aesthetic values. Environmental impact of pest control activities and proper use of pesticides are important considerations. Many extension entomologists carry responsibilities in several areas. According to the January 1976 roster of extension entomologists, the assignments were as follows: general assignment, 122; pest management, 81; crops, 56; 4-H youth, 56; ornamentals/shade tree/turf, 49; household/structural, 42; fruits, 42; man/animals, 41; pesticides coordinator/pesticides safety/pesticides training, 38; vegetables, 36; administration, 30; apiculture, 21; survey, 14; cotton, 10; stored products, 9; taxonomy, 4; nematology, 3; forest insects, 3; and one each for environment, biological control, food processing, game management, parasitology, pesticide residues, ants, field testing, and rodent control.

Forest insects are serious pests and most forest land is in small holdings. It is surprising that only three extension entomologists are listed for forest insects, although many listed under general assignment probably do extension work in the forest insect area.

WHY EXTENSION ENTOMOLOGISTS?

Farm people and county extension agents had problems with insects, whether or not there was an extension entomologist to help them. Of 32 responses to the questionnaire, 19 said that such problems had been entirely or largely handled by experiment station entomologists. In four of these cases, the department head was also the state entomologist and apparently carried extension and regulatory responsibilities without direct or identifiable remuneration. One state had had an extension entomologist from the beginning of agricultural extension work. The remaining 12 states reported such problems had been handled by state entomologists or by state departments of agriculture (5 states), by USDA research entomologists (3 states), by anybody available (2 states), or not handled (4 states).

The decision to first appoint an extension entomologist in a state was usually in response to need or demand expressed by farmers and county agents, often centered around a specific insect problem. Relief of experiment station entomologists from the distractions of extension-type work was also a strong consideration.

THE EARLY DAYS

Traditionally, extension programs were locally oriented. County staffs existed to serve the needs of rural people in agriculture, homemaking, and youth (4-H clubs). Local leaders participated actively in planning and carrying out these programs on a voluntary basis. People participated because it provided a more comfortable life style, gave an opportunity to help friends and neighbors, and was a lot of fun. Administrators and specialists at the state level served as support personnel for the county staffs. The Federal Extension Service served as support for the several state programs and has always had only a small staff. This upward flow from the grass roots level has been a unique hallmark of extension work.

Most other programs directed toward the same clientele were centrally planned at the federal or state level, and many were compulsory to a considerable degree. Farmers and farm families did not always welcome such programs with enthusiastic cooperation; in fact they sometimes resisted bitterly.

The original mission of much of the Agricultural Extension Service program was devoted toward living at home with emphasis on home food production. Cash crop and commercial livestock production were also encouraged, to provide needed income, but the live-at-home program was designed to lessen dependence on them. The original synthesis of this approach stemmed from the devastation wrought by the cotton boll weevil, often leaving the one-crop cotton farmers completely poverty-stricken. With a good live-at-home program, cotton farmers could survive a cotton crop failure and hope to compensate with good crops in years of lesser weevil outbreaks. This emphasis on live-at-home was a central theme of extension work through the 1930s.

WORLD WAR II AND ITS AFTERMATH

World War II, with the slogan "Food will win the war and write the peace," placed emphasis on commercial production of food, fiber, and livestock. Even with a greatly decimated labor force, farmers responded nobly. The extension services changed their focus to help meet this great need, oftentimes reacting quickly to emergency situations. Concurrently, great emphasis was placed on home gardening and home food production and preservation by nonfarmers. The extension services began to serve an urban clientele, at least temporarily.

Following World War II, prices for farm products were generally good for a few years. Mechanical power rapidly replaced horse and mule power. A new generation of agricultural pesticides gave farmers potent new tools. Chemical fertilizer became both abundant and relatively cheap, replacing manure, cover crops, and crop rotation for maintaining soil fertility. New varieties made great contributions. Farmers put these tools together to become commercial farmers with greatly increased output per man-hour. Usually the acreage farmed per man increased, but some farmers turned to more intensive land and resource use. An agricultural revolution was born.

Commercial farmers purchased many production inputs: machinery, tractor fuel, pesticides, certified seed, etc. They needed credit to make these purchases and large, well-organized markets. Agribusiness greatly expanded to fill these needs.

The extension services maintained a strong position of leadership in this revolution. To do so required intensive training and some specialization of county staffs, more and better-trained specialists at the state level, changed and improved teaching techniques, and development of a new set of relationships with agribusiness and certain state and federal agencies.

Extension entomology led this agricultural revolution. Thanks largely to the new insecticides, control of many pests became a practical farm practice for the first time. Farmers were responsive, often led by World War II veterans who were free of a depression-oriented philosophy. Successful insect control increased yields and profits directly. Furthermore, the greatly reduced risks of crop failures from insect pests inspired farmers to make investments in fertilizer, machinery, and other production inputs, leading to still higher yields and profits.

The new insecticides were powerful tools for controlling pests, but with them came risks of induced outbreaks of nontarget pests. Extension entomologists had the great responsibility of teaching county extension agents and farmers to use these new tools correctly. They welcomed the challenge and discharged the responsibility to the limits of their numbers and abilities. In retrospect, many based their programs too much on insecticidal control at the expense of alternative control measures, but considering the times they did extremely well. To have failed to take the lead with the new insecticides would have resulted in farmers adopting, or attempting to adopt, general use of these potent new insecticides without a source of unbiased information from knowledgeable entomologists. To have permitted this might well have jeopardized the position of leadership of the entire Land Grant system.

Many rural people did not join in this agricultural revolution, partly from choice and partly from shortage of available land for everyone to expand operations. Many of these people moved to cities and temporarily lost contact with the extension service. Others continued to farm their small acreages, some living in poverty compared to the norm in the ever-more affluent society and some becoming parttime farmers with off-farm work for supplementary income. Others rented their land to commercial farmers but continued to live on the farm, either retired or devoting all of their energies to nonfarm work.

The extension services have had problems in adjusting to working with small and parttime farmers. Traditional channels of neighborhood and community leadership and communications had been disrupted. Lack of leadership, limited education, and limited income among this clientele presented difficult problems. The number of small farmers was too great for each to receive intensive personal attention from extension agents, and it proved difficult to develop new channels of leadership to replace the traditional ones that had been disrupted.

ADMINISTRATIVE AND TECHNICAL RELATIONS

Extension entomologists have in-line administrative responsibilities to extension administrations. In addition, they are dependent on research entomologists for technical information on which to base recommendations. They also apprise researchers of problems and needed research. Questions were asked about relations of extension entomologists with Entomology departments and with individual researchers and about where the extension entomologists were officed in relation to the Department of Entomology from the state land grant university. Relations with the department and with researchers were informal at the discretion of the extension entomologist in 21 states, formal in 6 states, and both in 5 states; there was no extension entomologist in one state, according to the 33 respondents.

Extension entomologists are on-campus and in the Department of Entomology in 26 states. All except two preferred this arrangement. Nine of these also had extension entomologists at off-campus locations and another wished they had.

In six states extension entomologists were on-campus but officed separately from the Department of Entomology. Three considered this the preferable arrangement, two did not, and one expressed no opinion. One of these six states also had extension entomologists at a branch experiment station.

One state was quite a mixture: one extension entomologist was on-campus but separate from the Department of Entomology, one was off-campus at the state extension headquarters, one area entomologist was at a branch experiment station, and three other area extension entomologists were at various locations. All of these were fulltime extension. In addition, two entomologists, a 50% and a 33% joint appointment with extension, respectively, were in the Department of Entomology.

Extension entomologists work with and through county agents, home demonstration agents, and 4-H agents. The average distribution of extension entomologists time showed 79.9% devoted to agriculture, 10.4% to home economics, and 6.0% to 4-H, with ranges of 45–95%, 0–35%, and 2–20%. The total is only 96.3% because 4 of the 27 respondents did not report a full 100%, presumably feeling that certain of their activities fall outside of these categories. The distribution between agriculture and home economics reflects a weakness in the questionnaire with respect to urban entomology. Those we have traditionally called county agents and home demonstration agents may both be involved in urban entomology, and the questionnaire had no place for this category. Putting all identifiable urban entomology under agriculture is manifestly unfair, but we had no choice because of the questionnaire weakness. It does appear that most urban entomology at the county level is carried out by county agents rather than by home demonstration agents.

It is the responsibility of extension entomology to develop recommendations and programs based on research done primarily by experiment station and USDA researchers. These are carried to farmers and other extension clientele through county extension staffs. The clientele may choose to bypass the county staff or the extension entomologist and go directly to the researchers for information and recommendations. Only 4 of the 27 respondents to the questionnaire considered that bypassing was more than an occasional problem. Suggestions for preventing such bypassing emphasized a well-trained staff at county and state levels with adequate numbers to accomplish extension's mission. One respondent suggested that the extension entomologist should locate the problem and get involved. Another suggested bringing top farmers into extension program planning. In one state where bypassing had been a problem, administrative action solved it, but an extension entomologist in another state questioned the likelihood of effective administrative action. One state with a problem suggested that researchers located in the field be officed with the county extension staffs. One reported bypassing varies by county, undoubtedly a universal circumstance based on the attitudes and capabilities of the county staffs.

Relationships with experiment station researchers were considered to be good to excellent or improving in 21 states, unchanged in 7, and little or no improvement in 2. Problems were insufficient backup research to meet present or expanding needs and a need for experiment station researchers to become more involved with or more sympathetic to extension needs. One thought that taxpayers are demanding more relevance, which would mean extension would be favored over research in the future. Another commented that as pest management grows, consultants would become so numerous that they must work through extension entomology to avoid overloading researchers with their problems.

Relationships with the US Department of Agriculture were good to excellent or improving in 14 states, unchanged in 8, and little or no improvement in 6. Reaction to the Agricultural Research Service reorganization was mixed. One suggestion was to make more use of federal entomologists in their areas of expertise in extension programs and meetings.

Relations with extension administrations are generally good, but there are too many layers of administration and too many bureaucratic channels. These restrict the freedom of action and decision-making of extension entomologists.

SPECIALIZATION AND PROFESSIONAL STATUS AND TRAINING

There is a strong trend toward more specialization throughout the profession of entomology. Extension entomology has peculiar problems in adapting to specialization. When a county extension agent contacts an extension entomologist for information, he wants an answer, not a referral to another specialist. However, the need for specialization cannot be denied.

One common device to effectuate specialization is through joint appointments. Many entomologists hold joint appointments between extension and research and a few between extension and resident teaching or all three areas. As noted earlier, 113 of the 286 extension entomologists held joint appointments. Opinions expressed in the questionnaire were about evenly divided on the desirability of joint appointments. Those opposing joint appointments felt that extension entomology was a fulltime responsibility that should not be diluted. At the other extreme was the expressed need to involve researchers more directly with insect problems of the state and its people. Joint appointments probably profit the institution but place great conflicting demands on entomologists who hold them.

Another trend in extension entomology approaches specialization from a different angle. Area and county extension entomologists are on the increase. Narrowing the geographical area of responsibility of an extension entomologist permits specialization on the problems of the area. All too often it also removes him from close contact with research entomologists, good library facilities, and rapid insect identification service. This can be most frustrating to a dedicated entomologist.

Additional comments were solicited in the questionnaire. One response complained about filling out questionnaires. Professional status of extension entomologists within entomology, agriculture, and administration received 11 comments, ranging from full acceptance of extension on the one hand to (*a*) the academic viewpoint that extension entomologists are "spraying freaks" and (*b*) that agriculture is low in the pecking order and extension entomology is low priority in agriculture. Most agreed that progress was being made where actual problems existed.

Training for extension entomologists received seven comments. One reply stated that the quality and vigor of extension entomologists was below that of their research and teaching counterparts. Another noted a past tendency of extension administrators to promote specialists from the ranks of county staffs with little regard for academic preparation. Two noted that present and future demands require more formal education and specialization, and another predicted a future shortage of trained extension entomologists. One thought experience and knowledge was more important than the degree held, and another concluded that more resource entomologists will be required to service the extension entomologists who must remain generalists.

Professional pride and need for full professional recognition were quite evident on the part of the respondents. The need for adequate training to do high-quality extension entomology was recognized. How to achieve this training within the framework of our research-oriented graduate schools was a source of some frustration. The extension part of extension entomology is largely learned on the job. Not all entomologists have the personal and mental attributes to become good extension workers. An internship in extension as part of or interspersed with graduate work could contribute toward training of potential extension entomologists. It would also give the individual the opportunity to find out first-hand if he were really interested in extension and had the temperament for it. Work in pest management gives a young agriculturist or entomologist practical experience of great value to potential extension entomologists.

A great deal of research is done by para- or subprofessionals working under the guidance of qualified research entomologists. Extension entomologists have made little direct use of such personnel. Possibly they should not. After all, the county extension agents fill the role of paraprofessionals within extension entomology. In many states extension entomologists are seriously overloaded, but at present there is no clear alternative to adding more extension entomologists where the need exists.

CURRENT TRENDS IN EXTENSION ENTOMOLOGY

It is difficult to summarize current trends in extension entomology because so much is happening on so many fronts and each state is reacting somewhat differently. Generally speaking, extension entomologists are more comfortable when dealing with production agriculture. The need to service a much broader clientele is recognized and accepted but with varying degrees of enthusiasm. Such programs as pesticide safety and applicator training have long been part of extension entomology programs, and recent federal dictation in these areas is not entirely welcome. If left alone and adequately funded, extension entomologists would have done these jobs well in their own way.

With extension's proven success, coupled with a changing social, economic, and political scene, new directions have emerged and new programs have been added. Urban sprawl has changed many formerly agricultural areas into urban-suburban areas, bringing extension into close contact with a new, nonrural clientele. In recognition of its broader mission, extension's name has widely been changed to the Cooperative Extension Service. Of a sample of 22, only 2 retained the older title of Agricultural Extension Service.

The reaction of extension entomologists to the sum of these changes was about evenly divided, according to responses to the questionnaire. One entomologist optimistically stated that extension now serves all of the people and can do much more than was ever dreamed possible. This particular state is adequately staffed for the increased opportunities and demands, and 12 extension entomologists are expected by January, 1976. Many extension entomologists praised the pest management and survey programs as effective and unifying influences. One pointed out that entomology can change from serving an agricultural clientele to an urban-suburban clientele, whereas specialists in some other fields cannot. (Imagine keeping a cow in an apartment!) The negative reactions centered around two themes: (*a*) increased work load and responsibilities without increased funds and staffing, and (*b*) federal- or occasionally state-initiated programs that lack practicality or have too little grass roots understanding. A particular complaint was pesticide safety education and applicator training for restricted-use pesticides—both programs were handed down from the federal level with limited grass roots input and inadequate funding.

On specific programs, most extension entomologists attributed increasing importance and effort to the insect pest management, urban entomology, pesticide usage and safety, and applicator training program. There was much less increase in extension entomology for the low-income group, apparently accompanied by some frustration over the difficulties of reaching this audience. There are in general fewer farm visits or one-on-one contacts with farmers, but considerable increase by working through groups and with mass media. Four-H entomology showed little, if any, increase.

The extension services have added on many programs but have dropped few, if any; this results in frustration for many county extension agents and specialists. Continual adding of programs without adding staff is likely to result in less training time for extension at the county and area levels, which in turn may result in loss of leadership at the local level, the very basis of extension's strength and effectiveness.

Pest management programs in many areas provided new personnel and allowed more regular personal contact with producers than had been the case for two decades. It appears that this personal contact between extension and producers in many areas of the United States has resulted in rapidly regaining of lost leadership and has once again placed extension entomologists in the forefront as program leaders. The pest management project leaders have generated additional support at the state level and in most projects have generated direct financial support from producers. That people and government have been willing to supply additional support during a period of economic recession must be viewed as a favorable evaluation of one kind of extension program.

Enthusiasm for the pest management and survey programs shows that extension entomologists want well-rounded programs of insect control, with the best possible assessment of current pest problems. Clearly, they are not wedded to the idea of insecticidal control at the expense of alternative methods.

Developments in Educational Techniques

In recent developments in educational techniques, 11 extension entomologists reported use of improved visual aids, some quite sophisticated: television, closed circuit television, Telenet for reaching county offices in emergencies, video-cassette tapes, etc. Leaflets and handouts are still very important, and efforts are being made to improve them. Three respondents mentioned plant health clinics for urban-suburban clientele; one uses malls of city shopping centers for these. More interdisciplinary cooperation was evident, even to training area crop specialists and area livestock specialists to handle extension entomology problems in the field. Need for regional extension publications and/or more exchange of recommendations between states to reduce duplication of extension and research effort was recognized. There was concern for teaching aids specifically designed for the diversity of clientele that is so rapidly emerging; more help from information specialists and faster turnaround time of information from and to the field were desired.

One state reported that use of aide help has strengthened training of applicators, dealers, and commercial fieldmen. Their Master Gardener Program—retreading old gardeners and training young people to assist county agents in horticultural programs—has been very helpful in educating these people to some of the basic principles of entomology and pest management.

In-service training, short courses, field tours, and group meetings are being used to offset the fewer county visits and farmer contacts. It is difficult to get a group of farmers to attend an indoors subject matter meeting. Field tours with discussions along the way are becoming a much better teaching device. None of these are fully adequate substitutes for county visits and one-on-one producer contacts, but they are necessitated by the too-frequent overloading and dilution of effort forced on extension entomologists and county extension staffs.

Relationships Outside of Entomology

In what manner, if at all, have relationships changed between extension entomologists and farmers, farm organizations, industry and cooperatives that supply inputs, industry and cooperatives that process and/or market, consumer groups, county extension staffs, state agencies, and federal agencies?

Farmers want more specialized assistance, are more receptive, or just have good, improving relations with extension entomologists in 15 states that responded to the questionnaire. Five reported no change, and eight reported less direct farmer contact.

Relationships with farm organizations were improved or good in 15 states, about the same in 9 states, and less contact to none at all in 6 states. Relationships with industry or cooperatives that supply inputs were improved or good in 23 states, about the same in 6 states, and less contact in one. Relationships with industry or cooperatives that process and/or market were improved or good in 17 states, about the same in 7, and little or less contact in 5. Relationships with consumer groups were improved or good in 16 states, about the same in 6, and little contact in 6.

Relationships with county extension staffs were improved or good to excellent in 16 states. Six more reported better servicing of county staffs, with fewer visits by utilizing in-service training, more modern methods of communication, etc. Two reported about the same and 2 less or a little less contact. One reported great variation from county to county with more emphasis on programs and less on everyday problems. One expressed the need for more extension entomologists, which suggests the difficulty of maintaining adequate rapport with county staffs in the face of other pressures.

Relationships with other state agencies were good to excellent or improving in 24 states and the same in 5. In one state the antipesticide attitude of the Commission on Environmental Protection was creating problems for extension entomology. In another, the State Department of Agriculture was carrying on extension-type work instead of limiting itself to its regulatory function.

Relationships with federal agencies other than the US Department of Agriculture were mixed. Six states reported good relationships. Fifteen reported closer but not necessarily good relationships; the Environmental Protection Agency (EPA) was the target of considerable criticism. Three reported good relationships with some federal agencies and conflict with others. Four reported about the same, and one reported little contact.

Suggestions were solicited for improvement of relationships with all of the above or other relevant matters. Eight offered no suggestions, while some others offered more than one. Three suggested no change, and eight suggested continuing along present lines but trying harder. Six saw a need for increased staffing; apparently they could not try harder because of understaffing. Need for better communications between individuals in similar program areas were noted by two. More timely information for specific problems in extension's traditional areas of responsibility was an expressed need by two states. Others mentioned by one state were (*a*) to emphasize servicing county staff clientele rather than to be so involved in state and federally initiated programs; (*b*) to assume greater leadership in pest management and applicator certification; (*c*) to not knuckle under to outside forces on matters of principle injurious to agriculture; (*d*) to staff EPA with people who have grass roots experience; and (*e*) to become involved with mosquito abatement districts.

Relationships with agriculture in these changing times drew several varied comments: (*a*) working closely with the agricultural chemicals industry would be productive; (*b*) commodity specialists are supported by pressure groups whereas entomologists are not; (*c*) emphasis on food production would help; (*d*) it is hard to sell the problems faced by agriculture in an industrialized urban state; and (*e*) more pest management and survey programs are needed.

The essentiality of good insect identification service was expressed. Two main concerns were for more and better insect collections to be available within the state and convincing county agents to submit specimens (even though vials and mailing tubes are furnished, this remains a problem).

Comments on adjusting to the changing times mentioned "our greatest opportunity to serve all the people," increasing urban needs, desirability of less governmental control and interference, risk of direct involvement in regulatory aspects rather than teaching. It was also felt that extension entomology, which is needed more than ever, has lost some identity in the proliferation of bureaucracy. Miscellaneous comments included the following: "Will extension entomologists retain freedom in decision making?" and "One must have solutions to be a hero." The need for assistants or subprofessionals was mentioned, and extension entomology was called effective considering the understaffing problem.

One commented that the Federal Extension Service (FES) leadership is weak. This is a two-sided coin: traditionally, extension has been strong at the grass roots level, with state and federal extension workers providing support. This has not required a large federal staff, although an energetic and competent one is needed. By contrast, straight-line agencies have power centered at the federal level, and state and local personnel served in lesser or subservient roles. It naturally follows that the workers in the Federal Extension Service have been greatly outnumbered by those in other agencies at the national level. To compensate for this requires not only a highly competent FES staff but also development of better mechanisms of getting a "power flow" from counties and states to the national level.

World Hunger

Worldwide shortages of food have evoked strong public and political reaction. States were asked whether this was being manifested in emphasis on more extension work in production agriculture. Nineteen respondents replied no or not yet, but most were ready and waiting for the opportunity. Of these, two noted that the United States is no longer criticized for adding to surpluses. One noted that higher costs of pesticides were encouraging farmers to treat only as needed, which is a boon to pest management. Of the 12 others, 5 replied yes; 3 yes for home gardens, roadside stands, and small scale beekeeping; 3 that it had always been their goal; and one reported some increase. In all cases extension entomology appeared deeply involved where there was increasing emphasis on production agriculture.

It appears that the powers-that-be nationally and internationally have not yet really faced up to the specter of present and imminent food shortages despite some present horrible examples and the continued increase in world population. If and when they do, extension entomologists are eager to get on with it.

THE FUTURE OF EXTENSION ENTOMOLOGY

The commitment of the Extension Services of the several states to extension entomology varies greatly. It is fervently hoped that those states which presently do not adequately support extension entomology will soon correct this deficiency. By and large, extension entomologists need more time for visits with county extension staffs and producers and more planning time with the people who have problems.

Despite current foot-dragging, an all-out effort in production agriculture is viewed as a certainty in the near future. Extension entomologists have demonstrated their competence and readiness to play a major role in the necessary educational effort to make increased production of food, feed, fiber, and livestock a reality. Greater use of extension entomology is foreseen to help increase agricultural productivity worldwide through training of foreign entomologists visiting the United States and by serving in foreign assignments.

Insect pest management is a burgeoning field. Extension entomologists are already heavily committed and the commitment will increase. Current programs are largely to assess pest populations and treat as needed, based on the best available, but usually inadequate, economic threshold information. Alternative practices to incorporate into pest management programs are eagerly awaited from research.

The operational aspects of pest management at the county and farm level are in a state of flux. Several types of programs will very likely emerge and work in harmony: insect scouts paid by farmers but recruited and supervised by county extension agents, independent pest management consultants, pest management districts, and others. Areawide forecasting will replace regular (weekly) inspection of all fields for certain pests. Regardless of the operational format, extension entomologists will play key roles in the intensified educational effort to make pest management work. To cope with the vast amounts of incoming and outgoing information necessary to successful pest management programs, computers will become widely used.

Survey entomologists will play an increasing role as pest management grows. Statewide pest assessments and predictions will be indispensable in forewarning local pest management units of developing problems. At present, survey entomologists are attached to the extension service, the agricultural experiment stations, the state regulatory agency (the Plant Board or the Department or Commission of Agriculture), or a combination of the above. Survey entomology greatly benefits all three areas. An increase in survey entomology and a clarification of its role and relationships is badly needed. It has a great potential as a unifying force between extension, research, and regulatory entomology; this potential is often overlooked.

A source of prompt and correct insect identification is a must in all aspects of extension entomology. Insect pest management programs will greatly increase the work load of insect identification. To meet this need, each state should have one or more comprehensive collections of the state's more frequently encountered insect fauna. They could be used by entomologists and others not specialized in taxonomy. Such collections may be distinct from the larger research collections replete with large numbers of specimens of many species and many exotic and obscure species. Specialized taxonomists and research collections should be readily available to provide needed backup identifications that cannot be made by using the more limited state faunal collections by nontaxonomists. To that end, all state extension services should contribute to the support of taxonomists, as only a few states now do.

The field of urban entomology will continue to grow. Extension entomologists are moving into this field as rapidly as available time permits. In general this is much too slow. The extension effort in this area should be accelerated with adequate funds and personnel. Research backup for urban entomology is often limited.

Reaching the low-income groups, both urban and rural, is a vexing problem. Possibly personnel drawn from these groups and trained as paraprofessionals would help to solve this. Such an approach is underway in human nutrition.

No end is seen to federal intervention into pesticide safety, applicator training for licensing, and other fields. This will continue to erode the grass roots strength of extension entomology. Participation in extension education is voluntary on the part of the students, be they farmers, aerial applicators, housewives, or whatever. They are free to attend meetings, read leaflets, and participate or not participate in any way in extension education. If they do participate, they are free to accept or reject what is taught and recommended. Exercise of compulsion is completely foreign to extension education. Federal programs and sometimes state programs that use extension often border on compulsion or regulation. Unless this can be prevented, the credibility of extension entomologists will suffer.

Relationships with agribusiness, farm organizations, and environmental groups will become more common and complex. It is essential for extension entomologists to defend vigorously their proper role in education and the integrity of their sources of research information, while fully cooperating with these groups in carrying out their legitimate functions.

Within extension, the extension entomologists will cooperate with extension specialists in other fields to develop interdisciplinary programs. Pest management will become a common meeting ground for extension specialists in entomology, plant pathology, nematology, weed science, economics, and several commodity areas. Greater use will be made of new teaching and communication techniques to maintain adequate liaison with county extension staffs by centrally located extension entomologists. Decentralization of extension entomology will continue through increasing the number of area and county entomologists.

Contacts with farmers and other extension clientele will continue to be largely through county extension staffs. This will be accomplished with fewer visits by state-based extension entomologists. While the county staffs can be adequately serviced with fewer visits because of better teaching techniques, the county extension agents must accept greater responsibility for local programming in entomology at a time when greater and more diverse demands are being made on their time. Better training and motivation of county extension staffs in the field of entomology is a must, but in many situations there appears to be no way to accomplish this. In others county staffs are or will become large enough to permit specialization, even to accommodating a specialist in entomology.

The role of area entomologists will be clarified, but how is not yet clear. An area extension entomologist is a great relief to a state-level extension entomologist, but does he offer any relief to overburdened county extension staffs? To bypass the county agent and deal directly with farmers strikes at the very base of cooperative extension work. Alternatives that make efficient use of an area entomologist to relieve a county extension staff of part of its work load have not been fully explored to the mutual satisfaction of everyone.

Closer relations will develop between extension entomologists and research entomologists of the agricultural experiment stations and the US Department of

Agriculture. To do so will not be easy. In such fields as insect pest management, researchers will more and more leave the experiment stations and laboratories to carry out large-scale field experiments. With more specialization and higher educational levels of extension entomologists, county extension staffs, and their clientele, the traditional result demonstrations will become as sophisticated as many field research experiments. As field research and extension demonstrations merge and overlap, a new set of relationships between research and extension must emerge. It will be beset with many problems of personal relationships and administrative decisions on divisions of responsibility. Joint appointments between extension and research offer only a partial solution to this problem.

Extension entomologists will achieve full professional status, with academic training and degrees comparable to their research and teaching counterparts. Specialization will grow, but extension entomologists will not and should not become as narrowly specialized as many of their research counterparts. Specialized degree programs for training of extension entomologists are not envisioned, but present programs should be broadened rather than being so narrowly research-oriented.

SUMMARY

The Agricultural Cooperative Extension Services originated and developed through an evolutionary process within the Land Grant University System. The original role was "to aid in diffusing among the people of the United States useful and practical information on subjects relating to Agriculture and Home Economics and to encourage the application of the same." It is a grass roots approach with power centered at the county level. State administrators and specialists serve in support of county extension staffs and they, in turn, are supported by a few federal administrators and specialists.

Since extension services are locally oriented and few extension administrators are entomologists, there was and is great variation from state to state in initiation and expansion of extension entomology. One state has 25 fulltime extension entomologists, whereas 7 have only one, and one state has none. Insect problems do not wait for a state to employ an extension entomologist. Experiment station entomologists and any others available filled in as best they could in the absence of an extension entomologist. Need and demand were the principal reasons for employing or adding extension entomologists, but relief of experiment station entomologists from extension demands was also an important consideration.

As of January 1976, there were 286 extension entomologists, of whom 60.5% were fulltime. The rest held joint appointments with research and/or teaching. Joint appointments are useful in involving researchers and teachers in insect problems of the state and in servicing specialized areas, but they dilute the commitment to fulltime extension responsibility.

Prior to World War II, the extension services placed strong emphasis on a live-at-home program. In this framework and with the limited available tools for insect pest control, extension entomology was not very effective. Specialists were too often drawn from the ranks of county agents and had little formal training in entomology. Most farmers were not ready to accept and follow sound insect control programs.

During and following World War II, production agriculture came to the fore. Aided by a new generation of insecticides, extension entomologists led a far-reaching agricultural revolution. In retrospect, many based their programs too much on insecticidal control at the expense of alternative control measures, but considering the times they did extremely well.

Many farmers who did not join in this agricultural revolution came to live in poverty, both rural and urban. Traditional leadership channels within communities and neighborhoods had been shattered. Extension is struggling with this problem and making only limited progress in reaching these have-nots.

With the changing social, economic, and political scene, new directions for extension emerged and new programs have been added. The reaction of extension entomologists to the sum of these recent and current changes is about evenly divided. One optimistically stated that extension entomologists now serve all the people and can do much more than was ever dreamed possible. Another pointed out that entomology can change from serving an agricultural clientele to an urban-suburban clientele, whereas specialists in some other fields cannot. The negative reactions centered around increased work load and impractical programs. Many extension entomologists are seriously overloaded.

Extension entomology services agribusiness and many organizations in its educational role. Maintaining good working relations with the many businesses, organizations, and public agencies requires diplomacy, flexibility, and tolerance. Success is usual but not universal.

As educators, extension entomologists are naturally interested in improved communications and visual aids. Some techniques being used are quite sophisticated.

Extension entomologists are usually most comfortable when dealing with insect problems of production agriculture. The need to serve a broader clientele, as in urban entomology, is accepted.

Such programs as pesticide safety and applicator training have long been part of extension entomology programs, and recent federal dictation in these areas is not entirely welcome. If left alone and adequately funded, extension entomologists would have done these jobs well in their own way.

Enthusiasm for pest management and survey show that extension entomologists want well-rounded programs of insect control based on the best available information. Pest management is proving to be a great unifying force within entomology and between related professions. Survey and insect identification services should be expanded, especially to support pest management. To that end the extension services need to pay their share of the costs of survey and insect identification services.

Extension entomologists are ready and able to make great contributions toward improving agricultural production to alleviate world hunger. Few programs with these goals are underway because of decisions made outside of entomology.

Extension entomologists are administratively responsibilities to extension administrators. They are also professional entomologists dependent upon research entomologists for subject matter information. Considering this awkward but necessary arrangement, relationships among the factions are quite good. Extension entomologists work primarily with county agents, even in urban entomology, rather than with home demonstration or 4-H agents.

Extension entomologists want and usually receive professional status and recognition on par with their research and teaching counterparts. Specialization is necessary and inevitable for many extension entomologists, but it does create problems in selling a balanced extension entomology program to county extension staffs. Partial solutions to problems of specialization lie in the use of joint appointments, in area and county entomologists, and in research entomologists becoming more involved in insect problems of the state and its people.

An extension entomologist deals with a varied clientele: research entomologists, specialists in other fields, various organizations and agencies, agribusiness interests, and above all with county extension staffs and farmers. The techniques used are education and gentle persuasion, never force or regulation. An extension entomologist has great influence but almost no direct power. Considering the opportunities and restraints, extension entomologists do a very creditable job.

Ann. Rev. Entomol. 1977. 22:157–76

BIOSYSTEMATICS OF *TRICHOGRAMMA* AND *TRICHOGRAMMATOIDEA* SPECIES[1]

❖6125

Sudha Nagarkatti and H. Nagaraja
Commonwealth Institute of Biological Control, Indian Station, Bangalore 560 006, India

The Trichogrammatidae represent a large group of minute parasitic wasps which attack eggs of various insects, many of which are of economic importance. The small size of trichogrammatids and the relatively uniform morphology of species within each genus have made taxonomic studies difficult, and this has resulted in many nomenclatorial problems. Descriptions of species have often been rather vague, frequently based on one sex, and, in the absence of good type specimens, verification of descriptions has not always been possible. The most recent and comprehensive revisionary study of the family has been made by Doutt & Viggiani (6), who not only developed an excellent key for separation of the 70-odd genera and subgenera, but have also synonymized many of these groups. These authors have observed that the minuteness of these winged insects has apparently increased their chances of dispersion by wind, and as a result they are even found on remote isolated oceanic islands. Moreover, they are unique in that they occupy all kinds of habitats, from swampy marshlands to hot dry deserts, and occur on low-lying vegetation or in strictly arboreal habitats.

PHYLOGENY

Trichogrammatids that closely resemble modern species have been found in Oligocene amber from Chiapas, Mexico (6). According to Imms (20), who quoted Ashmead (1), the Trichogrammatidae are related to the Eulophidae, which connects the latter to the Mymaridae. Among the members of the family it is difficult to construct a phylogenetic tree. However, the comparative study of male genitalia of trichogrammatids by Viggiani (51) suggests the possibility of a gradual evolution of the phallus towards a simplified tubular structure.

[1]The survey of the literature pertaining to this review was concluded in 1975.

Much of the earlier work on the Trichogrammatidae was done by Girault. He erected 52 genera, of which 37 are now considered valid by Doutt & Viggiani (6). Girault (14–16) classified the family into several subfamilies and genera with tabular listings of species and individual references. Nowicki (30, 31) studied the trichogrammatids from the Palaearctic region, while Risbec (41, 42) dealt with the African trichogrammatids. In the revision of the family, Doutt & Viggiani (6) classified the Trichogrammatidae into 64 genera comprising 347 species. However, in a subsequent detailed study of the morphology of the male genitalia, Viggiani (51) found considerable variation between genera and suggested a regrouping of the family into two subfamilies, the Trichogrammatinae and the Oligosetinae, each subdivided as follows:

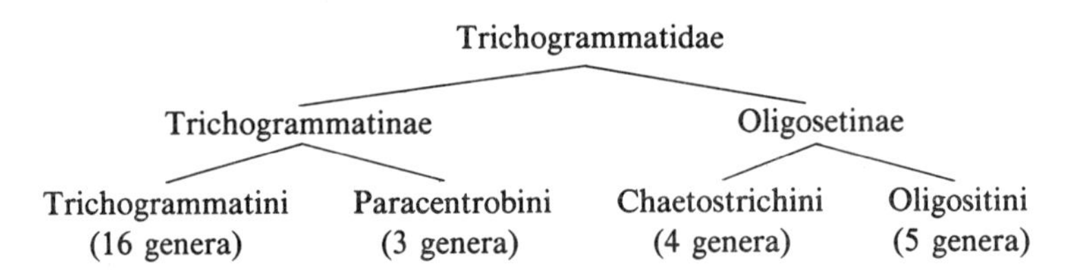

In this study, Viggiani showed that the phallus in the Trichogrammatinae has a distinct aedeagus with accessory structures like a phallobase, parameres, and volsellar digiti, whereas that in the Oligosetinae is simplified and tubular with the accompanying structures being undifferentiated from the aedeagus. This seems to be the most rational classification proposed so far for the Trichogrammatidae.

The best known genera of the Trichogrammatidae are *Trichogramma* Westwood and *Trichogrammatoidea* Girault. The ease with which they can be reared and the economic importance of the lepidopterous pests which they generally attack have given them a special status in the family. Since relatively little biosystematic work has been done on other genera of the Trichogrammatidae, this review is necessarily confined to these two genera. Both *Trichogramma* and *Trichogrammatoidea* are true representatives of the Trichogrammatidae in that they are exclusively egg parasites (primarily of Lepidoptera, but also attacking eggs of Hymenoptera, Neuroptera, Diptera, Coleoptera, and Hemiptera). They bear the characteristic three-segmented tarsi without a strigil on the fore tarsus and wings with long marginal cilia; the cilia on the hind wings in *Trichogramma* are considerably longer than in *Trichogrammatoidea.* The two genera can be differentiated by the longer wing fringes in *Trichogrammatoidea,* the presence of a two-segmented funicle in the antennae of both males and females and a three-segmented antennal club in males, the long stigmal vein, and the absence of vein track RS_1 in the forewing.

GENUS *TRICHOGRAMMA* WESTWOOD

The genus *Trichogramma* was erected by Westwood (54) with *T. evanescens* designated as the type species. This was followed by descriptions of *T. embryophagum* by Hartig (18), *T. minutum* and *T. pretiosum* by Riley (43, 44), and *T. semblidis* by Aurivilius (4). Early in the twentieth century, a description of *T. japonicum* Ashmead (1) appeared, followed by a flood of publications on *Trichogramma* by Girault (14, 15), in which *T. australicum, T. euproctidis, T. perkinsi,* and *T. retorridum* were described. About the same time, Perkins (36, 37) described *T. semifumatum* and *T. fasciatum.* The current status of *T. vitripenne* Walker, *T. koehleri* Blanchard, *T. erosicornis* Westwood, *T. minutissimum* Packard, *T. intermedium* How., *T. jezoensis* Ishii, and *T. odontotae* How. remains undetermined because of the unavailability of type specimens for study. A few specimens of *T. chilonis* Ishii obtained by the authors from Japan were found to be identical to *T. australicum* Gir.

The taxonomy of this genus cannot be studied using conventional methods since morphological differences between species are often difficult to discern and because biological and ecological differences do not appear to follow any regular pattern. Telenga (48) observed differences in the phragma of some European *Trichogramma* spp., but it has since been found that these differences are not consistent enough for use in species diagnoses. Among the characteristics used earlier for species diagnosis were pigmentation and wing and antennal trichiation, but these features are unstable and can be influenced either by temperature or by the host on which the insects are reared. Extensive studies on this aspect have been reported by Flanders (10, 11) and Quednau (38, 39).

In a later paper, Quednau (40), who studied the effect of temperature on pigmentation, developmental time, life cycle, fecundity, parthenogenesis, etc, has presented illustrations of adults of *Trichogramma australicum, T. embryophagum, T. evanescens, T. fasciatum, T. japonicum, T. minutum, T. retorridum, T. semblidis,* and *T. semifumatum* reared at 30°, 25°, 20°, and 15°C. He reported that males were always more darkly pigmented than females.

Other biological differences do not appear to be of significant taxonomic value. Whereas some species show habitat specificity—e.g. *T. semblidis,* which exclusively inhabits swampy areas, and *T. japonicum,* which occurs only in rice-fields—others such as *T. australicum, T. euproctidis,* and *T. evanescens* occur in field or arboreal habitats.

Flanders & Quednau (12) claimed that only six species were structurally identifiable, adding that "in the future, the validity of *Trichogramma* species requires that they be characterised by the mode of reproduction, the colour variation, the length of life cycles and the differences in setation, the descriptions to be based on specimens which developed at a constant high temperature of 30°C." These six species of *Trichogramma* were *embryophagum, evanescens, japonicum, minutum, retorridum,* and *semblidis,* and they considered *australicum* and *fasciatum* to be forms of *evanescens.* In the same year Quednau (40) published a comprehensive study including the following nine species of *Trichogramma* in a key: *australicum, embryophagum, evanescens, fasciatum, japonicum, minutum, retorridum, semblidis,* and *semifumatum.* Owing to difficulties in identification, numerous erroneous records are present in the literature. For example, Nagarkatti & Nagaraja (28) pointed out that what had been referred to in India as *Trichogramma minutum* Riley or *T. evanescens* Westwood for nearly 50 years is actually *T. australicum.*

Among the first to report the usefulness of male genitalia of *Trichogramma* for species diagnosis were Hintzelman (19) (who sketched the male genitalia of *T. evanescens*), Ishii (21), and Tseng (49). Viggiani (50), working independently on the Trichogrammatidae, had also recognized the importance of the male genitalia in classification. However, Nagarkatti & Nagaraja (28, 29) were probably the first to make a comprehensive study of interspecific differences among the better-known species of *Trichogramma* and to homologize the various structures. In addition, they compared the relative lengths of the ovipositor and hind tibia in females. All the descriptions were based on individuals reared on eggs of *Corcyra cephalonica* at 26°C (± 1°) and 50% relative humidity. They have prepared keys using male genitalia as the key characters (23, 24, 29). This has led to the discovery of many new species and helped to correct concepts regarding synonymy. Since then, other authors have also started including detailed descriptions and illustrations of the male genitalia of species that they describe (7, 35, 52).

Thelytokous forms (excluded from this review) and sibling species further complicate taxonomy of *Trichogramma.* In *Trichogramma* literature, terms such as forms, races, ecotypes, biotypes, etc, have been so commonly used that the biological control worker is often confused. Attempts have therefore been made to study the evolutionary relationships of some *Trichogramma* populations by crossing experiments (17, 26, 32, 33, 34).

Present knowledge suggests the existence of 36 biparental species, which can be differentiated using morphological and/or biological characters. On the basis of the male genitalia, these species can be grouped broadly as follows:

1. The Australicum group (Figure 1*a*) is characterized by the broad genitalia, broad dorsal expansion of gonobase (DEG), with two prominent lateral lobes and broad median ventral projection (MVP) (except in *T. nubilalum,* in which it is relatively narrow). This group includes *australicum, poliae* Nagaraja, *dendrolimi* Mats., *nubilalum* Ertle & Davis, *ivelae* Pang & Chen, and *closterae* Pang & Chen.
2. The Minutum group (Figure 1*b*) is characterized by somewhat narrow, triangular DEG with distinct constrictions at base, lacking lateral lobes and with a narrow MVP. It is the largest group and includes *minutum, evanescens, perkinsi, fasciatum, semifumatum, pretiosum chilotraeae* Nagaraja & Nagarkatti, *brasiliensis* Ashmead, *papilionidis* Viggiani, *papilionis* Nagarkatti, *platneri* Nagarkatti, *californicum* Nagaraja & Nagarkatti, *semblidis* (Auriv.), *retorridum* (Gir.), *embryophagum* Hartig, *leucaniae* Pang & Chen, and *ostriniae* Pang & Chen. Since this group has the largest number of members, it might be considered as representative of the ancestral species, with the other groups having evolved from it.
3. The Euproctidis group (Figure 1*c*) is a monotypic group comprising only *euproctidis,* which is characterized by the linear and very narrow male genitalia, the distal part being somewhat constricted near the gonoforceps, with a very short and narrow MVP and short aedeagus.
4. The Flandersi group (Figure 1*d*) is characterized by a spatulate extension of the DEG distally, minute MVP, and a short aedeagus (except in *lingulatum*); the median ridge on the ventral aspect extends up to the anterior margin of the

genitalia. This group includes *flandersi* Nagaraja & Nagarkatti, *hesperidis* Nagaraja, *bennetti* Nagaraja & Nagarkatti, *beckeri* Nagarkatti, and *lingulatum* Pang & Chen.

5. The Japonicum group (Figure 1*e*) is characterized by a horseshoe-shaped DEG, inconspicuous MVP, and a long aedeagus with short apodemes. This group includes two species: *japonicum* and *pallidiventris* Nagaraja.
6. The Agriae group (Figure 1*f*) is characterized by short DEG which does not reach the chelate structures. The MVP is minute. This group includes three species: *agriae* Nagaraja, *rojasi* Nagaraja & Nagarkatti, and *plasseyensis* Nagaraja.
7. The Maltbyi group (Figure 1*g*) is monotypic, comprised only of *maltbyi* Nagaraja & Nagarkatti. The DEG appears cup-shaped when viewed caudally and is constricted at the base. The chelate structures are very close to the tips of the gonoforceps.
8. The Parkeri group (Figure 1*h*) is characterized by its extremely linear appearance and the elongate DEG. This group is also monotypic and includes only *parkeri* Nagarkatti.
9. The Achaeae group (Figure 1*i*) is characterized by a simple and narrow DEG without lateral lobes and a blunt posterior extremity. The MVP is inconspicuous. This group includes *achaeae* Nagaraja & Nagarkatti and *raoi* Nagaraja.

Females of *Trichogramma* spp. show few interspecific differences, and as the relative lengths of the hind tibia and ovipositor frequently overlap, positive identification using these characteristics alone is not possible. The authors have observed that only in *T. japonicum* and *T. pallidiventris* females is the ventral genital plate more angular than is found in other *Trichogramma* spp. and has surface reticulations that are absent from other species. Positive identification of thelytokous species is therefore difficult unless males can be obtained by rearing at higher temperatures.

More recently, Voegele et al (53) have attempted to use the scanning electron microscope for taxonomic separation. They have reported the presence of nine different kinds of sensillae on the antennae of female *Trichogramma,* and they consider only those localized in the ventral part of the club to be of taxonomic value. The mean number of these sensillae appears to be statistically specific in the three species studied: *T. maltbyi* (20.23), *T. evanescens* (27.77) and *T. brasiliensis* (21.71). Moreover, the structure of the sensillae in *T. brasiliensis* and *T. evanescens* is different. Extensive studies of this nature would be desirable because they could provide information that it is impossible to obtain using an ordinary light microscope. It would also be interesting to see if taxonomic separation of species based on such studies coincides with classification based on male genitalia.

Some species of *Trichogramma* exhibit sexual dimorphism of the wings and antennae particularly when reared under certain conditions and on certain hosts. This is best seen in *T. semblidis,* which commonly parasitizes eggs of tabanids, sialids, and a few Lepidoptera in Europe, North America, and India. Salt (45) reported that all males of *T. semblidis* emerging from field-collected *Sialis lutaria* eggs were apterous. Also, of 757 *T. semblidis* males reared from *Sialis* eggs in the laboratory, all except 2 were apterous, whereas all those reared from *Ephestia*

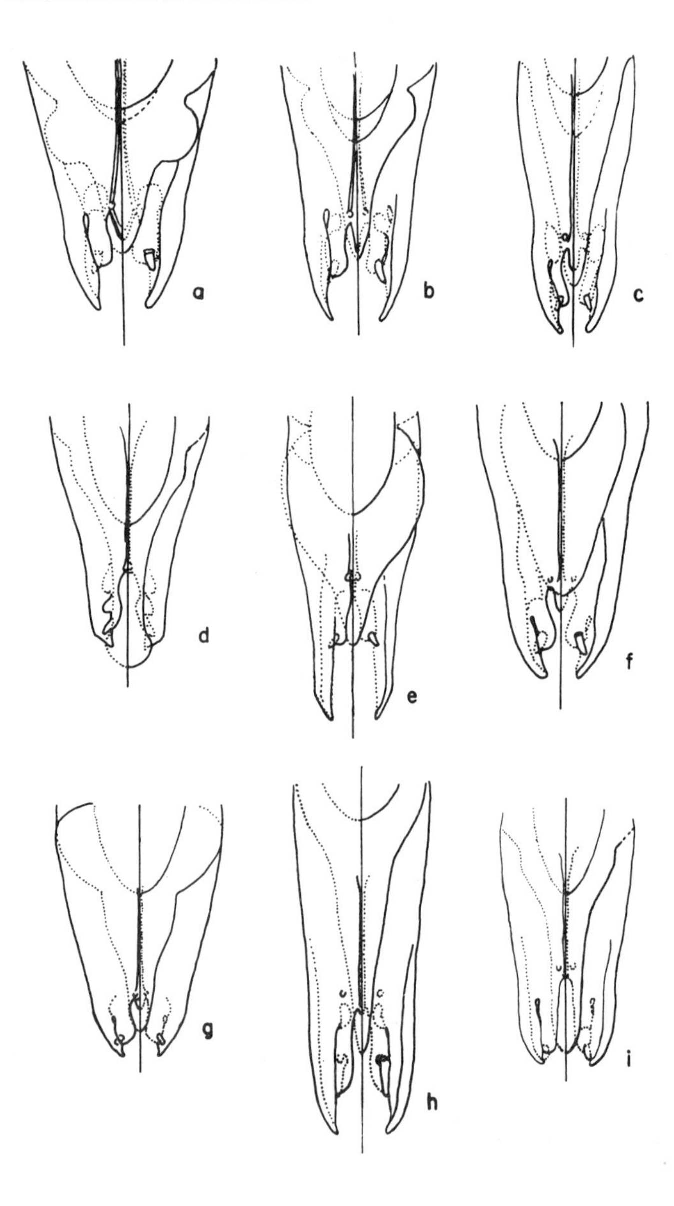

Figure 1 Diagrammatic sketches of male genitalia of *Trichogramma* showing only the terminal portion, excluding aedeagus: (*a*) Australicum group, (*b*) Minutum group, (*c*) Euproctidis group, (*d*) Flandersi group, (*e*) Japonicum group, (*f*) Agriae group, (*g*) Maltbyi group, (*h*) Parkeri group, and (*i*) Achaeae group.

kuhniella eggs in the laboratory were normally winged. Salt rightly concluded that the quality of food controls dimorphism.

Although apterous males of *T. semblidis* are otherwise perfectly normal [mating and insemination are normal, as shown by Nagarkatti & Jeyasingh (27)], the antennae are gynecoid, that is, they bear femalelike clubbed antennae. Although sexual dimorphism in itself is not of any taxonomic value, it is interesting to note that it is most evident in *T. semblidis.*

While studying parthenogenesis, Birova (5) observed that in populations of *T. embryophagum* (whose natural host is *Choristoneura fumiferana*) reared from *Scotia segetum* gynandromorphs occasionally appeared. These showed secondary signs of the female sex on the abdomen, whereas the head had male signs, that is, the left antenna bore a normally long-haired clavus.

GEOGRAPHIC DISTRIBUTION

Since *Trichogramma* spp. are small enough to be carried by wind and have also been intentionally transported from one country into another from the very early days, it is difficult to determine their natural ranges of distribution. Table 1 gives a list of *Trichogramma* species with their geographic distribution, hosts, and host plants. This list is based largely on material actually examined by the authors, rather than on records in the literature, where inaccuracies are likely. For this reason, species from mainland China are not included in the list. A few remarks on the distribution pattern appear necessary.

Recently, Pang & Chen (35) have reported 12 species of *Trichogramma* from mainland China: *australicum, closterae, dendrolimi, euproctidis, evanescens, ivelae, japonicum, leucaniae, lingulatum, ostriniae, raoi,* and *sericini.* Of these, *T. euproctidis, T. evanescens,* and *T. dendrolimi* are known to occur in the Palaearctic region, while *T. australicum, T. japonicum,* and *T. raoi* occur in other areas of the Orient. Six of the species were previously undescribed (35). The present authors suggest that a critical comparison and cross-breeding (crossing) studies between *T. ostriniae* and *T. chilotraeae* are needed because both species appear very similar and may be synonymous. The latter has been reared from *Ostrinia furnacalis* (= *salientalis*) in Thailand, while *T. ostriniae* is reported (35) to have been reared from *O. nubilalis,* which does not occur in Asia. The *Trichogramma* fauna of Japan also comprises some Oriental (*japonicum*) and some Palaearctic species (*dendrolimi, evanescens*).

The Nearctic and Neotropical regions are well represented with 14 species, while the European fauna includes at least 5. Several "species" of unconfirmed taxonomic status have also been reported, but until these populations are actually studied and crossed with others in the region, their identity will remain dubious. These are *T. turkestanica* Meyer, *T. latipennis* Haliday, *T. carpocapsae* Schreiner, and *T. barathrae* Skriptschinsky, all of which have been considered synonymous with *T. evanescens* by Quednau (40), and *T. piniperdae* Wolff, *T. cacoeciae pini* Telenga, and *T. pini* Meyer, which the same author considers synonymous with *embryophagum.* There is need for confirming the identity of a number of European populations through crossing tests, but this has not been possible because of various limitations.

Table 1 Distribution of *Trichogramma* by country, with records of hosts and host plants wherever known

Country	Species	Host	Host plant
Angola	*papilionidis*	*Papilio demodocus*	?
Argentina	*fasciatum*	?	?
	minutum	?	?
Australia	*australicum*	*Cactoblastis cactorum*	*Opuntia* spp.
Barbados	*fasciatum*	*Spodoptera* sp.	corn
Chile	*rojasi*	*Rachiplusia ou* and *Tatochila* sp.	mint
	semifumatum	?	rose
Colombia	*semifumatum*	*Diatraea saccharalis*	sugarcane
		Scrobipalpula absoluta	tomato
	perkinsi	*D. saccharalis*	sugarcane
		Alabama argillacea	cotton
Costa Rica	*pretiosum*	*Hypsipyla grandella*	*Cedrela* sp.
	semifumatum	*H. grandella*	*Cedrela* sp.
		Heliothis zea	corn
	beckeri	*Hypsipyla grandella*	*C. tonduzii*
	fasciatum	*H. grandella*	*C. odorata*
Czechoslovakia	*embryophagum*	*Choristoneura fumiferana*	spruce
	evanescens	*Ostrinia nubilalis*	corn
Egypt	*evanescens*	*Sesamia cretica*	sugarcane
France	*evanescens*	*Pieris* spp.	cabbage
Ghana	*Trichogramma* sp.	*Diopsis thoracica*	rice
Greece	*cacoeciae*	*Prays oleae*	olive
Guatemala	*perkinsi*	*Diatraea* sp.	corn
	fasciatum	*Diatraea* sp.	corn
	Trichogramma sp.	*Diatraea* sp.	corn
Hawaii	*semifumatum*	*Herpetogramma liscarsisalis*	grass
		Achaea janata	
	japonicum	*Sepedon sauteri*	—
	australicum	*Papilio xuthus*	citrus
India	*australicum*	*Chilo* spp.	sugarcane
		A. janata	castor
		Heliothis armigera	*Cicer arietinum*, etc
		Agrius convolvuli	*Colocasia antiquorum*
	japonicum	*Tryporyza incertulas*	rice
		T. nivella	sugarcane
		S. sauteri	—
	semblidis	*Tabanus macer*	grass
		Chilo infuscatellus	sugarcane
	flandersi	*C. infuscatellus*	sugarcane
	achaeae	*A. janata*	castor
	chilotraeae	*C. infuscatellus*	sugarcane
	agriae	*A. convolvuli*	*Colocasia antiquorum*

Table 1 *(Continued)*

Country	Species	Host	Host plant
India	*pallidiventris*	*T. incertulas*	rice
	hesperidis	unidentified hesperiid	rice
	plasseyensis	*C. infuscatellus*	sugarcane
	poliae	*C. infuscatellus*	sugarcane
	raoi	unidentified lepidopteran	sugarcane
Indonesia	*Trichogramma* sp.	*Milionia basalis*	pine
	australicum	*C. infuscatellus* and *C. sacchariphagus*	sugarcane
Israel	*evanescens*	noctuids	tomato
	euproctidis	noctuids	tomato
	pretiosum[a]	noctuids	tomato
	semifumatum[a]	noctuids	tomato
Japan	*australicum*	*P. xuthus*	citrus
		Mamestra brassicae	cabbage
		Heliothis assulta	?
	evanescens	*H. assulta*	?
	dendrolimi	*Dendrolimus spectabilis*	pine
		Dictyoploca japonica	?
		M. brassicae	cabbage
		Lampides boeticus	cabbage
	papilionis	*P. xuthus* and *P. memnon thunbergi*	citrus
Kenya	*Trichogramma* sp.	*Sesamia calamistis*	corn
Malagasy	*Trichogramma* sp.	*C. sacchariphagus*	sugarcane
Malaysia	*japonicum*	*C. polychrysa*	rice
		T. incertulas	rice
	chilotraeae	*C. suppressalis*	rice
Mexico	*fasciatum*	*H. zea*	corn
	semifumatum	*H. zea*	corn
	perkinsi	*H. zea*	corn
Mozambique	*Trichogramma* sp.	*Chilo partellus*	corn
Netherlands	*Trichogramma* sp.	*Cydia pomonella*	apple
	embryophagum	*Cydia pomonella*	apple
New Guinea	*Trichogramma* sp.	*Chilo terrenellus*	sugarcane
Pakistan	*australicum*	*Bactra* sp.	*Cyperus rotundus*
		C. partellus	corn
Peru	*fasciatum*	*D. saccharalis*	sugarcane
	perkinsi	unidentified lepidopteran	apple
	brasiliensis	*H. zea*	cotton
Philippines	*australicum*	*C. suppressalis*	rice
	japonicum	*T. incertulas* and *C. suppressalis*	rice
Poland	*embryophagum*	*Acantholyda pinivora*	pine
	evanescens	*Dendrolimus pini*	pine
	euproctidis	*A. pinivora*	pine
	dendrolimi	*Orgyia antiqua*	pine

Table 1 *(Continued)*

Country	Species	Host	Host plant
Sri Lanka	*australicum*	*Homona coffearia*	tea
Taiwan	*australicum*	borers	sugarcane
	Trichogramma sp.	borers	sugarcane
Thailand	*japonicum*	*C. polychrysa*	rice
	chilotraeae	*Ostrinia furnacalis*	corn
Trinidad	*bennetti*	*Hypsipyla ferrealis*	*Carapa guianensis*
	Trichogramma sp.	*Calpodes ethlius*	canna
Turkey	*Trichogramma* sp. (*fasciatum* ?)	*Calpodes ethlius*	canna
Uganda	*Trichogramma* sp.	*C. partellus*	corn
United States	*minutum*	*Argyrotaenia velutinana*	apple
		Limenitis archippus	willow
		Rhyacionia buoliana	pine
	retorridum	*Faronta albilinea*	cotton
		Papilio india pergamus	cabbage
	perkinsi	*Heliothis zea*	corn
		Vogtia malloi	*Alternanthera philoxeroides*
	fasciatum	*D. saccharalis*	sugarcane
	semblidis	*Sialis* spp.	—
	platneri	*Cydia pomonella*	apple
	parkeri	*H. zea*	corn
	maltbyi	*Oulema melanopus*	grasses
	californicum	*Hemerocampa pseudotsugata*	forest trees
	nubilalum	*H. zea*	corn
	pretiosum	*A. argillacea*	cotton
	semifumatum	*Colias eurytheme*	alfalfa
		Pieris rapae	corn
USSR	*euproctidis*	*M. brassicae*	cabbage
	cacoeciae pallida	*C. pomonella*	apple
West Germany	*euproctidis*	*Euproctidis chrysorrhoea*	forest trees

[a] Probably introduced from the United States.

The African continent has relatively few representatives of *Trichogramma.* With the exception of a few individuals reared from *Chilo partellus* in Uganda and Mozambique, a few specimens from *Sesamia calamistis* in Kenya, and another undetermined species occurring on *Diopsis thoracica* in West Africa, the authors have encountered no other *Trichogramma* from this region. The Ugandan and Kenyan populations contained no males, making positive identification difficult, and the genitalia of the single male in the West African population were damaged. It is certain, however, that the West African and East African populations are different. Specimens from Mozambique likewise could not be positively determined because they had been preserved in alcohol for a long time, making clearing and identification difficult. On examination by the authors, dead specimens of a *Tricho-*

gramma sp. reared in Malagasy from *Chilo sacchariphagus* appeared nearly identical to those of *T. minutum,* but the specific status could not be confirmed without crossing tests. If this is indeed *minutum,* the chances are that it has been introduced from the New World.

Some of the oceanic islands, such as Hawaii, also show an interesting combination of species from adjoining continental areas.

EXPERIMENTAL HYBRIDIZATION

Harland & Atteck (17) were the first to recognize that solutions to taxonomic problems involving *Trichogramma* could be found only by making extensive studies of their biological characteristics, including crossing relationships. Telenga (47) mentioned that *T. evanescens* in Europe is represented by many forms which are geographically separated by 200–300 km, and which do not normally cross between themselves, but which paired when reared under the same conditions for four to six months. However, it is not known whether or not fertile progeny were produced.

There is considerable merit in conducting crossing studies; they are helpful both in correcting misconceptions concerning synonymy and in determining the evolutionary proximity or distance between different populations, as has been shown by Fazaluddin & Nagarkatti (8) and Nagaraja (23). Nagarkatti & Nagaraja (28) conducted crosses between *T. minutum, T. evanescens,* and *T. australicum.* No hybrids were produced, which proves their distinct identity.

Crossing experiments have also been conducted by Oatman et al (33) between strains of *Trichogramma* occurring in southern California. These strains were reared from hosts such as *Heliothis zea, Spodoptera exigua,* and *Trichoplusia ni.* All the crosses proved to be interfertile, indicating the presence of a single species, but whether this species (referred to as a reference stock) was *T. pretiosum* or *T. semifumatum* was uncertain. In a later study, Oatman et al (34) crossed this reference stock with populations reared from Los Mochis and Tapachula (Mexico), as well as with *T. semifumatum* from Hawaii. The reference stock was found to interbreed readily with the Mexican population but not with the Hawaiian *semifumatum.* This led them to conclude that (*a*) the reference stock and the Mexican populations are *pretiosum* and not *semifumatum,* and (*b*) the latter two, as used by Stern & Bowen (46), are synonymous. These results, however, are not in agreement with those obtained by Nagarkatti & Fazaluddin (26), who worked with a *T. pretiosum* culture supplied by the Rincon Insectary in southern California. Details of crosses conducted by the above authors between various New World *Trichogramma* are included in Figure 2. The *Trichogramma* species which exhibited partial isolation, with hybrids produced only in one direction, and which were therefore considered full species with respect to each other were the following: *fasciatum* and *perkinsi, fasciatum* and *rojasi, maltbyi* and *fasciatum, minutum* and *perkinsi, perkinsi* and *californicum, semifumatum* and *perkinsi, minutum* and *californicum, minutum* and *semifumatum, maltbyi* and *californicum,* and *fasciatum* and *semifumatum.* The only crosses in which at least limited female progeny were obtained in both directions were *T. fasciatum* X *T. pretiosum, T. minutum* X

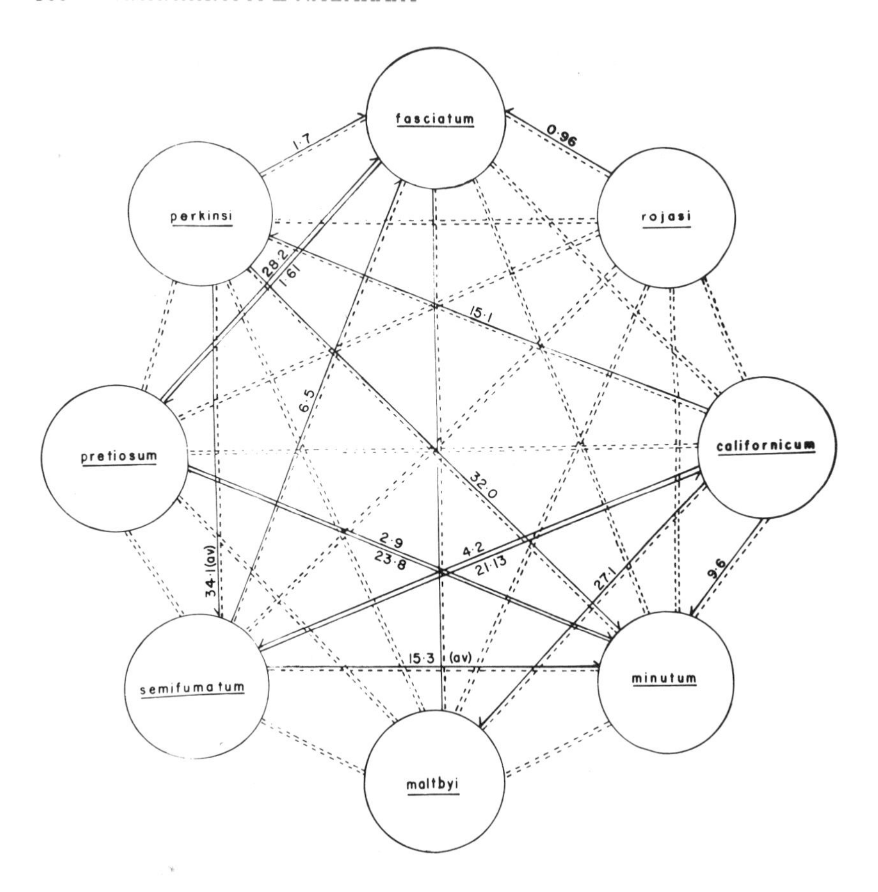

Figure 2 Crossing relationships between some New World species of *Trichogramma*. Solid lines indicate production of fertile hybrids; hatched lines indicate failure to cross; arrows point to female parent; numbers represent percentage of female progeny in the F_1 generation, and those followed by (av) are the average of two replicates.

T. pretiosum, and *T. semifumatum* X *T. californicum*. These species are therefore considered to be semispecies with respect to each other; not only are they morphologically distinguishable from each other, but it would appear that in course of time, with restricted gene flow between them, reproductive isolation would be strengthened and possibly become complete.

Oatman & Platner (32) studied stocks of *Trichogramma* spp. obtained from Missouri (United States). In all 15 stock cultures were involved, including 2 from California (*T. pretiosum*, a reference stock originally reared from *H. zea*, and *Trichogramma* sp., reared from codling moth eggs on apple). They reported the results of 42 crosses and concluded that 4 species of *Trichogramma* were present in the material from Missouri, including *T. pretiosum*, and also that two species from Missouri and one from California, which crossed with neither *T. pretiosum* nor *T. minutum,* were probably new species.

Isolating mechanisms in *Trichogramma* are rather interesting. By and large, ethological isolation is weak and males appear somewhat aggressive. In many of the laboratory crosses, mating occurred fairly readily and even insemination was frequently successful. However, fertilization appeared to be prevented in the majority of crosses, or when it did occur the sperms were probably killed, resulting in few or no hybrids. A discussion of the high degree of successful insemination in interspecific crosses between New World *Trichogramma* species is presented by Nagarkatti & Fazaluddin (26), together with a comparison between female (hybrid) production and insemination. Crossing success was scarcely ever related to geographical proximity or morphological similarity of the two species. In most of the crosses in which hybrids were produced, the latter were viable and fertile. Many species showed unilateral compatibility, with fertile hybrids being produced in one direction but not the other. In a few crosses, females that had successfully mated with alien males died shortly after insemination, possibly from mechanical injury. When dissected shortly after death, such females showed abundant sperm in the spermathecae. Hybrids produced in interspecific crosses were usually fertile and were almost invariably indistinguishable from the maternal parent.

An interesting cross was reported by Nagarkatti (25) between *T. perkinsi* females and *T. californicum* males; it resulted in 17 hybrid females in the F_1 generation. Of these, a single female proved to be thelytokous, and the seven females that could be tested proved to be arrhenotokous.

GENUS *TRICHOGRAMMATOIDEA*

Girault (14) erected the genus with *Trichogrammatoidea nana* as the genotype. Ferriere (9) did not at first consider the genus valid, because he considered the characteristics described for the genus to be variants of those of *Trichogramma.* In 1933, however, he conceded to Girault.

Phylogenetically, *Trichogrammatoidea,* which has a much simplified phallus, may have evolved from a *Trichogramma*-like ancestor. This theory is supported by the fact that a trichogrammatid reared from *Epiphyas postvittana* in New Zealand and Australia is somewhat intermediate, having *Trichogramma*-like male genitalia and fore wings but a segmented male antennal funicle as in *Trichogrammatoidea.* This may represent a transition between *Trichogramma* and *Trichogrammatoidea.*

The genus is characterized by the presence of (*a*) a segmented flagellum (except the ring segment) in the male antenna, and (*b*) long fringe setae in the fore wing, the longest (e.g. *T. simmondsi*) measuring about half the width of the wing, and by the absence of (*a*) oblique line of setae (R_s) in the fore wing below the stigma and (*b*) the DEG in the male genitalia. Furthermore, the aedeagus together with the apodemes is invariably shorter than the entire genital capsule.

Because the DEG is absent, the male genitalia of *Trichogrammatoidea* are not of as great diagnostic value as in *Trichogramma,* but useful characteristics include (*a*) the relative levels of chelate structures and gonoforceps, (*b*) presence or absence

Table 2 Distribution of *Trichogrammatoidea* spp. by country

Country	Species	Host	Host plant
Argentina	*signiphoroides*	*Pseudaulacaspis pentagona*	?
Australia	*flava*	?	?
	rara	*Cosmophila* sp.	?
Costa Rica	*hypsipylae*	*Hypsipyla grandella*	*Cedrela odorata*
Ethiopia	*lutea*	*H. armigera*	cotton
Germany	*stammeri*	?	?
Ghana	*simmondsi*	*Diopsis thoracica* and *Chilo* sp.	rice
Guam	*guamensis*	*Lampides boeticus*	*Cassia grandis*
India	*armigera*	*H. armigera*	*Polianthus tuberosa*
		unidentified lepidopteran	*Cajanus cajan*
	fulva	*Cryptophlebia ombrodelta*	*Acacia concinna*, *Cassia fistula*, *Cassia* sp., *Poinciana regia*, *Tamarindus indicus*
	bactrae	*Bactra venosana*	*Cyperus rotundus*
		Chilo indicus and *C. infuscatellus*	sugarcane
		Agrius convolvuli	*Colcocasia antiquorum*
		Emmalocera depressella	sugarcane
		Sepedon sauteri and *Pelopidas mathias*	rice
		Nymphula depunctalis	rice
	bactrae ssp. *fumata*	unidentified lepidopteran	*Achyranthus aspera*
	nana	*Bactra venosana*	*Cyperus rotundus*
		Sylepta derogata	cotton
	prabhakeri	*Achaea janata*	castor
	robusta	*Hypsipyla robusta*	*Cedrela toona*
Indonesia	*armigera*	*Etiella zinckenella*	unknown
	bactrae	*Chilo sacchariphagus stramineellus*	sugarcane
	nana	*Chilo sacchariphagus stramineellus*	sugarcane
Ivory Coast	*lutea*	*Earias biplaga*	cacao
Kenya	*lutea*	*Heliothis* sp.	cotton
Malaysia	*bactrae*	unidentified borer	sugarcane
Mauritius	*Trichogrammatoidea* sp.	*Sesamia calamistis* and *Argyroploce schistaceana*	sugarcane
		Plusia chalcytes	tomato
		Achaea trapezoides	*Acalypha* sp.

Table 2 *(Continued)*

Country	Species	Host	Host plant
Mozambique	*lutea*	*Chilo partellus*	corn
Pakistan	*bactrae*	*C. partellus*	corn
Philippines	*nana*	*A. schistaceana*	sugarcane
Senegal	*combretae*	unidentified pierid	*Combretum aculeatum*
South Africa	*lutea*	*H. armigera*	cotton
		Aspidiotus hederae	?
Sri Lanka	*nodicornis*	?	?
Taiwan	*bactrae*	*Eucosma* sp.	sugarcane
Trinidad	*hypsipylae*	*Hypsipyla ferrealis*	*Carapa guianensis*
Venezuela	*hypsipylae*	*H. grandella*	?

of the MVP, (*c*) the shape and length of the MVP, and (*d*) the distinctiveness of the connective membrane in place of the dorsal expansion of gonobase. In addition, the antennae, body color, the relative lengths of wing fringe, wing trichiation, the extent of basal infuscation in the fore wing, and to some extent the relative lengths of the aedeagus, hind tibia, and ovipositor also aid in identification.

Until recently, only six species were known: *T. nana* (Zehnt.), *T. lutea* Gir., *T. rara* Gir., *T. flava* Gir., *T. stammeri* Nowicki, and *T. signiphoroides* Breth. Doutt & Viggiani (6) included *T. combretae* (Risbec) and *T. nodicornis* (West.). In addition, an unidentified species was also known to occur in Mauritius (22), which, unlike *T. nana,* did not parasitize *Chilo sacchariphagus stramineellus.* Recent investigations have proven that there are additional species.

In a study of the genus, Nagaraja (unpublished data[2]) has described ten new species and a subspecies. A total of 20 species and a subspecies with hosts and host plants are listed in Table 2.

With a view to determining the phenetic relationships, a numerical taxonomic study using both morphological and biological data was undertaken on ten *Trichogrammatoidea* species and one subspecies. These species were *T. armigera* Nagaraja, *T. cryptophlebiae* Nagaraja, *T. guamensis* Nagaraja, *T. hypsipylae* Nagaraja, *T. lutea, T. nana, T. prabhakeri* Nagaraja, *T. robusta* Nagaraja, *T. simmondsi* Nagaraja, *T. thoseae* Nagaraja, and *T. bactrae bactrae* Nagaraja. Two species groups, Lutea and Nana, were evident (Figure 3). The Lutea group was characterized by a short fore wing fringe, whereas those in the Nana group possess a long fore wing fringe.

EXPERIMENTAL HYBRIDIZATION

Crosses could be conducted between the following five species of *Trichogrammatoidea*: *lutea, armigera, robusta, prabhakeri,* and *bactrae,* since live cultures of

[2]These *Trichogrammatoidea* have been described by Nagaraja in a doctoral dissertation as yet unpublished.

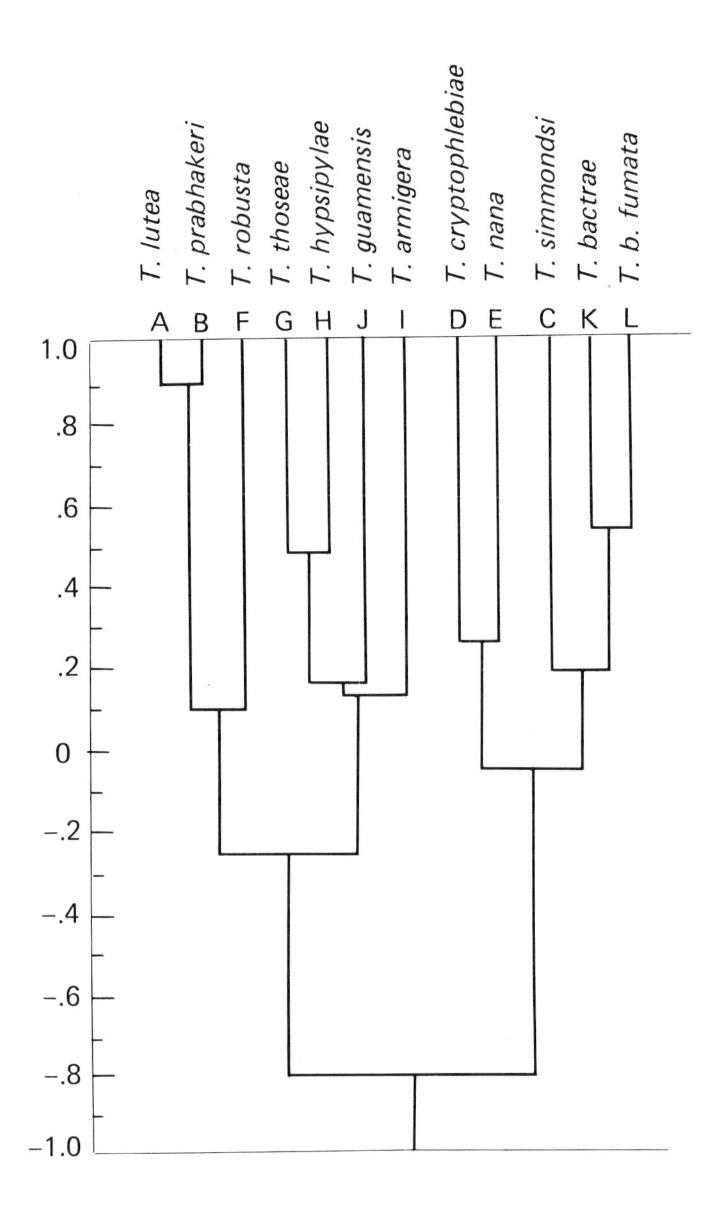

Figure 3 Dendrogram showing the phenetic relationships between various species and subspecies of *Trichogrammatoidea*.

these species were available. Of these, *bactrae* belongs to the Nana group, while the other four belong to the Lutea group. In these crosses, only a few hybrid progeny resulted in the F_1 generation in some crosses within the Lutea group. This can probably be accounted for by the inability of sperms to fertilize alien eggs to the normal extent observed in *Trichogramma* spp. Crosses between *T. lutea* and *T. prabhakeri,* which are morphologically identical, resulted in small numbers of fertile female progeny in both directions. These could therefore be considered semispecies with respect to each other. Results of the crossing studies within the Lutea group appear to conform with the clustering cycle.

No crosses could be conducted among species of the Nana group because, except for *T. bactrae,* no live cultures of the others were obtained. Since *T. bactrae* is widespread and polyphagous, five geographic populations reared from different hosts were used in intraspecific crossing experiments. Results of these experiments indicated a diverging trend between some populations that could be a precursor to speciation.

CONCLUSION

The genera *Trichogramma* and *Trichogrammatoidea* appear to be in a state of evolutionary flux, as is evident from the presence of so many populations that are phenotypically identical (or nearly so), but possibly genetically different, and others that conform to the definition of semispecies. The pattern of evolution also appears to differ, so much so that, although physiological and ecological needs diversify, the morphology of the insects remains relatively unchanged, leading to the production of sibling species. No specialization of anatomical structures appears to be necessary, even in the immature stages, because they feed passively in host eggs. The need to exercise caution in Chalcid taxonomy has been aptly discussed by Askew (2), who says that intraspecific variation in response to different hosts and seasonal dichroism must be thoroughly understood.

Limited knowledge of the cytogenetics of *Trichogramma* is available (13), and the extent to which interspecific differences occur at the chromosomal or genic levels is unknown. Although cytogenetic studies may not solve all the problems in trichogrammatid taxonomy, there is an obvious need for shedding more light on this aspect. Perhaps such studies would also lead to a better understanding of thelytokous species.

The origin of reproductive isolation in the two genera, and particularly in *Trichogramma,* is most perplexing. In *Trichogramma,* although there is ample evidence of gametic isolation, there is hardly any evidence of selection for sexual isolation to prevent the waste of gametes, as indicated by the high degree of successful insemination in many interspecific crosses. Some of the species that are separated by wide geographical barriers would obviously be prevented from crossing in nature, which would leave no opportunity for natural selection to strengthen sexual isolation. However, in some of the *Trichogrammatoidea* and New World *Trichogramma,* which occur in close proximity or whose distribution ranges overlap, there is little evidence of sexual isolation. Why natural selection has not brought about reinforcement of sexual isolation in such cases is a matter for debate. This brings us to the question: how do species arise in these genera?

Askew (3) has given the following reasons for rapid speciation in the Chalcidoidea: (*a*) sibmating and consequent inbreeding in isolated populations, combined with rapid decline in fertility of males; (*b*) haplodiploid method of reproduction; and (*c*) adaptation to different hosts. These reasons and the theory of sympatric race-formation are generally applicable to the Trichogrammatidae. For further discussions on speciation in Hymenoptera in general, the reader is referred a chapter in this volume (5a).

One can hardly overemphasize that taxonomic studies on *Trichogramma* and *Trichogrammatoidea* cannot be undertaken in the absence of good type specimens. Type specimens available at the present time are in poor condition or have been badly mounted, obscuring critical morphological characters such as male genitalia. Specimens on card points are of little value because the genitalia cannot be examined. The need for designating type specimens and preserving them adequately was stressed at a workshop on *Trichogramma* organized by the U.S. Department of Agriculture in Columbia, Missouri, in 1970. It was agreed that more permanent mounting media should be used to prevent deterioration of the specimens. [The technique developed by Wirth & Marston (55) for mounting small insects has been found to be excellent for Trichogrammatidae.] It was also agreed that types and paratypes of species described in the future should be deposited in museums in the United States, the United Kingdom, the Soviet Union, and India to facilitate access by other workers.

Type specimens are not available for several described species, and, although efforts have been made to obtain *Trichogramma* specimens from the type host and locality (e.g. 32) to enable redesignation of neotypes, this practice has not found favor with some taxonomists. There is, however, little doubt that, with genera such as *Trichogramma* and *Trichogrammatoidea,* correct identification is of utmost importance, as is true of other parasitoids, and this is facilitated when types are present and when facilities are available for studying live material.

Acknowledgments

The authors are grateful to Dr. F. J. Simmonds, Director, Commonwealth Institute of Biological Control, and Dr. T. Sankaran, Entomologist-in-Charge, Commonwealth Institute of Biological Control, Indian Station, for critically reviewing the manuscript and making useful suggestions.

Literature Cited

1. Ashmead, W. H. 1904. Descriptions of new Hymenoptera from Japan II. *J. Entomol. Soc. NY* 12:146–65
2. Askew, R. R. 1965. Host relations in the Chalcidoidea and their taxonomic significance. *Proc. Int. Congr. Ent. 12th London, 1964,* p. 94
3. Askew, R. R. 1968. Considerations on speciation in Chalcidoidea. *Evolution* 22:642–45
4. Aurivilius, C. 1897. En ny svensk äggparasit. *Ent. Tidskr.* 18:249–55
5. Birova, H. 1970. A contribution to the knowledge of the reproduction of *Trichogramma embryophagum* (Hartig). *Acta Entomol. Bohemoslov.* 67 (2): 70–82
5a. Crozier, R. H. 1977. Evolutionary genetics of the Hymenoptera. *Ann. Rev. Entomol.* 22:263–88
6. Doutt, R. L., Viggiani, G. 1968. The classification of the Trichogrammatidae. *Proc. Calif. Acad. Sci.* 35:477–586
7. Ertle, L. R., Davis, C.P. 1975. *Trichogramma nubilale* new species, an egg parasite of *Ostrinia nubilalis* (Hübner). *Ann. Entomol. Soc. Am.* 68(3):525–28
8. Fazaluddin, M., Nagarkatti, S. 1971. Reproductively incompatible crosses of *Trichogramma cacoeciae pallida* with *T. minutum* and *T. pretiosum. Ann. Entomol. Soc. Am.* 64(6):1470–71
9. Ferriere, C. 1930. On some egg-parasites from Africa. *Bull. Entomol. Res.* 21 (1):33–44
10. Flanders, S. E. 1931. The temperature relationships of *Trichogramma minutum* as a basis for racial segregation. *Hilgardia* 5:395–406
11. Flanders, S. E. 1935. Host influence on the prolificacy and size of *Trichogramma. Pan-Pac. Entomol.* 11:175–77
12. Flanders, S. E., Quednau, W. 1960. Taxonomy of the genus *Trichogramma. Entomophaga* 4 (4):285–94
13. Fukada, H., Takemura, M. 1943. Genetical studies of *Trichogramma. Jpn. J. Genet.* 19:275–81
14. Girault, A. A. 1911. Synonymic and descriptive notes on the chalcidoid family Trichogrammatidae with descriptions of new species. *Trans. Am. Ent. Soc.* 37:43–83
15. Girault, A. A. 1912. The chalcidoid family Trichogrammatidae. I. Tables of the subfamilies and genera and revised catalogue. *Bull. Wisc. Natl. Hist. Soc.* 10:81–100
16. Girault, A. A. 1913. The chalcidoid family Trichogrammatidae. II. Systematic history and completion of the catalogue and table. *Bull. Wisc. Natl. Hist. Soc.* 11:150–79
17. Harland, S. C., Atteck, O. M. 1933. Breeding experiments with biological races of *Trichogramma minutum. Z. Indukt. Abstamm. Vererbungsl.* 64: 54–76
18. Hartig, T. 1938. *Jahrb. Forsch. Forstwirtsch.* 1:250
19. Hintzelman, U. 1925. Beiträge zur Morphologie von *Trichogramma evanescens* Westw. *Arb. Biol. Reichsanst. Land. Forstwirtsch.* 14:225–30
20. Imms, A. D. 1961. *A General Textbook of Entomology,* p. 714. New Delhi: Asia Publ. House. 886 pp.
21. Ishii, T. 1941. The species of *Trichogramma* in Japan, with descriptions of two new species. *Kontyu* 14:169–76
22. Moutia, L. A., Courtois, C. M. 1952. Parasites of the mothborers of sugarcane in Mauritius. *Bull. Entomol. Res.* 43 (2):325–59
23. Nagaraja, H. 1973. On some new species of Indian *Trichogramma. Orient. Insects* 7 (2):275–90
24. Nagaraja, H., Nagarkatti, S. 1973. A key to some New World species of *Trichogramma,* with descriptions of four new species. *Proc. Entomol. Soc. Wash.* 75 (3):288–97
25. Nagarkatti, S. 1970. The production of a thelytokous hybrid in an interspecific cross between two species of *Trichogramma. Curr. Sci.* 39 (4):76–78
26. Nagarkatti, S., Fazaluddin, M. 1973. Biosystematic studies on *Trichogramma* species. II. Experimental hybridization between some *Trichogramma* spp. from the New World. *Syst. Zool.* 22 (2):103–17
27. Nagarkatti, S., Jeyasingh, S. 1974. Studies on *Trichogramma semblidis* (Auriv.), egg parasite of *Tabanus macer* Bigot in India. *Proc. Ind. Acad. Sci.* 80B (5):299–39
28. Nagarkatti, S., Nagaraja, H. 1968. Biosystematic studies on *Trichogramma* species: I. Experimental hybridization between *Trichogramma australicum* Girault, *T. evanescens* Westwood and *T. minutum* Riley. *Tech. Bull. CIBC.* 10:81–96
29. Nagarkatti, S., Nagaraja, H. 1971. Redescriptions of some known species of *Trichogramma,* showing the impor-

tance of the male genitalia as a diagnostic character. *Bull. Entomol. Res.* 61: 13–31
30. Nowicki, S. 1935. Descriptions of new genera and species of the family Trichogrammatidae from the Palaearctic region, with notes. I. *Z. Angew. Entomol.* 21:566–96
31. Nowicki, S. 1936. Descriptions of new genera and species of the family Trichogrammatidae from the Palaearctic region, with notes. II. *Z. Angew. Entomol.* 23:114–48
32. Oatman, E. R., Platner, G. R. 1973. Biosystematic studies of *Trichogramma* species: 1. Populations from California and Missouri. *Ann.Entomol. Soc. Am.* 66 (5):1099–1102
33. Oatman, E. R., Greany, P. D., Platner, G. R. 1968. A study of the reproductive compatibility of several strains of *Trichogramma* in Southern California. *Ann. Entomol. Soc. Am.* 61:956–59
34. Oatman, E. R., Platner, G. R., Gonzalez, D. 1970. Reproductive differentiation of *Trichogramma pretiosum, T. semifumatum, T. minutum* and *T. evanescens* with notes on the geographical distribution of *T. pretiosum* in the Southwestern United States and in Mexico. *Ann. Entomol. Soc. Am.* 63:633–35
35. Pang, X.-F., Chen, T.-L. 1974. *Trichogramma* of China. *Acta Entomol. Sinica* 17 (4):441–54
36. Perkins, A. 1910. Supplement of Hymenoptera (Previously treated in Vol. 1). In *Fauna Hawaiiensis,* ed. D. Sharp, 2 (6):600–86
37. Perkins, A. 1912. Parasites of insects attacking sugar cane. *Entomol. Ser. Exp. Stn. Hawaiian Sugar Planters Assoc. Bull.* 10:27
38. Quednau, W. 1956. Die biologischen Kriterien zur Unterscheidung von *Trichogramma*-Arten. *Z. Pflanzenke. Pflanzenschutz* 63 (6):333–44
39. Quednau, W. 1956. Der Wert des physiologischen Experiments für die Artsystematik von *Trichogramma. Ber. Hundertjahrefeier Dtsch. Entomol. Ges. Berlin, Sept. 30–Oct. 5, 1956, Akad.—Verlag, Berlin,* pp. 87–92
40. Quednau, W. 1960. Über die Identität der *Trichogramma* arten und einiger ihrer Ökotypen. *Mitt. Biol. Bundesanst. Land Fortwirtsch. Berlin-Dahlem* 100: 11–50
41. Risbec, J. 1955. Mymaridae et Trichogrammatidae des malagaches. *Bull. Mus. natl. Hist. Paris Ser. 2* 27 (4): 311–13
42. Risbec, J. 1957. I. Chalcidoides, D'A.O.F. II. Les Microgasterinae D'A.O.F. *Mem. Inst. Fr. Afr. N.* 13: 397–99
43. Riley, C. V. 1871. Third annual report on the noxious, beneficial and other insects of the State of Missouri. *Ann. Rep. Mo. Board Agric.* 3:57–58
44. Riley, C. V. 1879. Parasites of the cotton worm. *Can. Ent.* 11:161–62
45. Salt, G. 1937. The egg parasite of *Sialis lutaria:* a study of the influence of the host upon a dimorphic parasite. *Parasitology* 29:539–53
46. Stern, V. M., Bowen, W. 1963. Ecological studies of *Trichogramma semifumatum,* with notes on *Apanteles medicaginis,* and their suppression of *Colias eurytheme* in southern California. *Ann. Entomol. Soc. Am.* 56:358–72
47. Telenga, N. A. 1958. Biologische Schädlings Bekämpfung an den landwirtschaftlichen Kulturen und Forstpflanzen in den USSR. *Int. Konf. Quarantāne, 9th. Pflzanzenkr. Pflanzenschutz, Moskau,* pp. 3–17.
48. Telenga, N. A. 1959. Taxonomic and ecological characteristics of species from the genus *Trichogramma. Trans. Int. Conf. Insect Pathol. Biol. Contr. 1st, Prague, 1958,* pp. 355–59
49. Tseng, S. 1965. On the identification of *Trichogramma* insects. *Acta Entomol. Sinica* 14 (4):404–8
50. Viggiani, G. 1968. The significance of the male genitalia in the Trichogrammatidae. *Proc. Int. Congr. Entomol., 13th* 1:313–15
51. Viggiani, G. 1971. Ricerche sugli Hymenoptera Chalcidoidea XXVIII. Studie morfologico comparativo dell'armature genitale esterna maschile dei Trichogrammatidae. *Boll. Lab. Entomol. Agraria "Filippo Silvestri" Portici* 29:181–222
52. Viggiani, G. 1972. Ricerche sugli Hymenoptera Chalcidoidea XXXVI. Nuovi trichogrammatidi africani. *Boll. Lab. Entomol. Agraria "Filippo Silvestri" Portici* 30:156–64
53. Voegele, J., Cals-usciati, J., Pihan, J. P., Daumal, J. 1975. Structure de l'antenne femelle des trichogrammes. *Entomophaga* 20 (2):161–69
54. Westwood, J. O. 1833. Descriptions of several new British forms amongst the parasitic hymenopterous insects. *Phil. Mag. J. Sci.* 2:444
55. Wirth, W. W., Marston, N. 1968. A method for mounting small insects on microscope slides in Canada Balsam. *Ann. Entomol. Soc. Am.* 61 (3):783–84

Ann. Rev. Entomol. 1977. 22:177–92

PESTICIDES AND POLLINATORS[1]

❖6126

Carl A. Johansen
Department of Entomology, Washington State University, Pullman, Washington 99163

Bee poisoning, the accidental killing of bees through the use of insecticides, first became a problem in the United States during the 1870s. The problems remained restricted and localized until the surge in modern agricultural methods following World War II. Highly effective synthetic organic insecticides became feasible for use not only on all types of crops but also on tremendous acreages of rangelands and forests. Development of efficient herbicides led to severe reduction in bee forage plants on both cultivated and wild lands. This not only involves direct removal, such as elimination of sweetclover from wheat fields, but also treatments which cause displacement of bee plants by herbicide-tolerant, non-bee plants, such as dalmatian toadflax, yarrow, and nightshade. There can also be a conflict of interests when noxious weeds—e. g. yellow starthistle, Russian knapweed, hounds-tongue, and Canada thistle—are excellent pollen and nectar sources. Ever-increasing monocultures of crops and the associated removal of fencerows and wild strips have reduced both bee forage and nest sites for wild bees. These developments have caused greater dependency upon honey bees for pollination of crops and simultaneously forced beekeepers to pasture their bees on insecticide-treated fields.

Increasing the acreages of crops has a definite effect on pollination requirements. Wild bees and hobbyist colonies of honey bees normally provide for good yields on 10 acres of raspberries, but a grower with 50 acres usually requires pollination service from a beekeeper. Development of hybrid varieties often increases the need for pollination. Hybrid onion and hybrid cabbage seed production pose special problems in transferring pollen from male-fertile rows to male-sterile rows or from compatible pollen rows to self-incompatible rows. Intensive plantings of dwarf fruit trees normally require two colonies of honey bees per acre, compared to one per acre in standard orchards. Many trends in modern agriculture have led to a vicious circle

[1]Survey of the literature pertaining to this review mainly covers the period 1967 to 1975 and follows the comprehensive article by Anderson & Atkins (2).

of increasing pollination requirements and increasing dependency of beekeepers on insecticide-contaminated bee pasturage. Bee poisoning has become the number one problem for beekeepers on a worldwide scale.

SYMPTOMS

The most common symptom of bee poisoning is the appearance of excessive numbers of dead bees in front of the hives. Use of Todd dead bee traps (11) on honey bee colonies has shown that up to 100 dead bees per day is a normal die-off, 200–400 is a low kill, 500–900 is a moderate kill, and 1000 or more is a high kill. Aggressiveness in bees may be caused by such materials as lindane and organophosphorus compounds. Stupefaction, paralysis, abnormal jerky or rapid movements, and spinning on the back are commonly caused by DDT, other organochlorine materials, and organophosphorus compounds. Bees are often observed performing abnormal communication dances on the horizontal landing board outside the hive while under the influence of chemical poisoning. Schricker and Stephen (82–84) showed that sublethal doses of parathion cause mistakes in communicating distance and direction of feeding sites and in time-sense.

Many bees poisoned with carbaryl or dieldrin slow down and appear as though they had been chilled; such bees may take two to three days to die. Beekeepers familiar with carbaryl poisoning quickly learn to recognize the crawlers that move about in front of the hive but are unable to fly. These are older foragers which have contacted carbaryl residues in the field. Dead and dying newly emerged worker bees are the result of feeding on carbyl-contaminated pollen (64, 87). Such pollen can remain toxic to bees after storage in the combs for up to eight months (43). Dead brood inside or in front of the hive is typical of both carbaryl and arsenical poisoning. Brood is mainly killed by chilling or starvation due to the lack of hive bees for proper care.

Queens may be affected by contaminated pollen and behave abnormally; for example, they might lay eggs in a poor pattern. Colonies may become extremely agitated by the return of foragers that have become contaminated in the field. Workers can carry lethal concentrations of insecticides back to the hive in their honey stomachs. Queens are often superceded. We suspect this is caused by reduction in secretion of queen substance, since supercedure and presence of queen substance are closely linked (22). Queenlessness has been associated with the use of a wide variety of insecticides, including arsenicals, dieldrin, carbaryl, malathion, and parathion. Severely weakened and queenless colonies do not survive the following winter.

FORMULATION

Dust formulations of pesticides usually are more hazardous to bees than sprays. Wettable powders often have a longer residual effect than emulsifiable concentrates. This typical sequence probably results because of differential pickup of toxic residues by the bee (48).

Table 1 DDT residues and mortality to honey bees[a]

	Alfalfa foliage	Bees	Mortality
50% wettable powder	133 ppm	33 ppm	74%
2 lb gal^{-1} emulsifiable concentrate	211 ppm	11 ppm	12%

[a] 20 hours after spraying.

More than 25 years ago, Pradhan (81) showed that pickup is a major factor in the surface contact action of insecticides. Pickup may be associated with the branched or otherwise modified body hairs of bees, which are adapted for holding minute pollen grains. Differential sorption into plant tissues does not appear to be a major factor in pickup. Analysis of insecticide residues in surface washings, wax strippings, and tissue extracts of foliage treated with various formulations showed no significant differences in profiles of sorption (54). We had hoped microencapsulated methyl parathion might be less hazardous to bees, but found it to be more hazardous than the emulsifiable concentrate. Since the polymeric microcapsules are about the same size (30–50μ) as pollen grains, we believe they adhere readily to the bees. Currently (1976), microencapsulated methyl parathion is implicated in a number of severe bee poisoning problems involving kill of honey bee brood and newly emerged workers. This clearly indicates contamination of the pollen being consumed inside the hive. Granule formulations tend to be quite low in hazard to bees, as might be expected (32, 44).

Addition of solvents and oily substances to spray materials tends to make them safer to bees. For example, the 2 lb gal^{-1} demeton formulation, which contains about seven times as much xylene as the 6 lb gal^{-1} formulation, is safer for this reason (40). Acidifiers which increase the effectiveness of trichlorfon against pest insects do not increase the hazard to bees, except at excessive rates (46).

Both Standifer (86) and Moffett & Morton (66) found that low dosages of surfactants were not toxic to bees. However, Moffett & Morton showed that a nonionic surfactant, in low concentration such as 25 ppm in small ponds and puddles, caused extensive drowning of bees collecting water. They also found that several surfactants were quite repellent to bees when added to pond water at a concentration of 500 ppm (67). Foam additives that have been used to reduce the problem of spray drift do not increase the residual hazard to bees of insecticides tested.

SELECTIVITY

Many insecticides which are highly toxic to warm-blooded animals are also highly toxic to bees. However, there are many exceptions, such as malathion and carbaryl. Endrin is an outstanding insecticide for control of major pests on bee-pollinated crops, and it presents a short residual hazard to honey bees. Unfortunately, it has been necessary to severely restrict the use of this chemical on food and forage crops because of its hazard to other animals and man.

El-Aziz, Metcalf & Fukuto (30) showed that the probable source of extreme susceptibility to six carbamate insecticides was the very low titer of phenolase enzymes in the honey bee. In other insects, these enzymes reduce the effects of carbamates by detoxifying them. Carbamates are consistently more toxic to honey bees than to house flies (93).

Nazer et al (79) found that greater brain acetylcholinesterase (AChE) concentration in young honey bees was associated with greater tolerance to malathion, as compared to older bees. Investigators (19, 34) have demonstrated species differences in AChE specificity, and a reduction in AChE correlated with development of organophosphorus insecticide poisoning symptoms. Honey bee tolerance of isopropyl parathion has been well documented (18, 19, 29). Correctly timed applications of isopropyl parathion are relatively safe to honey bees, alkali bees, and alfalfa leafcutting bees (42). We investigated the unusually high tolerance of alfalfa leafcutting bees to trichlorfon. Only the relatively high pH of the body fluids as compared to that of honey bees correlated with this specificity (1).

Tahori, Sobel & Soller (89) studied the variability in tolerance to four insecticides in 18 colonies of honey bees. They concluded that selection based on existing polygenic variation could only provide minor increases in resistance and that useful developments would require major gene mutations.

FACTORS AFFECTING POISONING

No wide-scale bee poisoning catastrophes have occurred without the contamination of blooming plants by insecticidal chemicals. This provides an important direct approach for reducing the problem on short-blooming crops such as tree fruits. Insecticide applications simply are not recommended during the blooming period.

Residual action of a chemical is of paramount concern because it largely determines whether an insecticide can be safely used on a blooming crop. A material such as emulsifiable naled can be applied with relative safety in late evening because it has a short residual toxic effect on honey bees, even though the initial hazard at application time is high.

Temperature has a significant modifying effect on residual action. For example, Benedek (15) found low night temperatures greatly increased the residual toxicity of mevinphos to bees. Unusually cold nights following hot summer days cause condensation of copious dew on the foliage. Under these conditions, the residual action of insecticides is increased and many more bees may be killed the following day. Tests with an experimental insecticide in 1975 showed that an average 9°F lower mean temperature resulted in a four to five day increase (approximately 2X) in residual toxicity to alfalfa leafcutting bees. Regional differences in the hazard of a given pesticide to bees can often be explained in terms of differences in climate. For example, malathion often has a fumigant effect on bees in warm California, but does not in cooler Washington. Mevinphos normally has a short residual effect in California and can be used when bees are not in flight (3). Conversely, it sometimes continues to cause significant mortalities through one full day in Washington and cannot be safely applied to blooming crops (42).

Timing of applications is obviously related to the previous factors. Proper timing during late evening, night, or early morning provides relative safety to bees against short-residual chemicals. Honey bees have a specific time for foraging on corn, typically between 8 AM and noon each day. We found that 85% of the corn anthers are exserted from the male flowers between 4 AM and noon, following the lowest temperature and highest humidity period. As temperatures rise, the pores at the bottom of the double anther tubes open and the pollen begins to sift out. Bees enter the cornfields and actively collect the pollen as it begins to be shed. Thus, short-residual insecticides can be applied to corn between 1 PM and midnight with minimal hazard to bees.

Strength of the honey bee colony has a definite effect on toxicity. Populous colonies always suffer greater losses than weak colonies because greater numbers of foragers are exposed to the insecticidal residues.

Distance of colonies from treated fields is inversely proportional to the amount of mortality that occurs. Moffett & Stith (70) documented this effect in relation to applications of parathion on cotton in Arizona. Conversely, when there is a dearth of available pollen and nectar, bees may be severely poisoned by treatments applied at a distance from the apiary. Investigators (38, 58, 73) have noted such occurrences at distances up to three to four miles.

Lack of suitable pollen and nectar plants has severely aggravated bee poisoning problems in the United States since the late 1940s. We tried to develop bee forage preserves to counteract this trend but found it simply unfeasible (39). Todd & Reed found pollen gathering was reduced by insecticide treatments (90). Moeller (62) noted pollen availability on other crops reduced the poisoning problem on corn and (63) showed provision of pollen cakes prevented bee mortality in Wisconsin. Taber, Mills & Coe (88) used several management practices, including feeding pollen-soybean cakes and water to reduce the effects of poisoning.

Age of bees affects their tolerance to insecticides; newly emerged bees are most susceptible to DDT, dieldrin, and carbaryl, and older bees are most susceptible to malathion and methyl parathion (52, 60).

We have tested many two-material and three-material combinations of insecticides for their toxicity to honey bees and wild bees. In all cases the hazard is enhanced beyond what is expected from the individual materials used alone. Even such relatively innocuous materials as dicofol, tetradifon, and propargite cause a significant increase in hazard when added to mixtures.

Many factors cause differences in poisoning problems in different localities, such as the availability of pollen and nectar plants, the timing of major nectarflow in relation to pest control programs, the severity of pest infestations, the presence of crops requiring insect pollination, and the specific conditions of crop use and methods and materials of pest control.

TECHNIQUES

Laboratory methods of determining speed of action, residual contact toxicity, and residual fumigant toxicity of insecticides to honey bees have been devised by Clinch and co-workers (23, 24, 27). They note possible difficulties in interpreting results of field trials when fast-acting compounds are tested; for example, poisoned bees may not return to the hives. Clinch also notes direct contact toxicity tests are of little value as guides to residual hazard.

Beran (16, 17) classifies the hazard to honey bees of 200 pesticides as danger sum indexes, based on laboratory studies which correlate well with field tests. He also calculates a selectivity index from the topical mean lethal dose (LD_{50}) for the bee divided by the topical LD_{50} for the house fly. Atkins, Greywood & MacDonald (8) have developed an ingenious way to determine the expected kill from different dosages of an insecticide. They note the LD_{50} in micrograms per bee can usually be directly converted to the equivalent dosage in pounds per acre. Known LD_{50} values can be substituted in their prepared table to obtain the anticipated percent mortalities for an appropriate range of dosages. They emphasize that a few exceptions do not conform to this rule-of-thumb method.

Anderson et al (3) use a combination of colony strength, dead bees at the colony, bee visitation, and caged bees to assess the hazard of pesticides in the field. Clinch (26) prefers to collect bees from the treated crop with a vacuum device and to hold them for mortality determination. We have developed standardized procedures for quickly assessing the residual hazard of new chemicals (54). Bees are exposed in the laboratory to field-weathered residues on foliage samples taken from treated plots. Our comparative data indicate reasonable validity in transposing results obtained on alfalfa plots for use by orchardists. The dosage rate per acre on alfalfa is approximately equivalent to the amount per 100 gallons in a dilute spray application to fruit trees.

Table 2 presents a classification of selected insecticides currently in commercial use on crops in the United States. Categories are arranged in terms of correct timing of applications on blooming plants for minimal hazard to honey bees. In this way, the tabulation can be directly translated into preferred field usage (42).

BEE POISONING PROBLEMS

A variety of limited-scale problems have developed in recent years: several workers have shown honey bees and bumble bees are attracted to trap lures for Japanese beetles (21, 53). Vapors from dichlorvos slow-release strips are absorbed by beeswax and remain as a toxic hazard to honey bees for some time (12, 25). Nye (80) reported on bee losses associated with foragers collecting mineral feed meal containing ronnel. We found proprietary sugar-base fly spray materials containing trichlorfon or dichlorvos and ronnel were hazardous to bees. Our remedy was to recommend mixing the commercial material with molasses instead of more sugar. Glancey, Roberts & Spence (33) found mirex bait used for imported fire ant control was not hazardous to honey bees.

A number of workers have investigated the potential problem of nectar contamination by plant systemic insecticides. However, most have used high dosage rates and/or other unusual test conditions. Mizuta & Johansen (61) showed that a number of systemic materials were not hazardous to bees at standard field dosage rates. However, recent investigations have proven that aldicarb granules injected into the

Table 2 Poisoning hazard of selected insecticides to the honey bee

1. Hazardous at any time on blooming crops		
azinphosmethyl	dimethoate	mevinphos
carbaryl	fensulfothion	monocrotophos
carbofuran	malathion D, ULV	naled D
chlorpyrifos	methamidophos	parathion
diazinon	methidathion	phosmet
dicrotophos	methomyl D	phosphamidon
2. Minimal hazard, if applied during late evening on blooming crops		
malathion EC	naled WP	phorate EC
3. Minimal hazard, if applied during late evening, night, or early morning on blooming crops		
acephate	endosulfan	naled EC
carbophenothion	ethion	oxydemetonmethyl
chlordimeform D	formetanate	phosalone
demeton	leptophos	stirofos
dioxathion	methomyl LS	toxaphene
disulfoton EC	methoxychlor	trichlorfon
4. Minimal hazard on blooming crops		
carbofuran G	fensulfothion G	propoxur G
chlordimeform SP, EC	lime-sulfur	pyrethrum
cyhexatin	malathion G	rotenone
dicofol	oil sprays (superior)	ryania
dinocap	oxythioquinox	sulfur
disulfoton G	propargite	tetradifon

soil at 3 lb of active ingredient (ai) acre^{-1} can reduce reproduction by alfalfa leafcutting bee females. Conditions of dosage, timing, and irrigation required to eliminate this hazard are quite restrictive, so we no longer recommend the material for use on alfalfa grown for seed. Current studies (1976) show that dimethoate granules injected into the soil at 10 lb ai or more acre^{-1} kills alfalfa leafcutting bees through contamination of nectar.

Anderson et al (4) have studied the effects of smog on honey bees. They found that high levels of ozone (1–5 ppm) were toxic but noted that plants are more sensitive and would be destroyed before the bees would be injured. Atkins, Anderson & Greywood (6) reported that fluorine at 4–5 ppb was somewhat toxic to worker bees. Again, many plants tested were more seriously affected than the bees.

Gunther & Dallman (36) found arsenic oxide industrial pollutants could be dangerous to bees up to 7.45 miles downwind from the site of production. Muller & Worseck (77) noted that arsenic was the most damaging industrial discharge to bees from 1956 to 1966, but that fluorine was more important from 1967 to 1969. They (78) found that industrial effluents containing cyanide also affected bees.

Tong et al (91) investigated the presence of 47 elements in honey samples produced near highway, industrial, and mining areas. They concluded honey from polluted areas may acquire certain indicator elements.

LARGE-SCALE INSECT CONTROL OPERATIONS

Grasshopper control (GC) programs in the West often involve the treatment of millions of acres during a single season. By 1965, the federal Animal and Plant Health Inspection Service (APHIS) had established ultra-low-volume (ULV) malathion sprays as the standard method of control. Although water-diluted malathion sprays at 1 lb ai acre^{-1} are relatively low in hazard to bees, ULV applications of 8 fl oz (0.6 lb ai) acre^{-1} proved to be highly hazardous (50, 58). Currently, bee poisoning problems are minimized by an organized, cooperative program involving APHIS, state personnel, and beekeepers. Months before GC treatments are to be applied, beekeepers are alerted as to specific areas where APHIS expects to conduct campaigns during the coming season. As grasshopper populations develop and control details are finalized, beekeepers operating in localities to be sprayed are again notified concerning details of time and place. In this way, 1.08 million acres of rangeland were treated in Washington in 1973 without significant bee losses.

Mosquito abatement (MA) operations involving thousands or even millions of acres have been conducted in the United States for 30 years. Atkins (5) and Womeldorf, Atkins & Gillies (95) showed that ULV sprays of chlorpyrifos, fenthion, dichlorvos, and propoxur at 0.05–0.1 lb ai acre^{-1} are not likely to cause significant kills if applied when bees are not actively foraging. Six additional materials used for mosquito control would likewise cause negligible problems. We tested propoxur, fenthion, and chlorpyrifos at mosquito abatement rates and found that all lost their bee hazard within a few hours. One of the common causes of honey bee losses in MA programs is use of incorrect dosage. If ULV malathion is used at the GC rate of 8 fl oz acre^{-1} instead of at the MA rate of 3 fl oz, considerable bee poisoning may occur.

Thirteen and one-half million acres of land along the Gulf Coast of Louisiana and Texas were treated for control of the mosquito *Culex tarsalis* during an epidemic of Venezuelan equine encephalitis (VEE) in 1971. Three fluid ounces of ULV malathion per acre were applied by airplane between sunrise and 10 AM on each treatment date. Minimal bee kills were recorded during this entire campaign. Greater problems have occurred in smaller municipal operations against St. Louis encephalitis (SLE) in the Southwest. Such MA projects have often involved daytime applications and more hazardous materials such as naled. Colburn & Langford (28) found mist sprays of naled, pyrethrum, and malathion combinations at 2.4–10 lb ai 100 gal^{-1} used as mosquito adulticides were quite hazardous to bees.

Large-scale forest defoliator control programs have been conducted periodically in the United States since 1947. Early programs involved DDT without hazard to bees. Carbaryl first caused widespread losses of honey bees in New York gypsy moth control programs in 1959 (72). More recently, fenitrothion has been used against spruce budworm in the East and carbaryl against hemlock looper in the West. DDT was used one last time for Douglas fir tussock moth control on 460,000 acres in the Pacific Northwest in 1974. The most promising newer material for future defoliator control is Dimilin® (diflubenzuron, a urea compound), which apparently is low in hazard to bees. Dimilin fed to honey bee colonies in sugar syrup can cause supercedure of the queen, death of larvae and production of undersized workers. However, our 1976 studies have shown no adverse effects when bees collected pollen and nectar from plants treated in large-scale field trials. Low-hazard, viral and bacterial, microbial insecticides are also being tested for tussock moth control. Several studies have indicated that pollen traps reduce bee poisoning by removing contaminated pollen from foragers returning to the hives. However, Martin (59) found this method did not prevent bee kills from gypsy moth sprays in Michigan.

Simmons & Wilson (85) observed twice as much killing of honey bees in areas with normal boll weevil control programs as compared to areas following an eradication program. However, overwintering losses were much greater in the eradication areas because repeated applications of insecticides from early July through mid-October resulted in continuous low mortality of worker bees and minimal storage of honey.

Integrated pest management programs can be expected to reduce pesticide damage to pollinators. Both use of more selective chemicals and use of lower amounts of chemicals per acre will help diminish the problem. Our pest management programs on alfalfa grown for seed were developed through a search for insecticides that would be low in hazard to the alfalfa leafcutting bee (47, 49). In the process, we were successful in establishing programs with minimal hazard to beneficial predators and parasites.

MICROBIAL INSECTICIDES

Cantwell, Lehnert & Fowler (20) reviewed the effects of microbial insecticides on honey bees. Insect viruses, both polyhedrosis and granulosis types, seem to be completely harmless to both bee larvae and adults. Spores and crystals of the bacterium *Bacillus thuringiensis* are harmless also; its exotoxin would have to be present in abnormally high amounts to cause mortality. Four species of microsporidans that normally attack other insects were not infective when fed to adult honey bees. Eleven exotoxins of *B. thuringiensis* only shortened honey bee life at extreme dosage rates in tests by Haragsim & Vankova (37). They concluded this microbe was not one of the causative agents of bee paralysis during nectarflows.

HERBICIDES AND FUNGICIDES

King (51) noted that elimination of valuable nectar and pollen plants was the major effect of herbicides on honey bees. Direct poisoning of wild bee pollinators of alfalfa by pesticides was less important in Hungary than was previously assumed, according to Benedek (14). He stated that food shortage (especially of nectar) caused by mechanical and chemical weed control has a massive impact on bees and that weed killers are probably not justified on wastelands, roadsides, or railway embankments.

Herbicides have generally been assumed to be low in hazard to bees, with a few exceptions such as the arsenicals, dinoseb, and endothal. However, beekeepers continue to ask questions about the toxicity of various weed-killer treatments. J. O. Moffett, H. L. Morton, and other researchers at the Tucson, Arizona, laboratory attempted to answer some of these questions during the period from 1969 to 1974. The most toxic materials tested were paraquat and the organic arsenicals MAA, MSMA, DSMA, hexaflurate, and cacodylic acid. Phenoxy materials (2,4-D, 2,4, 5-T) were relatively nontoxic, except at unusually high dosages in feeding studies (68, 69, 74–76).

However, phenoxy herbicides may exert adverse effects on bees in the field that are not detected in most bee poisoning experiments (2). It has been suggested that since phytohormones alter plant secretions and metabolism, a change in nectar composition of plants sprayed with hormone weed killers would be expected (R. J. Barker, personal communication). Barker further notes that normal plant sap sometimes contains galactosides toxic to honey bees.

Use of blossom thinners, dinitrocresol, Amid-Thin® (naphthalene acetamide), NAA (naphthaleneacetic acid), Alar®, and ethephon in orchards has not been a hazard to honey bees. Use of xylene, diquat, and Acrolein for control of aquatic weeds in irrigation ditches and canals has not been a problem with respect to bees collecting contaminated water (55). Even excessive dosages (up to 1200 lb $acre^{-1}$) of soil sterilants, usually containing mixtures of sodium borates and sodium chlorate, have not been hazardous to either adults or larvae of the alkali bee in soil nesting sites. Desiccants applied to alfalfa seed crops in Washington or cotton in Arizona (65), usually formulated with dinitrocresol, endothal, or sodium chlorate, have not caused bee poisoning problems. No fungicides tested to date have proven hazardous to honey bees under field conditions.

REPELLENTS

Use of repellents to reduce the hazard of insecticide sprays to honey bees has been investigated since the early 1900s. Recently, E. L. Atkins at Riverside, California, has led an intensive search to find an effective material to deter bees from treated fields. Laboratory screening tests of 143 chemicals showed compounds containing nitrogen, short sidechain-substituted phenyl acetates, and tolyl derivatives had greatest promise (10). Field tests during 1971 to 1975 indicated that several compounds should be studied further (7, 9). Aromatic five-, six-, and seven-membered ring structures containing nitrogen in the ring, straight-chain amides, and phenyl ring structures with short-chain-length amide substitutions were among the most repellent compounds tested.

CHEMICAL IMMUNIZERS

Barker (13) tested atropine sulfate and several other materials as antidotes for bees against carbaryl, parathion, and monocrotophos poisoning. None reduced the mortality rate, regardless of method of application or whether administered before or after treatment. Moradeshaghi, Brindley & Youssef (71) and Lee & Brindley (56) tested several enzyme-inducing materials, including chlorcyclizine, for protection of alkali bees from parathion and alfalfa leafcutting bees from carbaryl. None of the drugs significantly decreased the susceptibility of the bees to the insecticides.

WILD BEES

Many insecticides have been tested against the alfalfa leafcutting bee, *Megachile pacifica,* and the alkali bee, *Nomia melanderi,* and a few against bumble bees, *Bombus* spp. Torchio (92) found that the leafcutting bee had the highest tolerance to most of the insecticides he tested with a topical drop technique in the laboratory. However, the reverse appears to be true in the field, since the leafcutting bee sustains much greater mortalities from exposure to most insecticides (45). We noted that the typical trend in susceptibility is directly correlated to the surface/volume ratios of the wild bee species (41). The major factor is obviously pickup, with the susceptibility sequence: alfalfa leafcutting bee>alkali bee>honey bee>bumble bees.

Waller (94) reported that several insecticides with long residual toxicity were hazardous to the alfalfa leafcutting bee by leaf-piece contamination. Chemicals normally applied to alfalfa seed crops were not hazardous in this way, according to our studies (31). Several investigators (35, 94) have shown carbamate materials to be much less toxic than organophosphorous materials to either larvae or adults of the alfalfa leafcutting bee. However, of the compounds tested (carbaryl, propoxur, Landrin®, methomyl, DDT, stirofos, monocrotophos, and azinphosmethyl), only stirofos EC can be applied at standard dosage rates to blooming alfalfa with relative safety to alfalfa leafcutting bees (42).

LOSSES

Carbaryl first became a severe problem in 1959 when it was used against certain orchard pests. As soon as it was registered for use on other crops, it became even more devastating. During a peak problem year, 1967, it caused the destruction of an estimated 70,000 colonies of honey bees in California from use on cotton and an estimated 33,000 colonies in Washington from use on corn. The estimated national loss from all pesticide poisoning for the same year was 500,000 colonies. Carbaryl still ranks as the most destructive bee-killing chemical to emerge since the near-complete removal of the older arsenical materials.

Lesser (57) conducted a comprehensive economic survey of 30 commercial beekeepers of Washington. Nineteen of these apiarists were full-time operators owning a total of 28,000 colonies and providing 60% of the commercial pollination service within the state. A composite income statement revealed that these 19 operators suffered a $204,000 loss from insecticide poisoning in 1967. They sustained a 3.2% loss on investment instead of the 11.2% gain that would have accrued without the adverse effects of insecticides.

Misuse of diazinon for aphid control on alfalfa hay fields in partial bloom near Touchet, Washington, in 1973 caused a 95% reduction in alkali bee larvae ft^{-3} in three nearby soil nesting sites. Losses in potential seed production and in pollinators totaled $287,000, and the alkali bees had only regained 25% of their initial populations by the end of 1975.

Immediate monetary losses caused by bee poisoning are considerable, but the long-term losses in yield of insect-pollinated crops can be even greater. This is particularly true of crops that are dependent upon wild bee pollinators; these may require three years or more to return to original population levels or may be eliminated from a locality indefinitely. Beekeepers often are unable to continue to supply strong colonies of honey bees for pollination service after suffering chronic poisoning losses over a period of years.

INDEMNITY

The federal Bee Indemnity Program was established in 1970 to help insure a continuing supply of honey bees for pollination service. Beekeepers who through no fault of their own suffered losses after January 1, 1967, stemming from the use of economic poisons registered and approved by the federal government, could apply for and receive indemnity payments. Total payments to beekeepers throughout the nation for the period 1967–1974 was $18 million. This legislation has been extended twice and is currently in effect through 1977.

The main objective of this program has been achieved: it has curtailed disintegration of the beekeeping industry. Fewer beekeepers are being driven out of business by poisoning losses, and most have been able to replace worn-out equipment and increase both quantity and quality of their honey bee stocks.

CONCLUSION

Modern agricultural methods have aggravated bee poisoning and pollination problems considerably. Herbicide destruction of bee forage plants is a major factor in this situation.

Removal of persistent organochlorine insecticides from agricultural crop uses by the Environmental Protection Agency often forces substitution of organophosphorus and carbamate materials, resulting in much greater hazard to bees. Formerly, properly timed treatments of DDT and endrin could be applied to blooming fruit or seed crops with relative safety to bees. Current nonpersistent sprays also allow development of various secondary pest problems, which often must be controlled during the blooming period. Development of longer persistence in organophosphorus compounds can cause additional problems. Formulations such as ULV malathion and microencapsulated methyl parathion are much more hazardous than the standard formulations. Bee losses associated with many crop pest control programs have greatly increased for these reasons.

Development of selective insecticides, effective bee repellents, and integrated control programs can help alleviate the problems. Formulations of a given insecticide can vary considerably in their hazard to bees because of differences in pickup. Therefore, choice of correct formulation as well as choice of material and timing can often eliminate bee poisoning conditions. Educational programs help by showing the beekeeper, the farmer, and the applicator how to avoid or minimize problems. Federal bee indemnity payments, although not a solution to the difficulties have aided commercial beekeepers in recuperating from severe losses and enabled them to continue essential pollination services.

Literature Cited

1. Ahmad, Z., Johansen, C. 1973. Selective toxicity of carbophenothion and trichlorfon to the honey bee and the alfalfa leafcutting bee. *Environ. Entomol.* 2:27–30
2. Anderson, L. D., Atkins, E. L. 1968. Pesticide usage in relation to beekeeping. *Annu. Rev. Entomol.* 13:213–38
3. Anderson, L. D., Atkins, E. L., Nakakihara, H., Greywood, E. A. 1971. Toxicity of pesticides and other agricultural chemicals to honey bees field study. *Calif. Agric. Ext. Serv. AXT-251.* 8 pp.
4. Anderson, L. D., Atkins, E. L., Todd, F. E., Levin, M. D. 1968. Research on the effect of pesticides on honey bees 1966–67. *Am. Bee J.* 108:277–79
5. Atkins, E. L. 1972. Rice field mosquito control studies with low volume Dursban sprays in Colusa county, California. V. Effects on honey bees. *Mosq. News* 32:538–41
6. Atkins, E. L., Anderson, L. D., Greywood, E. A. 1970. Research on the effect of pesticides on honey bees 1968–69. part 1. *Am. Bee J.* 110:387–89
7. Atkins, E. L., Kellum, D., Neuman, K. J., Ferguson, D. T. 1975. Repellent additives to reduce pesticide hazards to honey bees. *Ann. Rep. Proj. 3565-RR, Univ. Calif., Riverside.* 18 pp.
8. Atkins, E. L., Greywood, E. A., Macdonald, R. L. 1973. Toxicity of pesticides and other agricultural chemicals to honey bees laboratory studies. *Calif. Agric. Ext. Serv. M-16.* 38 pp.
9. Atkins, E. L., Macdonald, R. L., Greywood-Hale, E.A. 1975. Repellent additives to reduce pesticide hazards to honey bees: field tests. *Environ. Entomol.* 4:207–10
10. Atkins, E. L., Macdonald, R. L., McGovern, T. P., Beroza, M., Greywood-Hale, E. A. 1975. Repellent additives to reduce pesticide hazards to honey bees: laboratory testing. *J. Apic. Res.* 14:85–97
11. Atkins, E. L., Todd, F. E., Anderson, L. D. 1970. Honey bee field research aided by Todd dead bee hive entrance trap. *Calif. Agric.* 24(10):12–13
12. Baribeau, M. F. 1969. Warning against using Vapona strips in the honey house. *Am. Bee J.* 109:187
13. Barker, R. J. 1970. Cholinesterase reactivators tested as antidotes for use on poisoned honey bees. *J. Econ. Entomol.* 63:1831–34
14. Benedek, P. 1972. Possible indirect effect of weed control on population changes of wild bees pollinating lucerne. *Acta Phytopathol. Acad. Sci. Hung.* 7:267–78
15. Benedek, P. 1975. Effect of night temperature on the toxicity of field-weathered mevinphos residues to honeybees. *Z. Angew. Entomol.* 79:328–31
16. Beran, F. 1968. Bienen und Pflanzenschutz. *Besseres Obst* 13:36–37
17. Beran, F. 1969. Zur Kenntnis der Bienentoxizität chemischer Pflanzenschutzmittel. *Sonderdruck Vierjahresbericht Bundesanstalt Pflanzenschutz* 1966–1969: 215–53
18. Camp, H. B., Fukuto, T. R., Metcalf, R. L. 1969. Selective toxicity of isopropyl parathion: effect of structure on toxicity and anticholinesterase activity. *J. Agric. Food Chem.* 17:243–48
19. Camp, H. B., Fukuto, T. R., Metcalf, R. L. 1969. Selective toxicity of isopropyl parathion: metabolism in the house fly, honey bee, and white mouse. *J. Agric. Food Chem.* 17:249–54
20. Cantwell, G. E., Lehnert, T., Fowler, J. 1972. Are biological insecticides harmful to the honey bee? *Am. Bee J.* 112:255–58, 294–96
21. Caron, D. M., Morse, R. A. 1972. Attraction of Japanese beetle traps to honey bees, bumble bees, and other Apoidea. *Environ. Entomol.* 1:272–74
22. Chaudhry, M., Johansen, C. A. 1971. Management practices affecting efficiency of the honey bee, *Apis mellifera. Melanderia* 6:1–31
23. Clinch, P. G. 1967. The residual contact toxicity to honey bees of insecticides sprayed on to white clover (*Trifolium*

repens L.) in the laboratory. *NZ J. Agric. Res.* 10:289–300
24. Clinch, P. G. 1969. Laboratory determination of the residual fumigant toxicity to honey bees of insecticide sprays on white clover (*Trifolium repens* L.) *NZ J. Agric. Res.* 12:162–70
25. Clinch, P. G. 1970. Effect on honey bees of combs exposed to vapour from dichlorvos slow-release strips. *NZ J. Agric. Res.* 13:448–52
26. Clinch, P. G. 1971. A battery-operated vacuum device for collecting insects unharmed. *NZ Entomol.* 5:28–30
27. Clinch, P. G., Ross, J. G. M. 1970. Laboratory assessment of the speed of action on honey bees of orally dosed insecticides. *NZ J. Agric. Res.* 13:717–25
28. Colburn, R. B., Langford, G. S. 1970. Field evaluation of some mosquito adulticides with observations on toxicity to honey bees and house flies. *Mosq. News* 30:518–22
29. Dauterman, W. C., O'Brien, R. D. 1964. Cholinesterase variation as a factor in organophosphate selectivity in insects. *J. Agric. Food Chem.* 12:318–19
30. El-Aziz, S. A., Metcalf, R. L., Fukuto, T. R. 1969. Physiological factors influencing the toxicity of carbamate insecticides to insects. *J. Econ. Entomol.* 62:318–24
31. Eves, J. D., Johansen, C. A. 1974. Population dynamics of larvae of alfalfa leafcutting bee, *Megachile rotundata,* in eastern Washington. *Wash. Agric. Exp. Stn. Tech. Bull. 78.* 13 pp.
32. Free, J. B., Needham, P. H., Racey, P. A., Stevenson, J. H. 1967. The effect on honeybee mortality of applying insecticides as sprays or granules to flowering field beans. *J. Sci. Food Agric.* 18:133–38
33. Glancey, M., Roberts, W., Spence, J. 1970. Honey-bee populations exposed to bait containing mirex applied for control of imported fire ant. *Am. Bee J.* 110:314
34. Guilbault, G. G., Kuan, S. S., Sadar, M. H. 1970. Purification and properties of cholinesterases from honeybees - *Apis mellifera* Linnaeus—and boll weevils—*Anthonomus grandis* Boheman. *J. Agric. Food Chem.* 18:692–97
35. Guirguis, G. N., Brindley, W. A. 1974. Insecticide susceptibility and response to selected pollens of larval alfalfa leafcutting bees, *Megachile pacifica* (Panzer). *Environ. Entomol.* 3:691–94
36. Gunther, O., Dallman, H. 1968. Verbreitung und Wirkung von Rauchschaden, insbesondere der Arsenstaube auf die Honigbiene. *Garten Kleintierz. Imker* 7(4):10–11
37. Haragsim, O., Vankova, J. 1973. Effet pathogene des exotoxines de 11 varietes appartenant au groupe de *Bacillus thuringiensis* Berliner, sur l'abeille domestique. *Apidologie* 4:87–101
38. Johansen, C. 1959. The bee-poisoning hazard. *Wash. State Hort. Assoc. Proc.* 55:12–4
39. Johansen, C. A. 1969. Bee forage preserves. *Am. Bee J.* 109:96–97
40. Johansen, C. A. 1972. Spray additives for insecticidal selectivity to injurious vs. beneficial insects. *Environ. Entomol.* 1:51–54
41. Johansen, C. A. 1972. Toxicity of field-weathered insecticide residues to four kinds of bees. *Environ. Entomol.* 1: 393–94
42. Johansen, C. A. 1976. How to reduce poisoning of bees from pesticides. *Wash. Agric. Ext. Serv. EM 3473.* 6 pp.
43. Johansen, C. A., Brown, F. C. 1972. Toxicity of carbaryl-contaminated pollen collected by honey bees. *Environ. Entomol.* 1:385–86
44. Johansen, C. A., Eves, J. D. 1967. Systemic insecticides as lygus bug controls compatible with bee pollination on alfalfa. *J. Econ. Entomol.* 60:1690–96
45. Johansen, C., Eves, J. 1967. Toxicity of insecticides to the alkali bee and the alfalfa leafcutting bee. *Wash. Agric. Exp. Stn. Circ. 475.* 15 pp.
46. Johansen, C. A., Eves, J. D. 1972. Acidified sprays, pollinator safety and integrated pest control on alfalfa grown for seed. *J. Econ. Entomol.* 65:546–51
47. Johansen, C. A., Eves, J. D. 1973. Development of a pest management program on alfala grown for seed. *Environ. Entomol.* 2:515–17
48. Johansen, C. A., Kleinschmidt, M. G. 1972. Insecticide formulations and their toxicity to honeybees. *J. Apic. Res.* 11:59–62
49. Johansen, C. A., Klostermeyer, E. C., Retan, A.H., Madsen, R. R. 1975. Integrated pest management on alfalfa grown for seed. *Wash. Agric. Ext. Serv. EM 3755.* 11 pp.
50. Johansen, C. A., Levin, M. D., Eves, J. D., Forsyth, W. R., Busdicker, H. B., Jackson, D. S., Butler, L. I. 1965. Bee poisoning hazard of undiluted malathion applied to alfalfa in bloom. *Wash. Agric. Exp. Stn. Circ.* 455. 10 pp.
51. King, C. C. 1961. *Effects of herbicides on honey bees and nectar secretion.* PhD

thesis. Ohio State Univ., Columbus, Ohio. 176 pp.
52. Ladas, A. 1972. Der Einfluss verschiedener Konstitutions und Umweltfaktoren aud die Anfalligkeit der Honigbiene (*Apis mellifica* L.) gegenüber zwei insektizidin Pflanzenschutzmitteln. *Apidologie* 3:55–78
53. Ladd, T. L., McGovern, T. P., Beroza, M. 1974. Attraction of bumble bees and honey bees to traps baited with lures for the Japanese beetle. *J. Econ. Entomol.* 67:307–8
54. Lagier, R. F., Johansen, C. A., Kleinschmidt, M. G., Butler, L. I., McDonough, L. M., Jackson, D. S. 1974. Adjuvants decrease insecticide hazard to honey bees. *Wash. Agric. Res. Cent. Bull. 801.* 7 pp.
55. Langridge, D. F. 1965. Effects of acrolein weedicide on honey bees. *J. Agric. (Victoria, Aust.)* 63:349–51
56. Lee, R. M., Brindley, W. A. 1974. Synergist ratios, EPN detoxication, lipid, and drug-induced changes in carbaryl toxicity in *Megachile pacifica. Environ. Entomol.* 3:899–907
57. Lesser, R. K. 1969. *An investigation of the elements of income from beekeeping in the state of Washington.* MBA thesis. *Gonzaga Univ.*, Spokane, Washington. 119 pp.
58. Levin, M. D., Forsyth, W. B., Fairbrother, G. L., Skinner, F. B. 1968. Impact on colonies of honey bees of ultra-low-volume (undiluted) malathion applied for control of grasshoppers. *J. Econ. Entomol.* 61:58–62
59. Martin, E. C. 1974. Pollen trapping did not prevent bee kill from gypsy moth spray. *Glean. Bee Cult.* 102:339–40
60. Mayland, P. G., Burkhardt, C. C. 1970. Honey bee mortality as related to insecticide-treated surfaces and bee age. *J. Econ. Entomol.* 63:1437–39
61. Mizuta, H. M., Johansen, C. A. 1972. The hazard of plant-systemic insecticides to nectar-collecting bees. *Wash. Agric. Exp. Stn. Tech. Bull. 72.* 8 pp.
62. Moeller, F. E. 1971. Effect of pollen availability on poisoning of honey bees by carbaryl applied to sweet corn. *J. Econ. Entomol.* 64:1314–15
63. Moeller, F. E. 1972. Honey bee collection of corn pollen reduced by feeding pollen in the hive. *Am. Bee J.* 112:210–12
64. Moffett, J. O., MacDonald, R. H., Levin, M. D. 1970. Toxicity of carbaryl-contaminated pollen to adult honeybees. *J. Econ. Entomol.* 63:475–76
65. Moffett, J. O., Morton, H. L. 1971. Toxicity of airplane applications of 2,4-D, 2,4,5-T, and a cotton desiccant to colonies of honey bees. *Am. Bee J.* 11: 382–83
66. Moffett, J. O., Morton, H. L. 1973. Surfactants in water drown honey bees. *Environ. Entomol.* 2:227–31
67. Moffett, J. O., Morton, H. L. 1975. Repellency of surfactants to honey bees. *Environ. Entomol.* 4:780–82
68. Moffett, J. O., Morton, H. L. 1975. How herbicides affect honey bees. *Am. Bee J.* 115:178–89, 200
69. Moffett, J. O., Morton, H. L., Macdonald, R. H. 1972. Toxicity of some herbicidal sprays to honey bees. *J. Econ. Entomol.* 65:32–36
70. Moffett, J. O., Stith, L. S. 1972. Bee losses from parathion decreased as distance from sprayed field increased. *Am. Bee J.* 112:174–75
71. Moradeshaghi, M. J., Brindley, W. A., Youssef, N. N. 1974. Chlorcyclizine and SKF 525A effects on parathion toxicity and midgut tissue structures in alkali bees, *Nomia melanderi. Environ. Entomol.* 3:455–63
72. Morse, R. A. 1972. Honey bees and gypsy moth control. *NY Fd. Life Sci.* 5(2):7–9
73. Morse, R. A., Gunnison, A. F. 1967. Honey bee insecticide loss—an unusual case. *J. Econ. Entomol.* 60:1196–98
74. Morton, H. L., Moffett, J. O. 1972. Ovicidal and larvicidal effects of certain herbicides on honey bees. *Environ. Entomol.* 1:611–14
75. Morton, H. L. Moffett, J. O., Macdonald, R. H. 1972. Toxicity of herbicides to newly emerged honey bees. *Environ. Entomol.* 1:102–4
76. Morton, H. L., Moffett, J. O., Martin, R. D. 1974. Influence of water treated artificially with herbicides on honey bee colonies. *Environ. Entomol.* 3:808–12
77. Muller, B., Worseck, M. 1970. Bienenschaden durch arsen- und fluorhaltige Industrieabgase. *Mh. Vet Med.* 25: 554–56
78. Muller, B., Worseck, M. 1970. Bienenschaden durch zyanidhaltige Abwasser. *Mh. Vet Med.* 25:557–58
79. Nazer, I. K., Archer, T. E., Gary, N. E., Marston, J. 1974. Honeybee pesticide mortality: intoxication versus acetylcholinesterase concentration. *J. Apic. Res.* 13:55–60
80. Nye, W. P. 1975. Effect of ronnel on the honey bee. *Am. Bee J.* 115:12–13

81. Pradhan, S. 1949. Studies on the toxicity of insecticide films. II. Effect of temperature on the toxicity of DDT films. *Bull. Entomol. Res.* 40:239–65
82. Schricker, B. 1974. Der Einfluss subletaler Dosen von Parathion (E 605) auf das Zeitgedächtnis der Honigbiene. *Apidologie* 5:385–98
83. Schricker, B. 1974. Der Einfluss subletaler Dosen von Parathion (E 605) auf die Entfernungsweisung bei der Honigbiene. *Apidologie* 5:149–75
84. Schricker, B., Stephen, W. P. 1970. The effect of sublethal doses of parathion on honeybee behavior. I. Oral administration and the communication dance. *J. Apic. Res.* 9:141–53
85. Simmons, C. L., Wilson, C. A. 1975. Studies to determine the effects of the boll weevil eradication experiment on the honey bee, *Apis mellifera* L. *Am. Bee J.* 115:356–72
86. Standifer, L. N. 1972. Toxicity of Triton X-100 to honey bees. *J. Econ. Entomol.* 65:306
87. Strang, G. E., Nowakowski, J., Morse, R. A. 1968. Further observations on the effect of carbaryl on honey bees. *J. Econ. Entomol.* 61:1103–4
88. Taber, S., Mills, J., Coe, E. 1974. Colonies of honey bees: survival in insecticide-treated Arizona cotton fields through colony management. *J. Econ. Entomol.* 67:41–43
89. Tahori, A. S., Sobel, Z., Soller, M. 1969. Variability in insecticide tolerance of eighteen honey-bee colonies. *Entomol. Exp. Appl.* 12:85–98
90. Todd, F. E., Reed, C. B. 1969. Pollen gathering of honey bees reduced by pesticide sprays. *J. Econ. Entomol.* 62:865–67
91. Tong, S. S. C., Morse, R. A., Bache, C. A., Lisk, D. J. 1975. Elemental analysis of honey as an indicator of pollution. *Arch. Environ. Health* 30:329–32
92. Torchio, P. F. 1973. Relative toxicity of insecticides to the honey bee, alkali bee, and alfalfa leafcutting bee. *J. Kans. Entomol. Soc.* 46:446–53
93. Vinopal, J. H., Johansen, C. A. 1967. Selective toxicity of four O-(methylcarbamoyl) oximes to the house fly and the honey bee. *J. Econ. Entomol.* 60:794–98
94. Waller, G. D. 1969. Susceptibility of an alfalfa leafcutting bee to residues of insecticides on foliage. *J. Econ. Entomol.* 62:189–92
95. Womeldorf, D. J., Atkins, E. L., Gillies, P. A. 1974. Honey bee hazards associated with some mosquito abatement aerial spray applications. *Calif. Vector Views* 21:51–55

Ann. Rev. Entomol. 1977. 22:193–218

SORGHUM ENTOMOLOGY[1,2,3] ❖6127

W. R. Young
The Rockefeller Foundation, G.P.O. Box 2453, Bangkok, Thailand

G. L. Teetes
Department of Entomology, Texas A&M University, College Station, Texas 77843

Sorghum, *Sorghum bicolor,* ranks fifth in acreage and production among the world's major cereal crops, following wheat, rice, corn, and barley. World production of sorghum grain is currently about 52 million metric tons produced on some 42 million hectares (40). Thought to have originated in eastern Africa, sorghum is presently grown on all six continents in a zone extending about 40° on either side of the equator. It is used directly as a food for humans and as feed for livestock in the form of grain and fodder. Management of the crop varies considerably from small subsistence plots to immense monocultures; the latter occur most commonly in New World, but are also found in parts of Africa, Asia, and Australia. In the Old World, as well as in parts of Central and South America, multipurpose varieties are commonly used, while grain sorghum hybrids dominate in North America, where they produce slightly more than 50% of the world's sorghum grain. The introduction of hybrids during the 1950s resulted in greatly expanded grain production for the livestock feedlot industry and a concomitant intensification of insect and mite pest problems and pesticide use. In Africa and Asia, where about three fourths of the world's sorghum acreage currently produces one third of the grain crop, insect and mite pests have taken their toll. In these areas the use of insecticides has generally not been economically feasible.

Insect pests continue to compete with humans for the sorghum crop, and knowledge of both old and new pests has accumulated at a faster rate in recent years as the crop has received increasing attention. It is the purpose of this paper to review the current situation regarding sorghum pests and approaches to their control. Pest management, in contrast to imbalanced pesticide use, is emphasized, and the interrelationships of sorghum pests with appropriate components for their management are

[1]The survey of literature pertaining to this review was concluded September 30, 1975.
[2]Paper No. 205, The Agricultural Journal Series, The Rockefeller Foundation.
[3]Approved as TA 12497 by Director, Texas Agriculture Experiment Station.

described where possible. The pest management approach is most advanced in North America where higher sorghum yields, economics, and advanced technology have allowed a more sophisticated approach. In Africa and Asia, much more information is needed on the biology, ecology, and economic injury levels of the various pests. Unique pest management systems are required, of course, for each agroecosystem.

Reviews dealing solely with sorghum entomology did not exist until the last decade. Several recent reviews and bibliographies on sorghum have included either references or a chapter on the subject (31, 44, 106, 112, 151, 160). Such reviews were relied upon heavily in preparing this article.

PESTS OF SORGHUM PRODUCTION

As with most crops, sorghum is usually attacked by only one or two key pests in each agroecosystem. *Key pests* are serious, perennially occurring, persistent species that dominate control practices; in the absence of deliberate human intervention, the pest populations commonly exceed the economic-injury level each year, often over wide areas. The sorghum midge, *Contarinia sorghicola,* shoot fly, *Atherigona soccata,* stem borer, *Chilo partellus,* and greenbug, *Schizaphis graminum,* are examples of key pests of sorghum in various geographic areas in different agroecocystems.

Secondary pests may be present in sorghum fields or surrounding areas; generally they occur in numbers below damaging levels. These pests increase to levels that exceed the economic injury level largely as a result of changes in cultural practices or crop varieties or because of injudicious use of insecticides applied for a key pest. Spider mites, *Oligonychus* spp., may be secondary pests in sorghum.

A third group, *occasional pests,* cause economic damage only in localized areas or at certain times. Such pests are usually under natural control and exceed the economic injury level only sporadically. Most pests of sorghum are occasional pests and include armyworms, *Spodoptera* spp., and ear head caterpillars, *Heliothis* spp., *Stenachroia elongella,* and others,

A summary of information on the more common sorghum pests worldwide is given in Table 1. It includes the pest's scientific name, geographic distribution, pest status (key, secondary, occasional), nature of damage, and economic threshold (if known) and a key to pertinent literature (except for those pests discussed in greater detail in the text). The more important pests covered in the text are presented generally according to the time or stage of crop growth when damage occurs.

SOIL PESTS

The underground parts of the sorghum plant are injured by white grubs, wireworms, and rootworms, causing loss in stand or lodging resulting from root pruning.

White Grubs

Species of the genus *Phyllophaga* are injurious to a wide range of agricultural crops in North America (82). Damage by the white grub, *Phyllophaga crinita,* to grain

Table 1 Most common insect and spider mite pests of sorghum

Common name	Scientific name	Geographical distribution[a]	Pest status[b]	Nature of damage	Economic threshold	References
Soil pests						
White grub	*Phyllophaga crinita*	NW	Occ	seedling death, stunting and/or lodging	prior to planting: mean of 1 grub per cubic foot	see text
White grub	*Schizonycha* sp.	AF	Occ	seedling death, stunting and/or lodging	—	(92)
White grub	*Holotrichia consanguinea*	AS	Occ	seedling death, stunting and/or lodging	—	(111)
Wireworms	several species of true and false wireworms (*Eleodes*, *Conoderus*, *Aeolus*)	NW	Occ	destroy planted seed	2 or more larvae per linear foot of row	(103)
Rootworms	*Diabrotica* spp.	NW	Occ	pruning of roots	—	(103)
Aphids						
Greenbug	*Schizaphis graminum*	C	Key	suck plant sap, inject toxin that kills leaves, virus vector, disease predisposer	seedling—visible damage with colonies on lower leaf surface; preboot—before any entire leaves are killed; headed—when numbers are sufficient to cause death of two lower normal-sized leaves	see text
Yellow sugarcane aphid	*Sipha flava*	NW	Occ	suck plant sap, inject toxin that kills leaves	emergence to preboot, at first sign of damage prior to stand loss	see text

Table 1 *(Continued)*

Common name	Scientific name	Geographical distribution[a]	Pest status[b]	Nature of damage	Economic threshold	References
Sugarcane aphid	*Aphis sacchari*	AF, AS	Occ	suck plant sap	—	see text
Corn leaf aphid	*Rhopalosiphum maidis*	C	Occ	suck plant sap, virus vector	seedling—visible damage prior to stand loss; near-harvest—prior to accumulation of honeydew in heads	see text
Shoot fly	*Antherigona soccata*	AF, AS	Key	injure growing point, causing dead heart	—	see text
Armyworms						
Fall armyworm	*Spodoptera frugiperda*	NW	Occ	leaf feeder in whorl or destruction of seed in head	headed—mean of 2 larvae per head	see text
Armyworm	*Mythimna separata*	AS	Occ	leaf feeder in whorl and on leaf margins	—	see text
Nutgrass armyworm	*Spodoptera exempta*	AF	Occ	leaf feeder in whorl and on leaf margins	—	see text
Beet armyworm	*Spodoptera exigua*	AF, NW	Occ	leaf feeder in whorl and on leaf margins	—	see text
Armyworm	*Pseudaletia convecta*	AF, O	Occ	leaf feeder in whorl and on leaf margins	—	see text
Stem borers						
Sorghum borer	*Chilo partellus*	AF, AS	Key	some leaf feeding; boring in stalk may cause stalk lodging	—	see text
Southwestern corn borer	*Diatraea* (= *Zeadiatraea*) *grandiosella*	NW	Occ	some leaf feeding; boring in stalk may cause stalk lodging	most stages—25% of plants infested with eggs or small larvae	(24, 45)

Table 1 (Continued)

Sugarcane borer	*Diatraea saccharalis*	NW	Occ	some leaf feeding; boring in stalk may cause stalk lodging	—	(104, 145)
Sorghum maize borer	*Eldana saccharina*	AF	Occ	some leaf feeding; boring in stalk may cause stalk lodging	—	(31, 55, 65, 87)
Maize stalk bòrer	*Busseola fusca*	AF	Occ	some leaf feeding; boring in stalk may cause stalk lodging	—	(31, 55 65, 87)
Sorghum borer	*Sesamia cretica*	EE	Occ	some leaf feeding; boring in stalk may cause stalk lodging	—	(1, 9)
Sugarcane rootstock weevil	*Anacentrinus deplanatus*	NW	Occ	boring in stalk above and below soil surface causes lodging	—	(48)
Corn planthopper	*Peregrinus maidis*	C	Occ	suck sap from leaves in plant whorl	—	(17, 107)
Chinch bug	*Blissus leucopterus*	NW	Occ	such sap from leaves and stems	seedling—1–2 adults per 5 plants; 6–18-inch-tall plants—75% of plants infested	(82)
Spider mites						
Banks grass mite	*Oligonychus pratensis*	NW	Sec	suck plant sap, causing discoloration and death of leaves	after heading, rapid population buildup away from leaf midrib	see text
Sorghum mite	*Oligonychus indicus*	AS	Sec	suck plant sap, causing discoloration and deatn of leaves	—	see text

Table 1 *(Continued)*

Common name	Scientific name	Geographical distribution[a]	Pest status[b]	Nature of damage	Economic threshold	References
Sorghum midge	*Contarinia sorghicola*	C	Key	destroy developing seed	when 25–30% of the heads have begun to bloom—mean of 1 adult per head	see text
Head caterpillars						
Sorghum webworm	*Celama sorghiella*	NW	Occ	destruction of seeds in head	headed—mean of 5 larvae per head	see text
Webworm	*Stenachroia elongella*	AS	Occ	destruction of seeds in head	—	see text
Webworms	*Eublemma* spp.	AS	Occ	destruction of seeds in head	—	(23)
Yellow peach moth	*Dichocrosis punctiferalis*	AS, O	Occ	destruction of seeds in head	—	(23, 89)
Head caterpillar	*Cryptoblabes adoceta*	O	Occ	destruction of seeds in head	—	(89)
Corn earworm	*Heliothis zea*	NW	Occ	destruction of seeds in head	headed—mean of 2 larvae per head	see text
American bollworm	*Heliothis armigera*	AF, AS, O	Occ	destruction of seeds in head	2 larvae per head	see text
Head bugs						
Jowar earhead bug	*Calocoris angustatus*	AS	Key	feed on developing seed, causing smaller, lighter, distorted seed	—	see text
Lygaeid bug	*Nysius raphanus*	NW	Occ	feed on developing seed, causing smaller, lighter distorted seed	headed—mean of 140 bugs $head^{-1}$	see text

Table 1 *(Continued)*

Stink bugs	Pentatomidae	C	Occ	feed on developing seed, causing smaller, lighter, distorted seed	headed—mean of 2 bugs $head^{-1}$	(20, 168)
Pyrrhocorid bug	*Dysdercus superstitiosus*	AF	Occ	feed on developing seed, causing smaller, lighter, distorted seed	—	(42)
Leaffooted bug	*Leptoglossus phyllopus*	NW	Occ	feed on developing seed, causing smaller, lighter, distorted seed	—	(103)
Stored grain pests						
Rice weevil	*Sitophilus oryzae*	C	Occ	consume whole grain in storage	—	see text
Maize weevil	*Sitophilus zeamais*	C	Occ	consume whole grain in storage	—	see text
Angoumois grain moth	*Sitotroga cerealella*	C	Occ	consume whole grain in storage	—	see text
Lesser grain borer	*Rhyzopertha dominica*	C	Occ	consume whole grain in storage	—	(154)
Indian meal moth	*Plodia interpunctella*	C	Occ	feed on cracked grain, a secondary feeder	—	(154)
Grain mite	*Acarus siro*	C	Occ	feed on cracked grain, a secondary feeder	—	(154)
Confused flour beetle	*Tribolium confusum*	C	Occ	feed on cracked grain, a secondary feeder	—	(154)

[a] Distribution key: C = cosmopolitan, AF = Africa, EE = Eastern Europe, NW = New World, AS = Asia, and O = Oceania.
[b] Occ = occasional pest, Sec = secondary pest, and Key = Key pest.

sorghum has increased in severity in some areas. Approximately 40,000 acres are damaged annually on the Texas High Plains.

Adults are brown to brownish-black, 13–19 mm long, and are commonly referred to as May or June beetles. Larvae are C-shaped with brown heads and white bodies. Digested food can be seen through the shiny and transparent tip of the abdomen.

Damage to grain sorghum may occur in several different ways. The most obvious damage, and perhaps the most significant, is death of seedling plants from larvae feeding on the roots. Seed germination may occur and a satisfactory stand may be established, but within a short period, when plants are 4–6 inches tall, seedlings begin to die. Stand loss can occur within one week to ten days in severely infested areas. One grub is able to destroy all plants within 1–2 feet of row. Plants that are not killed as seedlings are severely stunted and in many cases never produce seed. A third type of damage results from root pruning by overwintered as well as current season larvae. Injured plants, although able to produce seedheads after such damage, frequently do not have sufficient roots to prevent lodging. Occasionally lodging is increased by secondary stalk rot organisms.

Seasonal field data indicate delayed planting is a possible means to escape seedling damage by overwintering larvae (150). However, the crop remains susceptible to attack later in the season. These studies have also revealed that some individuals have a life cycle longer than one year, in contrast to the one-year life cycle reported by Reinhard (110). Light trap data have shown only one major peak in adult abundance and activity.

Damage assessment studies show that the economic injury level is two grubs per cubic foot and the economic threshold has been set at one grub per cubic foot (139). Because of the nature of the grub infestations, a preplant broadcast incorporated application of insecticide is required for control when the economic threshold is exceeded. Effective insecticides include diazinon, carbofuran and fensulfothion (139), but only diazinon is currently labeled.

Injury by other species of white grubs, *Schizonycha* sp. (92) in the Sudan and *Holotrichia consanguinea* in India (111), is similar to that caused by *P. crinita* in the United States. Although some information is available on the biology and habits of these species and soil insecticides have been evaluated for their control, a reasonable approach to their management in the future appears to be modification of cultural practices. Limited use of soil insecticides in heavily infested areas may also be practical for high-yielding hybrids.

SEEDLING PESTS

Sorghum seedlings are sporadically attacked by leaf-feeding beetles of the family Chrysomelidae. However, the sorghum shoot fly in Africa and Asia is by far the most important key seedling pest.

Shoot Fly

The muscid fly, *Atherigona soccata,* is presently considered to be the predominant shoot fly species attacking *Sorghum bicolor* in the Old World (26, 95). Although first described before the turn of the century, it was not recognized as a pest until the 1920s (98, 99). Extensive research in its control was not started until after World War II, concurrent with the intensified sorghum improvement programs in Africa and India. This research effort has been reported in the proceedings of an international symposium on the shoot fly held in Hyderabad, India, in 1971 (75).

The adult fly is gray in color, about 5 mm in length with three pairs of black spots on the dorsum of its pale pink abdomen. The fly is attracted to seedlings from first to seventh leaf stage and deposits rod-shaped white eggs singly on the underside of the leaves. The small maggot usually hatches within 48 hr and tunnels toward the base of the plant behind or through the leaf sheath, ultimately cutting through the growing point. A dead heart results above the cut and the yellow larva completes its development on the decaying plant tissue above the cut. Plants infested after the fifth leaf stage usually escape injury. Pupation may occur either in the plant or in the soil. Developmental time from egg to adult is reported to be 16–24 days. Small seedlings may be killed by the fly, while larger seedlings continue to produce tillers, which may in turn be attacked. Losses result from reduced stand and a reduction in tiller size. Delayed tillers may escape fly attack. Complete loss of the crop often occurs (8, 87, 134).

Approaches to control of the shoot fly have included a combination of cultural practices, systemic insecticides, and resistant varieties. Early control recommendations in India suggested the removal of dead-heart plants, but this practice is of little help under severe and continuous fly attack on susceptible varieties. It has long been known that crops that were sown early, at the beginning of the rainy season, escape fly attack (93, 155). Adjustment of the planting date in Israel has reduced shoot fly attack to below economically damaging levels (A. Blum, personal communication).

The systemic insecticides phorate, disulfoton, and carbofuran, applied as granules in seed furrows, have given effective control but have been too costly for general use (6, 124, 156). Some reduction in cost has been possible by coating seed with carbofuran. This treatment is presently being used in some areas of India (64, 74). Foliar sprays have not given effective control.

Differential reaction of sorghum varieties to fly attack was first reported by Ponnaiya (93, 94). In India, screening of the world collection of sorghum germ plasm has identified a number of varieties that also exhibited resistance to the fly in Nigeria, Uganda, Israel, and Thailand (177). The mechanism of resistance has been nonpreference for oviposition plus a low level of antibiosis (10, 32, 59). Isolated planting studies in both India (M. G. Jotwani, personal communication) and Thailand have confirmed that nonpreference for oviposition is maintained where flies have no choice of hosts.

Resistance through rapid tillering and recovery from fly attack also has been observed and used in sorghum improvement programs (32). Methods developed for fly rearing (126) and for increasing infestation in field plots (128) are helping to increase development of bioassay techniques to develop agronomically acceptable resistant varieties. At present, a combination of planting date adjustment, limited use of systemic insecticides and continued efforts to improve fly resistant varieties seem to be the most practicable approaches to shoot fly management.

FOLIAGE PESTS

Aphids and armyworms are the most important pests attacking sorghum foliage during the vegetative and flowering stages.

Aphids

Recent outbreaks of the greenbug, *Schizaphis graminum,* on sorghum in the United States have spurred a considerable renewed interest in this insect, which Rondani (119) first described in Italy in 1852. The species for years was primarily a pest of small grains, although earlier references indicate an association with sorghum (60). Since 1968, the insect has caused widespread damage to sorghum in the United States and has become well established on the crop. Harvey & Hackerott (57) designated the sorghum greenbug as biotype C. Piper sundangrass, *Sorghum sudanese,* is susceptible to biotype C but resistant to biotype B. Wood (172) separated the biotypes A and B by the reaction of Dickinson Sel 28A and C.I. 9058 wheats; both are resistant to biotype A but are susceptible to B. Biotype C has apparently largely replaced the previous biotypes and infests sorghums as well as small grains.

The greenbug is approximately 1.6 mm long, light green in color, with a darker green dorsal abdominal stripe. The distal leg segments and tips of the cornicles are black. Alate and apterous forms may be present in the same colony. Females produce living young parthenogenetically; under optimum conditions, the young begin reproduction in about 7 days, and produce about 80 offspring during a 25-day period (2).

Infestations are detectable by reddish spots on the leaves caused by toxins injected into the plant as aphids feed in colonies on the underside of leaves. The reddened areas enlarge as greenbug numbers increase. The leaf finally begins to die, turning brown from the outer edges toward the center (2). Greenbugs also transmit maize dwarf mosaic virus (MDMV) and may predispose sorghum to charcoal rot (22, 41, 146).

Greenbug is presently a key pest of sorghum in most areas of the United States where the crop is grown, especially the Great Plains. Small grains, primarily wheat, provide a winter host, and where the growing season of this crop does not overlap that of sorghum, grasses such as johnsongrass, *Sorghum halepense,* serve as interim hosts. The greenbug may be a pest during the seedling stage of sorghum, although often it does not reach damaging proportions until after heading. In either case, sorghum generally becomes infested soon after emergence. Spring rains and predators suppress the aphid's rate of increase, which has been recorded in the field to be as high as 20-fold per week, with an average of 5- to 6-fold per week throughout the season (11).

As is so often the case, high rates of persistent systemic insecticides were initially relied on as the sole controlling agent of the greenbug in sorghum. These treatments were highly effective, but at the same time they were broadly toxic and ecologically disruptive.

Several direct control tactics have been determined which are applicable to population management of the greenbug. Seasonal population profiles of the pest in sorghum in Texas have shown peak population levels to occur in late July to early August (11, 136). Abundance of natural enemies has shown a characteristic lag time of about 1–2 weeks. However, native aphid predators have not always held greenbug populations in check. The major seasonal mortality factor of the greenbug has been parasitism by *Lysiphlebus testaceipes,* which usually causes a rapid decline in greenbug population levels by mid-August (159). Consequently, only about a 2-week period exists in which greenbug numbers may exceed the damage tolerance level of sorghum.

Based on the characteristics of greenbug population dynamics, a system of integrated control was developed. Ecological selectivity was achieved by dosage rate manipulation of several approved organophosphorous insecticides (16). Extremely low dosage rates ($\leq$0.1 lb of actual insecticide per acre) provided greenbug control yet spared most of the naturally occurring beneficial species (137, 141). Treatments with selective insecticide rates had the effect of throwing the balance back in favor of natural control by preserving natural enemies. Such selective treatments are applied when the aphid population reaches the economic threshold (142, 143). When properly timed, only one insecticide application is required and resurgence of the greenbug is prevented by the natural control afforded by parasites and predators.

Recent developments of insecticide resistance by the greenbug greatly hamper the use of selective insecticide rates. Laboratory tests confirmed field observations of resistance development at levels of about 30-, 10-, 3-, and 2-fold for disulfoton, dimethoate, phorate, and parathion, respectively (91, 149).

Soon after 1968, sources of greenbug-resistant germ plasm were found in sorghums such as SA 7536-1, KS 30, IS 809, and PI 264453 (28, 121, 131, 132, 147, 148, 173, 174). Releases of resistant breeding materials have recently been made to commercial seed companies (50, 51, 72, 167). Laboratory and field experiments have identified the mechanisms of resistance in these sorghums as moderate levels of nonpreference and antibiosis. Antibiosis is expressed as an increase in the duration of the developmental stages and decreases in progeny per adult, adult longevity, and duration of the reproductive period. The primary resistance mechanism has been shown to be tolerance. Greenbug-resistant sorghum acts to reduce greenbug numbers as a result of nonpreference and antibiosis and to increase the economic injury level through tolerance (49, 56, 58, 130, 144). Resistant sorghums are complemented by greenbug mortality caused by natural enemies (123, 129).

Biological and ecological studies have been conducted on native and exotic species of the major predators and parasites of the greenbug (3, 21, 33, 39, 61, 66, 69, 76, 105, 122, 159, 166). Indigenous primary parasites include *Lysiphlebus testaceipes, Aphelinus nigritus,* and *Diaeretiella rapae* and secondary parasites include *Charips* sp., *Pachyneuron siphonophorae, Asaphes lucens, Aphidencyrtus aphidovorus,* and *Tetrastichus minutus.*

Because indigenous beneficial insects have not always held the greenbug below damaging levels, the introduction of exotic parasites such as *Aphelinus asychis, A. varipes, Ephedrus plagiator, Praon gallicum,* and *Aphidius avenae* and predators such as *Propylea 14-punctata* and *Menochilus sexmaculata* has been attempted (37).

Basic studies have led to biological and ecological knowledge and provided methods of mass production (4, 5, 67–69, 113–117).

Other aphids that sometimes infest grain sorghum include the corn leaf aphid, *Rhopalosiphum maidis,* and the yellow sugarcane aphid, *Sipha flava.* The corn leaf aphid is greenish-blue in color and generally feeds within the plant whorl. Larger plants can tolerate large numbers of this insect without suffering serious damage (53). However, large populations may damage seedling plants, causing plant death and stand loss. Large populations infesting plants during the booting stage may cause poor head exertion (2). Heavy plant head infestations prior to harvest have resulted in harvesting difficulties because of the sticky honeydew deposited by the insects. The corn leaf aphid is also a vector of MDMV.

Since 1968, the presence of the corn leaf aphid in sorghum has attracted increased attention as a result of confusion with the greenbug. The species is generally not considered a serious pest of sorghum, but since some producers apply insecticides to control it, it does become an important consideration in sorghum pest management.

Some commonly used sorghum parental lines are more susceptible to corn leaf aphid damage than others. For example, B Redlan, a common parent of sorghum hybrids, is extremely susceptible. However, several sorghum lines, especially some Zera Zera types and certain yellow endosperm pollinators, are highly resistant to damage by corn leaf aphid.

The yellow sugarcane aphid is lemon yellow and covered with setae and has two double rows of dark tubercles down the dorsum. Feeding aphids secrete a plant toxin, and relatively light populations have been known to kill sorghum in the seedling stage and to cause severe yellowing of more mature plants (2). In the United States this aphid has only been observed damaging sorghum in the Gulf Coast and central counties of Texas. The species also occurs in the Texas Panhandle, Kansas, Nebraska, Illinois, Missouri, and Oklahoma where damage to seedling sorghum may occur. The insect has been reported to be a serious pest of sugarcane and sorghum in tropical areas such as Puerto Rico and has been recorded from many other areas in the warmer parts of the Western Hemisphere (13, 81, 84). In Asia the corn leaf aphid occurs on sorghum, although it commonly is held in check by coccinellid beetles and syrphid flies.

The sugarcane aphid, *Aphis sacchari* is commonly found sucking the sap from the underside of the lower leaves. It typically occurs on plants after panicle emergence through the stage of seed development. Very large colonies may nearly cover the undersurface of the leaf, and large amounts of honeydew are produced. Little work has been done to determine possible yield loss resulting from this injury. It is reported to be a serious pest in South Africa (79) and China (162). In India the coccinellid beetle, *Menochilus sexmaculata,* is quite effective in reducing populations late in the season (176). More research is needed to determine economic thresholds for the species.

Armyworms

The fall armyworm, *Spodoptera frugiperda,* is one of several lepidopterous insects known as armyworms which attack sorghum in North, South and Central America.

Armyworms often cause extensive leaf ragging in sorghum and commonly feed within the plant whorl. The leaves unfolding from the whorls are perforated by the feeding of the insect. The corn earworm, *Heliothis zea,* may cause similar damage. Damage rarely justifies control of these insects except with heavy infestations on small plants.

Similar leaf injury has been reported from Africa by *Spodoptera exempta* and *S. exigua,* (14); from Asia by *Mythimna separata* (100); and from Australia by *Pseudaletia convecta* (89). The effect of crop defoliation by armyworms has been studied (14), and although considerable recovery of plants is possible from early season attack, sustained defoliation considerably reduces yields of grain. Foliar application of insecticides usually gives effective control of larvae, but more research is needed to determine thresholds of economic injury with the several species in varied agro-ecosystems. Both hymenopterous and dipterous parasites have been observed to drastically reduce armyworm populations. Maximum use should be made of these natural enemies. In Thailand, differences in varietal susceptibility of sorghum to *Mythimna separata* have been observed. The use of varietal resistance should be explored more intensively.

STEM BORERS

Stem-boring Lepidoptera are one of the larger groups of injurious insects in sorghum throughout the world. The wide host range of borers and overlapping of borer species has complicated their study. Jepson (70) prepared an extensive review of the world literature on some 46 species of borers of graminaceous crops. Other extensive papers include those by Tams & Bowden (135), Bleszynski & Collins (7), Ingram (65), Nye (87), and Harris (55). Tunneling of borer larvae in the stalks and peduncles of sorghum and the leaf feeding of some species in certain seasons and regions may cause serious reduction in yield of grain and fodder. However, borer occurrence and injury tends to be sporadic over large areas and below levels requiring direct action by man. Natural enemies are certainly of importance in the suppression of borer populations (83, 125). Although detailed coverage of the several borer species attacking sorghum is beyond the scope of this review, a brief treatment of the current status of the sorghum stem borer, *Chilo partellus,* is exemplary of the group.

Sorghum Stem Borer

The pyralid *Chilo partellus* [= *zonellus*] is a key pest of sorghum in the Indian subcontinent and in East Africa. The adults are buff-colored nocturnal moths that deposit their scalelike, overlapping eggs on the underside of the leaves. The cream-colored, spotted larvae migrate to the plant whorl and then penetrate downward, where they feed on the leaves and tunnel in the stalk, thereby stunting the plants and sometimes killing the main shoot. Pupation occurs in the stalk. The life cycle is completed in 30–40 days, so two generations may attack a single crop. Principal cultivated hosts of the insect are sorghum, maize, and pearl millet (31, 65, 87, 152). Cultural control measures, such as destruction of trash, stubbles, and volunteer plants, are helpful but may not be very practical. Use of insecticides has not been economically feasible on the traditional varieties (65, 152). On the new hybrids and improved varieties in India, application of a number of insecticides in spray and granular formulations to the plant whorls has given effective control and significant yield increases in experimental plots (74). This approach is not commonly used. Natural enemies play an important role in suppressing pest population levels of *C. partellus* (83, 125). The use of host plant resistance also appears to be of promise in borer management. Resistant varieties have been reported from both Africa (31) and India (64), and this approach certainly warrants continued intensive research.

More information is needed on the numerous borers attacking sorghum to develop appropriate component practices for their management. There is certainly ample opportunity for an international network of research activity on this problem with collaboration of the International Research Institutes and national research programs. The International Research Institute for the Semi-Arid Tropics (ICRISAT) near Hyderabad, India, should probably coordinate this activity.

SPIDER MITES

Several species of spider mites, *Oligonychus* spp. and *Tetranychus* spp., infest grain sorghum in the United States. The Banks grass mite, *O. pratensis,* is the most frequently encountered species (36).

Adult Banks grass mites exhibit marked sexual dimorphism (78). After feeding, both sexes become a deep green, with the exception of the palpi and first two pairs of legs, which remain light salmon. The female, which is much larger than the male, reaches an overall body length of about 0.40–0.45 mm. Malcolm (78) found the life cycle to require 11 days at 78–80°F and as many as 61 days under less favorable conditions.

Spider mites are usually along the midrib on the underside of the lower functional leaves. The infested areas of the leaves are pale yellow initially and later take on a reddish color on the top side of the leaves. If mite numbers continue to increase, the entire leaf may turn brown. As mite numbers increase on the lower leaves, the infestation spreads upward through the plant. The underside of heavily infested leaves has a dense deposit of fine webbing spun by the mites. In the final stages of infestation, the mites may invade and web sorghum heads. Subsequently, plant lodging may occur.

Based on distribution records, Ehler (36) concluded that the Banks grass mite is native to North America, although this conclusion remains tentative until the tetranychid fauna of South America, Africa, and Asia are more thoroughly described. The pest is generally restricted to monocotyledonous plants, particularly grasses. Over 80 species of grasses in 17 genera have been recorded as hosts (36, 78, 80, 96, 153).

Spider mite outbreaks are closely correlated with reproductive maturity of the host plant (36). Rapid population increases generally begin after heading (15). Infestations of mites appear to be separated temporally and spatially from populations of effective natural enemies (36). Also, there is a positive correlation between mite density and hot, dry climatic conditions.

Natural enemies of spider mites include several species of general predators as well as some that are prey-specific for phytophagous mites, i.e. *Scolothrips sexmaculatus, Stethorus punctum, S. atomus, Amblyseius fallacis, A. mesembrinus, Pronematus ubiquitis,* and cecidomyiids (25, 34, 35). However, this predator complex has not provided adequate control of the tetranychid mite pests in grain sorghum (34).

Presently, chemical control is the only available method of suppressing outbreaks of mites, but grower experience and research data indicate varying degrees of success (63, 90, 138, 163, 165). Insectide resistance has accounted for control failures in some areas (C. R. Ward, unpublished information).

In some areas of Texas the Banks grass mite sometimes acts as a secondary pest where insecticidal treatments, especially parathion, that were applied for aphid control increases the severity of the mite (D. H. Kattes and G. L. Teetes, unpublished information). Parathion treatments resulted in the dispersal of mite populations, consequently releasing the reproductive inhibitory effect of crowded mite colonies.

Sorghum germ plasm resistant to mites has been identified from the Texas sorghum breeding program (J. W. Johnson and G. L. Teetes, unpublished information). SC 599-6, a partially converted Rio selection from the sorghum conversion program, appeared especially promising during several years of testing at Pecos, Texas (D. G. Foster, G. L. Teetes, and J. W. Johnson, unpublished information). This line is a nonsenescing type which maintains green leaves and healthier stalk much longer than most lines. Interestingly, it continues to maintain green leaves even when infested with mites; consequently, the resistance mechanism appears to be of the tolerance type. The line is higher in total sugars than standard grain sorghums, and this may be involved in the resistance mechanism.

The common mite on sorghum throughout India is *Oligonychus indicus.* Especially in dry seasons, the injury may be severe and webbing from this mite may cover the leaves and panicles. Normally it is controlled by coccinellid beetles, but populations have been observed to increase in large numbers following the application of insecticides in experimental sorghum plots (27, 158). This mite will undoubtedly increase in importance if insecticide use should increase on high-yielding varieties. Those involved with future pest management programs need to be aware of this problem.

EARHEAD OR PANICLE PESTS

The most important insect pests of the sorghum panicle and developing grain are the sorghum midge, head caterpillars, and sucking bugs that attack the developing seeds or grain.

Sorghum Midge

The cecidomyiid midge, *Contarinia sorghicola,* is probably the most widely distributed of all sorghum pests; it occurs in nearly all the regions of the world where the crop is grown. Its major hosts are all members of the genus *Sorghum,* and it is thought to have originated with the crop in Africa and spread with it around the world. The midge is probably first in importance of all the sorghum insect pests. Losses in grain production amounting to millions of dollars have been reported from Texas, Argentina, Nigeria, India, and Australia. The injury is caused by the larval feeding on the ovary, which prevents normal seed development.

The small, orange-colored female midges, less than 2 mm in length, oviposit in the flowering spikelets. These small flies live for only a day or two and may not be noticed in the field. The eggs hatch in about 2 days and the orange larvae complete their development in 9–11 days, pupating beneath the glume. Adults emerge some 3 days later. A generation may be completed in 14–16 days. This rapid developmental cycle permits 9–12 generations during a season and permits the buildup of high midge populations where flowering times are extended by a wide range of planting dates.

The midge overwinters in cooler climates as a larva in a state of facultative diapause within a cocoon. In this stage some individuals may resist cold and desiccation for 3–4 years. A few individuals of each generation enter diapause, which is later broken by exposure to warm, humid conditions. Periods of peak midge activity are thus controlled by the climatic conditions of a particular area, principally temperature and humidity. The widespread midge distribution can probably be associated with the transport of diapausing larvae with poorly cleaned seed (12, 30, 43, 52, 54, 88, 161, 169, 170).

The management of midge by cultural practices requires a good knowledge of the ecological relationships in a particular region. Since midge buildup early in the growing season takes place on wild sorghums, early planting of the main crop over a large area to shorten the flowering period has been found to reduce midge injury. This management practice has been very successful in the High Plains of Texas, where it is now the established practice to plant the main crop early so that flowering is completed by early August (62). Intermixing of plantings of new early hybrid sorghum with late traditional varieties in India has created a severe midge problem on the later-flowering local varieties.

Insecticidal control of sorghum midge has been difficult since applications need to be closely timed to flowering of the crop and peak emergence of ovipositing adults. Phytotoxicity of insecticides has also been a problem on sorghum. Where insecticides are needed on late-planted crops, aerial and ground spraying with toxaphene, diazinon, carbaryl, and carbophenothion have been partially effective in Texas (127) and Argentina (164). Passlow (89) in Australia has recommended two ultralow volume applications of diazinon or malathion where midge populations exceed six ovipositing females per head. The economic threshold in Texas is considered to be one adult per head (11).

The use of host plant resistance in midge management appears promising for the future. Screening of sorghum germ plasm for midge resistance has been underway for some time (73, 120). Sources of resistance in sorghums adapted for temperate zones have been identified and released (71, 171).

Efforts to identify host resistance to midge and to incorporate this resistance in plant breeding programs have been handicapped by a lack of techniques for rearing the insect for use in uniform artificial infestation of sorghum varieties. Segregating populations that flower at different times also presents difficulties.

Planting date management appears to be the most effective method of midge control at present. In Israel, where some 10,000 hectares of hybrid sorghum is planted annually within a period of about 3 weeks, midge has not been a problem (A. Blum, personal communication).

Head Caterpillars

A number of species of Lepidoptera feed upon the maturing seeds of sorghum in the panicle. The most important are the sorghum webworm, *Celama sorghiella,* the Old World webworm, *Stenachroia elongella,* the corn earworm, *Heliothis zea,* fall armyworm, *Spodoptera frugiperda,* and American bollworm, *Heliothis armigera,* although a number of other species cause similar injury. The head caterpillars are considered to be occasional pests and for the most part have not warranted the use of insecticides or other control measures.

The two New World species, *C. sorghiella* and *H. zea,* frequently feed on the seeds of headed sorghums. Cannibalism by *H. zea* larvae is an important limiting factor, and this plus control by naturally occurring predators is often sufficient to keep the species below economic levels. Chemical control is recommended when populations exceed an average of two small larvae per head in the absence of natural predators and parasites. The open-headed sorghums are seldom damaged to the extent of tight-headed sorghums.

The sorghum webworm occurs primarily in the more humid areas of North America (109). It frequently occurs in large numbers in sorghum heads where it eats circular holes in the seed and feeds on the starchy contents (108). Observations indicate that each larva may consume as many as 12 seeds in 24 hr, resulting in severe crop losses (103). Moths deposit from 100 to 300 eggs singly on the flowering parts or seeds of the host plant. There may be up to six generations annually. For making decisions for chemical control, heads should be inspected beginning in the bloom stage and continuing until the hard dough stage has been reached. Chemical control is justified when heads are infested with an average of five larvae per head (29, 77, 102).

In Africa and Asia, the American bollworm, *H. armigera,* tends to injure legumes and cotton more than cereal crops. However, at times it is found to feed on the soft grains of sorghum. Open-type sorghum panicles are less affected since the larvae on them are exposed to predaceous insects and birds (31). Although it is smaller in size than *Heliothis,* the larvae of the Old World webworm, *S. elongella,* probably causes more damage. As with the sorghum webworm, the eggs are deposited on the panicle, but each small larvae starts its development within a single seed. Later, as they become larger, a protective webbing is produced which binds the panicle together and protects the larvae from predators. Practically no detailed research has been done on the Old World sorghum head pests, and direct human intervention in their control is of doubtful economic value at present (23, 38, 89).

Head Bugs

The jowar or sorghum earhead bug, *Calocoris angustatus,* has long been considered to be a key pest of sorghum in the southern states of India. The yellowish green adult mirid, about 1 cm in length, deposits eggs on the panicles soon after their emergence from the boot, and both adults and nymphs suck sap from the developing grain, distorting and shrinking it and reducing yield. Although insecticide, mainly benzene hexachloride dust, has been recommended and used for the control of *Calocoris,* the economics of the practice and economic thresholds have not been worked out in any detail (18, 86). Little is known about its natural enemies, and little work has been done on varietal resistance to this species. The earhead bug problem seems to be spreading northward into central India with the introduction of high-yielding varieties and research on its biology and control needs to be intensified.

A lygaeid bug, *Nysius raphanus,* sporadically infests sorghum in the United States and causes considerable concern because of its abundance. Populations of this species are concentrated in small areas and general infestations over an entire sorghum field are rare. Reproduction has not been observed on sorghum.

Damage results from bugs sucking juices from the immature developing grain. Often the damaged seed are infected with a fungus (*Alternera* sp.) that causes the seed to turn black and results in further deterioration of quality. Damaged seed rarely develop fully and are considerably smaller, softer, and lighter in weight than the undamaged seed and are subject to loss during harvesting (175). This species also feeds in clusters on the leaves of sorghum; however, little leaf damage results.

Damage assessment studies made by confining different populations in cages on sorghum heads have shown in general that for each additional ten bugs per head there was 1% increase in damaged seed (140). However, actual yield loss occurred only when 200 bugs per head damaged 23% of the seed. Increase in weight of undamaged seed apparently compensates for yield loss to some extent because in most treatments the percentage of damaged seed was below the percentage increase in seed weight. Only when percentage of damaged seed was greater than percentage increase in seed weight was there evidence of reduced yields. The point of equal compensatory effect occurred at 14% damaged seed or approximately 140 bugs per head. Spot treatment with insecticides may be required to prevent yield loss.

A number of other bugs have been reported from sorghum heads both in the Old and New Worlds, but little research has been done on their management since they occur only occasionally and at low population levels (176). Sorghum workers need to be alert to possible changes in the status of these insects.

PESTS OF SORGHUM IN STORAGE

Sorghum grain, like the seeds of other cereals, is subject to the attack of a number of insect pests after harvest. The exposed sorghum seed is more susceptible than other cereals to field infestation of the rice weevil, *Sitophilus oryzae,* the maize weevil, *Sitophilus zeamais,* and the Angoumois grain moth, *Sitotroga cerealella*.

These three cosmopolitan species are probably the most important because they are capable of attacking whole healthy seeds. Weevils oviposit within the seeds, whereas moth adults place eggs on the seed surface. Principal damage is caused by the larvae feeding and developing within the seeds. In warm tropical climates where sorghum is grown, these insects complete a life cycle in less than one month, which means large populations can develop and severe losses may occur in 4–6 months of storage. Over 60% kernel damage and 30% loss in weight of sorghum grain has resulted from rice weevil infestations after 5 months of storage in a controlled study under tropical conditions in India (157). In India and Africa, losses are severe in village storage facilities, which are often primitive and where climate favors storage pests. In North America, where storage facilities are more adequate, better management of these pests is generally accomplished.

Several publications discuss in detail the more common insects injuring stored grain and effective methods of controlling them. These include those by Cotton (19) in the United States, Munro (85) in the United Kingdom, Ramirez Genel (101) in Mexico, Giles (46, 47) and Doggett (31) in Africa and Pruthi & Singh (97) in India.

The management of storage pests in commercial grain can be accomplished by maintaining the grain dry (less than 12% moisture) and under good sanitation conditions in properly constructed bins or silos that can be fumigated and aerated when necessary. In the tropics this is more difficult to achieve.

Seeds for planting can be protected by direct treatment with residual insecticides. Two of the more effective materials, DDT and dieldrin, will probably not be available for use in the future so substitutes need to be found. Malathion, though less hazardous to humans, is not as effective as a protectant as chlorinated hydrocarbon insecticides for long-term storage of seeds.

In recent years there has been increasing interest in breeding sorghums with flinty corneous endosperm that have some resistance to storage pests (31). Techniques that should facilitate screening sorghum varieties for resistance to weevils are being developed (118, 133). Varieties that resist attack in storage will be especially useful in the tropics.

CONCLUSION

Sorghum and its associated pests are in a dynamic state. The progress that has been made in sorghum entomology will undoubtedly continue at an accelerated rate, as it must to meet the ever-increasing pest problems. The area of greatest need lies in the establishment and refinement of economic threshold levels in order to eliminate needless insecticide treatments and to gain maximum benefits from the other components of pest management. This will require interdisciplinary, problem-oriented, integrated control approaches that are based on sound ecological principles.

Insecticides, at least in the foreseeable future, will remain a major management tool, since they are generally effective and economical and can be administered quickly to curb pest populations in emergency situations. They must be used judiciously on the basis of the potential positive values weighed against possible negative values occurring from hazards to nontarget organisms. Management components such as cultural practices, biological control, and resistant plant varieties must be given greater consideration in both monoculture production and subsistence farming.

The current literature on sorghum entomology reflects the increasing attention being given to the crop as the world faces the human food supply problem of the future. There is a tremendous potential in sorghum for increased production of food, feed, and fodder in the tropics that can be achieved with intensified research input on all aspects of its production, including better management of pests. This research effort has expanded during the past decade and should continue to improve: an international network of research on the crop is being organized through the collaboration of the International Centers, supported by the Consultative Group on International Agricultural Research and national and state sorghum improvement programs. During the next decade or two, much of the information now lacking for effective pest population management on sorghum, particularly in the tropics, should be accumulated.

Literature Cited

1. Abul-Nasr, S. E., El-Nahal, A. K. M., Shahoudah, S. K. 1968. Some biological aspects of the corn stem-borer, *Sesamia cretica* Led. *Bull. Soc. Entomol. Egypte* 52:429–44
2. Almand, L. K., Bottrell, D. G., Cate, J. R. Jr., Daniels, N. E., Thomas, J. G. 1969. Greenbugs on sorghum and small grains. *Tex. Agric. Ext. Serv. L-819.* 4 pp.
3. Archer, T. L., Cate, R. H., Eikenbary, R. D., Starks, K. S. 1974. Parasitoids collected from greenbugs and corn leaf aphids in Oklahoma in 1972. *Ann. Entomol. Soc. Am.* 67:11–14
4. Archer, T. L., Eikenbary, R. D. 1973. Storage of *Aphelinus asychis,* a parasite of the greenbug. *Environ. Entomol.* 2:489–90
5. Archer, T. L., Murray, C. L., Eikenbary, R. D., Starks, K. J., Morrison, R. D. 1973. Cold storage of *Lysiphlebus testaceipes* mummies. *Environ. Entomol.* 2:1104–8
6. Barry, D. 1972. Chemical control of sorghum shoot fly on a susceptible variety of sorghum in Uganda. *J. Econ. Entomol.* 65:1123–25
7. Bleszynski, S., Collins, R. J. 1962. A short catalogue of the world species of the family Crambidae. *Acta Zool. Cracoviensia* 7:197–389
8. Blum, A. 1963. The penetration and development of the shoot fly in susceptible sorghum plants. *Hassadeh* 44:23–25 (In Hebrew)
9. Blum, A. 1966. Evaluation of some exotic sorghum cultivars for resistance to stalk infestation by *Sesamia cretica* Led. *Isr. J. Agric. Res.* 18:95–98
10. Blum, A. 1972. See Ref. 75, pp. 180–91
11. Bottrell, D. G. 1971. See Ref. 151, pp. 35–37
12. Bowden, J. 1966. Sorghum midge, *Contarinia sorghicola* (Coq.) and other causes of grain sorghum loss in Ghana. *Bull. Entomol. Res.* 56:169–89
13. Box, H. E. 1953. List of sugarcane insects. *Commonw. Inst. Entomol. London* 101 pp.
14. Brown, E. S., Mohamed, A. K. A. 1972. The relation between simulated armyworm damage and crop-loss in maize and sorghum. *East Afr. Agric. For. J.* 37:237–57
15. Cate, J. R. Jr., Bottrell, D. G. 1971. Evaluation of foliar sprays for controlling the Banks grass mite on grain sorghum in the Texas High Plains, 1969. *Tex. Agric. Exp. Stn. PR-2870.* 6 pp.
16. Cate, J. R. Jr., Bottrell, D. G., Teetes, G. L. 1973. Management of the greenbug on grain sorghum. 1. Testing foliar treatments of insecticides against greenbug and corn leaf aphids. *J. Econ. Entomol.* 66:945–51
17. Chelliah, S., Basheer, M. 1965. Biological studies of *Peregrinus maidis* (Ashmead) on sorghum. *Indian J. Entomol.* 27:466–71
18. Cherian, M. C., Kylasam, M. S., Krishnamurti, P. S. 1941. Further studies on

Calocoris angustatus. *Madras Agric. J.* 29:66–69
19. Cotton, R. T. 1963. *Pests of Stored Grain and Grain Products.* Minneapolis: Burgess. 318 pp.
20. Dahms, R. G. 1942. Rice stinkbug as a pest of sorghums. *J. Econ. Entomol.* 35:945–46
21. Daniels, N. E., Chedester, L. O. 1972. Developmental studies of the convergent lady beetle. *Tex. Agric. Exp. Stn. PR-3109.* 3 pp.
22. Daniels, N. E., Toler, R. W. 1971. Transmission of maize dwarf mosaic by the greenbug. *Tex. Agric. Exp. Stn. Consol. PR-2869,* pp. 20–22
23. David, B. V., David, S. K. 1961. Lepidopterous larvae injurious to sorghum earheads. *Madras Agric. J.* 48:93–97
24. Davis, E. G., Horton, J. R., Gable, C. H., Walter, E. V., Blanchard, R. A. 1933. The southwestern corn borer. *US Dept. Agric. Tech. Bull. B-388* 62 pp.
25. Dean, H. A. 1957. Predators of *Oligonychus pratensis* (Banks). *Ann. Entomol. Soc. Am.* 50:164–65
26. Deeming, J. C. 1971. Some species of *Atherigona* Rondani from northern Nigeria, with special reference to those injurious to cereal crops. *Bull. Entomol. Res.* 61:133–90
27. Desai, M. K., Chavda, D. K. 1955. Mite (*Oligonychus* sp.) as a causal agent for "ratada" disease on sorghum. *Poona Agric. Coll. Mag.* 45:138–41
28. Dickson, R. C., Laird, E. F. Jr. 1969. Crop host preferences of greenbug biotype attacking sorghum. *J. Econ. Entomol.* 62:1241
29. Doering, G. W., Randolph, N. M. 1960. Field methods to determine the infestation of the sorghum webworm and the damage by the sorghum midge in grain sorghum. *J. Econ. Entomol.* 54:749–50
30. Doering, G. W., Randolph, N. M. 1963. Habits and control of the sorghum midge, *Contarinia sorghicola,* on grain sorghum. *J. Econ. Entomol.* 56:454–59
31. Doggett, H. 1970. *Sorghum,* Chap. 13. London: Longmans & Green. 403 pp.
32. Doggett, H. 1972. See Ref. 75, pp. 190–207
33. Dureaseau, L., Rivet, E., Drea, J. J. 1972. *Ephedrus plagiator,* a parasite of the greenbug in France. *J. Econ. Entomol.* 65:604–5
34. Ehler, L. E. 1972. Preliminary studies of spider mites on corn and sorghum in West Texas. *Ann. Tex. Conf. Insect, Plant Diseases, Weed Brush Control, 5th,* College Station, Texas
35. Ehler, L. E. 1973. Spider-mites associated with grain sorghum and corn in Texas. *J. Econ. Entomol.* 66:1220
36. Ehler, L. E. 1974. A review of the spider-mite problem on grain sorghum and corn in West Texas. *Tex. Agric. Exp. Stn. B-1149.* 15 pp.
37. Eikenbary, R. D., Rogers, C. E. 1973. Importance of alternate hosts in establishment of introduced parasites. *Proc. Tall Timbers Conf.* 5:119–34
38. Entomology Branch Officers. 1974. Sorghum pest control at a glance. *Queensl. Agric. J.* 100:478–81
39. Esmaili, M., Wilde, G. 1972. Behavior of the parasite *Aphelinus asychis* in relation to the greenbug and certain hosts. *Environ. Entomol.* 1:266–68
40. Food and Agricultural Organization of the United Nations 1975. *Mon. Bull. Agric. Econ. Stat.* 9:24. Rome: FAO
41. Frederiksen, R. A., Daniels, N. E. 1970. The influence of greenbugs on stalk rots of sorghum. *Tex. Agric. Exp. Stn. PR-2772.* 7 pp.
42. Geering, Q. A. 1952. A cotton stainer (*Dysdercus superstitiosus* Fabr.) *Emp. J. Exp. Agric.* 20:234–39
43. Geering, Q. A. 1953. The sorghum midge, *Contarinia sorghicola* (Coq.) in East Africa. *Bull. Entomol. Res.* 44: 363–66
44. George Washington University 1967. *Sorghum—A Bibliography of the World Covering the Years 1930–1963.* Metuchen, NJ: Scarecrow. 301 pp.
45. Gerhardt, P. D., Moore, L., Armstrong, J. F., Kaspersen, L. J. 1972. Southwestern corn borer control in grain sorghum. *J. Econ. Entomol.* 65:491–94
46. Giles, P. H. 1964. The insect infestation of sorghum stored in granaries in northern Nigeria. *Bull. Entomol. Res.* 55: 573–88
47. Giles, P. H. 1965. Control of insects infesting stored sorghum in northern Nigeria. *J. Stored Prod. Res.* 1:145–58
48. Goode, J. P., Randolph, N. M. 1961. Biology and control of the sugarcane rootstock weevil on grain sorghum in Texas. *J. Econ. Entomol.* 54:301–3
49. Hackerott, H. L., Harvey, T. L. 1971. Greenbug injury to resistant and susceptible sorghums in the field. *Crop Sci.* 11:641–43
50. Hackerott, H. L., Harvey, T. L., Ross, W. M. 1969. Greenbug resistance in sorghums. *Crop Sci.* 9:656–58
51. Hackerott, H. L., Harvey, T. L., Ross, W. M. 1972. Registration of KS41,

KS42, KS43 and KS44 greenbug resistant sorghum germ-plasm. *Crop Sci.* 12:720

52. Harding, J. A. 1965. Ecological and biological factors concerning the sorghum midge during 1964. *Tex Agric. Exp. Stn. Bull. M.P. 773.* 10 pp.
53. Harding, J. A. 1965. Effect of insecticidal phytotoxicity and aphids on grain sorghum yields. *Tex. Agric. Exp. Stn. PR-2350.* 6 pp.
54. Harris, K. M. 1961. The sorghum midge, *Contarinia sorghicola* (Coq.) in Nigeria. *Bull. Entomol. Res.* 52:129–46
55. Harris, K. M. 1962. Lepidopterous stem borers of cereals in Nigeria. *Bull. Entomol. Res.* 53:139–71
56. Harvey, T. L., Hackerott, H. L. 1969. Plant resistance to a greenbug biotype injurious to sorghum. *J. Econ. Entomol.* 62:1271–74
57. Harvey, T. L., Hackerott, H. L. 1969. Recognition of a greenbug biotype injurious to sorghum. *J. Econ. Entomol.* 62:776–79
58. Harvey, T. L., Hackerott, H. L. 1974. Effects of greenbugs on resistant and susceptible sorghum seedlings in the field. *J. Econ. Entomol.* 67:377–80
59. Harwood, R. R., Granados, Y., Jamornmarn, S., Granados, G. 1972. See Ref. 75, pp. 208–17
60. Hays, W. P. 1922. Observations on insects attacking sorghums. *J. Econ. Entomol.* 15:349–56
61. Hight, S. C., Eikenbary, R. D., Miller, R. J., Starks, K. J. 1972. The greenbug and *Lysiphlebus testaceipes. Environ. Entomol.* 1:205–9
62. Huddleston, E. W., Ashdown, D., Maunder, B., Ward, C. R., Wilde, G., Forehand, C. E. 1972. Biology and control of the sorghum midge. 1. Chemical and cultural control studies in West Texas. *J. Econ. Entomol.* 65:851–54
63. Huddleston, E. W., Ward, C. R., Hills, T. M., Owens, J. C. 1968. Evaluation of selected insecticides for control of mites on grain sorghum in West Texas. *Tex. Tech. Coll. Res. Farm Rep. ICASALS Spec. Rep. 4* pp. 23–25
64. Indian Council of Agricultural Research 1975. *Progress Report of the All India Co-ordinated Sorghum Improvement Project: Part III Sorghum Entomology 1973–1975.* New Delhi. 33 pp.
65. Ingram, W. E. 1958. The lepidopterous stalk borers associated with Graminae in Uganda. *Bull. Entomol. Res. 49–367–83*
66. Jackson, H. B., Coles, L. W., Wood, E. A. Jr., Eikenbary, R. D. 1970. Parasites reared from the greenbug and corn leaf aphid in Oklahoma in 1968 and 1969. *J. Econ. Entomol.* 63:733–36
67. Jackson, H. B., Eikenbary, R. D. 1971. Bionomics of *Aphelinus asychis,* an introduced parasite of the sorghum greenbug. *Ann. Entomol. Soc. Am.* 64:81–85
68. Jackson, H. B., Rogers, C. E., Eikenbary, R. D. 1971. Colonization and release of *Aphelinus asychis,* an imported parasite of the greenbug. *J. Econ. Entomol.* 64:1435–38
69. Jackson, H. B., Rogers, C. E., Eikenbary, R. D., Starks, K. J., McNew, R. W. 1974. Biology of *Ephedrus plagiator* on different aphid hosts and at various temperatures. *Environ. Entomol.* 3:618–20
70. Jepson, W. F. 1954. *A Critical Review of the World Literature on the Lepidopterous Stalk Borers of Graminaceous Crops.* London: Commonw. Inst. Entomol. 127 pp.
71. Johnson, J. W., Rosenow, D. T., Teetes, G. L. 1973. Resistance to the sorghum midge in converted exotic sorghum cultivars. *Crop Sci.* 13:754–55
72. Johnson, J. W., Rosenow, D. T., Teetes, G. L. 1974. Response of greenbug-resistant grain sorghum lines and hybrids to a natural infestation of greenbugs. *Crop Sci.* 14:442–43
73. Jotwani, M. G., Singh, S. P., Chaudhari, S. 1971. Relative susceptibility of some sorghum lines to midge damage. *Investigations on Insect Pests of Sorghum and Millets (1965–1970). Final Technical Report,* pp. 123–30. New Delhi: Div. Entomol., IARI
74. Jotwani, M. G., Young, W. R. 1972. See Ref. 106, p. 381
75. Jotwani, M. G., Young, W. R., eds. 1972. *Control of Sorghum Shoot Fly. Proc. Int. Symp. Hyderabad, India, Nov. 1–3, 1971.* New Delhi: Oxford & IBH. 324 pp.
76. Kelley, E. O. G. 1909. How *Lysiphlebus* fastens its aphid host to the plant. *Proc. Entomol. Soc. Wash.* 11:64–6
77. Kinzer, H. G., Henderson, C. F. 1967. Effect of sorghum webworm on yield of grain sorghum in Oklahoma. *J. Econ. Entomol.* 60:118–21
78. Malcolm, D. R. 1955. Biology and control of the timothy mite. *Wash. Agric. Exp. Stn. B-17.* 35 pp.
79. Matthee, J. J. 1962. Guard against aphids on kaffircorn. *Farming S. Afr.* 37:27–29

80. McGregor, E. A., Stickney, F. 1965. Distribution and host plants of *Oligonychus pratensis* (Banks). *Proc. Entomol. Soc. Wash.* 67:28
81. Medina-Gand, S. L., Martorell, L. F., Bonilla-Robles, R. 1967. Notes on the biology and control of the yellow aphid of sugarcane, *Sipha flava* Forbes, in Puerto Rico. *Proc. Congr. Int. Soc. Sugar Cane Technol., 12th, Puerto Rico, 1965,* pp. 1278–86
82. Metcalf, C. L., Flint, W. P., Metcalf, R. L. 1962. Destructive and useful insects, their habits and control. New York: McGraw-Hill. 1071 pp. 4th ed.
83. Milner, J. E. D. 1967. *Final Report on a Survey of the Parasites of Graminaceous Stem Borers in East Africa.* Kawanda, Uganda: Commonw. Inst. Biol. Control East Afr. Stn. 159 pp.
84. Miskimen, G. W. 1970. Population dynamics of the yellow sugarcane aphid, *Sipha flava,* in Puerto Rico, as affected by heavy rains. *Ann. Entomol. Soc. Am.* 63:642–45
85. Munro, J. W. 1966. *Pests of Stored Products.* London: Hutchinson. 234 pp.
86. Nagarajan, K. R., Edwards, J. J. D. 1959. Control of the Cholam earhead bug with BHC in Coimbatore. *Plant Protect. Bull. (India)* 8:12–13
87. Nye, I. W. B. 1960. *The Insect Pests of Graminaceous Crops in East Africa. Colon. Res. Study 31.* London: Colon. Office. 47 pp.
88. Passlow, T. 1965. Bionomics of sorghum midge *(Contarina sorghicola)* (Coq). in Queensland with particular reference to diapause. *Queensl. J. Agric. Anim. Sci.* 22:150–67
89. Passlow, T. 1973. Insect pests of grain sorghum. *Queensl. Agric. J.* 99:620–28
90. Pate, T. L., Neeb, C. W. 1971. The Banks grass mite problem in the Trans-Pecos area of Texas. *Tex. Agric. Exp. Stn. PR-2871.* 3 pp.
91. Peter, D. C., Wood, E. A. Jr., Starks, K. J. 1975. Insecticide resistance in selections of the greenbug. *J. Econ. Entomol.* 68:339–40
92. Pollard, D. G. 1956. The control of chafer grubs, *Schizonycha* sp., in the Sudan. *Bull. Entomol. Res.* 47:347–60
93. Ponnaiya, B. W. X. 1951. Studies in the genus sorghum. I. Field observations on sorghum resistance to the insect pest, *Atherigona indica* M. *Madras Univ. J.* 21:203–17
94. Ponnaiya, B. W. X. 1951. Studies in the genus *Sorghum.* II. The cause of resistance in *Sorghum* to the insect pest, *Atherigona indica. Madras Univ. J.* 21:203–17
95. Pont, A. C. 1972. See Ref. 75, pp. 27–102
96. Pritchard, A. E., Baker, E. W. 1955. A revision of the spider mite family Tetranychidae. *Mem. Pac. Coast Entomol. Soc.,* Vol. 2. 472 pp.
97. Pruthi, H. J., Singh, M. 1950. Pests of stored grain and their control. *Indian J. Agric. Sci.* 18 (4):1–88
98. Ramakrishna Ayyar, T. V. 1932. Entomology of the sorghum plant in south India. *Madras Agric. J.* 20:50–56
99. Ramakrishna Ayyar, T. V. 1950. *Handbook of Economic Entomology for South India.* Govt. Madras, India. 516 pp. Printed 1963
100. Ramamani, S., Subba Rao, B. R. 1965. On the identity and nomenclature of the paddy cutworm commonly referred to as *Cirphis unipuncta* Haworth. *Indian J. Entomol.* 27:363–65
101. Ramirez Genel, M. 1966. *Storage and Conservation of Grains and Seeds.* Mexico, DF: Comp. Edit. Cont., S.A. 300 pp. (In Spanish)
102. Randolph, N. M., Doering, G. W., Buckholt, A. J. 1960. The sorghum webworm and sorghum midge on grain sorghum. *Tex. Agric. Exp. Stn. PR-2130.* 5 pp.
103. Randolph, N. M., Garner, C. F. 1961. Insects attacking forage crops. *Tex. Agric. Ext. Serv. B-975.* 26 pp.
104. Randolph, N. M., Teetes, G. L., Jeter, B. E. Jr. 1967. Insecticide sprays and granules for control of the sugarcane borer on grain sorghum. *J. Econ. Entomol.* 60:762–65
105. Raney, H. G., Coles, L. W., Eikenbary, R. D., Morrison, R. D., Starks, K. J. 1971. Host preference, longevity, developmental period and sex ratio of *Aphelinus asychis* with three sorghum-fed species of aphids held at controlled temperatures. *Ann. Entomol. Soc. Am.* 64:169–76
106. Rao, N. G. P., House, L. R., eds. 1972. *Sorghum in Seventies,* Chap. 24. New Delhi: Oxford & IBH. 636 pp.
107. Rawat, R. R., Saxena, D. K. 1967. Studies on the bionomics of *Peregrinus maidis* (Ashmead) *JNKVV Res. J.* 1:64–67
108. Reinhard, H. J. 1937. Sorghum webworm studies in Texas. *J. Econ. Entomol.* 30:869–72
109. Reinhard, H. J. 1938. The sorghum webworm (*Celama sorghiella* Riley). *Tex. Agric. Exp. Stn. B-559.* 35 pp.

110. Reinhard, H. J. 1940. The life history of *Phyllophaga lanceolata* (Say) and *Phyllophaga crinita* Burmeister. *J. Econ. Entomol.* 33:572–78
111. Ritcher, P. O. 1961. Descriptions of some common North Indian scarabeid larvae. *Indian J. Entomol.* 23:15–19
112. Rockefeller Foundation 1973. *Sorghum—A Bibliography of the World Literature, 1964–1969.* Metuchen, NJ: Scarecrow. 393 pp.
113. Rogers, C. E., Jackson, H. B., Eikenbary, R. D. 1972. Voracity and survival of *Propylea 14-punctata* preying upon greenbugs. *J. Econ. Entomol.* 65:1313–16
114. Rogers, C. E., Jackson, H. B., Angalet, G. W., Eikenbary, R. D. 1972. Biology and life history of *Propylea 14-punctata,* an exotic predator of aphids. *Ann. Entomol. Soc. Am.* 65:648–50
115. Rogers, C. E., Jackson, H. B., Eikenbary, R. D. 1972. Response of an imported coccinellid, *Propylea 14-punctata,* to aphids associated with small grains in Oklahoma. *Environ. Entomol.* 1:198–202
116. Rogers, C. E., Jackson, H. B., Eikenbary, R. D., Starks, K. J. 1972. Host-parasitoid interaction of *Aphis helianthi* on sunflowers with introducted *Aphelinus asychis, Ephedrus plagiator,* and *Praon gallicum,* and native *Aphelinus nigritus* and *Lysiphlebus testaceipes. Ann. Entomol. Soc. Am.* 65:38–41
117. Rogers, C. E., Jackson, H. B., Eikenbary, R. D., Starks, K. J. 1971. Sex determination in *Propylea 14-punctata,* an imported predator of aphids. *Ann. Entomol. Soc. Am.* 64:957–59
118. Rogers, R. A., Mills, R. B. 1974. Reaction of sorghum varieties to maize weevil infestation under three relative humidities. *J. Econ. Entomol.* 67: 692
119. Rondani, C. 1852. *Aphis graminum* n. sp. *Nuove Ann. Sci. Nat. Bologna* 6:9–11
120. Rossetto, C. J., Banzatto, N. V. 1967. Resistencia de variedades de sorgo a *Contarinia sorghicola* (Coquillet). VII. Reunion Latino Americana de fitotecnia, Maracay, Venezuela 17–23 Setiembre. *Resume Trabajos Cient.,* pp. 292–93
121. Schuster, D. J., Starks, K. J. 1973. Greenbugs: components of host-plant resistance in sorghum. *J. Econ. Entomol.* 66:1131–34
122. Schuster, D. J., Starks, K. J. 1974. Response of *Lysiphlebus testaceipes* in an olfactometer to a host and a non-host insect and to plants. *Environ. Entomol.* 3:1034–35
123. Schuster, D. J., Starks, K. J. 1975. Preference of *Lysiphlebus testaceipes* for greenbug resistant and susceptible small grain species. *Environ. Entomol.* 4: 887–88
124. Sepsawadi, P., Meksongsee, G., Knapp, F. W. 1971. Effectiveness of various insecticides against a sorghum shoot fly. *J. Econ. Entomol.* 64:1509–11
125. Sharma, A. K., Saxena, J. D., Subba Rao, B. R. 1966. A catalogue of the hymenopterous and dipterous parasites of *Chilo zonellus* (Swinhoe). *Indian J. Entomol.* 28:510–42
126. Soto, P. E., Laxminarayana, K. 1971. A method for rearing the sorghum shoot fly. *J. Econ. Entomol.* 64:553
127. Stanford, R. L., Huddleston, E. W., Ward, C. R. 1972. Biology and control of the sorghum midge. 3. Importance of stage of bloom and effective residual of selected insecticides. *J. Econ. Entomol.* 65:796–99
128. Starks, K. J. 1970. Increasing infestations of the sorghum shoot fly in experimental plots. *J. Econ. Entomol.* 63: 1715–16
129. Starks, K. J., Muniappan, R., Eikenbary, R. D. 1972. Interaction between plant resistance and parasitism against the greenbug on barley and sorghum. *Ann. Entomol. Soc. Am.* 65:650–55
130. Starks, K. J., Wood, E. A. Jr. 1974. Greenbugs: damage to growth stages of susceptible and resistant sorghum. *J. Econ. Entomol.* 67:456–57
131. Starks, K. J., Wood, E. A. Jr., Teetes, G. L. 1973. Effects of temperature on the preference of two greenbug biotypes for sorghum selections. *Environ. Entomol.* 2:351–54
132. Starks, K. J., Wood, E. A. Jr., Weibel, D. E. 1972. Nonpreference of a biotype of the greenbug for a broomcorn cultivar. *J. Econ. Entomol.* 65:623–24
133. Stevens, R. A., Mills, R. P. 1973. Comparison of techniques for screening sorghum grain varieties for resistance to rice weevil. *J. Econ. Entomol.* 66: 1222–23
134. Swaine, G., Wyatt, C. A. 1954. Observations on the sorghum shoot fly. *E. Afr. Agric. For. J.* 20:45–48
135. Tams, W. H. T., Bowden, J. 1953. A revision of the African species of *Sesamia* Guenee and related genera. *Bull. Entomol. Res.* 43:645–78
136. Teetes, G. L. 1971. Research results on grain sorghum pests in the High Plains.

Proc. Ann. Tex. Conf. Insect, Plant Disease, Weed and Brush Control, 4th, College Station, Tex: Tex. A&M Univ., pp. 89–98

137. Teetes, G. L. 1972. Differential toxicity of standard and reduced rates of insecticides to greenbugs and certain beneficial insects. *Tex. Agric. Exp. Stn. PR-3041.* 6 pp.
138. Teetes, G. L. 1973. Insecticidal control of a spider mite in grain sorghum on the Texas High Plains. *Tex. Agric. Exp. Stn. PR-3178.* 4 pp.
139. Teetes, G. L. 1973. *Phyllophaga crinita:* damage assessment and control in grain sorghum and wheat. *J. Econ. Entomol.* 66:773–76
140. Teetes, G. L., Johnson, J. W., Rosenow, D. T. 1974. Damage assessment of a false chinch bug, *Nysius raphanus* (Howard), in grain sorghum. *Tex. Agric. Exp. Stn. PR-3260.* 4 pp.
141. Teetes, G. L., Brothers, G. W., Ward, C. R. 1973. Insecticide screening for greenbug control and effect on certain beneficial insects. *Tex. Agric. Exp. Stn. PR-3116.* 6 pp.
142. Teetes, G. L., Johnson, J. W. 1973. Damage assessment of the greenbug on grain sorghum. *J. Econ. Entomol.* 66:1181–86
143. Teetes, G. L., Johnson, J. W. 1974. Assessment of damage by the greenbug in grain sorghum hybrids of different maturities. *J. Econ. Entomol.* 67: 514–16
144. Teetes, G. L., Johnson, J. W., Rosenow, D. T. 1975. Response of improved resistant sorghum hybrids to natural and artificial greenbug populations. *J. Econ. Entomol.* 68:546–48
145. Teetes, G. L., Randolph, N. M. 1968. Resistance of certain corn and grain sorghum varieties to attack of the sugarcane borer, *Diatraea saccharalis* (F). *Tex. Agric. Exp. Stn. PR-2580.* 5 pp.
146. Teetes, G. L., Rosenow, D. T., Frederiksen, R. D., Johnson, J. W. 1973. The predisposing influence of greenbugs on charcoal rot of sorghum. *Tex. Agric. Exp. Stn. PR-3173.* 6 pp.
147. Teetes, G. L., Schaefer, C. A., Johnson, J. W., Rosenow, D. T. 1974. Resistance in sorghums to the greenbug: field evaluation. *Crop Sci.* 14:706–8
148. Teetes, G. L., Schaefer, C. A., Johnson, J. W. 1974. Resistance in sorghums to the greenbug; laboratory determination of mechanisms of resistance. *J. Econ. Entomol.* 67:393–96
149. Teetes, G. L., Schaefer, C. A., Gipson, J. R., McIntyre, R. C., Latham, E. E. 1975. Greenbug resistance to organophosphorous insecticides on the Texas High Plains. *J. Econ. Entomol.* 68: 214–16
150. Teetes, G. L., Wade, L. 1974. Seasonal biology and abundance of the white grub, *Phyllophaga crinita* Burmeister, in the Texas High Plains. *Tex. Agric. Exp. Stn. PR-3261.* 4 pp.
151. Texas Agricultural Experiment Station 1971. *Grain Sorghum Research in Texas, Consol. PR–2938–2949.* (PR–2940 by D. G. Bottrell). College Station, Texas. 120 pp.
152. Trehan, K. N., Butani, D. K. 1949. Notes on the life history, bionomics and control of *Chilo zonellus* Swinhoe in Bombay Province. *Indian J. Entomol.* 11:47–59
153. Tuttle, D. M., Baker, E. W. 1968. *Spider mites of the southwestern United States and a revision of the Tetranychidae.* Tucson: Univ. Ariz. Press. 143 pp.
154. US Department of Agriculture 1955. Stored grain pests. *Farmer's Bull. 1260.* 46 pp.
155. Usman, S. 1968. Preliminary studies on the incidence of shoot fly on hybrid jowar under differential sowings. *Mysore J. Agric. Sci.* 2:44–48
156. Veda Moorthy, G., Thobbi, V. V., Matai, B. H., Young, W. R. 1965. Preliminary studies with seed and seed-furrow applications of insecticides for the control of sorghum stem maggot, *Atherigona indica* Malloch. *Indian J. Agric. Sci.* 35:14–28
157. Venkatarao, S., Nuggehalli, R. N., Swaminathan, M., Pingale, S. V., Subrahmanyan. V. 1956. Effect of insect infestation on stored grain. III. Studies of Kafir corn *(Sorghum vulgare) J. Sci. Food Agric.* 9:837–39
158. Vijayaraghavan, S., Vasudava Menon, P. P. 1962. The Cholam mite (*Oligonychus indicus* Hirst) and its control. *Madras Agric. J.* 49:254–57
159. Walker, A. L., Bottrell, D. G., Cate, J. R. Jr. 1973. Hymenopterous parasites of biotype C greenbug in the High Plains of Texas. *Ann. Entomol. Soc. Am.* 66:173–76
160. Wall, J. S., Ross, W. M., eds. 1970. *Sorghum Production and Utilization,* Chap. 7. Westport, Conn.: Avi. 702 pp.
161. Walter, E. V. 1941. *The Biology and Control of the Sorghum Midge. U.S. Dept. Agric. Tech. Bull. 778.* 26 pp.

162. Wang, Y. S. 1961. Studies on the sorghum aphid, *Aphis sacchari* Zehntner. *Acta Entomol. Sinica.* 10:363–80
163. Ward, C. R., Huddleston, E. W., Owens, J. C., Hills, T. M., Richardson, L. G., Ashdown, D. 1972. Control of the Banks grass mite attacking grain sorghum and corn in West Texas. *J. Econ. Entomol.* 65:523–29
164. Ward, C. R., Huddleston, E. W., Parodi, R. A., Ruiz, G. 1972. Biology and control of the sorghum midge. 2. Chemical control in Argentina. *J. Econ. Entomol.* 65:817–18
165. Ward, C. R., Richardson, L. G., Ashdown, D., Huddleston, E. W., Gfeller, R. 1971. Development of pesticide resistance in Banks grass mite indicated in field studies. *Tex. Tech. Univ. Entomol. Spec. Rep. 71-1.* 8 pp.
166. Webster, F. M. 1909. Investigations of *Toxoptera graminum* and its parasites. *Ann. Entomol. Soc. Am.* 2:67–87
167. Weibel, D. E., Starks, K. J., Wood, E. A. Jr., Morrison, R. D. 1972. Sorghum cultivars and progenies rated for resistance to greenbugs. *Crop Sci.* 12:334–36
168. Whitfield, F. G. S. 1929. The Sudan millet bug, *Agonoscelis versicolor. Bull. Entomol. Res.* 20:209–24
169. Wiseman, B. R., McMillian, W. W. 1970. Preference of the sorghum midge among selected sorghum lines, with notes on overwintering midges and parasite emergence. *US Dept. Agric. Prod. Res. Rep. 122.* 8 pp.
170. Wiseman, B. R., McMillian, W. W. 1973. Diapause of the sorghum midge, and location within the sorghum spikelet. *J. Econ. Entomol.* 66:647–49
171. Wiseman, B. R., McMillian, W. W., Widstrom, N. W. 1973. Registration of SGIRL-MR-1 sorghum germ plasm (Reg. No. GP19). *Crop Sci.* 13:398
172. Wood, E. A. Jr. 1971. Designation and reaction of three biotypes of the greenbug cultured on resistant and susceptible species of sorghum. *J. Econ. Entomol.* 64:183–85
173. Wood, E. A. Jr., Chada, H. L., Weibel, D. E., Davies, F. F. 1969. A sorghum variety highly tolerant to the greenbug, *Schizaphis graminum* (Rond.) *Okla. Agric. Exp. Stn. Prog. Reg. P-618.* 7 pp.
174. Wood, E. A. Jr., Starks, K. J. 1972. Effect of temperature and host plant interaction on the biology of three biotypes of the greenbug. *Environ. Entomol.* 1:230–34
175. Wood, E. A. Jr., Starks, K. J. 1972. Damage to sorghum by a lygaeid bug, *Nysius raphanus. J. Econ. Entomol.* 65:1507–8
176. Young, W. R. 1970. See Ref. 160, pp. 235–87
177. Young, W. R. 1972. See Ref. 75, pp. 168–79

Ann. Rev. Entomol. 1977. 22:219–40

❖6128

THE PERITROPHIC MEMBRANES OF INSECTS[1]

A. Glenn Richards and Patricia A. Richards
Department of Entomology, Fisheries and Wildlife, University of Minnesota, St. Paul, Minnesota 55108

A membrane surrounding the food bolus in the midgut (MG) of various arthropods was described by a number of authors in the nineteenth century, dating at least as far back as 1843. Balbiani (5) appropriately gave it its present name, *peritrophic* (surrounding the food), in 1890. Peritrophic membranes (PM) have been reviewed a number of times—first by Vignon (84) and subsequently by others (32, 56, 65, 88, 91, 92). There are also reviews for Crustacea (29, 30) and all animal phyla (55) and for some special topics such as penetration by parasites (38). In this review, with few exceptions, we cite only papers since 1950.

DEFINITIONS

We thought we were fairly familiar with the literature, but until we began compiling abstracts we did not realize that there obviously is no agreed-upon criterion or test for saying that a PM is present or absent. This is the same as saying that there is no single generally accepted definition for a PM. In this review we limit ourselves to just five possible characteristics:

1. A positive chitin test (usually the chitosan color test) from fragments associated with the food bolus or feces (e.g. 85, 86), provided the species is not cannibalistic, predaceous, or inclined to consume its own exuvium (safe indeed with adult Lepidoptera, etc). This criterion is commonly combined with one or more of the following characteristics.
2. A membrane or set of membranes forming a tube or sack around the food and capable of being dissected off or washed free (e.g. 52, 72).
3. A line separating the food from the epithelium in histological sections. Goodchild (31) commented that he has frequently seen lines suggestive of a PM in serial

[1]Paper No. 9448, Scientific Journal Series, Minnesota Agricultural Experiment Station, St. Paul, Minn. 55108.

sections of Heteroptera but has been unable to locate a membrane upon dissection. We concur and might add that we have seen sets of serial sections where some of the sections showed no trace of a line that might represent a PM, others showed fragments, and yet others showed a line around the entire food bolus (light microscopy). Perhaps such lines are artifacts. Other current difficulties deal with individual variability (86), differences in possible artifact production for light vs electron microscopy, and questions about membranes that topographically fit PMs but otherwise seem more characteristic of a glycocalyx (41, 50).

4. Any recognizable layer around the food produced by the MG cells (27, 54, 55). There are difficulties: How does one recognize them? How does one distinguish between PM, mucous, and glycocalyx? Are they all PMs? Are those species in which an evident PM has been shown to disperse in water (25) to be said to have or to lack a PM?

5. Any membranous or filamentous secretion of MG cells, whether or not it is concerned with food (e.g. cocoon precursors) (75, 81, 82). This is most extreme with the beetle *Murmidius,* where no PM was detected in feeding larvae, but after the cessation of feeding the MG begins to secrete chitin-containing threads that are used for cocoon formation. Peritrophic, then, is a misnomer, but probably few would object to calling the material modified PM.

For our purposes in this review we consider any solid or seemingly solid secretion of the MG epithelium to be a PM. Most, although not all, of these contain some form of chitin.

OCCURRENCE OF PERITROPHIC MEMBRANES

Species reported to possess or to lack PMs are listed in the various reviews cited (see 29, 55, 56, 86). To summarize those reviews, most insects have PMs in both larval and adult instars. PMs are found in insects whose diets contain harsh particles, as well as in many insects, such as adult butterflies and bloodsucking flies, that have diets with no rough particulate matter. PMs are usually reported as absent in insects that suck plant juices, especially Hemiptera and Homoptera [for recent references see (13, 22)]. Because in a considerable number of insects (Diptera) the PM is secreted only after distension of the gut, negative reports are to be viewed with caution, unless it is known that food is present and has been for at least a day. In tabanids, PMs were found in only 14 of 17 specimens examined (86). In most adult mosquitoes a definite membrane is present, but in one species (*Anopheles atroparvus*) it is reported that there is no definite membrane but only a viscous material (25). Waterhouse (86) reports that, even within single families of moths, adults of some species possess a PM, whereas other species are negative.

ORIGIN OF PERITROPHIC MEMBRANES

There is no question among PM workers that PMs are formed by secretion from the MG epithelium. The only question is whether the MG is of endodermal, ectodermal, or of some other embryonic origin. Not many people today worry about this old controversy (see 11, 26, 33, 47, 64), but one still occasionally sees a reiteration of the old, and never proven, dictum that only cells of ectodermal origin can secrete chitin (36, 48).

In terms of cell differentiation, the one obvious line of demarcation between foregut and MG is the cell boundary where the cuticle of the foregut ceases and the cells become microvillate (see Figure 5; see also 69). The material condensing into PM is secreted by these microvillate cells, which we consider to be MG (see Figure 4). Cells secreting PM precursors may be localized at the anterior end in the proventricular region, distributed throughout the MG, or both (see reviews cited). In the first case, a certain group of cells are presumably specialized for producing PM (59, 71); in the second case, PM secretion seems to be done by cells that also secrete enzymes and absorb the products of digestion. Studies to discover different cell types in these MG epithelia have revealed either only one type (9, 50) or several types, with no evident correlation of any particular type with PM production (2, 89).

Time of Appearance

Most embryologic papers stop with the formation of the major organ systems and therefore do not state when PMs first appear. In the few cases where a PM is mentioned for an embryo, it is not clear whether this PM is the same as at later stages (11, 26). For the embryo of *Drosophila,* Rizki (71) reported that a PM equivalent to that of the larva begins to be secreted by a ring of cells (four cells wide) at the anterior end of the MG at 80% of developmental time to hatching. This same ring of cells, characterized by heavy staining with the PAS reaction, continues to secrete PM throughout larval life. A PM has been reported in just-hatched larvae of *Aedes aegypti* (69). In contrast, several authors report that the first-instar larva of the honey bee lacks a PM at hatching and does not develop one until several days later (6). Presumably any species that does not produce a PM except after the stimulus of feeding would not have a PM at hatching, but this has not been reported, perhaps because all of the reported cases of stimulated production of PMs involve adult bloodsucking flies.

THE SO-CALLED TYPES OF PERITROPHIC MEMBRANES

Much of the literature on the classification of PMs refers to two types (91, 92). In type I, which includes most insects, the PM is produced by the entire MG epithelium; in type II, which shows best development in Diptera, production is restricted to a belt of cells at the anterior end of the MG. Commonly, the term delamination is used for production of type I PMs and the term secretion for type II; however, in terms of phenomena of cell biology there is no reason to think there is any real difference—both are secretions and neither is truly a delamination. In several orders of insects there is production at the anterior end of the MG, which is supplemented more towards the posterior end (33, 56, 101). There are also cases where PM production is limited to a belt of cells in the middle third of the MG (the beetle *Ptinus*) or a belt at the posterior end of the MG (the weevils *Cionus* and *Cleopus*) (75). These latter cases seem to be as distinct from type I as the situation in Diptera. One could list all of the above as separate types or could recognize that many insects produce PMs from all MG cells (which may all look alike), whereas other insects, as a byproduct of mosaic differentiation of the MG, have restricted PM production to one or another part of the MG.

A more useful distinction might be to distinguish between the continuous production of PM in contrast to a response to the stimulus of a distended MG. This distinction was originally proposed by Stohler (79) for biting mosquitoes and subsequently has been studied in detail by a number of investigators (9, 24, 25, 27, 70, 73, 95). Similar reports have been made for biting females of various species of *Simulium* (21, 39), *Culicoides* (42), and *Phlebotomus* (28). All of these reports have much in common, except for considerable variation in the amount of time reported to elapse between feeding and the appearance of a PM (20 min to 8 hr for various mosquitoes, 5 hr for *Culicoides,* and 24 hr for *Phlebotomus*). It is also reported that at least 0.5 mg of blood must be ingested by an adult female of *Aedes aegypti* to induce formation of a complete PM, but that minute amounts of blood suffice in *Anopheles stephensi* (24). Incidentally, a second blood meal in adult mosquitoes becomes surrounded by a second PM, which can also enclose the first blood meal and its PM (86).

On the basis of sequential histological appearances, Gänder (27) has suggested that the mosquito MG may exhibit a temporal separation of synthetic activities, i.e. on stimulation produce PM precursors, then stop and start producing digestive enzymes. This might be compared to epidermis first producing epicuticle, then procuticle. Such a temporal separation is presumably not found in insects that produce PM continuously.

Although a sugar meal may not induce a PM in adult mosquitoes (it may go into the crop instead of the MG), Freyvogel & Jaquet (24) report that a PM can be induced by an enema of saline solution or even air! These enema membranes are formed at about the same speed as those induced by a blood meal (4–6 hr in *Aedes aegypti,* 15–20 hr in *Anopheles stephensi*) and seem to have the same chemical composition (27). This supports the suggestion that the stimulus for PM production is physical distension.

Here under the heading of types may be a good place to summarize recent, surprising findings: certain beetles use chitinous materials from the MG to form cocoons. This was discovered by Streng (81, 82) and has been further analyzed by Rudall & Kenchington (75). Briefly, (*a*) the MG or some portion of it may secrete a tubular PM that surrounds the food and that may, with no detectable modification, be used for forming a cocoon (*Prionomerus*); (*b*) the PM may be flattened into a ribbonlike thread for use as cocoon material (*Gibbium*); (*c*) the MG may shift from secreting tubular lamellae to producing thick fibrous assemblages that become organized into slender threads (*Ptinus*); (*d*) a MG that seems to produce no PM around the food bolus may commence secreting ribbons when it is time to make a cocoon (*Murmidius*); or (*e*) a MG, part of which delaminated typical PMs during feeding stages, may reorganize its epithelium with the development of deep crypts where ribbons (70 × 10–20 μm) of leaflike or feathershaped assemblages of microfibers (6 nm wide) are produced as cocoon material (*Cionus*).

The Endo- and Ectoperitrophic Spaces

The endoperitrophic space is where the food is located; the ectoperitrophic space is that between the PM and MG epithelium. With continuously delaminated membranes the ectoperitrophic space may be small and possibly of no functional significance. However, when the PM is secreted as a tube by a band of cells near the anterior end of the MG, the ectoperitrophic space may be considerable (and at times may equal the volume of the endoperitrophic space).

The contents of these two spaces is of course different, since undispersed food does not go through the PM. There are two points of significance about this separation. (*a*) The rate of movement in the ectoperitrophic space is very much slower and may even be in the posterior to anterior direction (10, 35, 38, 62). Materials entering this space may then be acted upon for much longer than one would expect from measuring the rate of passage of food through the gut. (*b*) Some parasites that enter this space, either through a fluid anterior end of the PM or by forward migration from the open posterior end of the PM (even if in hindgut), seem to find it a haven where they can undergo further development without danger of being passed out through the anus (20, 35, 38, 39, 45, 80, 94).

SYNTHESIS AND SECRETION

The brilliant analysis of secretion in vertebrate cells by Palade and his colleagues (which led to his being elected a Nobel Laureate in 1974) can be summarized as secretion consisting of six steps: (*a*) synthesis by ribosomes on rough endoplasmic reticulum (ER); (*b*) segregation of the product into cisternae of the ER; (*c*) intracellular transport to Golgi areas; (*d*) concentration in vesicles associated with the Golgi; (*e*) intracellular storage in secretory granules or vacuoles that get moved to the cell surface; and (*f*) discharge through the surface. The above was worked out over many years by elaborate studies with tracer techniques, primarily on the exocrine cells of the mammalian pancreas (53).

Numerous workers, beginning with Bertram & Bird (9), have pointed out that the above sequence is not to be seen in MG cells of various insects. However, in the work cited above, Palade remarks briefly that some secretory cells shorten the sequence. Specifically, fibroblasts, chondrocytes, and plasma cells omit the concentration step, fail to form secretory granules visible in sections viewed with an electron microscope, store the product for only a short time or not at all, and discharge the product continuously. With such modifications, Palade's picture could be applied to what can be seen in insect MG cells (Figures 1–4), but we can never feel really safe about this until someone performs appropriate tracer tests.

What can be said is that insect MG cells, including ones differentiated for producing only PM, contain tremendous numbers of ribosomes on a well-developed ER (Figures 1–2); a considerable number of typical mitochondria; large numbers of rather diffuse areas, which sometimes definitely have the characteristics of Golgi (Figure 1) but at other times (or in other species) border on being amorphous (Figures 2–3); no secretory granules or secretory vesicles moving to the microvillate

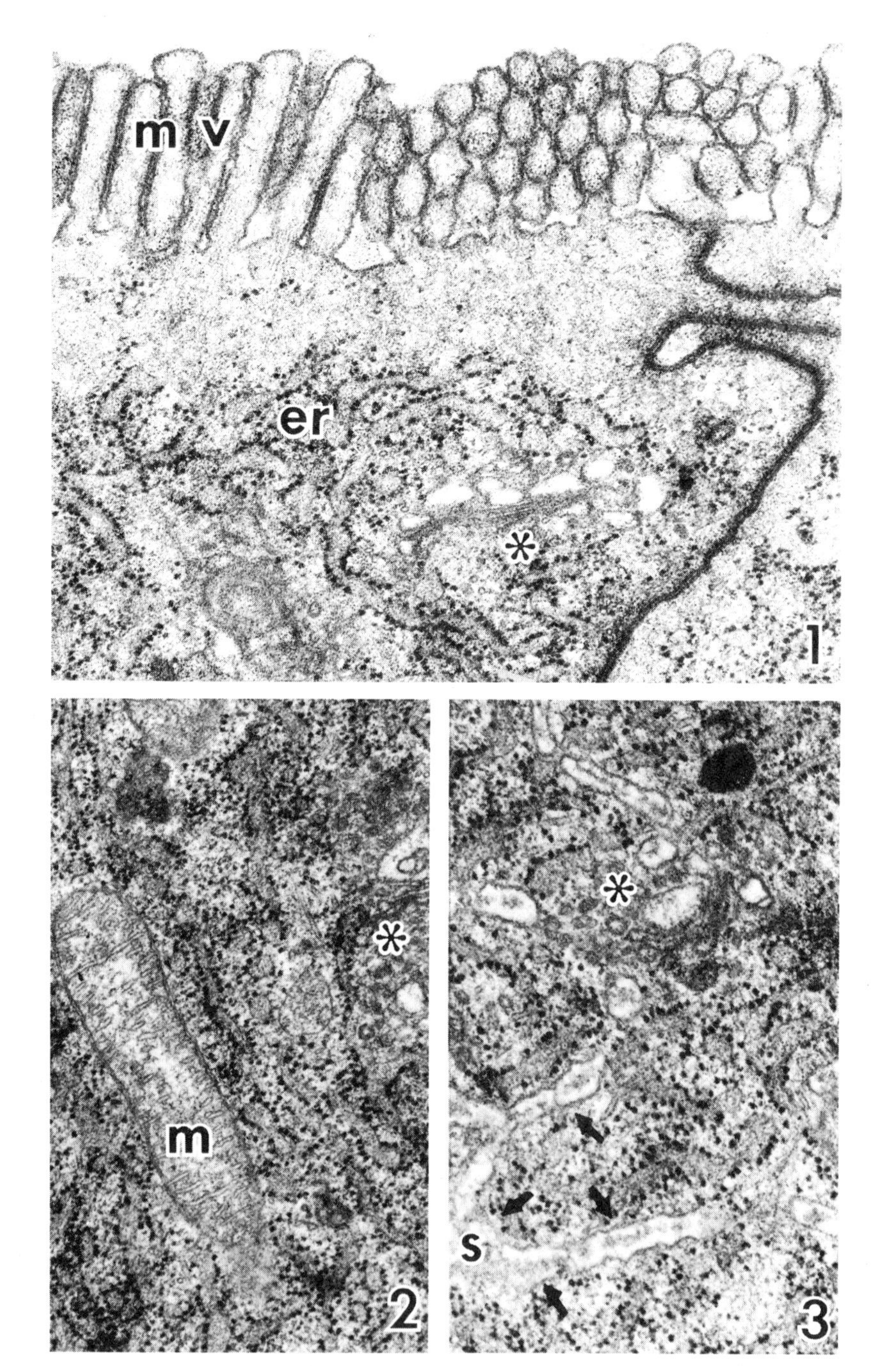

m v
er
*
1
*
m
2
*
s
3

surface for discharge; and a secretion that, whether continuous or discontinuous, appears in electron micrographs as an amorphous, finely granular product in which microfibers appear and aggregate as postsecretion phenomena (Figures 4, 6, 7). In the region where PM is secreted in mosquito larvae, there is an organelle-free region 0.25 to 0.5 μm thick at the base of the microvilli. Basal labyrinth is well developed throughout the MG (66).

Moloo et al (45) report three types of cells in the pad that secretes PM in *Glossina* adults, but their interpretation seems more subjective than well documented. Becker et al (8) give a better case for three components from three areas in adult *Calliphora.*

Just what is secreted remains to be determined, but it is presumably a mixture of precursors that, after secretion, condense into the definitive PM. Presumably the material is in fluid form at the time of secretion. This is consistent with the fact that in bloodsucking flies the PM is a blind sack closed posteriorly (44, 93; see also above references); it could hardly become a blind sack unless the material could flow around the blood meal. In some cases bacteria are seen in a PM under circumstances that lead one to suspect that the bacteria got into a still-fluid PM (A. G. Richards and S. Seilheimer, unpublished data), and a number of authors have postulated that certain parasites (especially trypanosomes) penetrate through a fluid anterior end of the PM (discussed below). Furthermore, it seems unlikely that the known rearrangements of microfibers into mats or laminae (Figures 6–7) could take place if the medium were not fluid.

Differentiation of PM After Secretion

In numerous dipterous larvae all stages of PM formation may be found in a single section. Longitudinal sections of the proventricular region of mosquito larvae show all stages in PM formation from an initial, rather diffuse granular layer at the anterior end to the more condensed, completed PM, which consists of the first-secreted granular layer (Figures 6*a* and 7*a*) followed by coarsely fibrous layers (Figures 6*b* and 7*b*), which differentiate within the granular layer and which are connected by some very fine but random microfibrils (Figures 6*c* and 7*c*) (68, 69). Studies involving a series of extractions show that all three states of aggregation (nonfibrous, microfibrous, and microfibrillar) contain chitin (69). More important, these studies showed that the formation of microfibrils and the subsequent aggrega-

---

Figure 1 (*opposite, top*) Section of midgut epithelium of fourth-instar larva of *Aedes aegypti* in pouch where peritrophic membrane is secreted. Shows microvilli (*mv*) in longitudinal and cross section, the 0.25- to 0.5-μm band of clear cytoplasm at base of microvilli, ribosomes on rough endoplasmic reticulum (*er*), an intercellular boundary, and a typical Golgi apparatus (asterisk). Fixed with glutaraldehyde followed by OsO_4, stained with uranyl acetate and lead citrate. Magnification 48,000X.

Figure 2 (*opposite, lower left*) Portion of another cell showing a mitochondrion (*m*), much rough ER, and a vague area that may represent Golgi (asterisk). Magnification 42,000X.

Figure 3 (*opposite, lower right*) Another area within same cell showing what is presumably secretory material (*s*) within the cisternae of smooth ER (arrows) and leading towards a presumed Golgi area (asterisk). Magnification 48,000X.

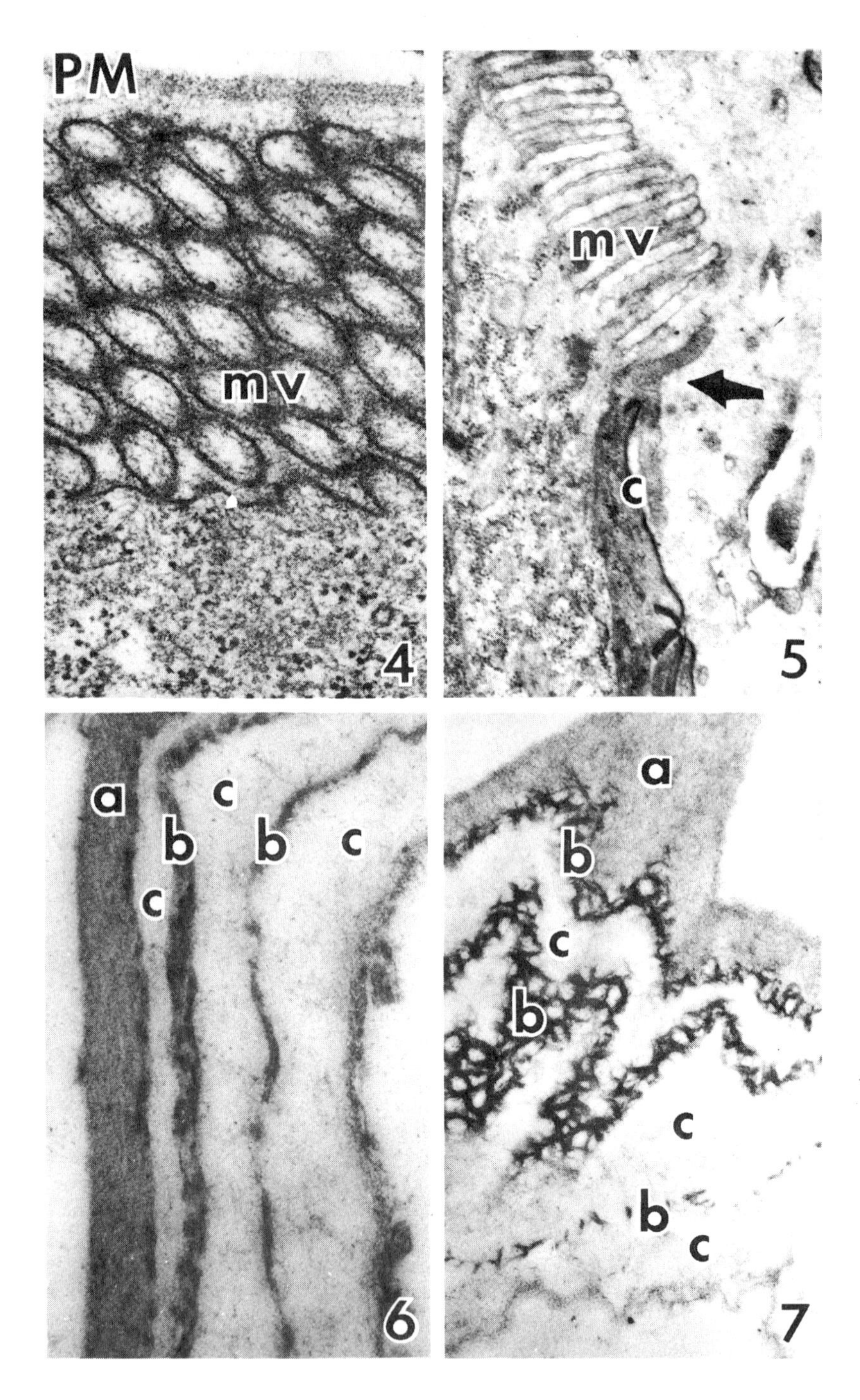
PM
mv
4
mv
c
5
a
c
b
b
c
c
6
a
b
c
b
c
b
c
7

tion into larger microfibers can be the result of postsecretion phenomena after the mixture is far removed from all cell surfaces.

In a favorable set of longitudinal sections of the proventricular region of a larva of *Anopheles freeborni,* the PM goes from trace thickness to full thickness over a distance of six to eight cells. After full thickness of the granular-appearing secretion is obtained, aggregation into fibrous mats begins. The first layer formed is at the tips of the microvilli, then a second one forms outside this, and finally a third mat forms outside the first two. There is no change in total thickness of the PM during formation of the mats (A. G. Richards and S. Seilheimer, unpublished data).

Sections of the MG of *Simulium* larvae reveal a much more complicated PM [see Figure 4 in reference (66)]. There are extremely regular fibrous meshes (orthogonal) of three different magnitudes (approximately 15, 35, and 120 nm), and these bear a relationship in terms of center-to-center spacing of fibrous aggregates of 1:2:8 (Figure 8). Only the largest of the spacings are in the size range of microvillar spacings. We had hoped to have the development of this PM fully worked out before writing this review, but the story is turning out to be too complicated for quick solution. It seems that fibrils for some of the meshes will be found to be formed at their final position in the membrane (Figure 9), but others will be aggregated as small units, which are then assembled (Figures 10–12). The last may bear similarity to the leaflike assemblages secreted as cocoon materials by the MG of the beetle *Cionus.* Further work on simuliids is needed, but since they do make cocoons the comparison may be fruitful. (However, the cocoons of *Simulium* are not only dispersed by concentrated KOH at 160° C but even by overnight in 1 N NaOH at 60° C. Presumably, then, they do not contain a significant amount of chitin.)

The reported diameters of the finest microfibrils in arthropod PMs are in the 5 to 10 nm range, usually closer to 5 nm. Synthetic chitin microfibrils are significantly larger (76). The microfibrils in PMs aggregate into larger microfibers with diameters up to 10 times the single fibril (19, 29, 44, 56, 69). The modified PM fibrils used for cocoon formation are reported to be even larger (10–50 nm) (see 82).

Figure 4 (*opposite, top left*) Oblique section through microvilli (*mv*) of a first-instar larva of *Aedes aegypti* showing granular secretion, which is condensing into a granular PM (no fibrous layers in early first-instar larvae). Preparation as in preceding figures. Magnification 60,000X.

Figure 5 (*opposite, top right*) Longitudinal section of proventricular pouch of last-instar larva of *Simulium,* subgenus *Simulium,* probably *venustum.* Shows end of esophageal cuticle (*c*) and appearance of microvilli (*mv*) at one particular cell boundary (arrow). Magnification 25,000X.

Figure 6 (*opposite, lower left*) Section of fully formed PM of last-instar larva of *Aedes aegypti.* The first secreted granular layer (*a*) shows no microfibrils. Material secreted subsequently condenses to give layers (usually three or four) of microfibers (*b*), which are separated but also are held together by less dense material containing randomly oriented microfibrils (*c*). Magnification 45,000X.

Figure 7 (*opposite, lower right*) Section of another fully formed PM of a last-instar larva of *A. aegypti.* A small fold in the PM results in an almost tangential section of one of the fibrous layers. Labeling as in Figure 6. Magnification 22,500X.

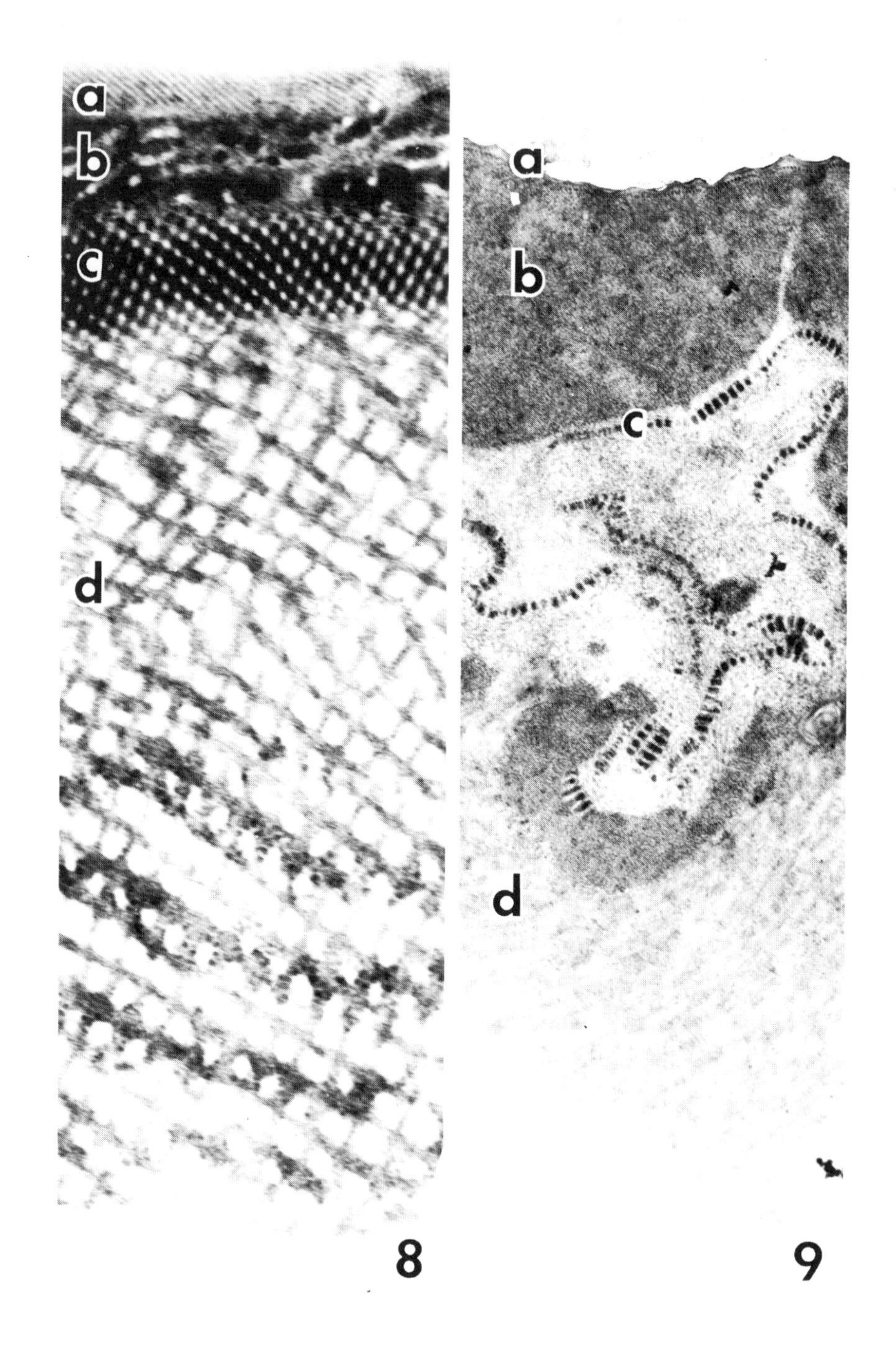
a
b
c
d
8
a
b
c
d
9

In many cases the microfibers are randomly distributed among the other components of the PM and form a feltlike texture. Commonly, however, the microfibers are aggregated both into successive layers or laminae (Figures 6–7) and into meshworks, which usually are either hexagonal (fibers set at a 60° angle to one another) or orthogonal (fibers at 90° angle), as beautifully illustrated in the papers by Peters (55, 56) and Georgi (29, 30). The shadowed whole mounts used by Peters and Georgi show these grids much better than they are ever seen in sections.

Since the mesh spacing of PM grids is commonly correlated with the center-to-center spacing of microvilli (mostly 150–200 nm), Mercer & Day (44) postulated that the mesh is templated in the depressions between tips of microvilli. Although this postulate has been accepted by several authors (48, 56), it is considered inadequate by others (69, 75) because meshworks can be formed in unorganized PMs after spatial separation from the microvilli. Presumably there is some explanation for the common but not universal correlation of mesh spacings and microvillar spacings, but it cannot be as simple and direct as envisioned by Mercer & Day. [In a chilopod, Rajulu (61) reports that grid spacings change from 23–38 nm in the anterior part of the MG, to 36–49 nm in the middle part, and to 46–58 nm in the posterior part. It would be interesting to know if the spacing of microvilli shows the same gradation.]

In connection with the hexagonal vs orthogonal vs random or feltlike meshes, it should be remembered that these can all be seen within relatively short distances in a single PM (e.g. in the beetle *Ptinus tectus*) (75). In another beetle, *Gibbium psylloides,* the same authors report that feeding larvae give PMs with only small areas of random orientation whereas spinning larvae produce predominantly the random type. There is nothing to indicate that the arrangements called hexagonal, orthogonal, and random represent fundamental or important differences.

Rates of Formation

A number of values are given in the literature as to either millimeters per hour or number of sheets delaminated per hour; however it does not seem possible to evaluate these.

For delaminated membranes, *Aeschna* larvae are reported to produce two per day (4), bee and wasp larvae produce a half-dozen per day (63), and damselfly larvae produce three per day (one per day when starving) (87).

Figure 8 (*opposite, left*) A slightly oblique section through the fully formed PM of larva of *Simulium,* subgenus *Simulium,* probably *venustum.* Layer *a* is the fine outer grid; *b,* amorphous chunk layer; *c,* larger inner grid; and *d,* largest grid with spots, suggesting that the fibers may be formed of linear aggregates of particles. Fixed with glutaraldehyde and OsO_4, stained with $Ba(MnO_4)_2$. Magnification 50,000X.

Figure 9 (*opposite, right*) From the same section as Figure 8 but up in pouch at level where the PM is less than half formed. Layer *a* is complete but faint; layer *b,* uniformly granular; layer *c,* forming but still an incomplete layer; and layer *d,* a vaguely granular layer not yet showing any trace of the large grid. Magnification 38,000X.

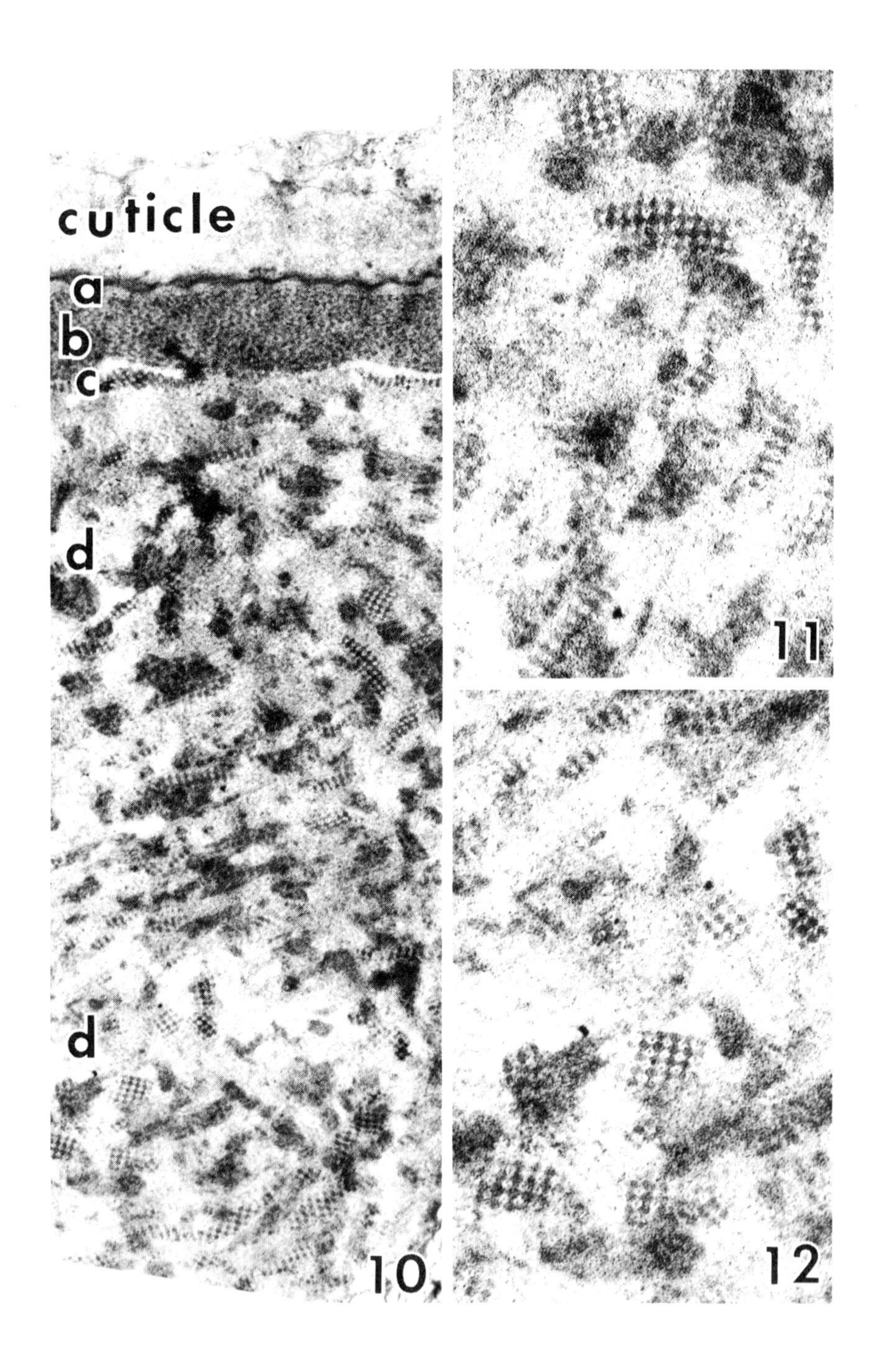
cuticle
a
b
c
d
d
10
11
12

For secreted tubes that protrude from the anus, *Eristalis* larvae produce 5–6 mm hr^{-1} dropping to 2 mm hr^{-1} when starving (3, 87), *Glossina* adults produce 1 mm hr^{-1} (93), *Calliphora* larvae produce 5–10 mm hr^{-1} (87), and earwigs produce 1.6 mm hr^{-1} at 30° C, falling to 0.25 mm hr^{-1} at 18° C (87). Mosquito larvae are said to produce a length of PM equal to the length of the MG every 30–40 min (14). Waterhouse (87) remarks that individual variation is considerable, commonly more than tenfold! Constancy of thickness of tubes of the dipterous type imply both constancy in production and backward movement. Correspondingly, an abrupt change in thickness and number of laminae might imply a rate change [see Figure 3 in reference (66)]. In the special case of a modified PM being used for cocoon material, Streng (82) says this is produced at the very high rate of about 10 mm min^{-1}. In another special case, proventriculi of *Calliphora* adults grown in vitro, the recorded rate is slower, i.e. 2–4 mm hr^{-1} (7, 8, 99).

Mosquito larvae stressed by disease or DDT may frequently show long streamers of PM tube protruding from the anus (1, 14). We, too, have occasionally seen mosquito larvae with long protruding PM tubes. It is not clear, however, whether this represents increased PM production or delayed PM deterioration.

One point does seem clear: the rate of movement of food through the gut is 10–12 times faster than the rate of production of PM in dronefly and blowfly larvae (87), but of course it is essentially the same in mosquito and tsetse fly adults where the PM forms a sack around a blood meal.

The Backward Movement of PM

For the tubular PMs of Diptera and honey bee larvae, it is commonly assumed that posterior movement is caused by pressure of continued secretion (37). Certainly it seems independent of the peristalsis of the MG (14). It has been suggested that minute spines or circumferential flanges in the proventriculus, coupled with muscular activity of this organ, help push the PM posteriorly (69, 91). In some cases, spines in the hindgut are reported to pull the PM posteriorly (60, 93).

The above can hardly be true for PMs delaminated by the whole MG. Presumably these would be affected by peristalsis and by further ingestion.

Changes with Age of the Insect

Changes that occur with age have seldom been recorded. Henson (33) remarked that the PM of first-instar *Vanessa urticae* larvae is too delicate for easy removal, but

Figure 10 (*opposite, left*) Section of the proventricular area of another last-instar larva of *Simulium*, subgenus *Simulium*, probably *venustum*. The outer part of the PM in this specimen is pressed tightly against the cuticle of the reflexed surface of the esophagus (transverse black line is epicuticle). Layers *a*, *b*, and *c* are similar to those in Figures 8 and 9; layer *d*, which has no continuous grid, has thousands of little pieces that are somewhat gridlike but resemble the basement membranes of the MG of fleas and mosquitoes. Fixed in glutaraldehyde followed by OsO_4; stained with uranyl acetate and lead citrate. Magnification 25,000X.

Figures 11 and *12* (*opposite, right*) Higher magnifications within region *d* of Figure 10. Magnification 50,000X.

that in older larvae it is readily removed as a colorless membrane investing the food. Similarly, in first-instar larvae of *Aedes aegypti, Anopheles albimanus,* and at least some other mosquito species, the PM, as seen in sections for the electron microscope, is thin and devoid of fibrous mats or grids (Figure 4), whereas in third- and fourth-instar larvae the PM is both thicker and has relatively thick fibrous layers (Figures 6–7; see 69). It seems quite likely that the above will be generally true for insects.

Differences when feeding larvae change to spinning larvae have already been stated.

Although there are only the slight changes mentioned above between larval instars and between larval and adult instars in hemimetabolous insects, there may be great differences between larval and adult PMs in holometabolous insects. This is best known for mosquitoes where larvae continuously secrete a tubular PM from a band of cells at the anterior end of the MG (69, 91), whereas adults delaminate an entirely different-appearing PM from the whole MG, but only on the stimulus of distension (9, 24). Similar differences occur for larval vs adult *Simulium;* differences are also recorded between larval and adult PMs in *Sarcophaga* (51) and *Apis* (34). All caterpillars have PMs, but adults of less than half the 128 species examined (38 families represented) have a PM (86).

Recently, PMs have been reported from pupae of various mosquitoes (72, 74). These determinations are from dissections and little data are given. The membranes are said to surround the meconium and to contain chitin, and they seem to resemble the PMs of adults rather than of larvae. We suspect that they are produced by pharate adults rather than pupae, but we cannot be sure from the published data.

Changes in the PM with Time

As already stated, it seems that the PM materials are secreted in fluid form and condense into a solid membrane within minutes or at most a few hours. However, it is not a highly durable membrane despite the common grids of chitin microfibers (the actual percentage of chitin is never high).

Some authors use terms such as viscous, elastic, hard, fragile, etc, without having precise data for those properties. What is certain is that an immature membrane matures (i.e. becomes stabilized and usually solid) and then begins to deteriorate. Since the PM exists in an environment with active digestive enzymes and since it has been shown that trypsin and pepsin remove some components from PMs (67), it is hardly surprising that the PM has a short and finite life. In fact, we suggest that the continual or repeated renewal of PM may be necessitated by the fact that PMs can survive the MG enzymes for only a few hours. [It has been suggested (61) that acid mucopolysaccharides in the PM give protection from digestive enzymes. Conceivably this is so, but we know of no data supporting the speculation.]

Microscopically (light or electron), one can commonly see in sections of various insects that the PM becomes thinner and more ragged and shows signs of beginning to disintegrate in the posterior part of the MG (28, 39, 60, 69, 90). In numerous reports it is said that mechanical factors, especially spines in the hindgut, assist in the breakup (60, 93).

Periodicities in some PMs

An anisotropic banding was reported for the PMs of some Diptera by Zhuzhikov (97) and was further analyzed by Peters et al (57, 99, 100). This is really a corresponding alternation of circumferentially and longitudinally oriented fibrous regions with a center-to-center spacing of 26–90 μm (there is much less variation if stated for only the PM of one species, i.e. the outer PM layer of *Lucilia* adults has spacings of 54 ± 10 μm). The periodicity is somewhat correlated with the diameter of the PM tube (300–500 μm).

We are tempted to compare this situation with tracheae where the spacing of taenidia is distinctly although not precisely related to tracheal diameter, where the ratios of periodicity to tube diameter are similar, and where it has long been known (67) that the microfibers are oriented circumferentially to the tube in taenidia but longitudinally between taenidia. However, in tracheae, the longitudinal components are really continuous and the circumferential elements are stronger. Richards (65) suggested and Locke (40) analyzed the possibility that the tracheal situation could fit engineering analyses where stress forces are viewed as inducing this sort of alternating orientation. Since electron microscope pictures of tracheae (from numerous different insect species) in our possession show ratios of periodicity (spacing of taenidia) to tube diameter of 0.1 to 0.5, and the PMs measured by Peters et al (57) give ratios of 0.06 to 0.6, we wonder if the same engineering equations might not be applicable.

CHEMICAL COMPOSITION

Other than the qualitative identification of chitin as one component, little is known about the composition of PMs. There are several reasons for this: usually only a small amount is available [however Japanese workers made analyses of 30,000 isolated and washed PMs weighing 18 g (49)]; there is always the possibility of contamination with food and other items of the gut contents; there is also the certain knowledge that the PM is deleteriously affected by digestive enzymes and hence will change in composition postsecretion. For persons interested in making analyses, two recent methods offer promise: (*a*) culturing cardia of adult Diptera in vitro, thereby getting usable amounts free from food and free from digestive enzyme effects (7, 8); and (*b*) inducing enema PMs in species that make PMs only when stimulated by distension (24, 27).

Chitin is present in the PMs of most insects but the percentages are much lower than the percentages in various cuticles. Using an enzymatic analytical method, de Mets & Jeuniaux (18) report 13% chitin in PMs of *Bombyx mori* larvae, 11% in *Aeschna* larvae, and only 3.7% in *Dixippus.* [A similar range of only 13–20% is reported for PMs of polychaete annelids (15).]

Everyone agrees that protein is also present. Quantitative estimates are 35–40% for *Bombyx* and *Antherea* larvae (52), 47% for *Dixippus,* and 42% for *Bombyx* larvae, but only 21% in *Aeschna* larvae (18). Several authors have given amino acid analyses (52, 61, 100), which show relatively high values for the diamino and dicarboxylic acids, even higher than the values recorded for these in proteins from cuticle. This suggests no relationship to collagen or silk fibroin but a general similarity to arthropodins.

Presumably the microfibers forming the PM grids are chitin-protein aggregations. In mosquito larvae they are stained by uranyl acetate and lead citrate, and, although they are not significantly affected by the first four treatments of Hackhan's series, they are by warm 1 N NaOH, which ruptures chitin-protein bonds and removes all protein (69).

Several investigators have reported acid mucopolysaccharides, neutral mucopolysaccharides, mucins, hyaluronic acid-like components, or just nonchitin polysaccharides (16, 18, 43, 61). Comparison to a glycocalyx is sometimes made (13, 41, 50). For larvae of *Bombyx mori* these components are said to account for 15% of the total dry weight (49).

Lipids have seldom been mentioned. Mello et al (43) say simply that lipids are present in the PM of honey bees. Gänder (27) says that, in enema membranes from mosquitoes, the PM lipids can be separated by thin-layer chromatography into fractions moving with the speeds of known lecithin, cholesterin, and sterol esters. No one gives quantitative values for any lipid and no one states any positive identifications.

It is disturbing that if one adds up the quantitative values reported one never obtains a total as high as 75%, and commonly it is less than 50%. Thus, for *Dixippus morosus* there is listed 3.7% chitin, 47% protein, 3.8% hexosamine, 8.0% glucose, and 3.4% glucuronic acid, a total of 66% (18). Where is the discrepancy?

We have used the term chitin as though it referred to a single known entity, which is not entirely true. Jeuniaux (36) has spent many years developing the concepts of free and bound chitin. Free chitin is digested by pure chitinase; bound chitin requires some kind of treatment before being attacked by chitinase. Most of the chitin in PMs is free.

Also, Rudall has elaborately developed the fact that X-ray diffraction analyses show that there are at least three different types of chitin in crystallographic terms. In other words, chitin chains can be associated in at least three ways (just as bricks can be put together in different ways for making a wall). We do not have space to explain this. Rudall and Kenchington (75) have recently given a beautifully detailed presentation of chitin types in PMs. In summary, although insect cuticle contains chitin in the α configuration, insect PMs may have chitin in either α, β, or γ configurations. Chitin in the α form is found in PMs of bees and wasps and of the weevil *Cionus* during feeding stages. Chitin in the γ form seems to be the most common for insect PMs (beetles, locusts, cockroaches, mantids, dragonflies, caterpillars, and sawflies). Chitin in the β form is produced by the MG when the secretion is to be used for making a cocoon (several beetles).

PERMEABILITY

There is little significant literature on permeability. Obviously, digestive enzymes and the products of digestion somehow get through, but in opposite directions. Zhuzhikov (98) says that digestive enzymes go through the PM of mosquito larvae but that food components of comparable size do not, and therefore the PM is a special type of membrane. We know of no good evidence for this statement.

There is some old literature (17, 46, 58) stating that various dyes usually penetrate PMs, though the reports differ as to whether large dye molecules of colloidal dimensions (e.g. congo red, trypan blue) do or do not penetrate. Other reports state that PMs are readily penetrated by water, salts, and small organic molecules (96), and one also says by radioactive barium^{140} and lanthanum^{140} (10). More definite data are given for mosquito larvae by Schildmacher (77), who reported that colloidal gold particles of 2 to 4 nm diameter penetrate, but that particles of 20 nm do not. Zhuzhikov (98) extended this study and reported that particles up to 9 nm penetrate, although ones less than 5 nm are much faster. For comparison, a hydrated glucose molecule has a diameter of 1 nm or a bit more, sucrose about 2 nm, and the various amino acids 1–3 or 4 nm.

Probably considerable differences will be found for different insects. For instance, Smith et al (78), without specifically mentioning the PM, record the uptake of ingested ferritin molecules (11 nm diameter) from the gut of a caterpillar, whereas A. G. Richards and S. Seilheimer (unpublished data) have found no penetration of ferritin through the PM of a mosquito larva. Also, Wigglesworth (90) reported the penetration of large hemoglobin molecules through the PM of tsetse flies, whereas Zhuzikov (96) reported no penetration of proteins for house flies and blowflies.

Numerous papers mention or attempt to treat the penetration of viruses and microorganisms; they are reviewed by Stohler (80) and LeBerre (38). Obviously, these must somehow get past the PM or transmission would be limited to regurgitation (conceivably nonpenetration could account for some insect species not being vectors, but we know of no documentation).

It seems clear that ookinetes of malarial parasites can wander along and wriggle through the PM of adult mosquitoes "by means of annular waves of contraction" (23, 79). But Trager (83) has warned against generalizing either from one protozoan to another or from one insect to another. And the reports do not all agree. Thus Hoare (35) concluded that trypanosomes in tsetse flies do not penetrate the PM but always migrate forward in the ectoperitrophic space from the hindgut, whereas Moloo et al (45) say they penetrate at the anterior end, where they assume the PM is fluid. Insect viruses have diameters of 15–50 nm (12) but direct evidence on how these invade insects is lacking.

THE FUNCTIONS OF PERITROPHIC MEMBRANES

Almost always it is said that PMs function in protecting the MG epithelium from abrasion by food particles. Certainly they do prevent such harsh particles from making contact with the MG cells, but this does not explain why a few insects (e.g. some adult Neuroptera and Mecoptera) seem to lack a PM or why numerous juice feeders possess one (some adult Lepidoptera, etc). Several authors have pointed out that there is no good correlation to phylogeny or nutrition (e.g. 56).

PMs do keep food out of the gastric caeca, but many insects lack caeca and Heteroptera lacking a PM may have elaborately developed caeca. PMs also may produce an ectoperitrophic space where movement may be slower than in the endoperitrophic space. In a few cases PMs, or material from the same source, are used for making cocoons. Any physiological significance to permeability values is uncertain. We have a feeling that some significance is eluding us.

CONCLUDING REMARKS

With the thousands of papers that mention PMs and the hundreds that treat them specifically, it seems surprising that there is so much vagueness and uncertainty. Doubtless there are a variety of reasons, including the fact that the PM is incidental to many studies (e.g. parasite transmission). More particularly though, there have been few attempts to apply modern analytical methods, except for some details (e.g. α, β, and γ chitin characterization). It is time for PM workers to become knowledgeable in current methods of cell biology and to apply such methods, particularly with the recently developed in vitro (7, 8) and enema membrane (24, 27) techniques, which avoid the complications of contaminating food and might even avoid the presence of digestive enzymes.

ACKNOWLEDGMENTS

This work was supported by grant No. AI09559 from the Parasitology and Medical Entomology Branch of the National Institutes of Health.

Acknowledgment is made to Mrs. Sandra Seilheimer, for preparing most of the electron micrographs used, and Dr. E. F. Cook, for identifying the *Simulium* larvae.

ADDENDUM Since this manuscript was submitted a monograph has appeared with two chapters dealing with certain PMs (see 36a, 56a).

Literature Cited

1. Abedi, Z. H., Brown, A. W. A. 1961. Peritrophic membrane as vehicle for DDT and DDE excretion in *Aedes aegypti* larvae. *Ann. Entomol. Soc. Am.* 54:539–42
2. Anderson, E., Harvey, W. R. 1966. Active transport by the cecropia midgut. 2. Fine structure of the midgut epithelium. *J. Cell Biol.* 31:107–34
3. Aubertot, M. 1932. Origine proventriculaire et évacuation continue de la membrane péritrophique chez les larves d'*Eristalis tenax*. *C. R. Soc. Biol.* 111:743–45
4. Aubertot, M. 1932. Les sacs péritrophiques des larves d'Aeschna; leur évacuation périodique. *C. R. Soc. Biol.* 111:746–48
5. Balbiani, E. G. 1890. Etudes anatomiques et histologiques sur le tube digestif des *Crytops. Arch. Zool. Exp. Gen. Ser. 2* 8:1–82
6. Bamrick, J. F. 1964. Resistance to American foulbrood in honey bees. 5. Comparative pathogenesis in resistant and susceptible larvae. *J. Insect Pathol.* 6:284–304
7. Becker, B., Peters, W., Zimmermann, U. 1975. *In vitro* synthesis of peritrophic membranes of the blowfly, *Calliphora erythrocephala. J. Insect Physiol.* 21:1463–70

8. Becker, B., Peters, W., Zimmermann, U. 1976. The fine structure of peritrophic membranes of the blowfly, *Calliphora erythrocephala,* grown in vitro under different conditions. *J. Insect Physiol.* 22:337–45
9. Bertram, D. S., Bird, R. G. 1961. Studies on mosquito-borne viruses in their vectors. 1. The normal fine structure of the midgut epithelium of the adult female *Aedes aegypti,* and the functional significance of its modifications following a blood meal. *Trans. R. Soc. Trop. Med. Hyg.* 55:404–23
10. Bowen, V. T., Rubinson, A. C., Sutton, D. 1951. The uptake and distribution of barium^{140} and lanthanum^{140} in larvae of *Drosophila repleta. J. Exp. Zool.* 118:509–30
11. Butt, F. H. 1934. The origin of the peritrophic membrane in *Sciara* and the honey bee. *Psyche* 41:51–56
12. Chamberlain, R. W., Sudia, W. D. 1961. Mechanism of transmission of viruses by mosquitoes. *Ann. Rev. Entomol.* 6:371–90
13. Cheung, W. W. K., Marshall, A. T. 1973. Studies on water and ion transport in homopteran insects: ultrastructure and cytochemistry of the cicadoid and cercopoid midgut. *Tissue Cell* 5:651–69
14. Christophers, S. R. 1960. *Aedes aegypti (L.), the Yellow Fever Mosquito.* London: Cambridge Univ. Press. 739 pp.
15. Dales, R. P., Pell, J. S. 1970. The nature of the peritrophic membrane in the gut of the terebellid polychaete *Neoamphitrite figulus. Comp. Biochem. Physiol.* 34:819–26
16. Day, M. F. 1949. The occurrence of mucoid substances in insects. *Aust. J. Sci. Res. Ser. B* 2:421–27
17. von Dehn, M. 1933. Untersuchungen über die Bildung der peritrophischen Membran bei den Insekten. *Z. Zellforsch. Mikrosk. Anat.* 19:79–105
18. de Mets, R., Jeuniaux, C. 1962. Sur les substances organiques constituant la membrane péritrophique des insectes. *Arch. Int. Physiol. Biochem.* 70:93–96
19. Edwards, G. A., Souza-Santos, P. 1953. A estrutura da membrana peritrófica dos insetos. *Cienc. Cult.* 5:195–96
20. Fairbairn, H. 1958. The penetration of *Trypanosoma rhodesiense* through the peritrophic membrane of *Glossina palpalis. Ann. Trop. Med. Parasitol.* 52: 18–19
21. Fallis, A. M. 1964. Feeding and related behavior of female Simuliidae. *Exp. Parasitol.* 15:439–70
22. Forbes, A. R. 1964. The morphology, histology and fine structure of the gut of the green peach aphid, *Myzus persicae. Mem. Entomol. Soc. Can.* 36:1–74
23. Freyvogel, T. A. 1966. Shape, movement in situ and locomotion of plasmodial ookinetes. *Acta Trop.* 23:201–22
24. Freyvogel, T. A., Jaquet, C. 1965. The prerequisites for the formation of a peritrophic membrane in Culicidae females. *Acta Trop.* 22:148–54
25. Freyvogel, T. A., Staubli, W. 1965. The formation of the peritrophic membrane in Culicidae. *Acta Trop.* 22:118–47
26. Gambrell, F. L. 1933. The embryology of the black fly, *Simulium pictipes. Ann. Entomol. Soc. Am.* 26:641–71
27. Gänder, E. 1968. Zur Histochemie und Histologie des Mitteldarmes von *Aedes aegypti* und *Anopheles stephensi* in Zusammenhang mit der Blutverdauung. *Acta Trop.* 25:132–75
28. Gemetchu, T. 1974. The morphology and fine structure of the midgut and peritropic membrane of the adult female *Phlebotomus longipes. Ann. Trop. Med. Parasitol.* 68:111–24
29. Georgi, R, 1969. Feinstruktur peritrophischer Membranen von Crustaceen. *Z. Morphol. Tiere* 65:225–73
30. Georgi, R. 1969. Bildung peritrophischer Membranen von Decapoden. *Z. Zellforsch. Mikrosk. Anat.* 99:570–607
31. Goodchild, A. J. P. 1966. Evolution of the alimentary canal in the Hemiptera. *Biol. Rev. Cambridge Philos. Soc.* 41:97–140
32. Gooding, R. H. 1972. Digestive processes of hematophagous insects. *Quest. Entomol.* 8:5–60
33. Henson, H. 1931. The structure and postembryonic development of *Vanessa urticae.* 1. The larval alimentary canal. *Q. J. Microsc. Sci.* 74:321–60
34. Hering, M. 1939. Die peritrophischen Hüllen der Honigbiene mit besonderer Berücksichtigung der Zeit während der Entwicklung des imaginalen Darmes. Ein Beitrag zum Studium der peritrophischen Membran der Insekten. *Zool. Jahrb. Anat.* 66:129–90
35. Hoare, C. A. 1931. The peritrophic membrane of *Glossina* and its bearing upon the life-cycle of *Trypanosoma grayi. Trans. R. Soc. Trop. Med. Hyg.* 25:57–64
36. Jeuniaux, C. 1963. *Chitine et Chitinolyse.* Paris: Masson & Cie. 181 pp.

36a. Kenchington, W. 1976. Adaptation of insect peritrophic membranes to form cocoon fabrics. In *The Insect Integument,* ed. H. R. Hepburn, pp. 497–513. New York: Elsevier
37. Kusmenko, S. 1940. Über die postembryonale Entwicklung des Darmes der Honigbiene und die Herkunft der larvalen peritrophischen Hüllen. *Zool. Jahrb. Anat.* 66:463–530
38. LeBerre, R. 1967. Les membranes péritrophique chez les arthropodes. Leur rôle dans la digestion et leur intervention dans l'evolution d'organisimes parasitaires. *Cah. ORSTOM Ser. Entomol. Med. Parasitol.* 5:147–204
39. Lewis, D. J. 1950. *Simulium damnosum* and its relation to onchocerciasis in the Anglo-Egyptian Sudan. *Bull. Entomol. Res.* 43:597–644
40. Locke, M. 1958. The formation of tracheae and tracheoles in *Rhodnius prolixus. Q. J. Microsc. Sci.* 99:29–46
41. Marshall, A. T., Cheung, W. W. K. 1970. Ultrastructure and cytochemistry of an extensive plexiform surface coat on the midgut cells of a fulgorid insect. *J. Ultrastr. Res.* 33:161–72
42. Megahed, M. M. 1956. Anatomy and histology of the alimentary tract of the female of the biting midge *Culicoides nubeculosus* Meigen. *Parasitology* 46: 22–47
43. Mello, M. L., Vido, B. C., Valdrighi, L. 1971. The larval peritrophic membrane of *Melipona quadrifasciata. Protoplasma* 73:349–65
44. Mercer, E. H., Day, M. F. 1952. The fine structure of the peritrophic membrane of certain insects. *Biol. Bull.* 103:384–94
45. Moloo, S. K., Steiger, R. F., Hecker, H. 1970. Ultrastructure of the peritrophic membrane formation in *Glossina. Acta Trop.* 27:378–83
46. Montalenti, G. 1931. Sulla permeabilità della membrana peritrofica dell' intestino degli insetti. *Boll. Soc. Ital. Biol. Sper.* 6:89–94
47. Mori, H. 1969. Normal embryogenesis of the waterstrider, *Gerris paludum insularis,* with special reference to midgut formation. *Jpn. J. Zool.* 16:53–67
48. Neville, A. C. 1975. *Biology of the Arthropod Cuticle.* Berlin: Springer. 448 pp.
49. Nishizawa, K., Yamaguchi, T., Honda, N., Maeda, M., Yamazaki, H. 1963. Chemical nature of a uronic-acid-containing polysaccharide in the peritrophic membrane of the silkworm. *J. Biochem. Tokyo* 54:419–26
50. Noirot, C., Noirot-Timothée, C. 1972. Structure fine de la bordure en brosse de l' intestin moyen chez les insectes. *J. Microsc. Paris* 13:85–96
51. Nopanitaya, W., Misch, D. W. 1974. Developmental cytology of the midgut in the flesh fly, *Sarcophaga bullata. Tissue Cell* 6:487–502
52. Ono, S., Kato, S. 1968. Amino acid composition of the peritrophic membrane in the silkworm, *Bombyx mori. Bull. Seric. Exp. Stn. Gov. Gen. Chosen* 23:1–8
53. Palade, G. 1975. Intracellular aspects of the process of protein synthesis. *Science* 189:347–58
54. Peters, W. 1966. Zur Frage des Vorkommens und der Definition peritrophischer Membranen. *Verh. Dtsch. Zool. Ges.* 30:142–52
55. Peters, W. 1968. Vorkommen, Zusammensetzung und Feinstruktur peritrophischer Membranen im Tierreich. *Z. Morphol. Tiere* 62:9–57
56. Peters, W. 1969. Vergleichende Untersuchungen der Feinstruktur peritrophischer Membranen von Insekten. *Z. Morphol. Tiere* 64:21–58
56a. Peters, W. 1976. Investigations on the peritrophic membranes of Diptera. See Ref. 36a, pp. 515–543
57. Peters, W., Zimmermann, U., Becker, B. 1973. Anisotropic cross-bands in peritrophic membranes of Diptera. *J. Insect Physiol.* 19:1067–77
58. Platania, E. 1938. Richerche sulla struttura del tubo digerente di *Reticulitermes lucifugus,* con particulare riguardo alla nature, origine e funzione della peritrofica. *Arch. Zool. Ital. Torino* 25: 297–328
59. Platzer-Schultz, I., Reiss, F. 1970. Zur Histologie der Bildungszone der peritrophischer Membran einiger Chironomiden larven. *Arch. Hydrobiol.* 67: 396–411
60. Puchta, O., Wille, H. 1956. Ein parasitisches Bakterium im Mitteldarmepithel von *Solenobia triquetrella. Z. Parasitenkd.* 17:400–18
61. Rajulu, G. S. 1971. An electron microscope study on the ultrastructure of the peritrophic membrane of a chilopod, *Ethmostigmus spinosus,* together with observations on its chemical composition. *Curr. Sci.* 40:134–35
62. Ramsay, J. A. 1950. Osmotic regulation in mosquito larvae. *J. Exp. Biol.* 27:145–57

63. Rengel, C. 1903. Über den Zusammenhang von Mitteldarm und Enddarm bei den Larven des aculeaten Hymenopteren. *Z. Wiss. Zool.* 75:221–32
64. Richards, A. G. 1932. Comments on the origin of the midgut in insects. *J. Morphol.* 53:433–41
65. Richards, A. G. 1951. *The Integument of Arthropods.* Minneapolis: Univ. Minn. Press. 441 pp.
66. Richards, A. G. 1975. The ultrastructure of the midgut of hematophagous insects. *Acta Trop.* 32:83–95
67. Richards, A. G., Korda, F. H. 1948. Studies on arthropod cuticle. 2. Electron microscope studies of extracted cuticle. *Biol. Bull.* 94:212–35
68. Richards, A. G., Richards, P. A. 1969. Development of microfibers in the peritrophic membrane of a mosquito larva. *Ann. Proc. Electron Microsc. Soc. Am.* 27:256–57
69. Richards, A. G., Richards, P. A. 1971. The origin and composition of the peritrophic membrane of the mosquito *Aedes aegypti. J. Insect Physiol.* 17: 2253–75
70. Richardson, M. W., Romoser, W. S. 1972. The formation of the peritrophic membrane in adult *Aedes triseriatus. J. Med. Entomol.* 9:475–500
71. Rizki, M. T. M. 1956. The secretory activity of the proventriculus of *Drosophila melanogaster. J. Exp. Zool.* 131:203–21
72. Romoser, W. S. 1974. Peritrophic membranes in the midgut of pupal and preblood meal adult mosquitoes. *J. Med. Entomol.* 11:397–402
73. Romoser, W. S., Cody, E. 1975. The formation and fate of the peritrophic membrane in adult *Culex nigripalpus. J. Med. Entomol.* 12:371–78
74. Romoser, W. S., Rothman, M. E. 1973. The presence of a peritrophic membrane in pupal mosquitoes. *J. Med. Entomol.* 10:312–14
75. Rudall, K. M., Kenchington, W. 1973. The chitin system. *Biol. Rev. Cambridge Philos. Soc.* 48:597–636
76. Ruiz-Herrera, J., Bartnicki-Garcia, S. 1974. Synthesis of cell wall microfibrils in vitro by a "soluble" chitin synthetase from *Mucor rouxii. Science* 186:357–59
77. Schildmacher, H. 1950. Darmkanal und Verdauung bei Stechmuckenlarven. *Biol. Zentralbl.* 69:390–438
78. Smith, D. S., Compher, K., Janners, M., Lipton, C., Wittle, L. W. 1969. Cellular organization and ferritin uptake in the midgut epithelium of a moth, *Ephestia kühniella. J. Morphol.* 127:41–72
79. Stohler, H. 1957. Analyse des Infektionsverlaufes von *Plasmodium gallinaceum* im Darme von *Aedes aegypti. Acta Trop.* 14:302–52
80. Stohler, H. 1961. The peritrophic membrane of blood-sucking Diptera in relation to their role as vectors of blood parasites. *Acta Trop.* 18:263–66
81. Streng, R. 1969. Chitinhaltiger Spinnfaden bei der Larve des Buchenspringrusslers (*Rhynchaenus fagi* L.). *Naturwissenschaften* 56:333–34.
82. Streng, R. 1973. Die Erzeugung eines chitinigen Kokonfaden aus peritrophischer Membran bei der Larve von *Rhynchaenus fagi* L. *Z. Morphol. Tiere* 75:137–64
83. Trager, W. 1974. Some aspects of intracellular parasitism. *Science* 183: 269–71
84. Vignon, P. 1901. Recherches sur les epitheliums. *Arch. Zool. Exp. Gen. Ser. 3* 9:371–715
85. Waterhouse, D. F. 1953. Occurrence and endodermal origin of the peritrophic membrane in some insects. *Nature* 172:676
86. Waterhouse, D. F. 1953. The occurrence and significance of the peritrophic membrane with special reference to adult Lepidoptera and Diptera. *Aust. J. Zool.* 1:299–318
87. Waterhouse, D. F. 1954. The rate of production of the peritrophic membrane in some insects. *Aust. J. Biol. Sci.* 7:59–72
88. Waterhouse, D. F. 1957. Digestion in insects. *Ann. Rev. Entomol.* 2:1–18
89. Waterhouse, D. F., Wright, M. 1960. The fine structure of the mosaic epithelium of blowfly larvae. *J. Insect Physiol.* 5:230–39
90. Wigglesworth, V. B. 1929. Digestion in the tsetse fly: a study of structure and function. *Parasitology* 21:288–321
91. Wigglesworth, V. B. 1930. The formation of the pertitrophic membrane in insects, with special reference to the larvae of mosquitoes. *Q. J. Microsc. Sci.* 73:593-616
92. Wigglesworth, V. B. 1974. *The Principles of Insect Physiology.* London: Methuen. 546 pp. 7th ed.
93. Willett, K. C. 1966. Development of the peritrophic membrane in *Glossina* and its relation to infection with trypanosomes. *Exp. Parasitol.* 18:290–95
94. Yorke, W., Murgatroyd, F., Hawking, F. 1933. The relation of polymorphic trypanosomes, developing in the gut of

Glossina, to the peritrophic membrane. *Ann. Trop. Med. Parasitol.* 27:347–54

95. Zhuzhikov, D. P. 1962. Formation of a peritrophic membrane in *Aedes aegypti* L. mosquitoes *Sci. Rep. Higher Sch. Biol. Sci.* 4:25–27 (In Russian)
96. Zhuzhikov, D. P. 1964. Function of the peritrophic membrane in *Musca domestica* L. and *Calliphora ertythrocephala* Meig. *J. Insect Physiol.* 10:273–78
97. Zhuzhikov, D. P. 1966. Investigations on the peritrophic membranes of some Diptera in polarized light. *Vestn. Mosk. Univ.* 1966:37–41 (In Russian)
98. Zhuzhikov, D. P. 1970. Permeability of the peritrophic membrane in the larvae of *Aedes aegypti. J. Insect Physiol.* 16:1193–1202
99. Zimmermann, U., Hallstein, H. 1970. Untersuchungen über den Transport von Stoffen durch peritrophische Membranen. 2. Regelmässige Bandenstruktur in peritrophischen Membranen. *Z. Naturforsch. Ser. B* 25:1155–57
100. Zimmermann, U., Mehlan, D., Peters, W. 1973. Periodic incorporation of glucose, methionine and cysteine into the peritrophic membranes of the blowfly, *Calliphora erythrocephala,* in vivo and in vitro. *Comp. Biochem. Physiol.* 45(B):683–94
101. Zimmermann, U., Peters, W., Hallstein, H. 1969. Struktur und Bildungsgeschwindigkeit peritrophischer Membranen von *Calliphora erythrocephala. Z. Naturforsch. Ser. B* 24:1456–60

Ann. Rev. Entomol. 1977. 22:241–61

MODEL ECOSYSTEM APPROACH TO INSECTICIDE DEGRADATION: A CRITIQUE

❖6129

Robert L. Metcalf
Department of Entomology and Institute for Environmental Studies, University of Illinois, Urbana-Champaign, Urbana, Illinois 61801

INTRODUCTION

Laboratory model ecosystems or microcosms play an increasingly important role in the research and teaching of both ecological effects and the disposition and fate of environmental contaminants. As developed over the past decade, such model ecosystems provide for the simplification of real life situations and for the modeling of particular environmental compounds. These systems offer the advantages of compactness, standardizing and replicating conditions in environmental chambers, rapid evaluation, and convenient use of radiolabeled molecules, without which many important problems of environmental toxicology cannot be studied.

The essence of model ecosystem evaluations is comparative, both (*a*) among ecological, behavioral, toxicologic and degradative processes in organisms of different phyla, class, and order, and (*b*) between a wide range of contaminants or pollutants in which environmental effects can be related to physical-chemical properties. Thus if the relative model ecosystem behavior of a variety of organic compounds is compared to that of such widely research pollutants as DDT, dieldrin, and polychlorobiphenyl isomers, it is possible to make meaningful judgments regarding the ultimate environmental fate of new compounds that have not yet been produced commercially or about older compounds whose real world fate has not yet been scrutinized.

Model ecosystem studies have been most effectively applied to environmental toxicology of insecticides. Insecticides are environmental contaminants purposefully applied based on the concept that benefit/risk ratios are tilted in favor of returns to the user over damage to environmental quality. Most entomologists are painfully aware, however, that this is not always the case, as demonstrated by laws forbidding or regulating the use of persistent insecticides (e.g. DDT, aldrin, dieldrin, heptachlor, chlordane) in the United States, Canada, most countries of Western Europe, the USSR, China, and Japan. These laws are based on the belated realization that (*a*) trace amounts of these micropollutants can permeate the entire biosphere, (*b*) these nearly water-insoluble and fat-soluble compounds can be stored in the tissues of living organisms of all descriptions, (*c*) storage levels in living organisms can be biomagnified from 10^3 to 10^7 by food-chain assimilations and by partitioning from water into aquatic organisms, and (*d*) by increasing knowledge that degradation products of insecticidal micropollutants can sometimes be more persistent environmentally than the parent substance and/or more directly toxic to organisms in the environment. The environmental conversion of aldrin to dieldrin and photodieldrin, of chlordane to oxychlordane, and of DDT to DDE (dichlorodiphenyl ethylene) are good examples (48). Defining and realizing these problems has required at least a generation and has resulted in environmental damage that may require many years to repair. Thus 28 years elapsed between the introduction of DDT in 1945 and the decision by the Environmental Protection Agency (EPA) to ban it for agricultural use in January 1973. Chlordane has been widely used for control of household pests since 1948; however, knowledge of its persistent environmental degradation product, oxychlordane, now present in virtually all human tissues, has existed only since 1970 (54, 64). An early warning technology must be devised, not only to give rapid advance notice about potential environmental problems that may result from specific uses of insecticides, but also to serve as a screening tool for the quantitative evaluation of proposed new insecticidal molecules to ensure that substitue and newly devised insecticides have more benign environmental properties than the materials they replace.

There is no better example of these problems than chlordecone (Kepone®) or decachlorooctahydro-1,3,4-methano-2H-cyclobuta [cd] pentalene-2-one, which was introduced in the United States in 1958 (U.S. Patent 2,616,825, reissue 24,435) and has been widely used as bait for cockroaches and ants, including the imported fire ant (*Saevissima solenopsis richteri*). It is also used in Latin America to control banana pests. There was, however, almost no knowledge of its degradative fate (34, 35) or environmental effects (56) until 1975 when chlordecone wastes from a manufacturing plant in Hopewell, Virginia, were found to have poisoned more than 70 plant workers and to have contaminated the James River so severely that 100 miles of it, from Richmond, Virginia, to Chesapeake Bay, has been closed to fishing (1). Model ecosystem technology could have rapidly developed the information necessary to institute proper control of this pollutant.

This review brings laboratory model ecosystem technology up to date and analyzes the value and limitations of this approach in studying the evaluation of toxic substances in the environment. Model ecosystems can be used to study the toxicology of all types of environmental contaminants such as radionuclides (69), trace metals (31), petroleum spills (28), heavy organic chemicals (49, 51, 60), plasticizers (38, 62), industrial chemicals (30), herbicides (48, 73, 74), and insecticides (48, 50, 63). Because of severe space limitations we have considered only studies involving insecticides; thus the large variety of model ecosystems used to evaluate radionuclides and trace metals, which have been thoroughly reviewed by Taub & Pearson (69), will not be considered further here.

Use of Model Ecosystems

Potential areas of usefulness of model ecosystems are as follows: (*a*) chemical degradation pathways in the environment (i.e. the chemical and biochemical transformations that occur in air, water, and soil and in the organisms of the model system under standardized conditions); (*b*) fate of chemicals in the environment (i.e. transport, distribution, and accumulation in various organisms); (*c*) toxicological effects of chemicals in the environment and the effects of their transformation products on the organisms of the model system (i.e. the identification of potentially hazardous chemicals); (*d*) evaluation of fate of chemicals after various modes of application and entry into the environment (i.e. directly to water, to air, to various soil types, on plant or seed, on organisms, etc); (*e*) behavioral effects on organisms, especially in food chains (e.g. predatorism, feeding behavior, mating, etc); (*f*) identification and quantification of ecosystem processes (e.g. correlation of physical-chemical parameters of chemicals with bioaccumulation and degradation); (*g*) evaluation of the interactions between chemicals in the environment (e.g. insecticide and synergist, insecticide and herbicide, qualitatively and quantitatively); (*h*) evaluation of biochemical mechanisms involved in comparative toxicology in a variety of organisms; and (*i*) screening a variety of analogues or derivatives of a new insecticide to determine their relative environmental behavior and safety.

Limitations of Model Ecosystems

At the same time model ecosystems have well-defined limitations, including limited applicability to long-term effects (i.e. multigeneration studies, time for ecosystem stability), uncertain predictability from one model system to another, limited predictability as to vital biological components and vulnerable biological processes, and the fact that replicatability and reproducibility need careful evaluation for each type of system.

The validity of model ecosystem studies can be evaluated by measuring replicatability and reproducibility of evaluation between model ecosystem units, by the degree of similarity of results with standard chemicals between model systems and natural ecosystem processes for known pollutants, and by the relation of model ecosystem results and natural ecosystem processes to mathematical simulation based on physical-organic properties of chemicals, i.e. water solubility, lipid/water partition, electron density, molecular orbital calculations, etc.

Types of Model Ecosystems

Laboratory model ecosystems or microcosms are potentially almost as diversified as the natural environment whose components are being modeled. Such systems range in complexity from petri dishes containing soil microflora and flasks containing microorganisms in water or nutrient medium to elaborately constructed and instrumented terrestrial chambers, e.g. the National Ecological Research Laboratory world in miniature (9, 14) and model streams such as the computer-operated and instrumented plexiglass system of the Southeastern Water Research Laboratory, Athens, Georgia. The most standardized feature of all these microcosms is the general choice of radiolabeled contaminants to produce the maximum amount of information with the minimum expenditure of analytical effort.

There is no such thing as a standard model ecosystem anymore than there is a standard environment, and carefully conducted studies of contaminants in any microcosm can yield useful information about their effect on environmental quality. Many types of model ecosystems are described in the literature (69). A variety of these, especially selected because of relevance to studies with insecticides, are described in Table 1.

RADIOLABELED COMPOUNDS AND RADIOCHEMICAL TECHNIQUE

The majority of the complex problems of degradative fate of pesticides and their environmental interactions with living organisms cannot be solved without the use of radiolabeling and radiochemical techniques. However, radiolabeled compounds are often expensive and there are increasing restrictions on their use as environmental pollutants. Thus some sort of model system, in which minimal quantities of the radiolabeled insecticide can be confined and exposed to environmental conditions of light, temperature, moisture, soil, plant life, and a variety of target and nontarget animals, is imperative to successful use of the informative radiotracer methodology. Such model systems can be as simple as cultures of bacteria or alga (7, 11, 52, 55, 67) or fish in flasks of water (57, 60). However, each molecule of radiolabeled compound contains its own intrinsic analytical detector and, for optimum information, it is generally useful to use the radiolabeled pesticide in a model system sufficiently complex to provide a variety of information about degradative pathways, absorption, translocation, bioconcentration, detoxication mechanisms, comparative pharmacology, etc. Waste disposal problems from such confined systems using microgram-to-milligram quantities, i.e. 1–100 μCi of radioactivity, are minimal.

Radioisotopes

A large variety of ^{14}C-labeled compounds, including most widely used insecticides, are available from commercial suppliers.[1] Others are often generously supplied by the manufacturers who use them in animal metabolism studies. Lesser known compounds can be synthesized on a milligram scale from commercially available radiolabeled precursors, e.g. heptachlor from [^{14}C]hexachlorocyclopentadiene (32), carbaryl from ^{14}C ring-labeled (UL)1-naphthol (48), fonofos (Dyfonate®) from [^{14}C]ethanol (48), methylchlor or 2,2-bis(p-tolyl)-1,1,1-trichloroethane from ^{14}C ring UL toluene (20), and Prolan® or 1,1-bis(p-chlorophenyl)-2-nitropropane from ^{14}C ring UL chlorobenzene (17). ^{14}C has a half-life of about 5360 years and the β-particle has an energy of 0.156 Mev.

Tritium radiolabeling is particularly useful for aromatic compounds and a simple and inexpensive method involves exchange between the compound and the

[1]For example, Amersham-Searle Corporation, New England Nuclear, California Bionuclear Corporation.

Table 1 Model ecosystems used in insecticide research

Type	Organisms	Size	Pollutant	Exposure	Data	Reference
Aquatic systems — single species						
Flask	*Daphnia magna*	100-ml beakers	DDT (GLC)	26 hr	bioconcentration, distribution	(6)
Flask	*Platichthys flesus*	Oral	^{14}C-labeled DDT	7.14 days	bioconcentration, distribution, degradation	(10)
Flask	*Chlamydotheca arcuata*	500 ml	^{14}C-labeled aldrin ^{14}C-labeled dieldrin	48 hr	bioconcentration, epoxidation	(23)
Flask	microorganisms	100 ml	^{14}C-labeled organochlorines	1–16 days	bioconcentration	(27)
Flask	*Euglena gracilis*	70 ml	^{14}C-labeled DDT	6 days	effects on cells bioconcentration	(7)
Flask	*Lepomis cyanellus*	2000 ml	^{14}C-labeled DDT ^{14}C-labeled DDE ^{14}C-labeled PCB's	2–16 days	bioconcentration	(60)
Continuous flow	*Chlorella* sp.	100 ml	^{14}C-labeled DDT	1–6 days	bioconcentration	(65)
Aquatic systems — multiple species						
Flask added in water	*Oedogonium cardiacum* *Daphnia magna* *Culex pipiens* *Physa spp.* *Gambusia affinis*	2000 ml	^{14}C-labeled DDT ^{14}C-labeled aldrin ^{14}C-labeled benzene derivatives	3 days	degradation bioconcentration volatile products food-chain effects comparative detoxication	(30)
Flask added in water, alga, or daphnia	*Daphnia* protozoa ostracoda	500 ml	DDT PCB		variable bioconcentration	(51)
Flask added in water or daphnia	*Daphnia magna* *Physa* sp. *Lepomis cyanellus* *Tilapia mossambica* *Lebistes reticulatus*	5000 ml	^{14}C-labeled DDT $[^{3}H]$methoxychlor	3–31 days	bioconcentration food chain effects degradation	(57)

Table 1 *(continued)*

Type	Organisms	Size	Pollutant	Exposure	Data	Reference
Flask	*Chlamydomonas* *Tetrahymena* rotifers *Daphnia magna* Ostracoda *Serratia*	500 ml	toxaphene PCB's Hg	7–35 days	growth of organisms in size and numbers, bioconcentration replication	(68)
Aquarium continuous flow added to water, alga, fish	algae *Daphnia* *Poecilia*	7400 ml	dieldrin	6 days	bioconcentration	(58)
Flask continuous flow added in water	*Daphnia magna* *Gammarus fasciata* *Oronectes nais* *Palaemonetes kadiakensis* *Hexagenia bilineata* *Siphlonurus* *Isnura verticalis* *Libellula* *Chironomus* *Culex pipiens*	1000 ml	^{14}C-labeled DDT ^{14}C-labeled aldrin	1.5 days	bioconcentration	(19)
Terrestrial—aquatic systems						
Aquarium—10 gal with sand or soil shelf, insecticide applied to plant or soil	sorghum, rice *Estigmene acrea* plankton *Daphnia magna* *Culex pipiens* *Gambusia affinis* *Oedogonium cardiacum* *Physa* spp.	13,000 g soil 7,000 ml water	30 radiolabeled insecticides	33 days	distribution, toxicity, degradation, bioconcentration, food-chain effects, leaching, soil types	(36, 37, 50) (48)

Table 1 *(continued)*

Type	Organisms	Size	Pollutant	Exposure	Data	Reference
Flask	microorganisms *Arthrobacter* *Hydrogenmonas*	30–2000 ml water, sewage, sediment	^{14}C-labeled DDT diphenyl methane	4–168 days	bioconcentration degradation degradation	 (55) (11)
Terrestrial systems						
Flask—19.1 with vapor traps bottom drain, insecticide applied to plant or soil	corn, soybean, cotton; *Lumbricus terrestris* *Armadilidium vulgare* *Limax maximus* *Estigmene acrea* *Microtus ochregaster*	400 g soil 2000 ml water	^{14}C-labeled aldrin dieldrin DDT methoxychlor fonofos parathion methyl parathion	20 days	distribution toxicity, degradation, bioconcentration, volatilization leaching soil types, food-chain effects	(4, 5)
Box—plastic with vapor traps, surface to ground water flow and rain system	corn, sorghum, johnson-grass, rye grass, blue grass alfalfa *Pristionchus inheriteira* *Cephalobus perseghis* *Lumbricus* *Helix pomata* *Armadilidium* *Pordellia* *Tenebrio molitor* *Achaeta domestica* *Anoleus* *Excalfactoria chinensis* *Microtus canicaudus*	1.0 × 0.75 × 0.61 m 40,000 g soil	^{14}C-labeled dieldrin parathion methyl parathion	30–45 days	distribution, bioconcentration, volatilization leaching degradation food-chain effects reproductive effect	(9, 14)
Stream systems						
Trough and paddle wheel	fresh water rocks, periphylon		dieldrin	60–120 days	bioconcentration	(59)
Model streams	review				possibilities, constraints	(29, 71)

$^{3}H_3PO_4 \cdot BF_3$ complex (15). This technique has provided milligram-to-gram quantities of very high-specific-activity compounds (20–100 mCi mmole^{-1}) such as parathion, methoxychlor, carbofuran, and a variety of DDT analogues (15, 16, 20–22) at a cost of 10% or less of ^{14}C labeling. With aromatic compounds at least there is little or no problem with isotopic exchange; results in relatively complex experiments are equivalent to those obtained with ^{14}C labeling (22, 46). ^{3}H has a half-life of 12 years but the β-energy of 0.018 Mev is relatively very low so that radioautography is very difficult if not impractical.

^{32}P labeling is very useful for organophosphorus insecticides and provides a marker for the reactive P atom (13, 18, 33, 41–43). O-P insecticides of very high specific activity can be produced, which along with the high energy of the β-particle, 1.701 Mev, is an advantage in radioautography. The 14.3-day half-life of radioactive decay can be both advantageous and disadvantageous.

^{35}S labeling has been used for phosphorothionates and other S-containing insecticides. ^{35}S has a half-life of 87.1 days and the energy of the β-particle is 0.168 Mev, so that ^{35}S-labeled compounds behave similarly to ^{14}C-labeled products.

^{36}Cl, with a half-life of about one million years and a β-energy of 0.714 Mev, has been used for labeling organochlorine insecticides such as DDT, lindane, and dieldrin. Because of the very long half-life, however, specific activity is relatively low.

POSITION OF RADIOLABEL The position of the radiolabel in the insecticide molecule is of obvious importance in degradation studies, in which substantial portions of the molecule may be cleaved. For example, [^{14}C]propoxur, [2-^{14}C]isopropoxyphenyl *N*-methylcarbamate is readily cleaved to 2-hydroxyphenyl *N*-methylcarbamate by in vivo *O*-dealkylation, and the latter compound can no longer be identified in further degradation pathways (37, 47). Carbamates, e.g. carbofuran, radiolabeled in the $^{14}C{=}O$(36, 72) or in the $NH^{14}CH_3$ moieties (24, 39), can be followed only as long as the carbamoyl ester moiety is intact or until *N*-demethylation has occurred. Such relatively labile radiotracers can give valuable information, especially when compared with radiolabeling on other parts of the molecule (24, 36, 37, 72). For example, the degradative fate of aldicarb or 2-methyl-2-(methylthio)-propionaldehyde *O*-(methylcarbamoyl)-oxime has been evaluated by using $^{14}CH_3S$, *tert*-^{14}C, and $NH^{14}CH_3$ labels in separate experiments (40, 48); the model ecosystem fate of 3,5-xylyl *N*-methylcarbamate was studied with ring $^{14}CH_3$, $^{14}C{=}O$, and $NH^{14}CH_3$ moieties (24). For aromatic compounds, the best labeling position is the aromatic ring either with ^{14}C or ^{3}H (48). Where there are two aromatic rings that may be cleaved through degradation processes, e.g. dimilin or 1-(2',6'-difluorobenzoyl)-3-(4'-chlorophenyl)-urea, the fate of both aromatic moieties was conveniently followed in separate experiments by using [2,6-$^{14}C{=}O$]difluorobenzoyl ^{14}C ring UL, *p*-chlorophenyl, and [^{3}H]difluorobenzoyl radiotracers (46).

Organophosphorus esters are conveniently labeled with ^{32}P, which is easy to trace because of its high energy β-particles and has rapid decay. This ^{32}P label makes it convenient to follow all the reactions of the biologically active phosphorus moiety, e.g. P=S to P=O activation, *O*-dealkylation, and hydrolysis, through the environment (18, 33, 36, 41–43, 48). Environmental concern, however, is also directed at the fate of leaving groups that may be persistent, and ^{14}C or ^{3}H ring labeling yields valuable information, e.g. ^{14}C ring-labeled 3,5,6-trichloro-2-pyridinol from chlorpyrifos (36, 48), ^{14}C ring-labeled *p*-nitrophenol from parathion (48), and the ^{3}H ring-labeled 4,4'-dihydroxydiphenyl sulfide moiety from temephos (48). With phosphonothionates $O^{14}C_2H_5$ has been used to trace the environmental fate of fonofos (48) and $O^{14}CH_3$ to evaluate the fate of leptophos (48). In these insecticides these *O*-alkoxy moieties are extraordinarily stable.

Essentially parallel information about various biological oxidation states of the P=S and S-alkyl moieties of phorate and its analogue, terbufos (Counter®), have been gained from ^{32}P (41) and $^{14}C(CH_3)_3$ labeling (48).

RADIOPURITY This is easily evaluated by TLC and radioautography or by counting 0.5-cm serial sections. Compounds used in model ecosystem research should be of 98–99+% radiopurity, which can be readily attained by preparative column or TLC.

DETECTION OF RADIOACTIVITY Radioactivity is commonly detected by liquid scintillation counting, and ^{14}C, ^{35}S, and ^{32}P labels can be easily determined in tissue homogenates and extracts added directly to scintillation vials. The low energy of ^{3}H, however, leads to problems of quenching in the scintillation medium; accurate results are best obtained by total combustion of the sample in O_2 to give $^{3}H_2O$. The Schöniger oxygen flask combustion technique (25) is simple, inexpensive, and used routinely in our laboratory (20–22, 36, 48).

For investigating insecticide degradative pathways in model ecosystems it is essential to extract the samples of water, sand, soil, and biological material with organic solvents, such as diethyl ether, acetone, or acetonitrile, to concentrate them to small volumes, and to subject them to analysis by TLC, by using solvents of varying polarities to separate the variety of environmentally produced degradation products. These are then identified quantitatively by radioautography on X-ray film[2] (usually with exposures of 1 week to 2 months), followed by scraping the areas containing radioactivity (blackened on the film) into liquid scintillation vials, or by scraping 0.5-cm serial sections directly from the TLC plates (22, 36, 37, 46, 49, 50). Using fluorescent silica gel (e.g. E. Merck GF-254) is of great value as aromatic compounds readily quench the fluorescence and can be observed as dark spots under UV light.

OPERATION OF MODEL ECOSYSTEMS

Operating various types of model ecosystems has been adequately described in individual reports (Table 1) and reviews (36, 37, 66, 69, 71). For precise work the individual units should be held at constant or controlled temperatures with constant or controlled lighting (4, 5, 36–38, 50). Environmental plant growth chambers with high-intensity fluorescent lighting provide the most convenient means for maintain-

[2]Eastman No-Screen or Royal Blue X-ray film.

ing constant conditions to improve replicatability and also provide optimal conditions for isolating experiments with radiolabeled compounds. Using standard substrates, such as vermiculite (4, 5), white quartz sand (36, 37, 50), and standard reference water (12), is important in providing uniform backgrounds for comparing various pesticides (48). Conversely studying individual pesticides with a variety of soil types (5) has almost endless possibilities in explaining the complexities of insecticide volatility, codistillation, soil binding, leaching, and run-off. Selective application of the radiolabeled insecticides to the model ecosystems can be made to simulate practical usage: directly to plants, directly to soil, seed treatment, directly to water, or as animal excreta.

TERRESTRIAL AQUATIC MODEL ECOSYSTEMS

The movement of pesticides applied to plants or soil into an aquatic environment is an important consideration because of the contamination of water supplies (as in the lower Mississippi River) and the phenomena of bioconcentration. For example, the dieldrin and DDT present in Lake Michigan at average concentrations of 2 and 6 ppt are found stored in the tissues of large lake trout, *Salvelinus namaycush,* to average values of 0.26 ppm dieldrin (bioconcentration 1.2×10^5) and 19.23 ppm DDT (bioconcentration 3.2×10^6) (8). Many of these fish have tissue residues that exceed the safe limits for human consumption established by the Food and Drug Administration.

The terrestrial aquatic model ecosystem (Figure 1*A*) was devised to study the transfer of insecticides from crops to lake, e.g. from field to farm pond, and to evaluate the degradation and fate and food chain transfer in a variety of organisms: plankton⟶daphnia⟶mosquito larva⟶fish, and alga⟶snail. More than 50 radiolabeled insecticides have been carefully evaluated in this type of model ecosystem, and a considerable amount of data is available on the individual compounds describing degradative pathways, comparative toxicity, persistence, bioaccumulation, and quantitative parameters such as ecological magnification, biodegradability index, and percentage of unextractable radioactivity (21, 48, 49).

The composite results from the evaluation of 49 pesticides show some interesting correlations with basic properties of the individual chemicals. Perhaps the most interesting correlations are those between biomagnification in the fish *Gambusia affinis* and water solubility of the pesticide ($r = -0.76$) (Figure 1*D*), and between the percentage of extractable radioactivity and biomagnification ($r=0.74$) (48). These results indicate the possibilities of predicting ecological effects from the basic physical-chemical properties of the individual compounds. Thus from Figure 1D we suggest (48) that pesticides with water solubilities of <0.5 ppm are likely to show objectionable biomagnification and those with solubilities >50 ppm are unlikely to be magnified. The pesticides in the in-between range should be regarded with caution. This relationship is clearly predicated upon the lipid/water partition coefficients of these pesticides (30, 49), and when accurate values of this parameter are available for all the pesticides, even more impressive and useful relationships can be developed from model ecosystem studies.

Replicatability of Model Ecosystems

Quantitative parameters of model ecosystem technology, i.e. bioconcentration and ecological magnification (EM), biodegradability index (BI),[3] unextractable radioactivity, etc, demand a reasonable precision of statistical reliability. Several studies have been made with the terrestrial-aquatic model ecosystem (Figure 1*A*) to indicate the reliability of the data. With the insect growth regulator dimilin or 1-(2', 6'-difluorobenzoyl)-3-(4'-chlorophenyl)-urea, three replicate model ecosystems treated with dimilin and having a different position of radiolabeling gave very comparable results (46) (e.g. EM fish, 19, 14, 80; BI fish, 4.99, 6.64, 3.38). The EM values for *Culex* mosquito larvae, which incorporate dimilin into the cuticle, were 779, 596, and 1099 with BI values of 0.035, 0.005, and 0.004, respectively. In similar study with the herbicide bifenox or 2-nitro-5-(2',4'-dichlorophenoxy)-benzoic acid, methyl ester, radiolabeled in three different positions, the replicate values were 58, 85, and 79 for EM fish and 3.2, 9.5, and 7.4 for BI fish (48) (R. L. Metcalf, unpublished data). Values for other organisms were of the same degree of precision, which is considered highly satisfactory, especially since degradative changes involving loss of one or more of the radiolabels could alter BI values.

Sanborn & Chio (61) have designed studies evaluating the replicatability of this same terrestrial-aquatic model ecosystem, by using five replicates of [^{14}C]atrazine. EM values in 15 individual fish, *Gambusia* (three in each replicate system), gave a mean value of 4.38 ± 0.31 SE. Values for unextractable ^{14}C in the five systems were 0.052, 0.031, 0.027, 0.042, and 0.047 ppm, 0.039 ± 0.0053 SEM. Similar precision in replication was obtained with the other organisms.

It can be concluded from these and other studies that the statistical reliability for these quantitative parameters between individual systems is highly significant. Indeed, since the distribution and biodegradation of any pollutant are controlled by its intrinsic physico-chemical parameters, and the individual systems are run in precisely controlled environmental chambers, a high degree of consistency in results is to be expected.

TERRESTRIAL MODEL ECOSYSTEMS

Although the more dramatic effects of environmental pollution seem to have occurred in the aquatic environment where astonishing bioconcentrations have been observed in creatures of the Great Lakes and the oceans, perhaps the more subtle problems are those involving soil plant interrelationships. The proliferation of pre-emergent treatments of crops with insecticides and herbicides, often in a bewildering number of combinations, has focused scientific attention on physical absorption of insecticides in soil, volatility and codistillation, degradation by soil microorganisms, absorption and translocation by plants, and the results of leaching and erosion. Terrestrial model ecosystems are well suited to exploit the values of radiolabeled insecticides in producing the answers to these and other questions. As outlined in

[3]These terms are defined in Table 3.

A

B

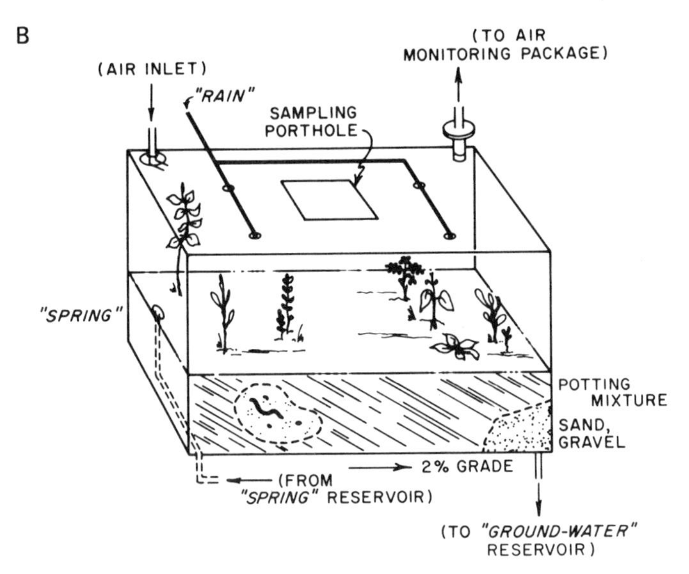

Figure 1 Model ecosystem studies of the fate of insecticides. (*A*) Terrestrial-aquatic model ecosystem (36, 50). Reprinted from *Chemtech,* Feb. 1972, p. 104. Copyright (1972) by the American Chemical Society, reprinted by permission. (*B*) Diagram of terrestrial microcosm

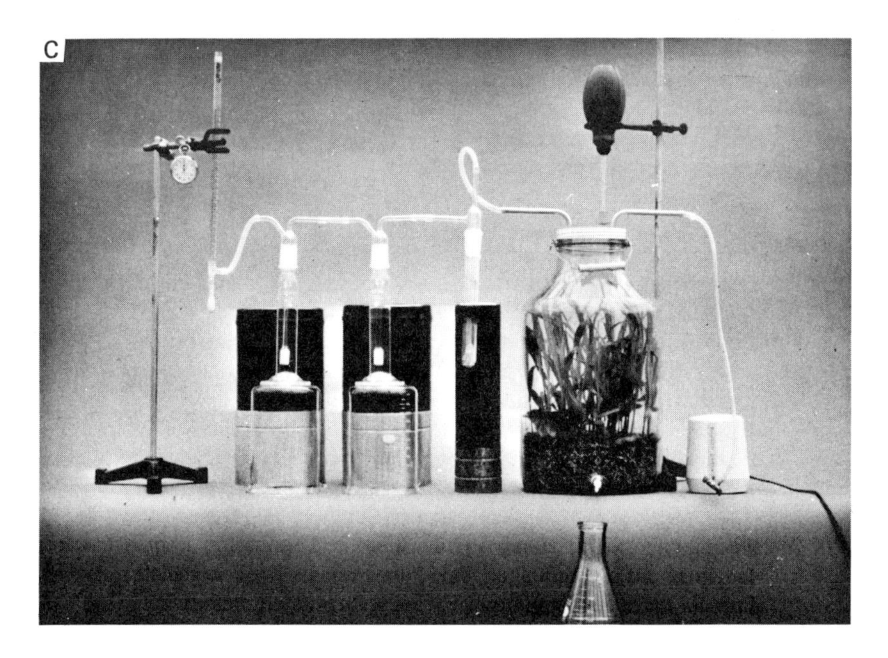

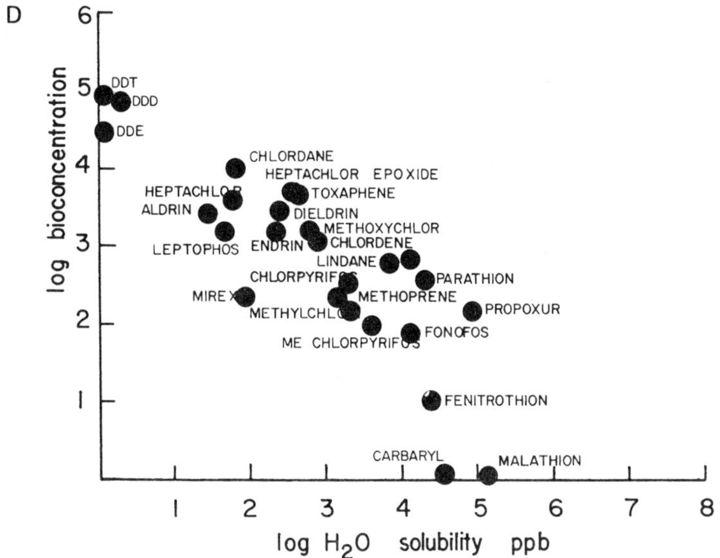

chamber. Courtesy of J. W. Gillette and J. D. Gile (9, 14). (*C*) Terrestrial model ecosystem showing arrangements for moritoring volatile components and leaching (4, 5). (*D*) Plot of data from model ecosystem studies with insecticides showing relation of bioconcentration in *Gambusia affinis* to water solubility (48).

Table 1, modern systems (such as those being used in the study of substitute pesticides developed to replace the organochlorines) contain arrangements to monitor the physical aspects of insecticide-soil interaction, such as volatility and leaching, incorporate crop plants and weeds as desired, and have a wide variety of food chain organisms, earthworms, snails, insects, the vole, and the Japanese quail. A variety of pesticide applications may be simulated: to seeds, soil, or foliage (4, 5, 9, 14).

The information that can be obtained with such a system is shown in Table 2. This table clearly shows that the substitute insecticides methoxychlor and fonofos (Dyfonate) have great advantages over DDT and aldrin when applied to corn foliage or to soil in regard to uptake and accumulation by the prairie vole *Microtus ochregaster* at the top of the food chain. The amount of data that can be obtained by these terrestrial systems is virtually inexhaustable (4, 5, 9, 14). These terrestrial model ecosystems are particularly suited to evaluating the effects of various soil types on the fate of radiolabeled pesticides through volatility, leaching, run-off, and mobilization by living organisms (Figure 1*B* and 1*C*).

SCREENING OF NEW COMPOUNDS

Model ecosystem technology is particularly suited to screening prospective new insecticidal compounds in the research and development stage. Not only is such information important in the process of Federal Registration (53), it is essential to demonstrate biodegradability (21). The quantitative aspects of biodegradability are particularly important when a number of analogues of a promising insecticidal nucleus are under consideration, as for example synthetic pyrethroids, organophosphorus esters, *N*-methylcarbamates, or DDT analogues; the data gained from such studies are a vital part of the decision-making process for further selection and development.

The study of the DDT biodegradable analogues provides a good example of the value of this information. The problems of bioconcentration (26) of DDT in living

Table 2 Residues of insecticides in prairie voles feeding in soil-corn terrestrial model ecosystems treated with ^{14}C-labeled insecticides at 1.0 lb per A[a]

	Fresh weight (ppm)					
	Total body		Liver		Brain	
Determinants	Total ^{14}C	Parent	Total ^{14}C	Parent	Total ^{14}C	Parent
DDT foliar	11.9	1.37	14.8	0.48	10.5	1.99
Methoxychlor foliar	0.47	0.041	0.13	trace	0.02	trace
Aldrin preemergent	3.81	3.48[b]	2.95	2.62[b]	0.57	0.55[b]
Fonofos preemergent	0.27	0	0.07	0	0.017	0

[a] From (4).
[b] Aldrin and dieldrin.

organisms, due to its extreme water insolubility (1.2 ppb) and high lipid/water partition coefficient, are well known (8) and easily demonstrable by model ecosystems (4, 20, 44, 50). These are compounded by the ready transformation of DDT to the more environmentally recalcitrant DDE (8, 49, 50). This compound has been shown to have the lowest percentage of unextractability (0.25% in fish) of all pesticides presently examined in a model ecosystem (48, 49). Studies of the degradative pathways of [^{14}C]- and [^{3}H]methoxychlor (22, 50) indicated the CH_3O groups as degradophores, i.e. molecular groupings that are substrates for attack by microsomal oxidases. These enzymes seek to convert xenobiotic molecules into more water-partitioning derivatives that are excreted rather than stored in lipid tissues. *O*-demethylation in vivo converts methoxychlor (water solubility, 0.62 ppm) into 2,2-bis-(*p*-hydroxyphenyl)-1,1,1-trichloroethane (water solubility, 76 ppm) (22). As a result methoxychlor, as demonstrated by model ecosystem study, is much more biodegradable and less bioconcentrated than DDT (Table 3). The CH_3 groups of methylchlor or 2,2-bis-(*p*-methylphenyl)-1,1,1-trichloroethane are also effective degradophores and in vivo oxidation converts methylchlor (water solubility, 2.21 ppm) to 2,2-bis-(*p*-carboxyphenyl)-1,1,1-trichloroethane (water solubility, 50 ppm) (20, 21). Model ecosystem study with [^{14}C]methylchlor (20) demonstrated its biodegradability (Table 3). A considerable range of alkyl-, alkoxy-, and alkylthio-degradophores has been developed for the DDT-type compound and, when judiciously combined, has been shown to produce effective DDT-like insecticides with greatly enhanced biodegradability (21), such as the asymmetric 2-(*p*-ethoxyphenyl)-2-(*p*-methylphenyl)-1,1,1-trichloroethane (Table 3). Model ecosystem data demonstrating biodegradability were an important part of patent application (45).

Table 3 Environmental properties of DDT analogues evaluated in laboratory model ecosystems[a]

$R^1C_6H_4CH(R^3)C_6H_4R^2$			Ecological magnification[b]		Biodegradability index[c]	
R^1	R^2	R^3	Fish	Snail	Fish	Snail
Cl	Cl	CCl_3	84,500	34,500	0.015	0.045
CH_3O	CH_3O	CCl_3	1,545	120,000	0.94	0.13
C_2H_5O	C_2H_5O	CCl_3	1,536	97,645	2.69	0.39
CH_3	CH_3	CCl_3	140	120,270	7.14	0.08
CH_3S	CH_3S	CCl_3	5.5	300	47	0.77
CH_3O	CH_3S	CCl_3	310	3,400	2.75	105
CH_3	C_2H_5O	CCl_3	400	42,000	1.20	0.25
Cl	CH_3	CCl_3	1,400	21,000	3.43	2.0
CH_3O	CH_3O	$C(CH_3)_3$	1,636	23,300	1.04	0.23
Cl	Cl	$CH(NO_2)CH_3$	112	31,392	3.27	0.009

[a] Information taken from (3, 17, 21).
[b] Ratio of concentration in organism/concentration in water.
[c] Ratio of polar/nonpolar metabolites.

More recent model ecosystem studies of the effects of substituting degradophores into the aliphatic portion of the DDT-type molecule have shown the value of CH_3 and NO_2 degradophores in Prolan or 1,1-bis-(*p*-chlorophenyl)-2-nitropropane (17) (Table 3) and in DDT analogues that contain only C, H, and O, such as dianisyl neopentane or 2,2-bis-(*p*-ethoxyphenyl)-dimethylpropane (3).

Degradative Pathways

Carefully conducted model ecosystem studies with radiolabeled insecticides provide the simplest and most complete way to develop information on environmental fate and degradation of a candidate insecticide, not only in the overall ecosystem but also on a comparative basis in water, soil, plants, and individual components of food chains. The importance of such studies is gauged by a recent critique of 32 pesticides registered with the EPA, stating that data on soil degradation were lacking for 11, on water degradation for 17, on microbial degradation for 16, and in leaching and run-off for 24 (2). A recent survey (70) has shown that for 35 of the most widely used pesticides there was inadequate information about the nature of the environmental degradation products and their effects on environmental quality.

Degradative pathways have been studied in model ecosystems for a number of pesticides (3, 5, 10, 16, 17, 20–22, 24, 30, 32, 44, 46, 48–50, 55, 57, 63, 72). The model ecosystem evaluation of the fate of the insect growth regulator dimilin or 1-(2',6'-difluorobenzoyl)-3-(4'-chlorophenyl) urea (46) is a good example of the value of such studies. Three radiolabeled preparations available, [^{14}C=O]benzoyl, ring *p*-[^{14}C]chlorophenyl, and ring [^{3}H]benzoyl, were studied individually in three terrestrial-aquatic model ecosystems, and it was shown that dimilin is cleaved into the smaller fragments *p*-chloroaniline and 2,6-difluorobenzamide. The individual fate of these moities was followed further by their specific radiolabeling; it was shown that dimilin, although relatively persistent, was not bioaccumulative (46). The model ecosystem evaluation also gave a picture of the photochemical and hydrolytic reactions occurring in the water phase.

Model ecosystem study of the DDT-type compound ring α-[^{3}H]trichloromethyl-*p*-ethoxybenzyl-*p*-ethoxyaniline (16) showed that this insecticide is highly degradable environmentally and undergoes *O*-deethylation at both *p*-ethoxybenzyl and *p*-ethoxyaniline moieties to form mono- and bis-phenols. In addition the molecule is dehydrochlorinated to form α-dichloromethyl-*p*-ethoxybenzylidine-*p*-ethoxyaniline, which is rapidly cleaved to form *p*-ethoxyaniline and *p*-ethoxybenzoic acid.

Similar model ecosystem evaluation of ^{14}C ring-labeled Prolan® or 1,1-bis-(*p*-chlorophenyl)-2-nitropropane (17) showed for the first time the degradative pathways of this 20-year-old insecticide, which is degraded by elimination to 1,1-bis-(*p*-chlorophenyl)-1-propene and by oxidation to 1,1-bis-(*p*-chlorophenyl)-2-propanone and 1,1-bis-(*p*-chlorophenyl)-pyruvic acid, giving ultimately *p,p*'-dichlorobenzophenone and bis-(*p*-chlorophenyl)acetic acid.

Comparative Metabolism

Model ecosystem studies with radiolabeled insecticides are well suited to provide data about the comparative metabolism of the compounds in a variety of organisms of different phyla. This has become an integral part of the evaluation with our model aquatic ecosystem (30). Studies with [^{14}C]hexachlorocyclopentadiene, -chlordene, -heptachlor, and -heptachlor epoxide provide a good example. These four compounds represent stages in the environmental history of an insecticide from raw material (hexachlorocyclopentadiene) through intermediate and principal impurity (chlordene), to product (heptachlor), and to environmentally produced pollutant (heptachlor epoxide). In individual model aquatic ecosystems (32) chlordene was epoxidized to chlordene-2,3-epoxide and hydrolyzed to form 1-hydroxy-2,-3-epoxychlordene. Heptachlor was epoxidized to form heptachlor-2,3-epoxide, which comprised 22% of the extractable ^{14}C in alga, 37% in snail, 49% in mosquito, and 79% in fish. Heptachlor was also hydrolyzed at C_1 to 1-hydroxychlordene, but heptachlor epoxide without the activating allylic structure (CH=CHCHCl) was relatively inert, was stored in all the organisms, and comprised 91% of the extractable ^{14}C in alga, 82% in snail, and 69% in fish. Hydrolysis of C_1-Cl to form 1-hydroxy-2,3-epoxychlordene formed only 3.5% of the extractable ^{14}C in alga, 8.7% in snail, and 19% in fish. These data provide an excellent illustration of the source of environmental problems with heptachlor. A variety of data is now available under comparable controlled conditions to demonstrate the comparative metabolism of insecticides (21, 44, 48).

INSECTICIDE INTERACTIONS The biochemical importance of the microsomal or mixed function oxidases in the environmental fate of insecticides in a variety of organisms has been demonstrated by model ecosystem evaluations with [^{3}H]methoxychlor comparing systems with methoxychlor alone and in the presence of five times the concentration of the microsomal oxidase inhibitor, the synergist piperonyl butoxide (36). The ready *O*-demethylation of methoxychlor to produce the water-partitioning mono- and bis-phenols was severely inhibited by the presence of the piperonyl butoxide. As a result the organisms in the model ecosystem with the piperonyl butoxide contained two to four times as much intact methoxychlor, along with increased storage of methoxychlor ethylene and methoxychlor dichloroethane or 2,2-bis-(*p*-methoxyphenyl)-1,1-dichloroethane produced by reductive dechlorination. As a net result the presence of piperonyl butoxide reduced the BI value in the fish *Gambusia* from 0.33 for methoxychlor alone to 0.081 for the combination. This study illustrates the applicability of the model ecosystem technology to the immensely complicated problem of the interaction of pesticides in mixtures.

CONCLUSIONS

Insecticides are considered the most generally troublesome of all environmental contaminants because of their purposeful introductions into the environment, their lipophilicity, and their generally similar toxic action to target pests and other animals. For most of the insecticides in use today there is a dearth of information about environmental fate and ecological effects.

Model ecosystem technology, with the use of radiolabeled insecticides, is the most informative and convenient tool for studying the fate and environmental effects of insecticides. This technology is rapid and inexpensive and can be used in virtually every laboratory. Through its expanded use it will be possible to fill the many lacunae in our knowledge of pesticide interactions in the environment, to quantify the principles of environmental degradability, and most importantly to design new pesticide molecules for greater environmental compatibility.

Acknowledgments

The author expresses his great appreciation to the many co-workers named in the bibliography who participated in model ecosystem development.

The work described from our laboratory has been supported by grants from the Herman Frasch Foundation, the Rockefeller Foundation, the Department of the Interior through the Illinois Water Resources Center (Project B-050 Illinois), the National Science Foundation (Grant ESR 74-22760), the Environmental Protection Agency (Grant R-80-3249), and the World Health Organization.

Literature Cited

1. Bray, J. J. 1975. Health Hazard. *Wall Street J.* Dec. 2, p. 18
2. Chem. Eng. News. 1976. EPA runs into more flak over pesticides. Feb. 16, pp. 19–20
3. Coats, J. R., Metcalf, R. L., Kapoor, I. P. 1974. Metabolism of the methoxychlor isostere, dianisyl neopentane, in mouse, insects, and a model ecosystem. *Pestic. Biochem. Physiol.* 4:201–11
4. Cole, L. R., Metcalf, R. L., Sanborn, J. R. 1976. Environmental fate of insecticides in terrestrial model ecosystems. *Int. J. Environ. Stud.* In press
5. Cole, L. R., Sanborn, J. R., Metcalf, R. L. 1976. Inhibition of corn growth by aldrin and the insecticide's fate in the soil, air, crop, and wildlife of a terrestrial model ecosystem. *Environ. Entomol.* 5:583–89
6. Crosby, D. G., Tucker, R. K. 1971. Accumulation of DDT by *Daphnia magna. Environ. Sci. Technol.* 5:715–16
7. de Koning, H. W., Mortimer, D. C. 1971. DDT uptäke and growth of *Euglena gracilis. Bull. Environ. Contamin. Toxicol.* 6:244–48
8. Environmental Protection Agency 1972. An evaluation of DDT and dieldrin in Lake Michigan. *Ecol. Res. Ser. R3-72-003, Aug.*
9. Environmental Protection Agency 1975. Ecosystem 'boxed up' to help evaluate pesticides. *Terr. Ecol. Res. Highlights, March.* Natl. Ecol. Res. Lab. EPA, Corvallis, Oregon
10. Ernst, W. 1970. Metabolism of pesticides in marine organisms. 2. Biotransformations and accumulation of DDT carbon -14 in flatfish, *Platichthys flesus. Veroeff Inst. Meeresforsch. Bremerhaven* 12(3):353–60
11. Focht, D. D., Alexander, M. 1970. Bacterial degradation of diphenyl methane, a DDT model substrate. *Appl. Microbiol.* 20:608–11
12. Freeman, L. 1953. A standardized method for determining the toxicity of pure compounds to fish. *Sewage Ind. Wastes* 25:845, 1331
13. Fukuto, T. R., Metcalf, R. L. 1954. Isomerization of *O,O*-diethyl *O*-beta-mercaptoethyl ethyl thionophosphate (Systox). *J. Am. Chem. Soc.* 76:5103–06
14. Gillett, J. W., Gile, J. D. 1975. Progress and status report on terrestrial in-house system. *Proc. Substitute Chem. Progr.—First Year Progr.* EPA, Fredericksburg, Va.
15. Hilton, B. D., O'Brien, R. D. 1964. A simple technique for tritiation of aromatic insecticides. *J. Agric. Food Chem.* 12:236–39
16. Hirwe, A. S., Metcalf, R. L., Kapoor, I. P. 1972. α-Trichloromethylbenzyl anilines and α-trichloromethylbenzyl phenyl ethers with DDT-like insecticidal action. *J. Agric. Food Chem.* 20: 818–24
17. Hirwe, A. S., Metcalf, R. L., Lu, P.-Y., Chio, L.-C. 1975. Comparative metabolism of 1,1-bis-(*p*-chlorophenyl)-2-nitro-propane (Prolan) in mouse, insects, and in a model ecosystem. *Pestic. Biochem. Physiol.* 5:65–72

18. Hollingworth, R. M., Metcalf, R. L., Fukuto, T. R. 1967. Selectivity of Sumithion compared with methyl parathion. Metabolism in the white mouse. *J. Agric. Food Chem.* 15:242–49
19. Johnson, B. T., Saunders, C. R., Sanders, H. D., Campbell, R. S. 1971. Biological magnification and degradation of DDT and aldrin by fresh water invertebrates. *J. Fish. Res. Board Can.* 28:705–09
20. Kapoor, I. P., Metcalf, R. L., Hirwe, A. S., Lu, P. Y., Coats, J. R., Nystrom, R. 1972. Comparative metabolism of DDT, methylchlor, and ethoxychlor in mouse, insects, and in a model ecosystem. *J. Agric. Food Chem.* 20:1–6
21. Kapoor, I. P., Metcalf, R. L., Hirwe, A. S., Coats, J. R., Khalsa, M. S. 1973. Structure-activity correlations of biodegradability of DDT analogs. *J. Agric. Food Chem.* 21:310–15
22. Kapoor, I. P., Metcalf, R. L., Nystrom, R. F., Sangha, G. K. 1970. Comparative metabolism of methoxychlor, methiochlor, and DDT in mouse, insects, and in a model ecosystem. *J. Agric. Food Chem.* 18:1145–52
23. Kavatski, J. A., Schmulbach, J. C. 1971. Epoxidation of aldrin by a fresh water ostrocod. *J. Econ. Entomol.* 64:316–17
24. Kazano, H., Asakawa, M., Tomizawa, C. 1975. Fate of 3,5-xylyl methylcarbamate (XMC) in a model ecosystem. *Appl. Ent. Zool.* 10:108–15
25. Kelly, R. G., Peets, E. A., Gordon, S., Buyske, D. A. 1961. Determination of C^{14} and H^3 in biological samples by Schöniger combustion and liquid scintillation techniques. *Anal. Biochem.* 2:267–73
26. Kenaga, E. E. 1972. Guidelines for environmental study of pesticides: determination of bioconcentration potential. *Residue Rev.* 44:73–113
27. Kokke, R. 1971. Radioisotopes applied to environmental toxicity research with microbes. *Delft Univ. Technol. Rep. IAEA PL-469* 2:15–23
28. Ladner, L., Hagstrom, A. 1975. Oil spill protection in the Baltic sea. *J. Water Pollut. Control Fed.* 47:796–809
29. Lauff, G. H., Cumins, K. W. 1964. A model stream for studies in lotic ecology. *Ecology* 45:188–91
30. Lu, P.-Y., Metcalf, R. L. 1975. Environmental fate and biodegradability of benzene derivatives as studied in a model aquatic ecosystem. *Environ. Health Perspect.* 10:269–84
31. Lu, P.-Y., Metcalf, R. L., Furman, R., Vogel, R., Hassett, J. 1975. Model ecosystem studies of lead and cadmium and of urban sewage sludge containing these elements. *J. Environ. Qual.* 4:505–09
32. Lu, P.-Y., Metcalf, R. L., Hirwe, A. S., Williams, J. W. 1975. Evaluation of environmental distribution and fate of hexachlorocyclopentadiene, chlordene, heptachlor, and heptachlor epoxide in a laboratory model ecosystem. *J. Agric. Food Chem.* 23:967–73
33. March, R. B., Fukuto, T. R., Metcalf, R. L., Maxon, M. G. 1956. Fate of P^{32}-labeled malathion in the laying hen, white mouse, and American cockroach. *J. Econ. Entomol.* 49:185–95
34. Menzie, C. M. 1969. Metabolism of pesticides. U.S. Dept. Interior Special Scientific Report. Wildlife No. 127, Washington, D.C. 485 pp.
35. Menzie, C. M. 1974. Metabolism of pesticides an update. U.S. Dept. Interior Special Scientific Report. Wildlife No. 184, Washington, D.C. 486 pp.
36. Metcalf, R. L. 1974. A laboratory model ecosystem to evaluate compounds producing biological magnification. *Essays Toxicol.* 5:17–38
37. Metcalf, R. L. 1974. A laboratory model ecosystem for evaluating the chemical and biological behaviour of radiolabeled micropollutants. In *Comparative Studies of Food and Environmental Contamination.* International Atomic Energy Agency. Vienna. SM-175/52. pp. 49–63
38. Metcalf, R. L., Booth, G. M., Schuth, C. K., Hansen, D. J., Lu, P.-Y. 1973. Uptake and fate di-2-ethylhexyl phthalate in aquatic organisms and in a model ecosystem. *Environ. Health Perspect.* 4:27–34
39. Metcalf, R. L., Fukuto, T. R., Collins, C., Borck, K., El-Aziz, S. A., Munoz, R., Cassil, C. E. 1968. Metabolism of 2,2-dimethyl-2,3-dihydrobenzofuranyl-7 *N*-methylcarbamate (Furadan) in plants, insects, and mammals. *J. Agric. Food Chem.* 16:300–11
40. Metcalf, R. L., Fukuto, T. R., Collins, C., Borck, K., Burk, J., Reynolds, H. T., Osman, M. F. 1966. Metabolism of 2-methyl-2-(methylthio)-propionaldehyde *O*-(methylcarbamoyl)-oxime in plant and insect. *J. Agric. Food Chem.* 14:579–84
41. Metcalf, R. L., Fukuto, T. R., March, R. B. 1957. Plant metabolism of dithio-Systox and Thimet. *J. Econ. Entomol.* 50:38–45

42. Metcalf, R. L., Fukuto, T. R., March, R. B. 1959. Toxic action of Dipterex and DDVP to the house fly. *J. Econ. Entomol.* 52:44–49
43. Metcalf, R. L., Fukuto, T. R., Reynolds, H. T., March, R. B. 1955. Schradan residues in cotton and cotton seed products. *J. Agric. Food Chem.* 3: 1101–13
44. Metcalf, R. L., Kapoor, I. P., Lu, P.-Y., Schuth, C., Sherman, P. 1973. Model ecosystem studies of the environmental fate of six organochlorine pesticides. *Environ. Health Perspect.* 4:35–44
45. Metcalf, R. L., Kapoor, I. P., Hirwe, A. 1974. Insecticidal biodegradable analogues of DDT. U.S. Pat. 3,787,505 Jan. 22.
46. Metcalf, R. L., Lu, P.-Y., Bowlus, S. 1975. Degradation and environmental fate of 1-(2,6-difluorobenzoyl)-3-(4-chlorophenyl)-urea. *J. Agric. Food Chem.* 23:359–64
47. Metcalf, R. L., Osman, M. F., Fukuto, T. R. 1967. Metabolism of C^{14}-labeled carbamate insecticides to $C^{14}O_2$ in the house fly. *J. Econ. Entomol.* 60:445–50
48. Metcalf, R. L., Sanborn, J. R. 1975. Pesticides and environmental quality in Illinois. *Bull. Ill. Nat. Hist. Surv.* 31(9): 381–436
49. Metcalf, R. L., Sanborn, J. R., Lu, P.-Y., Nye, D. 1975. Laboratory model ecosystem studies of the degradation and fate of radiolabeled tri-, tetra-, and pentachlorobiphenyl compared with DDE. *Arch. Environ. Contamin. Toxicol.* 3:151–65
50. Metcalf, R. L., Sangha, G. K., Kapoor, I. P. 1971. Model ecosystem for the evaluation of pesticide biodegradability and ecological magnification. *Environ. Sci. Technol.* 5:709–13
51. Morgan, J. R. 1972. Effects of Aroclor 1242 (a polychlorinated biphenyl) and DDT on cultures of an alga, protozoan, daphnia, ostracod, and guppy. *Bull. Environ. Contamin. Toxicol.* 8(3):129–37
52. Mosser, J. L., Fisher, N. S., Wurster, C. G. 1972. Polychlorinated biphenyls and DDT alter species composition in mixed cultures of algae. *Science* 176:533–35
53. National Academy of Sciences 1975. Simulated Systems. In *Principles for Evaluating Chemicals in the Environment,* Chap. 14. Washington, DC: NAS.
54. National Academy of Sciences 1976. *Contemporary Pest Control: An Assessment of Present and Alternative Technologies,* vol. 1. Washington, DC: NAS
55. Pfaender, F. K., Alexander, M.A. 1972. Extensive microbial degradation of DDT *in vitro* and DDT metabolism by natural communities. *J. Agric. Food Chem.* 20:842–46
56. Pimentel, D. 1971. *Ecological effects of Pesticides on Non-Target Species.* Washington DC: Off. Sci. Technol. 220 pp.
57. Reinbold, K., Kapoor, I. P., Childers, W. F., Bruce, W. N., Metcalf, R. L. 1971. Comparative uptake and biodegradability of DDT and methoxychlor by aquatic organisms. *Bull. Ill. Nat. Hist. Surv.* 30(6):405–15
58. Reinert, R. E. 1972. Accumulation of dieldrin in an alga (*Scenedesmus obliquus*), *Daphnia magna,* and the guppy (*Poecilia reticulata*). *J. Fish. Res. Board Can.* 29:1413–18
59. Rose, F. L., McIntire, C. D. 1970. Accumulation of dieldrin by benthic algae in laboratory streams. *Hydrobiologia* 35(3–4):481–93
60. Sanborn, J. R., Childers, W. F., Metcalf, R. L. 1975. Uptake of three polychlorinated biphenyls, DDT, and DDE by the green sunfish, *Lepomis cyanellus* Ref. *Bull. Envir. Contamin. Toxicol.* 13(2):209–17
61. Sanborn, J. R., Chio, L.-C. 1976. *Validation Studies Using a Terrestrial Aquatic Model Ecosystem. EPA Contract Res. Progr. Rep.*
62. Sanborn J. R., Metcalf, R. L., Yu, C.-C., Lu, P.-Y. 1975. Plasticizers in the environment: the fate of di-*n*-octyl phthalate (DOP) in two model ecosystems and uptake and metabolism of DOP by aquatic organisms. *Arch. Environ. Contamin. Toxicol* 3:244–55
63. Sanborn, J. R., Yu, C.-C. 1973. The fate of dieldrin in a model ecosystem. *Bull. Environ. Contamin. Toxicol.* 10:340–46
64. Schwemmer, B., Cochrane, W. P., Polen, P. B. 1970. Oxychlordane, animal metabolite of chlordane: isolation and synthesis. *Science* 169:1087
65. Sodergren, A. 1968. Uptake and accumulation of C^{14} DDT by Chlorella sp. *Oikos Copenhagen* 19:126–38
66. Taub, F. B. 1974. Closed ecological systems. *Ann. Rev. Ecol. Syst.* 5:139–60
67. Taub, F. B. 1969. A biological model of a freshwater community: a gnotobiotic ecosystem. *Limnol. Oceanogr.* 14:136–42
68. Taub, F. E., Pearson, N. 1973. *Toxic Effects on Aquatic Communities. Final Rep. EPA*

69. Taub, F. B., Pearson, N. 1974. Use of laboratory microcosms to determine environmental fate of chemicals. *Oak Ridge Nat. Lab. Environ. Sci. Div. ORNL IIX-49403V, Aug. 1,* pp. 1–153
70. Von Rumker, R., Horay, F. 1972. *Pesticide Manual. U.S. Dept. State Agency Int. Dev. AID/csd 3296*
71. Warren, C. E., Davis, G. E. 1971. Laboratory stream research: objectives, possibilities and constraints. *Ann. Rev. Ecol. Syst.* 2:111–44
72. Yu, C.-C., Booth, G. M., Hansen, D. J., Larsen, J. R. 1974. Fate of carbofuran in a model ecosystem. *J. Agric. Food Chem.* 22:431–34
73. Yu, C.-C., Booth, G. M., Hansen, D. J., Larsen, J. R. 1975. Fate of pyrazon in a model ecosystem. *J. Agric. Food Chem.* 23:309–11
74. Yu, C.-C., Hansen, D. J., Booth, G. M. 1975. Fate of dicamba in a model ecosystem. *Bull. Environ. Contamin. Toxicol.* 13:280–83

Ann. Rev. Entomol. 1977. 22:263–88

EVOLUTIONARY GENETICS OF THE HYMENOPTERA

❖6130

R. H. Crozier

School of Zoology, University of New South Wales, Kensington,
New South Wales 2033, Australia

INTRODUCTION

The order Hymenoptera is of great size and possesses a number of notable characteristics: wide diversity among life patterns of its members, extraordinary repeated evolution of social forms, arrhenotoky (origin of males from unfertilized eggs), and male haploidy. These characteristics interact with manifold aspects of the genetic biology of the group, including karyotype evolution, speciation, and population genetics. These various topics are reviewed below; other genetical and evolutionary areas are covered elsewhere (30, 90, 115, 130, 158–160, 169).

SEX DETERMINATION

The basic mode of sex determination in the order is arrhenotoky, except in those species, particularly parasitoids, that are secondarily thelytokous. All normal hymenopteran males are haploid and females diploid (30). These characteristics make the question of genic-level sex determination in the order a complex one upon which there is still disagreement (30, 90), even though the mechanism in two divergent species is well known: heterozygotes at a single sex locus are female, hemi- and homozygotes are male. Genic balance is the mechanism applied to many organisms with both sexes diploid, and interlocus dosage differences between genotypes determine sexual differences. With every hymenopteran chromosome normally present once in male eggs and twice in eggs destined to be females, no interlocus dosage differences are possible, and genic balance is eliminated as an explanation of hymenopteran sex determination.

Nucleocytoplasmic Balance Model

A *genic-balance* model, with sex determination by the balance between male- and female-determining loci, was proposed by da Cunha & Kerr (34). The female-tendency loci are additive, with increased activity as a group in diploids compared with haploids, whereas male-determining loci do not show increased activity in diploids. The male-determining loci outweigh the others in a haploid, leading to development as a male, whereas in a diploid the increased relative activity of female-determiners yields a female. Kerr (90) has modified this model proposing that the male-determiners are also additive, but less so than the female-determiners. The single sex-locus of some species is a major female-determining locus that is not additive except when heterozygous. According to Chaud-Netto (20), the diploid males of single-locus species should be hypermasculine (super- or metamales), if anything, because of the increased activity of the male-determiners; however, feminization would be equally plausible under the genic-balance model, because of activity of minor, fully additive female-determiners.

No molecular-level mechanism has been proposed for the genic-balance model. Without introducing allelic diversity, no mechanism seems possible, except one that relies on relative chromosomal/cytoplasmic dosage differences between haploids and diploids. For the sake of brevity and consistency with previous usage, this nucleocytoplasmic balance model is referred to below as the genic-balance model.

Allelic Diversity Models

Extensive genetic work over several decades firmly established that sex in the parasitoid *Bracon hebetor* is determined by the genotypes at a single locus: heterozygotes are female, homo- and hemizygotes are male (161). Demonstration of the *single-locus* mechanism hinges ideally on the production of diploid males, which proved hard to obtain in the honey bee *Apis mellifera* because here these drones are eaten by the workers (173); nevertheless, the evidence for this model applying to this species too is unequivocal (39, 109, 129, 174). Diploid males are known in species of seven other taxonomically divergent genera; the single-locus model is plausible for one of these cases (144). However, the single-locus model is not generally applicable to Hymenoptera because close inbreeding does not produce diploid males in expected proportions for some other species where detection would seem likely.

The difficulty in producing diploid males is understandable if sex is generally determined by a number of loci, with individuals heterozygous for one or more of these loci being female (28, 30). The single-locus case results either from homozygosity at all sex loci except one, or through one sex locus becoming predominant in effectiveness. The diploid males of single-locus species should, if anything, be feminized in comparison to normal haploid males, because of the activity of the residual sex loci; supermales are not expected. A molecular mechanism for this *multiple-locus* model is suggested by gynandroids: males mosaic for different haploid genotypes that, at least in *B. hebetor* (164), often have feminized borders between genetically different tissue patches. An allele at a sex locus produces a peptide that forms active female-determining molecules only as heteropolymers with the peptides produced by other alleles (28). Another suggestion (92) is that each sex allele at a locus produces activator mRNA (13) defective in one or more cistrons: in heterozygotes, with defective regions in one allele compensated for in the other, a further female-determining locus is activated.

In the copepod *Tisbe reticulata,* inbreeding increases the proportion of males, as also expected in Hymenoptera under the multiple-locus model. Although he referred to two sets of loci, Battaglia's (8) model for this male-diploid species reduces to sex being determined by the balance between male- and female-tendency effects at a number of loci scattered over various chromosomes. Female- and male-determining genes are allelic, with the female-determiners dominant to the male-determiners. Likelihood of development as a male increases with increasing homozygosity for male-determiners, which is in turn increased by inbreeding. Hemizygosity (as in haploid males) would also lead to a higher proportion of the recessive male-determiners being active, since none would then be masked by (dominant) female-determiners. This mechanism cannot be universal, even among copepods, because many have female heterogamety (131).

Comparing Models and Data

The single-locus model holds well for only two genera, and so it does not provide a general model for hymenopteran sex determination. A general model should adequately explain sex determination in other cases, encompass the single-locus case with a minimum of special pleading, be testable, and preferably admit of a plausible molecular-level explanation. I will concentrate below on comparing the genic-balance and multiple-locus models because these are the paradigms for current research and because the *Tisbe* model, although close to the multiple-locus scheme in some key predictions, cannot readily encompass the single-locus case. I attempt to avoid a Procrustean treatment of the data, but my remaining biases (28, 30) can be judged against other treatments (20, 90, 175).

Smith & Wallace (144) found that inbreeding produces diploid males in the sawfly *Neodiprion nigroscutum,* and from pedigree analysis they concluded that the single-locus mechanism applies to this species. Noting a deficiency of diploid males in one class of crosses, Kerr (90) suggests that a two-locus multiple-locus model might apply. However, Smith & Wallace presented only lumped data, not individual results. In view of similar deficiencies found in *Bracon* crosses (162), the best-known single-locus case, I tentatively favor the single- over the multiple-locus model for *N. nigroscutum.*

A mother-son mating in the bumblebee *Bombus atratus* produced 14 workers (presumably diploid), 10 males that were probably all diploid, and 17 males of unknown ploidy (44). Garófalo (44) assumes the unclassified males were all diploid and concluded that a two-locus variant of the multiple-locus model fits the data, with only double heterozygotes becoming female. This mechanism would be inefficient, leading to the production of large numbers of diploid males (effectively semisterile), unless there were a number of such loci, heterozygosity at any two of which leads to development as a female. Garófalo's cross would then involve heterozygosity at only two such loci. Although the model is plausible, the analysis supporting it ignores the likely production by the queen of some haploid males from unfertilized eggs. The upper 95% confidence limit (hypergeometric distribution) for haploid males in Garófalo's sample is about six; when this is taken into account the data do not distinguish between the single-locus and Garófalo's models. Similar and other difficulties occur with the progeny analysis for a virgin triploid *B. atratus* (45), compounded by the occurrence of three types of known males—diploid, haploid, and aneuploid—plus unkaryotyped males and intersexes. Plausible though Garófalo's model may be in general terms, tentative acceptance of the single-locus model for *Bombus* seems advisable.

Two of the *Bombus* aneuploids had almost the triploid number ($3n = 60$) of chromosomes, indicating a somewhat chaotic maternal meiosis with all three homologs in many trivalents segregating together. Assuming the intersexes were aneuploid, they can be understood under the single-locus model as resulting from hemi- or homozygosity at the major sex locus, with many minor sex loci heterozygous (28), and under the genic-balance model from an increased chromosome/cytoplasm ratio for loci other than those mutated to dependence on heterozygosity for additivity.

Chaud-Netto (20), in an interesting approach, measured 22 characters in diploid and haploid drones and their diploid and triploid worker sisters and calculated Mahalanobis' generalized distances, D^2, between all types. The resulting plot of 2D values is a tilted quadrilateral: both diploid males and triploid workers are "masculinized" (i.e. displaced away from the female end along the normal male-female axis). This result would be compatible with the genic balance but not the allelic-diversity models if statistically significant, which was not tested for. While the distances between the same-sex castes are indeed significant ($P < .05$) (145, p. 403), the positions of the aberrant castes on the normal drone-worker axis are not significantly distant from those of the corresponding normals.

The apparent resemblance between same-sex adults of different ploidy when nonsexual characters are considered (20) contrasts with the reduced testis size of diploid drone adults (175) and similar feminization of larvae for three characters (96). Woyke (175) found that testis size in diploid drones with differing proportions of African and Italian ancestry is governed polygenically by these proportions and not by degree of background homozygosity ($F \approx .08$–$.27$) nor by which sex locus allele they are homozygous for. Under the multiple-locus paradigm, Woyke's results indicate a threshold—heterozygosity at a few minor sex loci is as effective as heterozygosity at many in causing feminization. No genetic background effect is expected under the genic-balance model. Testis measurements have not been reported for *Bracon* males, but *Bombus atratus* diploid males have smaller testes than haploids (91).

Inbreeding produces diploid males in the bees *Melipona quadrifasciata* and *Trigona quadrangula* (90), but the mechanisms involved cannot be determined since no data have yet appeared. Cytologically verified diploid males have been found in the ants *Pseudolasius* nr *emeryi* (73) and *Solenopsis invicta* (multi-queened form); the latter males lack functional testes (75; B. M. Glancey, personal communication) and include allozyme marker heterozygotes (74).

In the parasitoid *Nasonia vitripennis,* polyploidy arose three times in laboratory stocks (163), and crosses yielded tetraploid females as well as the original diploid males and triploid females (108). The progeny of virgin females of any ploidy level are all sons, which, considering the preferential segregation of parental markers known in *Nasonia* (22, 108), suggests close linkage of the sex-determining loci to their centromeres (28).

Diploid males are thus known in nine divergent genera, ruling out nucleocytoplasmic balance models for these. In terms of both testability and plausibility, the multiple-locus paradigm seems favored over the genic-balance model, but remains to be firmly demonstrated. Why inbreeding bisexual species do not die out under a sterile diploid male surplus is explicable with closer examination of their life pattern (see below). Why there are not more reports of diploid males (90) probably stems from the special conditions needed for detection: genetic markers, good cytologists, or serendipity. The *Neodiprion* diploid males were discovered fortuitously through being heavier than haploids, which is apparently not so for other diploid males [although among *Bracon* males, diploids have larger cells (55)].

A difficulty with the supposed evolutionary derivation of single-locus from genic-balance species via the mutational production of complementary defective alleles is that these—being effectively recessive sterility alleles—are selectively eliminated.

One definitive test (e.g. see 30) between the genic-balance and multiple-locus models would be to produce completely homozygous diploids in species lacking the single-locus mechanism. Such individuals would be female under the genic-balance model and male under the alternative model. Completely homozygous *Apis* males should have larger testes than the incomplete homozygotes known so far. How might such individuals be produced and recognized? One way would be through colchicine-induced genome doubling, but a better method utilizes the genetic uniformity of the sperm from a single haploid male. Heavy irradiation can destroy the egg nucleus, permitting the development of "motherless" males from sperm nuclei (43, 157); when such eggs are fertilized by more than one sperm, diploid tissue can result (102). Simple chromosome techniques are now available for screening large numbers of males, so this is a manageable problem for hymenopteran geneticists, especially for those working on *Apis* and *Nasonia.*

CASTE DETERMINATION

Caste in most social insects investigated is determined environmentally (see 169), but a strong genetic influence on caste determination has been widely accepted for the bee genus *Melipona* because of the strongly suggestive evidence of Kerr and his group (86, 89, 92, 97). Unlike those of other eusocial apids, *Melipona* combs, which are mass-provisioned, have no different cell types, and queens, workers, and males are distributed apparently at random. A maximum of 25% of females become queens; this 3:1 ratio of workers to queens leads Kerr to plausibly suggest that caste in *Melipona* is determined by two loci and that only double heterozygotes become queens. An initial (86) hypothesis of a third caste-determination locus in some species has been abandoned. The proportion of queens drops under unfavorable conditions, which Kerr initially (86) proposed indicates the induction of meiotic thelytoky (boosting the proportion of homozygotes). Recent evidence (89) indicates that pupae weighing less than 72 mg in *M. quadrifasciata* become workers irrespective of caste locus genotype, indicating a strong environmental effect. *Melipona quadrifasciata* queens do, in fact, mate but once (98), as required by the model. *Marginata* queens invariably have four abdominal ganglia, while workers have either four or five, suggesting to Kerr & Nielsen (95) that the four-ganglion workers are nutritionally deprived potential queens: the fit of five-ganglion to four-ganglion females to a 3:1 ratio is better than that for workers to queens. However, in *M. quadrifasciata anthidiodes,* ganglion number variation occurs in both castes, requiring the postulate of a third, ganglion-number-mediating locus that interacts complexly with the caste loci (89).

Without new queens, a *Melipona* colony is mortal and its reproductive performance impaired. However, if 25% of the females emerging are queens, they are greatly in excess of requirements and are killed by the workers (89); evolution of mechanisms limiting the excess would be expected, such as additional loci requiring heterozygosity for development as a queen. The dynamics of the postulated two locus system are interesting: two alleles are selected for at each locus, but additional alleles (which would boost wasteful queen production) are selected against. Nevertheless, bottleneck effects could lead to establishment of different pairs of alleles in different populations; interpopulation crosses might then yield high proportions of queens and support the theory.

The Darchens (36, 37) have attacked Kerr's genetically mediated caste-determination hypothesis, obtaining experimental ratios showing an exclusive environmental role in *M. beechei.* However, the data upon which the ratios are based remain unpublished, preventing assessment of the Darchens' report.

Kerr (89, 92) believes that the caste- and sex-determination systems are functionally intergrated in bees and that the caste loci act by increasing juvenile hormone titer. A role for juvenile hormone is supported by experiment for *Melipona quadrifasciata* (18) and the ant *Myrmica rubra* (12). *Melipona,* but not *Apis,* workers are markedly displaced in a male direction from queens in terms of generalized distance (91), leading Kerr to regard *Melipona* workers as intersexual. Analyzing these data as for Chaud-Netto's shows that the malewards displacement of *Melipona* but not that of *Apis* workers is indeed statistically significant ($P < .01$). However, similar analysis of distance measures made on the wasps *Vespula germanica* and *V. rufa* (10) shows that these workers, too, are strongly "masculinized" ($P < .01$), and caste-determination in vespids seems to be strongly envirionmental (77).

Harpagoxenus sublaevis has both normal queens, which are rare, and ergatoid females. Buschinger (17), on the basis of various crosses involving the two female types, plausibly suggests a marked genetic influence on caste determination in this ant. According to Buschinger, there is a locus at which recessive homozygotes (*ee*) become queens if adequately cared for during development, but at which the other genotypes (*Ee, EE*) become ergatoids irrespective of their envirionmental milieu. Buschinger failed, however, to induce any putatively 100% *ee* broods to all become queens. The *Harpagoxenus* model differs markedly from that proposed for *Melipona,* as the latter involves balancing selection, but, as it stands says little about normal ant caste-determination because the *E* allele should there be deleterious.

This model (if confirmed) would, on the other hand, be a great potential aid to understanding various ponerine ants lacking true queens, and especially those *Rhytidoponera* species where true queens occur, but sporadically (65). Geographical surveys of the frequencies of the postulated *e* allele in *Harpagoxenus* (and perhaps in *Rhytidoponera*) might yield key insights into the selective pressures involved in this apparent evolutionary loss of morphologically differentiated queens.

In at least one ant species (11), the faster-growing larvae are the most likely to become queens. In *Polistes* wasps (153) and the ant *Leptothorax gredleri* (15), dominance struggles between potential queens determine which becomes the egg layer. In these and similar cases, queen potential could increase with heterozygosity; this hypothesis is testable with electrophoretic techniques.

KARYOTYPE EVOLUTION

Haploid chromosome numbers in the order range from 3 to 42; the spread is caused by the best-known group, the ants (30; H. T. Imai, R. H. Crozier, and R. W. Taylor, in preparation). Distributions of haploid numbers have been plotted under the *genus-karyotype* convention (30), which minimizes the effects of differences in intensities of sampling between genera by counting each number only once per genus. The discussion below is based on such plots. Many early references are listed elsewhere (30).

Symphytan numbers range from 6 to 26, with most at the lower end of the scale (30). *Diprion simile* once seemed a polyploid, with $n = 14$ as opposed to $n = 6$, 7, and 8 for other diprionids, but the describer of the karyotype (S. G. Smith) later attributed this increase to centric fission. Parasitican karyotypes once appeared highly uniform within superfamilies (30), but recent findings have eroded this uniformity. Although haploid numbers of five still predominate in chalcidoids, haploid numbers of four, six, and, surprisingly, ten are now known (48, 50, 52, 76). Cynipids remain uniform with $n = 10$, although here, as elsewhere, centromere positions differ between species (51; A. R. Sanderson, personal communication). The braconid *Phaenocarpa persimilis* has $n = 17$ (G. Prince and H. Stace, unpublished information), whereas other ichneumonoids have lower numbers, down to $n = 10$ (30). Robertsonian changes and pericentric inversions, rather than polyploidy, seem exclusively to have molded parasitican karyotype evolution (52); this conclusion draws support from *Nasonia vitripennis* ($n = 5$) having twice the DNA value of *Bracon hebetor* ($n = 10$) (125), instead of the reverse, which accords with a prediction based on chromosome sizes (30).

Among aculeates other than bees or ants, *Polistes* has a particularly wide range (6–26) of haploid numbers (30, 49), suggesting a species-level role here for cytotaxonomy. Goodpasture (49) tentatively concludes that the higher numbers are ancestral for five eumenid wasp species ($n = 6$–10).

Bee haploid numbers range from a possible 6 to 20 and are bimodally distributed, with most clustered around 17 and with a much smaller peak around $n = 8$ (30, 132). Bizarre phenomena during spermatogenesis occur in some species, including a clumping of the chromosomes that roughly simulates pairing. Kerr (92; also earlier papers) believes that the lower-numbered karyotypes are ancestral and that the far more numerous higher-numbered karyotypes were repeatedly derived from them by polyploidy. Kerr's conclusion should be regarded as highly tentative because of the small number of bees karyotyped (48 species), the general lack of published details on chromosome morphology, the apparent action of two caste-determination loci in both diploid and polyploid *Melipona* species (89), the lack of supporting DNA values (needed also in Hymenoptera generally), and the plausibility of mechanisms other than homologous pairing to explain the observed clumping (30).

More ants (over 300 species) have been karyotyped than all other hymenopterons combined (30, 32; H. T. Imai, R. H. Crozier, and R. W. Taylor, in preparation; B. M. Glancey, M. K. St. Romain, and R. H. Crozier, unpublished information). Marked differences occur between genera in the degree of karyotype variation. Thus for 30 *Formica* species $n = 26$ or 27, whereas $n = 5$–42 for 8 species of *Myrmecia,* in which there are two cases of probable sibling species differing greatly in chromosome number; karyotype change has also been rapid in genera such as *Camponotus* and *Aphaenogaster* (30; H. T. Imai, R. H. Crozier, and R. W. Taylor, in preparation). The predominant detectable rearrangements in ant karyotype evolution have been Robertsonian changes and pericentric inversions, with growth of pericentromeric or terminal heterochromatin also important; rarer events include simple and complex translocations and tandem fusions. Higher-numbered karyotypes have smaller chromosomes and higher proportions of acrocentrics than lower-numbered ones. That the ancestral ant chromosome number was close to the high end of the scale (30) now seems unlikely, not only because the highest *Myrmecia* numbers now known (41 and 42) greatly exceed the highest known number in other aculeates (26 in a *Polistes*) (49), but also because the generation of the lowest numbers from the highest would lead to much loss of functional genetic material. An intermediate number is more plausible, for example, 15, the median of the roughly normal ant genus-karyotype histogram, or, perhaps preferably, the modal number of 11. However, current reappraisals of basic chromosomal rearrangement processes suggest that centric fission, rather than centric dissociation, may often cause increases in chromosome number (e.g. 83). Various lines of evidence support a model, principally synthesized by H. T. Imai (H. T. Imai, R. H. Crozier, and R. W. Taylor, in preparation), in which chromosome numbers tend to increase rather than decrease, as is often thought. Testing this theory further in ants, other Hymenoptera, and diverse other animals awaits widespread phylogenetic analysis and also analysis of c-banding patterns. Preliminary c-banding analysis supports the model for ants. Phylogenetic analysis indicates marked increase of number for *Pheidole nodus,* for which $n = 17$–20 as against $n = 9$, 10 for congeners, marked decrease of number for the western form of *Rhytidoponera metallica,* with $n = 11$, 12 as against $n = 17$–26 for congeners, and repeated conversion of acrocentric chromosomes to metacentrics in *Camponotus* species.

Supernumerary (B–) chromosome polymorphism occurs in the ant *Leptothorax spinosior* (78), probably also in two *Aphaenogaster "rudis"* sibling species (30, 32), and possibly in another myrmicine ant, *Podomyrma adelaidae* (H. T. Imai, R. H. Crozier, and R. W. Taylor, in preparation). Robertsonian and other numerical polymorphisms are known in species of the genera *Myrmecia, Rhytidoponera, Ponera, Aphaenogaster,* and *Pheidole* (30, 79; H. T. Imai, R. H. Crozier, and R. W. Taylor, in preparation). *Iridomyrmex gracilis* and *Tapinoma sessile* are polymorphic for pericentric inversions, and the western form of *Rhytidoponera metallica* is polymorphic for complex rearrangements (30; H. T. Imai, R. H. Crozier, and R. W. Taylor, in preparation).

C-banding reveals that in some ants, notably *Myrmecia brevinoda,* the long arms of most chromosomes are heterochromatic (H. T. Imai, R. H. Crozier, and R. W. Taylor, in preparation), as in the bee *Megachile rotundata* (C. Goodpasture, personal communication) and the beetle *Chilocorus stigma* (143).

SIBLING SPECIES AND SPECIATION

In accord with a theoretical finding that male-haploid species should evolve one-third faster than male-diploids (64), sibling species are common in many hymenopteran groups, suggesting that speciation is rapid. Mating-choice experiments and exacting morphological analysis showed that the solitary pteromalid *Muscidifurax raptor,* probably introduced into the Americas with its housefly host in Columbian times, has diversified there to yield four new species. One of these new species is uniparental and another evolves towards gregariousness (5, 101, 103). The evolution of gregariousness implies a shift from outbreeding to sibmating. Hybridization experiments also detected sibling species in the scelionid genus *Gryon* where, unlike the *Muscidifurax* case, the siblings differ in host preferences (136). The enigmatic *Aphytis mytilaspidus* case is discussed below; hybridization experiments showed sexual isolation between other *Aphytis* strains and also produced some strains partially reproductively isolated from both parents (124). Hybridization or genetic evidence has frequently indicated, if not sibling species, then genetic divergence between allopatric populations consistent with incipient speciation (e.g. 70, 110, 138). Extensive analyses on the evolutionary relationships of *Trichogramma* are reviewed by Nagarkatti & Nagaraja (116a) elsewhere in this volume. The evidence leaves unsettled whether or not speciation is more rapid in groups with the inbreeding life pattern than in relative outbreeders, as suggested by Askew (3).

Genetic evidence has also proved important in determining symphytan sibling relationships. Smith (142) found that, whereas *Diprion polytomum* in Europe is biparental and has a haploid number of six, a form introduced into Canada is thelytokous and has seven chromosomes. The name *hercyniae* was later resurrected and applied to the thelytokous form (100). Speciation in *Neodiprion* frequently involves shifts in host plant preferences, as well as emergence time (100), allowing either sympatric speciation or swift establishment of coexistence. Various factors cooperate to maintain the integrity of sibling species placed under *Neodiprion "abietis";* thus female hybrids between the spruce and balsam strains preferentially oviposit on balsam, upon which their larvae do poorly (100). Knerer & Atwood

(100) point out that, as in other hymenopterans, all F1 hybrids are female (males being uniparental), so that introgression is likely to be limited, especially where there are emergence time differences between the parental populations.

The apparent ready derivation of thelytokous strains from biparental species is understandable when the very widespread occurrence of sporadic thelytoky in the Hymenoptera (30) is considered. This sporadic thelytoky is doubtless facilitated by oviposition rather than fertilization being the stimulus for embryonic development (see 152). An excess of homogametic matings during natural crossing between the honey bee subspecies *Apis mellifera adansonii* and *A. mellifera ligustica* in Brazil (94) has not prevented their widespread amalgamation (115).

Ant species can often be distinguished only on minor morphological details (168, 172); it now seems likely that many remaining widespread "species" are in fact clusters of siblings. Karyotype differences, some extreme, are known between disjunct populations of *Rhytidoponera metallica, Camponotus compressus, Myrmecia pilosula, M. fulvipes,* and *Myrmica sulcinodis* (30; H. T. Imai, R. H. Crozier, and R. W. Taylor, in preparation). *Conomyrma bicolor* (from Arizona) and *C. ? thoracicus* (from Peru) have morphologically indistinguishable workers but divergent karyotypes (30). In the absence of analysis of the zones of overlap for the above cases, significant intergradation there might be postulated (6), but this possibility is remote for two other cases. The *Iridomyrmex "purpureus"* complex comprises a number of contiguously allopatric sibling species differing in coloration, nest form, and isozyme gene frequencies (56). *Aphaenogaster "rudis"* comprises two forms differing in karyotype, habitat preferences, and (slightly) in coloration in upper New York State (30); in northeast Georgia there are three sibling species ($n = 18, 20, 22$), between which, on chromosomal and allozyme evidence, there is little or no introgression in sympatry. The 18- and 22-chromosome forms are montane species which show microhabitat differences but which are extremely similar in worker morphology; the 20-chromosome form is lighter in color and largely restricted to the Coastal Plains province. Genetic distances between populations are slight within these sibling species compared to distances between populations of different cytotypes (32). The variation with environmental factors of allozyme allele frequencies at three loci in *Pogonomyrmex barbatus* (84) and two in *P. badius* (148) may indicate that both of these are clusters of siblings.

Speciation may be occurring in the red imported fire ant, *Solenopsis invicta.* There are contradictions between published accounts as to whether colonies are single-queened and hostile to one another, or multi-queened and mutually tolerant (107). Diploid males (see section on sex determination) have so far been found only in highly multi-queened colonies. One possibility is that the populations in the United States are in transition from the single- to the multi-queened form. Another possibility is that a multi-queened form largely lacking in intercolony hostility is emerging and will coexist with the single-queened form. There are a number of cases in other ant genera of such unicolonial species (169), often coexisting with related normal multicolonial species, from which they probably arose. The evolution of unicolonial species is difficult to explain, as progressive loss of the worker caste would be expected.

The development of artificial mating techniques for some ant species at least (16, 17, 35, 53, 72) should allow the analysis of postmating (and sometimes premating) reproductive isolating mechanisms.

THE PARASITOID LIFE PATTERN

Many parasitic species, especially chalcidoids, have males with short or no wings and eggs laid in clutches that are usually isolated from each other, and they mate either close by or even inside the host. Hamilton (59) and Askew (3) rightly stress that such species are close inbreeders, with sibmating the rule. This life pattern raises a problem: if inbreeding is total, why do inbreeders not abandon males altogether? Sporadic thelytoky could well occur in every hymenopteran species (30), so that the continued, inefficient, production of males is puzzling if indeed there is no outbreeding.

The problem is simply resolved by the assertion that outbreeding occurs every few generations at least in all normally arrhenotokous species. A number of careful studies that support the above assertion have been made on species in which superficial examination would suggest no relief from inbreeding. Sibmating species tend to minimize male production in favor of female production (59); that they outbreed frequently is indicated by the fact that ovipositing females increase the proportion of eggs not fertilized when other such females are present on the host, thus increasing their share of the genetic output of the group (59). Such competitive sex-ratio shifts are known in the pteromalids *Nasonia vitripennis* (68, 176, 177) and *Lariophagus distinguendus* (4), the ichneumonid *Gregopimpla himalayensis* (141), and the eulophid *Dahlbominus fuscipennis* (151, cf 166). Increases in the percentage of males following multiparasitism are also known for many other species, being attributed, perhaps erroneously, to male larvae being more resistant to crowding stress (133). Intermale combat for possession of the emergence site and the virgin females (99, 111, 171) also indicates outbreeding and competition between sibships. The females of many species avoid previously parasitized hosts (134), recognition of which is often facilitated by pheromonal marking by the females following oviposition. The effectiveness of marking, however, is far from complete: Rabb & Bradley (122) found that only 77% of attacked host eggs are marked by *Telenomus sphingis* females, and inexperienced females of this species and others (80, 135) tend to lay even in marked hosts. Hidaka (67) reports that *Telenomus gifuensis* females fight for posssession of the host egg mass, inferring that only one finally oviposits in it, but in another scelionid, *Trissolcus basalis,* inter-female combat occurs only after almost all the host eggs have been attacked (171).

The oviposition of inexperienced females in previously parasitised hosts suggests that outbreeding increases with parasite density, because the frequency of such hosts should increase with parasite numbers. Another factor favoring outbreeding in *Nasonia vitripennis* at high parasite and host densities is that males displaced from one host seek other parasitized puparia and compete for them (99). When females are emerging rapidly and occupying the attention of several males, they are likely to prefer males from puparia other than their own (54).

Even many thelytokous species may occasionally outbreed, if recent findings for the aphelinid *Aphytis mytilaspidus* (128) are any guide. Actually *mytilaspidus* comprises two species, one arrhenotokous and one reproducing by meiotic thelytoky. The latter form produces occasional males, thus resembling many other thelytokous species in which these males have been considered functionless. However, females of both *mytilaspidus* species mate with the males of either form. If mated, even thelytokous females lay fertilized eggs. Variability can thus pass between the species and from one thelytokous lineage to another. Of course, even in the absence of such outbreeding, selection can maintain genetic variation in thelytokous species (2). Parallels to the *mytilaspidus* case could be widespread, but have not been looked for.

Genetic systems similar to those of inbreeding parasitoids occur elsewhere in the order as well. For example, the queen of the socially parasitic ant *Anergates atratulus* penetrates a colony of its host, *Tetramorium caespitum,* leading to the host queen's death. The *Anergates* males are flightless and mate within the parental nest before the young queens disperse (169). The vespid *Stenodynerus miniatus* lays eggs in mixed-sex pairs; the resulting sibs usually mate together (81).

Only a little direct evidence exists yet to support the above assertion that outbreeding occurs even in sibmating Hymenoptera. In *Anergates,* Gösswald (see 61) found that sometimes more than one queen is established in a host nest, and Ramirez (123) found that although often only one fig wasp (Agaonidae) colonizes a fig, up to four may do so. Some *Stenodynerus* individuals leave the emergence site unmated (81) and thus could mate elsewhere. Allozyme analyses of the level and distribution of heterozygosity in inbreeding parasitoid populations would provide the missing evidence. Such studies should show that the probability of heterozygosity is highly correlated between loci, with most heterozygosity caused by a few multiple heterozygotes, the results of recent outbreeding.

POPULATION GENETICS

Hymenopteran population genetics theory is sex-linked theory—any deviations would probably result from the absence in male haploids of interactions between autosomal and sex-linked loci. Oscillations in gene frequency occur readily at sex-linked (and hence hymenopteran) loci in populations with discrete, non-overlapping generations; the behavior of neutral alleles that initially differ in frequencies between males and females is well known (see 24, 104), and oscillations also occur under a range of selective conditions (19). Dyson (40) observed such fluctuations in the frequencies of two deleterious genes in *Bracon hebetor.* Populations with overlapping generations lack oscillations for neutral genes (117), and the same result seems likely for the (unstudied) case of selective models. These results imply significant differences between annual and perennial hymenopterons in terms of the behavior of allele frequencies.

Curious allele-frequency behavior is also expected in other ways at the form-determination loci in those cynipids which have bizarre alternating-generation life cycles which prevent interbreeding among the descendants of a mating until the F4 generation. The precise genetic mechanism determining the forms is unknown, but whichever of the possibilities (30, 42) it turns out to be, sampling error acts in one sex but not the other to eliminate one or more genotypes (30).

Li (105) and Hartl (63) found measures of the average fitness of male-haploid populations that are maximized under weak selection. Hartl (64) derived an extension of Fisher's well-known fundamental theorem of natural selection that applies fairly well to male-haploids under the additional constraints of additive fitness relationships in females and identity of male fitnesses with those of corresponding female homozygotes. Under these assumptions, male haploids evolve one-third faster than similar male diploids, but one-third slower than total haploids (64).

There are few data on the modes of natural selection operating in Hymenoptera. One sample with a marked heterozygote excess was found in *Aphaenogaster* "*rudis*" (29, 32), and several with excess numbers of homozygotes in *Pogonomyrmex barbatus* (84) and *P. badius* (148). Heterozygote excess is certainly not general in *Aphaenogaster* populations (32), and the *Pogonomyrmex* result says little about modes of selection because the individuals concerned were workers and thus not part of the Mendelian population (queens and males). However, *badius* possibly sibmate frequently (150), so that a nest-by-nest analysis of this species at the level of the reproductive population would be worthwhile. Grant, Snyder & Glessner (54) found frequency-dependent selection (rare-type male advantage) in laboratory trials with *Nasonia vitripennis,* but their procedures do not distinguish between the basis of female choice being genotype per se or culture odor (66). Judicious use of multiparasitism by females carrying different markers to yield genetically mixed cultures as the source of the experimental wasps would solve this problem.

Is the a priori likelihood of balanced polymorphism in Hymenoptera the same as in male diploids? Two approaches to this question have unfortunately yielded different conclusions. Hartl (63) found that if males have the same fitness values as the corresponding female homozygotes, then male haploids should have less balanced polymorphism than male diploids, but how much less is unknown. However, if there is no correlation between male and female fitness values, then the probability of balanced polymorphism in the two genetic systems is about the same (26, 30), but whether or not the probabilities are identical is uncertain. However, male haploidy definitely leads to a decrease in the equilibrium frequencies of deleterious recessives (26).

What data are there on the levels of polymorphism in Hymenoptera? The general impression until very recently was that Hymenoptera have equivalent stocks of genetic variation to male diploid animals, in that there is ready response to selection in many species (90, 139, 155, 165), and polymorphic isozyme loci were readily found in the ants *Aphaenogaster rudis* (32), *Pogonomyrmex barbatus* (84), *P. badius* (148), *Solenopsis invicta* (74), and *S. geminata* (A.C.F. Hung, personal communication) and in the bees *Apis mellifera* (112; cf 14) and *Melipona subnitida* (23). Multilocus studies, however, indicate that these general impressions may be faulty, with low levels now reported in a number of species. Thus, little variation was found in ants of the *Formica rufa* (119) and *Rhytidoponera chalybaea* (P.S. Ward, personal communication) groups or in a variety of solitary aculeates (113), and no variation at all was found in three bee species (146). If further work supports the view that male haploids have less genetic variation than male diploids, then this would indicate that most variation is caused by deleterious alleles, that most polymorphism involves substantial correlation between the fitnesses of males and corresponding female homozygotes, or that effective population size is critical in governing levels of genic diversity (see below).

There is also some evidence from *Drosophila robusta* that male haploids may have reduced amount of genetic variability compared with male diploids, in that Prakash (121) estimated that sex-linked loci have much lower rates of polymorphism than autosomal loci in the same population. This finding is not definitive, however, because Prakash estimated the proportion of all potentially polymorphic loci that are sex-linked strictly on the basis of the relative euchromatic lengths of the X-chromosome and the autosomes; the relationship between length of chromosome segment and the number of included loci is not necessarily constant from one chromosome to another. The *Drosophila robusta* data raise the point of effective population size differences between autosomal and sex-linked loci in the same organism. Considering the formulas for effective population size in sex-linked and autosomal loci (104, p. 321), it can readily be shown that when the numbers of males and females are equal, the sex-linked population size is only 75% that of the autosomal population size. The disparity increases if males outnumber females and lessens if females outnumber males; in fact, if there are more than seven females per male, the sex-linked effective population size is greater than the autosomal one! Hymenopterous parasitoids with strongly female-biased sex ratios thus have larger effective population sizes than they would if not arrhenotokous.

Loci limited in effect to females in Hymenoptera follow male diploid conditions for genetic polymorphism, not those for male haploids. Sex-limitation might therefore be a significant factor in hymenopteran evolution (25, 87, 156), on the tacit assumption that it is selectively advantageous at the population level to inactivate genes in males. Apparently in confirmation of this view, almost 40% of the mutant alleles known in *Nasonia vitripennis* are female-sterile or female-lethal (137), and a substantial portion of the genetic load (about 14%) in *Apis mellifera* that is detectable through inbreeding results from sex-limited gene effects (88–90, 92, 93). However, consideration of the formulas for mutation-selection balance of deleterious recessives at sex-limited and at non-sex-limited loci (31) shows that such alleles reach higher frequencies at sex-limited loci (and hence are detected more readily by inbreeding) than the others, so that the assumption of a high proportion of hymenopteran loci being sex-limited is unwarranted. Furthermore, when the sexes are weighted according to their contributions to succeeding generations, the hymenopteran genetic load due to deleterious recessives is the same at both sex-limited and other loci, which negates the suggestion that sex limitation is advantageous in the long term.

The Causes of Eusociality

The evolution of sterile or semisterile worker castes poses a problem under usual selectionist reasoning: how can reduction of reproductive ability be selected for?

However, there are many insect species with such castes, especially in the Hymenoptera. Such species are termed *eusocial* if, in addition to reproductive specialization and cooperative brood care, they exhibit generation overlap among adults, and *semisocial* if they do not (see 114, 169, 170). There is not enough space here for a thorough analysis of the torrent of recent papers, but a brief overview seems called for.

The approach to the conundrum of sociality is now entering a transitional stage, from a games-theoretic period to an allele-frequency-theoretic one. The games-theoretic approach involves prediction of the behavior of individuals in the light of their supposed best interests, whereas we are now seeing papers tackling the conditions under which alleles conferring altruistic behavior become fixed. This transition is unlikely to be complete for some time because of the complexity of the systems and logic involved.

Hamilton (57) began the games-theoretic period with his notion of *inclusive fitness* (which shifted the emphasis from strictly individual selection to the fitness of all carriers of an allele as a group) and the *kin-selection* formula, giving conditions for the spread of altruism under selection:

$$K > 1/r$$

where K is the ratio of the gain to the beneficiary to the altruist's loss and r is the coefficient of relationship between them [this formula has been significantly modified (61, 154) to deal with the relationships between the various potential offspring, but the original form suffices here to introduce concepts]. Under this formula, potential altruists sacrifice their own lives only to save those of at least two sibs ($r = .5$), or at least eight full cousins ($r = .125$), and so forth, to use a crude example. Special significance for the Hymenoptera arises because, under male haploidy, full sisters have $r = .75$ (27, 58); aid between sisters, or of mother by daughter to produce more sisters in lieu of the daughter's own offspring, would seem especially favored under male haploidy. In line with the prediction of Hamilton's general formula, namely, that high degrees of relationsip favor the evolution of sociality, individuals that form associations leading to eusocial colonies seem to be more related than expected by chance, in some wasps at least (41, 82, 153).

West-Eberhard (154) points out that emphasis should not be placed exclusively on the r in Hamilton's formula; consideration of K aids understanding of phenomena previously inexplicable. Thus, for sterile workers, such as those of army ants, K is infinite because they have no potential reproduction to lose. If the relationship of workers in a dequeened army ant colony to the queen of another colony is higher than the population average, by however slight an amount, then selection favors the fusion of the queenless group with the complete colony. Although army ant males are strong fliers, the queens lack wings, so two sympatric colonies should on average be more highly related than two from different areas.

Other paradigms competing with the kin-selection model for Hymenoptera are the *mutualistic* (106) and the *parental-manipulation* (1, 116) hypotheses. The mutualistic hypothesis cannot explain the repeated origins of eusocial species with nonlaying workers and seems to have been tacitly abandoned by one of its authors (116). The parental-manipulation hypothesis holds that workers are essentially specially produced offspring who, through environmental deprivation and possibly because of continuing maternal suppression, have no option but continued assistance to their mother in the "hope" of taking over if she dies. Clearly, the association of (mainly) sisters to found colonies in *Polistes* (153), *Trigonopsis* (41), and *Mischocyttarus* (82) is hard to explain under this model, although it has been suggested (1) that these sisters were preprogrammed to associate by their dead mother!

That Hymenoptera, but not termite, workers are all females is explicable under the male-haploid kin-selection hypothesis, because of the lower relationship of sibs to any male ($r = .5$) than to any female [$r > .5$, assuming sex-ratio bias (see below)]. The parental-manipulationist argument (1) that males are unsuited to worker duties and so are underproduced is countered (154) by observations that *Polistes* males can perform such duties.

A difficulty with the games-theoretic approach, particularly in the case of the parental-manipulation model, is the tendency to consider different castes and sexes almost as though they were competing species, whereas of course they are drawn from the same gene pool. Alleles causing carriers to disadvantage other individuals have to face those disadvantages as well as benefit by them. Modeling the parental-manipulation system should also take into account the likely effect of such a selective scheme: the whole population will become manipulators, and there will be selection, if manipulation is highly successful, for any allele that leads to escape.

Eusociality in Hymenoptera is restricted to aculeates, suggesting that further conditions are required for its evolution. These include the ability to manipulate objects and a life pattern in which food is stored in one place for the larvae and spatial and temporal coexistence of mother and offspring is possible. These conditions are not met in the Symphyta, Parasitica, or other male-haploid orders. Paternal care (subsociality) is extremely rare among the Parasitica and lower aculeates, although not unknown (21, p. 30, 38), because the parasitoid life pattern does not facilitate accumulation of larval food at a central site. Subsociality is widespread in other insect groups, however, so that Wilson's (169, 170) question as to why eusociality has evolved so very rarely (once) outside the Hymenoptera remains a vital point that those who deny a key role to male-haploidy in the development of eusociality have not yet squarely faced.

Despite the above arguments, the special male-haploidy kin-selection argument is currently in difficulties. One such difficulty stems from the observation of Trivers & Hare (149) that a hymenopteran female will have an average of half her genes in common with siblings just as in the male diploid case! This previously unnoticed relationship, which depends upon a 1:1 sex ratio, exists because the proportions of genes identical by descent to those in a female held by a sister (0.75) are counterbalanced by those of a brother (0.25). However, workers in established colonies can still boost the average degree of relationship between themselves and the colony brood by two means: by laying the male-producing eggs themselves or by adjusting the sex ratio of reproductives produced. Trivers & Hare (149) found that expenditure on reproductives in many eusocial species examined deviate from a 1:1 male-female ratio in the direction of the 1:3 ratio expected under sex-ratio theory. Exceptions include termites (male diploids) and slave-making ants (where the host workers have no interest in biasing the sex ratio of the slave-makers). These data are incompatible with the parental-manipulation model because the queen should favor a 1:1 expenditure ratio. The ability of eusocial Hymenoptera to detect the sex of immature stages is well-documented (see 30). Male production by workers is widespread but not universal in eusocial Hymenoptera (30) and is most easily understood under the kin-selection model rather than the others. Worker production of males is in fact somewhat maladaptive (from the population viewpoint): an examination of the appropriate formulas for loss of genetic variation shows that it decreases effective population size (cf 9). For worker oviposition to be in the interests of the queen (1), egg production would have to be a factor controlling colony size: this is a possibility, but most evidence seems to indicate that queen oviposition is in excess of needs and that the amount of worker care available for the brood is a greater restriction on colony size. The work by Trivers & Hare (149) explains the continued cohesion of already eusocial societies, but does not explain their origin since their model requires the ability on the part of the protoworkers to distinguish the sex of the immatures. Although the ability of hymenopteran females to determine the sex of eggs during oviposition is well-known [see especially (46)], the ability of nonsocial females to tell the sex of already laid eggs has not been reported and does not seem selectively advantageous. Workers therefore need to have worker abilities before they can develop them. However, if eusociality arises sporadically in many insect groups, under the model (149) it is much more likely to persist in male haploids than elsewhere.

Recent papers by Scudo & Ghiselin (140) and by Orlove (118) herald the transition to the allele-frequency-modeling phase (cf 60). Scudo & Ghiselin find that Hamilton's results are correct for male haploids with potential female workers and for male diploids with both sexes as potential workers, but paradoxically they report that kin selection should result in eusociality much more readily for male haploids with male workers and for male diploids with only one sex as workers. They therefore reject male haploidy as a significant factor in the evolution of eusociality. If further work supports Scudo & Ghiselin's findings, then a most meticulous examination of relevant natural history details is called for to explain the extraordinary concentration of eusocial species in the Hymenoptera. However, these findings should be regarded as preliminary on a number of counts. Only one mode of inheritance (complete dominance of the altruism allele in potential altruist heterozygotes) was examined, and it is not certain how much checking through simulation was made of the various results. A further unexpected point, not commented upon (140), is that whereas Scudo & Ghiselin (140) find $K > 1/0.75$, as did Hamilton, for altruism to develop among arrhenotokous females, inspection of Orlove's Figure 6 (118) shows that at intermediate gene frequencies (and they are always intermediate) K can be less than $1/0.75$ under the inheritance model which they consider. The differences in results could reflect differences between semisociality (118) and eusociality (140) or could exemplify the warning (47, p. 229) that even mathematical models can err. Even given Scudo & Ghiselin's (140) results as definitive, however, those of Trivers & Hare (149) suggest that male haploidy could still be the key factor in the relative explosion of occurrence of eusociality in the Hymenoptera. Contrary to Scudo & Ghiselin (140), Levitt (103a) found, through extensive analysis, that male haploidy is the genetic system most favorable to the evolution of eusociality.

The Consequences of Eusociality

If hymenopteran colonies are bound together by kinship ties—or parental intrigue—and are in competition with other such units, then mechanisms to maintain their integrity and separate identity are expected. Such a mechanism is colony odor, widespread but not universal among social Hymenoptera (169), whereby individuals from other colonies are recognized and excluded from the nest and sometimes from the foraging area. Although there has been a tendency to suggest either external environmental or internal genetic sources of this colony odor, very likely both are involved (127, 169), possibly to different degrees in different species. Environmental sources of colony odor have been found easier to demonstrate, but recent studies firmly implicate a genetic component. Thus in two ant species the workers can distinguish their nest soil from samples of either unnested soil or that from other colonies, even when the soil in the samples is either extremely similar (62) or from the same mixed-up batch (71). The evidence is particularly clear-cut for the bee *Lasioglossum zephyrum,* where workers can similarly distinguish aliens and nestmates, even when all nests are made in the same batch of mixed soil, and males can individually recognize females through pheromones (7). Workers of the ant *Pogonomyrmex badius* distinguish their nestmates' trail secretions from those of strangers (126), which is additional evidence suggesting a genetic component to colony distinction. Our present knowledge implies that environmentally mediated odor differences might be the most important sources of colony distinction, but that in very uniform environments genetic differences between colonies play a major part in determining intercolony odor differences.

Two simple models can readily be constructed for the genetic determination of colony odor (R. H. Crozier and M. W. Dix, unpublished information). Both models assume that odor is determined by one or more pheromone loci and that workers who share at least one allele per locus show no hostility to each other. The individualistic model supposes that each worker maintains its pheromonal integrity, whereas the gestalt model supposes that pheromones are intermixed and that each worker assumes an odor determined by colony composition. Under both models, in the case of species with colonies with single, once-mated queens, fusion of colonies can occur on occasion. Species fitting the individualistic model would also often show amity between some workers of a pair of colonies, but hostility between other workers of the same colonies. Only the gestalt model could apply to multiqueened colonies, or those with multiply mated queens, because intracolony hostility would occur under the individualistic model. The models are being developed to assess their plausibility in terms of numbers of alleles and loci required for adequate colony distinction.

There is a marked trend in social Hymenoptera, contrary to that among parasitoids, towards multiple-mating females (queens) with males mating only once (see 169). Multiple insemination of the queen is indicated when her spermatheca contains far more sperm than can be carried by a single male. However, the sperm-counting technique cannot demonstrate single insemination, because equivalance of male-carried and spermatheca amounts may indicate that the males do not deliver the full load to any one female, which might happen if they also mate multiply. Male polygamy is known in some ants (85, 120). However, the plugging of the vagina by the male's genitalia suggests enforced queen (and male) monogamy in *Melipona quadrifasciata* (98). Genetic analysis of the workers revealing a simple familial distribution of genotypes in many colonies does demonstrate female monogamy. Mixing of the sperm following multiple mating, although not complete, is, judging from three species studied (69, 147, 167), sufficiently complete that significant distortion of genotype frequencies within colonies should not occur from this cause. This genetic technique has the opposite bias from that of sperm counting, as discovery of unexpected worker genotypes can result from multiple queens as well as from single insemination. Such progeny analysis has demonstrated female monogamy in the bee *Melipona marginata* (89) and the ant *Aphaenogaster rudis* (29,32) and indicated it in the fire ant *Solenopsis invicta* (74), although the number of fire ant colonies studied was somewhat low.

Selection on colonies is not necessarily equivalent to selection on their queens (cf 154). Though the queen is a vital and initially the only living member of the colony, the mature colony consists mostly of a large force of workers, the composition of which determines the efficiency and survival of the colony, including the queen. The same queen genotype can have different genotypic mixes of worker offspring, depending upon the male mated with. Balanced polymorphism assuming constant fitness values can result from selection on colonies treated as worker groups, and the conditions for polymorphism are not the same as queen-level overdominance (33). Some extreme colony-fitness values lead to oscillation of male, female, and weighted-average gene frequencies. Such oscillations might occur in annual wasp species, although it is unlikely, but not in perennial species such as ants and honey-bees because generation overlap eliminates the oscillations. Potential tests for colony-level selection are easily devised.

CONCLUSIONS

There are interesting problems in every area of hymenopteran evolutionary genetics. However, there are a few key points arising from the topics covered above that should be stressed.

Sex determination remains a problematic and difficult area, despite signal successes for some species. Satisfactory resolution of the problem of sex determination in the great majority of the bisexual species will require clearer recognition of the operational nature of the hypotheses involved. As I attempt to show, the genic-balance hypothesis is indistinguishable from a nucleocytoplasmic-balance hypothesis. Fortunately, a test to distinguish between the genic-balance and multiple-locus hypotheses is at hand: totally homozygous diploids under genic-balance are female, but male under the alternative scheme. Methods for producing such diploids are well understood in principle.

The genetic analysis of caste determination is a potentially exciting area. Perhaps in some species genetic methods can be used to dissect the developmental pathways involved, much as genetic methods clarified biochemical pathways.

Karyotype information varies greatly in quantity and quality from group to group. Hopefully, c-banding techniques will be applied widely, extending the analysis begun on ants throughout the order. One major lack is that comparative DNA values are known for only two species. Discussions about the possible occurrence of polyploidy remain highly tentative in the absence of such measurements or indeed of any indications of relative genome sizes of the species involved.

Sibling species seem to be widespread in the groups in which they have been sought. However, speciation mechanisms are likely to differ markedly from one group to another, and there has been little comparative work to confirm or refute this likelihood. In particular, work is needed to compare speciation rates in inbreeders and outbreeders. Furthermore, it should be relatively easy to make tests of the assertion of this paper that all bisexual hymenopterons outbreed at least occasionally.

Population genetics is a field in which theory traditionally outstrips the gathering of data. It is thus surprising that significant gaps in basic theory remain for hymenopteran population genetics, especially regarding the effects of various levels of correlation of fitnesses between the sexes on the likelihood of balanced polymorphism. Data are needed to test the preliminary impression of lower levels of genic diversity in hymenopterons than in other animals, and also to see if the response to selection fits the predictions of present theory on evolutionary rates in male haploids.

The question of the origins of eusociality in the order remains a knotty problem. Work in this area is now entering a gene-frequency-modeling stage, and the attendant greater potential rigor of the analyses may eventually yield definitive results. The first results from the new approach include significant discrepancies with predictions based on the previous games-theoretic approach. These contradictions will doubtless be expanded or resolved in the future—the production of papers on this general topic is still in the early exponential phase.

One unfortunate consequence of the great interest in the origins of eusociality is that the field seems almost obsessed with this one problem. Social insects live in colonies. Although denied by some, the level of the colony is one at which selection can act, and this level is ideally studied in social insects. Integrating the many levels at which selection can act in social insects may lead us closer to a basic understanding of what is happening on any one level. Such work has begun; it has a long way to go, both in theory and experiment.

ACKNOWLEDGMENTS

I thank H. T. Imai, J. James, R. W. Poole, and P. J. Staff for valuable discussions; K. Brown, P. I. Dixon, and P. S. Ward for critical comments on the manuscript; and F. Romer for translating a reference. Responsibility for the views expressed in the review and for any remaining errors remains mine.

Literature Cited

1. Alexander, R. D. 1974. The evolution of social behaviour. *Ann. Rev. Ecol. Syst.* 5:325–83
2. Asher, J. H. 1970. Parthenogenesis and genetic variability. II. One locus models for various diploid populations. *Genetics* 66:369–91
3. Askew, R. R. 1968. Considerations on speciation in Chalcidoidea. *Evolution* 22:642–45
4. Assem, J. van den. 1971. Some experiments on sex ratio and sex regulation in the pteromalid *Lariophagus distinguendus. Neth. J. Zool.* 21:373–402
5. Assem, J. van den, Povel, G. D. E. 1973. Courtship behaviour of some *Muscidifurax* species: a possible example of a recently evolved ethological isolating mechanism. *Neth. J. Zool.* 23:465–87
6. Baker, R. J., Bleier, W. J., Atchley, W. R. 1975. A contact zone between karyotypically-characterized taxa of *Uroderma bilobatum. Syst. Zool.* 24:133–42
7. Barrows, E. M., Bell, W. J., Michener, C. D. 1975. Individual odor differences and their social functions in insects. *Proc. Natl. Acad. Sci. USA* 72:2824–28
8. Battaglia, B. 1961. Rapporti tra geni per la pigmentazioni e la sessualita in *Tisbe reticulata. Atti Assoc. Genet. Ital.* 6:439–47
9. Beig, D. 1972. The production of males in queenright colonies of *Trigona* (*Scaptotrigona*) *postica. J. Apic. Res.* 11:33–39
10. Blackith, R. E. 1958. An analysis of polymorphism in social wasps. *Insectes Soc.* 5:263–72
11. Brian, M. V. 1956. Studies of caste differentiation in *Myrmica rubra* L. 4. Controlled larval nutrition. *Insectes Soc.* 3:369–94
12. Brian, M. V. 1974. Caste differentiation in *Myrmica rubra:* the rôle of hormones. *J. Insect Physiol.* 20:1351–65
13. Britten, R. J., Davidson, E. H. 1969. Gene regulation for higher cells: a theory. *Science* 165:349–67
14. Brückner, D. 1974. Reduction of biochemical polymorphisms in honey bees (*Apis mellifica*). *Experientia* 30:618–19
15. Buschinger, A. 1968. Mono- und Polygynie bei Arten der Gattung *Leptothorax* Mayr. *Insectes Soc.* 15:217–25
16. Buschinger, A. 1972. Kreuzung zweier sozialparasitischer Ameisenarten, *Doronomyrmex pacis* Kutter und *Leptothorax kutteri* Buschinger. *Zool. Anz.* 189:169–79
17. Buschinger, A. 1975. Eine genetische Komponente im Polymorphismus der dulotischen Ameise *Harpagoxenus sublaevis. Naturwissenschaften* 62:239–40
18. Campos, L. A. de Oliveira, Velthuis-Kluppell, F. M., Velthuis, H. H. W. 1975. Juvenile hormone and caste determination in a stingless bee. Sex determination in bees, VII. *Naturwissenchaften* 62:98–99
19. Cannings, C. 1967. Equilibrium, convergence and stability at a sex-linked locus under natural selection. *Genetics* 56:613–18
20. Chaud-Netto, J. 1975. Sex determination in bees. II. Additivity of maleness genes in *Apis mellifera. Genetics* 79:213–17
21. Clausen, C. P. 1940. *Entomophagous Insects.* Reprinted 1962. New York: Hafner. 688 pp.
22. Conner, G. W. 1966. Preferential segregation in *Mormoniella. Genetics* 54:1041–48.
23. Contel, E. P. B., Mestriner, M. A. 1974. Esterase polymorphism at two loci in the social bee. *J. Hered.* 65:349–52
24. Crow, J. F., Kimura, M. 1970. *An Introduction to Population Genetics Theory.* New York: Harper & Row. 591 pp.
25. Crozier, R. H. 1969. Chromosome number polymorphism in an Australian ponerine ant. *Can. J. Genet. Cytol.* 11:333–39
26. Crozier, R. H. 1970. On the potential for genetic variability in haplo-diploidy. *Genetica* 41:551–56
27. Crozier, R. H. 1970. Coefficients of relationship and the identity of genes by descent in the Hymenoptera. *Am. Nat.* 104:216–17
28. Crozier, R. H. 1971. Heterozygosity and sex determination in haplo-diploidy. *Am. Nat.* 105:399–412
29. Crozier, R. H. 1973. Apparent differential selection at an isozyme locus between queens and workers of the ant *Aphaenogaster rudis. Genetics* 73: 313–18
30. Crozier, R. H. 1975. *Animal Cytogenetics 3 Insecta 7 Hymenoptera.* Berlin: Gebrüder Borntraeger. 95 pp.
31. Crozier, R. H. 1976. Why male-haploid and sex-linked genetic systems seem to have unusually sex-limited mutational genetic loads. *Evolution.* In press

32. Crozier, R. H. 1976. Genetic differentiation between populations of the ant *Aphaenogaster "rudis"* in the southeastern United States. *Genetica.* In press
33. Crozier, R. H., Consul, P. C. 1976. Conditions for genetic polymorphism in social Hymenoptera under selection at the colony level. *Theor. Popul. Biol.* In press
34. da Cunha, A. B., Kerr, W. E. 1957. A genetical theory to explain sex-determination by arrhenotokous parthenogenesis. *Forma Functio* 1:33–36
35. Cupp, E. W., O'Neal, J., Kearney, G., Markin, G. P. 1973. Forced copulation of imported fire ant reproductives. *Ann. Entomol. Soc. Am.* 66:743–45
36. Darchen, R. 1973. Essai d'interprétation du déterminisme des castes chez les Trigones et les Mélipones. *C.R. Acad. Sci. Ser. D* 276:607–9
37. Darchen, R., Delage-Darchen, B. 1974. Nouvelles expériences concernant le déterminisme des castes chez les Mélipones (Hyménoptères apidés). *C. R. Acad. Sci. Ser. D* 278:907–10
38. Doutt, R. L. 1973. Maternal care of immature progeny by parasitoids. *Ann. Entomol. Soc. Am.* 66:486–87
39. Drescher, W., Rothenbuhler, W. C. 1964. Sex determination in the honey bee. *J. Hered.* 55:91–96
40. Dyson, J. G. 1965. *Natural selection of the two mutant genes honey and orange in laboratory populations of Habrobracon juglandis.* PhD thesis. North Carolina State Univ., Raleigh, NC. 87 pp.
41. Eberhard, W. G. 1972. Altruistic behavior in a sphecid wasp: support for kin-selection theory. *Science* 175: 1390–91
42. Folliot, R. 1964. Contribution a l'étude de la biologie des cynipides gallicoles. *Ann. Sci. Nat., Zool. Biol. Anim. Ser. 12* 6:407–564
43. Friedler, G., Ray, D. T. 1951. Androgenesis in the wasp *Mormoniella. Anat. Rec.* 111:475
44. Garófalo, C. A. 1973. Occurrence of diploid drones in a Neotropical bumblebee. *Experientia* 29:726–27
45. Garófalo, C. A., Kerr, W. E. 1975. Sex determination in bees. I. Balance between femaleness and maleness genes in *Bombus atratus* Franklin. *Genetica* 45:203–9
46. Gerber, H. S., Klostermeyer, E. C. 1970. Sex control by bees: a voluntary act of egg fertilization during oviposition. *Science* 162:82–84
47. Ghiselin, M. T. 1974. *The Economy of Nature and the Evolution of Sex.* Berkeley, Calif: Univ. Calif. Press. 346 pp.
48. Goodpasture, C. 1974. *Cytological data and its uses in the classification of the Hymenoptera.* PhD thesis. Univ. Calif., Davis. 178 pp.
49. Goodpasture, C. 1974. Karyology and taxonomy of some species of eumenid wasps. *J. Kans. Entomol. Soc.* 47:364–72
50. Goodpasture, C. 1975. Comparative courtship behavior and karyology in *Monodontomerus. Ann. Entomol. Soc. Am.* 68:391–97
51. Goodpasture, C. 1975. The karyotype of the cynipid *Callirhytis palmiformis* (Ashmead). *Ann. Entomol. Soc. Am.* 68:801–2
52. Goodpasture, C., Grissell, E. E. 1975. A karyological study of nine species of *Torymus. Can. J. Genet. Cytol.* 17:413–22
53. Gösswald, K., Schmidt, G. H. 1960. Untersuchungen zum Flugelabwurf und Begattungsverhalten einiger *Formica*-arten (Ins. Hym.) im Hinblick auf ihre systematische Differenzierung. *Insectes Soc.* 7:297–321
54. Grant, B., Snyder, G. A., Glessner, S. F. 1974. Frequency-dependent mate selection in *Mormoniella vitripennis. Evolution* 28:259–64
55. Grosch, D. S. 1945. The relation of cell size and organ size to mortality in *Habrobracon. Growth* 9:1–17
56. Halliday, R. B. 1975. Electrophoretic variation of amylase in meat ants, *Iridomyrmex purpureus,* and its taxonomic significance. *Aust. J. Zool.* 23:271–76
57. Hamilton, W. D. 1963. The evolution of altruistic behavior. *Am. Nat.* 97:354–56
58. Hamilton, W. D. 1964. The genetical evolution of social behavior. I & II. *J. Theor. Biol.* 7:1–52
59. Hamilton, W. D. 1967. Extraordinary sex ratios. *Science* 156:477–88
60. Hamilton, W. D. 1971. Selection of selfish and altruistic behavior in some extreme models. In *Man and Beast: Comparative Social Behavior,* ed. J. F. Eisenberg, W. S. Dillon, pp. 59–91. Washington: Smithsonian Inst. Press. 401 pp.
61. Hamilton, W. D. 1972. Altruism and related phenomena, mainly in social insects. *Ann. Rev. Ecol. Syst.* 3:193–232
62. Hangartner, W., Reichson, J. M., Wilson, E. O. 1970. Orientation to nest material by the ant, *Pogonomyrmex badius* (Latreille). *Anim. Behav.* 18:331–34

63. Hartl, D. L. 1971. Some aspects of natural selection in arrhenotokous populations. *Am. Zool.* 11:309–25
64. Hartl, D. L. 1972. A fundamental theorem of natural selection for sex linkage or arrhenotoky. *Am. Nat.* 106:516–24
65. Haskins, C. P., Whelden, R. M. 1965. "Queenlessness," worker sibship, and colony versus population structure in the formicid genus *Rhytidoponera. Psyche* 72:87–112
66. Hay, D. A. 1972. Recognition by *Drosophila melanogaster* of individuals from other strains or cultures: support for the role of olfactory cues in selective mating? *Evolution* 26:171–76
67. Hidaka, T. 1958. Biological investigation on *Telenomus gifuensis* Ashmead, an egg-parasite of *Scotonophara lunda* Burmeister in Japan. *Acta Hymenopt.* 1:75–93
68. Holmes, H. B. 1972. Genetic evidence for fewer progeny and a higher percent males when *Nasonia vitripennis* oviposits in previously parasitized hosts. *Entomophaga* 17:79–88
69. Holmes, H. B. 1974. Patterns of sperm competition in *Nasonia vitripennis. Can. J. Genet. Cytol.* 16:789–95
70. Hoy, M. A. 1975. Hybridization of strains of the gypsy moth parasitoid, *Apanteles melanoscelus,* and its influence upon diapause. *Ann. Entomol. Soc. Am.* 68:261–64
71. Hubbard, M. D. 1974. Influence on nest material and colony odor on digging in the ant *Solenopsis invicta. J. Ga. Entomol. Soc.* 9:127–32
72. Hung, A. C.-F. 1973. Induced mating in *Formica* ants. *Entomol. News* 84:310–13
73. Hung, A. C.-F., Imai, H. T., Kubota, M. 1972. The chromosomes of nine ant species from Taiwan, Republic of China. *Ann. Entomol. Soc. Am.* 65:1023–25
74. Hung, A. C.-F., Vinson, S. B. 1976. Biochemical evidence for queen monogamy and sterile male diploidy in the fire ant *Solenopsis invicta. Isozyme Bull.* 9:In press
75. Hung, A. C.-F., Vinson, S. B., Summerlin, J. W. 1974. Male sterility in the red imported fire ant, *Solenopsis invicta. Ann. Entomol. Soc. Am.* 67:909–12
76. Hunter, K. W., Bartlett, A. C. 1975. Chromosome number of the parasitic encyrtid *Copidosoma truncatellum* (Dalman). *Ann. Entomol. Soc. Am.* 68:61–62
77. Ishay, J. 1975. Caste determination by social wasps: cell size and building behaviour. *Anim. Behav.* 23:425–31
78. Imai, H. T. 1974. B-chromosomes in the myrmicine ant, *Leptothorax spinosior. Chromosoma* 45:431–44
79. Imai, H. T., Kubota, M. 1975. Chromosome polymorphism in the ant, *Pheidole nodus. Chromosoma* 51:391–99
80. Jackson, D. J. 1966. Observations on the biology of *Caraphractus cinctus* Walker, a parasitoid of the eggs of Dytiscidae. III. The adult life and sex ratio. *Trans. R. Entomol. Soc. London* 118:23–49
81. Jayakar, S. C., Spurway, H. 1966. Reuse of cells and brother-sister mating in the Indian species *Stenodynerus miniatus* (Sauss.). *J. Bombay Nat. Hist. Soc.* 63:378–98
82. Jeanne, R. L. 1972. Social biology of the neotropical wasp *Mischocyttarus drewseni. Bull. Mus. Comp. Zool., Harv. Univ.* 144:63–150
83. John, B., Freeman, M. 1975. Causes and consequences of Robertsonian exchange. *Chromosoma* 52:123–36
84. Johnson, F. M., Schaffer, H. E., Gillaspy, J. E., Rockwood, E. S. 1969. Isozyme genotype-environment relationships in natural populations of the harvester ant, *Pogonomyrmex barbatus,* from Texas. *Biochem. Genet.* 3:429–50
85. Kannowski, P. B. 1963. The flight activities of formicine ants. *Symp. Genet. Biol. Ital.* 12:74–102
86. Kerr, W. E. 1950. Genetic determination of caste in the genus *Melipona. Genetics* 35:143–52
87. Kerr, W. E. 1951. Bases para o estudo da genética dos Hymenoptera em geral e dos Apinae sociais em particular. *Ann. Esc. Super. Agric. Luiz de Queiroz* 8:220–354
88. Kerr, W. E. 1967. Genetic structure of the populations of Hymenoptera. *Cienc. Cult. (São Paulo)* 19:39–44
89. Kerr, W. E. 1969. Some aspects of the evolution of social bees. *Evol. Biol.* 3:119–75
90. Kerr, W. E. 1974. Advances in cytology and genetics of bees. *Ann. Rev. Entomol.* 19:253–68
91. Kerr, W. E. 1974. Sex determination in bees. III. Caste determination and genetic control in *Melipona. Insectes Soc.* 21:357–68
92. Kerr, W. E. 1975. Evolution of the population structure in bees. *Genetics* 79:73–84

93. Kerr, W. E. 1976. Population genetic studies in Hymenoptera. 2. Sex limited genes. (1). *Evolution.* In press
94. Kerr, W. E., Bueno, D. 1970. Natural crossing between *Apis mellifera adansonii* and *Apis mellifera ligustica. Evolution* 24:145–48
95. Kerr, W. E., Nielsen, R. A. 1966. Evidences that genetically determined Melipona queens can become workers. *Genetics* 54:859–66
96. Kerr, W. E., Nielsen, R. A. 1967. Sex determination in bees. *J. Apic. Res.* 6:3–9
97. Kerr, W. E., Stort, A. C., Montenegro, M. J. 1966. Importância de alguns fatôres ambientais na determinãço das castas do gênero *Melipona. Ann. Acad. Bras. Cienc.* 38:149–68
98. Kerr, W. E., Zucchi, R., Nakadaira, J. T., Butolo, J. E. 1962. Reproduction in the social bees. *J. NY Entomol. Soc.* 70:265–76
99. King, P. E., Askew, R. R., Sanger, C. 1969. The detection of parasitized hosts by males of *Nasonia vitripennis* (Walker) and some possible implications. *Proc. R. Entomol. Soc. London Ser. A* 44:85–90
100. Knerer, G., Atwood, C. E. 1973. Diprionid sawflies: polymorphism and speciation. *Science* 179:1090–99
101. Kogan, M., Legner, E. F. 1970. A biosystematic revision of the genus *Muscidifurax* with descriptions of four new species. *Can. Entomol.* 102:1268–90
102. Laidlaw, H. H., Tucker, K. W. 1965. Diploid tissue derived from accessory sperm in the honeybee. *Genetics* 50:1439–42
103. Legner, E. F. 1969. Reproductive isolation and size variation in the *Muscidifurax raptor* complex. *Ann. Entomol. Soc. Am.* 62:382–85
103a. Levitt, P. R. 1975. General kin selection models for genetic evolution of sib altruism in diploid and haplodiploid species. *Proc. Natl. Acad. Sci. USA* 72:4531–35
104. Li, C. C. 1955. *Population Genetics.* Chicago: Univ. Chicago Press. 366 pp.
105. Li, C. C. 1967. The maximization of average fitness by natural selection for a sex-linked locus. *Proc. Natl. Acad. Sci. USA* 57:1260–61
106. Lin, N., Michener, C. D. 1972. Evolution of sociality in insects. *Q. Rev. Biol.* 47:131–59
107. Lofgren, C. S., Banks, W. A., Glancey, B. M. 1975. Biology and control of imported fire ants. *Ann. Rev. Entomol.* 20:1–30
108. Macy, R. M., Whiting, P. W. 1969. Tetraploid females in *Mormoniella. Genetics* 61:619–30
109. Mackensen, O. 1951. Viability and sex determination in the honey bee (*Apis mellifera* L). *Genetics* 36:500–9
110. Martin, A. 1948. Genetic evidence for speciation in *Habrobracon juglandis. Proc. P. Acad. Sci.* 22:64–67
111. Matthews, R. W. 1975. Courtship in parasitic wasps. In *Evolutionary Strategies of Parasitic Insects and Mites,* ed. P. W. Price, pp. 66–86. New York & London: Plenum. 224 pp.
112. Mestriner, M. A., Contel, E. P. B. 1972. The *P-3* and *Est* loci in the honeybee *Apis mellifera. Genetics* 72:733–38
113. Metcalf, R. A., Marlin, J. C., Whitt, G. S. 1975. Low levels of genetic heterozygosity in Hymenoptera. *Nature* 257:792–94
114. Michener, C. D. 1974. *The Social Behavior of the Bees. A Comparative Study.* Cambridge, Mass: Harvard Univ. Press. 404 pp.
115. Michener, C. D. 1975. The Brazilian bee problem. *Ann. Rev. Entomol.* 20:399–416
116. Michener, C. D., Brothers, D. J. 1974. Were workers of eusocial Hymenoptera initially altruistic or oppressed? *Proc. Natl. Acad. Sci. USA* 71:671–74
116a. Nagarkatti, S., Nagaraja, H. 1977. Biosystematics of *Trichogramma* and *Trichogrammatoidea* species. *Ann. Rev. Entomol.* 22:157–76
117. Nagylaki, T. 1975. A continuous selective model for an X-linked locus. *Heredity* 34:273–78
118. Orlove, M. J. 1975. A model of kin selection not invoking coefficients of relationship. *J. Theor. Biol.* 49:289–310
119. Pamilo, P., Vepsäläinen, K., Rosengren, R. 1975. Low allozymic variability in *Formica* ants. *Hereditas* 80:293–96
120. Petersen, M., Buschinger, A. 1971. Das Begattunsverhalten der Pharaoameise, *Monomorium pharaonis* (L). *Z. Angew. Entomol.* 68:168–75
121. Prakash, S. 1973. Patterns of gene variation in central and marginal populations of *Drosophila robusta. Genetics* 75:347–69
122. Rabb, R. L., Bradley, J. R. 1970. Marking host eggs by *Telenomus sphingis. Ann. Entomol. Soc. Am.* 63:1053–56
123. Ramirez, B. W. 1970. Taxonomic and

biological studies of Neotropical fig wasps. *Univ. Kans. Sci. Bull.* 49:1–44
124. Rao, S. V., DeBach, P. 1969. Experimental studies on hybridization and sexual isolation between some *Aphytis* species. III. The significance of reproductive isolation between interspecific hybrids and parental species. *Evolution* 23:525–33
125. Rasch, E. M., Cassidy, J. D., King, R. C. 1975. Estimates of genome size in haplo-diploid species of parasitoid wasps. *J. Histochem. Cytochem.* 23:317
126. Regnier, F. E., Nieh, M., Holldöbler, B. 1973. The volatile Dufour's gland components of the harvester ants *Pogonomyrmex rugosus* and *P. badius*. *J. Insect Physiol.* 19:981–92
127. Ribbands, C. R. 1965. The role of recognition of comrades in the defence of social insect communities. *Symp. Zool. Soc. London* 14:159–68
128. Rössler, Y., DeBach, P. 1973. Genetic variability in a thelytokous form of *Aphytis mytilaspidus* (Le Baron). *Hilgardia* 42:149–76
129. Rothenbuhler, W. C. 1957. Diploid male tissue as new evidence on sex determination in honey bees. *J. Hered.* 48:160–68
130. Rothenbuhler, W. C., Kulinčevič, J. M., Kerr, W. E. 1968. Bee genetics. *Ann. Rev. Genet.* 2:413–38
131. Rüsch, M. E. 1960. Untersuchungen über Geschlechtsbestimmungsmechanism bei Copepoden. *Chromosoma* 11:419–32
132. Rust, R. W., Thorp, R. W. 1973. The biology of *Stelis chlorocyanea,* a parasite of *Osmia nigrifrons*. *J. Kans. Entomol. Soc.* 46:548–62
133. Salt, G. 1941. The effects of hosts upon their insect parasitoids. *Biol. Rev. Cambridge Philos. Soc.* 16:239–64
134. Salt, G. 1961. Competition among insect parasitoids. *Symp. Soc. Exp. Biol.* 15:96–119
135. Samson-Boshuizen, M., Lenteren, J. C. van, Bakker, K. 1974. Success of parasitization of *Pseudeucoila bochei* Weld: a matter of experience. *Neth. J. Zool.* 24:67–85
136. Sankaran, T., Nagaraja, H. 1975. Observations on two sibling species of *Gryon* parasitic on Triatominae in India. *Bull. Entomol. Res.* 65:215–19
137. Saul, G. B., Whiting, P. W., Saul, S. W., Heidner, C. A. 1965. Wild-type and mutant stocks of *Mormoniella*. *Genetics* 52:1317–27
138. Schmieder, R. G., Whiting, P. W. 1947. Reproductive economy in the chalcidoid wasp *Melittobia*. *Genetics* 32:29–37
139. Scossiroli, R. E., Von Borstel, R. C. 1963. Selection experiments after inbreeding in *Habrobracon*. *Proc. Int. Congr. Genet., 11th, The Hague* 1:152
140. Scudo, F. M., Ghiselin, M. T. 1975. Familial selection and the evolution of social behavior. *J. Genet.* 62:1–31
141. Shiga, M., Nakanishi, A. 1968. Variation in the sex ratio of *Gregopimpla himalayensis* Cameron parasitic on *Maracosoma neustria testacea* Motschulsky, with considerations on the mechanism. *Kontyu* 36:369–76
142. Smith, S. G. 1941. A new form of spruce sawfly identified by means of its cytology and parthenogenesis. *Sci. Agric.* 21:245–305
143. Smith, S. G. 1965. Heterochromatin, colchicine and karyotype. *Chromosoma* 16:162–65
144. Smith, S. G., Wallace, D. R. 1971. Allelic sex determination in a lower hymenopteran, *Neodiprion nigroscutum* Midd. *Can. J. Genet. Cytol.* 13:617–21
145. Sneath, P. H. A., Sokal, R. R. 1973. *Numerical Taxonomy. The Principles and Practice of Numerical Classification.* San Francisco: Freeman. 573 pp.
146. Snyder, T. P. 1974. Lack of allozymic variability in three bee species. *Evolution* 28:687–689
147. Taber, S. 1955. Sperm distribution in the spermathecae of multiple-mated queen honeybees. *J. Econ. Entomol.* 48:522–25
148. Tomaszewski, E. K., Schaffer, H. E., Johnson, F. M. 1973. Isozyme genotype-environment associations in natural populations of the harvester ant, *Pogonomyrmex badius*. *Genetics* 75:405–21
149. Trivers, R. L., Hare, H. 1976. Haplodiploidy and the evolution of the social insects. *Science* 191:249–63
150. Van Pelt, A. 1953. Notes on the aboveground activity and a mating flight of *Pogonomyrmex badius* (Latr.). *J. Tenn. Acad. Sci.* 28:164–68
151. Victorov, G. A., Kochetova, N. I. 1973. On the regulation of the sex ratio in *Dahlbominus fuscipennis* Zett. *Entomol. Rev.* 52:434–38
152. Went, D. F., Krause, G. 1974. Alteration of egg architecture and egg activation in an endoparasitic hymenopteron as a result of natural or imitated ovipo-

sition. *Wilhelm Roux' Arch. Entwicklungsmech. Org.* 175:173–84
153. West Eberhard, M. J. 1969. The social biology of polistine wasps. *Univ. Mich. Mus. Zool. Misc. Publ.* 140:1–101
154. West Eberhard, M. J. 1975. The evolution of social behavior by kin selection. *Q. Rev. Biol.* 50:1–33
155. White, E. B., DeBach, P., Garber, M. J. 1970. Artificial selection for genetic adaptation to temperature extremes in *Aphytis lingnanensis* Compere. *Hilgardia* 40:161–92
156. White, M. J. D. 1954. *Animal Cytology and Evolution.* Cambridge, Engl: Cambridge Univ. Press. 454 pp. 2nd ed.
157. Whiting, A. R. 1946. Motherless males from irradiated eggs. *Science* 103:219–20
158. Whiting, A. R. 1961. Genetics of *Habrobracon. Adv. Genet.* 10:295–348
159. Whiting, A. R. 1965. The complex locus *R* in *Mormoniella vitripennis* (Walker). *Adv. Genet.* 13: 341–58
160. Whiting, A. R. 1967. The biology of the parasitic wasp *Mormoniella vitripennis* (=*Nasonia brevicornis*) (Walker). *Q. Rev. Biol.* 42:333–406
161. Whiting, P. W. 1939. Sex determination and reproductive economy in *Habrobracon. Genetics* 24:110–11
162. Whiting, P. W. 1943. Multiple alleles in complementary sex determination of *Habrobracon. Genetics* 28:365–382.
163. Whiting, P. W. 1960. Polyploidy in *Mormoniella. Genetics* 45:949–970
164. Whiting, P. W., Greb, R. J., Speicher, B. R. 1934. A new type of sex-intergrade. *Biol. Bull. Woods Hole, Mass.* 66:152–165
165. Wilkes, A. 1964. Inherited male-producing factor in an insect that produces its males from unfertilized eggs. *Science* 144:305–307
166. Wilkes, A. 1965. Sperm transfer and utilization by the arrhenotokous wasp *Dahlbominus fuscipennis* (Zett.). *Can. Entomol.* 97:647–57
167. Wilkes, A. 1966. Sperm utilization following mutiple insemination in the wasp *Dahlbominus fuscipennis. Can. J. Genet. Cytol.* 8:451–61
168. Wilson, E. O. 1955. A monographic revision of the ant genus *Lasius. Bull. Mus. Comp. Zool. Harv. Univ.* 113:1–201
169. Wilson, E. O. 1971. *The Insect Societies.* Cambridge, Mass: Harvard Univ. Press. 548 pp.
170. Wilson, E. O. 1975. *Sociobiology. The New Synthesis.* Cambridge, Mass: Harvard Univ. Press. 697 pp.
171. Wilson, F. 1961. Adult reproductive behavior in *Asolcus basalis. Aust. J. Zool.* 9:737–51
172. Wing, M. W. 1968. Taxonomic revision of the Nearctic genus *Acanthomyops. Mem. Cornell Univ. Agric. Exp. Sta.* 405:1–173
173. Woyke, J. 1967. Diploid drone substance—cannibalism substance. *Proc. Int. Apicult. Congr., 21st, Maryland,* pp. 471–72
174. Woyke, J. 1969. A method of rearing diploid drones in a honeybee colony. *J. Apic. Res.* 8:65–74
175. Woyke, J. 1974. Genic balance, heterozygosity and inheritance of size of testes in diploid drone honeybees. *J. Apic. Res.* 13:77–9
176. Wylie, H. G. 1966. Some mechanisms that affect the sex ratio of *Nasonia vitripennis* (Walk.) reared from superparasitized housefly pupae. *Can. Entomol.* 98:645–53
177. Wylie, H. G. 1973. Control of egg fertilization by *Nasonia vitripennis* when laying on parasitized house fly pupae. *Can. Entomol.* 105:709–18

Ann. Rev. Entomol. 1977. 22:289–308

QUALITY CONTROL IN MASS REARING

◆6131

Derrell L. Chambers
Insect Attractants, Behavior and Basic Biology Research Laboratory, Agricultural Research Service, United States Department of Agriculture, Gainesville, Florida 32604

INTRODUCTION

The development of pest management techniques alternative to or augmentive of toxicant application requires the evaluation and implementation of systems more complex than those previously used. Among the most intricate is the production of insects in specialized factories and their release in massive numbers to achieve economic control or eradication. These applications include the inundative release of entomophagous insects and autocidal programs based on the release of pest species that have been genetically altered so their mates or progeny are unable either to reproduce or to enact their detrimental effect (e.g. vector a disease).

The imposition of control on the target population by the released insects requires physiological and behavioral performance factors adequate in kind and quality. However, it is now well understood that during the intricate process of production and release these factors can be affected detrimentally. This review considers the process of production and release as a system of interacting components with an interplay of effects and demands. The effects and demands are both biologic and logistic, requiring consideration not only of technical requirements for success but also of fiscal and social elements that demand an acceptable cost/benefit ratio. Attention is given the procedures available at each stage of the process that can be used to monitor its continuity and measure the acceptability of its product. Reviews and articles on this and related subjects are referred to in the text.

DEFINITIONS

"Quality Control in Mass Rearing" as a topic of discourse requires immediate attention to the definition of key terms. Mass rearing can have such broad interpretation as to imply producing insects in numbers ranging up to hundreds of millions per week. Indeed, production levels of a billion per week have been envisioned. The term is most often used with an implied definition, referring to the ease of rearing larger numbers of insects than previously possible (83), to the numbers needed to impose economic control per unit of area (49), or to production efficiency per unit of effort and/or space (27). However, massiveness of a production effort also depends on the biotic capacity of the reared insect; therefore, the definition of Mackauer (54) is attractive for its inclusion of this component. He defined *mass production* (of entomophagous insects) as the rearing per generation cycle of one million times the mean number of offspring per female. This is realistic for entomophagous insects, but it may be an order of magnitude or more too large for programs involving inundative release of sterile or genetically engineered pestiferous insects. A really comprehensive definition would also have to include specifications such as the level of domestication of the female, reference to the expected competence traits and levels of the produced insects, and the ease and efficiency of producing the numbers needed for program accomplishment. An inordinate portion of this paper could be used in an attempt to make my definition comprehensive, therefore I limit it as follows: *mass rearing* is the production of insects competent to achieve program goals with an acceptable cost/benefit ratio and in numbers per generation exceeding ten thousand to one million times the mean productivity of the native population female.

Quality is in more critical need of definition. Although it is of increasing concern to those rearing insects in large numbers and is the objective of considerable research, it often exists only in the eye of the beholder. As I have noted previously (19) quality has, by dictionary definition, three components—skill, relativity, and reference—that is, *quality* is the degree of excellence in some trait(s) or skill(s) relative to a reference. Thus, for the purposes of this paper, *skills* are the performance requirements for achieving the objective, *relativity* has to do with ranking (comparing) the degree of excellence in performing the skills, and *references* are the standards, needs, and constraints against which the skills are compared.

The dictionary definition of *control* includes checking, verification, comparison, regulation, periodicity, and restraint. In its congruence herein with quality there are also implications of input and feedback mechanisms, cross-check comparisons, decision making, and developing protocols for the imposition of control upon quality. Protocols are badly needed in insect production, and their development and application to decision making should receive major attention. However, developing the means for making comparisons has been an objective of the greater part of recent research, and such research has tended to stress the effects of inputs. I try to make the point herein that feedback mechanisms and information should form a critical portion of the data used in establishing and refining protocols because feedback effects occur throughout the process of production and have an impact upon its efficiency and effectiveness. Unfortunately, these impacts are often overlooked or accepted quid pro quo.

THE PROCESS

A program involving massive numbers of released insects, whether entomophagous or pestiferous, contains two major elements: (*a*) the production of the insects, and (*b*) the subsequent performance of the insects for accomplishing the objective of release. These elements can be visualized as a series of stages and events, each of which affects the subsequent events (input); moreover, the standards, needs, and constraints for each event impose requirements upon the preceding events (feedback). Figure 1 is a stylized diagram showing the principal elements of a typical mass release program of either entomophagous or pestiferous insects. The events and their input typically occur in series from the top in counterclockwise sequence, as indicated by arrows. Feedback information on needs and impact is shown by arrows indicating clockwise sequence. Much input and feedback information loops around intervening stages and events; these are included by implication only, in order to simplify the figure. In the rest of this paper I use the figure as a reference and an outline.

PRODUCTION

The capability of producing an insect species easily, efficiently, and in adequate numbers is a prerequisite for considering a program of release for that insect; thus, research and development are concentrated on this aspect initially. Unfortunately, product quantity often remains the overriding consideration in the subsequent program phases as well. The similarities between the insect factory and an industrial production line are obvious, and the philosophies and technologies of industry often are incorporated to great advantage in insect production facilities. However, in the industrial sector, a marketing division usually provides continuous feedback on the marketability and competitiveness of the product. The insect factory is putting out a product in the same context, and protocols for producing insects should also be influenced by the marketability (skills) and competitiveness of the product via comparison with standards and references. Although a program that considers productivity its sole or principal function can be criticized for lack of attention to the marketability of its product, there are very legitimate considerations relative to production line efficiency. Quality measurements also provide the input that ensures maintenance of an efficient output on schedule and in needed numbers. No program can be operated without attention to these pragmatic matters.

General treatises on insect rearing are few. Handy hints that are dated but useful are found in a compilation edited by Needham (63). A more recent text, edited by Smith (87), is more directly concerned with large-scale production technology, and Finney & Fisher (27) reviewed the culturing of entomophagous insects. Rodriguez (74) edited a current text on arthropod nutrition and Singh prepared a bibliography on diets (85), although a review of artificial diets for insects by Vanderzant (99) is the most recent information source. N. C. Leppla & T. R. Ashley of this laboratory are currently editing papers for a compendium on facilities for insect production that will provide useful guides. The reader is also referred to the Panel Proceedings Series of the International Atomic Energy Agency, Vienna. However, most of the information on technologies for rearing specific entomophages or pests is published in individual articles in the various journals and this paper does not pretend to list them comprehensively. It is dedicated rather to the quality of the insect produced

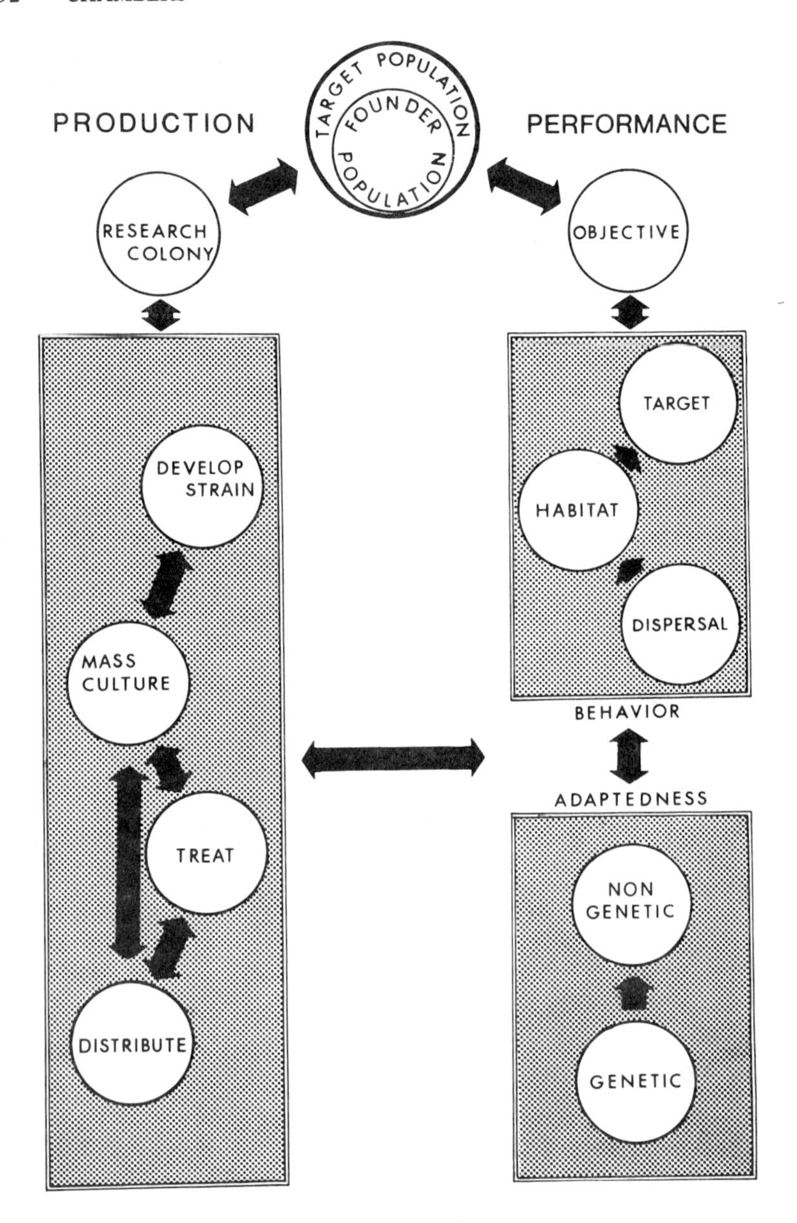

Figure 1 Schematic representation of the production process and performance of mass-reared insects.

and to the impact of the production process on that quality. Nevertheless, I emphasize the importance of critical studies on insect colonization to entomology. Rearing and rearing research must have the image of a credible discipline. As long as they are down-graded, or quality research pertinent to the subject is euphemistically called physiology, life history, or some other more esoteric term, the topic will not receive the attention it needs or deserves.

Founder Population

Mackauer (54, 55) recently discussed the effects of size and origin of the source colony on the ultimate genetic and, hence, potential behavioral profiles of the production colony. Clearly, both kind and range of variability of essential performance traits must be retained if the released insects are to be adequate for competent interaction with the target population. The difficulties in matching such factors as biotype, niche width, founder effects, and genetic variability with performance needs presently place precise or secure engineering of production colonies via discriminating selection of founder populations beyond the realm of possibility. Also, subsequent selection during research and colonizing stages may cause changes in the population counter to those envisaged in the engineering efforts. The present state of the art of genetic modeling is such that one can only take certain precautionary measures and then resort to testing for retained competence. These measures include colonizing with maximum possible numbers, selecting from a central portion of the population and from the biotype appropriate to target interaction, and other procedures that maximize genetic diversity and suitability.

Research Colony

Needham (63) defined the determinants of successful rearing as food, protection from enemies, a suitable physical environment, and fit conditions for reproduction. Methods of providing these are developed by using relatively small research colonies. These colonies also provide individual insects for conducting treatment investigations, such as irradiation, marking, handling, and distribution, and for making preliminary estimates on the feasibility of mass production and efficient successful control via mass release. Frequently, most of the estimates on the quality of the colonized insect, both in terms of program needs and of negative effects of production, are made at this time, and preconceived protocols are established that are fairly rigid. Also, production colonies are often established from the research colony without benefit of feedback that could improve selection of the founder colony.

The remainder of the production events are graphically isolated from prior and subsequent events (see Figure 1). They constitute the factory and distribution aspects of this model and include the rearing, treatment, and release phases. Desirable or undesirable attributes that arise during production or that are needed during performance can be directly manipulated or corrected (e.g. rearing procedures or treatments that affect vigor can be altered or eliminated). The events occurring on the performance side of Figure 1 are principally susceptible to indirect correction or manipulation only, and then solely during the production phase (e.g. reproductive isolation can only be corrected during establishment and culture).

Strain Development

The founding population from which production stocks are derived may be carefully selected, and the conditions of propagation carefully managed. Nevertheless there are requirements for economy, scheduling, and productivity that restrict the design and conditions of rearing. These, plus a lack of recognition of all components of the natural habitat and a lack of capacity to incorporate them, lead to developing an insect strain in the factory that differs to some degree from the target or source population. The processes most contributory to genetic decay have been identified as the founder effect, inbreeding, genetic drift, and selection (54, 55, 56). In the large populations used as factory breeding stock, selection can be considered to have the greatest impact (55) and the primary result is a quantitative rather than qualitative change in behavior (56). Thus, it may be expected that behavioral thresholds and frequencies are more likely to be altered than are the kinds of behavioral traits themselves. Also, selection and inbreeding tend to affect most deleteriously those traits with the highest adaptive value (13).

The managed or inadvertent selection that occurs in an insect factory rapidly and efficiently produces similar siblings; the result, then, is increasing homozygosity (72). It can be anticipated that the greatest amount of selection will occur during the initial stages of colonization (12, 72) when the population is pressed to accept the artificial and constrained environment and resources of the factory. The development of homozygosity is a legitimate concern; it has been indicated that it should be avoided in principle (12) and that it can be deterimental in engineered stocks, e.g. for optimum performance of key traits needed for program success (52, 55). However, the opinion is expressed in a recent review (55) that genetic models have great potential for, but are not yet capable of, relating theory to practice in these systems.

Situations exist where selection may be advantageous, at least in theory (55, 104). For example, selected strains could provide improved performance of uniparental parasite strains (5), improved sexual performance of released sterile insects (61), engineered strains for genetic pest control (9, 80), selected nondiapausing strains (37), color and morphological markers (16), and other qualities such as suitability for production (22). Selection for these may be anticipated to increase homozygosity, but one may hope that serious disadvantage can be avoided. Therefore, I am inclined to retain the opinion (19) that known inbreeding or indeed relative homozygosity should be warning signals rather than rejection criteria.

The genetic divergence of the production strain is susceptible to analysis, not only by classic techniques but now by electrophoretic display of allozymes and their comparison with those of the target or source populations (15). Also, specific performance traits such as attraction to a pheromone (84) indicate the effects of selection. Improving such modeling and testing should be stressed.

Mass Culture

A variety of values are routinely determined to monitor the quality of the rearing process and to ensure continuity in production facilities. These include the following: egg viability; density and yield of immature forms; survival, size, and yield of adults; adult density, sex ratio, and age; life history measurements; diet texture, composition, and microbial load; and environmental conditions. Since such factors are reasonably well understood I will not impose on limited space to list references on the subject, though they are many and significant. The general treatises mentioned previously will provide entry to the literature.

Treatment

Preparing mass-reared insects for release often involves applying one or more treatments, e.g. sterilizing with radiation or chemicals (in sterile insect release programs) and marking with internal (45) or external (77) dyes, tracers (89), or rare elements, individually or in combination (106) (in releases of sterile insects, genetically engineered pests, or entomophagous insects). Separating the sexes involves manipulation, as does packaging or other preparations for distribution. Anesthesia with cold, carbon dioxide, or anoxia is frequently used.

The suitability of such treatments is frequently analyzed by their impact on specific performance traits (20, 73, 78, 95). However, these considerations are more applicable to the performance side of the model. It is appropriate here to consider measuring the adequacy of each treatment for its purpose (degree of sterility, duration of markers, etc). Production protocols then routinely monitor the quantity and occasionally the quality of the treatment applied. Scheduled monitoring for continuity of treatment suitability and effectiveness can proceed with a generalized test that scans a spectrum of potential impact. The commonest evaluation for sterile releases is the so-called ratio test. For example, proportions of treated males, treated females (sometimes), untreated males, and untreated females are placed in small or moderate-size cages and the impact of the sterile population is assessed by measuring the fertility of the untreated females (3). Resulting data may be statistically manipulated to yield a value representing the relative competitiveness of the sterile insect and, hence, the adequacy of the treatment (29, 97). Because of the artificiality of the cage environment such tests cannot ensure assessment of all performance requirements nor can they identify wherein deficiencies occur. Thus, they are very useful as monitors of relative changes in treatment effects and effectiveness but should be used after other more specific tests have established the suitability of the treatment(s).

General trends in treatment effects are evaluated by life table measurements. Also, physiological tests can monitor treatment continuity once their relevance has been established. These potentially include tests for respiration (6), vision (1, 41), sound production (100), and metabolism (58, 86).

Future emphasis should include ensuring the relativity of monitoring efforts to essential performance traits.

Distribution

The production phase of the process terminates with transporting the insects to and releasing them in the target area. Techniques for handling, packaging, delivering, and releasing mass-reared insects are tailored to each program's needs; they vary from quick transport in simple containers from which the insects are released to complex sorting, suppression of activity, controlled environment transport, and mechanical distribution. Release may involve either the active life stage or a preceding stage in which the effects of distribution may be delayed. Protocols for monitoring the effects of distribution are much like those discussed previously. In many cases, distribution is considered a part of treatment and monitoring is conducted on a subsample of the population delivered for release. In cases where other than the simplest techniques are employed such evaluations should be conducted.

Pertinent topics, techniques, and evaluations include (*a*) packaging for transportation with low-temperature immobilization (91), controlled atmospheres (42), low oxygen pressure (98), or under ambient conditions (92); (*b*) ground release systems (92); (*c*) aerial package drop (33, 40) or free-fall release systems for pupae (23) and adults (62); and (*d*) nutritional support for packaged insects (47).

It is important that the mass production program manager be aware that subsequent to release the product of his effort is largely beyond his control. Therefore, by this time he should have taken cognizance of the input into his product and the feedback from all impact, needs, and constraints that will arise in later performance phases. Thorough analysis of the dynamics of his system, including performance, should enable creation of analytic procedures that will ensure detection of defects and indicate corrective measures within the production phase.

PERFORMANCE

In the model in Figure 1, performance is composed of two subelements (adaptedness and behavior) that individually and in combination affect the capacity of the factory product to achieve its objective. The first of these, adaptedness, represents the intrinsic genetic and physiological constitution and capacity of the insects as modified by the preceding processes. In programs of release of entomophagous insects and genetically engineered pest species, adaptability and fitness are both important. The former affects the range and kinds of performance characteristics required in response to diverse ecological conditions; the latter affects ability to contribute to subsequent generations (55). In programs of sterile release the impact of the released insects is expected to be felt over a shorter time span. Thus, neither adaptability nor fitness is as important as in colonizing releases. Therefore, the term adaptedness is used here, rather than adaptability, to convey meanings suitable to both colonizing and noncolonizing releases. Adaptedness provides the negative connotation pertaining to the selection of a factory ecotype, and also it implies preimposed characteristics that are positively or negatively correlated with the behavioral traits essential for success. Preimposition occurs through both genetic and nongenetic mechanisms; the latter includes behavioral conditioning and physiological effects.

These two subelements (adaptedness and behavior) are comparable in concept to the two types of quality discussed by Huettel (44), namely overall quality and particular traits or sets of traits. The behavioral traits shown in Figure 1 are those that can be identified as performance characteristics generally essential to program accomplishment. They include behavior related to locomotion, the habitat, and the target organism (host or mate). Other reviews have identified similar requisites (12, 19, 44, 54, 55).

Genetic Adaptedness

The causes and implications of genetic alteration as a result of production are considered in other reviews (54, 55), and relevant citations are provided. It is appropriate here to consider techniques for monitoring this aspect of quality.

Results with classical techniques exemplify the potential for analyzing the production impact, as in selecting for positive or negative phototactic behavior in *Drosophila melanogaster* (57) or for pleitropic effects of color mutation linked to disadvantageous traits in vision and longevity and advantageous traits in locomotory behavior in a blow fly (*Phaenicia sericata*) (17). Alterations in mating behavior in two strains of *D. melanogaster* were linked to the hybridization of color mutants and the subsequent selection (21). Since such selection probably occurs under any artificial factory regime, basic genetic studies of candidates might be considered a prerequisite to mass rearing, thus anticipating and allowing subsequent genetic monitoring.

Electrophoretic analysis of allozymes offers the relative ease of routine procedure required for quality monitoring. A spectrum of genetic loci can be surveyed to determine variation or shifts within production lines or in comparison with source or target populations (15). Genetic divergence and inbreeding during production of codling moths *(Laspeyresia pomonella)* have been demonstrated (15), and use of the technique for determining fitness components including mating competitiveness and dispersal of the plum curculio *(Conotrachelus nenuphar)* has been modeled in field tests (44).

Analyzing physiological processes can indicate variation within and between populations. Because thresholds of behavioral trait performance may be expected to change more rapidly or frequently than would specific kinds of traits, physiological tests may be more sensitive monitors of quality than behavioral tests. Thus, electroretinographic analysis of the visual acuity of wild and production line screwworm flies (*Cochliomyia hominivorax*) detected differences due to strain and selection (30). The physical characteristics of sounds associated with four behavioral patterns of the Caribbean fruit fly (*Anastrepha suspensa*) were studied (101). We proposed that acoustic monitoring could sensitively detect changes and differences in general and specific activity patterns in production insects with nearly total automation.

The adaptation of the Caribbean fruit fly to colonization was studied by comparing life history values and respiration (CO_2 production) of a colonizing wild strain and an established production strain (51). During five generations, adaptation by the colonizing strain was manifested primarily in increased fecundity and decreased developmental time. Measurements of carbon dioxide production indicated equivalent metabolic changes. Thus, CO_2 production or oxygen consumption (6) could serve as monitors of general production quality.

Life history values might be considered the most basic of generalized measurements that indicate adaptive changes during the production phase. For example, a colonizing strain of tobacco budworms (*Heliothis virescens*) showed gradual acclimatization to culture conditions over a period of seven generations until, in comparison with a standard strain, equilibrium in oviposition and mating patterns was reached (68). Genetic control of these attributes was demonstrated by crosses between strains. Also, the oviposition, premating, and mating patterns of the Mediterranean fruit fly (*Ceratitis capitata*) examined during twelve generations indicated adaptation to colonization procedures (75, 76).

The genetic status of the mass-produced insect is as important at the beginning of the performance phase as it is at the beginning of production. Routine genetic monitoring at this stage may provide very meaningful information to the production manager and avoid erosion of the adaptability, fitness, and competence of the product. Physiological, behavioral, or life history examinations are sensitive to genetic variations, and some are quite suitable for routine monitoring.

Corrective responses to positive evidence of genetic deviation in the production stock or of its lack of performance attributes would include correction of factors causing bottlenecking and selection. Appropriate steps might include incorporating token stimuli (12) or altering the strain development procedures, such as incorporating wild genotypes (31, 105). Selecting parasites for increased parasitization rates (5) and for temperature adaptability (103) have been demonstrated. Mackauer (55) and Whitten & Foster (104) discuss selective breeding, colony infusion, and colony replacement in detail.

Nongenetic Adaptedness

The other effects felt by the insect as a result of mass colonization arise through processes that are not genetic. They are either conditioned behavioral changes or alterations in physiological processes. Inadequate or unsuitable nutrition and disease and the effects of treatment and handling can result in altered, and probably deficient, execution of most necessary performance traits. Lack of exposure to appropriate sensory stimuli or exposure to inappropriate stimuli may result in subsequent altered behavior. Entrainment to factory photoperiod regimens (or to conditions without periodicity) or temperature conditions (extremes or variation) can result in intolerance to field conditions or asynchrony with the life cycles or behavior of field populations. Bush (14) recently proposed that genetic and conditioned effects may be mutually reinforcing through the process of reproductive isolation. This would seem particularly pertinent to systems of long-term artificial colonization.

Phenotypic deviants occasionally arise in a production program or are intentionally developed and found useful in marking insects for release and field study. An ebony phenotype of the boll weevil (*Anthonomus grandis*) has three color-marked deviations that were found to be expressions induced by stress occurring in the rearing phase (59). The food of the California red scale (*Aonidiella aurantii*) affects the ability of its parasites to successfully oviposit, as well as affecting their mortality, longevity, size, sex ratio, and fecundity (88). *Trichogramma evanescens* and *T. minutum* reared from eggs of the cabbage looper, *Trichoplusia ni,* were more motile than those from eggs of the Angoumois grain moth (*Sitotroga cerealella*) (11). Others have found that parasites (*Apanteles chilonis*) produced on hosts fed an artificial diet were detrimentally affected (46) and, conversely, that prerelease adult nutrition did not affect performance of *Trichogramma* (*pretiosum*?) (4).

Although treatment may be expected to affect performance, reduced competitiveness of a leafroller (*Adoxophyes orana*) after irradiation was attributed to effects of diet rather than radiation itself or genetics (25).

Laboratory-reared tobacco budworms released for sterile suppression were found to be asynchronous with their field counterparts by about 2 hr, and hence males were unable to obtain mates (69). In the codling moth, photophase entrainment in the factory seems to occur in late larval stages, and here again asynchrony with the target population is detrimental to male performance (L. D. White, personal communication). Stilt bugs (*Jalysus spinosus*) proposed to be produced for control of tobacco pests displayed locomotory activity conditioned by the rearing photoperiod; less than 13 hr of light resulted in more and longer flights than did longer photoperiods (26).

The methods for detecting and analyzing nongenetic input to performance should be selected for appropriateness for each program and target organism; they are among those mentioned in the previous section. Again, techniques amenable to routine application that can provide a performance evaluation of a general nature or that will permit detection of changes in thresholds may be the most suitable. For example, electroretinographic analysis, as developed for quality monitoring, made it possible to detect differences among the effects of various larval diets on the visual acuity and threshold of adult Caribbean fruit flies (1).

Dispersal Related Behavior

Appropriate locomotory behavior of adequate capability and propensity is identified as a primary performance trait in mass-release programs (12, 19, 44, 55). Locomotory deficiencies have been implicated in problems encountered in sterile release programs of tephritids with which I have been associated. Propensity to flight was considered equally as involved as capability of flight, as would be expected if detriment is expressed in rising thresholds. Locomotion has become a concern to most of those planning or employing massive releases of other insect species.

However, locomotion must not only be adequate, it must be appropriate. Sterile screwworm adults entering a diffuse population in hidden habitats might need to display considerable motive propensity and ability for wide dispersal and interaction. Sterile male boll weevils distributed adequately over infested cotton fields might need to disperse only to the nearest host plant and remain there while potently attracting females. Thus, aggressive dispersal away from points of distribution would be a program handicap, as it would in many releases of entomophagous insects.

Flight capability has often been evaluated in the laboratory by using flight mills. Test insects are fastened to the end of a horizontal arm that rotates on a vertical axis in response to insect flight. The rotation of the arm is used in quality control to analyze flight behavior in terms of frequency, duration, velocity, and distance (20, 70).

Mediterranean fruit fly males marked with a dye, as in a release campaign, were tested after they had been sterilized by irradiation during the pupal stage or as 2-day-old adults. Although affected by the treatments, flies treated as adults exhibited greater flight capacity than those treated as pupae, and performance was better at 5 days of age than at 10. In anticipation that released males have their greatest impact soon after release, it was suggested that treatment of adults was indicated (81). However, flight characteristics of cabbage looper moths (36) or house flies, *Musca domestica* (86), displayed on a mill were not adversely affected by sterilizing treatment with tepa. In studies with another tephritid, the olive fly (*Dacus oleae*), the influence of various biological attributes on flight capacity were evaluated (70). For example, distance flown increased in proportion to body weight and feeding adults with sucrose (vs starving them) increased flight distance nearly 100% in males and 50% in females.

Motive capacity is also measurable in stationary systems. Wingbeat frequency of tethered stationary insects is readily determined stroboscopically (82), and this technique appears able to detect irradiation effects not apparent in flight mill studies. Analysis of the physical properties of the sounds produced by tethered fruit flies provided information on sound pressure levels and harmonic content, as well as wingbeat frequency. The sensitivity of the analysis and specificity of the sounds indicated potential for discerning variations in locomotory capability (100, 101).

A tethered insect is necessarily less able to express propensity to flight than one not tethered because the strong tarsal reflex results in lower thresholds for wing beating, and ad libitum flight is not clearly expressed. Devices that test free movement in an arena (7) or actograph (19), through a runway (5), or from one chamber to another (17, 78) do discern locomotory propensity. In this way, laboratory stocks of Mediterranean fruit flies, for example, were shown to be 50% less motile than field stocks (78).

The most applicable measurements of the quality of locomotory performance of an insect produced for release are made in the field if possible. Two principal methods are employed. In the more frequently used method, released insects (that may be sterilized) are marked in some way, as with pigment (106), radioactive isotope (8), or phenotypic character (16), and are recovered after release at various distances from the release point. Thus, interisland movement of tobacco budworms was detected in a sterile-release test (32). Also, average and maximum migration of a black fly (*Simulium venustum*) was determined to be about 7 and 22 miles, respectively (8). Mean distance to site of recovery of released tsetse flies (*Glossina morsitans*) was about equal for laboratory (sterilized and untreated) and native strains (24). The recovery of released laboratory strain olive flies in traps arrayed over various distances was only about 58% of that of released field-collected flies and was attributed to deficiences in dispersal behavior (66). The second technique requires the release of fertile insects that are genetically marked and whose progeny can be identified by morphological or biochemical expression (44). The location of these progeny implies the presence of their parents. However, both of these measures include expression of factors other than simple dispersal, such as survival, host recognition (if that is where recoveries are made), and in the second case the complex behavior associated with reproduction. Also, field dispersion is affected by climatic conditions, especially wind, that along with a low proportion of insects recovered can complicate interpretation of the recovery data. Thus, a combination of occasional field and routine laboratory analyses may be desirable, if the latter tests can be related to field tests, because their relationship has been established.

Habitat-Related Behavior

The next event occurring in the process encompasses a complex of behavioral attributes related to the habitat in which the production insect is expected to interact with its target. These include seeking and accepting food, water, and shelter, and avoiding predators; the presence of appropriate rhythms, both circadian and seasonal; tolerance of environmental conditions; and response to stimuli of rendezvous sites. A number of these, such as nutrition, predation, environmental tolerance, and longevity, are often evaluated in a summed value presented as survival. For example, tsetse flies of laboratory (sterilized and untreated) and native strains released in Rhodesia survived equally well as measured by mean time elapsed between release and recovery (24). Longevity and temperature tolerance of radiolabeled *Culex pipiens quinquefasciatus* were evaluated by the presence of egg rafts containing radiotracers. Temperature reduction to a mean of 8°C and a low of –0.6°C extended time to oviposition by 13 days, but females survived these conditions for at least 38 days (90). Marked adult mosquitoes (*Anopheles albimanus*) were recorded in natural resting places for 11 days after release during stress (dry) conditions (38).

On the other hand, these factors may be partitioned and studied in the laboratory or field. *Trichogramma* hybridized for temperature tolerance were significantly more successful in parasitizing cabbage looper eggs in field cages (5). Honey was the most effective of a variety of foods for prefeeding adult *Trichogramma* based on effects on longevity in the laboratory (4). *Trichogramma* reared on *Sitotroga* eggs rather than *Heliothis* eggs had a mean longevity of 2 days rather than 3 in the laboratory, but differences were less distinct in field cage tests (96). I accumulated data (unpublished) indicating that Mexican fruit flies (*Anastrepha ludens*) cultured more than 10 years in the laboratory searched for, accepted, and utilized honeydew as successfully as the artificial food (hydrolyzed protein) in small cages.

Locomotory propensity is associated with predator avoidance and was implicated in ant predation and the reduction in flight propensity of a laboratory strain of Mediterranean fruit fly (78). The presence or absence of diapause in the life cycle of released insects expected to reproduce through several generations may be important. Entrance to diapause of *Apanteles melanoscelus* varied according to the origin and history of the colony strain and ranged from 8–96%, even in established and new colonies both from within France (43).

A requirement generally exists for appropriate interaction with the host's or target's substrate. Apple maggot flies (*Rhagoletis pomonella*) use the fruit of the larval host as an assembly site for mating (67), and laboratory-cultured tephritids may be less able to distinguish host plants than native counterparts (R. J. Prokopy, personal communication) (1).

Although these attributes may be measured in tests of summed habitat behavior, there appears to be a need for more routine techniques that are known to test for those specific traits that are critical to performance.

Target-Related Behavior

The interaction of the released insect with its target (host or mate) is not only the final behavioral event, it is a key event because it represents the function for which release is made. That this is recognized by the research and pest control community is attested to by the large numbers of references available concerning this portion of the model (especially for sterile releases). It is tempting, in quality monitoring, to conceptually simplify the mating process and test in systems that essentially measure copulatory activity. However, sexual behavior is much more than mating and includes many components that may be critical to success in the field. Monitoring must occasionally if not routinely assess these characters. To assess them they must be known via intensive behavioral studies.

Laboratory testing of mating competitiveness has been conducted in two principal ways. Sexual aggressiveness of selected or treated males was tested by enclosing female screwworm adults with relatively larger numbers of males. The mortality rate of females expressed the harassment by and hence aggressiveness of males (10). The second method, ratio tests and their analysis, was mentioned previously. Combining proportions of treated and untreated insects in large outdoor cages probably allows expression of more behavioral components, as in a test showing that chemosterilized male *Aedes aegypti* were equal to normal males and that heterosis enhanced competitiveness (79). Field cage tests of competitiveness of *Adoxophyes orana* indicated the reared male insect was about 50% as competitive in mating as native males (25). Rearing *Trichogramma* on natural host eggs resulted in improved host-finding ability in small containers (96). Field cage tests showed increased parasitization, and therefore more host interaction, by *Trichogramma* selected uniparentally for heat tolerance and locomotion (5). Placing female noctuid moths with one wing clipped on mating tables allows measurement, in the field, of the periodicity of calling and attractiveness of these females (69). The recovery of mating pairs tests strain interaction.

Field tests of insects not restrained or caged would be expected to provide the most complete evaluation of target interaction, though they are not generally capable of identifying deficient characteristics. However, capability for making field assessments of sterility varies greatly between species and depends upon the possibility of observing mating behavior or assessing fertility of captured females or deposited eggs. This difficulty can be overcome by measuring incorporation of morphological or biochemical genetic material into wild populations when fertile insects can be released (44, 104). In insects such as mosquitoes whose eggs are readily collected, fertility can be measured and competitiveness as well as other components of population dynamics can be determined (102).

Temporal behavior patterns of released insects must conform to those of the target, especially when mating occurs at discrete times. Large outdoor cage tests showed irradiated laboratory-reared male Mediterranean fruit flies were delayed in initial and peak mating times, and were 50% less effective in competing for mates (39). Males of a laboratory strain of tobacco budworms released into cotton fields were almost totally noncompetitive (69).

Reproductive isolation through strain development or physiological change such as size can affect interaction with the target. Mediterranean fruit flies of laboratory and field strains showed mating preferences, laboratory males having become partially isolated sexually from field-strain females (75). Size is reported to effect mate selection in screwworm flies: small males are at a disadvantage in competition with normal-sized males for normal-sized females (2).

The key role of chemical cues in host and mate interaction has received major recent attention. Clearly, if a released parasite does not adequately respond to the kairomones of its host its success will be impaired. Also, production of and/or response to pheromones can have a critical role in the mate interactions of released sterile or genetically engineered insects. Thus, no differences were found in pheromone production or response between colonized and field strains of the almond moth (*Cadra cautella*) (93), whereas strain differences have been found in screwworm flies (28), gypsy moths [*Lymantria* (=*Porthetria*) *dispar*] (71), and European corn borers (*Ostrinia nubilalis*) (53) among others. The impact of laboratory diet or sterilizing treatment on pheromone production has been documented. Chemosterilization results in reduced production of two of the components of the boll weevil pheromone, as did feeding on synthetic diet instead of cotton buds (34). Irradiation, however, did not seriously reduce pheromone production in boll weevils (60) or the attractiveness of female tobacco budworms (35) or gypsy moths (94).

Target-related behavior in most species is amenable to general analysis, tests of specific attributes, or both. The importance of this complex of behavioral traits to program success indicates need for their complete description and routine monitoring.

Objective

The released insect discharges its duty when it interacts successfully with the target organism. The objective of release is not achieved until the impact of that interaction is felt by the target organism, and the target population is thereby affected. The released parasite will cause the organic death of the target through the action of its progeny; the sterile or genetically altered insect will cause the genetic death of its target. The level of success achieved may be moderated by physiological factors.

The success of entomophagous insects can be directly evaluated through studies of egg fecundity and fertility, growth rate and activity of immature forms, host resistance or tolerance, and host mortality. Such studies seem promising for monitoring and ensuring the success of inundative releases. Thus, this strategy in classical biological control may be advanced beyond the release and hope approach often encountered.

Numerous physiological factors influence the success of sterile or other autocidal releases after target interaction. These have been previously discussed as postmating factors (12, 19). Such factors vary among species in their occurrence and importance. Among them are sperm numbers, motility, and longevity; sperm precedence; monocoitic factors; normal or induced female monogamy or polygamy; sperm depletion and mixing; hybrid inviability, sterility, or fertility; and partial sterility.

A number of reviews consider these matters, which are beyond the scope and capacity of this paper (48, 50, 64, 65, 104).

CONCLUDING REMARKS

Great progress has been made in recent years in understanding the events and mechanisms that may influence the effectiveness of a massive release program. Techniques have been developed to test most of these; routine monitoring is now conducted on some. Protocols exist for assessing quality and continuity of production. Resources need to be applied to understanding and incorporating behavioral requisites into monitoring protocols that realistically reflect program objectives.

Although the events in a program typically occur in counterclockwise order, with reference to Figure 1, quality priorities often appear to decrease in clockwise order beginning with accomplishment of the objective. The complex and distinctive characteristics of each program indicate that this judgment should not be applied without rigorous examination. Nevertheless, high quality in preceding attributes is fruitless if mortality or sterility does not ensue. Similarly, high performance in locomotion or habitat interaction will not overcome deficient target relationships. Also, logistic input can correct deficiencies in some stages. For example, insects can be physically distributed near their target; in monocultures, habitat interaction may be less critical than elsewhere. Target interaction might be enhanced through the use of pheromones or kairomones (18). However, physiological deficiencies at the objective level can only be countered by increasing the numbers released, if at all.

Each proposed insect release program must be preceded by studies of the physiology, ethology, and ecology of the production and target populations of such intensity that critical factors will be discerned and accounted for. The complexity of the process of production and release indicates need, not only for systematic analysis, but also systems design. The application of modeling, with computer analysis and forecasting, has promise. Massive releases for pest control have been undertaken by inflating research scale methods and by extrapolating laboratory or small field test results. For general application and success, mass rearing requires development of quality control as a distinct discipline complete with a cadre of trained and dedicated specialists.

Literature Cited

1. Agee, H. R., Park, M. L. 1975. Use of electroretinogram to measure the quality of insect vision. *Environ. Lett.* 10:171
2. Alley, D. A., Hightower, B. G. 1966. Mating behavior of the screwworm fly as affected by differences in strain and size. *J. Econ. Entomol.* 59:1499–1502
3. Anwar, M., Chatha, N., Ohinata, K., Harris, E. J. 1975. Gamma irradiation of the melon fly: laboratory studies of the sexual competitiveness of flies treated as pupae 2 days before eclosion or as 2-day-old adults. *J. Econ. Entomol.* 68:733–35
4. Ashley, T. R., Gonzalez, D. 1974. Effect of various food substances on longevity and fecundity of *Trichogramma. Environ. Entomol.* 3:169–71
5. Ashley, T. R., Gonzalez, D., Leigh, T. F. 1974. Selection and hybridization of *Trichogramma. Environ. Entomol.* 3:43–48
6. Bailey, P. 1973. Respiration and nutrition of radiation-sterilized female *Dacus*

cucumis. Entomol. Exp. Appl. 16: 433–44
7. Bailey, P. 1974. Behavioral changes induced by irradiating *Dacus cucumis* with gamma rays. *J. Insect Physiol.* 21:1247–50
8. Baldwin, W. F., West, H. S., Gomery, J. 1975. Dispersal pattern of black flies tagged with ^{32}P. *Can. Entomol.* 107: 113–18
9. Bartlett, A. C., Butler, G. D. Jr. 1975. Genetic control of the cabbage looper by a recessive lethal mutation. *J. Econ. Entomol.* 68:331–35
10. Baumhover, A. H. 1965. Sexual aggressiveness of male screw-worm flies measured by effect on female mortality. *J. Econ. Entomol.* 58:544–48
11. Boldt, P. E. 1974. Temperature, humidity, and host: effect on rate of search of *Trichogramma evanescens* and *T. minutum. Ann. Entomol. Soc. Am.* 67:706–08
12. Boller, E. 1972. Behavioral aspects of mass-rearing of insects. *Entomophaga* 17:9–25
13. Bruell, J. H. 1967. Behavioral heterosis. In *Behavior-Genetic Analysis,* ed. J. Hirsch, pp. 270–86. New York: McGraw-Hill. 522 pp.
14. Bush, G. L. 1974. The mechanism of sympatric host race formation in the true fruit flies. In *Genetic Mechanisms of Speciation in Insects,* ed. M. J. D. White, pp. 3–23. Sydney: Australia and New Zealand Book Co. 170 pp.
15. Bush, G. L. 1975. Genetic variation in natural insect populations and its bearing on mass-rearing programmes. In *Controlling Fruit Flies by the Sterile-Insect Technique,* pp. 9–17. Vienna: IAEA Panel Proc. Ser. 175 pp.
16. Butler, G. D. Jr., Hamilton, A. G., Bartlett, A. C. 1975. Development of the dark strain of cabbage looper in relation to temperature. *Environ. Entomol.* 4:619–20
17. Chabora, P. C. 1969. Mutant genes and the emigration behavior of *Phaenicia sericata. Evolution* 23:65–71
18. Chambers, D. L. 1974. Chemical messengers as tools in the management of insect pests. *Proc. Tall Timbers Conf. Ecol. Anim. Contr. Habitat Manage.* 5:135–45
19. Chambers, D. L. 1975. Quality in mass-produced insects. See Ref. 15, pp. 19–32
20. Chambers, D. L., Sharp, J. L., Ashley, T. R. 1976. Tethered insect flight: a system for automated data processing of behavioral events. *Behav. Res. Methods Instrum.* 8:352–56
21. Crossley, S. A. 1974. Changes in mating behavior produced by selection for ethological isolation between ebony and vestigial mutants of *Drosophila melanogaster. Evolution* 28:631–47
22. Crystal, M. M., Ramirez, R. 1975. Screwworm flies for sterile-male release: laboratory tests of the quality of candidate strains. *J. Med. Entomol.* 12: 418–22
23. Cunningham, R. T., Suda, D., Chambers, D. L., Nakagawa, S. 1971. Aerial broadcast of free-falling pupae of the Mediterranean fruit fly for sterile-release programs. *J. Econ. Entomol.* 64:1211–13
24. Dame, D. A., Birkenmeyer, D. R., Nash, T. A. M., Jordan, A. M. 1975. The dispersal and survival of laboratory-bred and native *Glossina morsitans morsitans* Westw. in the field. *Bull. Entomol. Res.* 65:453–57
25. Denlinger, D. L., Ankersmit, G. W., Noordink, J. Ph. W. 1973. Studies on the sterile male technique as a means of control of *Adoxophyes orana.* 3. An evaluation of competitiveness of laboratory-reared moths. *Neth. J. Plant. Pathol.* 79:229–35
26. Elsey, K. D. 1974. *Jalysus spinosus:* effect of age, starvation, host plant and photoperiod on flight activity. *Environ. Entomol.* 3:653–55
27. Finney, G. L., Fisher, T. W. 1964. Culture of entomophagous insects and their hosts. In *Biological Control of Insect Pests and Weeds,* ed. P. DeBach, pp. 328–55. New York: Reinhold. 844 pp.
28. Fletcher, L. S., Claborn, H. V., Turner, J. P., Lopez, E. 1968. Differences in response of two strains of screw-worm flies to the male pheromone. *J. Econ. Entomol.* 61:1386–88
29. Fried, M. 1971. Determination of sterile-insect competitiveness. *J. Econ. Entomol.* 64:869–72
30. Goodenough, J. L., Wilson, D. D., Agee, H. R. 1976. Electroretinogram used to measure and compare vision of wild and mass-reared screwworm flies. *J. Med. Entomol.* In press
31. Haeger, J. S., O'Meara, F. G. 1970. Rapid incorporation of wild genotypes of *Culex nigripalpus* into laboratory-adapted strains. *Ann. Entomol. Soc. Am.* 63:1390–91
32. Haile, D. G., Snow, J. W., Young, J. R. 1975. Movement by adult *Heliothis* re-

leased on St. Croix to other islands. *Environ. Entomol.* 4:225–26
33. Harris, E. J., Mitchell, W. C., Baumhover, A. H., Steiner, L. F. 1968. Mutilation and survival of sterile oriental fruit flies and melon flies emerging in drop boxes. *J. Econ. Entomol.* 61: 493–96
34. Hedin, P. A., Rollins, C. S., Thompson, A. C., Gueldner, R. C. 1975. Pheromone production of male boll weevils treated with chemosterilants. *J. Econ. Entomol.* 68:587–91
35. Hendricks, D. E. 1974. Tobacco budworm: effects of cobalt irradiation on female attractiveness and mating frequency. *J. Econ. Entomol.* 67:610–12
36. Henneberry, T. J., Kishaba, A. N., Iqbal, M. Z., Klingler, B. B. 1968. Reproduction, longevity, and flight of cabbage looper moths treated topically with tepa. *J. Econ. Entomol.* 61:1536–40
37. Herzog, G. A., Phillips, J. R. 1974. Selection for a nondiapause strain of the bollworm, *Heliothis zea. Environ. Entomol.* 3:525–27
38. Hobbs, J. H., Lowe, R. E., Schreck, C. E. 1974. Studies of flight range and survival of *Anopheles albimanus* Wiedemann in El Salvador. I. Dispersal and survival during the dry season. *Mosq. News* 34:389–93
39. Holbrook, F. R., Fujimoto, M. S. 1970. Mating competitiveness of unirradiated and irradiated Mediterranean fruit flies. *J. Econ. Entomol.* 63:1175–76
40. Holbrook, F. R., Steiner, L. F., Fujimoto, M. S. 1970. Holding containers for melon flies and Mediterranean fruit flies for use in sterile fly aerial releases. *J. Econ. Entomol.* 63:908–10
41. Holt, G. G. 1975. λ-Radiation studies of two strains of *Trichoplusia ni:* sensitivity of the compound eye and induced sterility. *J. Insect Physiol.* 21:1167–74
42. Hooper, G. H. S. 1970. Use of carbon dioxide, nitrogen, and cold to immobilize adults of the Mediterranean fruit fly. *J. Econ. Entomol.* 63:1962–63
43. Hoy, M. A. 1975. Hybridization of strains of the gypsy moth parasitoid, *Apanteles melanoscelus,* and its influence upon diapause. *Ann. Entomol. Soc. Am.* 68:261–64
44. Huettel, M. D. 1976. Monitoring the quality of laboratory-reared insects: a biological and behavioral perspective. *Environ. Entomol.* In press
45. Jones, R. L., Harrell, E. A., Snow, J. W. 1972. Three dyes as markers for corn earworm moths. *J. Econ. Entomol.* 65:123–26
46. Kajita, H. 1973. Rearing of *Apanteles chilonis* Munakata on the rice stem borer, *Chilo suppressalis* Walker, bred on a semi-artificial diet. *Jpn. J. Appl. Entomol. Zool.* 17:5–9
47. Keiser, I., Schneider, E. L. 1969. Need for immediate sugar and ability to withstand thirst by newly emerged oriental fruit flies, melon flies, and Mediterranean fruit flies untreated or sexually sterilized with gamma irradiation. *J. Econ. Entomol.* 62:539–40
48. Kilgore, W. W., Doutt, R. L., eds. 1967. *Pest Control Biological, Physical, and Selected Chemical Methods.* New York: Academic. 477 pp.
49. Knipling, E. F. 1966. Introduction. In *Insect Colonization and Mass Production,* ed. C. N. Smith, pp. 1–12. New York: Academic. 618 pp.
50. LaBrecque, G. C., Smith, C. N., eds. 1968. *Principles of Insect Chemosterilization.* New York: Appleton-Century-Crofts. 354 pp.
51. Leppla, N. C., Huettel, M. D., Chambers, D. L., Turner, W. K. 1976. Comparative life history and carbon dioxide output of wild and colonized Caribbean fruit flies. *Entomophaga* In press
52. Levins, R. 1969. Some demographic and genetic consequences of environmental heterogeneity for biological control. *Bull. Entomol. Soc. Am.* 15:237–40
53. Liebherr, J., Roelofs, W. 1975. Laboratory hybridization and mating period studies using two pheromone strains of *Ostrinia nubilalis. Ann. Entomol. Soc. Am.* 68:305–09
54. Mackauer, M. 1972. Genetic aspects of insect production. *Entomophaga* 17: 27–48
55. Mackauer, M. 1976. Genetic problems in the production of biological control agents. *Ann. Rev. Entomol.* 21:369–85
56. Manning, A. 1967. Genes and the evolution of insect behavior. See Ref. 13, pp. 44–60.
57. Markow, T. A. 1975. A genetic analysis of phototactic behavior in *Drosophila melanogaster.* II. Hybridization of divergent populations. *Behav. Genet.* 5: 339–50
58. Mayer, R. T., Bridges, A. C. 1975. Some effects of ionizing radiation on the lipid and glycogen content of adult horn flies, *Haemotobia irritans. Insect Biochem.* 5:387–88
59. McCoy, J. R. 1975. Phenotypic devi-

ants in the ebony genotype in the boll weevil. *J. Econ. Entomol.* 68:775–76

60. McGovern, W. L., McKibben, G. H., Gueldner, R. C., Cross, W. H. 1975. Irradiated boll weevils: pheromone production determined by GLC analysis. *J. Econ. Entomol.* 68:521–23
61. McKibben, G. H., McGovern, W. L., Cross, W. H., Lindig, O. H. 1976. Search for a super laboratory strain of boll weevils: a rapid method for pheromone analysis of frass. *Environ. Entomol.* 5:81–82
62. Nadel, D. J., Monro, J., Peleg, E. A., Figdor, H. C. F. 1967. A method of releasing sterile Mediterranean fruit fly adults from aircraft. *J. Econ. Entomol.* 60:899–902
63. Needham, J. G., ed. 1937 *Culture Methods for Invertebrate Animals.* New York: Dover. 590 pp.
64. North, D. T. 1975. Inherited sterility in Lepidoptera. *Ann. Rev. Entomol.* 20:167–82
65. Pal, R., Whitten, M. J., eds. 1974. *The Use of Genetics in Insect Control.* Amsterdam: Elsevier/North-Holland. 241 pp.
66. Prokopy, R. J., Economopolos, A. P. 1975. Attraction of laboratory-cultured and wild *Dacus oleae* flies to sticky-coated McPhail traps of different colors and odors. *Environ. Entomol.* 4:187–92
67. Prokopy, R. J., Bennett, E. W., Bush, G. L. 1971. Mating behavior in *Rhagoletis pomonella.* I. Site of assembly. *Can. Entomol.* 103:1405–09
68. Raulston, J. R. 1975. Tobacco budworm: observations on the laboratory adaptation of a wild strain. *Ann. Entomol. Soc. Am.* 68:139–42
69. Raulston, J. R., Graham, H. M., Lingren, P. D., Snow, J. W. 1976. Mating interaction of native and laboratory-reared tobacco budworms released in cotton. *Environ. Entomol.* 5:195–98
70. Remund, U., Boller, E. F. 1975. Qualitatskontrolle bei Insekten: Messung von Flugparametern. *Z. Angew. Entomol.* 78:113–26
71. Richerson, J. V., Cameron, E. A. 1974. Differences in pheromone release and sexual behavior between laboratory-reared and wild gypsy moth adults. *Environ. Entomol.* 3:475–81
72. Roberts, R. C. 1967. Some concepts and methods in quantitative genetics. See Ref. 13, pp. 214–57
73. Robinson, A. S. 1975. Influence of anoxia during gamma irradiation on the fertility and competitiveness of the adult male codling moth, *Laspeyresia pomonella. Radiat. Res.* 61:526–34
74. Rodriguez, J. G., ed. 1972. *Insect and Mite Nutrition.* Amsterdam: North-Holland. 702 pp.
75. Rossler, Y. 1975. The ability to inseminate: a comparison between laboratory-reared and field populations of the Mediterranean fruit fly *(Ceratitis capitata). Entomol. Exp. Appl.* 18: 225–60
76. Rossler, Y. 1975. Reproductive differences between laboratory-reared and field-collected populations of the Mediterranean fruit fly, *Ceratitis capitata. Ann. Entomol. Soc. Am.* 68:987–91
77. Schroeder, W. J., Cunningham, R. T., Miyabara, R. Y., Farias, G. J. 1972. A fluorescent compound for marking Tephritidae. *J. Econ. Entomol.* 65: 1217–18
78. Schroeder, W. J., Chambers, D. L., Miyabara, R. Y. 1973. Mediterranean fruit fly: propensity to flight of sterilized flies. *J. Econ. Entomol.* 66:1261–62
79. Seawright, J. A., Kaiser, P. E., Dame, D. A. 1975. Mating competitiveness of chemosterilized hybrid males of *Aedes aegypti* in outdoor cage studies. *Mosq. News* 35:308–14
80. Seawright, J. A., Kaiser, P. E., Willis, N. L., Dame, D. A. 1976. Field competitiveness of double translocation heterozygote males of *Aedes aegypti. J. Med. Entomol.* 13:208–11
81. Sharp, J. L., Chambers, D. L. 1976. Gamma irradiation effect on the flight mill performances of *Dacus dorsalis* and *Ceratitis capitata. Proc. Hawaii. Entomol. Soc.* In press
82. Sharp, J. L. 1972. Effects of increasing dosages of gamma irradiation on wing-beat frequencies of *Dacus dorsalis* Hendel males and females at different age levels. *Proc. Hawaii. Entomol. Soc.* 21:257–62
83. Shorey, H. H., Hale, R. L. 1965. Mass-rearing of the larvae of nine noctuid species on a simple artificial medium. *J. Econ. Entomol.* 58:522–24
84. Showers, W. B., Reed, G. L., Oloumi-Sadeghi, H. 1974. European corn borer: attraction of males to synthetic lure and to females of different strains. *Environ. Entomol.* 3:51–58
85. Singh, P. 1972. Bibliography of artificial diets for insects and mites. *N. Z. Dep. Sci. Ind. Res. Wellington Bull. 209.* 75 pp.
86. Skelton, T. E., Hunter, P. E. 1970. Flight and respiration responses in

house flies following topical application of sterilization levels of tepa. *Ann. Entomol. Soc. Am.* 63:770–73
87. Smith, C. N., ed. 1966. *Insect Colonization and Mass Production.* New York: Academic. 618 pp.
88. Smith, J. M. 1957. Effects of the food plant of California red scale, *Aonidiella aurantii* on reproduction of its hymenopterous parasites. *Can. Entomol.* 89:219–30
89. Smittle, B. J., Lowe, R. E., Ford, H. R., Weidhaas, D. E. 1973. Techniques for ^{32}P labeling and assay of egg rafts from field-collected *Culex pipiens quinquefasciatus* Say. *Mosq. News* 33:215–20
90. Smittle, B. J., Lowe, R. E., Patterson, R. S., Cameron, A. L. 1975. Winter survival and oviposition of ^{14}C-labelled *Culex pipiens quinquefasciatus* Say in northern Florida. *Mosq. News* 35:54–56
91. Smittle, B. J., Patterson, R. S. 1974. Container for irradiation and mass transport of adult mosquitoes. *Mosq. News* 34:406–08
92. Snow, J. W., Burton, R. L., Sparks, A. N., Cantelo, W. W. 1971. Attempted eradication of the corn earworm from St. Croix, U.S. Virgin Islands. U.S. Dep. Agric., Agric. Res. Serv. *Prod. Res. Rep. No. 125.* 12 pp.
93. Sower, L. L., Hagstrum, D. W., Long, J. S. 1973. Comparison of the female pheromones of a wild and a laboratory strain of *Cadra cautella,* and male responsiveness to the pheromone extracts. *Ann. Entomol. Soc. Am.* 66:484–85
94. Statler, M. W. 1970. Effects of gamma radiation on the ability of the adult female gypsy moth to attract males. *J. Econ. Entomol.* 63:163–64
95. Stinner, R. E., Ridgway, R. L., Kinzer, R. E. 1974. Storage, manipulation of emergence, and estimation of numbers of *Trichogramma pretiosum. Environ. Entomol.* 3:505–07
96. Stinner, R. E., Ridgway, R. L., Morrison, R. K. 1974. Longevity, fecundity, and searching ability of *Trichogramma pretiosum* reared by three methods. *Environ. Entomol.* 3:558–60
97. Szentesi, A., Jermy, T., Dobrovolszky, A. 1973. Mathematical method for the determination of sterile insect population competitiveness. *Acta Phytopathol. Acad. Sci. Hung.* 8:185–91
98. Tanaka, N., Ohinata, K., Chambers, D. L., Okamoto, R. 1972. Transporting pupae of the melon fly in polyethylene bags. *J. Econ. Entomol.* 65:1727–30
99. Vanderzant, E. S. 1974. Development, significance, and application of artificial diets for insects. *Ann. Rev. Entomol.* 19:139–60
100. Webb, J. C., Sharp, J. L., Chambers, D. L., Benner, J. C. 1976. Acoustical properties of the flight activities of the Caribbean fruit fly. *J. Exp. Biol.* In press
101. Webb, J. C., Sharp, J. L., Chambers, D. L., McDow, J. J., Benner, J. C. 1976. The analysis and identification of sounds produced by the male Caribbean fruit fly, *Anastrepha suspensa. Ann. Entomol. Soc. Am.* 69:415–20
102. Weidhaas, D. E., LaBrecque, G. C., Lofgren, C. S., Schmidt, C. H. 1972. Insect sterility in populations dynamics research. *Bull. WHO* 47:309–15
103. White, E. B., DeBach, P., Garber, M. J. 1970. Artificial selection for genetic adaptation to temperature extremes in *Aphytis lingnanensis* Compere. *Hilgardia* 40:161–92
104. Whitten, M. J., Foster, G. G. 1975. Genetical methods of pest control. *Ann. Rev. Entomol.* 20:461–76
105. Young, J. R., Snow, J. W., Hamm, J. J., Perkins, W. D., Haile, D. G. 1975. Increasing the competitiveness of laboratory-reared corn earworm by incorporation of indigenous moths from the area of sterile release. *Ann. Entomol. Soc. Am.* 68:40–42
106. Zehnder, H. J., Remund, U., Boller, E. F. 1975. Zur Markierung von *Rhagoletis cerasi* L. II. Indikatoraktivierungsanalyse. *Z. Angew. Entomol.* 79:390–98

Ann. Rev. Entomol. 1977. 22:309–31

FACTORS AFFECTING FEEDING BY BLOODSUCKING INSECTS

❖6132

W. G. Friend and J. J. B. Smith
Department of Zoology, University of Toronto, Toronto, Ontario, Canada M5S 1A1

In this review we attempt to present a critical assessment of current knowledge about physical and chemical factors that affect feeding by bloodsucking insects, emphasizing events that occur after tarsal contact. Such a review is timely because of an increasing number of reports on phagostimulatory substances in natural and artificial diets based on experiments in which the complexity and variability of the feeding response often may be overlooked.

The reader is referred to other reviews on the behavior, physiology, and morphology of blood-feeding insects concerning the following topics: general orientation and mouthpart morphology (3); phagostimulants (32, 38, 39, 75); regulation of feeding (6, 47, 48); evolution of blood feeding, general behavior and physiology, and orientation to host (23, 53); host-ectoparasite interactions (82, 105); structure of chemoreceptors (15, 97); nutrition (57); symbionts (14); digestion (49); feeding habits (24, 25, 26, 37); saliva of Hemiptera (79); and metabolism of blood meal (17).

The habit of blood feeding has arisen independently in a wide range of insects, sometimes within the same order (3). Differences in feeding mechanisms, sensory receptors, and host characteristics result in differences in the factors affecting feeding and in their interplay; however, there are also features common to the process. Dethier (23) suggests that feeding in mosquitoes is the culmination of a stepwise series of stimulus-response events, as shown in plant-feeding insects. This probably applies to blood feeders generally. The events include detecting the host, alighting on it, probing (active movement of the mouthparts towards the surface), piercing or penetrating (insertion of mouthparts into the surface), locating blood (either in a blood vessel or from a hemorrhage), taking up blood into part or parts of the gut, and ceasing feeding. The earlier events may be missing if the insect remains in contact with the host between meals. Factors affecting feeding may determine individual events in the sequence or may modify parts or all of it. The process may be distorted by isolating subsets of the sequence, as is done in many investigations by using forced presentation of food. While reviewing the literature we often found it difficult to determine the conditions under which experiments were performed. Because of the complexity of the response and the range of known and unknown factors, we feel it imperative to consider and report as many factors as possible in studies of blood feeding and to exercise extreme caution when generalizing.

We do not attempt to be encyclopedic; we restrict coverage to what we consider the more important and representative papers appearing before January 30, 1976, and we exclude some families of Diptera, the arachnids (mites and ticks), and some less-studied blood-feeding insects, such as the vampire wasps of British Columbia (85).

REDUVIIDAE

The most extensively studied blood-feeding hemipteran is *Rhodnius prolixus.* Feeding has also been studied in another reduviid, *Triatoma infestans* (75), as have stylet morphology and electrophysiology (9, 10, 64, 86, 87). The other group of blood-feeding bugs, represented by the bed bug, *Cimex lectularius,* has received relatively little attention concerning feeding behavior and its stimulation (109) and is included below in the discussion of other insects.

The following description of the feeding act applies to both *R. prolixus* and *T. infestans.* The orientation and feeding process in *R. prolixus* has been described (113) as a "chain of reflexes," in which the initiation of the response is conditioned by the physiological state of the insect. The insect is aroused from a state of akinesis by the antennae responding to air currents or heat. It is then attracted to the host, natural or artificial, by heat (and perhaps odor) receptors in the antennae. Probing and piercing has been described in some detail (30, 72, 98, 99). During probing the tip of the labium is applied to the surface. The mandibles penetrate the surface, and the maxillae, which form the food and salivary canals, are then inserted in a series of thrusts and withdrawals during which saliva is continuously ejected from between their tips. After one or more samplings of the available fluid, ingestion of the diet commences in the presence of suitable phagostimulants. Feeding stops when a critical abdominal volume is reached; this is controlled by abdominal stretch receptors (2, 76) rather than by back pressure on the pharyngeal pump (8). In larvae, one meal of up to nine times their body weight is taken per moult cycle; adults take meals of up to three times their body weight each time they feed (29). The gut is simple; blood meals go directly into an expanded portion of the midgut (the crop). *R. prolixus* and *T. infestans,* both males and females, are exclusively hematophagous under normal conditions.

Factors known or suspected to affect feeding behavior in *R. prolixus* include: (*a*) temperature gradients, carbon dioxide, odors, and visual stimuli (e.g. movement, vibration, the texture of the surface on which the insects are standing, and any free liquid on that surface), all of which affect the readiness to probe (28, 30, 32, 72, 98, 99, 101, 113, 114); (*b*) diet composition (e.g. the nature and concentration of phagostimulatory nucleotides), osmotic pressure, pH, ion species and concentrations, which determine whether or not gorging occurs (28, 31, 32, 33a, 100, 101, 101a); (*c*) length of time since previous meal (33); and (*d*) degree of abdominal stretch (2, 76).

Heat alone incites short-range orientation, probing, and salivation. Parts of rearing jars that have been warmed by recent handling attract *R. prolixus,* which probe and salivate at such spots (30, 113). The heat response is affected by the gradient of air temperature, rather than by the radiant heat emitted from the source of the stimulation (114). Insects whose eyes had been covered with paint respond somewhat more slowly to the heat stimulus but nonetheless can successfully orient and probe. When the proboscis is cut off near the base and the animal is exposed to a heat source, the insect orients in the usual way, although when it reaches the stage where the proboscis would normally be extended it becomes "confused and agitated" and soon stops trying to probe (113). Insects with one antenna removed orient to heat and probe normally although the response time is greatly extended. Complete antennectomy blocks the orientation and probing response (113) and feeding on guinea pigs (64). "The antennae are the chief and perhaps the sole sense organs concerned in locating the source of the stimulus" (113). Antennae were oriented toward odors produced by a fresh mouse skin at ambient temperatures, but very few of the insects approached the skin and none probed. Adult insects showed an obvious preference for skin-covered heat sources over chemically neutral sources at 35° to 38°C; first-instar larvae that had been deprived of food for 10 days showed no such preferences. Thus odor plays some role in orientation, but under certain circumstances it can be completely masked by the heat stimulus (113). Moisture evokes no attractant or repellant responses when presented to insects under conditions where the tarsi did not contact the moisture (113). When free moisture comes in contact with any of the tarsi during feeding, as sometimes happens when the insects feed on artificial diets through a rubber membrane, the feeding pattern is immediately broken off and the insects leave the membrane (W. G. Friend and J. J. B. Smith, unpublished data).

The mouthparts of *R. prolixus* and *T. infestans* are similar; both pierce and feed in a similar manner (4, 5, 9, 30, 64, 72, 98, 99). *R. prolixus,* with the terminal 0.25 mm of the maxillae removed, feeds both through a rubber membrane on artificial diets containing phagostimulants and on a rabbit (32; Friend and Smith, unpublished data). Removing the last segment of the labium prolongs the time taken to probe and pierce guinea pigs, but afterward they gorge normally (64). Although both maxillae and mandibles are innervated, their sensory nerve endings appear to be mechanoreceptive and there is no evidence that they contain chemoreceptors (10, 86, 87). Consequently response to phagostimulants must be from sensilla internal to the food canal, probably the eight peglike sensilla located in the epipharynx (32, 64).

The response to phagostimulants under our test conditions is usually an all-or-none phenomenon; partial engorgement rarely occurs even if the potency of the particular chemical being tested is low or the solution is very dilute (31, 101). Barton Browne (6) suggests three explanations for this: (*a*) that the relevant receptors respond in an all-or-none manner to stimulation depending on the threshold of response; (*b*) that adequate stimulation triggers CNS activity, which then proceeds independently of the stimulation; and (*c*) that although the level of CNS activity is independent of stimulus magnitude, continuous excitatory input is required. In experiments in which an ATP-containing diet was rapidly replaced during feeding by saline lacking ATP, we found that the presence of this gorging stimulant was necessary only during the first 2 min for complete engorgement to occur (33a). This result supports Barton Browne's second suggestion, i.e. that a phagostimulant triggers the CNS, although other signals, arising for example from the ionic or osmotic properties of the diet, may take over to provide the excitatory input required by the third scheme. The concept of stimuli acting as triggers to release the next phase of behavior is supported further by our data that a warm (37°C) diet chamber may be replaced by a chamber at ambient temperature once probing has commenced; as long as the insect kept its proboscis extended during the exchange it continued to probe, pierce, and show the normal gorging response to ATP solutions. Thus, a temperature differential appears necessary only to initiate probing and is unnecessary afterwards (W. G. Friend and J. J. B. Smith, unpublished data). It should be pointed out, however, that even with diets that do not normally induce gorging, such as 0.15 M NaCl, *R. prolixus* repeatedly samples the diet and thus imbibes a definite but variable amount of liquid (30, 98). This has proved useful in our recent experiments in that weight gain can be used to distinguish between insects that probed and tasted and those that did not. Sometimes the amount consumed in this way becomes an appreciable fraction of a maximum meal, even though electrical monitoring shows that this type of behavior is different from that of a steadily pumping insect feeding to engorgement. Thus the degree of engorgement that determines a positive response must be chosen with caution. Also lately we have found that as many as 20% of the test insects feeding on diets such as human blood plasma take less than complete meals. The feeding response may be more complex, even in this regard, than previously assumed.

R. prolixus apparently responds to a wider range of nucleotide phagostimulants than the other insects studied (28, 31, 32, 38, 39, 101, 101a). In descending order the potency of compounds tested are ranked A(TETRA)P $>$ ATP $>>$ deoxyATP $>$ CTP $=$ ADP $=$ GTP $\geq$ CDP $\geq$ ITP $>$ cAMP $=$ UTP $\geq$ deoxyADP $\geq$ IDP $\geq$ GDP $>>$ AMP. Potencies range from 3.2×10^{-6} M for A(TETRA)P and 3.8×10^{-6} M for ATP to 0.63×10^{-3} M for AMP; these values are doses that elicit 50% gorging (ED_{50}) in test populations (101). Large numbers of insects were tested to establish this order, usually a minimum of 27 insects per dose with four to six doses per compound. This ranking of potencies, which is roughly correlated with the degree of departure in molecular configuration from ATP, has led to a model for interaction between stimulant and receptor protein involving principally the NH_2 group on the C–6 position of the adenine moiety, the terminal phosphates, and the OH group on C–2 of the ribose moiety (101).

Several factors affect the real or apparent potency of such gorging stimulants, including test procedures, period of previous food deprivation, and other dietary components. The criterion used in our studies was engorgement, determined by visual inspection, to at least 75% of an average maximum meal after 30 min of exposure to the test diet; however, this depends on engorgement being an all-or-none phenomenon (see above). Since probing and piercing preceed gorging but are not necessarily followed by it, the fraction engorged may be artificially low if not all insects probe. Readiness to probe seems to be affected by a variety of factors, including chamber design, ambient light, sound, vibration, animal odors, and tobacco smoke. Consequently rigidly controlled and uniform conditions of rearing and handling were employed in our studies on *R. prolixus* (28, 31, 33, 100, 101, 101a). In common with many other insects, *R. prolixus* shows a sensitivity to gorging stimulants that increases linearly with the length of food deprivation. Thus, the period between the previous meal and testing is a critical variable in experimental design. The sensitivity change is not a decreasing discrimination since there is no appreciable increase in the numbers that gorge on the control saline. Water loss is not likely to be the cause of this sensitivity increase since insects kept in a dessicator for 3 days prior to testing (rather than in the normally humid incubator) showed no difference in sensitivity to ATP (33). The nature and amounts of other dietary components also can have a significant effect on apparent sensitivity to particular phagostimulants. The drugs theophylline and caffeine enhance the sensitivity to ATP two- or fourfold, which suggests cAMP involvement in the chemoreceptive mechanism (100). *R. prolixus* feeds on diets containing ATP but not on those without it, even when 0.15 M NaCl is replaced by isosmotic concentrations of KCl, sucrose, mannitol, sorbitol, LiCl, NaBr, KBr, NaI, and KI; in all cases, however, the ED_{50} of ATP is raised. These findings contradict those of Salama (94), who found a median threshold of rejection for KCl, for example, at 0.34 M by using 10^{-3} M ATP as a phagostimulant. Although he claims osmotic effects are an unlikely factor, the KCl in fact was added to 0.15 M NaCl, which gives a total osmolarity of approximately 520 mosmoles at his median rejection threshold. We found that osmolarity has a significant effect at such levels, which when combined with the higher ED_{50} of ATP seen in KCl may well explain these results (32, 100; W. G. Friend and J. J. B. Smith, unpublished data). Replacing NaCl by $CaCl_2$ or $MgCl_2$ inhibits feeding entirely (Friend and Smith, unpublished data). The effect of lower concentrations of Ca^{2+} is more subtle. We found that the ED_{50} for ATP increases from about 3.8×10^{-6} M to about 5×10^{-5} M when it is dissolved in insect or mammalian Ringer's rather than in 0.15 M NaCl. This was found to depend on Ca^{2+}. The most likely explanation for the apparent sensitivity decline is our recent finding that *R. prolixus* possesses a potent Ca-dependent ATPase in its saliva. Since saliva is injected into the diet prior to sampling, and probably mixes with the diet during sampling (30), the level of ATP might be appreciably reduced by the time it reaches the epipharyngeal sensilla in Ca-containing diets. The salivary ATPase may also be involved in the synergistic effect seen when combinations of nucleotides are tested (J. J. B. Smith and W. G. Friend, manuscript in preparation). The potencies of mixtures of ATP with deoxyATP or UTP were significantly higher by a factor of 1.25 to 1.9 than predicted by an additive effect; this slight synergism may be explained by competitive inhibition of the salivary ATPase. We found that phosphate ions in 0.15 M NaCl can elicit gorging, although the effect was highly variable. Since phosphate is sometimes phagostimulatory at 10 mM, its use in buffer test diets is obviously hazardous! The response to ATP is also modified by pH and osmolarity; optimal responses occur around pH 7.0 and with solutions isosmotic with blood (Friend and Smith, unpublished data).

MOSQUITOES

The feeding response of mosquitoes is complex. Females probably feed more often on nectar than on blood (24); in addition, females and males also drink water (84). Blood feeding, at least outwardly, differs from nectar feeding. Feeding on blood follows landing on the host, although sometimes penetration may be effected before landing (59). The tip of the proboscis, which comprises labium, labrum, hypopharynx, mandibles, and maxillae, is applied to the substrate, bringing the sensilla of the labellar lobes in contact with it. With alternating movements of the terminally serrated maxillae and mandibles, the fascicle (stylets minus labium) penetrates the host substrate. When a blood vessel is pierced or an adequate pool of blood from damaged capillaries is located, the blood is imbibed by action of the cibarial and pharyngeal pump muscles via the food canal formed by the labrum and hypopharynx. Blood then passes into the midgut and in some species also into the crop or diverticula (108). Feeding is terminated by input to the CNS from abdominal stretch receptors (51). When feeding on sugar solutions presented as free drops or on cotton pledgets, the proboscis is extended toward the sugar solution seconds after the tarsi have contacted it. When the labellar hairs touch the solution the labellar lobes open, bringing the ligular hairs into contact with it. The solution then is imbibed, again by the activity of the two pumps but without extending the fascicle beyond the proboscis sheath. The meal is directed mainly to the crop rather than the midgut. Water is imbibed similarly (19, 83, 108). We have been unable to find detailed descriptions of feeding on natural sources of sugars such as flowers.

The distinction between blood and sugar feeding is by no means always clear, particularly under many investigative conditions. This poses a problem in examining factors that affect blood feeding in mosquitoes. We may, however, list the following: temperature, olfactory, and visual signals are involved in locating the host (53); heat is involved in probing (19) although CO_2 (16) and moisture (16, 63) are probably cofactors; chemical and mechanical stimuli to the tarsi and proboscis affect probing (93); chemical and osmotic properties of the diet determine activation of pumping and destination of the meal (21, 36, 40, 44, 56); age and state of food and water deprivation and the nature of its previous meal affects the readiness to feed and the sensitivity to test diets (27, 61, 62, 77, 96, 102, 111), as does the state of egg development (27); and abdominal stretch determines the cessation of feeding (51).

Both blood and sugar feeding can be considered as temporal sequences of events, from initial seeking of, or response to, a source of food until final cessation of feeding. The simplest way to regard these sequences is as a series of isolated stimulus-response elements, with the output of each event becoming part of the input for the next. This view is implicit in many investigations, which often ignore the possible effect of earlier components of the sequence or any overall modification by factors, such as diurnal rhythms, state of food and water deprivation, etc. Accepting this description for a moment, however, an important question can be asked: are the two sequences different at all times or are blood and sugar feeding merely overlapping subsets of one process, feeding? It is generally assumed that blood ends up in the midgut and sugar solutions in the crop, and that the destination of the meal is solely controlled by chemical signals in the diet, which is independent of the mode of feeding (21, 39, 56). This implies that the two processes coincide at the level of fluid intake, at which point a switching mechanism diverts the food to either of two destinations, dependent only on the nature of that food. Several studies have provided evidence for the independence of this switching mechanism. In *Aedes aegypti,* blood, or red cells in saline, pass to the midgut whether they are imbibed through a membrane or from an open drop, and sugar solutions go mainly to the crop in both cases, although they are seldom imbibed through a membrane (11). Trembley (108), in a study involving nine different species, came to similar conclusions; however, she also found that species differed in terms of the destination of the meal, e.g. 77% of 96 *Anopheles quadrimaculatus* and 54% of 183 *Culex pipiens,* biting humans, had blood (an unreported amount) in the crop, whereas only one of 174 *A. aegypti* had even a partial distension of the crop. Similarly glucose solutions, although they almost invariably enter the crop, are also found in the midgut with a frequency that varies between species. She points out that " . . . one cannot arbitrarily group these insects and refer to reactions or responses of 'mosquitoes'."

Although these results support the idea that the destination of the meal is independent of the mode of feeding, at least in the species tested, it is still possible that the decision mechanism may be biased by the insect having started in a blood-feeding or sugar-feeding mode of behavior. The bias may take the form of elevated or reduced thresholds to phagostimulants, or of changes in the relative amounts needed to activate the switch. Unfortunately, studies on the dependence of the switching mechanism on the feeding mode did not investigate the response to different mixtures of sugar and blood or blood components (11, 108), and studies in which diet proportions varied only used one mode of feeding (21, 56). By using *Culiseta inornata* we recently demonstrated that at critical concentrations of sucrose and ATP the destination of the meal is affected by the temperature of the diet (Friend and Smith, manuscript in preparation). Thus, extreme care should be taken when attempting to compare the relative effects of dietary phagostimulants presented to different species under different conditions, with different criteria of response.

Information on other aspects of the feeding sequence is even more scanty and confusing. Too few experiments have been carried out on the same species under adequately controlled and reported conditions to know, for example, whether the amount of probing depends on whether the diet is presented in a cotton pledget or under a membrane, and whether the acceptance threshold of different diets is thereby affected. We are forced, therefore, to conclude that no clear distinction can be made at present between blood feeding and nectar or sugar feeding, and so any distinction is made arbitrarily according to the particular circumstances of the experiment.

Feeding starts with the location of a food source. Hungry *A. aegypti* are stimulated into locomotory, orientation, and probing activity by ethyl-ether-soluble honey odors (111), and olfaction probably plays a role in the location of nonblood sources. There have been an enormous number of studies on the stimuli involved in the initial orientation of mosquitoes to animal hosts. In 1957, Dethier (23) optimistically suggested that the pieces of this complex puzzle were beginning to fall into place and identified four principles: the importance of "internal drive"; the heirarchical nature of stimuli and responses; the "non-additive interaction of combined stimuli"; and the relative strength of the stimuli. Hocking (53), however, feels that "rarely has so much work yielded so little consensus of opinion." He identifies three main reasons for this: variations in behavior among taxa at all levels, including subspecies; variation in physiological state caused by nutritional conditions, environmental factors, stimulus history, and internal cycles and rhythm; and inadequate knowledge of sensory physiology.

Comments similar to Hocking's can be made about studies on the various phases of feeding after tarsal contact. Active probing at the host substrate must preceed the blood-feeding response, at least with unrestrained insects. The factors affecting probing once the mosquito has landed appear to include thermal gradients, humidity, and CO_2, as well as the state of hunger or dessication. *A. aegypti* lands on and probes at a plastic-wrapped arm only in the presence of slightly elevated CO_2 levels (16); even then probing activity varies over a 3- to 4-day cycle. This species probes at wet filter paper optimally at 34° to 36°C but not at dry paper at the same temperature (63). Water-starved insects probed more avidly at warm and moist artificial targets than did water-satiated insects (61); water deprivation induced probing in previously blood-fed *A. aegypti,* which do not normally probe (62). The readiness with which different species probe on a target is variable and can be affected by the presence of other mosquitoes (93). Heat may not always be necessary (59); however, all membrane feeding experiments used a warmed diet, and Davis & Sokolove (20) believe that for *A. aegypti* thermal stimuli are most important. They describe heat and cold receptors on the antennae of females that could respond to the thermal gradients from a human arm. Receptors responding to water vapor and CO_2 have also been described (60). Probing and/or piercing can also be affected by the type of membrane used in artificial feeding experiments. Natural membranes, such as chick skin and Baudruche membrane, allowed greater feeding responses in female *Aedes aegypti* and *Anopheles stephensi* than did Saran wrap or Parafilm (93); in these experiments probing was not distinguished from piercing and feeding, and the reasons for the differences in feeding response are unspecified. Considering the variability and complexity of the probing response, it seems necessary at least to distinguish lack of probing from lack of feeding after probing in studies of various factors, such as phagostimulants, a condition that is often not met. One way of eliminating this problem appears to be to present the test diet forcibly to the proboscis of a restrained insect (56, 84), but, as Hocking (53) points out, there is danger in starting in the middle of a sequence of reflexes.

Both acceptability and final destination of a diet, once it has been contacted, are affected by a number of factors. Upon probing, contact is made between the labellar lobes of the labium and the substrate, whether feeding on skin-covered surfaces or on a free liquid at ambient or above-ambient temperatures. Chemoreceptors on the labella are sensitive to sucrose and may initiate the next steps of sugar feeding in *Culiseta inornata,* but they are insensitive to blood in this species (71, 83). Diet does not contact these receptors during feeding through a membrane, however (93), although Galun speculates that some materials may diffuse through Silverlight or Baudruche membranes and can be detected by tarsal chemoreceptors (36). During membrane feeding the chemoreceptors on the labrum or internal to the food canal are exposed to the diet (73, 110). There is disagreement about the ability and role of labral chemoreceptors in detecting blood. A correlation exists between the blood-feeding habit and the presence of apical labral chemoreceptors (73, 110). There is experimental evidence both for (11, 95) and against (56, 83) blood stimulating the labrum; however, different species and different methods of stimulus presentation were used.

Because they are in an appropriate position to monitor the composition of the food, it is generally agreed that the cibarial chemoreceptive sensilla play a major role in determining the acceptability and destination of a meal; however there is little direct evidence for this (73, 83, 95, 110). Phagostimulants associated with blood, presumably detected by cibarial sensilla in mosquitoes, have been investigated (11, 21, 36, 40, 44, 56, 84, 93) and recently reviewed (32, 38, 39). ATP and other adenine nucleotides act as phagostimulants for *Aedes aegypti* (36, 40) and *Culex pipiens* (56) but possibly not for restrained *Culiseta inornata* (84). The average amount of nucleotide solution consumed by the latter species was similar to that of water and less than that of blood or sucrose solution. Whereas Galun (39) explains this apparently anomalous result by stating that the nucleotides in this study were dissolved in water and not in isotonic saline, and that positive responses can be obtained only in solutions isotonic with blood, Owen & Reinholz (84) state that these nucleotides were dissolved in Tris buffer (molarity unspecified), pH 7.4, and that similar results were obtained by using 10^{-2} M ATP dissolved in 0.15 M NaCl. It should be noted that test diets were presented at room temperature to restrained, water-deprived females, that the destination of the meal is unreported, and that although under these conditions blood was accepted (as was sucrose), no numbers or proportions gorging on blood or sucrose are given. Another possible explanation is that the nucleotides were presented at the high concentrations of 10^{-2} M and 5×10^{-3} M. These authors themselves point out that Hosoi (56) found 10^{-2} M AMP to be less acceptable to *C. pipiens* than 10^{-3} M AMP; the response of *A. aegypti* to ATP also declines at levels greater than 10^{-2} M (93). Current work in our own laboratory has shown that unrestrained *C. inornata* feeding through a Silverlight membrane do in fact gorge on warmed diets of 10^{-3} M ATP in Ringers solution, with the diet going to the mid-gut. Ringers without ATP results only in uptake of very small samples of diet (Friend and Smith, manuscript in preparation).

Galun (39) reports that, in *A. aegypti* fed through a membrane on warm isotonic diets, 50% feeding is induced by 1×10^{-3} M AMP, 5×10^{-4} M ADP, and 1×10^{-4} M ATP. Her criterion for feeding is the presence (an unreported quantity) of dye in the midgut. This conclusion does not appear to be supported by the data presented in her two original tables (36, 40). In 1963 it was reported that, at 10^{-3} M concentrations, 31% gorge on AMP, 17% on ADP, and 58% on ATP; at 10^{-2} M concentrations, 70% gorge on AMP, 76% on ADP, and 85% on ATP (40). Assuming a linear relation between response and log dose, this would give estimates of ED_{50} (effective dose for 50% response) for AMP, ADP, and ATP of approximately 3×10^{-3} M, 3.6×10^{-3} M, and 5×10^{-4} M, respectively [these values are all substantially higher than those stated by Galun (39)], and the compounds rank ATP $>>$ AMP $\geq$ADP. In a later paper she reports responses for similar doses of the same nucleotides (36). Making the same assumption as previously, we calculated that these give ED_{50} values of 3.8×10^{-3} M (AMP), 3×10^{-3} M (ADP), and 5×10^{-4} M (ATP). In this case the ranking is as reported in Galun's reviews (38, 39), but our estimates again are substantially higher than those she reported. Because the data presented for AMP and ADP appear so similar in the two earlier papers (36, 40) but are listed differently, some of the discrepancies may be the result of reporting errors. Of other nucleotides tested, A(TETRA)P appears somewhat more potent than the other adenine nucleotides, and the nonadenine nucleotides had little or no stimulatory power at the doses tested.

Culex pipiens gorged on adenine nucleotides in 0.15 M NaCl when warmed solutions were forcibly presented to the fascicle of restrained insects (56); however, the difference between responses to ATP, ADP, and AMP was small at 10^{-4} and 10^{-5} M, the two principal doses tested. Although the response to ATP was slightly lower at both doses than that to ADP or AMP, little can be concluded about relative potency. Although ATP has been used in diets presented to several other species of mosquito (93), the only clear-cut effect was seen in *A. aegypti,* which was induced to gorge on otherwise unacceptable bovine serum.

Thus, our knowledge of phagostimulatory potencies of nucleotides for mosquitoes is scanty and confusing, and it is premature to conclude that adenine nucleotides stimulate feeding in mosquitoes in the order ATP $>$ ADP $>$ AMP, as does Galun (38, 39).

The feeding of *A. aegypti* is affected by solutes other than nucleotides in artificial diets (36, 40). Lower percentage feeding resulted if NaCl was replaced by isotonic KCl, $CaCl_2$, $MgCl_2$, glucose, sucrose, and lactose in solutions of 10^{-2} M AMP. Sucrose alone produced no appreciable feeding through a membrane, a result consistent with that of Bishop & Gilchrist (11). Feeding was inhibited in *A. aegypti* by 2 mM Zn^{2+}, Ni^{2+}, and Cu^{2+} in 10^{-3} M ATP with 0.15 M NaCl, but not by 2 mM Ca^{2+}, Mg^{2+}, Co^{2+}, and Mn^{2+} (36), a finding used to support a hypothesis that ATP stimulates the receptors by removing Zn^{2+} from the receptor surface by chelation (36). These results, however, do not hold for the tsetse fly *Glossina austeni* (42); furthermore, other hypotheses can explain the same results (e.g. the receptor site may recognize the different ATP chelates). Given these two observations, and the limitations in the potency studies discussed earlier, further evidence is needed for this hypothesis. A large number of electrolytes have been tested for acceptance or rejection by restrained *A. aegypti* (94). Most were rejected; osmolarity was not controlled in this experiment, however. The osmotic concentration of diets seems to affect blood feeding (diet taken into the midgut) in *Culex pipiens* (56) but not feeding on sucrose (diet taken into crop). Galun (36) also found an effect of ionic concentration on *A. aegypti;* low NaCl concentrations produced less feeding. Owen & Reinholz (84) found an appreciable intake of water in thirsty *Culiseta inornata;* however, in this case the type of feeding appears to determine the importance of osmolarity in that the response to blood or nucleotides is affected by osmolarity more than the response to sugars or water. Although this may not be a surprising conclusion, it does highlight the importance of controlling osmolarity, and the nature of the osmotic particles, in studies on blood feeding.

Food deprivation is a factor that can influence food seeking and feeding. In *A. aegypti,* food deprivation lowered both the threshold of the orientation and probing response to odors associated with honey (111) and the acceptance threshold for sugars, while increasing the tolerance to electrolytes (96). Food deprivation for up to 5 days did not affect the avidity with which these insects fed on the shaved abdomen of a guinea pig (102). *Culex tarsalis,* deprived of sugar and water for 24 hr before being fed on human blood through natural membranes, fed better and ingested more blood than nondeprived insects (104). In *C. inornata,* food deprivation for 3 days resulted in maximum sensitivity to sucrose, indicated by a spreading of the labellum (77). In *Culex nigripalpus,* females that were interrupted before they had consumed half or less of a full meal usually attempted to refeed on a human forearm when it was offered within 6 hr after the first meal; fewer refed after a 12- or 24-hr period. The initiation of egg development caused a decline in feeding. The feeding response in partially fed individuals seems to be all or none (27).

The cessation of feeding in many mosquitoes is controlled by stretch receptors in the abdomen. Cutting the ventral nerve cord of adult female *Aedes aegypti, Aedes taeniorhynchus, Aedes triseriatus, Armigeres subalbatus, Culex pipiens fatigans,* or *Anopheles quadrimaculatus* anterior to the second abdominal ganglion causes massive hyperphagia. As the site of incision was moved posteriorly the degree of hyperphagia decreased (51). Manjra (77) believes that in *Culiseta inornata* the labellar threshold of response to sucrose is regulated by body wall stretch receptors in the abdomen acting via the recurrent nerve and the ventral nerve cord.

TSETSE FLIES

The feeding process in tsetse flies may be summarized as follows: the fly lands, exudes a small drop of saliva from its mouthparts, and then applies the haustellum to the surface (probing). The labellar lobes on the end of the labium evert, exposing the prestomal teeth and the sensilla on the inner surfaces of the labella. The surface is pierced by rapid movements of the labellar lobes. Upon penetration of a suitable blood vessel and active pumping of the cibarial and pharyngeal pumps, blood is taken up through the food canal formed by the labrum and the hypopharynx (22, 50, 58). The blood first fills the midgut and then enters the crop (81, 112). Feeding stops when a critical abdominal volume is reached (81, 90, 107). Certain factors are believed to affect this sequence of events: stimulation of the heat receptors on the antennae and tarsi of the prothoracic legs (22, 88); thickness and texture of the substrate (22, 68, 69, 78, 91); chemical and osmotic characteristics of substrate and diet (39, 41–43, 68, 70; B. K. Mitchell, personal communication); the previous history of the fly, i.e. wild or colony reared, whether it had previously fed or not, and the nature of the previous host (animal or artificial feeding); the age of the fly (68; B. K. Mitchell and H. A. Reinouts van Haga-Kelker, personal communication); the amount of food already present in the gut and the fly's reproductive state (81, 107); and the time elapsed since the previous meal (12, 13, 22).

Some flies probe and even pierce and salivate when exposed to a chemically neutral surface (bond paper) at ambient temperature, although the fraction responding varies with species from 22.7% probing and 4.5% piercing in *Glossina palpalis* to 14.4% probing and none piercing in *Glossina brevipalpis* (22). Smoother surfaces or surfaces that entangled the tarsi reduced this response. Dethier (22) claims that heat is the most effective factor for inducing probing and piercing in the three species of *Glossina* he studied, although the maximum response to either a finger or a paper-wrapped tube at 38°C was less than 50%, and a minimum difference of 14°C between ambient and surface temperatures was necessary to elicit appreciable responses. Antennectomy reduced the biting response of *G. palpalis* (either probing or piercing), although about 33% still probed or pierced. Since flies with antennae removed did not feed on guinea pigs, whereas removal of palps had no effect on feeding, it would appear that the heat receptors are mainly in the antennae (22). Antennectomy caused a 43% reduction in feeding frequency (feeds per fly per day) in 29 *Glossina morsitans,* but these flies did gorge often enough on rabbits to maintain the mean weight of pupae produced at the average normal level (68). The probing response of *G. morsitans* to an artificial feeding apparatus covered with an agar-Parafilm membrane and warmed to 37°C was reduced almost to zero by antennectomy; however, when the antennae were present removal of the prothoracic legs also reduced the probing response, an effect greater in teneral flies than in post-teneral flies. Electrophysiological techniques revealed the presence of temperature receptors on the tarsomeres of the prothoracic legs (88). From these results it appears that heat plays an important role in the feeding responses of *Glossina,* principally via heat receptors on the antennae, and that it acts as an incitant, initiating probing. However, other factors modify this heat response, and the interpretation of these experiments is further complicated by the use of different species of *Glossina* in different physiological states. Flies caught in the wild, such as those used by Dethier (22), behave differently from laboratory-reared flies (68). *G. morsitans* previously fed through artificial membranes gave a significantly lower response to rabbits than did insects previously reared on rabbits (B. K. Mitchell and H. A. Reinouts van Haga-Kelker, personal communication). The conditions under which responses were tested varied: some insects were tested tethered (22) and others free (68), some were tested singly (22) and others in groups (68; B. K. Mitchell and H. A. Reinouts van Haga-Kelker, personal communication). The criteria used for a positive response may also affect conclusions: for example, biting (probing and/or piercing) (22) always precedes but does not necessarily imply gorging, and thus results using these two criteria can be expected to differ.

The role of odor in tsetse fly feeding is unclear. Dethier (22) obtained no biting reactions to guinea pig or human odors with wild-caught flies of unknown previous history, whereas B. K. Mitchell and H. A. Reinouts van Haga-Kelker (personal communication) suggest that odors may be one of the clues enabling their post-teneral, laboratory-reared *G. morsitans* to distinguish between rabbits and artificial membranes. They note that clues such as odor may only be important after the fly has been fed a number of times on the same host, showing that the relative importance of feeding signals may change with time and/or physiological condition within any one species.

The process of piercing the host surface and locating blood involves a number of factors that may influence the overall feeding response. Adult tsetse flies have at least eight types of sensilla on the labium (91) and four on the labrum and cibarium (92). In contrast to *Phormia,* which has predominantly chemoreceptive sensilla on the labium (54), *Glossina* has mainly mechanoreceptors. Electrophysiological recordings indicated that mechanoreceptive sensilla on the shaft of the labium signal the depth to which the labium has been inserted (91). Tsetse flies feed optimally from artificial membranes more than 40 μm thick and this discrimination probably involves the same receptors (68, 69, 78). Labral chemoreceptors and mechanoreceptors are thought to monitor and control food and saliva flow and, together with cibarial receptors, control operation of the cibarial pump (89, 92).

Chemical signals may help guide the piercing labium to adequate sources of blood (92) and are important in inducing engorgement (39, 41, 42, 43, 68). The labellar lobes of *Glossina austeni* possess eight chemoreceptive sensilla, each innervated by three neurons, one mechanoreceptive and two chemoreceptive (91). *G. morsitans* has similar sensilla. Mitchell (80, 80a) has shown electrophysiologically that one neuron of each of these sensilla in *G. morsitans* responds to ATP and related nucleotides; their sensitivity characteristics match those of behavioral responses to these nucleotides in *G. austeni* (41; B. K. Mitchell and H. A. Reinouts van Haga-Kelker, personal communication). It is not clear whether they actually mediate the gorging response. There are chemoreceptive sensilla in the cibarium internal to the food canal that are believed to control whether or not fluid sucked into the cibarial region passes into the gut (92), and it has been suggested that the labellar receptors instead control the piercing action, which tends to lead the labium toward rich ATP sources (92). Such a source could be blood platelets activated by wounded blood capillaries. Thus nucleotides, primarily ATP, probably act as signals in two phases of feeding, piercing and engorging. Two sets of receptors may be involved, and there is no a priori reason why they should have identical sensitivity characteristics. Consequently the particular way test diets containing nucleotides are presented may be expected to affect the response, as may the criterion of response if biting rather than engorging were chosen.

All tsetse flies studied responded positively to ATP as a gorging stimulant (39, 41, 42, 43, 68). However, the sensitivity to ATP may vary with species; male teneral *G. morsitans* are significantly more sensitive to 10^{-3} and 10^{-4} M ATP in 0.15 M NaCl, measured as percentage feeding, than are *G. austeni,* which was attributed to smaller fat reserves in *G. morsitans* at emergence (68). In this experiment the fraction of insects gorging was reported, but there was no report of the fraction not probing. Thus the lower feeding response (percentage gorging) may reflect a reluctance to probe rather than a lower sensitivity to ATP. ATP is a phagostimulant for *Glossina tachinoides* and *Glossina fuscipes,* but the relative effects of other nucleotides have not yet been studied (39).

Information on the relative potencies of nucleotides in eliciting gorging in tsetse flies is very scanty. According to several papers (38, 39, 41, 43) only adenine nucleotides stimulate, and these are ranked in potency: ATP > ADP > AMP. These conclusions were based on a study involving only 11 to 22 insects per test, however, and no report of variability was given.

Salts present in the diet also affect feeding. In one study (68) 33% of teneral male *G. morsitans* and 13% of *G. austeni* gorged on a solution of 0.15 M NaCl alone through an agar and Parafilm membrane; in another (42) 5% of teneral unsexed *G. austeni* fed on 0.15 M NaCl through either rat skin or Baudruche (bovine intestine) membrane. Salts may affect feeding through osmotic effects, although only a slight reduction in numbers gorging occurred when feeding on pure water and on 0.3 M NaCl; the stimulatory affect of ATP may override any osmotic effect (70). Rice et al (91) report receptors that respond to tsetse fly saline on the labella of *G. austeni.* The gorging response is not affected significantly by 2 mM concentrations of the metal ions Mg^{2+}, Co^{2+}, Mn^{2+}, Zn^{2+}, Ni^{2+}, and Cu^{2+}, although Ca^{2+} reduces the response slightly (42). Sensitivity to ATP is affected by pH; values below 7.0 inhibit both feeding (42) and discharge in the labellar ATP receptors (80a) with an optimum in both cases around 7.5.

The amount of blood that a tsetse fly takes in a normal meal is regulated. Tobe & Davey (107) found that before ovulation, female *G. austeni* feed to a constant weight of about 50 mg. After ovulation they feed to a constant weight of about 80 mg irrespective of the size of the developing larva; that is, the weight of the meal equals a constant minus the weight of the larva. This may also be interpreted as feeding to a constant volume or pressure. In males of *G. brevipalpis* the size of the blood meal varied with the state of food deprivation. Flies deprived of food for 1 day took 80 mg on the average, and flies deprived for 4 days took 115 mg. The amount of weight lost by the flies due to food deprivation was of the same order as the increase in meal size, and so again the flies seem to be feeding to a constant weight or volume (81). Increasing the viscosity of 10^{-3} M ATP in saline diet by adding various amounts of dextran caused a concomitant increase in the feeding times of tsetse flies; on all such diets tested the volume ingested did not vary significantly, which demonstrates that the meal was terminated when a critical volume had been taken, regardless of the duration of the feeding time (39). Proprioceptors measuring the amount of abdominal distension may provide the control that stops feeding. If the gut of a tsetse fly is punctured in a way that prevents the abdomen from being stretched by the blood meal, feeding is prolonged (90).

The hunger or dehydration state can change the internal physiological state of tsetse flies in a way that affects their inclination to bite; the biting response increased as the flies became hungrier until they reached the final stages of starvation (22). Recently Brady (12) found that the probing responsiveness of *G. morsitans* males to stimuli provided by a warm foam rubber ball increased linearly during a 4-day food deprivation period. Spontaneous activity and visual responsiveness increased exponentially for about 5 days after a blood meal, at which time premoribund decline set in. Both abdominal and total body weight changed in parallel with this behavior, which suggests that the flies behavioral thresholds are regulated by some internal monitoring of weight or abdominal volume (13).

OTHER INSECTS

Probing by the bed bug, *Cimex lectularius,* can be elicited by temperature differentials of 1–2°C (1). In another experiment, female *C. lectularius* would feed through mouse skin membranes at temperatures between 34° and 42°C, with an optimum at 37–38°C (7). The ambient temperature during these tests was not reported. Moisture in the feeding chamber inhibited feeding. Bugs fed readily on citrated rabbit blood, either whole or diluted with insect Ringer's 1:1 or 1:9, but less readily on blood diluted with distilled water or with added sucrose (0.05 g per 3 ml of blood). Some feeding was obtained on blood fractions such as washed erythrocytes, rabbit plasma or calf serum, homogenized milk, and insect or mammalian Ringer's. The best responses were to calf serum and Ringer's solutions, but the numbers tested (10 on each diet) were too small to draw any firm conclusions. A blood meal of a certain critical size is required to induce moulting (106).

Little is known about the mechanism of, and factors affecting, blood-feeding in simuliids (103). As in mosquitoes, the simuliid females feed on sugars (nectar) and water as well as blood; males are exclusively sugar feeders (24). Thus problems of separating blood feeding from sugar feeding undoubtedly apply equally in this group. Wild-caught female *Simulium venustrum* tested on artificial diets covered by a rubber membrane require heat to elicit probing (103). At the optimum temperature of the saline control diet of 37°C, 39 out of 80 insects probed. The amino acids serine, leucine, alanine, and proline were slightly repellent rather than phagostimulatory at 10^{-3} M in phosphate-buffered 0.15 M NaCl at pH 7.2. ATP at 10^{-4} caused 46 out of 52 flies probing to gorge, and at 10^{-5} M, 8 out of 50. ADP, somewhat more potent, caused gorging in 50 out of 53 probing flies at 10^{-4} M, and 26 out of 56 at 10^{-5} M. Under the test conditions, about half of the tested insects probed. Later work has shown AMP and adenosine to be phagostimulatory (J. F. Sutcliffe and S. B. McIver, personal communication).

The functional morphology of mouthparts and the blood-sucking habits of New World phlebotomine sandflies have been recently reviewed (74). The great diversity of mouthpart structure in this otherwise rather uniform group suggests interesting special adaptations to different hosts. The passage of blood may be monitored by labral and cibarial sensilla. Factors affecting blood feeding remain virtually unknown.

Tabanids also feed on both sugars and blood, although blood feeding is restricted to females. As in mosquitoes, blood feeding in natural conditions involves a set of sensory clues and actions different from those in sugar feeding. We were unable to find reports of experiments on phagostimulants that may be associated with blood feeding, in which insects were fed through a warmed membrane. Thus the effect of feeding conditions remains unclear. A tabanid orientates to a vertebrate or plant host by the use of visual clues, odors, temperature, and humidity gradients (67). After landing the fly uses its foretarsi to palpate a wide area; appropriate stimulation of tarsal receptors results in applying the proboscis to the substrate, spreading the labellum, and sucking. The labellar chemoreceptors can play a major role in triggering the sucking action, at least when the fly is exposed to sucrose solutions (67). The labellar chemoreceptors of *Chrysops vittatus, Tabanus lineola,* and *Hybomitra lasiophthalama* females are triggered to mediate a response only when the foretarsi contact a suitable substrate. The labellar thresholds for sucrose acceptance are lower than the tarsal thresholds in these insects (67) and in *Tabanus sulcifrons* (34). As well as being sensitive to water and to sucrose, glucose, and fructose, the labellar sensilla could differentiate between salt and sugar. Lall (65) tested the feeding response of females of the tabanid *C. vittatus* both by applying various solutions to the labella of hungry, water-satiated flies and by measuring feeding responses to test solutions containing methylene blue, which were presented on a cotton pad. The presence of dye in the gut after a 30- to 45-min test period was used as a feeding index. Under these test conditions the flies fed on whole blood only when it was prewarmed to 37°C, and then only partial meals were taken. This poor response suggested that "factors that evoke ingestion of blood are probably missing" (65). Amino acid solutions tested at 10^{-2} and 10^{-3} M in water and in phosphate buffer at pH 7.4 failed to stimulate the labellar chemoreceptors significantly, as did ATP and ADP at 10^{-2} and 10^{-3} M in phosphate buffer, pH 7.4. ADP or AMP at 10^{-2} or 10^{-3} M in water, phosphate buffer (pH 7.4), or 0.15 M NaCl solution failed to evoke feeding at levels higher than the controls. ATP at 5×10^{-3} M in phosphate buffer or at 5×10^{-2} M in saline caused four out of ten flies to feed. Since blood feeding did not occur under equivalent test conditions, and the number of insects tested was very small, very little can be concluded about the role of nucleotides as phagostimulants for blood feeding in this species (65).

Blood normally enters the midgut and sugar solutions and water enter the crop in *Chrysops vittatus* (65) and *Hybomitra lasiophthalma* (66); the destination probably is controlled by cibarial receptors responding to glucose or other sugars. This does not seem to be the case in the tabanids *Chrysops mlokosieviczi, Hybomitra peculiaris, Tabanus autumnalis,* and *Haematopota pallens* (18). Blood or dyed sucrose solutions or water were accepted from pipettes placed over the haustellum of restrained flies. All fluids accepted this way went first into the gut and then into the crop only when larger doses were consumed. It is suggested that the crop filling normally described results from a rapid absorbtion of nonblood meals from the midgut, thus leaving fluid only in the crop when examined later.

Among the Muscidae, blood, sugar, and water feeding occur in both males and females of stable flies and horn flies. These insects have highly modified mouthparts in which the hard pointed and toothed labium acts as a stabbing organ; the mandibles and maxillae largely are lost (3). They also differ from mosquitoes and tsetse flies in normally taking several meals per day (52). Factors that affect probing in the stable fly *Stomoxys calcitrans* have been investigated and summarized by Gatehouse (45, 46). During probing the extended proboscis is repeatedly and forcefully applied to the skin. Piercing is usually initiated only after a hair follicle or wrinkle in the skin is located. Water-deprived flies will probe a moist cotton-wool plug. Rapid increases in relative humidity or in substrate temperature, approximating the changes that are encountered when the fly lands on mammalian skin, induced probing in hungry, water-satiated flies. Olfactory stimuli from fresh blood, sweat, ammonia, butyric or valeric acid vapors, or CO_2 all failed to elicit probing. This contrasts with earlier findings (45). Surfaces at 35°C with low reflectance induced more probing than did those of high reflectance, whereas rough surfaces of low reflectance increased probing times but did not influence the numbers that probed. No interaction was shown between humidity or substrate temperature increases and olfactory stimulation by CO_2 or the odor of horse sweat. Although ammonia at unnaturally high levels (1.5 mg liter^{-1}) synergized the stimulatory effect of rapid increases in relative humidity, Gatehouse concludes that olfactory stimuli play no part in the induction or probing on the live host (46). Probing can be induced by stimulation of tarsal chemoreceptors with whole blood or blood components (55), and *S. calcitrans,* which gorges on plasma, can be stimulated (under unstated conditions) by ATP as well as by leucine but not by GMP or methionine (39). Consistent variability has been found between successive samples of *S. calcitrans* in spite of a very high level of standardization of materials and techniques (45). Such variability in studies of feeding behavior probably reflects the complexity of the intrinsic and extrinsic factors that control responsiveness. Factors listed are sex and age (the reproductive state of the female can influence the results), endogenous rhythms, nutritional state, and environmental conditions (45).

Little work has been done on horn flies. Harris et al (52), by using an electronic biteometer, showed that the feeding activity of *Haematobia irritans* confined in relatively large cages on a steer was quite different from that of similar flies exposed to citrated bovine blood on a pad. The females took more than twice as long to complete a feeding on the steer (30 min maximum) as on the blood-soaked pad (11 min maximum). Insects feeding on the steer spent ten times as long feeding and took four times as many meals each day as did those feeding on the blood pad. Males fed for shorter periods and took less blood than did females. These results show, yet again, how many different factors can influence the act of feeding.

Fleas are allied to Diptera but their mouthparts are highly modified compared to those of the flies (3). The adults are also obligate blood feeders. Thus, the factors affecting feeding can be expected to differ from, and perhaps be simplier than, those affecting mosquitoes. Our limited knowledge of factors affecting feeding stems mainly from a paper by Galun (35). The weight change in groups of from 100 to 300 fleas after a 1-hr exposure to diets at 37°C by using a membrane-feeding technique was the criterion of response. On diets of whole blood, washed erythrocytes in saline, or 5×10^{-3} M ATP in plasma or in 0.15 M NaCl, the percentage of weight gain ranged from 30 to 38%. Plasma, serum, or hemolyzed blood gave a maximum of 21% weight gain. The response to plasma was not increased by the addition of 5×10^{-3} M AMP, ADP, GTP, CTP, ITP, creatine phosphate, or glutathione. Saline alone (0.15 M NaCl) caused an 18% weight gain; 0.3 M sucrose, 21%; 0.3 M sucrose plus 5×10^{-3} M ATP, 22%. Only an 8% weight gain was shown when distilled water was used as the diet, and addition of 5×10^{-3} M ATP to this raised the weight gain to 10%. Unfortunately, it is not reported whether differences in weight gain were due to differences in the average amount consumed, or to differences in the proportions gorging on, as opposed to rejecting, the diet. Thus, it is difficult to evaluate Galun's suggestion (35) that the amount of intake may be controlled by "adaptation of the osmoreceptors." Indeed, no clear effect of osmotic pressure was seen; the difference between responses to isosmotic doses of NaCl and lactose does not support her contention that high osmolarity is itself a phagostimulant. Isotonic solutions of sodium chloride, bromide, iodide, nitrate, and acetate were accepted at about the 20% level; sodium sulfate, carbonate and citrate, and potassium chloride were rejected. Thus the ionic content of the diet may be monitored. Since the only nucleotide that appeared to be phagostimulatory is ATP, Galun concludes that fleas are more specific in their requirements for phagostimulants than other insects that have been tested (39).

CONCLUSIONS

The blood-feeding response of hematophagous insects, although often treated as a simple stereotyped sequence of events, parts of which can be isolated for the purposes of a particular study, is in reality a complex behavioral system that may be modified by a large number of factors acting alone or in combination. In many Diptera, blood-feeding is only one of perhaps three types of feeding, between which distinctions are unclear. Too often the importance of studies on blood feeding is lessened by inadequate control or reporting of factors that can affect the results or interpretation. Generalizations are made difficult by a lack of parallel studies on different species of hematophagous insects.

We are tempted, however, to make two generalizations. A temperature gradient or differential appears to be the prime signal inciting probing in all hematophagus insects studied so far. Although there are cases reported of mosquitoes feeding on poikilothermic animals, and mites and ticks are well known to feed on animals such as snakes, too little is known of the signals eliciting probing and feeding to determine whether or not this generalization holds up here. The second generalization is that nucleotides, particularly ATP, may be, or at least can mimic, the signal that indicates to a hematophagous insect the successful location of a blood source. Both platelets and erythrocytes contain high concentrations of ATP, but it is not clear which, if either, act as the natural source of this phagostimulant; there is also no reason why different insects might not differ in the type and origin of signals to which they respond by gorging on blood.

Several areas for profitable future research suggest themselves. How different are the blood- and sugar-feeding modes of behavior seen in mosquitoes, tabanids, and black flies? Does the feeding mode affect the sensitivity of the gorging response to different stimulants? Do the apparent exceptions to the generalization that ATP is a phagostimulant for blood feeding disappear when the test conditions are modified? Many insects naturally show a diurnal rhythm in feeding behavior; is this an important factor in laboratory studies? It is not yet known how a hematophagous insect locates a blood vessel or distinguishes between penetrable and nonpenetrable surfaces. The function of saliva is only partly understood in a few species.

These are but a few of the questions that may be asked. Blood feeding is a highly specialized and relatively rare mode of existence among animals, and yet hematophagous insects have an enormous impact on humans by affecting their health, economics, and enjoyment of the outdoors. Because of the special requirements of this feeding method, study on a variety of species can be expected to produce useful and valid generalizations as long as the complexity of this behavioral system is fully understood and appreciated.

Acknowledgments

We gratefully acknowledge the support of the National Research Council of Canada; the aid of colleagues who sent reprints, allowed us to use unpublished material, and constructively criticized the manuscript; the diligent and careful assistance of Mrs. R. Hewson; and the help of Mrs. C. Smith in preparing the manuscript.

Literature Cited

1. Aboul-Nasr, A. E. 1967. On the behaviour and sensory physiology of the bedbug (*Cimex lectularius*). I. Temperature reactions. *Bull. Soc. Entomol. Egypte* 51:43–54
2. Anwyl, R. 1972. The structure and properties of an abdominal stretch receptor in *Rhodnius prolixus. J. Insect Physiol.* 18:2143–53
3. Askew, R. R. 1971. *Parasitic Insects.* London: Heinemann Educ. Books. 316 pp.
4. Barth, R. 1952. Estudos anatômicos e histológicos sôbre a subfamília Triatominae. I: a cabeça do *Triatoma infestans. Mem. Inst. Oswaldo Cruz* 50:69–196
5. Barth, R. 1953. Estudos anatômicos e histológicos sôbre a subfamília Triatominae. III: pesquisas sôbre o mecanismo da picada dos Triatominae. *Mem. Inst. Oswaldo Cruz* 51:11–94
6. Barton Browne, L. 1975. Regulatory mechanisms in insect feeding. *Adv. Insect Physiol.* 11:1–116
7. Bell, W., Schaefer, C. W. 1966. Longevity and egg production of female bed bugs, *Cimex lectularius,* fed various blood fractions and other substances. *Ann. Entomol. Soc. Am.* 59:53–6
8. Bennet-Clark, H. C. 1963. The control of meal size in the blood-sucking bug, *Rhodnius prolixus. J. Exp. Biol.* 40: 741–50
9. Bernard, J. 1974. Mécanisme d'ouverture de la bouche chez l'hémiptère hématophage, *Triatoma infestans. J. Insect Physiol.* 20:1–8
10. Bernard, J., Pinet, J. M., Boistel, J. 1970. Électrophysiologie des récepteurs des stylets maxillaires de *Triatoma infestans*—action de la température et de la teneur en eau de l'air. *J. Insect Physiol.* 16:2157–80
11. Bishop, A., Gilchrist, B. M. 1946. Experiments upon the feeding of *Aedes aegypti* through animal membranes with a view to applying this method to the chemotherapy of malaria. *Parasitology* 37:85–100
12. Brady, J. 1973. Changes in the probing responsiveness of starving tsetse flies (*Glossina morsitans* Westw.). *Bull. Entomol. Res.* 63:247–55
13. Brady, J. 1975. 'Hunger' in the tsetse fly: the nutritional correlates of behaviour. *J. Insect Physiol.* 21:807–29
14. Brooks, M. A. 1964. Symbiotes and the nutrition of medically important insects. *Bull. WHO* 31:555–59
15. Buerger, G. 1967. Sense organs on the labra of some blood-feeding Diptera. *Quest. Entomol.* 3:283–90
16. Burgess, L. 1959. Probing behaviour of *Aedes aegypti* (L.) in response to heat and moisture. *Nature* 184:1968–9
17. Bursell, E., Billing, K. C., Hargrove, J. W., McCabe, C. T., Slack, E. 1974. Metabolism of the blood meal in tsetse flies. *Acta Trop.* 31:297–320
18. Chirov, P. A., Alekseev, A. N. 1970. To the physiology of feeding of the horsefly. *Byull. Mosk. Ova. Ispyt. Prir. Otd. Biol.* 75:60–67
19. Christophers, S. R. 1960. *Aedes aegypti (L.) The Yellow Fever Mosquito. Its Life History, Bionomics, and Structure.* New York: Cambridge Univ. Press. 739 pp.
20. Davis, E. E., Sokolove, P. G. 1975. Temperature responses of antennal

receptors of the mosquito, *Aedes aegypti. J. Comp. Physiol.* 96:223–36

21. Day, M. F. 1954. The mechanism of food distribution to midgut or diverticula in the mosquito. *Aust. J. Biol. Sci.* 7:515–24
22. Dethier, V. G. 1954. Notes on the biting response of tsetse flies. *Am. J. Trop. Med. Hyg.* 3:160–71
23. Dethier, V. G. 1957. The sensory physiology of blood-sucking arthropods. *Exp. Parasitol.* 6:68–122
24. Downes, J. A. 1958. The feeding habits of biting flies and their significance in classification. *Ann. Rev. Entomol.* 3: 249–66
25. Downes, J. A. 1974. The feeding habits of adult Chironomidae. *Entomol. Tidskr.* 95:84–90
26. Downes, J. A., Downe, A. E. R., Davies, L. 1962. Some aspects of behaviour and physiology of biting flies that influence their role as vectors. *Proc. Int. Congr. Entomol., 11th, Vienna, 1960* 3:119–21
27. Edman, J. D., Cody, E., Lynn, H. 1975. Blood-feeding activity of partially engorged *Culex nigripalpus. Entomol. Exp. Appl.* 18:261–68
28. Friend, W. G. 1965. The gorging response in *Rhodnius prolixus* Stahl. *Can. J. Zool.* 43:125–32
29. Friend, W. G., Choy, C. T. H., Cartwright, E. 1965. The effect of nutrient intake on the development and egg production of *Rhodnius prolixus* Stahl. *Can. J. Zool.* 43:891–904
30. Friend, W. G., Smith, J. J. B. 1971. Feeding in *Rhodnius prolixus:* mouthpart activity and salivation, and their correlation with changes of electrical resistance. *J. Insect Physiol.* 17:233–43
31. Friend, W. G., Smith, J. J. B. 1971. Feeding in *Rhodnius prolixus:* potencies of nucleoside phosphates in initiating gorging. *J. Insect Physiol.* 17:1315–20
32. Friend, W. G., Smith, J. J. B. 1972. Feeding stimuli and techniques for studying the feeding of haematophagous arthropods under artificial conditions, with special reference to *Rhodnius prolixus.* In *Insect and Mite Nutrition,* ed. J. G. Rodriguez, pp. 241–56. Amsterdam: North Holland. 702 pp.
33. Friend, W. G., Smith, J. J. B. 1975. Feeding in *Rhodnius prolixus:* increased sensitivity to ATP during prolonged food deprivation. *J. Insect Physiol.* 21:1081–84

33a. Friend, W. G., Smith, J. J. B. 1976. An apparatus for exchanging diets rapidly during artificial feeding of sucking insects. *Can. Entomol.* 108:In press

34. Frings, H., O'Neal, B. R. 1946. The loci and thresholds of contact chemoreceptors in females of the horsefly, *Tabanus sulcifrons* Macq. *J. Exp. Zool.* 103: 61–80
35. Galun, R. 1966. Feeding stimulants of the rat flea *Xenopsylla cheopis* Roth. *Life Sci.* 5:1335–42
36. Galun, R. 1967. Feeding stimuli and artificial feeding. *Bull. WHO* 36:590–93
37. Galun, R. 1971. Recent developments in the biochemistry and feeding behaviour of haematophagous arthropods as applied to their mass rearing. In *Sterility Principle for Insect Control or Eradication,* pp. 273–82. Vienna: International Atomic Energy Agency. 542 pp.
38. Galun, R. 1975. Behavioural aspects of chemoreception in blood-sucking invertebrates. In *Sensory Physiology and Behavior,* ed. R. Galun, P. Hillman, I. Parnas, R. Werman, pp. 211–21. New York: Plenum. 357 pp.
39. Galun, R. 1975. The role of host blood in the feeding behavior of ectoparasites. In *Dynamic Aspects of Host-Parasite Relationships,* ed. A. Zuckerman, Vol. 2, pp. 132–62. New York: Wiley. 225 pp.
40. Galun, R., Avi-Dor, Y., Bar-Zeev, M. 1963. Feeding response in *Aedes aegypti:* stimulation by adenosine triphosphate. *Science* 142:1674–75
41. Galun, R., Margalit, J. 1969. Adenine nucleotides as feeding stimulants of the tsetse fly *Glossina austeni* Newst. *Nature* 222:583–84
42. Galun, R., Margalit, J. 1970. Some properties of the ATP receptors of *Glossina austeni. Trans. R. Soc. Trop. Med. Hyg.* 64:171–74
43. Galun, R., Margalit, J. 1970. Artificial feeding and feeding stimuli of the tsetse fly *Glossina austeni. Proc. Int. Symp. Tsetse Fly Breeding Under Lab. Cond. Pract. Appl., 1st, Lisbon, 1969,* pp. 211–20
44. Galun, R., Rice, M. J. 1971. Role of platelets in haematophagy. *Nature* 223:110–11
45. Gatehouse, A. G. 1970. The probing response of *Stomoxys calcitrans* to certain physical and olfactory stimuli. *J. Insect Physiol.* 16:61–74
46. Gatehouse, A. G. 1970. Interactions between stimuli in the induction of prob-

ing by *Stomoxys calcitrans*. *J. Insect Physiol.* 16:991–1000

47. Gelperin, A. 1971. Regulation of feeding. *Ann. Rev. Entomol.* 16:365–78
48. Gelperin, A. 1972. Neural control systems underlying insect feeding behavior. *Am. Zool.* 12:489–96
49. Gooding, R. H. 1972. Digestive processes of haematophagous insects. I. A literature review. *Quest. Entomol.* 8:5–60
50. Gordon, R. M., Crewe, W., Willet, K. C. 1956. Studies on the deposition, migration, and development to the blood forms of trypanosomes belonging to the *Trypanosoma brucei* group. I. An account of the process of feeding adopted by the tsetse-fly when obtaining a blood-meal from the mammalian host, with special reference to the ejection of saliva and the relationship of the feeding process to the deposition of the metacyclic trypanosomes. *Ann. Trop. Med. Parasitol.* 50:426–37
51. Gwadz, R. W. 1969. Regulation of blood meal size in the mosquito. *J. Insect Physiol.* 15:2039–44
52. Harris, R. L., Miller, J. A., Frazar, E. D. 1974. Horn flies and stable flies: feeding activity. *Ann. Entomol. Soc. Am.* 67:891–94
53 Hocking, B. 1971. Blood-sucking behavior of terrestrial arthropods. *Ann. Rev. Entomol.* 16:1–26
54. Hodgson, E. S. 1968. Taste receptors of arthropods. *Symp. Zool. Soc. London* 23:269–77
55. Hopkins, B. A. 1964. The probing response of *Stomoxys calcitrans* (L.) (the stable fly) to vapours. *Anim. Behav.* 12:513–24
56. Hosoi, T. 1959. Identification of blood components which induce gorging of the mosquito. *J. Insect Physiol.* 3:191–218
57. House, H. L. 1958. Nutritional requirements of insects associated with animal parasitism. *Exp. Parasitol.* 7:555–609
58. Jobling, B. 1933. A revision of the structure of the head, mouth-part and salivary glands of *Glossina palpalis* Rob.-Desv. *Parasitol.* 24:449–95
59. Jones, J. C., Pillitt, D. R. 1973. Blood-feeding behavior of adult *Aedes aegypti* mosquitoes. *Biol. Bull.* 145:127–39
60. Kellogg, F. E. 1970. Water vapour and carbon dioxide receptors in *Aedes aegypti*. *J. Insect Physiol.* 16:99–108
61. Khan, A. A., Maibach, H. I. 1970. A study of the probing response of *Aedes aegypti*. 1. Effect of nutrition on probing. *J. Econ. Entomol.* 63:974–76
62. Khan, A. A., Maibach, H. I. 1971. A study of the probing response of *Aedes aegypti*. 2. Effect of desiccation and blood feeding on probing to skin and an artificial target. *J. Econ. Entomol.* 64:439–42
63. Khan, A. A., Maibach, H. I. 1971. A study of the probing response of *Aedes aegypti*. 4. Effect of dry and moist heat on probing. *J. Econ. Entomol.* 64:442–43
64. Kraus, C. 1957. Versuch einer morphologischen und neurophysiologischen Analyse des Stechaktes von *Rhodnius prolixus* Stål. 1858. *Acta Trop.* 14:36–84
65. Lall, S. B. 1969. Phagostimulants of haematophagous tabanids. *Entomol. Exp. Appl.* 12:325–36
66. Lall, S. B. 1970. Feeding behavior of haematophagous tabanids. *J. Med. Entomol.* 7:115–19
67. Lall, S. B., Davies, D. M. 1968. Comparative studies of the labellar sensitivity of female tabanids to sucrose and sodium chloride. *Ann. Entomol. Soc. Am.* 61:222–25
68. Langley, P. A. 1972. The role of physical and chemical stimuli in the development of *in vitro* feeding techniques for tsetse flies *Glossina* spp. *Bull. Entomol. Res.* 62:215–28
69. Langley, P. A., Maly, H. 1969. Membrane feeding technique for tsetse flies (*Glossina* spp.). *Nature* 221:855–56
70. Langley, P. A., Pimley, R. W. 1973. Influence of diet composition on feeding and water excretion by the tsetse fly, *Glossina morsitans*. *J. Insect Physiol.* 19:1097–1109
71. Larsen, J. R., Owen, W. B. 1971. Structure and function of the ligula of the mosquito *Culiseta inornata* (Williston). *Trans. Am. Microsc. Soc.* 90:294–308
72. Lavoipierre, M. M. J., Dickerson, G., Gordon, R. M. 1959. Studies on the methods of feeding of blood-sucking arthropods. I. The manner in which Triatomine bugs obtain their blood-meal as observed in the tissues of the living rodent, with some remarks on the effects of the bite on human volunteers. *Ann. Trop. Med. Parasitol.* 53:235–50
73. Lee, R. 1974. Structure and function of the fascicular stylets, and the labral and cibarial sense organs of male and female *Aedes aegypti* (L.). *Quest. Entomol.* 10:187–215

74. Lewis, D. J. 1975. Functional morphology of the mouth parts in New World phlebotomine sandflies. *Trans. R. Entomol. Soc. London* 126:497–532
75. Lindstedt, K. J. 1971. Chemical control of feeding behavior. *Comp. Biochem. Physiol.* 39(A):553–81
76. Maddrell, S. H. P. 1963. Control of ingestion in *Rhodnius prolixus* Stål. *Nature* 198:210
77. Manjra, A. A. 1971. Regulation of threshold to sucrose in a mosquito *Culiseta inornata* (Williston). *Mosq. News* 31:387–90
78. Margalit, J., Galun, R., Rice, M. J. 1972. Mouthpart sensilla of the tsetse fly and their function. I. Feeding patterns. *Ann. Trop. Med. Parasitol.* 66:525–36
79. Miles, P. W. 1972. The saliva of Hemiptera. *Adv. Insect Physiol.* 9:183–255
80. Mitchell, B. K. 1976. ATP reception by the tsetse fly, *Glossina morsitans* Westw. *Experientia* 32:192–93
80a. Mitchell, B. K. 1976. Physiology of an ATP receptor in labellar sensilla of the tsetse fly *Glossina morsitans* Westw. *J. Exp. Biol.* 64:In press
81. Moloo, S. K., Kutuza, S. B. 1970. Feeding and crop emptying in *Glossina brevipalpis* Newstead. *Acta Trop.* 27:356–77
82. Nelson, W. A., Keirans, J. E., Bell, J. F., Clifford, C. M. 1975. Host-ectoparasite relationships. *J. Med. Entomol.* 12:143–66
83. Owen, W. B. 1963. The contact chemoreceptor organs of the mosquito and their function in feeding behavior. *J. Insect Physiol.* 9:73–87
84. Owen, W. B., Reinholz, S. 1968. Intake of nucleotides by the mosquito *Culiseta inornata* in comparison with water, sucrose, and blood. *Exp. Parasitol.* 22:43–49
85. Phipps, J. 1974. The vampire wasps of British Columbia. *Bull. Entomol. Soc. Can.* 6:134
86. Pinet, J. M. 1968. Données ultrastructurales sur l'innervation sensorielle des stylets maxillaires de *Rhodnius prolixus*. *C.R. Acad. Sci. Paris* 267:634–37
87. Pinet, J. M., Bernard, J. 1972. Essai d'interprétation du mode d'action de la vapeur d'eau et de la température sur un récepteur d'insecte. *Ann. Zool. Ecol. Anim.* 4:483–95
88. Reinouts van Haga, H. A., Mitchell, B. K. 1975. Temperature receptors on tarsi of the tsetse fly *Glossina morsitans* West. *Nature* 255:225–26
89. Rice, M. J. 1970. Cibarial stretch receptors in the tsetse fly (*Glossina austeni*) and the blowfly (*Calliphora erythrocephala*). *J. Insect Physiol.* 16:277–89
90. Rice, M. J. 1972. *Proc. East African Med. Res. Com. Conf.* See Ref. 39
91. Rice, M. J., Galun, R., Margalit, J. 1973. Mouthpart sensilla of the tsetse fly and their function. II. Labial sensilla. *Ann. Trop. Med. Parasitol.* 67:101–07
92. Rice, M. J., Galun, R., Margalit, J. 1973. Mouthpart sensilla of the tsetse fly and their function. III. Labiocibarial sensilla. *Ann. Trop. Med. Parasitol.* 67:109–16
93. Rutledge, L. C., Ward, R. A., Gould, D. J. 1964. Studies on the feeding response of mosquitoes to nutritive solutions in a new membrane feeder. *Mosq. News* 24:407–19
94. Salama, H. S. 1966. Taste sensitivity to some chemicals in *Rhodnius prolixus* Stål. and *Aedes aegypti* L. *J. Insect Physiol.* 12:583–89
95. Salama, H. S. 1966. The function of mosquito taste receptors. *J. Insect Physiol.* 12:1051–60
96. Salama, H. S. 1967. Factors affecting taste sensitivity in *Aedes aegypti* L. *Bull. Soc. Entomol. Egypte* 51:343–46
97. Slifer, E. H. 1970. The structure of arthropod chemoreceptors. *Ann. Rev. Entomol.* 15:121–42
98. Smith, J. J. B., Friend, W. G. 1970. Feeding in *Rhodnius prolixus:* responses to artificial diets as revealed by changes in electrical resistance. *J. Insect Physiol.* 16:1709–20
99. Smith, J. J. B., Friend, W. G. 1971. The application of split-screen television recording and electrical resistance measurement to the study of feeding in a blood-sucking insect (*Rhodnius prolixus*). *Can. Entomol.* 103:167–72
100. Smith, J. J. B., Friend, W. G. 1972. Chemoreception in the blood-feeding bug *Rhodnius prolixus:* a possible rôle of cyclic AMP. *J. Insect Physiol.* 18:2337–42
101. Smith, J. J. B., Friend, W. G. 1976. Further studies on potencies of nucleotides as gorging stimuli during feeding in *Rhodnius prolixus. J. Insect Physiol.* 22:607–11
101a. Smith, J. J. B., Friend, W. G. 1976. Potencies of combined doses of nucleotides as gorging stimulants for *Rhodnius prolixus. J. Insect Physiol.* 22:In press

102. Strauss, W. G., Maibach, H. I., Khan, A. A., Pearson, T. R. 1965. Observations on biting behavior of *Aedes aegypti* (L.). *Mosq. News* 25:272–76
103. Sutcliffe, J. F., McIver, S. B. 1975. Artificial feeding of simuliids (*Simulium venustum*): factors associated with probing and gorging. *Experientia* 31:694–95
104. Tarshis, I. B. 1959. Feeding *Culex tarsalis* on outdated whole human blood through animal-derived membranes. *Ann. Entomol. Soc. Am.* 52:681–87
105. Tatchell, R. J. 1969. Host-parasite interactions and the feeding of blood-sucking arthropods. *Parasitology* 59: 93–104
106. Tawfik, M. S. 1968. Effects of the size and frequency of blood meals on *Cimex lectularius* L. *Quest. Entomol.* 4: 225–56
107. Tobe, S. S., Davey, K. G. 1972. Volume relationships during the pregnancy cycle of the tsetse fly *Glossina austeni*. *Can. J. Zool.* 50:999–1010
108. Trembley, H. L. 1952. The distribution of certain liquids in the esophageal diverticula and stomach of mosquitoes. *Am. J. Trop. Med. Hyg.* 1:693–710
109. Usinger, R. L. 1966. *Monograph of Cimicidae.* College Park, Md: Thomas Say Foundation. 585 pp.
110. von Gernet, G., Buerger, G. 1966. Labral and cibarial sense organs of some mosquitoes. *Quest. Entomol.* 2:259–70
111. Wensler, R. J. D. 1972. The effect of odors on the behavior of adult *Aedes aegypti* and some factors limiting responsiveness. *Can. J. Zool.* 50:415–20
112. Wigglesworth, V. B. 1929. Digestion in the tsetse fly: a study of structure and function. *Parasitology* 21:288–321
113. Wigglesworth, V. B., Gillett, J. D. 1934. The function of the antennae in *Rhodnius prolixus* and the mechanism of orientation to the host. *J. Exp. Biol.* 11:120–39
114. Wigglesworth, V. B., Gillett, J. D. 1934. The function of the antennae in *Rhodnius prolixus:* confirmatory experiments. *J. Exp. Biol.* 11:408

Ann. Rev. Entomol. 1977. 22:333–53

ARTHROPOD PROBLEMS IN RECREATION AREAS

❖6133

H. D. Newson
Entomology Department and Department of Microbiology and Public Health, Michigan State University, East Lansing, Michigan 48824

INTRODUCTION

Arthropod problems have been present in many areas of the New World now used for recreational purposes since before the first settlement by European immigrants. The earliest recorded confrontation occurred on the Atlantic coast where, to this day, many of the same arthropod pests continue to harass both the local residents and the hapless tourists. Dove, Hall & Hull (43) described accounts of the sand fly (*Culicoides*) infestations present along the south Atlantic coast when only the Creek Indians resided there and, later, during the Civil War, and they concluded that "... we share the opinion that these biting insects were largely responsible for the lack of early development of the southern areas along the Atlantic Seaboard." Jamnback (61) quoted an early visitor (1771) to New York. "I never saw the muskitoes more plentiful in any part of *America* than they are here. They are so eager for our blood that we could not rest all the night, though we had surrounded ourselves with fire." There are numerous accounts of the enthusiasm with which swarms of black flies and mosquitoes surrounded early travelers in eastern Canada and the northeastern United States during the pre-Revolutionary War period, and explorers and early settlers in the Great Lakes regions of both countries wrote lurid accounts of the misery inflicted by the variety of biting insects they found there. The Forty-Niners and other westbound settlers in the mid-1800s encountered large numbers of mosquitoes in the Mississippi valley and, while in transit through that area, became ill with malaria transmitted by the indigenous *Anopheles.* At times the incidence and severity of the disease was so severe that wagon trains were forced to disband or turn back (T. B. Hall, personal communication). Upon their arrival in Utah the first Mormon pioneers found large numbers of mosquitoes breeding in the waterway flowing through the arid Salt Lake valley, and trappers and early settlers in the high valleys of the Rocky and Sierra Nevada Mountains also found these areas swarming with mosquitoes in early summer (D. M. Rees, personal communication).

Although some of these arthropod problems were eliminated or minimized as areas of North America were settled and developed, many of the regions with these problems were bypassed and only now, because of their natural attributes and the increased demand for real estate suitable for recreational purposes, are being developed or considered as likely locations for future recreation areas. Some of the arthropod problems in these areas are essentially unchanged from those encountered by the first explorers, whereas others have resulted from more recent human activities. Whatever the source, these arthropods now are significant barriers to the full use or future development of many otherwise suitable recreation areas. The magnitude of both the economic and public health implications of these problems is such that they are becoming a matter of serious concern to the general public, governmental agencies, and real estate developers.

The mechanisms and organizations that have been established in various parts of North America to cope with arthropod problems are quite varied, and it frequently is impossible to distinguish the true motivation involved in their origins. There is no question that some of the first efforts were primarily concerned with arthropod-borne disease control and only secondarily with the improvement of human comfort. In many instances, however, it was only after the health problems had been eliminated and the comfort levels improved that the localities could achieve their full potential as recreation areas.

The first state in the United States to enact specific legislation concerning the control of biting arthropods provided for the establishment of local mosquito abatement agencies (1). This law, passed by New Jersey in 1912, required the supreme court justice presiding over the courts of a county to appoint six persons to serve, without salary, as members of the County Mosquito Extermination Commission. In the original law the director of the state experiment station was designated as an ex officio member of each county commission and a 1915 amendment gave the same appointments to the state director of health. It is significant that the organization of these districts in New Jersey was on a county unit basis and was mandatory, and that the state could exercise control through the approval of each control commission's budgets and plans. California was the second state to pass legislation for the abatement of mosquitoes and the original 1915 law, with many subsequent amendments, now is included in the State Health and Safety Code (1). Unlike New Jersey, the California law is permissive and allows the formation of control districts that include parts of a single county, an entire county, or parts of two or more counties. In that state the initiative for the formation of control districts must come from the citizens of the localities involved, and the law clearly specifies, in detail, the powers of such districts. It also makes it a misdemeanor for anyone to interfere with authorized work being done by control districts. With these legal bases, both California and New Jersey have developed comprehensive arthropod control programs that encompass most of their established recreational facilities, and their successful programs involve close cooperation between the organized districts, the state health departments, and active supporting research activities in their state universities (1, 122).

The United States federal government has conducted both operational control programs and limited research concerned with arthropods in the National Parks and other recreation areas it operates, but these have been, for the most part, local efforts with little overall coordination. In recent years the Vector Control Committee of the Federal Water Resources Council has been concerned with critical evaluations of arthropod problems associated with stream modification projects, most frequently those conducted by the U.S. Army Corps of Engineers (5). Perhaps the most comprehensive of all the federally funded programs is that conducted by the Tennessee Valley Authority (TVA), the quasi-governmental agency that constructs and operates the extensive water impoundment complex in the Tennessee River Valley. The potential for this extensive river impoundment system to produce serious arthropod-related health and nuisance problems was recognized at the outset so, since its inception, TVA has had an active arthropod control and research program involving malariologists, biologists, and engineers. These groups have worked cooperatively with municipalities adjacent to the TVA complex to eliminate problems along the waterways that predated the water impoundments. By careful planning and appropriate corrective and control measures (118–120) they have been able to create a vast recreational complex that has minimal annoyance from pest arthropods and even less potential for the transmission of arthropod-borne diseases that otherwise might have been a major health problem.

An example of the degree of success that can be achieved in controlling arthropods in recreation areas can be seen in Florida, which undoubtedly has one of the best and most comprehensive programs now in existence. In 1845 when Florida was seeking statehood it was described by congressman John Randolph from Virginia as "a land of swamps, of quagmires, of frogs and alligators and mosquitoes." Mulrennan & Sowder (87) quoted a description of conditions in Tallahassee, Florida, in 1842. "But unfortunately in opposition to these numerous advantages there are the greatest plagues that can afflict a new settlement; an unhealthful climate; every year bilious fevers of a most dangerous nature spread consternation in the whole region.—However, although the climate is dangerous for strangers at all times, the most insalubrious months are August, September, October and November; then no one can be sure of escaping the plague,—. The comparative extent of the huge cemeteries is a sad warning for one who, charmed by the beauty of the sight, would want to establish himself in this region." In view of the current popularity of Florida it would be difficult to convince modern tourists that such conditions existed, essentially unchanged, until World War I when some mosquito control measures were conducted around an army camp near Jacksonville. Following this, in 1919, the Florida State Board of Health undertook its first malaria control project in cooperation with a lumbering plant in the city of Perry. The Florida Anti-Mosquito Association was formed in 1922 and from this has developed the arthropod control organizations that now operate throughout most of the state. In fiscal 1974–75, this included 59 districts or county arthropod control organizations that encompassed nearly every county in the state (J. A. Mulrennan, personal communication), as well as state-supported research laboratories at Vero Beach and Panama City (47). As presently operated it is not possible to differentiate between the public health and pest control aspects of these activities, but there is no question that the comprehensive arthropod control programs of Florida have enabled the state to develop its unique resources and become one of the most popular year-round recreation areas in the world.

In recent years there have been major changes in the recreation industry that have a definite bearing upon the arthropod problems encountered by tourists and vacationers and their reaction to them. Primitive camping has become very popular and large numbers of back packers now invade many primitive areas that have remained essentially unchanged since first discovered by early explorers and trappers. The biting arthropods and enzootic diseases indigenous to some of these areas now are encountered by more people each summer than formerly would have gone there in several years. Others also have joined the back to nature movement, but these campers prefer to enjoy the conveniences of home while vacationing so there has been a proliferation of campgrounds that provide electricity, hot showers, and flush toilets for their guests. Patrons of these facilities have a much lower discomfort threshold than the back packers and their expectations while camping do not include harrassment by pest arthropods. Since these modern campgrounds require major capital investment they must provide conditions that will maintain the occupancy levels needed to remain solvent. For those even more affluent there has been a marked increase in luxury resort hotels along the coast of the Caribbean and complete resort communities in many parts of the United States and Canada. Hotel guests and individuals who purchase or rent property in these exclusive communities have perhaps the highest comfort demands and lowest discomfort thresholds of any recreation area users. Unless an entomological evaluation is included in the site selection considerations, the commercial developers may find they have intractable pest arthropod populations that markedly reduce the attractiveness of the resort and are costly to control. Specific problems encountered in these and other recreation areas are described in the following sections.

MOSQUITOES

Mosquitoes undoubtedly are the most widespread and serious arthropod pests in recreation areas of the western hemisphere, extending from Alaska and Northern Canada on south to the tropics. They undoubtedly have delayed and complicated the development and operation of recreation facilities wherever they have been present in large numbers. Their medical and economic importance has long been recognized and has been the stimulus for developing and operating the organized mosquito control programs that now encompass many densely populated areas and high-use recreation and resort areas. This group of insects has been the object of extensive research, and the literature resulting from these studies is voluminous (24–27). Where adequate organized control programs are in operation mosquito problems usually are nominal, but there are many localities where these are not available. Many people now are seeking out new camping and fishing sites, and improved highways make it possible for them to reach remote locations formerly utilized only by local residents. These new people often come from localities where mosquitoes either are not a problem or adequate arthropod control measures are used, and they are not tolerant of mosquitoes or other biting insects that interfere with their full enjoyment of the recreation resources. Thus, increasing pressure is put upon the owners or operators to "do something about the mosquitoes."

Isolated recreation areas owned or operated by state or federal agencies often have effective mosquito control (112, 118, 119). Some privately owned resorts and campgrounds obtain competent technical assistance for their mosquito problems (100), but too often technical assistance is not available and the attempts at control are ineffective. Protection from mosquitoes then becomes an individual challenge in the use of topical repellents and protective clothing (51). In many cases, if the money spent for repellents and insecticides by individual users of recreation sites could be used collectively, it would be more than adequate to finance a very effective control program.

In attempts to develop effective control for mountain, snow, or early season *Aedes*, it has become apparent that biological information now available is inadequate. Although the species may be the same, available information indicates their behavior in the far western mountains, the northern midwest, and the northeastern United States may be quite different. Although some successful efforts to control the mountain *Aedes* were initiated a number of years ago (89, 90, 100), many other control projects and the biological studies to support them are of much more recent origin (95–97, 125, 126). A notable exception is the long series of studies made by Carpenter of the mountain *Aedes* in California and published in *California Vector Views* from 1961 through 1974 (24–26).

Mosquito-Borne Diseases

Mosquito-borne diseases are not usually considered important problems in the recreation areas of North America but there should be an awareness of their potential. These diseases are dynamic and their appearance, either in resort areas or their vicinity, can generate adverse publicity that often has a severe economic impact on recreation facilities (75, 77). In New Jersey the first reported epidemic of eastern equine encephalitis occurred in 1959. During that outbreak one hotel in Atlantic City had 1000 reservation cancellations. It was estimated that the hotels in that city suffered a two-million-dollar revenue loss because tourists were afraid to visit this popular resort city. In the same period the Garden State Parkway, a toll road running through the state, also lost approximately $60,000 in revenue (61). The extensive St. Louis encephalitis epidemic that occurred in the central United States and Canada during the summer of 1975 caused many people to avoid rustic vacation areas with known mosquito problems, not realizing that this disease usually is transmitted in urban and suburban areas. Estimates of the revenue losses to these resorts is not available but almost certainly they were extensive.

Transmission of malaria by indigenous mosquitoes is a rarity in the United States even though large populations of *Anopheles* mosquitoes still are present in many recreation areas. Focal outbreaks are a definite possibility if these mosquitoes have the opportunity to feed upon individuals with *Plasmodium* infections. Such occurrences, involving military returnees from Korea and Vietnam, have happened (16, 48; R. Singer, personal communication) and, with jet travel and the frequent movement of tourists between malarious areas and parts of North America with large populations of *Anopheles*, it is quite likely to happen again. Even though there are very effective drugs for treating most malaria infections, the index of suspicion is very low in most parts of North America and it is quite possible that the diagnosis of the disease, should it involve transmission by indigenous *Anopheles*, would be missed or delayed, with perhaps serious consequences to the patient.

Diseases of the California encephalitis virus group are enzootic in many parts of Canada and the United States and some are known to be involved in human infections (75, 116). Two of these, the LaCrosse and trivittatus subtypes, are known to overwinter in the eggs of their mosquito vectors, *Aedes triseriatus* and *A. trivittatus* respectively (3, 128, 129). These mosquito species are common in the woodland camping areas of the central United States. Studies in these areas support the contention that campers in these localities may be exposed to the bites of infected mosquitoes (13, 39, 40, 95–97, 103, 104, 117) and that summer recreation activities could be a mechanism for the dissemination of California encephalitis virus infections.

Dog heartworm, *Dirofilaria immitis,* has been a major problem along the eastern seaboard and in the southeastern United States for many years, and now it has become widespread throughout many parts of the midwest (91). Recent reports indicate that it also may be established in some locations in the far western part of the country (130). One or more of the many mosquito species thought to be capable of transmitting this parasite are present in most geographical areas of North America (27, 71). Many campgrounds with large mosquito populations provide an excellent means for transmitting dog heartworm infections when infected dogs are brought to the campsites and become an infection source for the local mosquitoes.

BLACK FLIES

Judged by the degree of annoyance and irritation they create, black flies are second only to mosquitoes as major arthropod pests in recreation areas. Some fortunate individuals are immune, but black fly bites usually produce allergic responses ranging from transitory local irritation to severe systemic illness that may require hospitalization. In unsensitized individuals the bite is characterized by a hyperemic wheal, formed while the fly is feeding, and by a small hemorrhage of varying extent that occurs when the mouthparts are withdrawn from the skin. The anesthetic bite usually is not perceived at the time it is inflicted and the localized swelling and intense itching do not appear until several hours afterwards. More severe reactions may occur with multiple bites or in sensitized individuals. These may range from localized edema with blister formation to systemic reactions that include headache, fever, nausea, and adenitis. Symptoms usually are transitory except for glandular swelling, which subsides slowly, and pruritis, which may persist for several weeks.

Relatively small numbers of black flies can discourage most individuals from remaining in infested areas, whatever the attractions, and the massive numbers present in many parts of Canada and the northern United States during the spring and early summer drive away even the most avid outdoor enthusiasts (46, 62, 63). Localized outbreaks also may occur wherever there are streams suitable for black fly breeding. Many of the water management and flood control projects developed in river systems of the United States have produced ideal breeding habitats, and the parks, campgrounds, and other recreation facilities that invariably develop in association with such projects often have black fly problems of varying severity. Serious black fly problems now are recognized in Alaska, much of Canada, the United

States, and many parts of Mexico and Central America (2, 41, 42, 54, 61, 84, 94, 107, 131). Black fly populations usually are greatest in late spring and early summer in the north but may exist throughout most of the year in southern localities (84, 131). It also has been noted that the feeding habits of some species appear to change in late summer and early fall when normally ornithophilic species begin to feed on humans (40, 42, 107, 133). Whether or not this is a significant factor in recognized problem areas is not known.

Some of the earliest black fly control projects, involving the use of DDT, were developed in eastern Canada (58) and resort areas in the Adirondack Mountains of New York State (63), but because of changing attitudes concerning the use of persistent pesticides DDT is no longer allowed. Methods now available for black fly control and the prospects for their improvement were recently discussed in some detail (62). The most effective black fly control in recreation areas now is obtained using both adulticides and larvicides, and the best results are obtained where the treated areas are surrounded by natural barriers that limit the inward migration of adults from surrounding untreated localities. In the absence of these barriers, or when stream treatments are less than complete, the degree of control obtained is reduced proportionately. Black fly problems resulting from the modification of existing streams and rivers often are easier to eliminate than those in more pristine habitats, since the larvae usually are localized near the water impoundments or in dam spillways and constructed water courses, and are more easily found and treated. In areas where black fly breeding is both intensive and widespread, such as eastern Canada and the northeastern United States, comprehensive control is economically feasible only in those localities that have closely spaced and intensively used recreation facilities. The more remote and isolated sites in these areas usually do not have either the technical personnel or the financial resources needed for complete control programs, so must rely if, indeed, any control is attempted, on procedures that provide something less than completely satisfactory results. For the future, the increasing anxiety about the deliberate introduction of pesticide chemicals into surface water and the proliferation of federal, state, and local laws and regulatory agencies concerned with environmental pollution make it quite probable that the use of pesticides for black fly larval control will meet with increasing opposition and that alternate control methods will be needed. Of equal importance, site selection for the development of new campgrounds, golf courses, tourist lodges, and hotels should include consideration of existing black fly populations and every attempt should be made to locate new facilities as far away as possible from known problem areas.

STABLE FLIES

Although the stable fly, *Stomoxys calcitrans* (also called the dog fly or beach fly), is widely acknowledged as a severe pest, there are surprisingly few published accounts of studies made of this insect in the various recreation areas where it occurs. Most publications are concerned with its role as an agricultural pest. It appears that the most severe and prolonged difficulties with this insect occur along the Gulf Coast from the Florida-Alabama border approximately 400 miles east to Aldon Keys,

Florida (110). In this coastal area very large numbers usually are present from the middle of August to mid-October of each year and can make beach areas untenable during this time. King & Lenert (66) found the source of stable fly production on these beaches was in tidal drifts of marine vegetation that had accumulated along the shorelines of certain types of inlets. In later, more extensive studies in this area it was found that stable fly breeding occurred primarily in bay grasses and marine algae deposited on the beaches of shallow, narrow inlets and bays by abnormally high storm tides (109, 110), and that only green grass accumulations underwent fermentation to provide suitable breeding media for the flies. After the seasonal die-off of sea grass in mid-September, beach accumulations of the dead grass were not utilized as breeding sites and the problem abated over a period of a few weeks. Only vegetation above the normal daily tide produces stable flies since inundation by salt water prevents fermentation and larval development in the submerged parts of the media. Simmons' continued studies of stable flies in this area resulted in the publication of what is perhaps the definitive work on the biology of this species (109).

Blakeslee (14) developed the first practical chemical control technique for *Stomoxys,* spraying the sea grass accumulations on the beaches with DDT. In successive years a variety of laboratory and field testing programs were concerned with attempts to improve control levels to the point that stable flies would no longer prevent or reduce the use of these beach areas during the tourist season (78, 80). More recently, extensive work has been done to incorporate sterile male release techniques into an integrated pest management program for stable flies (67, 93). Despite this, stable flies still are a significant problem at some coastal resort areas in northwest Florida.

Sporadic outbreaks of stable flies have been reported along Atlantic coast beaches but these usually are much less severe and of shorter duration than those that occur regularly along the Gulf Coast of Florida. Even though the populations are smaller they still cause severe economic losses since relatively few flies can cause bathers to leave the beaches and cut short their stay at a resort. Hansens (55) studied a typical *Stomoxys* problem on New Jersey beaches and found that breeding occurred in decaying marine vegetation on the beach above the normal daily high tide level. Satisfactory control was achieved by using Blakeslee's control technique (14).

Less publicized but of equal importance in the disruption of recreation activities are the *Stomoxys* outbreaks that occur with some regularity at inland resorts and recreation areas. It is probable that the potential for the production of large numbers of stable flies exists in most parts of North America and those recreation facilities that maintain horses or have other activities that produce large amounts of waste organic material could have problems if proper disposal or management procedures are not followed. In addition to these, however, there are many recreation sites with major *Stomoxys* population that develop quite independently of the facility operations. Most of these are near fresh water lakes or reservoirs but some also occur in the vicinity of fresh water marshes.

For a number of years stable flies have plagued campers and fishermen in certain areas of the Kentucky and Pickwick Reservoirs, part of the TVA complex on the

Tennessee River. The pain and discomfort caused by this biting fly have become so acute at times that visitors have been forced to abandon camping and recreational facilities operating in these areas. A study in the Kentucky Reservoir side of the Land Between The Lakes indicated that the *Stomoxys* were breeding in a moist windrow of fine flotage mixed with mayfly bodies left undisturbed when water levels were lowered during scheduled draw downs. Nuisance populations of stable flies in this area followed large-scale mayfly emergences with regularity throughout the season as residual accumulations of dead mayflies were left along the shoreline by continued water level recessions. Additional sources of stable fly breeding occurred in accumulations of dead mayfly bodies under security lights in recreation areas adjacent to the reservoirs (92). In this situation the conventional control methods, treating the breeding media with insecticide, is not feasible, since this would almost certainly result in the insecticide getting into the river water, and this type of contamination is prohibited (W. W. Barnes, personal communication).

A quite different situation exists along the shores of Lake Superior where periodic appearances of *Stomoxys* force visitors to leave the beach areas. Although these occasions occur periodically rather than continuously during the summer, the large numbers of flies that may be present (up to 3000 per day have been collected in a single Malaise trap) make life intolerable for swimmers, campers, and fishermen (G. D. Gill, personal communication). Unlike the coastal areas and the Tennessee River Valley, it appears unlikely that the aquatic vegetation and organic debris that accumulate along the shores of Lake Superior are the principal breeding sites. Gill and associates conducted a series of studies extending over a period of several years and determined that in most cases the shoreline debris was unable to support the full development of stable fly larvae, either in the field or the laboratory. Their hypothesis, for which supportive but not conclusive evidence was obtained, is that the primary sources of the flies are conventional breeding sites in the farms and stables of the area, and that a combination of weather conditions periodically causes the flies to gravitate to the lake shoreline in large numbers (124, 127, 134). Flight range studies of *Stomoxys* (Gill, personal communication; 44) also support this hypothesis but further studies are needed to establish, with certainty, the source of these flies.

Accurate estimates of the economic losses caused by *Stomoxys* outbreaks in the recreation areas of this country are not available, but Patterson, LaBrecque & Williams (93) quoted estimates that the infestations in the Gulf beach resort areas alone cost the tourist industry of Florida a million dollars a day in lost revenue.

BITING MIDGES (CERATOPOGONIDAE)

Biting midges, variously called sand flies, punkies, or no-see-ums, were recognized as major insect pests in North America and the Caribbean in the first recorded accounts of those areas, and they continue to be major problems and to deter the development of many locations otherwise ideal for recreation facilities (43, 61, 69). The blood-feeding species of this group breed in a variety of habitats, including coastal marshes, bottoms of fresh water ponds and streams, accumulations of moist leaves and other organic material, and clay-type soils (61, 69, 101). The irritation associated with their bites far exceeds that expected from insects of their minute size. The overwhelming numbers that often occur can be intolerable. Up to 15,000 females have been collected in a one-night light trap operation and over 300 in a 2-min landing collection made in the Adirondack Mountains of New York state (61). Problems that occur with these insects in the inland recreation areas usually are localized, whereas those in the coastal areas, because of the extensive and often contiguous breeding sites present, are often much more extensive (11, 43, 69, 132). Inland species usually can be controlled by using chemical and cultural methods or personal protective measures (53, 60, 101), but the coastal species often present a more difficult and complex control problem.

On the Caribbean islands, in Florida, and to a lesser extent in some of the coastal areas of the southeastern United States sand flies have been a major obstacle in the growth of tourism, an industry that is of great economic importance there. Florida established a Medical Entomology Laboratory in 1947 and since that time a significant portion of the laboratory efforts has been concerned with sand fly problems (47). In 1959 the Jamaican government organized a Sandfly Research and Control Laboratory at Montego Bay to study the difficult sand fly pest problems at the beach areas and resort hotels on the northern coast of Jamaica, and later a similar laboratory was established on Grand Cayman island. Since the early 1930s the Bahaman government has utilized foreign entomologists to work on the biology and control of sand flies, and through the efforts of all these groups effective methods have been developed for controlling sand flies in coastal areas. The ability to control these insects has made it possible to develop many new tourist facilities that have become increasingly important to the economy of the Caribbean region. Linley & Davis (69) provide an excellent and very comprehensive review of the sand fly problems in the Caribbean and the methods available for their solution. One of their comments is pertinent to many recreation areas. "Though great imagination has almost invariably been shown in the provision of man-made facilities for the tourist, frequently considerably less skill has been demonstrated in entomological matters. . . . Unless an entomological evaluation is made, the owners of a magnificent new facility may soon find that they have a monumental problem which can be brought under control only after considerable difficulty and substantial expenditures."

NONBITING FLIES AND MIDGES

Relatively few insects occur in such massive profusion as to be pests simply because of their numbers, but Trichoptera (caddisflies), Ephemeroptera (mayflies, shadflies, or eelflies), and the chironomid and chaoborid midges are among those that do (6). Although these are not blood feeders, the nuisances they can create are noteworthy. Adults may emerge from fresh water habitats by the millions and can totally disrupt human activities in nearby areas. They enter the eyes, ears, nose, and mouth, fall into food and drinks, stain walls and windows, and at the ends of their relatively short lives the dead bodies often accumulate in large piles that become putrid and odoriferous and may serve as breeding media for stable flies and other Diptera (92).

When these emergences occur in or near recreation facilities the result can be totally disruptive, as indicated by a quotation from an Associated Press dispatch (61). "Shadflies . . . [were] . . . in some places piled to a depth of four feet [and] blocked traffic over the . . . highway bridge for nearly two hours . . . Fifteen men in hip boots used shovels and a snowplow to clear a path . . . Trucks without chains were unable to operate until most of the flies had been shovelled into the Mississippi River." Because of a similar mayfly outbreak at Verona Beach, New York, the general manager of the Central New York State Parks Commission estimated an attendance loss of 50,000 people during the infestation period (61). Similar problems involving caddisflies have been reported from Canada (88).

In the western United States perhaps the best known nonbiting midge is the Clear Lake gnat, *Chaoborus astictopus,* the target of control efforts in California since the 1930s (35). From the arrival of the first settlers in Lake County this insect was a severe pest until 1949 when the lake was treated with TDE (DDD), with spectacular success. The residents and visitors were confident that the problem had been forever solved. Their high expectations for the future did not materialize, however, and the periodic resurgences of *C. astictopus* populations and the attempts to control them have provided a classic example of the complications that can develop when insecticide resistance and environmental pollution problems coincide with high expectations for insect control programs that will maintain pest insect populations below the nuisance threshold (35). This insect has developed resistance to methyl parathion, the most recent in a series of insecticides that have been used, and the application rates now needed to achieve 95% control, which still is not adequate to eliminate the nuisance factor, poses serious toxicological hazards to nontarget organisms in the lake. Abate, a substitute insecticide that offers an acceptable safety margin, is prohibitively costly when applied at effective rates, with a projected price of over one million dollars per treatment (C. S. Apperson, personal communication). Local resort owners still are adamant in their demand for gnat control.

In recent years there has been a marked trend towards developing recreational lakes as focal points for new housing subdivisions, so-called complete recreation communities, and seasonal resort areas. Since porous soils can be sealed with a variety of materials and the lakes can be filled with water from wells, natural runoff, springs, surplus agricultural water, diverted rivers and streams, and even reclaimed sewage effluent, these lakes can be constructed almost anywhere the builders and developers choose to place them. Lakes of this type may have relatively large surface areas but typically are shallow and have inordinately long shorelines designed to provide the maximum number of waterfront building lots. Property values in these developments are relatively high and the owners and prospective buyers are attracted by the variety of outdoor recreation opportunities provided by the lake. Typically they are from metropolitan areas and are not kindly disposed to accept any conditions that reduce the use of the available recreation area facilities. Although most of the published accounts of insect problems associated with these types of lakes have been concerned with locations in southern California, it is quite probable that similar lakes in other parts of the country will develop the same difficulties, so a brief review of the California experiences is in order.

The problem manifestations in California resort communities with man-made lakes, essentially the same as those in Canada and the northeastern United States described earlier, have been massive numbers of midges, in this case chironomids and chaoborids, that swarm into the yards and business property adjacent to the lakes, disrupting outdoor activities and entering homes and business establishments in large numbers. The water used to fill these lakes frequently has been rich in nutrients that are often supplemented by organic material in runoff water sometimes diverted into the lakes to maintain water levels. The southern California climate is mild through most of the year and many of the artificial lakes there have become eutrophic after only one season and have produced huge midge populations (72, 81). A number of research projects have been conducted to develop effective chemical and biological control measures and water management techniques that will reduce the insect productivity of the lakes to acceptable levels (36, 52, 82, 83, 85, 86). These have shown that proper design and water management operations in these lakes can extend the pre-eutrophication period, but the ultimate development of algal blooms and large midge populations is nearly inevitable unless some type of chemical and/or biological control measures are employed. The design, water sources, soil characteristics, and other features make each lake unique, as are the phytoplankton and complex of midge species they produce. Because of this, attempts to develop general control recommendations have not been particularly successful and it has been necessary to study each lake individually to determine the procedures that will provide an acceptable level of control (82).

TICKS AND TICK-BORNE DISEASES

The tick species that may feed on humans vary considerably in different North American recreation areas, and their distribution usually is spotty and correlated with the populations of wild or domestic animals that normally serve as their hosts. Experienced hikers and campers minimize the potential for tick bites by using repellent and by making frequent personal inspections to remove ticks before they begin feeding. In the absence of suitable wild mammalian hosts, tick infestations in and around campgrounds and recreation areas can be controlled or eliminated by using chemical controls or management techniques (74, 79, 111). Of increasing concern at this time, however, are three tick-borne diseases that are endemic in many vacation localities in the United States, Rocky Mountain spotted fever, Colorado tick fever, and tick-borne relapsing fever.

Rocky Mountain Spotted Fever

Rocky Mountain spotted fever was first reported in areas of the intermountain west that now are frequented by back pack hikers and advocates of primitive style camping, and it still persists in some of these localities in enzootic transmission cycles involving species of ticks that include *Dermacentor andersoni,* the Rocky Mountain wood tick. This disease has been recognized in the eastern United States since 1930 and the primary tick vector involved in human infections there is *Dermacentor variabilis,* the American dog tick. In recent years the preponderance of reported human infections has been in the middle Atlantic and southeastern part of the country, and the numbers have increased progressively from a low point of just under 200 in 1959 to over 800 in 1975 (17, 30). Many high-use recreation areas are located in this part of the country and the potential for the transmission of Rocky Mountain spotted fever to the many visitors of these areas has been a matter of concern (37, 38). The changing epidemiology of this disease and its apparent increase have been reported extensively (18, 19, 50, 56, 57, 70, 98, 108, 123) and were discussed in detail by Burgdorfer (17). Sonenshine and his associates have conducted extensive studies on the epidemiology of this disease in Virginia (4, 33, 102, 113–115), but there still are many informational gaps that must be filled before the significance of the changing incidence and distribution of Rocky Mountain spotted fever in the eastern United States and its importance to the recreation areas there are understood.

Colorado Tick Fever

Colorado tick fever has been known to exist in localized areas of the Rocky Mountains for many years, and a number of studies have been made to determine its distribution and the arthropod and vertebrate hosts involved in its normal transmission cycle (15, 20, 21, 32, 45, 49). Because of the relative mildness of this virus disease in humans and its apparently limited distribution, research efforts have been, until recently, at a low level. With increased surveillance during the past few years it now appears that Colorado tick fever is more widespread and has a much higher incidence in humans than was formerly appreciated (98). It should be considered a potential disease hazard to individuals who camp or travel through some areas of the Rocky Mountains.

Tick-Borne Relapsing Fever

Tick-borne relapsing fever is endemic, although of low incidence, throughout the western United States and occasional infections are incurred by campers and hikers in the mountains of the west. Cases usually are sporadic and involve only a few individuals in any given outbreak (31). In 1973, however, 45 confirmed cases were contracted at North Rim, Grand Canyon National Park, the largest single focus outbreak ever to occur in the United States (73). This was a classic example of the complexities between arthropod-borne diseases and the difficulty in making epidemiological investigations when these diseases involve populations of transient tourists. In July two epidemiologically related cases of relapsing fever were reported, one in Arizona and the other in Georgia. Both were apparently contracted at the Grand Canyon North Rim and, subsequently, other confirmed infections were diagnosed in employees working at that location. In the investigation that followed it was found that the rodent population, which had been extremely high in 1972, apparently had been decimated by an undetected epizootic of sylvatic plague prior to the relapsing fever outbreak. With the rodent die-off, the soft ticks (*Ornithodorus hermsi*) present in rodent nests in the area apparently sought out and fed on humans since their normal rodent hosts were no longer available, and in the course of feeding they transmitted the disease. These soft ticks feed rapidly and cause little irritation so their bites were largely undetected. Retrospective reports of relapsing fever cases traced to this outbreak came from several states within the United States and as far away as West Germany. The large number of visitors to this park (over 600,000 between June and October) and the unusually high rodent population greatly enhanced the potential for human involvement in what otherwise might have remained strictly an enzootic disease cycle. Similar potentials also exist in many recreation areas, particularly when their management practices encourage the development of unusually large rodent populations (A. M. Barnes, personal communication).

SYLVATIC PLAGUE

Measured in terms of morbidity and mortality, plague has never been a major public health problem in the United States, even though there have been some outbreaks in urban areas and several hundred human cases, mostly sporadic in both locality and time distribution, are known to have occurred (10, 28, 29, 68). Despite the potential for the introduction of plague into densely populated urban areas (23, 68), the chemotherapeutic measures and techniques for rodent and flea control now available make it rather unlikely that outbreaks in the densely populated areas of the United States, should they occur, will involve major portions of the human population (7–9, 12, 64, 76). On the other hand sylvatic plague has occurred in all of the Rocky Mountain and far western states, as well as parts of Texas, Oklahoma, Kansas, and North Dakota (28). Enzootic foci persist in many localities and, periodically, both localized and widespread epizootics decimate the resident populations of prairie dogs, ground squirrels, and other susceptible rodent species. The recreation areas in the western United States, having either enzootic plague foci or susceptible rodent populations, present a latent danger to their residents and visitors that has required the development of suppressive measures and control techniques that are suitable for use in these areas. Although effective control measures are available, the remoteness of some of the localities in which epizootic outbreaks may begin make comprehensive surveillance impossible, and some sylvatic outbreaks are so extensive that control measures for the entire area involved are not economically feasible.

Extensive field tests and laboratory experiments have shown that treating rodent burrows with appropriate insecticides and using bait to entice rodents into boxes containing insecticides will effectively control the flea vectors of plague (7–9, 12, 64, 106). These measures, used individually or in combination, now are standard plague control techniques. Recent work by Miller and associates (76) indicated that rodent bait treated with a systemic insecticide may offer advantages in flea vector control over the procedures now used most extensively.

Depending upon the local conditions and circumstances, sylvatic plague control measures may include the reduction of rodent populations as well as the methods outlined above. When flea/rodent control programs are initiated in these plague outbreaks, a major decision, usually made by the health authorities conducting the program, is whether the recreation facility should be closed or allowed to continue its operations while the control measures are employed. These decisions are based upon a variety of factors, but when a recreation facility is closed, particularly those that are privately owned or operated, there may be severe economic repercussions to the operators and surrounding communities. However desirable it may be to close a recreation site this course of action may not be possible in situations where plague outbreaks coincide with major tourist events. In 1965 an extensive epizootic plague outbreak was in progress in New Mexico at the same time as the annual Gallup Inter-Tribal Indian Ceremonial, which each year attracts thousands of tourists from all parts of the country (B. E. Miller, personal communication). At this time six Navajo Indian children became ill with plague they had contracted on their reservation, the same area visited and camped in by many of the visiting tourists (34, 65, 121). Because of the large number of tourists involved it was not possible to control their movements within the plague area and cancelling the ceremonies was not an acceptable alternative. Fortunately this outbreak of human plague was limited, but it is apparent that the potential for major future problems will continue as long as prairie dogs and other plague-susceptible rodent hosts coexist with a highly mobile tourist population.

That the mobility of the tourist population further complicates the problem of proper diagnosis and treatment of plague has been amply demonstrated. In 1957 an individual acquired plague infection in Colorado but traveled to Texas before the onset of symptoms; another person exposed to plague in New Mexico (1961) returned to his home in Massachusetts before becoming sufficiently ill to require hospitalization; and in 1974 a child became infected while visiting his grandparents near Alburquerque and then returned to San Francisco (29; B. E. Miller, personal communication). All three died despite the availability of very effective drugs for the treatment of plague infections. Outside enzootic areas the index of suspicion for plague is not high, and attending physicians often overlook this possibility in making their differential diagnosis and initiating treatment, which must be started early in the course of the disease. Delayed treatment, an incorrect diagnosis, or the use of the wrong antibiotic can be fatal.

Available evidence indicates that there may be basic differences in the ecology of plague in the Rocky Mountain areas and that which is present in Pacific coast states. There is a critical need for more qualitative and quantitative background data on the mammal populations, particularly in the Rocky Mountains, to evaluate the enzootic plague activity in that area. A better understanding of plague ecology in the entire intermountain western states will be needed to develop effective measures for the prevention of human plague. More effort also is needed to determine the potential of fox squirrels *(Sciurus niger)* as reservoirs of plague in certain environments. Only one epizootic involving this species has been reported (59), but its widespread distribution throughout the eastern United States and the serological evidence that it has a high plague infection survival rate opens many questions as to whether or not it might become involved in plague infection outside the currently known enzootic areas of the west. The known involvement of wild carnivores and domestic cats and dogs (99, 104, 105) in plague outbreaks also make them suspect as possible amplifying hosts in the western recreation areas, particularly as a link between rodents in enzootic localities and the tourists and residents who visit or reside there.

CONCLUSIONS

Space limitations have limited this review and discussion to some of the more important arthropod problems that are present in recreation areas. The recreation industry is becoming increasingly important because it is a major economic factor in many areas and also because the natural resources available for new recreational development are becoming progressively harder to obtain, even though the demand for recreation sites is increasing sharply. Arthropod problems in these facilities, whether pest or health related, can diminish their utilization and cause major economic problems to the owners and businesses that are dependent upon tourism. Although the technical expertise needed to control arthropod problems in recreation areas is well developed, there is a definite shortage of the technically qualified people and economic resources needed to develop and use satisfactory control programs in some areas where they are badly needed. There also is a critical need for additional biological studies of many of the arthropods involved to ensure that available control techniques can be utilized most effectively. Finally, entomological considerations should be an integral part of planning new resort facilities to avoid undesirable site selections and to preclude or minimize future arthropod problems.

ACKNOWLEDGMENTS

The preparation of this article was made possible only through the cooperation and assistance of many people who provided reprints and information. I am especially indebted to the following individuals whose helpfulness greatly exceeded the usual limits of professional cooperation: R. C. Axtell; A. M. Barnes; W. W. Barnes; G. R. DeFoliart; G. D. Gill; G. Grodhaus; G. C. LaBrecque; B. E. Miller; M. S. Mulla; L. T. Nielsen; and D. E. Sonenshine.

Literature Cited

1. American Mosquito Control Assoc. 1961. *Organization for Mosquito Control. Am. Mosquito Control Assoc. Bull. No. 4,* ed. H. F. Gray. Oakland: Abbey. 54 pp.
2. Anderson, D. R., DeFoliart, G. R. 1961. Feeding behavior and host preferences of some black flies in Wisconsin. *Ann. Entomol. Soc. Am.* 54:716–29
3. Andrews, W. N., Rowley, W. A., Wong, Y. W., Dorsey, D. C., Housler, W. J. Jr. 1976. Isolation of trivittatus virus (California group) from larvae and adults reared from field-collected larvae of *Aedes trivittatus. J. Med. Entomol.* 13: In press
4. Atwood, E. L., Lamb, J. T. Jr., Sonenshine, D. E. 1965. A contribution to the epidemiology of Rocky Mountain spotted fever in the eastern United States. *Am. J. Trop. Med. Hyg.* 14:831–37
5. Bagley, J. R. Jr. 1974. A study of mosquito prevention and control problems associated with stream modification projects. Coll. Vet. Med. Biomed. Sci. Fort Collins: Colorado State Univ. 50 pp.
6. Balyeat, R., Stemen, T., Taft, C. 1932. Comparative pollen, mold, butterfly and moth emanation content of the air. *J. Allergy* 3:227–34
7. Barnes, A. M., Kartman, L. 1960. Control of plague vectors on diurnal rodents in the Sierra Nevada of California by use of insecticide bait boxes. *J. Hyg.* 58:159–67
8. Barnes, A. M., Ogden, L. J., Archibald, W. S., Campos, E. 1974. Control of plague vectors on *Peromyscus maniculatus* by use of 2% carbaryl dust in bait stations. *J. Med. Entomol.* 11: 83–87

9. Barnes, A. M., Ogden, L. J., Campos, E. G. 1972. Control of the plague vector *Opisocrostis ludovicianus,* by treatment of prairie dog (*Cynomys ludovicianus*) burrows with 2% carbaryl dust. *J. Med. Entomol.* 4:330–33
10. Beadle, L. D., ed. 1972. *Plague Activity of Major Concern - 1971. Vector Control Briefs,* Feb. 1972, p. 11. Atlanta: CDC.
11. Beck, E. C. 1958. A population study of the *Culicoides* of Florida. *Mosq. News* 18:6–11
12. Bennett, W. C., Graves, G. N., Wheeler, J. R., Miller, B. E. 1975. Field evaluations of dichlorvos as a vapor toxicant for control of prairie dog fleas. *J. Med. Entomol.* 12:354–58
13. Berry, R. L., Parsons, M. A., LaLonde, B. J., Stegmiller, H. W., Lebio, J., Jalil, M., Masterson, R. A. 1975. Studies on the epidemiology of California encephalitis in an endemic area in Ohio in 1971. *Am. J. Trop. Med. Hyg.* 24: 992–98
14. Blakeslee, E. B. 1945. DDT surface sprays for control of stable flies breeding in shore deposits of marine grass. *J. Econ. Entomol.* 38:548–52
15. Bowen, G. S., McLean, R. G., Francy, D. B., Shriner, R. B., Strasser, K. S. 1975. Colorado tick fever in Rocky Mountain National Park. Part II. Field and laboratory studies of mammalian hosts. Presented at Ann. Meet. Am. Soc. Trop. Med. Hyg., 24th, New Orleans.
16. Brunetti, R., Fritz, R. F., Hollister, A. C. Jr. 1953. An outbreak of malaria in California, 1952–1953. *Am. J. Trop. Med. Hyg.* 3:779–88
17. Burgdorfer, W. 1975. A review of Rocky Mountain spotted fever (tick-borne typhus), its agent, and its tick vectors in the United States. *J. Med. Entomol.* 12:269–78
18. Burgdorfer, W., Atkins, T. R. Jr., Priester, L. E. 1975. Rocky Mountain spotted fever (tick-borne typhus) in South Carolina: an educational program and tick/rickettsial survey in 1973 and 1974. *Am. J. Trop. Med. Hyg.* 24:866–72
19. Burgdorfer, W., Cooney, J. C., Thomas, L. A. 1974. Zoonotic potential (Rocky Mountain spotted fever and tularemia) in the Tennessee Valley region. II. Prevalence of *Rickettsia rickettsi* and *Francisella tularensis* in mammals and ticks from Land Between the Lakes. *Am. J. Trop. Med. Hyg.* 23:109–17
20. Burgdorfer, W., Eklund, C. M. 1959. Studies on the ecology of Colorado tick fever in western Montana. *Am. J. Hyg.* 69:127–37
21. Burgdorfer, W., Eklund, C. M. 1960. Colorado tick fever. I. Further ecological studies in western Montana. *J. Infect. Dis.* 107:379–83
22. Burks, B. 1953. The mayflies or Ephemeroptera of Illinois. *Bull. Ill. Natl. Hist. Surv.* Vol. 26, 216 pp.
23. Campos, E. G., Barnes, A. M. 1972. Rediscovery of *Nosopsyllus londiniensis* (Rothschild, 1903) in San Francisco, California. *Calif. Vector Views* 19:87
24. Carpenter, S. J. 1968. Review of recent literature on mosquitoes of North America. *Calif. Vector Views* 15:71–98
25. Carpenter, S. J. 1970. Reviews of recent literature on mosquitoes of North America. Supplement I. *Calif. Vector Views* 17:39–65
26. Carpenter, S. J. 1974. Review of recent literature on mosquitoes of North America. Supplement II. *Calif. Vector Views* 21:73–99
27. Carpenter, S. J., LaCasse, J. W. 1955. *Mosquitoes of North America (North of Mexico).* Berkeley: Univ. Calif. 360 pp.
28. Center for Disease Control. 1970. *Plague Surveillance, Report No. 1.* July 1970. 20 pp.
29. Center for Disease Control. 1971. *Plague Surveillance, Report No. 2.* July 1971. 14 pp.
30. Center for Disease Control. 1975. Cases of specified notifiable diseases - United States. *Morbid. Mortal. Weekly Rep.* 24:429
31. Center for Disease Control. 1975. Tick-borne relapsing fever—California. *Morbid. Mortal. Weekly Rep.* 24:419–20
32. Clark, G. M., Clifford, C. M., Fadness, L., Jones, E. K. 1970. Contributions to the ecology of Colorado tick fever virus. *J. Med. Entomol.* 7:189–97
33. Clifford, C. M., Sonenshine, D. E., Atwood, E. L., Roberts, C. S., Hughes, L. E. 1969. Tests on ticks from wild birds collected in the eastern United States for rickettsiae and viruses. *Am. J. Trop. Med. Hyg.* 18:1057–61
34. Collins, R. N., Martin, A. R., Kartman, L., Brutsché, R. L., Hudson, B. W., Doran, H. G. 1967. Plague epidemic in New Mexico, 1965. I. Introduction and description of the cases. *Public Health Rep.* 82:1077–84
35. Cook, S. F. Jr. 1965. The Clear Lake gnat: its control, past, present, and future. *Calif. Vector Views* 12:43–48

36. Cook, S. F. Jr. 1968. The potential role of fishery management in the reduction of chaoborid midge populations and water quality enhancement. *Calif. Vector Views* 15:63–69
37. Cooney, J. C., Burgdorfer, W. 1974. Zoonotic potential (Rocky Mountain spotted fever and tularemia) in the Tennessee Valley region. I. Ecologic studies of ticks infesting mammals in Land-Between the Lakes. *Am. J. Trop. Med. Hyg.* 23:99–108
38. Cooney, J. C., Pickard, E. 1972. Comparative tick control field tests–Land Between the Lakes. *Down Earth* 28: 9–11
39. DeFoliart, G. R., Anslow, R. O., Hanson, R. P., Morris, C. D., Papadopaulos, O., Sather, G. E. 1969. Isolation of Jamestown Canyon serotype of California encephalitis virus from naturally infected *Aedes* and tabanids. *Am. J. Trop. Med. Hyg.* 18:440–47
40. DeFoliart, G. R., Anslow, R. O., Thompson, W. H., Hanson, R. P., Wright, R. E., Sather, G. E. 1972. Isolations of trivittatus virus from Wisconsin mosquitoes, 1964–1968. *J. Med. Entomol.* 9:67–70
41. DeFoliart, G. R., Rao, M. R. 1965. The ornithophilic black fly *Simulium meridionale* Riley feeding on man during autumn *J. Med. Entomol.* 2:84–85
42. DeFoliart, G. R., Rao, M. R., Morris, C. D. 1967. Seasonal succession of bloodsucking diptera in Wisconsin during 1965. *J. Med. Entomol.* 4:363–73
43. Dove, W. E., Hall, D. G., Hull, J. B. 1932. Salt marsh sandfly problem. *Ann. Entomol. Soc. Am.* 25:505–27
44. Eddy, G. W., Roth, A. R., Plapp, F. W. Jr. 1962. Studies on the flight habits of some marked insects. *J. Econ. Entomol.* 55:603–7
45. Eklund, C. M., Kohls, G. M., Brennan, J. M. 1955. Distribution of Colorado tick fever and virus-carrying ticks. *J. Am. Med. Assoc.* 157:335–37
46. Fallis, A. M. 1964. Feeding and related behavior of female Simuliidae. *Exp. Parasitol.* 15:439–70
47. Florida State Bureau of Entomology. 1975. *Florida Medical Entomology Laboratory. Informational Brochure.* Vero Beach: Florida Dept. Health Rehab. Surv. 10 pp.
48. Fontaine, R. E., Gray, H. F., Aarons, T. 1953. Malaria control at Lake Vera, California in 1952–53. *Am. J. Trop. Med. Hyg.* 3:789–92
49. Francy, D. B., Jacob, W. L., Eads, R. B., Trimble, J. M., McLean, R. G., Barnes, A. M., Bowen, G. S., Calisher, C. H. 1975. Colorado tick fever in Rocky Mountain National Park. Part I. Studies on ticks. Presented at Ann. Meet. Am. Soc. Trop. Med. Hyg., 24th, New Orleans.
50. Fuller, H. S. 1956. Veterinary and medical acarology. Rocky Mountain spotted fever. *Ann. Rev. Entomol.* 1:353–56
51. Gorham, J. R. 1974. Tests of mosquito repellents in Alaska. *Mosq. News* 34: 409–15
52. Grodhaus, G. 1968. Considerations in controlling chironomids. *Proc. 36th Ann. Conf. Calif. Mosq. Control Assoc.* 36:37–39
53. Grothaus, R. H., Gouck, H. K., Weidhaas, D. E., Jackson, S. C. 1974. Wide mesh netting, an improved method of protection against blood-feeding Diptera. *Am. J. Trop. Med. Hyg.* 23:533–37
54. Hall, F. 1972. Observations on black flies of the genus *Simulium* in Los Angeles County, California. *Calif. Vector Views* 19:53–58
55. Hansons, E. J. 1951. The stable fly and its effect on seashore recreational areas in New Jersey. *J. Econ. Entomol.* 44:482–87
56. Hattwick, M. A. W. 1971. Rocky Mountain spotted fever in the United States, 1920–1970. *J. Infect. Dis.* 124:112–14
57. Hattwick, M. A. W., Peters, A. H., Gregg, M. R., Hanson, B. 1973. Surveillance of Rocky Mountain spotted fever. *J. Am. Med. Assoc.* 225:1338–43
58. Hocking, B., Richards, W. 1952. Biology and control of Labrador blackflies. *Bull. Entomol. Res.* 43: 237–57
59. Hudson, B. W., Goldenberg, M. I., McCluskie, J. D., Larson, H. E., McGuire, C. D., Barnes, A. M., Poland, J. D. 1971. Serological and bacterological investigations of an outbreak of plague in an urban tree squirrel population. *Am. J. Trop. Med. Hyg.* 20:255–63
60. Jamnback, H. 1961. The effectiveness of chemically treated screens in killing annoying punkies, *Culicoides obsoletus. J. Econ. Entomol.* 54:578–80
61. Jamnback, H. 1969. *Bloodsucking Flies and Other Outdoor Nuisance Arthropods of New York State.* Albany: N.Y. State Museum Sci. Serv. 90 pp.
62. Jamnback, H. 1973. Recent developments in the control of black flies. *Ann. Rev. Entomol.* 18:281–304

63. Jamnback, H., Collins, D. L. 1955. The control of black flies in New York. *New York State Bull. No. 350,* 113 pp.
64. Kartman, L. 1958. An insecticide-bait-box method for the control of sylvatic plague vectors. *J. Hyg. Cambridge* 56:455
65. Kartman, L., Martin, A. R., Hubbert, W. T., Collins, R. N., Goldenberg, M. I. 1967. Plague epidemic in New Mexico, 1965. II. Epidemiologic features and results of field studies. *Public Health Rep.* 82:1084–94
66. King, W. V., Lenert, L. G. 1936. Outbreaks of *Stomoxys calcitrans* (L.) ("dog flies") along Florida's northwest coast. *Fla. Entomol.* 19:33–39
67. LaBrecque, G. C., Meiffert, D. W., Weidhaas, D. E. 1975. Potential of the sterile-male technique for the control or eradication of stable flies, *Stomoxys calcitrans* Linnaeus. In *Sterility Principle for Insect Control,* pp. 449–59. Vienna: I. A. E. A. IAE-SM-186/55.
68. Link, V. B. 1955. *A History of Plague in the United States.* Public Health Monograph No. 26. Washington, DC: U.S. Dept. H.E.W. 120 pp.
69. Linley, J. R., Davies, J. B. 1971. Sandflies and tourism in Florida and the Bahamas and Caribbean area. *J. Econ. Entomol.* 64:264–78
70. Linnemann, C. C. Jr., Jansen, P., Schiff, G. M. 1973. Rocky Mountain spotted fever in Clermont County, Ohio. Description of an endemic focus. *Am. J. Epidemiol.* 97:125–30
71. Ludlam, K. W., Jachowski, L. A., Otto, G. F. 1970. Potential vectors of *Dirofilaria immitis. J. Am. Vet. Med. Assoc.* 157:1354–59
72. Magy, H. I. 1968. Vector and nuisance problems emamating from man-made recreational lakes. *Proc. 36th Ann. Conf. Calif. Mosq. Control Assoc.* 36:36–37
73. Maupin, G. O. 1974. An outbreak of tick-borne relapsing fever at Grand Canyon National Park. *Proc. 10th Biennial Public Health Vector Control Conf.* pp. 31–32
74. McDuffy, W. C., Smith, C. N. 1955. Recommended current treatments for tick control. *Public Health Rep.* 70:327–30
75. McGowan, D. E., Bryan, C. A., Gregg, M. B. 1973. Surveillance of arboviral encephalitis in the United States 1955–71. *Am. J. Epidemiol.* 97:199–207
76. Miller, B. E., Bennett, W. C., Graves, G. N., Wheeler, J. R. 1975. Field studies of systemic insecticides. I. Evaluation of phoxim for control of fleas on cotton rats. *J. Med. Entomol.* 12:425–30
77. Morris, C. D., Caines, A. R., Woodall, J. P., Bast, T. F. 1975. Eastern equine encephalomyelitis in upstate New York, 1972–1974. *Am. J. Trop. Med. Hyg.* 24:986–91
78. Mount, G. A., Gahan, J. B., Lofgren, C. S. 1968. Toxicity of insecticides to stable flies. *U.S. Dept. Agric., Agric. Res. Serv.* ARS 33–123. 27 pp.
79. Mount, G. A., Hirst, J. M., McWilliams, J. G., Lofgren, C. S., White, S. A. 1968. Insecticides for control of the lone star tick tested in the laboratory and as high- and ultra-low-volume sprays in wooded areas. *J. Econ. Entomol.* 61:1005–7
80. Mount, G. A., Lofgren, C. S., Gahan, J. B. 1966. Malathion, naled, fenthion and Bayer 39007 thermal fogs for control of the stable fly (dog fly) *Stomoxys calcitrans. Fla. Entomol.* 49:169–73
81. Mulla, M. S. 1974. Chironomids in residential-recreational lakes. An emerging nuisance problem—measures for control. *Entomol. Tidskr.* 95(Suppl.): 172–76
82. Mulla, M. S., Barnard, D. R., Norland, R. L. 1975. Chironomid midges and their control in Spring Valley Lake, California. *Mosq. News* 35:389–95
83. Mulla, M. S., Khasawinah, A. M. 1969. Laboratory and field evaluations of larvicides against chironomid midges. *J. Econ. Entomol.* 62:37–41
84. Mulla, M. S., Lacey, L. A. 1976. Biting flies in the lower Colorado River basin: economic and public health implications of *Simulium. Mosq. News* 36:In press
85. Mulla, M. S., Norland, R. L., Fanara, D. M., Darwazeh, H. A., McKean, D. W. 1971. Control of chironomid midges in recreational lakes. *J. Econ. Entomol.* 64:300–7
86. Mulla, M. S., Norland, R. L., Westlake, W. E., Dell, B., St. Amant, J. 1973. Aquatic midge larvicides, their efficiency and residues in water, soil, and fish in a warm-water lake. *Environ. Entomol.* 2:58–65
87. Mulrennan, J. A., Sowder, W. T. 1954. Florida's mosquito control system. *Public Health Rep.* 69:613–18
88. Munroe, E. 1951. Pest Trichoptera at Fort Erie, Ontario. *Can. Entomol.* 83:69–72

89. Nielsen, L. T. 1958. Control of snow mosquitoes in the mountains of Utah. *Proc. 10th and 11th Ann. Meet. Utah Mosq. Abatement Assoc.* pp. 33–34
90. Nielsen, L. T. 1959. Seasonal distribution and longevity of Rocky Mountain snow mosquitoes of the genus *Aedes. Proc. Utah Acad. Sci.* 36:83–87
91. Otto, G. F. 1972. Epizootiology of canine heartworm disease. In *Canine Heartworm Disease: The Current Knowledge,* ed. R. E. Bailey, pp. 1–15. Gainesville: Univ. Florida
92. Packard, E. 1968. *Stomoxys calcitrans* (L.) breeding along TVA reservoir shorelines. *Mosq. News* 28:644–46
93. Patterson, R. S., LaBrecque, G. C., Williams, D. F. 1975. Use of the sterile-male technique to control stable flies on St. Croix. Study in progress Jul 74–77 at USDA ARS Federal Exp. Sta., St. Croix, U.S. Virgin Islands
94. Peterson, B. V. 1970. The *Prosimulium* of Canada and Alaska. *Mem. Entomol. Soc. Canada No. 69.* 216 pp.
95. Pinger, R. R. Jr., Rowley, W. A. 1972. Occurrence and seasonal distribution of Iowa mosquitoes. *Mosq. News* 32: 234–41
96. Pinger, R. R., Rowley, W. A. 1975. Host preferences of *Aedes trivittatus* in central Iowa. *Am. J. Trop. Med. Hyg.* 24:889–93
97. Pinger, R. R., Rowley, W. A., Wong, Y. W., Dorsey, D. C. 1975. Trivittatus virus infections in wild mammals and sentinal rabbits in central Iowa. *Am. J. Trop. Med. Hyg.* 24:1006–9
98. Poland, J. D. 1974. Rocky Mountain spotted fever and Colorado tick fever. *Proc. 10th Biennial Public Health Vector Control Conf.* pp. 29–30
99. Poland, J. D., Barnes, A. M., Herman, J. J. 1973. Human plague from exposure to a naturally infected wild carnivore. *Am. J. Epidemiol.* 97:332–37
100. Rees, D. M., Nielsen, L. T. 1952. Control of *Aedes* mosquitoes in two recreational areas in the mountains of Utah. *Mosq. News* 12:43–49
101. Rees, D. M., Smith, J. V. 1950. Effective control methods used on biting gnats in Utah during 1949. *Mosq. News* 10:9–15
102. Rothenberg, R., Sonenshine, D. E. 1970. Rocky Mountain spotted fever in Virginia: clinical and epidemiological features. *J. Med. Entomol.* 7:663–69
103. Rowley, W. A., Wong, Y. W., Dorsey, D. C., Hansler, W. J. Jr. 1973. Field studies on mosquito-arbovirus relationships in Iowa, 1971. *J. Med. Entomol.* 10:613–17
104. Rust, J. H. Jr., Cavanaugh, D. C., O'Shita, R., Marshall, J. D. Jr. 1971. The role of domestic animals in the epidemiology of plague. I. Experimental infection of dogs and cats. *J. Infect. Dis.* 124:522–26
105. Rust, J. H. Jr., Miller, B. E., Bahmanyar, M., Marshall, J. D. Jr., Pevenaveja, S., Cavanaugh, D. C., Hla, U. S. T. 1971. The role of domestic animals in the epidemiology of plague. II. Antibody to *Yersinia pestis* in sera of cats and dogs. *J. Infect. Dis.* 124:527–31
106. Ryckman, R. E., Ames, C. T., Lindt, C. C., Lee, R. D. 1954. Control of plague vectors on the California ground squirrel by burrow dusting with insecticides and the seasonal incidence of fleas present. *J. Econ. Entomol.* 47:604–7
107. Sailer, R. I. 1953. The blackfly problem in Alaska. *Mosq. News* 13:232–35
108. Sexton, D. J., Burgdorfer, W. 1975. Clinical and epidemiologic features of Rocky Mountain spotted fever in Mississippi, 1933–1973. *South. Med. J.* 68:1529–35
109. Simmons, S. W. 1944. Observations on the biology of the stable fly in Florida. *J. Econ. Entomol.* 37:680–86
110. Simmons, S. W., Dove, W. E. 1941. Breeding places of the stable fly, or "dog fly," *Stomoxys calcitrans* (L.) in northwestern Florida. *J. Econ. Entomol.* 34:457–62
111. Smith, C. N., Gouck, H. K. 1944. Sprays for the control of ticks about houses or camps. *J. Econ. Entomol.* 37:85–87
112. Smith, G. E., Pickard, E., Hall, T. F. 1969. Tree plantings for mosquito control. *Mosq. News* 29:161–66
113. Sonenshine, D. E., Atwood, E. L., Lamb, J. T. Jr. 1966. The ecology of ticks transmitting Rocky Mountain spotted fever in a study area in Virginia. *Ann. Entomol. Soc. Am.* 59:1234–62
114. Sonenshine, D. E., Peters, A. H., Levy, G. F. 1972. Rocky Mountain spotted fever in relation to vegetation in the eastern United States, 1951–1971. *Am. J. Epidemiol.* 96:59–69
115. Sonenshine, D. E., Stout, I. J. 1970. A contribution to the ecology of ticks infesting wild birds and rabbits in the Virginia-North Carolina piedmont. *J. Med. Entomol.* 7:645–54
116. Sudia, W. D., Newhouse, W. F., Calisher, C. H., Chamberlain, R. W. 1971. California group arboviruses: isolations

from mosquitoes in North America. *Mosq. News* 31:576–600
117. Taylor, D. J., Lewis, A. L., Edman, J. D., Jennings, W. L. 1971. California group arboviruses in Florida: host vector relations. *Am. J. Trop. Med. Hyg.* 20:139–45.
118. Tennessee Valley Authority. 1965. Survey of insects and other arthropods of medical importance. Land Between the Lakes. Progress Report, Reservoir Ecology Branch, Division of Health and Safety, July 1, 1965.
119. Tennessee Valley Authority. 1974. Holston river gets a "debug." *TVA Today* 5:3&6
120. Tennessee Valley Authority. 1975. Plan for the investigation and control of arthropod pests and vectors - Fy 1975–1976. (Revised 4-11-75)
121. Tirador, D. F., Miller, B. F., Stacy, J. W., Martin, A. R., Kartman, L., Collins, R. N., Brutsché, R. L. 1967. Plague epidemic in New Mexico, 1965. III. An emergency program to control plague. *Public Health Rep.* 82:1094–99
122. University of California. 1975. *Research on Mosquitoes - 1974.* Special Publ. 3014, Div. Agric. Sci. 28 pp.
123. Vianna, N. J., Hinman, A. R. 1971. Rocky Mountain spotted fever on Long Island. Epidemiologic and clinical aspects. *Am. J. Med.* 51:725–30
124. Voegtline, A. C., Ozburn, G. W., Gill, G. D. 1965. The relation of weather to biting activity of *Stomoxys calcitrans* (Linnaeus) along Lake Superior. *Papers Mich. Acad. Sci. Arts Lett.* L:107–14
125. Wagner, V. E., Newson, H. D. 1975a. Mosquito biting activity in Michigan state parks. *Mosq. News* 35:217–22
126. Wagner, V. E., Newson, H. D. 1975b. Field investigations on *Aedes fitchii* mosquito populations in a woodland pool ecosystem. *Mosq. News* 35:518–22
127. Waldbillig, R. C. 1968. Color vision of the female stable fly, *Stomoxys calcitrans. Ann. Entomol. Soc. Am.* 61: 789–91
128. Watts, D. M., Morris, C. D., Wright, R. E., DeFoliart, G. R., Hanson, R. P. 1972. Transmission of La Crosse virus (California encephalitis group) by the mosquito *Aedes triseriatus. J. Med. Entomol.* 9:125–27
129. Watts, D. M., Pantuwatana, S., DeFoliart, G. R., Yuill, T. M., Thompson, W. H. 1973. Transovarial transmission of La Crosse virus (California encephalitis group) in the mosquito, *Aedes triseriatus. Science* 182:1140–41
130. Weinmann, C. J., Garcia, R. 1974. Canine heartworm in California with observations on *Aedes sierrensis* as a potential vector. *Calif. Vector Views* 21:45–50
131. WHO. 1971. Blackflies in the Americas. WHO/VBC/71. 283, 32 pp.
132. Wirth, W. W., Blanton, F. S. 1974. *The West Indian Sandflies of the Genus Culicoides.* U.S. Dept. Agr. Tech. Bull. No. 1474, 98 pp.
133. Wright, R. E., DeFoliart, G. R. 1970. Some hosts fed upon by ceratopogonids and simuliids. *J. Med. Entomol.* 7:600
134. Yu-Hwa, E. W., Gill, G. D. 1970. Effect of temperature and relative humidity on mortality of adult stable flies. *J. Econ. Entomol.* 63:1666–68

Ann. Rev. Entomol. 1976. 22:355–76

GRAPE INSECTS

❖6134

Alexandre Bournier
Chaire de Zoologie, Ecole Nationale Supérieure Agronomique, 9 Place Viala, 34060 Montpellier-Cedex, France

The world's vineyards cover 10 million hectares and produce 250 million hectolitres of wine, 70 million hundredweight of table grapes, 9 million hundredweight of dried grapes, and 2.5 million hundredweight of concentrate. Thus, both in terms of quantities produced and the value of its products, the vine constitutes a particularly important cultivation.

THE HOST PLANT AND ITS CULTIVATION

The original area of distribution of the genus *Vitis* was broken up by the separation of the continents; although numerous species developed, *Vitis vinifera* has been cultivated from the beginning for its fruit and wine producing qualities (43, 75, 184). This cultivation commenced in Transcaucasia about 6000 B.C. Subsequent human migration spread its cultivation, at first around the Mediterranean coast; the Roman conquest led to the plant's progressive establishment in Europe, almost to its present extent. Much later, the Western Europeans planted the grape vine wherever cultivation was possible, i.e. throughout the temperate and warm temperate regions of the world: North America, particularly California; South America, North Africa, South Africa, Australia, etc.

Since the commencement of vine cultivation, man has attempted to increase its production, both in terms of quality and quantity, by various means including selection of mutations or hybridization. A wide variety of cultivars have been obtained and then spread by vegetative multiplication (74).

Viticulturists, in wishing to spread the cultivation of the vine, have had to chose among the numerous varieties (clones) those that were best adapted to the soil or climate of each particular region. Moreover, a variety of ecological conditions (e.g. soil and climate) have led the viticulturists to carry out many different cultivation practices. Principally because of pruning the vine has become an extraordinary polymorphic plant. Low pruning results in rootstocks bearing short shoots that are not more than 50–60 cm high (bush, cordon, or fan pruning), whereas the tall vines (trellised, arbor, trained vines, and vines on trees) can reach a height of several meters. Moreover, modern methods of cultivation, particularly mechanized harvesting, have caused the creation of *wide and tall vines*—rows of widely separated rootstocks (2.5 m or more apart), as planned vegetation.

This polymorphism and the diversity of plantation methods creates ecological conditions which inhibit or favor the action of the animal or vegetable pests of the vine. It is highly probable that the modern procedures of noncultivation (suppression of ploughing and use of herbicides) will soon have an influence on the action of pests.

THE PESTS[1]

Originating from a rather limited region, grapes are now intensively cultivated, sometimes to the extent of being a monoculture. When growing over many tens of thousands of hectares, the vine has attracted diverse pests indigenous to the ecosystems into which it has been introduced (4). These pests have adapted so well to the plant that occasionally biological races infesting the vine have developed from indigeneous species. Less commonly, monophagous insects from other *Vitis* species have been transported on plant samples and have infested *V. vinifera.* This is the case with Phylloxera, an insect whose appearance in Europe considerably disturbed and subsequently modified the cultivation of the vine there.

An important note must also be added on the subject of the adaptation of indigeneous insects during the introduction of the vine into a given country. The ecological niches offered by the biocenosis of the vine has been occupied in each country by different species. For example, the vine chrysomelids (*écrivains*) in Europe are different species from those in North America. It is the same for the so-called grape worms. However, these species quite often belong to the same genus —*Lobesia botrana* in Europe, *Lobesia viteana* in the USA. Consequently, pests of the vine will be considered in relation to the part of the plant attacked (Table 1).

On account of the large number of pests attacking the plant throughout the world, papers that provide an exhaustive list are rare; in fact, the only one available is that by Stellwaag in 1928 (174). However, monographs on vine pests are easily found in each vine-growing country: Spain (48), Italy (23, 168), Algeria (62), Switzerland (41), Romania (121), Moldavia (166, 201), Turkey (80, 88), Azerbaidjan (100), USSR (114), California (170), eastern USA (66, 123), Chile (137), Japan (89), Australia (18, 125, 182), and France (19, 20, 32, 43, 64, 75, 110, 122).

Root Pests

Formerly the most damaging pest of the vine was, without doubt, Phylloxera, *Dactylosphaera* (*Viteus*) *vitifolii,* which in a few years completely destroyed 2,500,000 ha of vine in western Europe. The species was described in 1854 in New York State and subsequently recorded at several sites in the eastern United States.

[1]Only insects and mites are studied here; nematodes are not, even though they cause direct damage to the vine roots and even though certain species of the genus *Xiphinema,* for example, are redoubtable vectors of virus (82, 198).

Table 1 Grape insects

Plant organ attacked	Pest: Order	Pest: Region	Ref.
Roots	Homoptera		
	Dactylosphaera (*Viteus*) *vitifolii*	World	12, 33–36, 42, 58, 98, 99, 107, 108, 117, 118, 122, 134, 135, 170, 175–179, 186, 187
	Cicada spp.	S. Europe	174
	Tibicen haematodes	S. Europe	29
	Cicada septemdecim	California	170
	Rhizoecus falcifer	California	170
	Margarodes meridionalis	California	170
	Margarodes vitis	Chili	174
	Margarodes capensis	S. Africa	174
	Margarodes greeni	S. Africa	174
	Eurhizoecus brasiliensis	Brasil	72
	Coleoptera		
	Bromius obscurus	Europe, N. America	19
	Fidia viticida	N. America	66
	Scelodonta strigicollis	India	109
	Vesperus spp.	France, Spain, Italy	63, 146
	Pentodon spp.	S. Europe	19, 32, 122
	Phyllognatus excavatus	S. Europe	85
	Opatrum sabulosum	Europe	32
	Lepidoptera		
	Vitacea polistiformis	Missouri	59, 155
Wood			
Trunk	T. Isoptera		
	Calotermes flavicollis	S. Europe	71
	Reticulitermes lucifugus	S. Europe	71
	Reticulitermes hesperus	California	170
	Lepidoptera		
	Cossus cossus	S. Europe	70
	Paropta paradoxus	Israel, Egypt	141
	Coleoptera		
	Anaglyptus mysticus	Bulgaria	163
Shoots	S. Homoplera		
	Eulecanium corni	S. Europe	20, 32
	Pulvinaria vitis	S. Europe	20, 32
	Diaspidiotus uvae	California	170
	Divers		
	Ceresa bubalus (Membracidae)	E. United States	143, 168, 192
	Bostrychidae	S. Europe	122, 129
	Macrophya strigosa (Tenthredinidae)	S. Europe	122
	Polycaon confertus (Coleoptera)		170
Buds and very young shoots	Lepidoptera		
	Arctia caja	S. Europe	19, 20, 27, 32
	Noctuidae	World	19, 20, 26, 32, 53, 151, 170
	Coleoptera		
	Peritelus sphaeroides	S. Europe	19, 32, 84
	Peritelus noxius	S. Europe	19, 32, 84

Table 1 *(Continued)*

Plant organ attacked	Pest: Order	Pest: Region	Ref.
	Limonius canus	California	170
	Glyptoscelis squamulata	California	170
	Pocalta ursina	California	170
	Phlyctinus callosus	S. Africa	24, 25
	Eremnus cerealis	S. Africa	24, 25
	Eremnus stulosus	S. Africa	24, 25
	Acarina		
	Calepitrimerus vitis	S. Europe	120, 148
	Eriophyes oculivitis	Egypt	11
	Eriophyes vitineusgemma	Moldavia	119
Leaf	Lepidoptera		
	Sparganothis pilleriana	S. Europe	142, 145, 149, 169
	Celerio lineata	Europe, N. America	19, 20, 32, 170
	Pholus achemon	California	170
	Antispila rivillei	Georgia	57
	Harrisina brillans	Mexico, California	84a, 123, 140, 170
	Sylepta lunalis	India	136
	Coleoptera		
	Haltica lythri subsp. *ampelophaga*	Europe	122
	Haltica chalybea	California	170
	Haltica torquata	California	170
	Byctiscus betulae	Europe	19
	Desmia funeralis	N. America	5, 69, 93
	Homoptera		
	Aphis illinoisensis	E. United States	123
	Cicadoidea		
	Philaenus spumarius	N. America, Europe	83, 144, 170, 200
	Scaphoideus littoralis	N. America, France, Germany, Switzerland, Italy	45, 50, 51, 76, 79
	Empoasca flavescens	Europe	13, 14, 30, 31, 52, 57, 128, 131, 160, 162, 189, 190, 193
	Empoasca lybica	Spain, S. Italy, Maghreb, Tanganyka	191
	Flata ferrugata	Punjab	171
	Unnata intracta	Punjab	171
	Zygina rhamni	France	
	Erythroneura adanae vitisuga	Bulgaria	102
	Erythroneura comes	California	96, 170, 173
	Erythroneura variabilis	California	96, 170, 173
	Erythroneura elegantulae	California	96, 170, 173
	Erythroneura ziczac	British Columbia	124
	Heteroptera		
	Nyzius senecionis	France	
	Nyzius ericae	Europe, America	22
	Acarina		
	Eotetranychus carpini	S. Europe	65, 81
	Panonychus ulmi	S. Europe	47
	Brevipalpus lewisi	Bulgaria	21

Table 1 *(Continued)*

Plant organ attacked	Pest		Ref.
	Order	Region	
	Tetranychus pacificus	California	104, 105, 111, 173
	Tetranychus flavus	California	170, 173
	Tetranychus atlanticus	France	
	Oligonychus mangiferae	India	167
Gall makers	Acarina		
	Eriophyes vitis	S. Europe	77, 126, 148
	Eriophyes vitigenusgemma	Moldavia	119
	Eriophyes oculivitis	Egypt	11, 202
	Homoptera		
	Dactylosphaera (Viteus) vitifolii	World	179, 180
	Diptera		
	Janetiella oenophila	France, Italy	129
	Lasioptera vitis	E. United States	66
	Dasyneura vitis	E. United States	66
	Schizomyia pomorum	E. United States	66
Fruits	Lepidoptera		
	Eupoecila ambiguella	Europe	3, 4, 17, 44, 56, 73
	Lobesia botrana	Europe	78, 91, 92, 96, 101, 112, 133, 145, 153, 165, 175, 183, 188, 197
	Lobesia viteana	E. United States	18, 66, 87, 181, 185
	Argyrotaenia politana	France	32
	Argyrotaenia velutinana	E. United States	95, 185
	Platynota stultana	California	7, 8, 115, 173
	Phalaenoides glycine	Australia	18
	Epiphyas postvittana	Australia	18
	Serrodes partitus	S. Africa	132, 199
	Coleoptera		
	Craponius inaequalis	E. United States	123
	Lopus sulcatus	France, Italy	32
	Heteroptera		
	Euchistus conspersus	California	170
	Chalcidoidea		
	Prodecatoma cooki	Florida	2, 55, 164
	Cecidomyidae		
	Contarinia viticola	France	122
	Thysanoptera		
	Drepanothrips reuteri	N. America, S. Europe	15, 16, 38–40, 139
	Anaphothrips vitis	Bulgaria, Romania, Greece, Turkey	203
	Haplothrips globiceps	Turkey	54
	Retithrips aegyptiacus	N. Africa, Middle East	152
	Rhipiphorothrips cruentatus	India	10
	Scirtothrips dorsalis	Japan	89
	Heliothrips haemorrhoidalis	World	
	Scirtothrips citri	California	170
Honeydew producers	Homoptera		
	Planococcus citri	S. Europe	6, 60
	Planococcus maritimus	California	6, 60
	Planococcus ficus	France	138

Table 1 *(Continued)*

Plant organ attacked	Pest		Ref.
	Order	Region	
	Eulecanium corni	S. Europe	20, 32
	Eulecanium persicae	S. Europe	20, 32
	Pulvinaria vitis	S. Europe	20, 32
	Pulvinaria betulae	Romania	127
	Trialeurodes vittatus	California	170
Aerial polyphagous insects	Orthoptera		
	Barbitistes fischeri v. *berenguieri*	France, Spain, Italy	122
	Ephippiger spp.	S. Europe	122
	Miogryllus convolutus	S. America	113
	Locusta migratoria	S. Europe, N. Africa	122
	Schistocerca peregrina	S. Europe, N. Africa	122
	Dociostaurus maroccanus	S. France	
	Coleoptera		
	Macrodactylus subspinosus	E. United States	
	Popilia japonica	California	123
	Anomala spp.	France	85
	Hymenoptera		
	Vespidae	World	156, 195
Polyphagous soil insects	*Melolontha melolontha*	Europe	85, 121
	Polyphylla fullo	Europe	85, 121
	Anoxia villosa	Europe	85, 121
	Otiorrhynchus sulcatus	Bulgaria	86
	Otiorrhynchus turca	Bulgaria	86
	Agriotes obscurus	Romania	90

One of the areas of penetration into Europe occurred in France in the Gard region (in 1863), where the insect was introduced on some American vinestocks. Westwood recorded it in England in 1867. Little by little the pest spread throughout Europe (186) and then, despite precautions, throughout the whole world. At present, regions that are still untouched are rare. The biology of this species has been studied in each country where it has been introduced. In some places the full cycle of both aboveground and underground stages have been observed, in others only the underground stage is constant. Sometimes rare swarms of winged forms unable to reproduce are found (108, 117, 118, 170, 176). These variations in the insect's life cycle are usually explained by ecological conditions, temperature, and humidity (108, 117, 118). Some have thought also that the isolation of diverse populations of the underground stage and their adaptation to the biotope has permitted the creation of races with particular morphologies or biologies (33, 118, 135). For example, on *Vitis vinifera,* the radicicoles produce enormous tuberosities on the roots, thereby causing the death of the rootstock. On American vines they cause only very slight damage because they produce only shallow wounds and because many of them swarm before winter. The origin of this resistance has been the subject of many studies (34–36, 58, 107, 134) because it is the basis for the control of Phylloxera.

The cultivars of *Vitis vinifera* are therefore grafted onto American *Vitis* stocks. Attempts have also been made to hybridize them to avoid grafting, which is a burdensome operation which causes a delay in production. But these resistant directly-producing hybrids, if allowed to spread again to the area of vine cultivation, give a disagreeable taste to the wine. Planting these hybrids has been prohibited in France. Another method of control for radicicole Phylloxera consists of injecting carbon disulfide (CS_2) into the soil, but this method is very costly and can only be applied over small areas and in vineyards yielding high returns (42). Other methods of chemical control have been recommended for the gall-making stages (12, 42, 98, 177, 179), as well as for the root-feeding stages (98, 99, 175, 178, 187). Note that the radicicoles may also be destroyed by flooding and submersion of the vines for 50 days.

Among the other root-sucking insects one should mention the cicadas, *Cicada plebeja, C. orni, C. atra* (174), and *Tibicen haematodes* in Europe, (29) and *C. septemdecim* in California (170). The larvae, which live for several years in the soil, pierce the cortical parenchyma and suck the exudate.

Certain coccids are also found on vine roots. The ground mealybug *Rhizoecus falcifer* in California is seldom damaging (170). On the other hand in *Margarodes meridionalis* in California (170), *Eurhizococcus brasiliensis* in Brazil (72), *M. vitis* Chile, and *M. capensis* and *M. greeni* in South Africa (174), the globular females fix themselves on the rootlets and thus weaken the stocks.

Coleopterous larvae, which have a subterranean life, cause damage by gnawing the roots. In Europe *Bromius obscurus,* which was previously a pest of some importance, has now become quite rare. The larvae destroy the cortical part and the superficial wood of the roots (19). They live with difficulty on the roots of American *Vitis,* which explains their rarity after the widespread use of American rootstocks following the crisis caused by Phylloxera. The adult bites and pierces the parenchyma of the leaves and the epidermis of the young grapes following a sinuous line, hence the name *writer* (*écrivain*) which has been given to it. The species is parthenogenetic and only the female is known. It has been introduced into North America, where another species, the grape rootworm beetle, *Fidia viticida,* causes similar damage (66). In India a third Eumolpine, *Scelodonta strigicollis,* behaves in the same way as the two above-named species. The wounding caused by the larvae on the roots often leads to the death of the rootstock. The adult gnaws the surface of the young leaves of the vine shoots and the young grapes (109).

Other oligophagous Coleoptera larvae are also injurious to the roots of the vine. Around the Mediterranean the genus *Vesperus,* with *V. xatarti, V. strepens,* and *V. luridus,* causes local but often serious damage (63, 146); in addition, *Pentodon punctatus, P. idiota, P. bispinosus* (19, 32, 122), *Phyllognathus excavatus* (85), and a Tenebrionidae, *Opatrum sabulosum* (32), have larvae which attack the collar of the young rootstocks and gnaw the graft calluses.

The caterpillar of *Vitacea polistiformis* also causes considerable damage to roots in the Missouri region, where certain vineyards have been completely destroyed (59). Their life cycle lasts three years. The first-instar larvae have very poor resistance to drought (155).

Wood-Damaging Pests

ON THE TRUNK Termites attack rootstocks that are in poor condition for any of several reasons such as lesions caused by frosting, by agricultural implements or poor circulation of the sap, due, for example, to an incorrect choice of a graft stock that is ill-adapted to the soil. Examples in France are the yellownecked termite *Calotermes flavicollis* and, more rarely, *Reticulitermes lucifugus* (71), and in California *Reticulitermes hesperus* (170).

Cossus cossus (70), whose large, red caterpillar sometimes destroys the rootstock in Europe, has its homolog in *Paropta paradoxus* (141) in Israel and in Egypt. Finally, in Bulgaria a cerambycidae, *Anaglyptus mysticus,* has recently been observed (163). The larvae destroy the cambial tissue, causing the death of 13% to 15% of the rootstock of four- to seven-year-old vines.

The control of trunk-attacking species is difficult. For the termites and *Cossus,* the only method of control is to maintain the vine in excellent condition. On the other hand, it seems that oleoparathion preparations or dichlorvos are effective for controlling *A. mysticus* (163).

ON THE SHOOTS *Scale insects* In Europe and America, two Diaspidae, *Eulecanium corni* and *Pulvinaria vitis,* attack the shoots, as does the grape scale *Diaspidiotus uvae* in California (170). They are damaging not only because they remove the sap, but also because of the toxicity of their saliva. Other species of Coccidae [*Planococcus citri* (20, 32) and *Planococcus ficus* (138) in Europe and *Planococcus maritimus* in America (170)] cause a similar nuisance, but they are mobile and damage principally the young grapes because of the coccids production of honeydew and the subsequent growth of the sooty mould fungi on the honeydew.

Others Other insects attack the vine shoots but the damage inflicted is never very important, as in the following examples:

1. Cicadas make fusiform egg-laying cicatrices.
2. *Ceresa bubalus* has rings of egg-laying punctures around the shoot which destroy the vascular system and produce a spectacular callus above the wound (168, 192). This species passes part of its life cycle on fruit trees and the vine and then part on herbaceous cultivated or adventitious plants. An attempt at biological control using the egg parasite *Polynema striaticorne* is underway (143).
3. The Bostrychids—*Apate sexdentatum* (129), *Apate muricata, Schistoceros bimaculatus*—attack decaying stocks and penetrate at bud level to lay eggs there. The larvae bore longitunal galleries in the pith (122).
4. The branch and twig borer *Polycaon confertus* is found in California. The adult feeds by chewing a hole in the base of the young shoots, which then may be broken by the wind. The larva lives in the main branches of the stock in decaying or dead wood (170).
5. The vine sawfly, *Macrophya strigosa,* lays its eggs in the green shoots; the larva descends into the pith and bores its gallery until it reaches the buds. The damage can be locally important (122).
6. Finally, numerous Hymenoptera species make notches in the pith to nest there.

Bud Pests

The damage caused by these insects always has a catastrophic effect on the crop. In Europe, the tiger moth, *Arctia caja,* or *chenille bourrue,* so named because of its hairiness, overwinters as a caterpillar on weeds and afterwards passes onto the vine devouring five to six buds a day at the time of bud break and later on attacks the young shoots and the floral buds. A method of chemical control has been perfected (27); in addition, an entomogenous fungus, *Entomophtora sphoeroderme,* produces fierce and deadly epidemics among the caterpillar populations.

Throughout the world numerous species of noctuids (cutworms) are known as bud destroyers. In Europe the most important pests are *Autographa gamma, Agrotis segetum, Euxoa nigricans, Triphaena pronuba,* etc (53). In America some of the most prominent pests are the variegated cutworm *Peridromia margaritosa,* the Greasy cutworm *Agrotis ipsilon* (170), and the cutworms, *Anagrotis barnesis* (134) and *Rhynchagrotis cupida* (151). The caterpillars of all these species pass the day a few centimeters below the soil at the foot of the rootstocks; they come out at night to feed on the buds. Although the viticulturist was for a long time defenseless against them, they can now be effectively controlled (26, 53).

Some Coleoptera in the adult stage may also attack buds, e.g. in Europe the vine grubs (*coupe bourgeons*), *Peritelus sphaeroides* and *P. noxius* (84), and in America the diurnal click beetle *Limonius canus,* the grape bud beetle, *Glyptoscelis squamulata,* whose adults penetrate into the buds without damaging the scales, and the little bear beetle, *Pocalta ursina,* which does only occasional damage (170). In South Africa three species of vine grub are found: *Phlyctinus callosus, Eremnus cerealis,* and *E. stulosus* (24, 25).

Finally, one must mention the mite, *Calepitrimerus vitis,* which hinders the development of the buds and young shoots by the damage caused by infestations. The internodes are short and sometimes the grapes abort (120, 148). *Eriophyes oculivitis* (11) in Egypt and *E. vitineusgemma* (119) in Moldavia have recently been described, and their damage is similar to that of the preceding species. Some tydeids are effective predators of these phytophages (157, 159, 202).

Leaf Pests

LEPIDOPTERA The pyralid *Sparganothis pilleriana,* once common throughout all French vineyards, appears to have declined sharply (169). It has only one generation per year, and the first-instar caterpillar, which hibernates in a cocoon under the bark of the vinestock, can be destroyed at that moment by sprayings of sodium arsenite. Use of this chemical is currently allowed, but it is being replaced by phosphoric acid esters (145). At the three-to-four-leaf stage the larvae are killed by toxaphene, which is also active against the hairy caterpillar (*chenille bourrue*) (149). Numerous indigenous parasites control the populations of the pyralid. The caterpillars have a marked preference for the plants whose leaves are rich in protein (142). Among the sphingid caterpillars, the whitelined sphinx moth, *Celerio lineata,* is the most damaging in France as well as in California. In the latter region, the achemon sphinx moth, *Pholus achemon,* is also found (170). The larvae of *Antispila rivillei* have been recorded as vine leafminers in Georgia (57).

Another Lepidoptera, the western grapeleaf skeletonizer, *Harrisina brillans,* which has passed from Mexico into California, has curious, gregarious caterpillars with black-and-red-striped, yellow bodies. The larvae destroy the leaf blade without damaging the upper epidermis (140, 170). There have been attempts at microbiological control (123) and biological control (84a). In India the caterpillars of *Sylepta lunalis* destroy just as completely the parenchyma of the leaves which they roll (136).

COLEOPTERA The halticids, *Haltica lythri* ssp. *ampelophaga* in Europe and *Haltica chalybea* and the grape flea beetle *Haltica torquata* (170) in California, have gregarious larvae which also eat the leaf blade without damaging the cuticle.

The leafrollers *(cigariers), Byctiscus betulae,* in France and the grape leaffolder, *Desmia funeralis,* in America have females which cut the leaves and roll them into "cigars" within which they lay their eggs and which serve as food for the larvae. Treatment trials against this last species have shown that a preparation based on *Bacillus thuringiensis* is as effective as some chemical pesticides (carbaryl) (5, 93). Attempts at biological control by the rearing and release of natural parasites have given encouraging results (69).

HEMIPTERA Only one species of aphid has been recorded on the vine, *Aphis illinoisensis.* It produces heavy infestations on young shoots and on the leaves in summer to the east of the Mississippi River. It is dioecious and migrates in autumn onto its principal host (123).

On the other hand, a number of cicadellid species are vectors of phytopathogenic agents. Pierce's disease, found in North America, has 20 species as vectors (83, 144, 170, 200). One of the most efficient among these is *Philaenus spumarius,* a cercopid, which also occurs in France where very fortunately Pierce's disease is not found. The golden yellows, a disease which appeared about 1950 in the southwest of France, is also caused by a molicute (79). The vector is *Scaphoideus littoralis,* a species of American origin. The disease has spread and at present affects Switzerland, Italy, Sardinia, Corsica, and perhaps Germany (45, 50, 51, 76). Furthermore, the disease of black wood (*bois noir*), which is rife in Burgundy and in the Jura, very probably has a cicadellid as vector also (49, 194).

The cicadellids can also cause direct damage by piercing the phloem into which their toxic saliva is injected (30, 31, 160, 162, 189, 190), e.g. *Empoasca flavescens,* which has multiplied in France during the last few years, perhaps because of the physiological condition of the vine after pesticide treatment (57). It causes a crinkling of the leaf, a hardening of the leaf blade, and a browning of the nerves (14). The leaves dry up from the periphery towards the center and then fall (131, 193). Quite often the treatment against the grape moths *(tordeuses)* is sufficient to control *Empoasca vitis* (= *E. flavescens)* (13, 52, 128). *Empoasca libyca* (191) causes the same damage in hotter regions, such as southern Spain, Sicily, Sardinia, North Africa, Israel, and Tanganyika. *E. decipiens,* though not a pest in France, is one in Baluchistan. Recently, two Flatidae, *Flata ferrugata* and *Unnata intracta,* have infested the vine in the Punjab (171). Finally, other species—*Zygina rhamni* in

France, *Erythroneura adanae vitisuga* in Bulgaria (102), *E. comes, E. variabilis,* and the grape leafhopper, *E. elegantulae,* in California (96, 173), and *E. ziczac* in British Columbia (124)—are also damaging to the vine. *Ceresa bubalus,* already mentioned for the damage it does on the shoots, also causes a crinkling and a reddening of the leaves.

Other Hemiptera also damage the leaves, such as *Nyzius senecionis,* which pierces at bud level and whose toxic saliva produces the drying-up of what is above the wound. The false chinch buy *Nysius ericae* lives in Europe and America where it causes damage similar to that of the above-named species (22). These insects migrate by walking from their plant host to the vine. They can thus be controlled by creating a barrier consisting of a ten-meter band of land powdered with lindane.

ACARINES The punctures of the cell feeders gravely damage the leaves, which dry up and fall. The cell feeders in Europe consist of two Tetranychidae: the yellow spider mite, *Eotetranychus carpini,* which lives on the underneath side of the leaves and whose punctures cause red or yellow stains, according to the vine cultivar, along the nerves (65, 81), and the red spider mite, *Panonychus ulmi,* which lives on both sides of the leaves, causing them to take on a grayish tint (47). Recently, heavy infestations of *Tetranychus atlanticus* have appeared in the vineyards of Mediterranean Languedoc, as a consequence of using herbicides (A. Rambier, 1975, personal communication). Other, less-damaging species exist on the vine in France and throughout Europe (21, 126). In California *Tetranychus pacificus* is the homolog of the French tetranychids on the vine (104, 105, 111, 173), and *Tetranychus flavus* dries up the buds and the young, unopened leaves. Their most effective predator is *Metaseiulus occidentalis* (9, 104, 105). *Oligonychus mangiferae* has recently been observed in India, where it produces damage similar to that from the tetranychids on the French vines (167). The mite problem did not exist before the application of synthetic insecticides against the grape moths (*tordeuses*). Many authors have tried to determine the causes inducing these multiplications (45, 56, 158, 173). They are multiple, principally including the selection of resistant races, the suppression of predators (typhlodromids, anthocorids, thrips, etc) (37, 65, 147, 157), and the effects of insecticides on the physiology of the vine and on the mites are in a sense favorable to the fecundity and longevity of the latter (56). Because many synthetic insecticides induce heavy mite infestations, the control of other insects damaging to the vine is often rendered more difficult (3, 17, 73).

GALLMAKERS Leaf blister is caused by the mite *Eriophyes vitis*. The gall occurs as a depression on the underneath part of the leaf. It is lined with white hairs, within which are found the minute mites (77, 126, 148). *E. vitigenus gemma* and *E. oculivitis,* both recently described [the first is from Moldavia (119), the second from Egypt (11, 202)], do not form a blister but instead dry up the buds.

Some other species can also be gallmakers. By its punctures the gall-living form of Phylloxera provokes the formation of galls on American vines and on certain Franco-American hybrids. The gallforming female lives in a globular gall, which opens onto the upper surface of the leaf. Only the young larvae can leave by passing through the hairs that obstruct the orifice. The sensitivity of different vine cultivars to attack by the gallforming aphids has been tested (180), and diverse insecticides have tried as controls against them (179).

The leaf cecidomyid *Janetiella oenophila* forms lenticular galls between the two cuticles of the leaf (129). One can mention also in America the grapevine tomato gallmakers, *Lasioptera vitis* and *Dasyneura vitis,* as well as another Diptera, *Schizomyia pomorum* (66).

Fruit Pests

The grape moths *(tordeuse de la grappe)* give the most trouble to French viticulturists. The grape tortrix *Lobesia botrana* and the grape tineid *Eupoecilia ambiguella* are often confused by the experts because they have fairly similar biologies and synchronous flight periods, and they cause nearly the same damage. In France *L. botrana* has three annual generations and lives principally in dry situations, whereas *E. ambiguella* has only two generations and lives in humid situations. Their biology (44, 92, 112, 133, 165) and methods of control have been the object of many studies in all countries (3, 4, 68, 91, 97, 101, 112, 145, 153). The treatments used are of two types: preventive, by applications before hatching of the eggs, or curative, by using control measures against the young caterpillars. The treatments are applied after announcements are broadcasted by the Plant Protection Service. In effect, the biology of the two species depends directly on meteorological conditions (175), and it is necessary to have recourse to trapping to determine with precision the date and importance of the flight of adult moths. The captures from food traps are often difficult to identify (183, 197), and confusions with non-pest species can occur (96). The latter can be avoided by the use of sexual traps, which moreover have a high efficiency (78, 188). Observation of egg laying in the vines is being used more and more to fix with certainty the necessity and date of treatment. The damage by the second and especially that done by the third generation of *Eupoecilia* is often severe. The caterpillars penetrate into the grapes, and the entry holes favor the establishment of the fungus of the gray mold *Botrytis cinerea* (3). It is noteworthy that the treatments against the grape moths, in particular those with phosphoric esters, have induced heavy infestations of mites in the vineyards (3, 17, 56, 73).

In the eastern United States, the grape berry moth, *Lobesia viteana,* has the same biology and causes the same damage as the French *Eudemis,* but because pupation occurs not under the bark of the vine stock but instead in a rolled-up leaf on the ground, efficient control by cultural means (66, 87, 181) and by using *Bacillus thuringiensis* (28) is possible. A confusion operation with the aid of sex pheromones has also given good results (185).

The small grape moth *Argyrotaenia politana* was discovered for the first time in Montpellier, France, in 1954. It is questionable whether it is a species newly adapted to the vine, coming from the numerous apple orchards planted a few years before in the region, or whether it has been confused with *Eudemis,* which the adult closely resembles. The damage inflicted by the two species is the same, only the treatment dates differ (32). In the eastern half of the United States, a related species, the redbanded leafroller, *Argyrotaenia velutinana* occurs (95, 185). In California

Platynota stultana (7, 8) has become an important pest during the last few years (115, 173); all of the nonligneous parts are attacked, but the greatest damage is to the grapes. The numerous wounds inflicted on the epidermis permit fungal spores to penetrate and cause rot. Crop losses can reach 15%. In Australia the vine moth, *Phalaenoides glycine,* and the light brown apple moth, *Epiphyes postvittana,* cause the same damage and are justifiably controlled by the same methods as the above-named species (18). A Coleoptera, the grape curculionid *Craponius inaequalis* lays its eggs in the grapes, where the larvae consume the seeds and the pulp (123). In California the consperse stink bug, *Euchistus conspersus,* which habitually lives on low plants, pierces the grapes and thereby greatly depreciates the crop of table grapes (170). A chalcid in Florida, *Prodecatoma cooki,* whose larva lives in the pulp and the seeds (2, 55, 164) and the fruit-piercing moth, *Serrodes partitus* (F.), in South Africa must also be noted (132, 199). Finally, two species whose very infrequent damage can be considered as negligible in France must be mentioned: the polyphagous capsid *Lopus sulcatus,* which sometimes passes onto the vine (32) and whose larvae and adults pierce the flower buds that blacken and fall, and the vine cecidomyid *Contarinia viticola,* cause similar damage (122).

Many species of thrips (Thysanoptera) attack the vine in the world. They are very rarely mentioned because on account of their small size they pass unnoticed and their damage is attributed to other pests. The grape thrips, *Drepanothrips reuteri,* probably originated from California where its biology and control have been studied (15). Introduced into Europe, it had been recorded as damaging solely American vines (139). During the last ten years it has been recorded as damaging hybrid vines first of all (38), then French vines (16, 39, 40). Over and above the damage it causes to the leaves (necrosis or holes in the blade), the punctures of the larvae and of the adults produce a toxic reaction which retards the development of the shoots and causes a certain amount of abortion at the flowering period. The most severe damage is recorded on the grapes where it causes necrosis and suberization of the epidermis; this considerably decreases the varieties of table grape. Its biology and methods of control have been studied in Europe (16, 38–40). The treatments against the second generation of the grape moths are also valuable against thrips. Some vicariant species in other climates occupy the same ecological niche as this species and cause similar damage to the grapes. These include, for instance, *Anaphothrips vitis* (203) in Bulgaria, Romania, and Greece; *Haplothrips globiceps* (54) in Turkey; *Retithrips aegyptiacus* in the Middle East (152), Egypt, and North Africa; *Rhipiphorothrips cruentatus* (10) in India; *Scirtothrips dorsalis* in Japan (89); and *Heliothrips haemorrhoidalis,* a polyphagous species of warm and warm temperate climates which also passes onto the vine in Chile, for example. Other species can also occasionally attack the vine: *Scirtothrips citri* (170) close to citrus orchards and *Hercothrips fasciatus,* as well as the grass thrips *Frankliniella moultoni, F. occidentalis,* and *F. minuta* in the United States.

Honeydew Producers

Certain insects are also a nuisance indirectly because of the production of a sweet honeydew. This secretion serves as a substrate for a black fungal growth, the sooty mould, which greatly depreciates the quality of table grapes.

Among the coccids, the females of *Planococcus citri,* mobile in all stages, have soft integument covered with a white bloom. The species can be effectively controlled by releases of a coccinellid, *Cryptolaemus montrouzieri* (13). *Pseudococcus maritimus* (6, 60) is the California homolog of the above species. One can also mention *Pulvinaria vitis, Eulecanium corni,* and *Eulecanium persicae,* which are controlled successfully by the entomogenous fungus *Beauveria bassiana,* and *Pulvinaria betulae* in Romania (127). These species are quite often controlled by the treatments against the grape moths. Furthermore, since 1969, perhaps because of the employment of organophosphorus insecticides, a new species that is locally very damaging has multiplied throughout the Mediterranean basin: *Planococcus ficus* (138). In California the aleurodid *Trialeurodes vittatus* also produces honeydew.

Polyphagous Aerial Insects

These are principally the Orthoptera. Among the Ensifera, the *Boudrague, Barbitistes fischeri* var. *berenguieri* (122) and three species of the genus *Ephippiger*—*E. ephippiger, E. bitterensis,* and *E. terrestris*—gnaw the foliage and the green or ripe grapes in the Mediterranean vineyards. The damage, although episodic, can be important (122). Among the acridians, *Locusta migratoria, Schistocerca peregrina,* and *Dociostaurus maroccanus* have gregarious bands that are particularly destructive. *Miogryllus convolutus* has recently been recorded as damaging vines in South America (113).

Some polyphagous Coleoptera, in particular the scarabs, can sometimes destroy the green parts of the vine. In the eastern United States the rose chafer, *Macrodactylus subspinosus,* and the Japanese beetle *Popilia japonica,* whose larvae also attack the roots (123), are found. In France there are the green vine chafer, which includes in reality a complex of three related species—*Anomala vitis, A. dubia,* and *A. ausonia* (85)—which are found only in the vines on the sands of the Mediterranean coast and whose larvae only feed on decomposing plant material. Finally, practically everywhere in the world (156, 164, 195) the wasps (Vespidae) attack the ripe grapes, causing serious damage to table varieties.

Polyphagous Soil Insects

Mostly polyphagous soil insects are cockchafer grubs, larvae of various species: the common cockchafer, *Melolontha melolontha,* the pine chafer, *Polyphylla fullo,* and the hairy chafer, *Anoxia villosa* (85, 121). In Bulgaria (86) *Otiorrhynchus sulcatus* and *O. turca* larvae are recorded damaging the roots. Larvae of *Agriotes obscurus* cause damage in Romania (90).

CONCLUSION

In France, the major arthropod grape pests are the grape leaf rollers, *Lobesia botrana* and *Eupoecilia ambiguella,* and the tetranichids, *Eotetranychus carpini* and *Panonychus ulmi.* The two mite species were not pests until 1950 when the ill-considered use of organophosphates in controlling leaf rollers caused catastrophic outbreaks of mites. To avoid such outbreaks, studies were undertaken to promote the concept of integrated control in vineyards (1, 73, 148).

To control leaf rollers knowledgeably, levels of economic tolerance were established; these vary in relation to the phenologic stages of the vine, the climate, and the value of the crop. Population dynamics were based on data from traps, by using light, attractants, and sex pheromones, as well as from observing egg laying by adult moths. The hibernating forms of mites were carefully counted to predict future outbreaks.

French workers have shown that the use of certain pesticides causes mite outbreaks in two ways. First, pesticides destroy mites' natural predators, *Scolothrips* spp., *Orius vicinus* (Anthocoridae), *Stethorus punctillum* (Scimniini), and *Typhlodromus* spp. (147, 148). Additionally, certain insecticides and fungicides alter plant metabolism to produce conditions more favorable to mite growth and reproduction. In this way it has been shown that soluble nitrogen (amino acids) and reducing sugars play a large role in the nutrition and reproduction of mites (1, 56, 57). Outbreaks of leafhoppers (*Empoasca flavescens* in particular) seem to follow an analogous process. For this reason, after detailed investigations, a list of pesticides and their effects on pests and auxiliaries was compiled (1). The problem of tetranychid outbreaks on grapes has also been studied in California (73). Ecological studies have made it possible to intensify the predation of *Metaseiulus occidentalis,* which attacks *Eotetranychus willamettei* (a minor pest) when *Tetranychus pacificus* is absent. As in France, pesticides that spare predators were selected.

To avoid the disadvantages of chemical control, recent experimental work has tested the effectiveness of *Bacillus thuringiensis* against larvae of *Lobesia botrana* (153). These are very susceptible but two difficulties are not yet resolved. The fact that eggs are laid and hatched successively over protracted periods and the behavior of the insect make contact between larvae and spores uncertain. Preventive treatment therefore cannot be based on defined infection thresholds.

Other projects, using entomophagous insects, have also been initiated. Two parasites of *Harrisina brillians,* the tachnid *Sturmia harrisinae,* and the hymenopteran *Apanteles harrisinae,* as well as a virus introduced from Arizona, were effective in lowering *H. brillians* populations in San Diego County, California.

Controlling the leafhopper *Erythroneura elegantula* by the mymarid *Anagrus epos* has been improved by providing refuges of *Rubus* sp. that harbor the nonpest species *Dikrella cruentata,* the eggs of which allow hibernation of the parasite. Other trials of entomophagous insects, e.g. against *Desmia funeralis,* are in progress.

There is no question that the continuing progress of ecological sciences will lead to a greater effectiveness of integrated control programs for grape pests as they have for other crops.

Literature Cited

1. A C T A. 1975. Lutte intégrée en vignoble. *Assoc. Coord. Tech. Agric. Spéc.* 5. 62 pp.
2. Adlerz, W. C. 1972. *Prodecatoma cooki* (How.), a seed Chalcid on Florida grapes. *J. Econ. Entomol.* 65(5):1530
3. Agulhon, R. 1973. Trois problÈmes phytosanitaires étudiés dans le vignoble en 1972: l'excoriose, les Tordeuses de la grappe, les acariens. *Vignes et Vins Mai 1973,* 219:15–22
4. Alexandri, A. A. 1973. Efficacy of some insecticides in the control of the grape moth (*Polychrosis botrana* Schiff.) (Rouman). *Ann. Inst. Cerc. Pentru. Protect. Plant.* 9:507–14
5. Aliniazee, M. T., Jensen, F. L. 1973. Microbial control of the grape leaf folder with different formulations of *Bacillus thuringiensis. J. Econ. Entomol.* 65(1):157–58
6. Aliniazee, M. T., Stafford, E. M. 1972. Control of the grape mealybug on "Thompson Seedless" grapes in California. *J. Econ. Entomol.* 65(6):1744
7. Aliniazee, M. T., Stafford, E. M. 1972. Notes on biology, ecology and damage of *Platynota stultana* WLSM on grapes. *J. Econ. Entomol.* 65(4):1042–44
8. Aliniazee, M. T., Stafford, E. M. 1973. Management of grape pests in Central California vineyards. 1. Cultural and chemical control of *Platynota stultana* on grapes. *J. Econ. Entomol.* 66:154–57
9. Aliniazee, M. T., Stafford, E. M., Kido, H. 1974. Management of grape pests in Central California vineyards:toxicity of some commonly used chemicals to *Tetranychus* pacificus and its predator *Metaseiulus occidentalis. J. Econ. Entomol.* 67(4):543–47
10. Ananthakrishnan, T. N. 1971. Thrips in Agriculture, Horticulture and Forestry-Diagnosis, Bionomics and Control. *J. Sci. Indust. Res.* 30(3):113–46
11. Attiah, H. H. 1969. *Eriophyes oculivitis* n. sp. a new bud mite infesting grapes in the U.A.R. *Bull. Soc. Entomol. Egypte* 51:17–19
12. Avdyshev, Sh. E. 1971. Prospects for chemical control of grape vine Phylloxera. *C.R. Int. Congr. Moscou, 13th* 1968:206–7
13. Aykac, M. K., Erguder, T. M. 1974. A study of control measures against *Planococcus citri* (RISSO) in the vineyards of Tokat province. *Sams. Böl. Zir. Mucad. Arast. Enstit.* 43:171–72
14. Baggiolini, M., Canevascini, V., Caccia, R. 1972. La Ciccadelle verte (*Empoasca flavescens* F.) cause d'importants rougissements du feuillage de la vigne. *Bull. OEPP* 3:43–49
15. Bailey, S. F. 1942. The grape or vine thrips *Drepanothrips reuteri* Uzel. *J. Econ. Entomol.* 35(3):382–86
16. Baillod, M. 1974. Dégâts de thrips sur vigne en Suisse romande. *Rev. Suisse Vitic. Arboric. Hortic.* 6(2):45–48
17. Baillod, M. 1974. La protection de la vigne contre l'araignée rouge *Panonychus ulmi* (Koch) et l'araignée jaune commune *Tetranychus urticae* (Koch). *Rev. Suisse Vitic. Arboric. hortic.* 6(1): 17–22
18. Baker, B. T. 1974. Practical pest and disease management in vineyard. *Aust. Grapegrower Winemaker April* pp. 1–10
19. Balachowsky, A. S. 1962–1972. *Entomologie Appliqué à l'agriculture,* Vol. 4 Paris:Masson
20. Balachowsky, A. S., Mesnil, L. 1935. *Les insectes nuisibles aux plantes cultivées,* Vol. 1–2. Paris:Masson. 1921 pp.
21. Balevski, A., Martinov, S., Nachev, P. 1970. The grape-vine mite (*Brevipalpus lewisi* (McGr.) the cause of dessication of grape vines in the Vidin district. *Gradinar. Lozar. Nauka,* 7(6):67–76
22. Barnes, M. M. 1970. Grape pests in Southern California. *Calif. Agric. Exp. Stn. Circ.* 553:1–10
23. Beffa, D. G. 1949. *Gli Insetti Dannosi All'agricoltura e i Moderni Metodi e Mezzi di Lotta,* ed. Miland: Hoepli. 978 pp.
24. Berg, H. C. Van den. 1971. The biology and control of vine snout beetle. *Decid. Fruit Grow.* 21:83–85
25. Berg, H. C. Van den, Giliomee, J. H. 1972. Aspects of the ecology and behaviour of *Eremnus cerealis* Marshall. I. The emergence and distribution of the adults and the distribution of the larvae in the vineyard. *J. Entomol. Soc. South. Afr.* 35(1):171–76
26. Berville, P. 1954. La lutte contre les vers gris de la vigne. *Phytoma* 56:26
27. Berville, P., Terral, A. 1958. Observations et essais de traitement sur les Chenilles bourrues. *Phytoma* 102:7–10
28. Biever, K. D., Hostetter, D. L. 1975. *Bacillus thuringiensis* against lepidopterous pests of vine grapes in Missouri. *J. Econ. Entomol.* 68(1):66–70
29. Blunck, H. 1956. See Ref. 172, pp. 190–95

30. Boller, E. 1971. Untersuchungen an der Rebzikade (*Empoasca flavescens* F.) und am embindigen Traubenwickler (*Clysia ambiguella* Hb.) in der Ostschweiz. *Schweiz. Z. Obst Weinbau* 106:651–60
31. Bonfils, J., Leclant, F. 1972. Reconnaissance et nuisibilité des Cicadelles sur la vigne. *Prog. Agric. Vitic.* 89(14):343–55
32. Bonnemaison, L. 1962. Les ennemis animaux des plantes cultivées et des forêts, Vol. 3 Paris:SEP. 1502 pp.
33. Börner, C. 1925. Die neuen Forschungen zur Reblausrassenfrage. *Dtsch. Weinbau* 1:5–8; 2:18–19; 3:29; 4:36–37; 5:48–49; 6:118–124
34. Börner, C. 1939. Anfalligkeit, Resistenz und Immunität der Reben gegen die Reblaus. *Z. Hyg. Zool. Schädlings be Kämpf.* 31:274
35. Boubals, D. 1966. Etude de la distribution et des causes de la résistance au Phylloxera radicicole chez les Vitacées. *Ann Amélior. Plant.* 16(2):145–84
36. Boubals, D. 1966. Hérédité de la résistance au Phylloxera radicicole chez la vigne. *Ann. Amélior. Plant.* 16(4): 327–47
37. Bournier, A. 1954. A propos des thrips de la vigne. *Prog. Agric. Vitic.* 142:104
38. Bournier, A. 1957. Le thrips de la vigne: *Drepanothrips reuteri. Ann. Ecole Natl. Agric. Montpellier* 30(1):145–57
39. Bournier, A. 1962. Dégâts de thrips sur vignes françaises. *Prog. Agric. Vitic.* 79(7):164–73
40. Bournier, A. 1965. Sur l'adaptation d'un Insecte nuisible à une culture *Drepanothrips reuteri* Uzel sur vignes françaises. *C.R. Congr. Int. Entomol. Londres 12th, 1964.* Son *9a, Agric. Entomol.* p. 537
41. Bovey, R. 1967. *La Défense des plantes Cultivées,* 176–218. Lausanne: Payot. 847 pp.
42. Branas, J. 1968. Le Phylloxera et les Insecticides. *Prog. Agric. Vitic.* 85(17): 401–9
43. Branas, J. 1974. *Viticulture.* Montpellier: ENSA. 990 pp.
44. Carbo Saguer, J., Ripolles Moles, J. L., Bricio Sanz, M., Fabregas Sole, C. 1973. Studies on the bionomics of the vine moths (*Lobesia botrana* and *Clysia ambiguella*). *Bol. Inf. Plag.* 112:49–51
45. Carle, P. 1965. Essais de pesticides en plein champ contre *Scaphoideus littoralis. Phytiatr. Phytopharm.* 14:29–38
46. Carles, P., Chaboussou, F., Harry, P. 1972. Influence de la nature du porte greffe de la vigne sur la multiplication de l'araignée rouge *Panonychus ulmi* Koch aux dépens d'un même greffon: le Merlot rouge. *C. R. Seances Acad. Agric. Fr.* 58(17):1403–15
47. Carmona, M. M. 1973. The presence of the mite *Panonychus ulmi* KOCH on vines. *Agricoltura* 4:16–21
48. Castro, A. R. 1965. Plagas y enfermedades de la vid. *Inst. Nac. Invest. Agric.* p. 757
49. Caudwell, A. 1961. Etude sur la maladie du bois noir de la vigne: ses rapports avec la flavescence dorée. *Ann. Epiph.* 12(3):241–68
50. Caudwell, A., Moutous, G., Brun, P., Larrue, J., Fos, A., Blancon, G., Schick, J. P. 1974. Les épidémies de flavescence dorée en Armagnac et en Corse et les nouvelles perspectives de lutte contre le vecteur par des traitements ovicides d'hiver. *Bull. Tech. Inf. Minist. Agric.* 294:1–12
51. Caudwell, A., Ottenwaelter, M. 1957. Deux années d'études sur la flavescence dorée nouvelle maladie grave de la Vigne. *Ann. Amélior. Plant.* 4:359–93
52. Caudwell, A., Schwester, D., Moutous, G. 1972. Variété des dégâts des Cicadelles nuisibles à la vigne. Les méthodes de lutte. *Prog. Agric. Vitic.* 89(24): 583–90
53. Cayrol, R. A. 1972. Famille des *Noctuidae.* See Ref. 19, pp. 1255–520
54. Cengiz, F. 1973. Recherches sur les Thysanoptères vivant sur vigne dans les territoires d'Izmir et de Mamisa: leurs caractères morphologiques, les plantes hôtes, leurs dégâts et leurs ennemis naturels. *Thèses Doc. Etat Bornova Izmir.* 112 pp.
55. Cermeli, L. M. 1973. The wasp attacking grapes (*Vitis vinifera*) a new pest of this crop in Venezuela. *Agron. Tropic.* 23(4):413–17
56. Chaboussou, F. 1969. Recherches sur les facteurs de la pullulation des Acariens phytophages de la vigne à la suite des traitements pesticides du feuillage. *Thèse Doc. Sci. Paris.* 238 pp.
57. Chaboussou, F. 1971. Le conditionnement physiologique de la vigne et la multiplication des Cicadelles. *Rev. Zool. Agric. Pathol. Vég.* 70(3):57–66
58. Chebotar, T. I. 1971. See Ref. 201, pp. 96–102
59. Clark, G. N., Enns, V. R. 1964. Life history studies of the grape root borer in Missouri. *J. Kans. Entomol. Soc.* 37(1):56–63
60. Cone, W. W. 1971. Grape mealybug control in Concord grape field trials in

central Washington. *J. Econ. Entomol.* 64(6):1552–53
61. Couturier, A. 1938. Remarques sur la biologie de *Ceresa bubalus* F. membracide d'origine américaine. *Rev. Zool. Agric.* 37(10):145–57
62. Delassus, M., Lepigre, A., Pasquier, R. 1930. Les ennemis de la vigne en Algérie et les moyens pratiques de les combattre.I. Les parasites animaux. Alger: Carbonnel. 249 pp.
63. Delmas, H. G. 1954. Le vespère de la vigne. Techniques de lutte. *Rev. Zool. Agric. Appl.* 53:110–20
64. Delmas, R. 1956. Lutte contre les ennemis animaux de la vigne. *C.R. Cong. Int. Vigne Vin,* 8th, pp. 57–71
65. Delmas, R., Rambier, A. 1954. L'invasion des araignées rouges sur la vigne. *Prog. Agric. vitic.* 142:34–35; 101–4
66. Demaree, J. B., Still, G. W. 1951. Control of grape diseases and insects in eastern United States. *Farm. Bull.* 1893: 36 pp.
67. Demetrashvili, M. I. 1971. The grape mining. *Zashch. Rast.* 16(8):53
68. Dirimanov, M., Kharizanov, A. 1972. Tests of preparations for the control of the larvae of variegated grape moth. *Zashch. Rast.* 20(10):23–26
68a. Doutt, R. L., Nakata, J., Skinner, F. E. 1966. Dispersal of grape leafhopper parasites from a blackberry refuge. *Calif. Agric.* 20:14–15
69. Doutt, R. L., Nakata, J., Skinner, F. E. 1969. Parasites for control of the grape leaf folder. *Calif. Agric.* 23(4):4
70. Feron, J., Audemard, H., Balachowsky, A. S. 1966. Superfamille des *Cossoidea.* See Ref. 19, 1:39–59
71. Ferrero, F. 1973. Les Dégâts de termites dans le cru de Banyuls. *Phytoma* 25(251):25–27
72. Figueiredo, E. R. de 1970. Nova praga da videir em Sao Paulo *Eurhizococcus braziliensis* (Hempel). *Biologico* 36(9): 229–34
73. Flaherty, D. L., Lynn, C. D., Jensen, F. L., Luvisi, D. A., 1969. Ecology and integrated control of spider mites in San Joaquin vineyards. *Calif. Agric.* 23(4):11
74. Galet, P. 1968. *Précis d'Ampelographie Pratique.* Montpellier:ENSA. 230 pp.
75. Galet, P. 1970. *Précis de Viticulture.* Montpellier:ENSA. 490 pp.
76. Gartel, W. 1965. Untersuchungen über das Auftreten und das Verhalten der "Flavescence dorée" in den Weinbaugebieten an Mosel und Rhein. *Weinbau Keller Wirtschaft.* 12:347–76
77. Gartel, W. 1972. Die Rebblattgallmilbe *Eriophyes vitis* PGST. der Erreger der Pockenkranheit (Erinose) als Knospensschädling und als Ursache starken Blattrolens. *Weinbau Kellerwirtschaft.* 19:589–614
78. Guennelon, G., d'Arcier, F. 1972. Piegeage sexuel de l'eudemis de la vigne (*Lobesia botrana* Schiff.) dans la région d'Avignon. *Rev. Zool. Agric. Pathol. Veg.*
79. Giannotti, J., Caudwell, A., Vago, C., Duthoit, J. L. 1969. Isolement et purification de micro-organismes de type mycoplasme à partir de vignes atteintes de flavescence dorée. *C.R. Acad. Sci. Paris, Sér. D* 268:845–47
80. Günyadin, T. 1974. A survey of vine pests in south-east and east Anatolia. *Dtyarb. Bölge Zirai Muc. Inst.* 42:170
81. Hatzinikolis, E. N. 1970. Neuf espèces d'acariens signalées pour la première fois en Grèce. *Ann. Inst. Phytopath. Benaki N.S.* 9(3):238–41
82. Hewitt, W. et al. 1958. Transmission of fanleaf virus by Xiphinema index. *Phytopathology* 48:293
83. Hewitt, W. B., Raski, D. J. 1967. Facteurs limitant la production 6:La vigne *Span* 10(1):56–59
84. Hoffman, A. 1962. See Ref. 19, Vol 1, 2:902–3
84a. Huffaker, C. B. 1971. *Biological Control,* pp. 273–274. New York: Plenum. 511 pp.
85. Hurpin, B. 1962. See Ref. 19, Vol. 1, 1:24–122
86. Ignator, B., Kirkov, K. 1972. The grape vine wevils. *Rastch. Zash.* 20(1):29–31
87. Iordanou, N. 1974. Chemical control of grape berry moth *Tech. Pap. Agric. Res. Inst. Cyprus* 5:7
88. Iren, Z. 1972. Investigations to determine the most important pests of viticulture in central Anatolia. *Anka Bölge Zorai Mucad. Enstit.* 40–41:168–69
89. Ishii, K. 1975. Control of grape disease and insect pests in Japan. *Jpn. Pestic. Inf.* 23:17–23
90. Ivan, A., Zahatia, V. 1969. New aspects of attack produced by *Agriotes lineatus* on vines *Rev. Hortic. Vitic.* 6:66–69
91. Ivanov, I. 1969. An experiment in forecasting and the control of the one-banded grape moth in the Burgas district. *Rastch. Zash.* 17(8/9):30–38
92. Ivanov, I. 1969. On the biology of the one-banded grape moth. *Rastch. Zash.* 17(12):19–23
93. Jensen, F. L. 1969. Microbiol insecti-

cides for control of grape leaf folder. *Calif Agric.* 23(4):5–6

94. Jensen, F. L., Flaherty, D. D., Chiarappa, L. 1969. Population densities and economic injury levels of grape leafhoppers. *Calif. Agric.* 23(4):9–10
95. Jubb, G. L., Cox, J. A. 1974. Catches of redbanded leafroller moths in liquid bait traps in Erie County, Pennsylvania vineyards:12 years summary. *J. Econ. Entomol.* 67(3):448–49
96. Jubb, G. L. 1973. Catches of *Episimus argutanus* (Clem) in grape berry moth sexpheromone traps in Pennsylvania. *J. Econ. Entomol.* 66(6):1345–46
97. Kara'ozova, A. 1971. The control of the grape moth in 1970. *Rastch. Zash.* 19(5):13–16
98. Kazas, I. A., Gorkavenko, A. S., Kiryukhin, G. A., Asriev, E. A. 1971. The protection of vineyards from *Phylloxera. Vzesay Nauch. Protcvolf. Stant. Odessa.* 264 pp.
99. Kazas, I., Gorkavenko, A., Kiryukhin, G. A., Poldenko, V. 1966. *Le Phylloxera de la Vigne Simferopol (Crimée).* 157 pp.
100. Khalilov, B. B. 1971. Vine entomofauna in Azerbaijan. *C.R. Int. Cong. Entomol. Moscow, 13th, 1968,* 2:344–45
101. Kharizanov, A. 1969. Conclusions from the control of the variegated vine moth in the Plovdiv district in 1968. *Rastch. Zash.* 17(6)20–22
102. Kharizanov, A. 1969. A new pest of grape vine in Bulgaria. *Rastch. Zash.* 17(11):21–23
103. Kharizanov, A., Stoilov, A. 1969. The grape-vine mite and its control in the Plovdiv district. *Rastch. Zash.* 17(4): 25–28
104. Kinn, D. N., Doutt, R. L. 1972. Initial survey of arthropods found in North Coast vineyards of California. *Environ. Entomol.* 1(4):508–13
105. Kinn, D. N., Doutt, R. L. 1972. Natural control of spider mites on wine grape varieties in Northern California. *Environ. Entomol.* 1(4):513–18
106. Kisakurek, O. R. 1972. Studies on the distribution, rate of infestation, parasites and predators of the grape cluster moth (*Lobesia botrana* Den. and Schiff.) in southern districts of Anatolia. *Bitki Kor. Bült.* 12(3):183–86
107. Kiskin, P. K. 1961. Regeneration des racines et résistance au Phylloxera. *Acad. Sci. Moldavie Kichinev. Inst. Zool. Publ.* 1:71–81
108 Klerk, C. A. de 1974. Biology of *Phylloxera vitifoliae* (Fitch) in South Africa. *Phytophylactica* 6:109–18
109. Kulkarni, K. A. 1971. Bionomics of the grape flea beetle *Sceledonta strigicollis* (Motsch.). *Mysore J. Agric. Sci.* 5: 308–16
110. Lafon, J., Couillaud, P., Hude, R. 1955. Maladies et parasites de la vigne. Paris: Bailliere. 364 pp.
111. Laing, J. E., Calvert, D. L., Huffaker, C. B. 1972. Preliminary studies of effects of *Tetranychus pacificus* McG. on yield and quality of grapes in the san Joaquin Valley California. *Environ. Entomol.* 1(5):658–63
112. Laurent, M. 1972. Lutte contre les Tordeuses de la Grappe. *Vignes Vins Janv. Fév.* 206:20–30
113. Liebermann, J., Espul, J. C., Mansur, P. S. 1971. On *Miogryllus convolutus* Johannson in the vineyards of San Carlos, La Consulta, Mendoza. *Idia* 281:55–61
114. Lipetskaia, A. D., Rouzaiev, K. C. 1958. Les ravageurs et les maladies de la vigne. Moscow: 280 pp.
115. Lynn, C. D. 1969. Omnivorous leafroller an important new grape pest in San Joaquin Valley. *Calif. Agric.* 23(4): 16–17
116. Mac Lellan, C. R. 1973. Natural enemies of the light brown apple moth *Epiphyas postvittana* (Wlk.) in the Australian capital territory. *Can. Entomol.* 105(5):681–700
117. Maillet, P. 1957. Contribution à l'étude de la biologie du Phylloxera de la vigne. *Ann. Sci. Natl. Zool.:* 283–410
118. Maillet, P. 1957. Le Phylloxéra de la vigne. Quelques faits biologiques et les problèmes qu'ils soulèvent. *Rev. Zool. Agric. Appl.* 7–9:1–19
119. Mal'chenkova, N.I. 1970. *Eriophyes vitigineusgemma* sp.n., a pest of grape vine. *Zool. Zh.* 49(11):1728–31
120. Mal'chenkova, N. I. 1971. On the mite fauna of the grape vine. See Ref. 201, pp. 107–21
121. Manolache, C., Pasol, P., Romascu, E., Iordan, P., Naum, A., Sadagorschi, D., Popescu, M. 1974. Ecological contributions to the study of grape vines on sandy soils of the Platonesti-Saveni-Suditi area (*Ialomita*). *Ann. Inst. Cerc. Pentru Prot. Plant.* 10:257–64
122. Mayet, V. 1890. Les insectes de la vigne. Paris:Masson. 466 pp.
123. McGrew, J. R., Still, G. W. 1972. Control of grape diseases and insects in the eastern United States. *Farm. Bull.* 1893. 24

124. McKenzie, L. M., Beirne, B. P. 1972. A grape leafhopper *Erythroneura ziczac* Walsh and its Mymarid egg parasite in Okanagan Valley, British Colombia. *Can. Entomol.* 104(8):1229–33
125. McLachlan, R. A. 1970. Grape pest control in Queensland. *Queensl. Agric. J.* 96(4):231–35
126. Mikhailhuk, I. B. 1970. The pests of grape vine. *Rastch. Zash.* 15(6):34–35
127. Mirica, A., Savescu, A., Mirica, I. 1969. Factors in forecasting for the control of the vine mealybug (*Pulvinaria betulae* L.). *Ann. Inst. Cerc. Pentru Protect. Plant.* 5:387–93
128. Moutous, G., Fos, A. 1971. Essais de lutte chimique contre la cicadelle de la vigne *Empoasca flavescens* Fabr. Résultats 1970. *Rev. Zool. Agric. Pathol. Vég.* 70(2):48–56
129. Moutous, G., Fos, A. 1971. Observations sur quelques ravageurs nouveaux ou occasionnels de la vigne. *Phytoma* 23(233):25–26
130. Moutous, G., Fos, A. 1972. La lutte contre la flavescence dorée et les cicadelles de la vigne. *Rev. Zool. Agric. Pathol. Veg.* 71(1):55–60
131. Moutous, G., Fos, A. 1973. Influence of populations levels of the vine Cicadellid (*Empoasca flavescens* Fab.) on the "scorching" symptoms on the leaves. *Ann. Zool. Ecol. Anim.* 5(2):173–85
132. Myburgh, A. C., Whitehead, V. B., Daiber, C. C. 1973. Pests of deciduous fruit, grapes and miscellaneous other horticultural crops in South Africa. *Entomol. Mem. Dep. Agric. Tech. Serv. Rep. S. Africa* 27. 38 pp.
133. Neamtu, I., Varna, P., Bahnareanus, M. 1969. Considerations regarding the attack of the vine moth on vines in vineyards at Husi. *Rev. Hortic. Vitic.* 6:63–65
134. Nedov, P. N. 1971. A study of the resistance of the grape vine to *Phylloxera vastatrix* Planch. *C.R. Int. Congr. Entomol. Moscow, 13th,* 2–9 Août 1968 2:366
135. Nikolaev, P. I., Ass, M. Ya. 1973. Details of the development of the vine phylloxera (*Dactylosphaera vitifolii* Fitch) on the Sochi coast of the Black Sea. *Byull. Mosk. Obs. Ispyt. Prir. Otdel Biol.* 78(2):5–20
136. Odak, S. C., Dhamdhere, S. V. 1970. New record of Hymenopterous parasites of *Sylepta lunalis* Guen, a pest of grapevines. *Indust. J. Entomol.* 32(4): 395
137. Olalquiaga Faure, G., Contesse Pinto, J. 1959. Pests of grape vine in Chile. *FAO Plant. Protect. Bull.* 7(6):73–77
138. Panis, A. 1975. Lutte contre la cochenille farineuse dans le vignoble méditerranéen. *Prog. Agric. Vitic.* 92(15/16):470–73
139. Pantanelli, E. 1911. Danni di thrips sulle vite americane. *St. Sperim. Agric. ital.* 44(7):469–514
140. Pinnock, D. E., Mrlstead, J. E., Coe, N. F., Stegmiller, F. 1973. Evaluation of *Bacillus thuringiensis* formulations for control of larvae of western grape leaf skeletonizer. *J. Econ. Entomol.* 66(1): 194–97
141. Plaut, H. N. 1973. On the biology of *Paropta paradoxus* (H.S.) on grape vine in Israel. *Bull. Entomol. Res.* 63(2): 237–45
142. Predescu, S., Petraneu, F., Lazar, S. 1969. Contributii la studiul componentelor biochimiche a insectei *Sparganothis pilleriana* Schiff. si a compositei chimiche a vitei de vi ca mediu de hrana. *Sucr. Stintif. Inst. Agric. Timis. Agron* 12:437–46
143. Prota, R. 1970. Un nuovo insetto dannoso ai frutti feri della Sardegna : la *Ceresa bubalus* F. *Studi Sassar.* 18(3): 48–56
144. Purcell, A. H. 1975. Role of the blue-green sharpshooter, *Hordnia circellata,* in the epidemiology of Pierce's disease of grapevines. *Environ. Entomol.* 4(5): 745–52
145. Pykhova, V. T. 1968. The control of the grape moth. *Rastch. Zash.* 13(11): 48–49
146. Rambier, A. 1951. A propos du *Vesperus xatarti* Muls. *Prog. Agric. Vitic.* No. 7–8, pp. 89–93
147. Rambier, A. 1969. Influence de traitements au DDT, au parathion et au carbaryl sur les ennemis naturels de *Panonychus ulmi* (Koch). *C. R. Symp. OILB Lutte Intég. Vignes, 4th,* pp. 173–78
148. Rambier, A. 1972. Les acariens dans le vignoble. *Prog. Agric. Vitic.* 89(16): 385–97
149. Richard, M. 1975. Compte rendu des essais de lutte contre la pyrale. *Vigne Champ.* 1:18–20
150. Rings, R. W. 1972. Contributions to the bionomics of climbing cutworms distribution and developmental biology of *Abgrotis barnesi.* *J. Econ. Entomol.* 65(2):397–401
151. Rings, R. W. 1972. Contributions to the bionomics of climbing cutworms; the

distribution and developmental biology of the brown climbing cutworm, *Rhynchagrotis cupida. J. Econ. Entomol.* 65(3):734–37

152. Rivnay, E. 1939. Studies in the biology and ecology of *Retithrips syriacus* Mayet with special attention to its occurrence in Palestine. *Bull. Soc. Fouad Ier Ent. Cairo,* pp. 150–82
153. Roehrich, R. 1970. Essais de deux produits commerciaux à base de *Bacillus thringiensis* Berl. pour la protection de la vigne contre l'Eudemis *Lobesia botrana* Schiff. *Rev. Zool. Agric. Pathol. Veg.* 69(4):74–78
154. Sampayo Fernandez, M., Hernandez Esteruelas, P. 1973. The vine moth (*Lobesia botrana* Schiff.). *Bol. Inf. Plagas* 108:7–11
155. Sarai, D. S. 1972. Seasonal history and effect of soil moisture on mortality of newly hatched larvae of the grape root borer in Southern Missouri. *J. Econ. Entomol.* 65(1):182–84
156. Saxena, D. K. 1970. Honey bees and wasps as pests of grape. *J. Bombay Nat. Hist. Soc.* 67(1):121–22
157. Schruft, G. 1972. Les Tydeidé (Acariens) sur vigne. *Bull. OEPP* 3:51–55
158. Schruft, G. 1972. Effects secondaires de fongicides agissant sur les Acariens sur vigne. *Bull. OEPP* 3:57–63
159. Schruft, G. 1972. Das Vorkommen von Milben aus der Familie *Tydeidae* (Acari) an Reben. VI Beitrag über Untersuchungen zur Faunistik und Biologie der Milben (Acari) an Kulturreben (*Vitis* sp.). *Z. Angew Entomol.* 71(2): 124–33
160. Schwester, D. 1972. Cicadelles de la vigne. *Bull. OEPP* 3:37–42
161. Schwester, D., Carle, P., Moutous, G. 1961. Sur la transmission de la flavescence dorée des vignes par une cicadelle. *C. R. Seances Acad. Agric. Fr.* 147(18):1021–24
162. Schwester, D., Moutous, G., Bonfils, J., Carle, P. 1962. Etude biologique des cicadelles de la vigne dans le sudouest de la France. *Ann. Epiph.* 13(3):205–37
163. Sengalevich, G. 1969. New pests on permanent plantations in the Plovdiv area. *Rastch. Zash.* 17(3):21–23
164. Servicio Para el Agricultor. 1973. La avispita de la uva. *Notas Agric.* 6(26): 105
165. Seryi, N. I. 1971. On the biology of the vine moth (*Clysia ambiguella* H. B.). See Ref. 201, pp. 86–90
166. Seryi, N. I. 1972. The biological principles of methods for the control of the pests of grape vines in Moldavia. See Ref. 201, pp. 95–103
167. Sidhu, A. S., Singh, G. 1972. Studies on the chemical control of *Oligonychus mangiferus* (Rahman and Sapra) on grape vine. *J. Res. Punjab Agric. Univ.* 8(4):462–65
168. Silvestri, F. 1939. Compendio di entomologia applicata. *Ed. Portici* Vol. 1, 2. 974., 685 pp.
169. Siriez, H. 1970. Un ravageur presque oublié : la pyrale de la vigne. *Phytoma* 22(215):41–47
170. Smith, L. M., Stafford, E. 1955. Grape pests in California. *Calif. Agric. Exp. Stn. Circ.* 445. 63 pp.
171. Sohi, A. S., Singh, S. 1970. New pests of grape vine. *Labdev. J. Sci. Technol. 8 B* 3:170
172. Sorauer, P. 1949–1958. *Handbuch der Pflanzenkrankheiten.,* Vol. 1–6. Berlin: Parey
173. Stafford, E. M., Kido, H. 1969. Newer insecticides for the control of grape insect and spider mite pests. *Calif. Agric.* 23(4):6–8
174. Stellwaag, F. 1928. *Die Weinbauinsekten der Kulturländer.* Berlin : Parey 884 pp.
175. Stevenson, A. B. 1962. Insecticide dips to control grape phylloxera on nursery stocks. *J. Econ. Entomol.* 55(5):804–5
176. Stevenson, A. B. 1964. Seasonal history of root-infesting *Phylloxera vitifoliae* (Fitch) in Ontario. *Can. Entomol.* 96(7):979–87
177. Stevenson, A. B. 1966. Seasonal development of foliage infestations of grape in Ontario by *Phylloxera vitifoliae* (Fitch). *Can. Entomol.* 98(12):1299–305
178. Stevenson, A. B. 1968. Soil treatments with insecticides to control the rooform of the grape Phylloxera. *J. Econ. Entomol.* 61(5):1168–71
179. Stevenson, A. B. 1970. Endosulfan and other insecticides for control of the leaf forme of the grape phylloxera in Ontario. *J. Econ. Entomol.* 63(1):125–28
180. Stevenson, A. B. 1970. Strains of the grape phylloxera in Ontario with different effects on the foliage of certain grape cultivars. *J. Econ. Entomol.* 63(1): 135–8
181. Still, G. W. 1962. Cultural control of the grape berry moth. *ARS Agric. Inf. Bull.* 256:1–8
182. Still, G. W., Rings, R. W. 1973. Insect and mite pests of grapes in Ohio. *Res. Bull. Ohio Agric. Res. Dev. Cent.* 1060. 30 pp.

183. Suire, J. 1954. Contribution à l'étude morphologique des Microlépidoptères de la vigne. *Ann. Ecole Natl. Agric. Montpellier* 24 (3 and 4):1–7
184. Tairov, V. E. 1967. *Questions Relatives à Laviticulture et à la Vinification.* (In Russian) Kiev : 386 pp.
185. Taschenberg, E. F., Carde, R. T., Roelofs, W. L. 1974. Sex pheromone mass trapping and mating disruption for control of redbanded leafroller and grape berry moth in vineyards. *Environ. Entomol.* 3(2):239–42
186. Troitzky, N. N. 1929. Beiträge zur kenntnis der Reblaus und der Reblauswiederstomdsfähigkeit der Weinrebe : Die Reblausfrage in Mittel-Europa Leningrad : Bur. Appl. Entomol. 191 pp.
187. Vega, E. 1972. Control de la filoxera de la vid con hexachlorobutadien. *Idia* 290:15–18
188. Vidal, J. P. 1974. Le piegeage de l'Eudemis (*Polychrosis botrana*) substances attractives. *Bull. Tech. Pyr. Or.* 72: 67–71
189. Vidano, C. 1957–58. Le Cicaline italiane della vite. *Boll. Zool. Agric. Bachic. Milan* 2(1):61–115
190. Vidano, C. 1959. Sulla identificazione specifica di alcuni *Erythroneurini europei. Ann. Mus. Civ. Stor. Natur. Giacomo Doria* 71:328–48
191. Vidano, C. 1962. La *Empoasca libyca* Bergevin nuovo nemico della vite in Italia. *Cent. Entomol. Alp. forest. CNR* 55:327–45
192. Vidano, C. 1963. Deviazione trofica ampelofila della *Ceresa bubalus* Fabricius e rispondenza reattiva del vegetale. *Atti. Accad. Sci. Torino* 98:193–212
193. Vidano, C. 1963. Appunti comparativi sui danni da cicaline alla vite. *Inf. Fitopatol.* 13:173–77
194. Vidano, C. 1965. Responses of Vitis to insect vector feeding. *Proc. Conf. Virus Vector, Univ. Calif.* Davis pp. 73–80
195. Viswanath, B. N., Nalawadi, U. G., Kulkarni, K. A. 1970. The yellow banded wasp, *Vespa cincta* Fabr. as a pest of grapes. *Agric. Res. J. Kerala* 8(1):53
196. Voigt, E. 1970. Influence of meteorological factors on the population dynamica of *Eupoecilia ambiguella* Hb and *Lobesia botrana* Den. and Schiff. *Növény. Korsz.* 4:63–78
197. Voigt, E., Bodor, J., Javor, A. 1973. Morphological descriptions of larvae of moth injurious to grapevines (Hungarian). *Novenynemesitesi Novenyter mesztesi Kut. Intez. Sopronhorpacs Kozl.* 6:51–64
198. Vuittenez, A. 1961. Les nématodes vecteurs de virus et le problème de la dégénérescence infectieuse de la vigne. In *ACTA, Les Nématodes,* Paris, pp. 55–77
199. Whitehead, V. B., Rust, D. J. 1972. Control of the fruit piercing moth *Serrodes partita* F. *Phytophylactica* 4(1): 9–12
200. Winkler, A. S. 1959. Pierce's disease investigations. *Hilgardia* 19:207–264
201. Yaroshenko, M. F. 1972. The insect fauna of Moldavia and its economic importance. In *Izdatel'stvo Shtüntsa.* 142 pp.
202. Yousef, A. T. A. 1970. Mites associated with vine trees in the UAR. *Z. Angew. Entomol.* 67(1):1–6
203. Zinca, N. 1964. Cercetari asupra morfologiei, biologiei, si combatterii tripsului vitei de vie *Anaphthrips vitis* Pr. *Inst. Cent. Cerc. Agric. Ann. Protect. Plant.* 2:299–305

Ann. Rev. Entomol. 1977. 22:377–405

RESPONSES OF LEPIDOPTERA TO SYNTHETIC SEX PHEROMONE CHEMICALS AND THEIR ANALOGUES[1]

❖6135

W. L. Roelofs
Department of Entomology, New York State Agricultural Experiment Station, Geneva, New York 14456

R. T. Cardé
Department of Entomology, Michigan State University, East Lansing, Michigan 48824

Chemical communication systems are being analyzed at an increasing rate by defining only the chemicals involved, or by describing the complex behavioral patterns elicited by some natural chemical mix of unknown composition. With insects, too often the former approach has been prevalent and the behavioral roles of many of the identified chemicals have not been pursued. Many times, however, the possible behavioral functions of a chemical can be inferred from indirect measurements, such as the compound's effect on trap catch. This chapter discusses field trapping and field observation techniques in elucidating sex pheromone responses of Lepidoptera, as well as the modulation of these reactions by various chemicals that are not constituents of the natural pheromone system.

Terminology

Pheromone researchers have debated repeatedly the semantics of chemicals that elicit various types of behavior. It has become obvious, however, that subclassifications of pheromonal responses developed for certain insect species are not universally applicable. The optimum, of course, would be to describe for each species the precise behavior elicited by each chemical rather than to be concerned with classification. However, the importance of comparative studies and of key word abstracting, as well as the paucity of observations defining the behavioral role of individual pheromone components, dictate some type of categorization. Therefore, we define

[1]The survey of the literature pertaining to this review was concluded on March 15, 1976.

several terms that are useful in classifying pheromone components of communication systems, including long-distance upwind anemotaxis to a chemical source. Multichemical sex pheromone systems, in these cases, are subdivided into primary and secondary components.

SEX PHEROMONE As generally accepted, the term *sex pheromone* is used to include all chemicals (pheromone components) that are released from one organism and that induce responses, such as orientation, precopulatory behavior, and mating, in another individual of the same species (69, 70). The term applies (136) to all such chemicals emitted by the organism, even though some or all of them may be produced by symbiotic microorganisms or acquired intact from outside sources. It is important, however, that a chemical be rigorously identified either from the organism or its effluvium prior to classification as a pheromone component. It is equally important to demonstrate that the chemical elicits behavioral activity definable as part of a natural sequence at natural pheromone concentrations.

PRIMARY SEX PHEROMONE COMPONENTS *Primary components* are chemicals emitted by an insect that elicit long-distance (distances over 1 m) upwind anemotaxis in the responding insect. In some species primary components alone may elicit the entire behavioral sequence leading to copulation. For lack of appropriate behavioral observations, these components are approximately defined as those emitted by the calling insect that are obligatory for trap catch in the field at component emission rates similar to that used by the insect.

SECONDARY SEX PHEROMONE COMPONENTS *Secondary components* are chemicals emitted by an insect that are not essential for eliciting upwind anemotaxis, but that in combination with the primary components evoke other aspects of the mating sequence. Generally close-range responses, such as landing, wing fanning, hair pencilling, and copulatory attempts, represent the type of responses induced by secondary components. In species analyzed to date the behavioral responses induced by the secondary components occur after the initial effect of its primary components. These reactions may be more subtle and difficult to delineate; nevertheless they can be approximately defined as those components that increase trap catches when combined with primary pheromone components but that do not elicit intrinsic upwind anemotatic activity by themselves, nor are they part of a critical combination requisite for anemotactic responses.

The following sections discuss synthetic pheromone components that fall into the above classes (although doubtless intermediate situations exist) and the effect of behavioral modifiers (inhibitors, masking agents, or antipheromones) as they might relate to the sensory input from primary and secondary components.

FIELD RESPONSES TO PHEROMONES

Behavioral Role of Pheromone Components

Although the chemistry of numerous sex attractant communication systems has been described, the behavioral elements of response to these chemicals has only infrequently been accorded more than a cursory examination. In the moth species that apparently utilize a single pheromone component (female emitter), the entire sequence of wild male behavior that would occur in the field has been assumed to be a hierarchy of responses with successive steps in the behavioral sequence evoked by increased concentrations of the pheromone. Of course, additional nonchemical cues such as visual (46), tactile (55, 56), and auditory (40) stimuli may be involved in the orientation and mating responses. The behavioral patterns involved in the mating of feral individuals would include upwind anemotaxis (long-range orientation), generally followed by close-range orientation and landing. Intrinsic limitations of most laboratory bioassay procedures, however, do not allow observation of some of the most critical steps (e.g. upwind flight and landing) in the behavioral sequence. Indeed, the behavioral mileau of laboratory trials is so altered from the natural field conditions that chemical components crucial to the full field response may be difficult to assay, especially when only a limited array of behaviors (often just one selected key response) can be monitored. Thus, the applicability of laboratory bioassays to the entire sequence of wild male behavior is somewhat inferential.

In moths the pheromone communication systems characterized chemically within the last several years have demonstrated that systems comprised of multiple components are nearly ubiquitous. The behavioral responses evoked by such blends also may be hierarchical with increasing concentrations of the entire blend requisite for each successive behavioral step (in such cases all the components would be considered primary). In other multicomponent systems the first steps of the behavioral sequence may be mediated by one or more primary components. This initial portion of the behavioral sequence also could proceed in a hierarchy, possibly with successive steps dependent on increases in concentration of primary components. Continuing the behavioral sequence past the initial phases governed by the primary components would require the presence of both the primary components and additional chemical stimuli, the secondary components.

PRIMARY AND SECONDARY COMPONENTS In the tortricid, redbanded leafroller, *Argyrotaenia velutinana,* the sex pheromone from the female effluvium is a 90:10:120 blend of (Z)-11-tetradecenyl, (E)-11-tetradecenyl, and dodecyl acetates (Z11-14:Ac; E11-14:Ac; 12:AC) (75, 110, 118, 121). This mixture elicits the maximum male trap catch (121). The feral male's sequence of behavior in response to this blend (Figure 1) involves long-range upwind anemotaxis to the vicinity of the chemical emitter, landing, walking to the emitter while wing fanning, and finally copulatory attempts (4). Observing wild males has revealed that initial long-range orientation is elicited by the combination of Z11-14:Ac and E11-14:Ac in a 91:9 ratio. This blend alone also has the lowest threshold for activation and upwind orientation in laboratory assays, and in the field it produces an appreciable trap catch, fulfilling the criteria of primary components.

The presence of 12:Ac in combination with the primary components elicits an increase in the frequency of landing, wind fanning, and walking to the chemical source when the in-flight male has oriented to within 60 cm of the chemical dispenser. In the field, however, 12:Ac alone induces no overt reaction and does not yield a trap catch. In laboratory assays 12:Ac will evoke stimulation and upwind

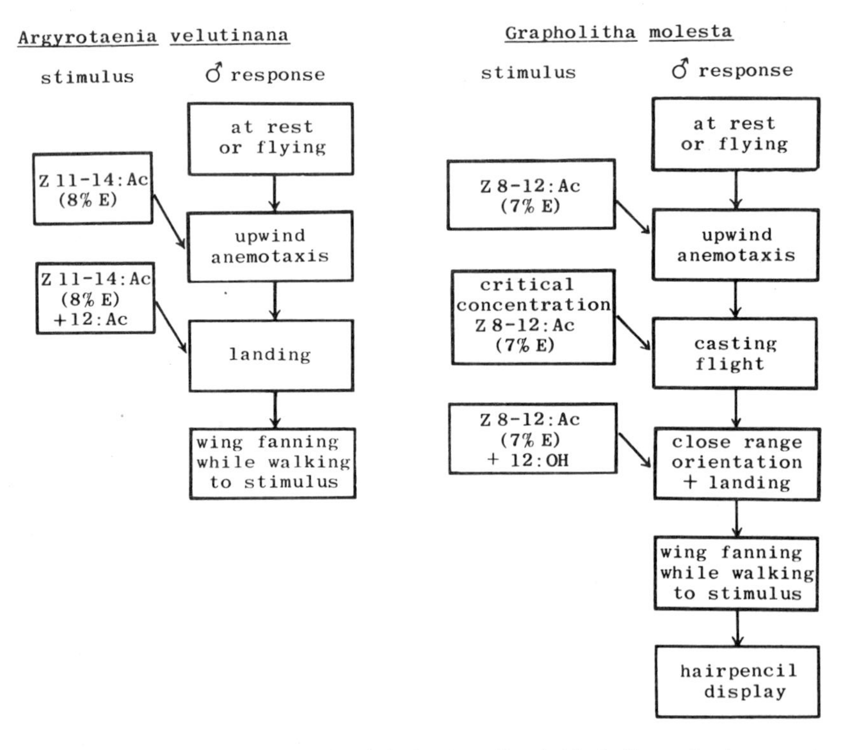

Figure 1 Stimulus-response reaction chain for the redbonded leafroller moth, *Argyrotaenia velutinana,* and the Oriental fruit moth, *Grapholitha molesta.*

orientation, but only at more than a 10^5-fold increase in stimulus dosage over the 91:9 Z11-14:Ac to E11-14:Ac combination (4).

A similar communication system exists in the Oriental fruit moth, *Grapholitha molesta.* One component of the pheromone system was identified as (*Z*)-8-dodecenyl acetate (Z8-12:Ac) (120), but trap catch requires emission of this component and its geometrical isomer (E8-12:Ac) in a 93:7 ratio (10, 13, 113). An additional component, dodecyl alcohol (12:OH), has been demonstrated to effect a twofold increase in trap catch over the optimum 93:7 blend alone, although 12:OH alone possesses no intrinsic attractiveness (117). Neither E8-12:Ac nor 12:OH have been characterized as yet from either the female or her effluvium.

Wild male behavior in the field (Figure 1) indicates that the blend of Z8-12:Ac and E8-12:Ac (93:7) is requisite for the long-range upwind anemotactic response (28, 29). A critical airborne concentration of these primary components elicits in-flight close approach, casting flight and a low incidence of landing, followed by walking to the chemical source while wing fanning (28). In observations of wild males the simultaneous emanation of 12:OH with the primary components evokes an increase in the frequencies of landing and close approach to the chemical emitter while wing fanning. The most spectacular behavioral effect of 12:OH is its inducement of an elaborate hair pencil and precopulatory display repertoire (28, 29).

Although the roles of attractant components in the redbanded leafroller and the Oriental fruit moth have been investigated in detail, a number of other moth species appear to possess similar attractant systems comprised of both primary and secondary components (Table 1). In these species the primary components produce male trap catch and the intrinsically nonattractive secondary components increase trap catch when emitted with the primary components. When the behavioral functions of these primary and secondary components are elucidated, their roles may be analogous to the communication systems in the redbanded leafroller and Oriental fruit moth. In some moth species for which only primary components have been characterized to date it is likely that secondary components remain to be discovered.

Just as partial elucidation (Figure 1) of the behavioral roles of secondary components has necessitated new assay techniques, unravelling the interactions of chemical cues with possible visual, tactile, or auditory stimuli will require novel and unambiguous tests. Indeed, in moths the exact processes involved in the perception and directed movement toward the chemical emitter are poorly understood. The most plausible mechanism appears to be optomotor anemotaxis (73), but this process has been experimentally verified in but a few Lepidoptera. Notwithstanding, additional reactions including chemotaxis (44, 45) could be involved, especially within a few centimeters of the pheromone source. The nature of the kineses and taxes potentially important in locating a pheromone source are considered in detail by Kennedy (72).

Most pheromone studies have described behaviors teleologically and have ascribed anthropomorphic characteristics to chemicals without attempting experi-

Table 1 Lepidoptera possessing primary components effecting trap catch and secondary components effecting an increase in trap catch when emitted simultaneously with primary components

Species	Primary components	Secondary components	Ratio[a]	Ref.
Antheraea polyphemus	(E,Z)-6, 11-16Ac	(E,Z)-6, 11-16:ALD	90:10	78
Archips mortuanus	Z11-14:Ac[b] E11-14:Ac[b] Z9-14:Ac[b]	12:Ac[b]	90:10:1:200	34
Argyrotaenia citrana	Z11-14:ALD	Z11-14:Ac	50:50	62
Argyrotaenia velutinana	Z11-14:Ac E11-14:Ac	12:Ac	90:10:150	75, 110, 118, 121
Diparopsis castanea	E9,11-12:Ac Z9,11-12:Ac	11-12:Ac	90:10:25	82, 97, 98
Grapholitha molesta	Z8-12:Ac E8-12:Ac[b]	12:OH[b]	93:7:300	10, 13, 113, 117, 120
Platynota idaeusalis	E11-14:OH	E11-14:Ac	50:50	61
Platynota stultana	E11-14:Ac Z11-14:Ac	E11-14:OH	94:6:1	3, 63
Spodoptera litura	(Z,E)-9, 11-14:Ac	(Z,E)-9, 12-14:Ac	90:10	172

[a]Optimum component ratio for male trap catch.
[b]Compounds have not been identified either from the female or the effluvium.

mental verification of the hypothesized behavioral mechanisms. Conventional laboratory assays usually select a single key response such as activation, wing fanning, or walking upwind. Some critical steps in the behavioral response sequence (e.g. landing) are impossible to ascertain in typical laboratory assays since the milieu is vastly different from the field, and the natural sequence of behavioral events is abbreviated.

Male Aphrodisiac Pheromones

The interchange of chemical messengers can extend beyond the male's perception of the primary and secondary components. In many moth species the male, after arrival within a few centimeters of the female, extrudes brush organs to disseminate an aphrodisiac pheromone. This is an olfactory stimulus that presumably induces female receptivity. In butterflies a male locating a female generally is dependent on visual rather than olfactory cues, but in numerous species a female accepting a mate evidently is influenced by an aphrodisiac. There have been numerous ethological studies of moths and butterflies demonstrating that scent glands and scales are displayed during courtship (18). However, experimental proof of behaviors mediated by an aphrodisiac may be very difficult to establish, especially since the only overt reaction may be acquiescence. In the Lepidoptera the evolutionary origin and diversity of aphrodisiac-disseminating structures is extraordinary, ranging from modified wing scales and tufts of hair to elaborate and often eversible organs located on the abdomen, legs, and thorax. Birch (18) has reviewed aphrodisiac pheromones in detail. We consider here only the types of behavior thought to be influenced by aphrodisiacs and the cases in which the role of either an extracted or synthetic component is established.

The most extensive and elegant analysis of an aphrodisiac-mediated courtship in the Lepidoptera is with *Danaus gilippus berenice,* the queen butterfly. The male's initial location of the female is apparently visual (21). Pliske & Eisner (105) have shown that the male overtakes the female in flight and induces her to land by extruding his abdominal hair pencils and dusting a pheromone onto her antennae. A female thus courted is not yet seduced and further bouts of hair pencilling by the hovering male are required for her acquiescence, which she signals by folding her wings, thereby allowing a landed male to copulate.

The hair pencil dust contains a crystalline ketone, 2,3-dihydro-7-methyl-1H-pyrrolizin-1-one, and a liquid diol, (*E,E*)-3,7-dimethyldeca-2,6-dien-1,10-diol (87). The ketone acts both as a landing stimulant (sometimes termed a flight arrestant) and an aphrodisiac, whereas the solvent diol imparts stickiness to the dust, thereby insuring adhesion to the antennae of the female. Males reared in the laboratory inexplicably possessed undetectable quantities of the ketone, but this deficiency afforded an opportunity to restore the male's sexual allurement by experimental addition to the hair pencils of either the hair pencil secretion, the ketone plus diol, or the ketone plus mineral oil, another compound imparting stickiness. The males, thus chemically restored, mated as frequently as wild individuals. The serendipitous circumstance of the laboratory-bred males lacking the aphrodisiac and the ability of the experimenters to restore mating competence allowed unequivocal verification of the ketone's behavioral role (105).

Male butterflies of the nymphalid subfamily Ithomiinae possess hair pencils located on the costal edge of their hindwings. Not surprisingly, initial ethological observations point to their dissemination of a courtship pheromone (106). The wing hair pencils of a number of species in the genera *Pteronymia, Ithomia, Hymenitis,* and *Godyris* contain identical lactones (43). Although the experimental proof of an aphrodisiac role for these components remains to be formally established, the lactones released from the hair pencils have been shown to act both as male territorial recognition pheromones and as allomones. The synthetic lactones tend to repel both conspecific males and males of other lactone-producing ithomiine species. Since many ithomiine species are quite similar in appearance, visual cues alone should be ineffective for species and sexual identification. The lactones function in males recognizing other males, thereby aiding the termination of fruitless male-male aerial pursuits. Hence, ithomiine hair pencil secretions apparently serve to elicit a wide spectrum of divergent behaviors (territorial marking, sex and species recognition, and female seduction), dependent on the context of presentation (106).

Essentially descriptive analyses of behavior cannot distinguish the effects of olfactory stimuli from those of visual (143), tactile (55, 56), and possibly auditory cues that may accompany and interact with the dissemination of an aphrodisiac. The reaction of acquiescence may be difficult to quantify. In moths there is no surfeit of studies demonstrating that the identified compounds elicit a behavioral response relevant to the mating sequence. In *Pseudaletia separata* the major constituent of the male hair pencils is benzaldehyde, to which a locomotory arrestant role has been ascribed (39); however, locomotion over 55 min analyzed with a cumulative ethogram is not clearly related to the rapid events of the mating sequence, so that the possible behavioral function of benzaldehyde is still enigmatic. Unhappily, in other Lepidoptera linking behavioral function with components identified from aphrodisiac glands also has not been definitive. Although numerous chemicals have been identified from supposed or known aphrodisiac structures, their behavioral roles have rarely been established. The chemical characterization of a major constituent of a pheromone gland cannot verify a behavioral role. In certain species male brush organ extracts (active components not identified) may provoke an overt response, such as close-range attraction of the female in *G. molesta* (28) and cessation of female calling behavior in *Heliothis virescens* (59).

Although the involvement of an apparent aphrodisiac structure in the mating ritual implies a requisite behavioral function (a teleological correlation), detailed experimentation may fail to establish even a rudimentary qualitative role. With cabbage loopers, *Trichoplusia ni,* males whose abdominal hair pencils have been surgically excised nonetheless mated in the laboratory as readily as control males splaying their hair pencils prior to coupling. Moreover, removing the female antennae did not affect mating success (53). With cabbage loopers no simple qualitative function for hair pencils is evident, although subtle quantitative aspects of mating success may yet be uncovered.

Experimental Design of Field Trapping Tests

The ability of the synthetic to lure males has been used both for verifying most moth attractant pheromone structures and for utilizing attractant-baited traps for programs of population monitoring and direct control via mass trapping. Since natural insect infestations are rarely distributed in a homogeneous fashion, the experimental design must compensate for the effect of trap placement. For example, tortricid moths within apple orchards may be distributed according to the foliage characteristics of individual trees (168), or they may immigrate into the test area from outside infestations (157).

Replicating treatments, if extensive enough, will tend to obviate the difficulties inherent in sampling scattered populations. However, the tactic of extensive replication by itself may be insufficient to overcome distributional uneveness, particularly where 10 to 20 replicates are plainly impractical. Another consideration is treatment interaction, e.g. in site placement within blocks where two very attractive treatments are in close proximity, competition will result in lower trap catches for these two treatments than for one similarly attractive treatment matched with a relatively unattractive one. In testing a series of treatments a degree of competition among traps (traps spaced about 10 m apart) may tend to increase the distinctiveness of treatments.

Replication in a randomized complete block design allows the choice of replicate blocks to reflect more homogeneous population groupings, with block choices based either on past trapping experience or on more subjective notions (e.g. relationship to foliage) as to the way in which the moths are likely to be dispersed. Since even within blocks heterogeneous dispersions are to be expected, one procedure to overcome their effect is frequent re-randomization of treatments within the blocks. In practice, counting catch followed by a re-randomization of treatments within blocks as often as possible (e.g. daily) is a simple and highly effective technique for compensating for uneven populations and treatment interaction.

Before statistical evaluation of the data is performed, transformation of the actual trap counts is usually advantageous. Many treatments (and hopefully the unbaited traps) will have no moths, whereas other treatments may lure appreciable numbers. Criteria, such as values less than ten and zeros, indicate the desireability of a transformation such as $\sqrt{(x + 0.5)}$ (149). The actual necessity of this manipulation and the most useful transformation could be verified by Tukey's test for additivity (158, 159). Typically the transformed data will be evaluated with an analysis of variance and with differences among means graded at $P = 0.05$, according to tests such as Duncan's new multiple range test (149).

Evaluating Synthetic and Organism Attractiveness

In determining the actual effectiveness of the synthetic attractant in luring males, the most important criterion of biological activity is the trap catch relative to that evoked by the virgin female. Some past evaluations of attractancy have contrasted the trap catch of synthetics with those of pheromone extracts, but this comparison can be misleading. Extracts may either contain compounds inhibitory to attraction (37, 166), possess suboptimal component ratios (3, 63), lack important components (60, 62, 121), or not emit the attractant bouquet at the same rate as the calling female. The most appropriate comparison is against the wild virgin female. For many species collecting field-reared insects is impractical, so using laboratory-reared insects is more feasible, although there are examples of reduced pheromone production in laboratory individuals (89, 108). In laboratory-bred *Danaus gilippus berenice,* the males possessed suboptimal quantities of the hair pencil aphrodisac pheromone (105). Extreme care should be exercised to insure that the diel rhythms of female calling of the laboratory and field population will be synchronous. Since subtle interactions of both temperature and light cues can modulate female attractiveness and male response (31, 67), conditioning laboratory-bred insects to natural thermo- and photoperiods prior to testing will produce the least ambiguous findings. For example, a 2-hr difference in photoperiod entrainment between laboratory and native *Heliothis virescens* apparently rendered these populations incompatible for mating (107).

Another mandatory basis of comparison is the catch of synthetic-baited traps relative to unbaited traps. Traps lacking attractant dispensers may nonetheless produce appreciable catches due either to accidental contamination of traps with pheromones or to the simple blundering of males and females into traps, the latter being a phenomenon particularly evident in population flushes. Without this critical standard, there is no objective basis for claiming attractancy.

In determining relative attractancy of synthetics a seminal but rarely weighed factor is the diel periodicity of female pheromone emission. For example, within the gypsy moth female calling period, the maxima of male attraction and female pheromone emission occur only during a relatively small interval (33, 108). In other species the rhythm of male pheromone responsiveness may be broader than the rhythm of female pheromone emission (36). Thus, a precise assessment of the attractiveness of any synthetic system relative to that of the natural organism requires a detailed knowledge of the natural periodicities of male attraction and female attractiveness.

The ability of a synthetic-baited attractant system to ensnare insects lured to the trap vicinity can be termed trapping efficiency. Direct observations of male red-banded leafroller and Oriental fruit moth behaviors showed that, after upwind anemotaxis to within 1 m of the traps, the optimal three component blends (Table 1) for these species captured 87 and 93%, respectively, of the males. However, in concurrent observations, binary attractant systems for these species comprised of only the primary components were only 20 and 56% effective in capturing males within 1 m of baited traps (4, 29). If the relative trap catch of synthetics is considered synonomous with trapping efficiency (a common usage), then important factors including multiple visits of single males to the trap before capture and the absence of components crucial to close-range orientation could be obscured, thereby giving a false impression of good trapping efficiency. An objective determination of trapping efficiency necessitates direct observations of male behavior.

DISRUPTION OF MALE ORIENTATION

Studies directed towards understanding pheromone perception involving sensory input of each component (acceptor specificity and transduction), integration, and finally CNS translation to possible behavioral effects have included single sensillum neurophysiology, investigating the physicochemical and biochemical processes with radiolabeled chemicals, and histochemical research with antennal receptors by using high-resolution electron microscopy (66, 130).

Although the olfactory processes are incompletely understood, much research has been conducted on modulating pheromone perception by effecting quantitative and qualitative changes in the chemical stimuli and determining the resulting behavioral alterations. Similar to the research on pheromone component activity, modulation effects usually are approximated by indirect means such as detecting differences in the number of male moths captured in traps and in the frequency of matings, or by determining the extent of larval damage by the ensuing generation. Research on behavior modification has received much attention because of the potential of using such sensory input manipulations for insect control. This section discusses various means of disrupting male moth orientation to pheromone traps.

Disruptants Used Within Pheromone Traps

PHEROMONE COMPONENTS Pheromone blends elicit the normal sequence of male responses in the field when the components are used in ratios and release rates approximating those of the natural system. Changes in either ratios or emission rates can affect the responses and prevent successful orientation to pheromone traps. Thus, in a sense, the pheromone components themselves represent some of the most potent disruptants.

Release rate It generally has been found that pheromone dispensers releasing the chemicals above a certain emission rate will catch fewer males. The optimum release rate or dispenser load for trap catch varies greatly among species (Table 2), ranging from those that exhibit attraction to lower emission rates, such as the larch bud moth and the grape berry moth, to species that are best lured to high release rates, such as the cabbage looper and the tufted apple bud moth. In many species males were captured in increasing numbers to increasing release rates well above those of the calling females. The highest average release rates recorded for females are about 0.8 $\mu g\ hr^{-1}$ and 0.6 $\mu g\ hr^{-1}$ for the gypsy moth (108) and the cabbage looper moth (146), respectively, whereas the lowest is 0.003 $\mu g\ hr^{-1}$ for the Indian meal moth (100, 145). Judging from the nanogram amounts of pheromone extracted from most females, it is tempting to speculate that for most species the natural emission rates are closer to the low release rate.

Component ratios Increasing the pheromone concentration to unnaturally high levels may disrupt male orientation, but a change in component ratios also can be very disruptive. Primary components that possess similar vapor pressures appear to

Table 2 Examples of the diversity of pheromone release rates for optimum trapping of male moths

Species	Approximate release rate (μg hr^{-1})			
	Optimum	Too high	Too low	Ref.
Red bollworm (*Diparopsis castanea*)	0.7	—	0.1	82
Codling moth (*Laspeyresia pomonella*)	1.3	3	0.5	81
European pine shoot moth (*Rhyacionia buoliana*)	3	6	0.1	144
Summerfruit tortrix moth (*Adoxophyes orana*)	5[a]	—	1	91
Cabbage looper (*Trichoplusia ni*)	60	180	2	139
	Load rate on rubber septa (μg)			
Larch bud moth (*Zeiraphera diniana*)	10	100	1	114
Grape berry moth (*Paralobesia viteana*)	30	100	1	154
Oriental fruit moth (*Grapholitha molesta*)	200	1000	10	117
Oak leafroller (*Archips semiferanus*)	1000	1×10^4	100	88
Tufted apple budmoth (*Platynota idaeusalis*)	12×10^3	—	—	61

[a] 10 mg of pheromone in a polyethylene cap, about 5 μg hr^{-1} release rate (79).

require a precise ratio for effective trapping; incorrect ratios lower male orientation to traps. Male moths of species using pheromones of this type may have a lower threshold of response for a specific range of component ratios, e.g. redbanded leafroller males respond to the natural ratio of 92:8 (Z11/E11-14:Ac) in the laboratory at concentrations much lower than required with ratios of 100:0 and 70:30 (4). This also was reflected in trap catch with a series of ratios (75, 121).

Trap catch with a series of component ratios has been used to determine the specificity range for many species. For example, the oak leafroller moth, *Archips semiferanus,* has a very restricted range with a 66:34 ratio of primary components, E11/Z11-14:Ac, catching significantly more males than ratios of 60:40 and 70:30 (88). No males were captured with 50:50 and 80:20 ratios. Similar requirements have been found with other species utilizing geometrical isomers (1, 3, 10, 13, 17, 35, 64, 68, 75, 77, 82, 90, 97–99, 104, 113, 124, 126, 129, 167) and positional isomers (34, 86, 90, 91, 115, 152, 153) as primary components. The optimum trapping ratios usually coincide with the ratios of the same components in the female moth. In some cases a critical range of ratios of three components was found (34).

The effectiveness in decreasing trap catch by selectively increasing the release rate of only one component of a multicomponent pheromone system is probably responsible for the results reported for the corn earworm, *Heliothis zea* (134). In this case a chemical (Z11-16:ALD), which was isolated from female moths (134) and found to elicit activation and wing fanning at 100 ng (122), was emitted at relatively high release rates along with calling females and caused a decrease in trap catch. This type of inhibition is not evidence that the chemical elicits behavior distinct from other moths' primary components, which also decrease trap catch when emitted at superoptimal rates. Most primary components would be classified as inhibitors in that respect.

The restricted ratios over which the primary components discussed previously are effective in luring males could well be related to evolutionary pressures exerted on the species for pheromone specificity (112). The fruittree leafroller moth, *Archips argyrospilus,* in British Columbia is captured with a relatively broad range of primary component (Z11/E11-14:Ac) ratios, whereas in New York this species is trapped with a comparatively narrow range of ratios, although the optimum blend for both populations appears similar (124). In New York other temporally synchronous sympatric leafroller species that are not present in British Columbia use the same primary components but in dissimilar proportions. In the former geographic region reproductive isolation could be aided by added specificity of response to precise component ratios. The European corn borer, *Ostrinia nubilalis,* populations optimally attracted to a 96:4 ratio of Z11/E11-14:Ac (75) are trapped by a narrow range of component ratios, whereas the European corn borer populations that are optimally attracted to the opposite ratio, 4:96 (77), are captured by ratios up to 50:50 in certain localities (76).

Males captured in traps containing a ratio different from the optimum ratio for that species do not necessarily represent divergent phenotypes or individuals optimally responsive to another ratio. Studies of the Oriental fruit moth (30) showed that males attracted in the field initially to various blends and marked with powdered dye color-coded to that blend were captured the following night by traps, representing the various blends, in similar distributions about the optimum blend. These data underscore the fact that males can orient to traps emitting a non-optimal ratio of components if the release rate is above the threshold level for response, and thus they can be trapped without further opportunities to seek the ratio to which they are the most sensitive.

A wide range of effective ratios is found if the primary components have dissimilar vapor pressures, such as with different carbon chain lengths (122, 160). In these cases, it would be difficult to maintain a specific ratio in the field because of the differing relative emission rates of the chemicals at varying environmental temperatures (34). A wide range of effective ratios also has been found with combinations involving primary components and components suspected of being secondary components (61, 62, 78, 82, 117, 121, 123, 124, 151, 172). These examples mainly involve dissimilar functional groups, such as acetate/aldehyde and acetate/alcohol combinations, dissimilar carbon chain lengths, such as 14-carbon/12-carbon chain compounds, and dissimilar numbers of double bonds. Secondary components, however, cannot be used in indiscriminate amounts in every case. With the omnivorous leafroller moth, *Platynota stultana,* which utilizes primary component acetates (E11/Z11-14:Ac), a secondary component alcohol (E11-14:OH) was found to increase trap catch if used in amounts from 0.2–2% of the primary components and to diminish trap catches at higher ratios (3, 63).

It is apparent from the foregoing considerations that the pheromone components that best disrupt male orientation at the lowest release rates when emitted from traps are those involved in critical, narrow-range component ratios. Trap catches are probably diminished with a non-optimal ratio of primary components since the threshold level of response in the males for that ratio is too high to evoke upwind anemotaxis. If the release rate of a non-optimal ratio is raised to its threshold level of response, the actual quantities involved may be so great that other inhibitory mechanisms, such as those found with high release rates of pheromones (previous section), come into play. It also is probable that close-range orientation near the trap is influenced by the precise proportion of components.

PHEROMONE-RELATED CHEMICALS The complex sequence of behavioral responses that occur before a male moth finally enters a pheromone trap can not only be disrupted by the pheromone components but also by certain related chemicals added to the attractant lure. Although very high release rates of many chemicals can decrease trap catches, the disruptants of interest are those that have an effect at the low emission values. It was first found with redbanded leafroller moths that some chemicals, such as positional isomers or the alkyne analogue of a pheromone component, could drastically reduce trap catch when present in a 1:1 ratio to the pheromone (118). A more extensive study (119) including a series of oxygen analogues confirmed definite structural requirements for the disruptant effect. Similar investigations with the summerfruit tortrix moth, *Adoxophyes orana* (164, 165), and the codling moth, *Laspeyresia pomonella* (2, 52), showed that the most potent disruptants were chemicals closely related to the pheromone components. Primary component geometric isomers, which are not found to be part of the pheromone system, are frequently very effective in decreasing trap catches at very low release rates (7, 26, 41, 42, 58, 61, 62, 68, 80, 82, 91, 111, 123, 150, 162–164). A dramatic example is the significant decrease in trap catch of lesser peach tree borer, *Synanthedon pictipes,* males when only 1% of one of the geometric isomers, (Z,Z)-3,13-18:Ac, is added to the pheromone dispenser (162).

In a number of cases the geometric isomer of the primary component is effective in decreasing trap catch at low concentrations, whereas the geometric isomer of the secondary component is not very disruptive (61, 62).

Unfortunately, discovering a decrease in trap catch with a certain chemical does not provide information on its neurophysiological and behavioral effects on the male moth. In cases where the disruptant must be structurally very similar to a primary component, the disruptant may be modulating the input of a primary component sensory cell (101–103). Disruptants of this sort may possibly affect the threshold for male activation and long-distance anemotaxis and may be very effective for disrupting orientation via the air permeation technique (discussed later).

Other chemicals that decrease trap catch may modify other portions of the behavioral sequence, such as close-range behavior, and have little or negligible effect on long-distance anemotaxis. With the Oriental fruit moth, 12:Ac was found to decrease trap catch to the attractant blend Z8-12:Ac/E8-12:OH (117). Further studies (125) showed that this disruptant did not eliminate the approach of males to the pheromone source, but rather it disrupted their in-flight behavior within 15 cm of the source. Behavioral studies had shown (28, 29) that the first two compounds are important for long-distance anemotaxis and that 12:OH (and also Z8-12:OH) is important in eliciting close-range behavior, including landing, wing fanning, and hair pencilling. One possibility is that the disruptant, 12:Ac, could modulate the sensory input (or alter thresholds) from the component(s) involved in evoking close-range behaviors and not affect long-distance anemotaxis.

The sensory input from close-range compounds is not understood, but in some instances these components could act in conjunction with one of the primary components upon the same sensory neuron. Sensory input modulation could be effected differently by primary component disruptants and secondary component disruptants. Behavioral (4), electroantennogram (5), and single-cell studies (101, 102) with the redbanded leafroller moth indicate that the close-range pheromone component, 12:Ac, acts in conjunction with a primary component, Z11-14:Ac. The single-cell studies indicate that 12:Ac can synergize the sensory cell response to Z11-14:Ac. This increase in nerve impulse activity may be necessary to induce the next stage of precopulatory responses, and chemicals interfering with this process could disrupt close-range behavior. Since 12:Ac alone does not seem to activate the cell, large quantities of this compound might not lessen the male's response to the primary components; the close-range compound or its disruptant might only affect the synergistic activity of the cell that influences close-range behavior.

It also is possible that the close-range disruptants are perceived via cells other than those specialized for pheromone perception, such as cells cued to detecting compounds emitted by other species possessing similar communication systems. With cabbage loopers, Z7-12:OH was found (161) to decrease trap catch when only 0.1% was added to the pheromone (Z7-12:Ac). Olfactometer studies (85) showed that this chemical evidently altered close-range behavior and not long-distance anemotaxis. Permeation of the air in a flight tunnel with the disruptor Z7-12:OH did not reduce the male responses to the pheromone, and so it was suggested that the disruptant is perceived by a separate sensory neuron. A similar conclusion had been reached in laboratory studies of two pyralid moths (148). These data by themselves, however, do not eliminate the possibility that the close-range components and their disruptants could affect only the synergistic activity of the primary component cell and not the activity associated with long-range anemotaxis.

It is not possible to know exactly the behaviors affected by disruptants from observing trap catch decreases; however, it may be possible to infer from their structural similarities to primary and secondary components whether the effect is likely to be on long-range anemotaxis or on close-range behavior. In practice these judgments could be useful for determining which compounds are the best candidates for disrupting long-range anemotaxis.

Disruptants Permeating the Air

PHEROMONES Pheromones (either single-component systems or the entire blend of multicomponent systems) can decrease trap catch when released at abnormally high release rates (see previous section). It was recognized (8) that this behavioral effect had potential in disrupting mating for insect control if large amounts of pheromone were released to permeate the air over the whole area of concern. To demonstrate this phenomenon, field tests with the gypsy moth, *Porthetria dispar,* were conducted (22) but the results were disappointing. Later research (15, 16, 65) showed that gyplure, the chemical used in the tests and thought to be an active analogue of the pheromone, was behaviorally inert to gypsy moth males and was structurally quite dissimilar from the correct pheromone structure. Speculation on the feasibility of the disruption concept continued (169, 170), and in 1967 a small-plot experiment (49, 141) with the cabbage looper demonstrated that the release of an artificially large amount of pheromone from evaporators spaced 3 m apart could eliminate trap catch by female moths. The pheromone in this test was permeating the atmosphere at a rate of about 100 mg hr^{-1} ha^{-1}. Apparently this pheromone concentration was above some threshold level needed to alter normal male stimulation or long-distance anemotaxis.

Further experiments with the cabbage looper showed that the most important factor in effecting disruption is the maintenance of a critical concentration of pheromone. Thus, evaporators releasing 0.2 μg of pheromone hr^{-1} and spaced 1 m apart were just as effective as evaporators releasing 4×10^4 μg hr^{-1} spaced 400 m apart (Table 3). When the total emission rate dropped to 0.007 mg hr^{-1} ha^{-1} (produced by evaporators releasing 7 μg hr^{-1} on 100-m spacings), there was little disruptive effect.

Other researchers (Table 3) also have found that male orientation of various species can be disrupted by maintaining a high level of pheromone in the air. Using the Oriental fruit moth, Rothschild (127) confirmed that there is no significant difference in disruption with release sites at various spacings, as long as the amount of pheromone emitted per unit area remains above a given level, 5 mg hr^{-1} ha^{-1}. Release sites (25 μg of pheromone hr^{-1}), spaced every 50 m^2 (total emission rate of 5 mg hr^{-1} ha^{-1}) provided no better disruption than release sites (200 μg hr^{-1}) spaced every 400 m^2 (total emission rate of 5 mg hr^{-1} ha^{-1}). Spacings can be separated too far, however, resulting in a heterogenous distribution of pheromone with areas of ineffective pheromone concentration. Thus, with release sites (400 μg hr^{-1}) every 800 m^2 (total emission rate of 5 mg hr^{-1} ha^{-1}), the disruptive effect was greatly diminished.

Research on maintaining an effective disruptive pheromone emission rate by a variety of techniques has received much attention (27, 92, 109, 137). One technique in particular that has provided promising results involves the application of microencapsulated pheromone. The method offers the advantages both of utilizing conventional pesticide application equipment and of distributing pheromone in a relatively uniform manner. Initial research with microencapsulated pheromone was conducted with the gypsy moth (11, 12, 14, 23, 24, 54, 132) and then with a number

Table 3 Pheromone emission rates that effected excellent male orientation disruption

Species	Pheromone released	Emission rate ($mg\ hr^{-1}\ ha^{-1}$)	Evaporators	Ref.
Cabbage looper (*Trichoplusia ni*)	Z7-12:Ac	0.07	Teflon cavities, 70 $\mu g\ hr^{-1}$, 100-m spacings	138
Grape berry moth (*Paralobesia viteana*)	Z9-12:Ac (4% E)	0.75	Hollow fibers, 0.083 $\mu g\ hr^{-1}$ 6 per vine; 9000 ha^{-1}	156
Cabbage looper	Z7-12:Ac	2	Thread, 0.2 $\mu g\ hr^{-1}$ 1-m spacings	138
Cabbage looper	Z7-12:Ac	2.5	Nylon mesh, 4×10^4 400-m spacings	47
European pine shoot moth (*Rhyacionia buoliana*)	E9-12:Ac	3.7	PVC strands 3.3 $\mu g\ hr^{-1}$ 3-m spacings	42
Oriental fruit moth (*Grapholitha molesta*)	Z8-12:Ac (3% E)	5	Closed polyethylene capillaries, 10 $\mu g\ hr^{-1}$, 2 per tree	127
Gypsy moth (*Porthetria dispar*)	Z7,8-2Me18:epoxy	5	Planchets 4 $\mu g\ hr^{-1}$ 3-m spacings	32
Grape berry moth	Z9-12:Ac (4% E)	14	Planchets	155
Redbanded leafroller (*Argyrotaenia velutinana*)	Z11-14:Ac (2% E)	14	100 $\mu g\ hr^{-1}$ 8- to 9-m spacings	
Red bollworm (*Diparopsis castanea*)	E9, 11-12:Ac (7% Z) 11-12:Ac	17	Polyethylene caps 0.85 $\mu g\ hr^{-1}$ 19,930 ha^{-1}	82
Plum fruit moth (*Grapholitha funebrana*)	Z8-12:Ac (2% E)	25-50	Closed polyethylene capillaries, 100 $\mu g\ hr^{-1}$ 2 per tree & supplemented	1

of other lepidopterous species, including the redbanded leafroller (38, 155), Oriental fruit moth (50, 51), codling moth (96), grape berry moth (155), and the spruce budworm (128). Again, disruption of male orientation appeared to occur when the total emission rate of pheromone per unit of land area was above a critical value for each species.

Microencapsulated pheromone for the redbanded leafroller moth emitted pheromone at the rate of about 4 $\mu g\ hr^{-1}\ g^{-1}$ for the first 6 days (38). Distribution at a rate of 22 g of pheromone ha^{-1} would provide a total emission rate of 3 $mg\ hr^{-1}\ ha^{-1}$, which is within the effective range for the other species (Table 3). Unfortunately the release rate for that formulation diminished rapidly after 6 days to ineffective levels, even though only about 4% of the pheromone had evaporated. A microencapsulated formulation used for the gypsy moth had a more constant release rate (9, 11), and 20 g of pheromone ha^{-1} appeared to be effective in disrupting male orientation and mating for at least 5 weeks (11, 54).

Although the microencapsulated pheromone applications provide a relatively uniform distribution of pheromone over the entire area, the inefficient release of the active ingredient from the capsules presents the difficulty of maintaining the required emission rate for long periods. Another microdispersable technique, hollow fiber dispensers applied with a modified spray procedure, appears to obviate this difficulty. Each fiber can release pheromone at a nearly constant rate for months, with the longevity of emission dependent on the length of fiber. A test with the grape berry moth (156) showed that the use of six fibers per vine (0.75 mg hr^{-1} ha^{-1}) gave excellent disruption of male orientation for over 2 months. The emission rate could be maintained at higher levels by using either fibers of greater diameter or more fibers per hectare.

Disruption of mating with stored product pests in enclosed environments should be achieved more easily than in the open field, but research (19) with the Indian meal moth, *Plodia interpunctella,* and almond moth, *Cadra cautella,* showed that mating was not disrupted with high emission rates of pheromone (7 mg hr^{-1} per 1000 m^3). Reasons presented for this failure include the high population density (close proximity of individuals) and additional mating stimuli that override the pheromone response. It was estimated that at the population density tested a pheromone emission rate of 0.6 kg of pheromone hr^{-1} per 1000 m^3 would have been required to decrease mating of the almond moth to about 25% of the controls (19, 20). Other investigations (147) also showed that disruption of the Indian meal moth is dependent on moth density and emission rate. In populations of 1 moth pair per m^2 of surface area, mating reduction of 20% with an emission rate of 4×10^{-4} mg hr^{-1} per 1000 m^3 increased to 60% mating reduction with an emission rate of 4 mg hr^{-1} per 1000 m^3. When the population density was decreased to only 0.1 pair per m^2, an emission rate of 1 mg hr^{-1} per 1000 m^3 effected a 96% reduction in mating. Behavioral observations revealed that males exposed to the pheromone were not stimulated by females to initiate orientation behavior, but males, habituated or not, would sometimes encounter females in spontaneous movements and copulate. The close-range behavior apparently was not affected by exposure to the pheromone component (Z,E)-9,12-14:Ac.

PHEROMONE COMPONENTS In 1965, Wright (171) predicted that the most efficacious modification of insect behavior would be achieved by raising the background concentration of one or several of the pheromone blend components. Recent field trapping studies with various pheromone blends show that slight increases in certain component concentrations sometimes lower trap catch drastically (see previous section).

Primary components As discussed previously, long-distance upwind anemotaxis can be expected to be disrupted by modulating the input of the primary component blends. Air permeation tests (74, 75) with the redbanded leafroller and European corn borer moths showed that E11-14:Ac, a pheromone component found in low ratios (8:92 and 3:97, respectively) to Z11-14:Ac (75, 121), could disrupt male orientation of both species when distributed in the field on cork granules or in microcapsules. In one test, the monitoring traps were encircled with microencapsulated E11-14:Ac applied at the rate of 9 mg m^{-2} in a 1-m wide circular strip with a 10-m radius. Trap catch of both species was almost eliminated in the 9-day test. An additional test, however, showed that the European corn borer required a higher level of compound emission for disruption than the redbanded leafroller. Applying microencapsulated E11-14:Ac at rates of 33.2 g ha^{-1} and higher disrupted male orientation to pheromone traps for 3 weeks with redbanded leafroller, but only for 12 days with European corn borer moths.

In this test the actual emission rate per hectare was not determined, but other experiments with the redbanded leafroller (116) showed that the natural blend (8:92) of primary components is a more effective disruptant than a 50:50 mix or the E11-14:Ac compound alone (100:0). Microencapsulated formulations (22 g ha^{-1}) with those ratios gave 98, 89, and 67% decreases in trap catch, respectively, compared to control plots. Similar results were obtained with the pink bollworm when the natural ratio (50:50) of primary components, (Z,Z)-7,11-16:Ac and (Z,E)-7,11-16:Ac, was compared with each component alone in disruption tests (140). With the smaller tea tortrix moth (150), each pheromone component (Z9-14:Ac and Z11-14:Ac) alone was as disruptive as the natural blend (70:30), but the very high emission rates generated with 10 mg of chemical on cotton wicks every 1.5-5 m might not allow discrimination among treatments. For each species, the level of emission necessary for disruption should be determined for each pheromone component and the natural blend to ascertain the system that disrupts at the lowest concentration.

Secondary components Due to the subtle, and usually undefined, role of secondary pheromone components, it is difficult to predict their precise behavioral effect in air permeation tests. Although it is suspected that at low concentrations secondary components would not affect long-distance upwind anemotaxis, at high concentrations they might disrupt close-range behavior obligatory to mating. With *Spodoptera litura,* evaporation (1 mg on cotton wicks at the corners of a 3.5-m square) of the secondary component, (Z,E)-9,12-14:Ac, was as effective in suppressing mating of tethered females as the primary component, (Z,E)-9,11-14:Ac (173). Curiously, a 10:1 combination of the two components was not very effective, although a possible increase in mating pressure caused by luring males specifically to these test plots (32) was not evaluated. Since only mating suppression was determined, it is not known if upwind anemotaxis or close-range behavior or both were disrupted by each component.

PHEROMONE-RELATED CHEMICALS When field trapping tests revealed that certain pheromone analogues could drastically reduce trap catch, it was suggested (118) that these chemicals might be effective disruptants. Studies in recent years have shown that a few chemicals that decrease trap catch at low rates (see previous section) are effective when used in the permeation technique, whereas others are not. Although the mechanism effecting disruption is not known, the chemicals known so far to be effective are closely analogous to the primary components.

Primary component analogues Some chemicals that are structurally similar to primary components have been found to be effective disruptants in air permeation tests. (*Z*)-9-tetradecenyl formate, a formate analogue, of a *Heliothis zea* and *Heliothis virescens* pheromone component, (*Z*)-11-hexadecenal, is very effective at a rate of 2 mg hr^{-1} ha^{-1} in disrupting male orientation of both species to their conspecific females (95). The two compounds are structurally identical except that the formate possesses an oxygen atom in place of the 2-position methylene group of the aldehyde. An advantage of this compound is that it does not attract males to the disruption area as is possible when using the pheromone. Further research (93) showed that the formate also is effective at the rate of 2 mg hr^{-1} ha^{-1} in disrupting mating of both species. Although concentration levels of effectiveness were not established, the data suggested that at the emission rate used the formate might be a better disruptant than the aldehyde pheromone component.

The alkyne analogue (11-tetradecynyl acetate) of the European corn borer and redbanded leafroller pheromone components (Z- and E11-14:Ac) disrupted male orientation of both species when distributed in microcapsules (74) at rates of 66.4 g of alkyne ha^{-1} and higher. At these high emission rates, the disruptive effect appeared to be similar to that of the pheromone component, E11-14:Ac. Distribution of microencapsulated acetate analogue, (E,E)-8,10-12:Ac, of the codling moth pheromone, (E,E)-8,10-12:OH, at a rate of 15 g ha^{-1} resulted in effective disruption for 2 days, but the level of disruption was less than that caused by equal amounts of the pheromone (96).

In the case of the pink bollworm, *Pectinophora gossypiella,* a monounsaturated pheromone analogue, Z7-16:Ac (hexalure), not only possessed intrinsic attractiveness to males (57, 71) but was an effective disruptant of male orientation at rates of 2 mg hr^{-1} ha^{-1} and higher (142). It was about 100-fold less attractive to male pink bollworms than the pheromone components (Z,Z)- and (Z,E)-7,11-16:Ac, and for equivalent disruption hexalure had to be emitted at 100 times the level of the pheromone components (140). A disruptant with an apparently analogous action was found with the red bollworm. In this case the monounsaturated analogue E9-12:Ac drastically decreased trap catches of the pheromone E9,11-12:Ac (10% Z) (82). In air permeation tests, the disruptant was as effective in disrupting male orientation as the pheromone components (emission rate was about 40 mg hr^{-1} ha^{-1}). Additionally field cage mating studies (82) showed that the analogue was more effective in disrupting mating than the pheromone.

With the fall armyworm, *Spodoptera frugiperda,* the diunsaturated compound (Z,E)-9,12-14:Ac was found to decrease greatly trap catches of the monounsaturated attractant (Z9-12:Ac) (94). It is not known how closely the disruptant resembles the pheromone since Z9-14:Ac was claimed to be the pheromone (133) and Z9-12:Ac was reported to be an attractant (94). Air permeation tests with (Z,E)-9,12-14:Ac showed it to be very effective in disrupting orientation and mating when released at a rate of 2 mg hr^{-1} ha^{-1} (93). In preliminary tests, it was as efficacious for disruption as the attractant Z9-12:Ac (94).

Pheromone analyses (162) of the lesser peachtree borer, *Synanthedon pictipes,* and the peachtree borer, *Sanninoidea exitiosa,* showed that the pheromones were (E,Z)-3,13-18:Ac and (Z,Z)-3,13-18:Ac, respectively. Each species was found to possess only a single isomer, but air permeation tests (83) showed that both chemicals were equally effective in disrupting male orientation of both species at the emission rate employed. It was not determined whether the pheromone or its geometrical isomer was better at lower concentrations.

The geometrical isomer (Z9-12:Ac) of the European pine shoot moth pheromone was very effective in decreasing trap catch when added to the pheromone (E9-12:Ac), but it was ineffective in disrupting male orientation to traps in air permeation tests (41) at concentrations (4 mg hr^{-1} ha^{-1}) that were very effective with the pheromone compound.

Secondary component analogues Some secondary components have been shown to mediate close-range behavior (discussed in previous sections), and therefore it is possible that closely related analogues of these components could modulate the behavioral thresholds, the sensory input, or both of the secondary components. A chemical exhibiting secondary component activity with the Oriental fruit moth is dodecan-1-ol (28, 29). A functional analogue, dodecyl acetate, was found to decrease trap catch when emitted with the attractant blend (117, 125), and field behavioral studies (125) revealed that the disruptant interfered with searching behavior close to the pheromone source. Since the long-distance upwind anemotaxis evidently was not modified, 12:Ac may be affecting behaviors mediated by the secondary component, 12:OH. In air permeation tests with 12:Ac (125), trap catches actually increased in the treated area and normal mating activity evidently was unaffected. Since the behavioral threshold for the secondary component, 12:OH, already is relatively high (117), an extremely high emission rate of either 12:Ac or 12:OH via the air permeation technique might be required to effect disruption of close-range behaviors. Subthreshold concentrations might alter behavioral thresholds but with unpredictable consequences.

Similarities can be seen with the effect of a functional analogue (Z7-12:OH) of the cabbage looper pheromone (Z7-12:Ac). The analogue is potent in decreasing trap catch when present in a pheromone trap (161), but it failed to disrupt male orientation to traps when used in an air permeation test at a rate of 0.3 mg hr^{-1} ha^{-1}. Laboratory behavioral studies (85) suggested that the chemical was affecting close-range behavior rather than upwind anemotaxis. The acetate is the only pheromone component reported for the cabbage looper, and consequently close-range behaviors are thought to be elicited simply by increasing the pheromone concentration close to the source or perhaps by additional, as yet undefined, secondary components. The alcohol disruptant could modify close-range behavior by modulating the input of the secondary components or by affecting the pheromone response thresholds for such behavior or both.

With other species, chemicals also have been found that are effective in decreasing trap catch when emitted from the trap but are ineffective in decreasing trap catch when used in air permeation tests. Examples include Z7-12:Ac with the fall armyworm (attractant in trap was Z9-12:Ac) (94), tetradecyl acetate with the pink bollworm (Z7-16:Ac, a pheromone analogue, was used in the attractant traps) (84), and 2-methyl-(*Z*)-7-octadecene with the gypsy moth (pheromone is *cis*-7,8-epoxy-2-

methyloctadecane) (25, 32). In the last case, behavioral studies (32) were conducted to ascertain male behavioral changes in air permeation tests. The ability of males to locate pheromone sources was decreased in all treated plots, but observations of male behavior within the plots revealed that the number of males searching within the plots was increased at least twofold over the number in check plots. The males terminated their searching behavior sooner ($\bar{x}$=34 sec) in the plots treated with both chemicals than with either chemical alone ($\bar{x}$= 64 sec). Although both the pheromone and the olefin increased male searching, there was a significant difference in the termination of this searching behavior. Males left the pheromone-treated plot by rapidly flying nearly straight upward 20 m or more, whereas males flew out of the olefin-treated plots at a height of 2 m or less. The olefin could affect behavior by input via sensory cells different from those for the pheromone (131), but alternatively the olefin could modulate the sensory input of the receptor cells as suggested above with other secondary components and their analogues. Single-cell studies (131) have shown that both the olefin and the pheromone elicit excitatory responses from the same cell. The olefin, in combination with the pheromone, could elicit differences in the temporal characteristics of the receptor neuron discharge to modify the resulting behavioral responses. Perhaps the combined input from both primary and secondary type chemicals would be the most disruptive in air permeation tests (32).

CONCLUDING REMARKS

Many inferences from trap catch approximations have been made concerning possible behavioral responses mediated by pheromone components and their analogues. Such categorization of the chemicals into broad groupings does not explain the mechanisms by which the chemicals actually elicit and modify behaviors. It is not yet clear what correlations precise laboratory studies on habituation and adaptation have to field situations. A number of species have been easily habituated in laboratory tests by pre-exposure to the pheromone. With the light brown apple moth, *Epiphyas postvittana,* a pulsed pre-exposure to pheromone reduced male responsiveness more than a single, prolonged stimulus (6). However, the opposite was found with cabbage looper moths, which exhibited lower pheromone-induced activity to a continuous exposure of pheromone than to repetitive pulses of pheromone (48). Notwithstanding, in some species males continuously exposed to pheromone in air permeation tests may nonetheless search for females during their normal diel rhythm of mating. Behavioral observations (82) of red bollworm males exposed to relatively high emission rates (40 mg hr^{-1} ha^{-1}) of pheromone while confined in a 0.2-ha field cage showed that as many males were searching for females in the treated sites as in the control. However, marked behavioral differences were evident as males approached the female positions, with males in the treated plots flying very briefly near the female and then departing. Males in these plots were not captured in the traps. The high concentration of pheromone in the air did affect behavior and, as stated by Marks (82), "the male response threshold may be increased to such an extent that emanations of pheromone from the female, once located, are insufficient to elicit the full behavioral sequence leading to successful mating." At high concentrations, habituation of later events of the searching and mating sequence could be altered.

Thus, behavioral thresholds, which are influenced by various environmental factors and raised by the pheromones or analogues used in disruption tests, in turn could alter the behavioral sequence in different ways. Males that have initiated the behavioral sequence, including upwind anemotaxis, may not be behaviorally affected by high pheromone concentrations in the same way as males exposed before stimulation. The male's stimulatory responses in olfactometers rapidly diminish (6, 135) when exposed to pheromone, but flight responses elicited by the same concentration of pheromone can be maintained for long periods in some species. In a sustained-flight chamber utilizing a moving floor for optomotor response (see 73), redbanded leafroller males have flown continuously without leaving the pheromone plume for over an hour (W. Roelofs and J. Miller, unpublished data). Also, red bollworm males were found to orient to females up to 143 times in 2 hr in a field cage, although remaining around the individual female cages no longer than 3 min each time (82). The behavioral consequences of any chemical system evaluated in the field may need to be deciphered in terms of both the atmospheric concentration and the length of exposure a responding moth has had both before and after flight initiation. Until these detailed field studies have been undertaken, approximations of behavior will need to be continued and may prove to be enlightening.

Literature Cited

1. Arn, H., Delley, B., Baggiolini, M., Charmillot, P. 1976. Communication disruption with sex attractant for control of the plum fruit moth, *Grapholitha funebrana:* a two-year field study. *Entomol. Exp. Appl.* 19:139–47
2. Arn, H., Schwarz, C., Limacher, H., Mani, E. 1974. Sex attractant inhibitors of the codling moth *Laspeyresia pomonella* L. *Experientia* 30:1142–4
3. Baker, T., Hill, A., Cardé, R., Kurokawa, A., Roelofs, W. 1975. Sex pheromone field trapping of the omnivorous leafroller, *Platynota stultana. Environ. Entomol.* 4:90–92
4. Baker, T., Cardé, R. T., Roelofs, W. 1976. Behavioral responses of male *Argyrotaenia velutinana* to components of its sex pheromone. *J. Chem. Ecol.* 2: 333–52
5. Baker, T., Roelofs, W. 1976. Electroantennogram responses of male *Argyrotaenia velutinana* to mixtures of its sex pheromone components. *J. Insect Physiol.* 22:In press
6. Bartell, R. 1976. Behavioral responses of Lepidoptera to pheromones. In *Chemical Control of Insect Behavior: Theory & Application,* ed. H. Shorey, J. McKelvey. New York: Wiley. In press
7. Benz, G., von Salis, G. 1973. Use of synthetic sex attractant of larch bud moth *Zeiraphera diniana* (Gn.) in monitoring traps under different conditions, and antagonistic action of *cis*-isomere. *Experientia* 29:729–30
8. Beroza, M. 1960. Insect attractants are taking hold. *Agric. Chem.* 15:37–40
9. Beroza, M., Bierl, B., James, P., DeVilbiss, D. 1975. Measuring emission rates of pheromones from their formulations. *J. Econ. Entomol.* 68:369–72
10. Beroza, M., Gentry, C., Blythe, J., Muschik, G. 1973. Isomer content and other factors influencing captures of Oriental fruit moth by synthetic pheromone traps. *J. Econ. Entomol.* 66:1307–11
11. Beroza, M., Hood, C., Trefrey, D., Leonard, D., Knipling, E., Klassen, W. 1975. Field trials with disparlure in Massachusetts to suppress mating of the gypsy moth. *Environ. Entomol.* 4:705–11
12. Beroza, M., Hood, C., Trefrey, D., Leonard, D., Knipling, E., Klassen, W., Stevens, L. 1974. Large field trial with

microencapsulated sex pheromone to prevent mating of the gypsy moth. *J. Econ. Entomol.* 67:659–64
13. Beroza, M., Muschik, G., Gentry, C. 1973. Small proportion of opposite geometric isomer increases potency of synthetic pheromone of Oriental fruit moth. *Nature New Biol.* 244:149–50
14. Beroza, M., Stevens, L., Bierl, B., Philips, F., Tardif, J. 1973. Pre- and post-season field tests with disparlure, the sex pheromone of the gypsy moth, to prevent mating. *Environ. Entomol.* 2:1051–57
15. Bierl, B., Beroza, M., Collier, C. 1970. Potent sex attractant of the gypsy moth: its isolation, identification, and synthesis. *Science* 170:87–89
16. Bierl, B., Beroza, M., Collier, C. 1972. Isolation, identification, and synthesis of the gypsy moth sex attractant. *J. Econ. Entomol.* 65:659–64
17. Bierl, B., Beroza, M., Staten, R., Sonnet, P., Adler, V. 1974. The pink bollworm sex attractant. *J. Econ. Entomol.* 67:211–16
18. Birch, M. 1974. Aphrodisiac pheromones in insects. In *Pheromones*, ed. M. Birch, pp. 115–34. New York: Elsevier. 495 pp.
19. Brady, U., Daley, R. 1975. Mating activity of *Cadra cautella* during exposure to synthetic sex pheromone and related compounds in the laboratory. *Environ. Entomol.* 4:445–47
20. Brady, U., Jay, E., Redlinger, L., Pearman, G. 1975. Mating activity of *Plodia interpunctella* and *Cadra cautella* during exposure to synthetic sex pheromone in the field. *Environ. Entomol.* 4:441–44
21. Brower, B., Brower, J., Cranston, F. 1965. Courtship behavior of the queen butterfly, *Danaus gilippus berenice. Zoologica* 50:1–39
22. Burgess, E. 1964. Gypsy moth control. *Science* 143:526
23. Cameron, E. 1973. Disparlure: a potential tool for gypsy moth population manipulation. *Bull. Entomol. Soc. Am.* 19:15–19
24. Cameron, E., Schwalbe, C., Knipling, E. 1974. Disruption of gypsy moth mating with microencapsulated disparlure. *Science* 183:972–73
25. Cameron, E., Schwalbe, C., Stevens, L., Beroza, M. 1975. Field tests of the olefin precursor of disparlure for suppression of mating in the gypsy moth. *J. Econ. Entomol.* 68:158–60
26. Campion, D., Bettany, B., Nesbitt, B., Beevor, P., Lester, R., Poppi, R. 1974. Field studies of the female sex pheromone of the cotton leafworm *Spodoptera littoralis* (Boisd.) in Cyprus. *Bull. Entomol. Res.* 64:89–96
27. Cardé, R. T., 1976. Use of pheromones in the population management of moth pests. *Environ. Health Persp.* 14:In press
28. Cardé, R. T., Baker, T., Roelofs, W. 1975. Ethological function of components of a sex attractant system for Oriental fruit moth males, *Grapholitha molesta. J. Chem. Ecol.* 1:475–91
29. Cardé, R. T., Baker, T., Roelofs, W. 1975. Behavioural role of individual components of a multichemical attractant system in the Oriental fruit moth. *Nature* 253:348–49
30. Cardé, R. T., Baker, T., Roelofs, W. 1976. Sex attractant responses of male Oriental fruit moths to a range of component ratios: pheromone polymorphism? *Experientia* In press
31. Cardé, R. T., Comeau, A., Baker, T., Roelofs, W. 1975. Moth mating periodicity: temperature regulates the circadian gate. *Experientia* 31:46–48
32. Cardé, R. T., Doane, C., Granett, J., Roelofs, W. 1975. Disruption of pheromone communication in the gypsy moth: some behavioral effects of disparlure and an attractant modifier. *Environ. Entomol.* 4:793–96
33. Cardé, R. T., Doane, C., Roelofs, W. 1974. Diel rhythms of male sex pheromone response and female attractiveness in the gypsy moth. *Can. Entomol.* 106:479–84
34. Cardé, R. T., Hill, A., Cardé, A., Roelofs, W. 1977. Sex attractant specificity as a reproductive isolating mechanism among the sibling species *Archips argyrospilus* and *mortuanus* and other sympatric tortricine moths. *J. Chem. Ecol.* 3:In press
35. Cardé, R. T., Kochansky, J., Stimmel, J., Wheeler, A. Jr., Roelofs, W. 1975. Sex pheromones of the European corn borer: *cis* and *trans* responding males in Pennsylvania. *Environ. Entomol.* 4:413–14
36. Cardé, R. T., Roelofs, W. 1973. Temperature modification of male sex pheromone response and factors affecting female calling in *Holomelina immaculata. Can. Entomol.* 105:1505–12
37. Cardé, R. T., Roelofs, W., Doane, C. 1973. Natural inhibitor of the gypsy

moth sex pheromone. *Nature* 241: 474–75

38. Cardé, R. T., Trammel, K., Roelofs, W. 1975. Disruption of sex attraction of the redbanded leafroller (*Argyrotaenia velutinana*) with microencapsulated pheromone components. *Environ. Entomol.* 4:448–50
39. Clearwater, J. 1972. Chemistry and function of a pheromone produced by the male of the southern armyworm, *Pseudaletia separata. J. Insect Physiol.* 19:19–28
40. Dahm, K., Meyer, D., Finn, W., Reinhold, V., Roller, H. 1971. The olfactory and auditory mediated sex attraction in *Achroia grisella. Naturwissenschaften* 58:265–66
41. Daterman, G., Daves, G. Jr., Jacobson, M. 1972. Inhibition of pheromone perception in European pine shoot moth by synthetic acetates. *Environ. Entomol.* 1:382–83
42. Daterman, G., Daves, G. Jr., Smith, R. 1975. Comparison of sex pheromone versus an inhibitor for disruption of pheromone communication in *Rhyacionia buoliana. Environ. Entomol.* 4:944–46
43. Edgar, J., Culvenor, C., Pliske, T. 1976. Isolation of a lactone, structurally related to the esterifying acids of pyrrolizidine alkaloids, from the costal fringes of male Ithomiinae. *J. Chem. Ecol.* 2:263–70
44. Farkas, S., Shorey, H. 1972. Chemical trail-following by flying insects: a mechanism for orientation to a distant odor source. *Science* 178:67–68
45. Farkas, S., Shorey, H. 1973. Odor-following and anemotaxis. *Science* 180: 1302
46. Farkas, S., Shorey, H. 1974. Mechanisms of orientation to a distant pheromone source. See Ref. 18, pp. 81–95
47. Farkas, S., Shorey, H., Gaston, L. 1974. Sex pheromones of Lepidoptera. The use of widely separated evaporators of looplure for the disruption of pheromone communication in *Trichoplusia ni. Environ. Entomol.* 3:876–77
48. Farkas, S., Shorey, H., Gaston, L. 1975. Sex pheromones of Lepidoptera. The influence of prolonged exposure to pheromone on the behavior of males of *Trichoplusia ni. Environ. Entomol.* 4:737–41
49. Gaston, L., Shorey, H., Saario, C. 1967. Insect population control by the use of sex pheromones to inhibit orientation between sexes. *Nature* 213:1155
50. Gentry, C., Beroza, M., Blythe, J., Bierl, B. 1974. Efficacy trials with the pheromone of the Oriental fruit moth and data on the lesser appleworm. *J. Econ. Entomol.* 67:607–9
51. Gentry, C., Beroza, M., Blythe, J., Bierl, B. 1975. Captures of the Oriental fruit moth, the pecan bud moth, and the lesser appleworm in Georgia field trails with isomeric blends of 8-dodecenyl acetate and air-permeation trials with the Oriental fruit moth pheromone. *Environ. Entomol.* 4:822–24
52. George, D., McDonough, L., Hathaway, D., Moffitt, H. 1975. Inhibitors of sexual attraction of male codling moths. *Environ. Entomol.* 4:606–8
53. Gothilf, S., Shorey, H. 1976. Sex pheromones of Lepidoptera: examination of the role of male scent brushes in courtship behavior of *Trichoplusia ni. Environ. Entomol.* 5:115–19
54. Granett, J., Doane, C. 1975. Reduction of gypsy moth male mating potential in dense populations by mistblower sprays of microencapsulated disparlure. *J. Econ. Entomol.* 68:435–37
55. Grant, G., Brady, U. 1975. Courtship behavior of phycitid moths. I. Comparison of *Plodia interpunctella* and *Cadra cautella* and role of male scent glands. *Can. J. Zool.* 53:813–26
56. Grant, G., Smithwick, E., Brady, U. 1975. Courtship behavior of phycitid moths. II. Behavioral and pheromonal isolation of *Plodia interpunctella* and *Cadra cautella* in the laboratory. *Can. J. Zool.* 53:827–32
57. Green, N., Jacobson, M., Keller, J. 1969. Hexalure, an insect sex attractant discovered by empirical screening. *Experientia* 25:682–83
58. Hathaway, D., McGovern, T., Beroza, M., Moffitt, H., McDonough, L., Butt, B. 1974. An inhibitor of sexual attraction of male codling moths to a synthetic sex pheromone and virgin females in traps. *Environ. Entomol.* 3:522–24
59. Hendricks, D., Shaver, T. 1975. Tobacco budworm: male pheromone suppressed emission of sex pheromone by the female. *Environ. Entomol.* 4:555–58
60. Hendricks, D., Tumlinson, J. 1974. A field cage bioassay system for testing candidate sex pheromones of the tobacco budworm. *Ann. Entomol. Soc. Am.* 67:547–52
61. Hill, A., Cardé, R., Comeau, A., Bode, W., Roelofs, W. 1974. Sex pheromones of the tufted apple bud moth (*Platynota ideausalis*). *Environ. Entomol.* 3: 249–52

62. Hill, A., Cardé, R., Kido, H., Roelofs, W. 1975. Sex pheromone of the orange tortrix moth, *Argyrotaenia citrana*. *J. Chem. Ecol.* 1:215–24
63. Hill, A., Roelofs, W. 1975. Sex pheromone components of the omnivorous leafroller moth, *Platynota stultana*. *J. Chem. Ecol.* 1:91–99
64. Hummel, H., Gaston, L., Shorey, H., Kaae, R., Byrne, K., Silverstein, R. 1973. Clarification of the chemical status of the pink bollworm sex pheromone. *Science* 181:873–75
65. Jacobson, M., Schwarz, M., Waters, R. 1970. Gypsy moth sex attractants: a reinvestigation. *J. Econ. Entomol.* 63: 943–45
66. Kaissling, K.-E. 1975. Sensorische Transduktion bei Riechzellen von Insekten, *Verh. Dtsch. Zool. Ges.* 67: 1–11
67. Karandinos, M. 1974. Environmental conditions and sex activity of *Synanthedon pictipes* in Wisconsin, monitored with virgin female pheromone traps. *Environ. Entomol.* 3:431–38
68. Karandinos, M., Tumlinson, J. H., Eichlin, T. D. 1977. Field evidence of synergism and inhibition in the Sesiidae sex pheromone system. *J. Chem. Ecol.* 3:In press
69. Karlson, P., Butenandt, A. 1959. Pheromones (ectohormones) in insects. *Ann. Rev. Entomol.* 4:39–58
70. Karlson, P., Lüscher, M. 1959. "Pheromones": a new term for a class of biologically active substances. *Nature* 183:55–56
71. Keller, J., Sheets, L., Green, N., Jacobson, M. 1969. *cis*-7-Hexadecen-1-ol acetate (hexalure), a synthetic sex attractant for pink bollworm males. *J. Econ. Entomol.* 62:1520–21
72. Kennedy, J. 1976. Olfactory responses to distant plants and other odor sources. See Ref. 6
73. Kennedy, J., Marsh, D. 1974. Pheromone-regulated anemotaxis in flying moths. *Science* 184:999–1001
74. Klun, J., Chapman, O., Mattes, K., Beroza, M. 1975. European corn borer and redbanded leafroller disruption of reproduction behavior. *Environ. Entomol.* 4:871–76
75. Klun, J., Chapman, O., Mattes, K., Wojtkowski, P., Beroza, M., Sonnet, P. 1973. Insect sex pheromones: minor amount of opposite geometrical isomer critical to attraction. *Science* 181: 661–63
76. Klun, J., Cooperators. 1975. Insect sex pheromones: intraspecific pheromonal variability of *Ostrinia nubilalis* in North America and Europe. *Environ. Entomol.* 4:891–94
77. Kochansky, J., Cardé, R., Liebherr, J., Roelofs, W. 1975. Sex pheromones of the European corn borer in New York. *J. Chem. Ecol.* 1:225–31
78. Kochansky, J., Tette, J., Taschenberg, E., Cardé, R., Kaissling, K.-E., Roelofs, W. 1975. Sex pheromone of the moth, *Antheraea polyphemus*. *J. Insect Physiol.* 21:1977–83
79. Kuhr, R., Comeau, A., Roelofs, W. 1972. Measuring release rates of pheromone analogues and synergists from polyethylene caps. *Environ. Entomol.* 1:625–27
80. Lange, R., Hoffmann, D. 1972. Essigsäure-cis-dodecen-(9)-yl-ester, ein Inhibitor des Sexualpheromons des Kiefernknospentriebwicklers. *Naturwissenschaften* 5:217
81. Maitlen, J., McDonough, L., Moffitt, H., George, D. 1976. Codling moth sex pheromone: baits for mass trapping and population survey. *Environ. Entomol.* 5:199–202
82. Marks, R. 1975. An interim report on investigations in Malawi into the sex pheromones of the red bollworm of cotton *Diparopsis castanea* HMPS with special reference to the feasibility of developing a noninsecticidal control method. *Rep. Agric. Res. Counc. Malawi*, pp. 1–56; *Bull. Entomol. Res.* 66:219–300
83. McLaughlin, J., Doolittle, R., Gentry, C., Mitchell, E., Tumlinson, J. 1976. Response to pheromone traps and disruption of pheromone communication in the lesser peachtree borer and the peachtree borer. *J. Chem. Ecol.* 2: 73–81
84. McLaughlin, J., Gaston, L., Shorey, H., Hummel, H., Stewart, F. 1972. Sex pheromones of Lepidoptera. XXXIII. Evaluation of the disruptive effect of tetradecyl acetate on sex pheromone communication in *Pectinophora gossypiella*. *J. Econ. Entomol.* 65:1592–93
85. McLaughlin, J., Mitchell, E., Chambers, D., Tumlinson, J. 1974. Perception of *Z*-7-dodecen-1-ol and modification of the sex pheromone response of male loopers. *Environ. Entomol.* 3: 677–80
86. Meijer, G., Ritter, F., Persoons, C., Minks, A., Voerman, S. 1972. Sex pheromones of summer fruit tortrix moth *Adoxophyes orana*: two synergistic isomers. *Science* 175:1469–70

87. Meinwald, J., Meinwald, Y., Mazzocchi, P. 1969. Sex pheromone of the queen butterfly: chemistry. *Science* 164:1174–75
88. Miller, J., Baker, T., Cardé, R., Roelofs, W. 1976. Reinvestigation of oak leafroller sex pheromone components and the hypothesis that they vary with diet. *Science* 192:140–43
89. Minks, A. 1971. Decreased sex pheromone production in an inbred stock of the summerfruit tortrix moth, *Adoxophyes orana*. *Entomol. Exp. Appl.* 14:361–64
90. Minks, A., Roelofs, W., Ritter, F., Persoons, C. 1973. Reproductive isolation of two tortricid moth species by different ratios of a two-component sex attractant. *Science* 180:1073–74
91. Minks, A., Voerman, S. 1973. Sex pheromones of the summerfruit tortrix moth, *Adoxophyes orana:* trapping performance in the field. *Entomol. Exp. Appl.* 16:541–49
92. Mitchell, E. 1975. Disruption of pheromonal communication among coexistent pest insects with multichemical formulations. *BioScience* 25:493–99
93. Mitchell, E., Baumhover, A., Jacobson, M. 1976. Reduction of mating potential of male *Heliothis* spp. and *Spodoptera frugiperda* in field plots treated with disruptants. *Environ. Entomol.* 5: 484–86
94. Mitchell, E., Copeland, W., Sparks, A., Sekul, A. 1974. Fall armyworm: disruption of pheromone communication with synthetic acetates. *Environ. Entomol.* 3:778–80
95. Mitchell, E., Jacobson, M., Baumhover, A. 1975. *Heliothis* spp.: disruption of pheromonal communication with (*Z*)-9-tetradecen-1-ol formate. *Environ. Entomol.* 4:577–79
96. Moffitt, H. 1973. Communication at 4th meeting of the IOBC working group on genetic control of codling moth and *Adoxophyes,* Wädenswil.
97. Nesbitt, B., Beevor, P., Cole, R., Lester, R., Poppi, R. 1973. Synthesis of both geometric isomers of the major sex pheromone of the red bollworm moth. *Tetrahedron Lett.,* pp. 4669–70
98. Nesbitt, B., Beevor, P., Cole, R., Lester, R., Poppi, R. 1975. The isolation and identification of the female sex pheromones of the red bollworm moth, *Diparopsis castanea*. *J. Insect Physiol.* 21:1091–96
99. Nielsen, D., Purrington, F., Tumlinson, J., Doolittle, R., Yonce, C. 1975. Response of male clearwing moths to caged virgin females, female extracts, and synthetic sex attractants. *Environ. Entomol.* 4:451–54
100. Nordlund, D., Brady, U. 1974. Factors affecting release rate and production of sex pheromone by female *Plodia interpunctella* (Hübner). *Environ. Entomol.* 3:797–802
101. O'Connell, R. 1972. Responses of olfactory receptors to the sex attractant, its synergist and inhibitor in the redbanded leaf roller. In *Olfaction and Taste IV,* ed. D. Schneider, pp. 180–86. Stuttgart: Wissenschaftliche GmbH. 400 pp.
102. O'Connell, R. 1975. Olfactory receptor responses to sex pheromone components in the redbanded leafroller moth. *J. Gen. Physiol.* 65:179–205
103. den Otter, C. 1974. Electrophysiology of sex pheromone sensitive cells on antennae of the summerfruit tortrix moth, *Adoxophyes orana. 1st Congr. Eur. Chemorec. Res. Org., Orsay/Paris.* p. 45
104. Persoons, C., Minks, A., Voerman, S., Roelofs, W., Ritter, F. 1974. Sex pheromones of the moth, *Archips podana:* isolation, identification and field evaluation of two synergistic geometrical isomers. *J. Insect Physiol.* 20:1181–88
105. Pliske, T., Eisner, T. 1969. Sex pheromones of the queen butterfly: biology. *Science* 164:1170–72
106. Pliske, T. 1975. Courtship behavior and use of chemical communication by males of certain species of Ithomiine butterflies. *Ann. Entomol. Soc. Am.* 68:935–42
107. Raulston, J., Graham, H., Lingren, P., Snow, J. 1976. Mating interaction of native and laboratory-reared tobacco budworms released in the field. *Environ. Entomol.* 5:195–98
108. Richerson, J., Cameron, E. 1974. Differences in pheromone release and sexual behavior between laboratory-reared and wild gypsy moth adults. *Environ. Entomol.* 3:475–81
109. Roelofs, W. 1975. Manipulating sex pheromones for insect suppression. *Environ. Lett.* 8:41–59
110. Roelofs, W., Arn, H. 1968. Sex attractant of the red-banded leaf roller moth. *Nature* 219:513
111. Roelofs, W., Bartell, R., Hill, A., Cardé, R. T., Waters, L. 1972. Codling moth sex attractant—field trials with geometrical isomers. *J. Econ. Entomol.* 65: 1276–77

112. Roelofs, W., Cardé, R. T. 1974. Sex pheromones in the reproductive isolation of lepidopterous species. See Ref. 18, pp. 96–114
113. Roelofs, W., Cardé, R. T. 1974. Oriental fruit moth and lesser appleworm attractant mixtures refined. *Environ. Entomol.* 3:586–88
114. Roelofs, W., Cardé, R. T., Benz, G., von Salis, G. 1971. Sex attractant of the larch bud moth found by electroantennogram method. *Experientia* 27:1438–39
115. Roelofs, W., Cardé, A., Hill, A., Cardé, R. T. 1976. Sex pheromone of the three-lined leafroller, *Pandemis limitata. Environ. Entomol.* 5:649–52
116. Roelofs, W., Cardé, R. T., Taschenberg, E., Weires, R. 1976. Pheromone research for the control of Lepidopterous pests in New York. *Advan. Chem. Ser.* 23:75–87
117. Roelofs, W., Cardé, R. T., Tette, J. 1973. Oriental fruit moth attractant synergists. *Environ. Entomol.* 2:252–54
118. Roelofs, W., Comeau, A. 1968. Sex pheromone perception. *Nature* 220:600–1
119. Roelofs, W., Comeau, A. 1971. Sex pheromone perception: synergists and inhibitors for the red-banded leaf roller attractant. *J. Insect Physiol.* 17:435–48
120. Roelofs, W., Comeau, A., Selle, R. 1969. Sex pheromone of the Oriental fruit moth. *Nature* 224:723
121. Roelofs, W., Hill, A., Cardé, R. T. 1975. Sex pheromone components of the redbanded leafroller, *Argyrotaenia velutinana. J. Chem. Ecol.* 1:83–89
122. Roelofs, W., Hill, A., Cardé, R. T., Baker, T. 1974. Two sex pheromone components of the tobacco budworm moth, *Heliothis virescens. Life Sci.* 14:1555–62
123. Roelofs, W., Hill, A., Cardé, A., Cardé, R. T., Madsen, H., Vakenti, J. 1976. Sex pheromone of the European leafroller, *Archips rosanus. Environ. Entomol.* 5:362–64
124. Roelofs, W., Hill, A., Cardé, R. T., Tette, J., Madsen, H., Vakenti, J. 1974. Sex pheromone of the fruittree leafroller moth, *Archips argyrospilus. Environ. Entomol.* 3:747–51
125. Rothschild, G. 1974. Problems in defining synergists and inhibitors of the Oriental fruit moth pheromone by field experimentation. *Entomol. Exp. Appl.* 17:294–302
126. Rothschild, G. 1975. Attractants for monitoring *Pectinophora scutigera* related species in Australia. *Environ. Entomol.* 4:983–85
127. Rothschild, G. 1975. Control of Oriental fruit moth [*Cydia molesta* (Busck)] with synthetic female pheromone. *Bull. Entomol. Res.* 65:473–90
128. Sanders, C. 1977. Disruption of mating behavior of the Eastern spruce budworm. *Environ. Entomol.* In press
129. Sanders, C., Weatherston, J. 1976. Sex pheromone of the Eastern spruce budworm: optimum blend of *trans*- and *cis*-11-tetradecenal. *Can. Entomol.* In press
130. Schneider, D. 1974. The sex-attractant receptor of moths. *Sci. Am.* 231:28–35
131. Schneider, D., Lange, R., Schwarz, F., Beroza, M., Bierl, B. 1974. Attraction of male gypsy and nun moths to disparlure and some of its chemical analogues. *Oecologia* 14:19–36
132. Schwalbe, C., Cameron, E., Hall, D., Richerson, J., Beroza, M., Stevens, L. 1974. Field tests of microencapsulated disparlure for suppression of mating among wild and laboratory-reared gypsy moths. *Environ. Entomol.* 3:589–92
133. Sekul, A., Sparks, A. 1967. Sex pheromone of the fall armyworm moth: isolation identification, and synthesis. *J. Econ. Entomol.* 60:1270–72
134. Sekul, A., Sparks, A., Beroza, M., Bierl, B. 1975. A natural inhibitor of the corn earworm moth sex attractant. *J. Econ. Entomol.* 68:603–4
135. Shorey, H. 1974. Environmental and physiological control of insect sex pheromone behavior. See Ref. 18, pp. 62–80
136. Shorey, H. 1976. Interaction of insects with their chemical environment. See Ref. 6
137. Shorey, H. 1976. Manipulation of insect pests of agricultural crops. See Ref. 6
138. Shorey, H., Gaston, L. 1974. Programs utilizing pheromones in survey or control: the cabbage looper. See Ref. 18, pp. 421–25
139. Shorey, H., Gaston, L., Jefferson, R. 1968. Insect sex pheromones. In *Advances in Pest Control Research,* ed. R. L. Metcalf, Vol. 8, pp. 57–126. New York: Interscience. 255 pp.
140. Shorey, H., Gaston, L., Kaae, R. 1976. Air-permeation with gossyplure for control of the pink bollworm. *Advan. Chem. Ser.* 23:67–74
141. Shorey, H., Gaston, L., Saario, C. 1967. Sex pheromones of noctuid moths. XIV. Feasibility of behavioral control

by disrupting pheromone communication in cabbage loopers. *J. Econ. Entomol.* 60:1541–45

142. Shorey, H., Kaae, R., Gaston, L. 1974. Sex pheromones of Lepidoptera. Development of a method for pheromonal control of *Pectinophora gossypiella* in cotton. *J. Econ. Entomol.* 67:347–50
143. Silberglied, R., Taylor, O. 1973. Ultraviolet differences between the sulphur butterflies, *Colias eurytheme* and *C. philodice,* and a possible isolating mechanism. *Nature* 241:406–8
144. Smith, R., Daterman, G., Daves, G. Jr., McMurtrey, K., Roelofs, W. 1974. Sex pheromone of the European pine shoot moth: chemical identification and field tests. *J. Insect Physiol.* 20:661–68
145. Sower, L., Fish, J. 1975. Rate of release of the sex pheromone of the female Indian meal moth. *Environ. Entomol.* 4:168–69
146. Sower, L., Gaston, L., Shorey, H. 1971. Sex pheromones of noctuid moths. XXVI. Female release rate, male response threshold, and communication distance for *Trichoplusia ni. Ann. Entomol. Soc. Am.* 64:1448–56
147. Sower, L., Turner, W., Fish, J. 1975. Population-density-dependent mating frequency among *Plodia interpunctella* in the presence of synthetic sex pheromone with behavioral observations. *J. Chem. Ecol.* 1:335–42
148. Sower, L., Vick, K., Ball, K. 1974. Perception of olfactory stimuli that inhibit the responses of male phycitid moths to sex pheromones. *Environ. Entomol.* 3:277–79
149. Steel, R., Torrie, J. 1960. *Principles and procedures of statistics.* New York: McGraw-Hill. 481 pp.
150. Tamaki, Y., Ishiwatari, T., Osakabe, M. 1975. Inhibition of the sexual behavior of the smaller tea tortrix moth by the sex pheromone and its components. *Jpn. J. Appl. Entomol. Zool.* 19:187–92
151. Tamaki, Y., Noguchi, H., Yushima, T. 1973. Sex pheromone of *Spodoptera litura* (F.): isolation, identification, and synthesis. *Appl. Entomol. Zool.* 8:200–3
152. Tamaki, Y., Noguchi, H., Yushima, T., Hirano, C. 1971. Two sex pheromones of the smaller tea tortrix: isolation, identification, and synthesis. *Appl. Entomol. Zool.* 6:139–41
153. Tamaki, Y., Noguchi, H., Yushima, T., Hirano, C., Sugawara, H. 1971. Sex pheromone of the summerfruit tortrix: isolation and identification. *Kontyu* 39:338–40
154. Taschenberg, E., Cardé, R., Hill, A., Tette, J., Roelofs, W. 1974. Sex pheromone trapping of the grape berry moth. *Environ. Entomol.* 3:192–94
155. Taschenberg, E., Cardé, R., Roelofs, W. 1974. Sex pheromone mass trapping and mating disruption for control of redbanded leafroller and grape berry moths in vineyards. *Environ. Entomol.* 3:239–42
156. Taschenberg, E., Roelofs, W. 1976. Pheromone communication disruption of the grape berry moth with microencapsulated and hollow fiber systems. *Environ. Entomol.* 5:688–91
157. Trammel, K., Roelofs, W., Glass, E. 1974. Sex pheromone trapping of males for control of redbanded leafroller in apple orchards. *J. Econ. Entomol.* 67:159–64
158. Tukey, J. 1949. One degree of freedom for non-additivity. *Biometrics* 5:232–42
159. Tukey, J. 1955. Queries. *Biometrics* 11:111–13
160. Tumlinson, J., Hendricks, D., Mitchell, E., Doolittle, R., Brennan, M. 1975. Isolation, identification, and synthesis of the sex pheromone of the tobacco budworm. *J. Chem. Ecol.* 1:203–14
161. Tumlinson, J., Mitchell, E., Browner, S., Mayer, M., Green, N., Hines, R., Lindquist, D. 1972. *cis*-7-Dodecen-1-ol, a potent inhibitor of the cabbage looper sex pheromone. *Environ. Entomol.* 1:354–58
162. Tumlinson, J., Yonce, C., Doolittle, R., Heath, R., Gentry, C., Mitchell, E. 1974. Sex pheromones and reproductive isolation of the lesser peachtree borer and the peachtree borer. *Science* 185:614–16
163. Vick, K., Sower, L. 1973. *Z*-9, *Z*-12-tetradecadien-1-ol acetate, an inhibitor of the *Plodia interpunctella* (Hübner) sex pheromone response. *J. Econ. Entomol.* 66:1258–60
164. Voerman, S., Minks, A. 1973. Sex pheromones of summerfruit tortrix moth, *Adoxophyes orana.* 2. Compounds influencing their attractant activity. *Environ. Entomol.* 2:751–56
165. Voerman, S., Minks, A., Houx, N. 1974. Sex pheromones of summerfruit tortrix moth, *Adoxophyes orana*: investigations on compounds modifying their attractancy. *Environ. Entomol.* 3:701–4
166. Weatherston, J., Maclean, W. 1974. The occurrence of (*E*)-11-tetradecen-1-ol, a known sex attractant inhibitor in the abdominal tips of virgin female Eastern spruce budworm, *Choris-*

toneura fumiferana. Can. Entomol. 106:281–84

167. Weatherston, J., Percy, J., MacDonald, L. 1976. Field testing of *cis*-11-tetradecenal as attractant or synergist in Tortricinae. *Experientia* 32:178–79
168. Willson, H., Trammel, K. 1975. Relationships between sex pheromone trapping of six tortricids and a foliage index of apple orchard canopies. *Environ. Entomol.* 4:361–64
169. Wright, R. 1964. After pesticides - what? *Nature* 204:121–25
170. Wright, R. 1964. Insect control by nontoxic means. *Science* 144:487
171. Wright, R. 1965. Finding metarchons for pest control. *Nature* 207:103–4
172. Yushima, T., Tamaki, Y., Kamano, S., Cyama, M. 1974. Field evaluation of a synthetic sex pheromone, "litlure," as an attractant for males of *Spodoptera litura* (F.) *Appl. Entomol. Zool.* 9:147–52
173. Yushima, T., Tamaki, Y., Kamano, S., Cyama, M. 1975. Suppression of mating of the armyworm moth, *Spodoptera litura* (F.), by a component of its sex pheromone. *Appl. Entomol. Zool.* 10:237–39

Ann. Rev. Entomol. 1977. 22:407–29

CYTOGENETICS OF MITES AND TICKS

❖6136

James H. Oliver Jr.
Department of Biology, Georgia Southern College, Statesboro, Georgia 30458

There are three reviews of the cytogenetics of certain mites and ticks (43, 48, 75). The number of acarines studied cytologically more than doubled during the 10 years between Sokolov's review and Oliver's first review. The number has now doubled again. Even with this increase in information our knowledge of the cytogenetics of acarines is still fragmentary. Over 30,000 species and 1700 genera have been described and it is believed that a half million more species are extant (35). Nevertheless, data from our small cytogenetic sample indicate that an enormous amount of cytological and genetic diversity exists among mites and ticks. Somatic chromosome numbers range from 2 to 36, various sex chromosome systems are operative, and different kinds of chromosomes and reproductive mechanisms exist. The purpose of this review is to consider some of the fascinating aspects of the cytogenetics of mites and ticks by examining karyotypes, sex determination, and chromosome types.

KARYOTYPES—ORDER PARASITIFORMES

Parasitiformes consists of the suborders Metastigmata (ticks), Mesostigmata, and Tetrastigmata. No cytogenetic data exist for the Tetrastigmata.

Metastigmata

Among the soft ticks (Argasidae), cytogenetic data are available on 11 species of *Ornithodoros,* 11 species of *Argas,* and 2 species of *Otobius* (Table 1). Cytogenetic data also exist from species assigned to 8 genera of hard ticks (Ixodidae), including 8 species of *Ixodes,* 2 species of *Boophilus,* 15 *Haemaphysalis,* 10 *Dermacentor,* 4 *Rhipicephalus,* 14 *Amblyomma,* 4 *Aponomma,* and 12 *Hyalomma.* The Ixodidae are composed of two major groups, the Prostriata (Ixodinae) and the Metastriata (Amblyommatinae). The former group consists of only *Ixodes* wheras the latter contains the other genera.

Table 1 Chromosome numbers and sex determination in Argasidae

Species	Sex	2*n*	Sex determination	Reference
Argas brumpti	♀	24	—	J. Oliver[a]
A. cooleyi	♂ ♀	26	XY, XX	J. Oliver[a]
A. hermanni	♂ ♀	26	XY, XX	17
A. japonicus	♂ ♀	26	XY, XX	K. Tanaka, J. Oliver[a]
A. persicus	♂	26	XY	17
A radiatus	♂ ♀	26	XY, XX	27
A reflexus	♂	26	XY	17, 64
	♀	26	XX	17
A. sanchezi	♂ ♀	26	XY, XX	27
A. tridentatus	♂ ♀	26	XY, XX	17
A. vespertilionis	♂ ♀	20	—	16
A. zumpti	♂ ♀	26	XY, XX	49
Ornithodoros alactagalis	♂ ♀	32, 34	—	18
O. asperus	♂ ♀	16	XY, XX	18
O. capensis	♂ ♀	20	XY, XX	18
	♀	20	—	K. Tanaka, J. Oliver[a]
O. gurneyi	♂ ♀	12	XY, XX	46
O. lahorensis	♂ ♀	26	—	18
O. macmillani	♂	16	—	46
O. moubata	♂ ♀	20	—	14, 84
	♂ ♀	20	XY, XX	18
O. nereensis	♂	24, 26	—	18
O. savignyi	♂ ♀	20	XY, XX	28
O. tartakovskyi	♂ ♀	16	XY, XX	18
	♂	26	—	75
O. tholozani	♂ ♀	16	—	18
	♂	16	—	75, 76
Otobius lagophilus	♂	20	—	J. Oliver, R. Osburn[a]
O. megnini	♂	20	—	J. Oliver, R. Osburn[a]

[a] Unpublished data.

ORNITHODOROS Diploid chromosome numbers range from 12 to 32, and occasionally to 34, among species of *Ornithodoros.* These extremes represent the least and greatest numbers yet recorded for any species of tick. One species has 12 chromosomes, four have 16, three species have 20, three have 26, and one has 32 (Table 1).

Oliver (48) reviewed the karyotypes of 10 of the 11 cytogenetically reported species of *Ornithodoros*; therefore, only summary statements are made relative to those species. Howell's excellent report (28) on the chromosomes of *O. savignyi* indicates that both sexes possess a diploid number of 20, and the presence of extremely large sex chromosomes is evident. The large, slightly submetacentric sex chromosomes are more than twice the length of the short group of autosomes. Morphologically the Y chromosome differs only in length from the X chromosome, with both the upper and lower arms slightly shorter; it is still longer than any of the autosomes. The autosomes can be arranged as follows. Group 1 has medium-sized chromosomes with submedian centromeres, and the three pairs are morphologically alike in size and shape. Group 2 has one pair of medium-sized isobrachial (metacentric) chromosomes of the same length as those in Group 1. Group 3 has the longest chromosomes in the heterobrachial (acrocentric) series, approximately the same length as those of the first two groups. Group 4 has three pairs of short heterobrachial (acrocentric) chromosomes of similar size and shape, and Group 5 has one pair of short chromosomes the same length as those in Group 4, but they are isobrachial and each chromatid possesses a satellite.

The chromosome number of *O.* (*Pavlovskyella*) *nereensis* is uncertain. Goroshchenko (18) claims $2n = 24$ in the spermatogonia and spermatocytes of three specimens and $2n = 26$ in the same type tissues in three other males; a fourth male had 13 bivalents, the longest connected to a nucleolus. He also reports variable numbers of chromosomes in *O.* (*Pavlovskyella*) *alactagalis.* A diploid number of 32 was found in spermatogonia, spermatocytes, and oocytes of six males and one female, but 34 in some spermatocytes of another male. The variation in chromosome numbers in these two species might be a case of natural chromosomal polymorphism unless some of the observed cells were abnormal. It was unclear whether all spermatocytes in the one male *O. alactagalis* had 34 chromosomes or whether some possessed 32.

The karyotype of *O.* (*Alectorobius*) *coniceps* was reported to consist of $2n = 20$ in both sexes (15), but later (18) it was noted that on the previous occasion *O.* (*Alectorobius*) *capensis* was identified as *O.* (*A.*) *coniceps.* Thus, presumably the cytogenetics of *O. coniceps* has not been recorded and data are available for only 11 species.

ARGAS Most *Argas* species that have been studied cytogenetically reveal a $2n$ number of 26 chromosomes, including two long sex chromosomes in the female and one long and one shorter sex chromosome in the male (Table 1). Two exceptions are *A.* (*Ogadenus*) *brumpti* with 24 (J. H. Oliver Jr., unpublished data) and *A.* (*Carios*) *vespertilionis* with 20 (16). Moreover, no outstanding differences in the sex chromosomes were observed in *A. vespertilionis.* An earlier report (64) of 25 chromosomes in male *A.* (*Argas*) *reflexus* (= *A. columbarum*) has been corrected (17, 18).

Subsequent to Oliver's review (48) additional information on the cytogenetics of *Argas* has become available. Analyses of spermatogonial and oogonial cells of *A.* (*Persicargas*) *zumpti* (49) reveal a diploid chromosome number of 26. The two longest chromosomes are assumed to be sex chromosomes; they are of equal length in the females (XX) but are heteromorphic in the males (XY). The mean length of the X chromosome at mitotic metaphase is 6.5 μm whereas the Y is 4.8 μm. Autosomes consist of two shorter chromosomes (2.0–3.0 μm) and 22 very short ones

(1.0–1.9 μm). The sex chromosomes appear to be isobrachial or slightly heterobrachial whereas the autosomes are probably cephalobrachial.

Two other species of the subgenus *Persicargas, A.* (*P.*) *radiatus* and *A.* (*P.*) *sanchezi,* have been studied cytologically (27). Both species possess a $2n$ number of 26 including an extremely long X chromosome (X = 9.0 μm in *A. radiatus* and 4.5 μm in *A. sanchezi* at mitotic metaphase). The Y chromosome was not always distinguishable from the autosomes and together they averaged 2.5 and 2.4 μm in *A. radiatus* and *A. sanchezi,* respectively. The X chromosome is probably isobrachial and the Y heterobrachial (submetacentric); six to eight autosomes are isobrachial and heterobrachial. *A.* (*Argas*) *japonicus* also possesses $2n = 26$, including two X chromosomes in females and one X and Y in males (K. Tanaka and J. H. Oliver Jr., unpublished data).

OTOBIUS No published data on chromosomes exist for species of *Otobius* but unpublished research (J. H. Oliver Jr. and R. L. Osburn) indicates that the chromosome complements of *O. megnini* and *O. lagophilus* consist of $2n = 20$ in the males. Presumably the males possess XY sex chromosomes and females XX chromosomes since this is the sex chromosome system in other argasids.

IXODES Oliver's review (48) discusses Nordenskiold's (39, 40) and Sokolov's (74) work on *I. ricinus* and Kahn's results (31) on *I. ricinus* and *I. hexagonus.* Subsequent to that review Oliver & Bremner (53) report on the chromosomes of *I. tasmani, I. holocyclus,* and *I. cornuatus* (Table 2). *I. tasmani* has a $2n$ number of 24, with 22 chromosomes that appear cephalobrachial. The location of the centromeres in the other two chromosomes, which are the longest of the complement, are uncertain although the chromosomes appear nearly isobrachial in some

Table 2 Chromosome numbers and sex determination in Ixodidae

Species	Sex	$2n$	Sex determination	Reference
Amblyomma americanum	♂ ♀	21, 22	XO, XX	J. Oliver, R. Osburn[a]
A. cajennense	♂ ♀	21, 22	XO, XX	J. Oliver[a]
A. darwini	♂	20	XY	J. Oliver[a]
A. dissimile	♂ ♀	21, 22	XO, XX	J. Oliver[a]
A. helvolum	♂ ♀	21, 22	XO, XX	K. Tanaka, J. Oliver[a]
A. inornatum	♂ ♀	21, 22	XO, XX	J. Oliver, R. Osburn[a]
A. limbatum	♂	21	X_1X_2Y	44, 53
A. maculatum	♂	21	XO	J. Oliver[a]
A. moreliae (Sydney)	♂ ♀	21, 22	X_1X_2Y, $X_1X_1X_2X_2$	44, 53
A. moreliae (Brisbane)	♂ ♀	20, 20	XY, XX	44, 53
A. testudinarium	♂ ♀	21, 22	XO, XX	K. Tanaka, J. Oliver[a]
A. triguttatum	♂ ♀	19, 20	XO, XX	53
A. tuberculatum	♂ ♀	21, 22	XO, XX	J. Oliver, R. Osburn[a]
Amblyomma sp. (Galapagos Is.)	♂	21	XO	J. Oliver[a]
Amblyomma sp. (Japan)	♂	21	—	K. Tanaka, J. Oliver[a]
Aponomma concolor	♂	19	XO	J. Oliver, B. Stone[a]
A. fimbriatum	♂	21	XO	53
A. hydrosauri	♂ ♀	17, 18	XO, XX	53
A. undatum	♂ ♀	19, 20	XO, XX	53
Boophilus annulatus	♂ ♀	21, 22	XO, XX	37, 38
	♂	21	XO	75
B. microplus	♂ ♀	21, 22	XO, XX	37, 38, 53

Table 2 *(Continued)*

Species	Sex	2*n*	Sex determination	Reference
Dermacentor albipictus	♂	21	XO	58
D. andersoni	♂ ♀	21, 22	XO, XX	31, 47, 51
Dermacentor sp. (*auratus* ?)	♂	21	XO	K. Tanaka, J. Oliver[a]
D. hunteri	♂ ♀	21, 22	XO, XX	47, 51
D. nitens	♂	21	XO	J. Oliver[a]
D. occidentalis	♂ ♀	21, 22	XO, XX	47, 51, 54
D. parumapertus	♂ ♀	21, 22	XO, XX	47, 51
D. silvarum	♂	21	XO	75
Dermacentor sp. (*taiwanensis* ?)	♂ ♀	20, 20	XY, XX	K. Tanaka, J. Oliver[a]
D. variabilis	♂ ♀	21, 22	XO, XX	47, 51
Haemaphysalis bancrofti	♂	21	—	53
H. bispinosa	♂ ♀	21, 22	XO, XX	62
H. bremneri	♂ ♀	21, 22	XO, XX	53
H. campanulata	♂ ♀	21, 22	XO, XX	62
H. flava	♂ ♀	21, 22	XO, XX	62
H. formosensis	♂ ♀	21, 22	XO, XX	62
H. hystricis	♂ ♀	20 or 21, 20	XY, XX	62
H. japonica	♂	21	—	62
H. kitaokai	♂	19	XO	62
H. lagrangei	♂	21	XO	J. Oliver[a]
H. leachii	♂	16 ?	—	86
H. leporispalustris	♂ ♀	21, 22	XO, XX	31, J. Oliver[a]
H. longicornis	♂ ♀	21, 22	XO, XX	61
	♂ ♀	30–33, 30–35	—	53, 61
H. megaspinosa	♂ ♀	21, 22	XO, XX	62
H. pentalagi	♂	21	XO	62
Hyalomma aegyptium	♂ ♀	21, 22	XO, XX	52
	♂	21	XO	12
H. anatolicum	♂ ♀	21, 22	XO, XX	75
H. anatolicum excavatum	♂ ♀	21, 22	XO, XX	31, 52
H. asiaticum	♂	21	XO	75
H. asiaticum excavatum × *plumbeum*	♀	21	—	75
H. detritum	♂	21	XO	75
H. dromedarii	♂ ♀	21, 22	XO, XX	31, 52
	♂	21	XO	75
H. franchinii	♂ ♀	21, 22	XO, XX	52
H. impeltatum	♂	21	XO	52
H. marginatum (= *plumbeum*)	♂ ♀	21, 22	XO, XX	31, 32, 75
H. rhipicephaloides	♂	21	XO	52
H. rufipes	♂ ♀	21, 22	XO, XX	31
Ixodes cornuatus	♀	24	—	53
I. hexagonus	♂ ♀	26, 26	XY, XX	31
I. holocyclus	♂ ♀	23, 24	XO, XX	J. Oliver, B. Stone[a]
	♀	24	XX	53
I. kingi	♂	26	XY	60
I. laysanensis	♂ ♀	28, 28	XY, XX	J. Oliver[a]
I. nipponensis	♀	28	—	K. Tanaka, J. Oliver[a]
I. ricinus	♂ ♀	28, 28	XY, XX	31
	♂	28	XY	39, 74
I. tasmani	♀	24, 25	—	53
Rhipicephalus bursa	♂	24 (?)	—	83
R. evertsi	♂	21	XO	J. Oliver[a]
R. sanguineus	♂ ♀	21, 22	XO, XX	31, 59
	♂	21	XO	12
	♂	24 (?)	—	77
R. secundus	♂ ♀	21, 22	XO, XX	31

[a] Unpublished data.

cells. Chromosomes range in length from 6.71–1.98 μm. An extra chromosome-like structure is present in some cells. Females of *I. holocyclus* and *I. cornuatus* contain a diploid number of 24, whereas unpublished research (J. H. Oliver Jr. and B. F. Stone) indicates that *I. holocyclus* males possess a diploid chromosome number of 23, not 24 as expected, made up of 22 autosomes and 1 sex chromosome. No data are available on the karyotype of male *I. cornuatus.* In males of *I. kingi* the diploid number is $2n = 26$, including heteromorphic sex chromosomes (XY) (60), with the X and Y approximately 5 and 4 μm, respectively. The Y is about the same length as the longest autosomes and the latter range downward to approximately 1 μm; centromeres appear to be terminal. Chromosome data exist for two other *Ixodes* species: *I. nipponensis* females possess a diploid chromosome complement of 28 (K. Tanaka and J. H. Oliver Jr., unpublished data); *I. laysanensis* males have 28 chromosomes including an XY pair, whereas the females have an XX pair (J. H. Oliver Jr., unpublished data).

BOOPHILUS Chromosome data exist for two of the four species in this genus. Sokolov (75) examined one male *B. annulatus* (*B. calcaratus* in Russia) and found 20 autosomes and 1 long sex chromosome; the autosomes differed only slightly among themselves and were considerably shorter than the sex chromosome. The karyotype of Australian male *B. microplus* is similar to that of *B. annulatus* (53). Females possess 29 autosomes and 2 long sex chromosomes (53). A careful comparison of karyotypes of the two species confirms earlier reports and emphasizes their similarities (37, 38). Karyotypic differentiation between these species would be difficult.

HAEMAPHYSALIS The karyotypes of *H. leporispalustris* and *H. leachii* have been reviewed recently (48) and summary information is presented in Table 2. Since the above review, cytogenetic data have been published on 12 additional species (53, 61, 62) and unpublished data exist for another species (Table 2). Although 12 species of *Haemaphysalis* show the usual $2n$ number of 20 autosomes plus 1 sex chromosome, there is an amazing amount of chromosomal diversity present (Table 2). Data available from females of seven of the 12 species show that all possess 20 autosomes plus 2 sex chromosomes. The sex chromosomes are 1.5 to 2.0 times the length of the longest autosome in some species (*H. campanulata, H. pentalagi, H. bremneri, H. formosensis*), approximately equal or slightly longer in another (*H. bispinosa*), and slightly shorter than the longest pair of autosomes in others (*H. flava, H. megaspinosa*). Variable numbers of chromosomes are present in some species, which in some cases are probably abnormal cells. For example, some testicular cells of *H. leporispalustris* have 22 instead of the normal 21 chromosomes (31) and irregular numbers are sometimes found in embryonic cells (J. H. Oliver Jr., unpublished data). A small, supernumerary chromosome was seen in *H. formosensis,* and in other cases it is almost certain that a variation in chromosome numbers is typical. Two of the most interesting examples are *H. hystricis* and *H. longicornis. H. hystricis* females have a $2n$ number of 20, including 2 long sex chromosomes. Two kinds of karyotypes are prevalent in males, one with 20 and the other with 21 chromosomes;

the 21 chromosome type was present in 23 males (53%) out of 43 collected from the Ryukyu Islands, Japan. Both the 21- and 20-chromosome ticks have a long X chromosome and a slightly shorter Y chromosome. In fact, the two kinds of males have identical karyotypes except for the presence of a supernumerary chromosome that is nearly the same size as the Y in the males with 21 chromosomes. Nine males from Mt. Sontra, Vietnam, possessed 20, including the long X and Y chromosomes. *H. longicornis,* probably the most interesting tick yet studied cytogenetically and reproductively, consists of diploid bisexual races (20+XX♀; 20+X♂), triploid obligatory parthenogenetic races (30–35 chromosomes), and an aneuploid race (22–28 chromosomes) capable of parthenogenetic and bisexual reproduction (61). Hybridization succeeds between bisexual diploid males and parthenogenetic aneuploid females but fails between diploid and triploid races. Another departure from the usual 20+XX♀ and 20+X♂ is found in *H. kitaokai*: females have 18 autosomes and 2 long sex chromosomes (twice as long as longest autosome). The karyotype of the male is the same except only one sex chromosome is present.

DERMACENTOR Cytogenetic data exist for approximately 30% of the described species of *Dermacentor. D. silvarum* males have a diploid chromosome complement of 20 autosomes and 1 sex chromosome; the latter is significantly longer than the autosomes, which vary in length. The chromosome number and sex chromosome system is similar for the six species of *Dermacentor* endemic to the United States (Table 2), and no major karyotypic differences are obvious, even in the one-host and winter-active *D. albipictus.* This is not surprising in view of the hybridization experiment results and the analyses of chromosome pairing in hybrids (*D. andersoni, D. variabilis, D. occidentalis*), which confirm a great degree of genetic similarity among these *Dermacentor* species (63). In *D. andersoni* the 20 autosomes range from approximately 1.8–4.5 μm, with the longest autosome slightly more than twice the length of the shortest and the others forming a gradual cline between the extremes. The mean length of all autosomes is 2.9 μm; the sex chromosome (two in females) is approximately 5.7 μm in length. There are chromosome data (K. Tanaka and J. H. Oliver Jr., unpublished data) on two other species of *Dermacentor.* One (presumably *D. auratus*) possesses 20 autosomes and 1 sex chromosome in males and the other (presumably *D. taiwanensis*) has 18+XX females and 18+XY males. The latter is the first instance of XY sex chromosomes and a departure in chromosome number in *Dermacentor.*

RHIPICEPHALUS Warren (85) erroneously describes spermatogenesis in *R. evertsi* and *R. appendiculatus* and gives the wrong chromosome number for the former species—*R. evertsi* males have 20 autosomes and 1 long sex chromosome (J. H. Oliver Jr., unpublished data). Possibly other errors were made when 12 bivalents were reported in *R. bursa* (83) and 12 tetrads in *R. sanguineus* (77). *R. sanguineus* has a diploid chromosome complement of 20+XX in females and 20+X in males, which was first described in males by Dutt (12) who claims the sex chromosome is the largest and is isobrachial. This is disputed by Kahn (31) who confirms the chromosome numbers of 21 and 22 in males and females, respectively. He states that the chromosomes differ little in size from each other and the sex chromosome is indistinguishable at mitosis. He further claims that the sex chromosomes are not the largest nor are they isobrachial, but that centromeres of all the chromosomes are terminal. These latter findings are confirmed from gonad tissues (59) and embryos of *R. sanguineus* (J. H. Oliver Jr., unpublished data) from California, Georgia, and Kansas, as well as from Egypt and Japan (J. H. Oliver, unpublished data, K. Tanaka and J. H. Oliver Jr., unpublished data). The karyotype of *R. secundus* is similar to that of *R. sanguineus* (31).

HYALOMMA Several earlier inaccurate reports (4, 70, 83) on the chromosomes of *H. aegyptium* were published before the correct number of $2n = 21$ for males was established (12). The latter paper erroneously reports the location of the centromere in the sex chromosome. Subsequently, it was found that the sex chromosome was cephalobrachial and that females possessed a $2n$ number of 20+XX (52). There is unanimous agreement (31, 32, 75) that $2n = 20+XX$ in females and 20+X in males of *H. marginatum* (= *plumbeum*). The cephalobrachial autosomes are similar in size, but the cephalobrachial sex chromosomes are much longer than the autosomes (31). The karyotypes of other species of *Hyalomma* are so similar to those described above that it is difficult to differentiate one from another (Table 2); indeed, it is impossible to differentiate among the karyotypes of *H. marginatum, H. dromedarii, H. excavatum,* and *H. rufipes* (31). Another study (52) indicates no major chromosomal differences among *H. dromedarii, H. franchinii, H. anatolicum excavatum, H. impeltatum, H. aegyptium,* and *H. rhipicephaloides.*

APONOMMA Only four species of *Aponomma* have been studied cytogenetically, yet it is clear that there is much chromosomal diversity present (Table 2). *A. fimbriatum* has confused taxonomists for years and the chromosomal data from this species are also confusing (53). A male tick (from an ornate population in Brisbane, Australia) may contain cells with 10 bivalents and 1 large sex univalent, and other cells of the same tick contain 10 bivalents, the sex univalent, and an additional univalent (shorter than the sex chromosome and larger than any autosome). This extra chromosome segregates to the same pole as the sex univalent (at least in some cases). Other irregularities occur in some specimens of an inornate and less punctuate population from Yea, Victoria, Australia. Meiosis in some males is normal and at anaphase I the 10 autosomal bivalents segregate to opposite poles and the sex univalent goes undivided to one pole; however, irregular, unequal divisions resulting in variable numbers of chromosomes per cell occur in some males. Postreductional cells frequently occur in which the chromosomes exceed the expected 10 autosomes. Some of these cells contain 10 autosomes plus 1 sex chromosome, 11, 11+X, 12, 12+X, 13, 13+X, 14, 14+X, 15, 15+X, 16, and 16+X. The source of these supernumerary chromosomes is unknown. The usual chromosome situation for this species is 20+X in the males with the sex chromosome approximately twice the length of the longest autosome. *A. hydrosauri* females contain oogonia with 16 autosomes and 2 sex chromosomes, 14 short cephalobrachial chromosomes, 1 pair of heterobrachials about twice the length of the longest of the 14 short cephalobrachials, and 1 long pair of isobrachial sex chromosomes approximately three times the length of the longest of the 14 short cephalobrachials. Males possess eight bivalents and one sex univalent resulting in a $2n$ number of 17. *A. undatum* males have 18 autosomes and 1 long sex chromosome and the females possess 2 sex chromosomes: the sex chromosome may be from four to eight times the length of the longest autosome, depending on the stage of division; one pair of autosomes is slightly longer than the four next longest pairs; and the remaining four pairs are decreasingly smaller. Although *A. concolor* was predicted to possess $2n = 20+XX$ females and 20+X males (53), males have 18 autosomes and 1 sex chromosome (J. H. Oliver Jr. and B. F. Stone, unpublished data). The female karyotype remains unknown.

AMBLYOMMA There are published cytogenetic data on only three species of *Amblyomma* (44, 53). *A. triguttatum* possesses a diploid number of 18 autosomes with 1 sex chromosome in males and 18+XX in females. The sex chromosomes are slightly heterobrachial, are more than three times the length (13 μm) of any autosome (all cephalobrachial), and range from 1.76–4.07 μm. *A. moreliae* males (Sydney, Australia, population) have 21 chromosomes, 18 autosomes plus X_1X_2Y sex chromosomes, not the 20 autosomes plus 1 sex chromosome most often found among species of Metastriata hard ticks. Nine autosomal bivalents and one sex trivalent are present during meiosis. Males from a Brisbane, Australia, population possess 20 chromosomes, 18 autosomes plus an X and Y sex chromosome. The autosomes are probably all cephalobrachial and the long X and medium length Y are probably isobrachial and heterobrachial, respectively. Females of the Sydney population have 11 bivalents ($2n = 22$), whereas females of the Brisbane group show 18 short autosomes plus 2 long sex chromosomes ($2n = 20$). There are unpublished chromosome data (J. H. Oliver Jr.; J. H. Oliver Jr. and R. L. Osburn; K. Tanaka and J. H. Oliver Jr.) on 11 other species of *Amblyomma* (Table 2). The $2n$ number of 20 autosomes plus 1 sex chromosome in males and the 20+XX in females occurs in 10 of them and 18+XY is present in male *A. darwini.*

Mesostigmata

There are data on chromosomes of mesostigmatid species assigned to eight families.

PHYTOSEIIDAE Almost all the phytoseiids thus far examined cytogenetically (*Amblyseius rotundus, A. brevipes, A. vazimba, A. bibens, A. masiaka, Typhlodromus gutierrezi, T. chazeaui,* and *Phytoseius amba,* all from Madagascar) are arrhenotokous: males have four short cephalobrachial chromosomes and females possess eight (3a; Table 3). The only exception to the four-eight chromosome situation is *Typhlodromus occidentalis,* in which females have six chromosomes and males have three (90). Thelytoky occurs in *Amblyseius elongatus* (= *A. quatemalensis*) (33), *Amblyseius deleoni* (89), and *Clavidromus jackmickleyi* (2). Females of the latter two species possess the expected $2n$ chromosome number of eight. The chromosome situation is unclear in *Paragignathus tamaricis* where nymphs showed eight chromosomes in three individuals, adult males had four chromosomes in six cases, and one female had four chromosomes whereas another female had eight (89).

Table 3 Chromosome numbers in Phytoseiidae

Species	2n	n	Reference
Amblyseius aberrans	8	4	89
A. barkeri	8	4	89
A. bibens	8	4	3a
A. brevipes	8	4	3a
A. chiapensis	8	4	89
A. chilensis	8	4	90
A. cucumeris	8	4	81, 89
A. deleoni	8	—	89
A. hibisci	8	4	90
A. judaicus	8	4	1
A. largoensis	8	4	90
A. masiaka	8	4	3a
A. messor	8	4	1
A. rotundus	8	4	3a
A. rubini	8	4	90
A. swirskii	8	4	90
A. vazimba	8	4	3a
Amblyseius sp. (VII)	8	4	89
Clavidromus aff. *jackmickleyi*	8	—	2
Iphiseius degenerans	8	4	90
Paragignathus tamaricis	8 ?	4 ?	89
Phytoseius amba	8	4	3a
P. finitimus	8	4	89
Phytoseiulus persimilis	8	4	20, 89, 90
Typhlodromus athiasae	8	4	90
T. caudiglans	8	4	20
T. chazeaui	8	4	3a
T. contiguus	8	4	89
T. drori	8	4	89
T. fallacis	8	4	20
T. gutierrezi	8	4	3a
T. occidentalis	6	3	90
T. phialatus	8	4	89
T. porathi	8	4	89
T. rhenanus	8	4	90
T. sternlichti	8	4	89
Typhlodromus sp.	8	4	89
Seiulus isotrichus	8	4	89

OTOPHEIDOMENIDAE It seems likely that a diploid number of six chromosomes is present in *Dicrocheles phalaenodectes,* but some cells only have four chromosomes (81; J. H. Oliver Jr., unpublished data).

ASCIDAE Unequivocal conclusions are impossible at present regarding the chromosome situation in *Blattisocius patagiorum*; more specimens need to be studied. It appears likely that the diploid number is six, but it might be eight (82).

PODOCINIDAE Chromosome data are available for two species of Podocinidae. *Podocinum pacificum* is thelytokous and the females have a diploid chromosome number of ten; however, *P. sagax* is arrhenotokous and females have a $2n$ number of ten whereas males possess five (88; J. H. Oliver Jr., unpublished data).

MACROCHELIDAE No published data exist on chromosomes of macrochelid species, but unpublished data (J. H. Oliver Jr.) indicate that males of *Macrocheles muscaedomesticae, M. pisentii, M. vernalis,* and *Areolaspis bifoliatus* contain five chromosomes whereas ten are present in females. Ten are also present in female *Macrocheles penicilliger.*

PARASITIDAE Diploid chromosome numbers for male *Pergamasus* (= *Gamasus*) *brevicornis, Amblyogamasus* (= *Gamasus*) *septentrionalis,* and *Eugamasus* (= *Gamasus*) *kraepelini* appear to be 12 and *Eugamasus* (= *Gamasus*) *magnus* seems to have 10 chromosomes (73).

DERMANYSSIDAE Contrary to an earlier report (87), *Dermanyssus gallinae* males have three chromosomes whereas the females possess six (45), and the same situation exists in *Dermanyssus prognephilus* (J. H. Camin, unpublished data; J. M. Pound and J. H. Oliver Jr., unpublished data).

MACRONYSSIDAE *Ophionyssus natricis* is arrhenotokous; males have 9 chromosomes and females have 18 (55). Identical numbers are present in *Ornithonyssus sylviarum,* whereas male *Ornithonyssus bacoti* have 8 chromosomes and females have 16 (45).

KARYOTYPES—ORDER ACARIFORMES

Acariformes consists of the three suborders—Astigmata, Cryptostigmata, and Prostigmata. Chromosome data exist for 8 species of Astigmata, 8 species of Cryptostigmata, and approximately 109 species of Prostigmata.

Astigmata

The Anoetidae contains three species on which chromosome data are available. *Anoetus laboratorium* is arrhenotokous: males have four chromosomes and females possess eight (29). Arrhenotokous and thelytokous strains (morphologically indistinguishable from each other) of *Histiostoma feroniarum* exist sympatrically and the females of both strains have 14 chromosomes and, as expected, the arrhenotokous males possess 7 (21). One embryo of *Histiostoma murchiei* contained 8 chromosomes, but more specimens need to be studied before we are certain of the karyotype of this species (J. H. Oliver Jr., unpublished data).

Chromosome data are available for six species of Acaridae. *Caloglyphus mycophagus* males have 14 autosomes plus 1 sex chromosome and females have 14 plus 2 sex chromosomes (22). *Caloglyphus berlesei* females possess a diploid chromosome number of 18 whereas males have 17, and *Caloglyphus michaeli* females and males possess 16 and 15 chromosomes, respectively (68a). Males of *Acarus siro* and *Tyrophagus putrescentiae* also show one sex chromosome present along with eight autosomes and seven autosomes, respectively, after meiosis in some cells. Other cells have only the autosomes. This indicates $2n$ numbers of 17 and 15 for males of these two species (74). Postreductional cells with five chromosomes were present in *Rhizoglyphus echinopus* and $2n$ cells had ten chromosomes (74). *Glycyphagus domesticus* (Glycyphagidae) also have sex chromosomes. Postreductional cells have either eight autosomes plus one sex chromosome or only eight autosomes (74).

Cryptostigmata

A diploid chromosome number of 18 and a haploid number of 9 is present in *Euzetes seminulum, Galumna* sp., *Parachipteria* (= *Notaspis*) *punctata, Damaeus* (= *Belba*) *verticillipes,* and *Hypochthonius rufulus,* whereas 36 chromosomes are present in *Nothrus silvestris* females. It is unknown whether the latter species is tetraploid (75). Eighteen seems to be the usual diploid number as it is also present in *Platynothrus peltifer* (78) and in *Trhypochthonious tectorum* (79). The latter two species are thought to be automictic and the term *mixokaryokinesis* (fusion of anaphase plates of the second maturation division resulting in restoration of the $2n$ condition) was applied. If automixis is functioning in the oribatid mites it makes an interesting contrast to the apomictic situation in the anoetid, *Histiostoma feroniarum* (21).

Prostigmata

Chromosome data are available on species assigned to several families of prostigs. Two species of Pyemotidae are represented. A diploid chromosome number of six is present in female *Siteroptes graminum* and males are haploid (8, 9). The same chromosome numbers and early presence of karyomerokinesis (individual chromosomes enclosed within a vesicle or karyomere) occurs in another arrhenotokous pyemotid, *Pyemotes ventricosus* (65). Low chromosome numbers and arrhenotoky are also prevalent among the families of cheyletoid mites. *Demodex caprae* (Demodicidae) females have four chromosomes and males possess two (R. R. Lebel and C. E. Desch, unpublished data). *Syringophiloidus minor* (Syringophilidae) males have three chromosomes and females have six (7). Two species of Harpyrhynchidae, *Harpyrhynchus brevis* and *Harpyrhynchus novoplumaris,* have females with four chromosomes and males with two (36, 57). The same chromosome numbers listed for the latter two species also exist in female and male *Cheyletus malaccensis,* in which two isobrachial and two heterobrachial chromosomes appear in most diploid cells (69). A diploid number of four is also present in the thelytokous *Cheyletus eruditus* (66, 69). A type of automixis occurs in *C. eruditus* in which reduction division occurs after two bivalents form in the first maturation division. The mature egg has the haploid complement and automixis occurs when the diploid state is restored by fusion of cleavage nuclei soon after the fifth cleavage (66). At the developmental stage where leg primordia are formed, some somatic cells are polyploid and 16 or more chromosomes are present.

The Eupodidae contains two species, *Linopodes* sp. and *Eupodes* sp., that possess a haploid chromosome number of nine (75). *Erythraeus* sp. (Erythraeidae) males have a $2n$ number of 16 and haploid number of 8 (75). There is controversy about the chromosomes of *Allothrombium fuliginosum* (Trombidiidae): one report indicates a $2n$ number of 24 for males and females (75), whereas another states 12 is the diploid number of chromosomes (5). Chromosome data on other trombidiids include a $2n$ number of 26 for female *Sericothrombium scharlatinum,* 22 (11 haploid) for a male *Sericothrombium* sp., and 18 (9 haploid) for males of another species of *Sericothrombium* (75).

The Hydrachnellae (water mites) are karyologically diverse (Table 4). Species of the genera *Limnochares, Eylais, Hydrachna, Hydryphantes, Thyas,* and *Hydrodroma,* with two exceptions (*Hyrachna leegei* and *Thyas dirempta*), have very low numbers of chromosomes (ranging from haploid numbers of two to six). In general, chromosomes of these species are relatively large and rod shaped (75). *Limnochares* (Limnocharidae) and *Eylais* (Eylaidae) are chromosomally different from most of the other water mites. The minimum number of chromosomes is characteristically $n = 2$ or 3. Moreover, the chromosomes are larger than those of most other Hydrachnellae and there is an absence of chiasmata. The homologous chromosomes at metaphase I are arranged in pairs at a considerable distance to each other (75). Later work on *Eylais setosa* and *Hydrodroma despiciens* indicates an achiasmatic meiosis in males of both species, whereas chiasmata appear to be formed during oogenesis of *E. setosa* (34). Chromosome behavior during mitotic anaphase suggests that the chromosomes are holokinetic in both species. During anaphase I, however, the chromosomes of *E. setosa* behave as if they might be telokinetic. Subsequent to treatment with X-irradiation, the chromosome fragments of *E. setosa* show an undisturbed capability to divide and migrate, which is excellent evidence for recognition of holokinetic chromosomes.

The Hydrachnidae have at least twice as many chromosomes as Limnocharae and the form of their chromosomes varies. Synapsis is normal with complete terminalization of chiasmata (75). Species of the superfamily Hygrobatoidea are characterized by greater numbers of chromosomes (haploid 7 to 13), except for *Neumania vernalis* ($n = 2$) and *Piona nodata* ($n = 4$), than many water mites. In size, however, the chromosomes are among the smallest.

The majority of species in the superfamily Tetranychoidea belong to the Tetranychidae (spider mites) and Tenuipalpidae (false spider mites). Cytogenetic data are available for species assigned to two subfamilies of Tetranychidae. Cytogenetic data on species of Bryobiinae indicate that four species are thelytokous with three of them possessing eight chromosomes and the other having four. Two of the three arrhenotokous species have a haploid number of four and one species has two (Table 5). All of the approximately 51 species from which cytogenetic data are available in the Tetranychinae are arrhenotokous and haploid chromosome numbers range from two to seven. The modal number of the family is three (Table 5).

Among the Tenuipalpidae, *Raoiella indica* males have two chromosomes and females have four. *Brevipalpus phoenicis, Brevipalpus californicus,* and *Brevipalpus*

Table 4 Chromosome numbers in Hydrachnellae

Species	Sex	2*n*	*n*	Reference
Arrhenuridae				
Arrhenurus bicuspidator	♂	20	10	75
A. caudatus	♂	—	13	75
A. maculata	♂	—	10	75
A. pustulator	♂	—	13	75
	♂	—	10	80
Eylaidae				
Eylais mutila	♂	—	3	75
E. rimosa	♂ ♀	4	2	75
E. setosa	♂	4	2	34, 75
Hydrachnidae				
Hydrachna globosa	♂	12	6	75
H. leegei	♂	20	—	75
H. uniscutata	♂	12	6	75
Hydrodromidae				
Hydrodroma despiciens	♂	6	3	34, 75
Hydryphantidae				
Hydryphantes bayeri	♂	—	5	75
H. clypeatus	♂	6	3	75
H. ruber	♀	12	—	75
Hydryphantes sp.	♂	—	5	75
Thyas dirempta	♂	—	9	75
Hygrobatidae				
Hygrobates calliger	♂	14	7	75
Lebertiidae				
Frontipoda musculus	♂	—	9	75
Lebertia porosa	♂	—	8	75
L. stackelbergi v. *saxicola*	♂	—	9	75
Limnesiidae				
Limnesia maculata	♂	18	9	75
L. undulata	♂	18	9	75
Limnocharidae				
Limnochares aquatica	♂ ♀	6	3	75
Pionidae				
Piona carnea	♂	—	11	75
P. coccinea coccinea	♂	20	10	75
P. nodata	♂	—	4	75
P. uncata uncata	♂	—	10	75
Unionicolidae				
Neumania vernalis	♂	—	2	75
Unionicola crassipes	♂	—	9	75

Table 5 Chromosome numbers in Tetranychoidea

Species	2*n*	*n*	Reference
Tetranychidae-Briobiinae			
Briobia praetiosa	8	—	26
B. rubrioculus	8	—	26
B. sarcothamni	8	4	23
Petrobia harti	4	2	26
P. latens	8	—	26
Porcupinchus insularis	8	4	26
Tetranycopsis horridus	4	—	23
Tetranychidae-Tetranychinae			
Anatetranychus tephrosiae	6	3	26
Eonychus curtisetosus	4	2	26
E. grewiae	4	2	26
Eotetranychus befandrianae	4	2	26
E. carpini	8	4	23
E. friedmanni	6	3	26
E. grandis	6	3	26
E. imerinae	6	3	26
E. paracybelus	6	3	26
E. ranomafanae	10	5	26
E. rinoreae	6	3	26
E. roedereri	6	3	26
E. sakalavensis	4	2	26
E. tiliarium	8	4	23
E. tulearensis	4	2	26
Eurytetranychus buxi	10	5	23
E. madagascariensis	6	3	26
Eutetranychus banksi	6	3	26
E. eliei	8	4	26
E. grandidieri	4	2	26
E. orientalis	6	3	26
E. ranjatori	6	3	26
E. sambiranensis	4	2	19
Neotetranychus rubi	14	7	23
Oligonychus andrei	4	2	26
O. bessardi	8	4	26
O. chazeaui	8	4	26
O. coffeae	6	3	26
O. gossypii	4	2	26
O. monsarrati	8	4	26
O. pratensis	8	—	26
O. quercinus	6	3	26
O. randriamasii	4	2	26
O. sylvestris	4	2	26
O. ununguis	6	3	23
O. virens	8	4	26

Table 5 *(Continued)*

Species	2*n*	*n*	Reference
Panonychus ulmi	6	3	23
Schizotetranychus australis	12	6	26
S. schizopus	6	3	23
Tetranychus atlanticus	6	3	26
T. cinnabarinus	6	3	23
T. hydrangeae	6	3	23
T. kaliphorae	6	3	26
T. ludeni	6	3	26
T. neocalendonicus	6	3	26
T. pacificus	6	3	23
T. panici	8	4	26
T. roseus	8	4	26
T. tumidus	12	6	26
T. urticae	6	3	23, 71
T. viennensis	6	3	26
Tenuipalpidae			
Brevipalpus californicus	2 (?)	—	25
B. obovatus	2 (?)	—	25
		2	68
B. phoenicis	2 (?)	—	25
Raoiella indica	4	2	25

obovatus are thelytokous and only two chromosomes per cell were seen in these embryos (25). These observations suggested that the species might have an *n* number of one. Subsequent work (68) proved the haploid number of *B. obovatus* to be two when two bivalents were observed during metaphase I instead of the expected one. Since early embryonic cells contain only two chromosomes and two bivalents are formed during meiosis, the question of when the diploid chromosome number is restored is unanswered. The most plausible suggestion could be that older cleavage nuclei might fuse and restore the 2*n* number. Further evidence substantiating that two is the haploid chromosome number in *B. obovatus* is the artificial induction of males via irradiation and the fact that these male embryos contain two chromosomes per cell (24).

SEX DETERMINATION

XX-XY, XX-XO, $X_1X_1X_2X_2$-X_1X_2Y chromosome systems and arrhenotokous and thelytokous types of parthenogenesis are operative among the Acari (43, 48, 50). Certain sex-determining mechanisms may be characteristic of many species in a particular taxon, but rarely is only one type represented. Thelytoky can be classified as two types: complete parthenogenesis and cyclical parthenogenesis. When complete parthenogenesis exists in a species, only females (or very rarely males) occur and each individual arises from an unfertilized egg. Cyclical parthenogenesis (heterogony) involves a situation in which one or more parthenogenetic generations alternate with a bisexual one, usually in an annual cycle. Many species of acarines exhibit the former type, but none have been reported yet to employ cyclical parthenogenesis.

XX-XY

XX females and XY males are present in many species of ticks, and this chromosome situation is the main type of sex chromosome system in the Argasidae (soft ticks); it is also present in some hard ticks. The X chromosome is usually several times longer than the autosomes; the Y, though shorter than the X, is also frequently longer than the autosomes. Heterogametic males are reported among *Argas* and *Ornithodoros* and although sex chromosomes have not been identified in all species from these genera [e.g. *O. tholozoni,* (75); *O. moubata,* (14, 84)] it is almost certain that the males are heterogametic in most species. Sex chromosomes have not been identified from *Otobius* yet, but it is probable that males are heterogametic although sex chromosomes may not be heteromorphic. Species of *Ixodes,* considered to be more closely related to the Argasidae than are other hard ticks, also have XY males (an exception is *I. holocyclus* males, which are XO). Most other hard ticks (Metastriata) do not have XY males, but there are exceptions. *Amblyomma darwini* males have 18 autosomes plus XY sex chromosomes (J. H. Oliver Jr., unpublished data) and males of the Brisbane, Australia, population of *Amblyomma moreliae* also possesses 18+XY. The latter is discussed under the topic of multiple sex chromosomes. *Dermacentor* sp. (*taiwanensis?*) males have 18 autosomes and XY sex chromosomes (K. Tanaka and J. H. Oliver Jr., unpublished data). Another hard tick with XY sex chromosomes in the male is *Haemaphysalis hystricis* from Japan (62) and Vietnam (J. H. Oliver Jr., unpublished data).

XX-XO

Almost all the hard ticks for which chromosome data are available, excluding species of *Ixodes* and the four species with XY males mentioned above, have XO males. The sex chromosome is usually much larger than the autosomes and may be heteropycnotic at certain stages. These generalizations hold for most *Amblyomma, Aponomma, Hyalomma, Rhipicephalus, Dermacentor, Haemaphysalis,* and *Boophilus.* Exceptions may be *Haemaphysalis leporispalustris, Rhipicephalus sanguineus,* and *Rhipicephalus secundus* (31).

The metastriatid hard ticks are not the only acarines possessing the XX-XO sex chromosome system. Certain species of Astigmata have XO males. Five species of Acaridae and one species of Glycyphagidae have males with only one sex chromosome, and contrary to the situation in ticks, the X chromosome is smaller than the autosomes. *Caloglyphus mycophagus, C. berlesei, C. michaeli, Acarus siro* (= *Tyroglyphus farinae*), and *Tyrophagus putrescentiae* (= *T. noxilus*) are the acarids (22, 74) and *Glycyphagus domesticus* is the glycyphagid (74).

Multiple Sex Chromosomes

At least two species of acarines possess multiple sex chromosomes. The Sydney, Australia, population of *Amblyomma moreliae* and *Amblyomma limbatum* have an $X_1X_1X_2X_2$-X_1X_2Y system (44, 53). Males of the Brisbane population of *A. moreliae* contain a diploid number of 20 chromosomes, 18 short autosomes plus 1 long isobrachial and 1 medium-sized heterobrachial sex chromosome. The Sydney population contains females with 22 and males with 21 chromosomes. The odd chromosome is much longer than all others. Cells from males of this population and *A. limbatum* show nine bivalents and one sex trivalent. Eleven bivalents are present in females. The chromosome situation found in most ticks is ten autosomal bivalents and one sex univalent. If an unequal reciprocal translocation occurred between the sex chromosome and one of the autosomes this would account for the observed sex trivalent in Sydney *A. moreliae* and *A. limbatum.* The translocation would result in two X chromosomes (X_1X_2) segregating from an unaltered autosome, the neo-Y. A large X chromosome is characteristic of most ticks, and in the case of *A. moreliae* (Sydney) and *A. limbatum,* the reciprocal translocation did not later significantly change its relative size.

Parthenogenesis

Parthenogenesis in mites and ticks has been reviewed recently (50) and so only a few summary statements are made here concerning it. Parthenogenesis may be of various types, i.e. arrhenotoky, thelytoky, deuterotoky, artificial parthenogenesis, gynogenesis. Arrhenotoky occurs in Mesostigmata, Astigmata, and Prostigmata and usually occurs in many (most) closely related species as the major type of reproduction (sex-determining mechanism) in certain genera, subfamilies, and families. For example, it is quite common among the Mesostigmata and is the predominant sex-determining mechanism in the Macrochelidae, Dermanyssidae, Macronyssidae, and Phytoseiidae. Much rearing data involving the separation of virgins have shown the prevalence of the arrhenotokous type of reproduction among the Macrochelidae (11, 13, 56; J. H. Oliver Jr., unpublished data) and Oliver's data (unpublished) on the chromosomes of *Macrocheles muscaedomesticae, Macrocheles pisentii, Macrocheles vernalis, Macrocheles penicilliger,* and *Areolaspis bifoliatus* prove the cytological haplodiploidy (five and ten chromosomes) of these species. Rearing data on species of Macronyssidae and Dermanyssidae (3, 6, 41, 72) and cytological proof of haplodiploidy (42, 45, 55) are available. Regular arrhenotoky occurs in *Ophionyssus natricis, Ornithonyssus bacoti,* and *Ornithonyssus sylviarum,* whereas a form of gynogenesis (male mates with female but spermatozoa do not penetrate the eggs) occurs in *Dermanyssus gallinae.* Nevertheless, males are haploid and females diploid in all these species. A similar type of gynogenesis that activates haplodiploidy in *D. gallinae* also occurs in the Phytoseiidae (1, 2, 20, 81, 89, 90). Males of most phytoseiids have a haploid chromosome number of four and females possess a diploid number of eight (Table 3).

Haplodiploidy is demonstrated among the Astigmata in the Anoetidae (*Anoetus laboratorium,* 8 and 4 chromosomes; *Histiostoma feroniarum,* 14 and 7 chromosomes). It is present in several families of Prostigmata. Indeed, it seems to be the dominant type of sex determination in the superfamily Cheyletoidea, e.g. Cheyletidae, Harpyrhynchidae, Syringophilidae, Demodicidae (7, 57, 66, 69; R. R. Lebel and C. E. Desch, unpublished data). Haplodiploidy also is the dominant type of sex-determining mechanism in the superfamily Tetranychoidea (Table 5).

Types of parthenogenesis mentioned above other than haplodiploidy may not qualify as separate types of sex-determining mechanisms; the mites merely employ one of the other sex-determining mechanisms and reproduce without mating. For example, thelytoky is not uncommon among acarines and it appears sporadically among bisexual diploid species. In these cases whichever sex-determining mechanism exists in the bisexual species (XX-XY, XX-XO, etc) continues to operate in the thelytokous individuals. The only difference is that the latter individuals do not require the stimulus of sperm penetration of their eggs.

CHROMOSOME TYPES

It is clear that at least two types of chromosomes are present in the acarines. Monokinetic chromosomes (one kinetochore or centromere per chromosome) are present in many species and are especially well documented among the ticks (see karyotype section of this paper). It is also obvious that holokinetic chromosomes (centromere or kinetochore activity diffused along the entire length of chromosome) are present, especially among some species of Prostigmata. The latter type of chromosomes are sometimes confused with polykinetic chromosomes in which the kinetic (centromeric) activity or loci are thought to alternate with akinetic loci along the length of the chromosome. As far as I am aware, no clear-cut unequivocal proof of polykinetic chromosomes exists, although this might be challenged. One of the interesting aspects of holokinetic (diffuse centric) chromosomes in mites is that no single centromere (primary constriction or kinetochore) can be demonstrated cytologically in them and the chromosomes behave holokinetically during somatic divisions, yet during one or both meiotic divisions the chromosomes orient differently. For example, in *Siteroptes graminum* during early cleavage divisions the chromosomes undergo parallel displacement at anaphase as though they were holokinetic, yet during the maturation divisions anaphase movements take place not by parallel displacement but by precession of an end of each chromosome at anaphase I or by both ends at anaphase II (9, 10). Chromosomal movements during somatic mitoses similar to those of *Siteroptes* are known in *Tetranychus* (67, 71), *Pyemotes* (65), *Eylais* (34), *Caloglyphus* (22), and in other species of mites (48). It is also well known in certain hemipteran insects (30).

Most of our knowledge of chromosome types has been accumulated as incidental observations made during studies of karyotypes and gametogenesis. Therefore, critical experiments on the nature of the kinetic activity of mite chromosomes have not generally been done. Pijnacker & Ferwerda (67) claim proof of holokinetic chromosomes in *Tetranychus urticae,* but their evidence is circumstantial and similar to earlier observations (71). Two investigations stand out as providing critical evidence of the holokinetic nature of chromosomes of two species of mites. The chromosomes of the water mites, *Eylais setosa* and *Hydrodroma despiciens,* were fractured by irradiation and the chromosome fragments continued to display kinetic activity. This strongly argues in favor of the holokinetic activity of the chromosomes (34). A more critical irradiation experiment on the fracturing of chromosomes of *Siteroptes graminum* proved their holokinetic nature (10).

It is presently impossible to delineate the distribution of monokinetic and holokinetic chromosomes among the taxa of acarines. It is safe to say that the former occur in the ticks (Metastigmata) and the latter occur in some mites of the Prostigmata. It is uncertain whether all prostigmatid mites possess holokinetic chromosomes even if they are present in all Tetranychoidea. *Cheyletus malaccensis* (Cheyletoidea) is said to have monokinetic chromosomes (69) and X-ray fragmentation experiments failed to characterize chromosomes of another Cheyletoidea species (*Harpyrhynchus brevis*) as to the chromosomal kinetic activity (57). It is clear that monokinetic chromosomes are present in some Mesostigmata (45) and this might be expected because of the presumed close phylogenetic relationship between the Metastigmata and Mesostigmata. Holokinetic chromosomes are presumed to occur among acarid mites (22) and they may be typical of all Astigmata. If in the future it is determined that all species of Acariformes possess holokinetic chromosomes and all species of Parasitiformes have monokinetic ones, then these data would give more credence to the theory of the diphyletic origin of the Acari.

Acknowledgments

The preparation of this review was aided in part by Research Grant 09556 from the NIAID (NIH).

I should like to thank Susan Oliver, R. L. Osburn, and J. M. Pound for their help in checking the draft typescript.

Literature Cited

1. Amitai, S., Wysoki, M. 1974. Two unknown males of genus *Amblyseius* Berlese and their karyotypes. *Acarologia* 16:45–51
2. Amitai, S., Wysoki, M., Swirski, E. 1969. A case of thelytoky in a phytoseiid mite with cytological studies. *Isr. J. Agric. Res.* 19:49–52
3. Bertram, D. S., Unsworth, K., Gordon, R. M. 1946. The biology and maintenance of *Liponyssus bacoti* Hirst, 1913, and an investigation into its role as a vector of *Litomosoides carinii* to cotton rats and white rats. *Ann. Trop. Med. Parasitol.* 40:228–54

3a. Blommers-Schlösser, R., Blommers, L. 1975. Karyotypes of eight species of phytoseiid mites from Madagascar. *Genetica* 45:145–48

4. Bonnet, A. 1907. Recherches sur l'anatomie comparee et le developpement des Ixodides. *Ann. Univ. Lyon* 20:1–180
5. Bottazzi, E. 1954. Numero cromosomico e ciclo gametogenetico in *Allothrombium fuliginosum. Boll. Zool.* 21: 207–17
6. Camin, J. H. 1953. Observations on the life history of the snake mite, *Ophionyssus natricis. Chicago Acad. Sci. Spec. Publ.* 10:1–75
7. Casto, S. D. 1974. Observations on the karyotype and maturation of the F_1 generation of *Syringophiloidus minor. Ann. Entomol. Soc. Am.* 67:136–37
8. Cooper, K. W. 1937. Reproductive behavior and haploid parthenogenesis in the grass mite, *Pediculopsis graminum. Proc. Natl. Acad. Sci. U.S.A.* 23:41–44
9. Cooper, K. W. 1939. The nuclear cytology of the grass mite, *Pediculopsis graminum,* with special reference to

karyomerokinesis. *Chromosoma* 1:51–103

10. Cooper, R. S. 1972. Experimental demonstration of holokinetic chromosomes, and of differential radiosensitivity during oogenesis, in the grass mite, *Siteroptes graminum*. *J. Exp. Zool.* 182:69–94
11. Costa, M. 1967. Notes on macrochelids associated with manure and coprid beetles in Israel. II. Three new species of the *Macrocheles pisentii* complex, with notes on their biology. *Acarologia* 9:304–29
12. Dutt, M. K. 1954. Chromosome studies on *Rhipicephalus sanguineus* Latreille and *Hyalomma aegyptium* Neumann. *Curr. Sci.* 23:194–96
13. Filipponi, A. 1964. Experimental taxonomy applied to the Macrochelidae. In *Proc. 2nd Int. Cong. Acarol.* ed. G. O. Evans, pp. 92–100. Budapest: Akademiai Kiado. 652 pp.
14. Geigy, R., Wagner, O. 1957. Ovogenese und Chromosomenverhaltnisse bei *Ornithodorus moubata*. *Acta Trop.* 14:88–91
15. Goroshchenko, Yu. L. 1959. *Coordination Conference on the Problems of Key Questions of Cytology*, Leningrad, 45 pp.
16. Goroshchenko, Yu. L. 1962. Chromosome complex of *Carios vespertilionis* in connection with the question of its generic relationship. *Dokl. Akad. Nauk USSR* 144:665–68
17. Goroshchenko, Yu. L. 1962. Karyological evidence for the systematic subdivision of the ticks belonging to the genus *Argas*, "reflexus" group. *Zool. Zh.* 41:358–63
18. Goroshchenko, Yu. L. 1962. The karyotypes of argasid ticks of the USSR fauna in connection with their taxonomy. *Tsitologiya* 4:137–49
19. Gutierrez, J., Helle, W. 1971. Deux nouvelles especes du genre *Eutetranychus* Banks vivant sur plantes cultivees a Madagascar. *Entomol. Ber.* 31:45–60
20. Hansell, R. I. C., Mollison, M. M., Putman, W. L. 1964. A cytological demonstration of arrhenotoky in three mites of the family Phytoseiidae. *Chromosoma* 15:562–67
21. Heinemann, R. L., Hughes, R. D. 1969. The cytological basis for reproductive variability in the Anoetidae. *Chromosoma* 28:346–56
22. Heinemann, R. L., Hughes, R. D. 1970. Reproduction, reproductive organs, and meiosis in the bisexual non-parthenogenetic mite *Caloglyphus mycophagus*, with reference to oocyte degeneration in virgins. *J. Morphol.* 130:93–102
23. Helle, W., Bolland, H. R. 1967. Karyotypes and sex-determination in spider mites (Tetranychidae). *Genetica* 38:43–53
24. Helle, W., Bolland, H. R. 1972. Artificial induction of males in a thelytokous mite species by means of X-rays. *Entomol. Exp. Appl.* 15:395–96
25. Helle, W., Bolland, H. R., Gutierrez, J. 1972. Minimal chromosome number in false spider mites (Tenuipalpidae). *Experientia* 28:707
26. Helle, W., Gutierrez, J., Bolland, H. R. 1970. A study on sex-determination and karyotypic evolution in Tetranychidae. *Genetica* 41:21–32
27. Homsher, P. J., Oliver, J. H. Jr. 1973. Cytogenetics of ticks. 11. Chromosomes of *Argas radiatus* Railliet and *Argas sanchezi* Duges (Argasidae) with notes on spermatogenesis and hybridization. *J. Parasitol.* 59:375–78
28. Howell, C. J. 1966. Studies on karyotypes of South African Argasidae. 1. *Ornithodoros savignyi*. *Onderstepoort J. Vet. Res.* 33:93–98
29. Hughes, R. D., Jackson, C. G. 1958. A review of the family Anoetidae. *Va. J. Sci.* 9:5–198
30. Hughes-Schrader, S., Schrader, F. 1961. The kinetochore of the Hemiptera. *Chromosoma* 12:327–50
31. Kahn, J. 1964. Cytotaxonomy of ticks. *Q. J. Microsc. Sci.* 105:123–37
32. Kahn, J., Feldman-Muhsam, B. 1958. A note on tick chromosomes. *Bull. Res. Counc. Isr.* 78:205–6
33. Kennett, C. E. 1958. Some predaceous mites of the subfamilies Phytoseiinae and Aceosejinae from central California with descriptions of new species. *Ann. Entomol. Soc. Am.* 51:471–79
34. Keyl, H. G. 1957. Zur Karyologie der Hydrachnellen. *Chromosoma* 8:719–29
35. Krantz, G. W. 1970. *A Manual of Acarology*. Corvallis: Oregon State Univ. 335 pp.
36. Moss, W. W., Oliver, J. H. Jr., Nelson, B. C. 1968. Karyotypes and developmental stages of *Harpyrhynchus novoplumaris* n. sp., a parasite of North American birds. *J. Parasitol.* 54:377–92
37. Newton, W. H., Price, M. A., Graham, O. H., Trevino, J. L. 1972. Chromosome patterns in Mexican *Boophilus annulatus* and *B. microplus*. *Ann. Entomol. Soc. Am.* 65:508–12

38. Newton, W. H., Price, M. A., Graham, O. H., Trevino, J. L. 1972. Chromosomal and gonadal aberrations observed in hybrid offspring of Mexican *Boophilus annulatus* X *B. microplus*. *Ann. Entomol. Soc. Am.* 65:536–41
39. Nordenskiold, E. 1909. Zur Spermatogenese von *Ixodes reduvius*. *Zool. Anz.* 34:511–16
40. Nordenskiold, E. 1920. Spermatogenesis in *Ixodes ricinus*. *Parasitology* 12:159–66
41. Ohmori, N. 1936. Studies on the tropical rat mite. Second report: on the sex ratio and parthenogenesis. *Zool. Mag.* (Tokyo) 48:627
42. Oliver, J. H. Jr. 1961. Sex determination and behavior in several dermanyssid species. *Bull. Entomol. Soc. Am.* 7:175
43. Oliver, J. H. Jr. 1964. Comments on karyotypes and sex determination in the Acari. *Acarologia* 6:288–93
44. Oliver, J. H. Jr. 1965. Cytogenetics of ticks. 2. Multiple sex chromosomes. *Chromosoma* 17:323–27
45. Oliver, J. H. Jr. 1965. Karyotypes and sex determination in some dermanyssid mites. *Ann. Entomol. Soc. Am.* 58: 567–73
46. Oliver, J. H. Jr. 1966. Cytogenetics of ticks. I. Karyotypes of the two *Ornithodoros* species restricted to Australia. *Ann. Entomol. Soc. Am.* 59:144–47
47. Oliver, J. H. Jr. 1966. Spermatogenesis in American species of *Dermacentor*. *Bull. Entomol. Soc. Am.* 12:301
48. Oliver, J. H. Jr. 1967. Cytogenetics of Acarines. In *Genetics of Insect Vectors of Disease,* eds. J. Wright, R. Pal, pp. 417–39. Amsterdam: Elsevier. 794 pp
49. Oliver, J. H. Jr. 1968. Cytogenetics of ticks. 4. Chromosomes of *Argas* (*Persicargas*) *zumpti*. *Ann. Entomol. Soc. Am.* 61:787–88
50. Oliver, J. H. Jr. 1971. Parthenogenesis in mites and ticks. *Am. Zool.* 11:283–99
51. Oliver, J. H. Jr. 1972. Cytogenetics of ticks. 6. Chromosomes of *Dermacentor* species in the United States. *J. Med. Entomol.* 9:177–82
52. Oliver, J. H. Jr. 1972. Cytogenetics of ticks. 8. Chromosomes of six species of Egyptian *Hyalomma*. *J. Parasitol.* 58:611–13
53. Oliver, J. H. Jr., Bremner, K. C. 1968. Cytogenetics of ticks. 3. Chromosomes and sex determination in some Australian hard ticks. *Ann. Entomol. Soc. Am.* 61:837–44
54. Oliver, J. H. Jr., Brinton, L. P. 1972. Cytogenetics of ticks. 7. Spermatogenesis in the Pacific Coast Tick, *Dermacentor occidentalis* Marx. *J. Parasitol.* 58:365–79
55. Oliver, J. H. Jr., Camin, J. H., Jackson, R. C. 1963. Sex determination in the snake mite *Ophionyssus natricis*. *Acarologia* V:180–84
56. Oliver, J. H. Jr., Krantz, G. W. 1963. *Macrocheles rodriguezi* a new species of mite from Kansas with notes on its life cycle and behavior. *Acarologia* V:519–25
57. Oliver, J. H. Jr., Nelson, B. C. 1967. Mite chromosomes: an exceptional low number. *Nature* 214:809
58. Oliver, J. H. Jr., Osburn, R. L. 1972. Cytogenetics of ticks. 10. Chromosomes of the winter tick, *Dermacentor albipictus*. *J. Parasitol.* 58:1182–84
59. Oliver, J. H. Jr., Osburn, R. L., Roberts, J. R. 1972. Cytogenetics of ticks. 9. Chromosomes of *Rhipicephalus sanguineus* and effects of gamma radiation on spermatogenesis. *J. Parasitol.* 58: 824–27
60. Oliver, J. H. Jr., Osburn, R. L., Stanley, M. A., Deal, D. 1974. Cytogenetics of ticks. 13. Chromosomes of *Ixodes kingi* with comparative notes on races east and west of the Continental Divide. *J. Parasitol.* 60:381–82
61. Oliver, J. H. Jr., Tanaka, K., Sawada, M. 1973. Cytogenetics of ticks. 12. Chromosomes and hybridization studies of bisexual and parthenogenetic *Haemaphysalis longicornis* races from Japan and Korea. *Chromosoma* 42: 269–88
62. Oliver, J. H. Jr., Tanaka, K., Sawada, M. 1974. Cytogenetics of ticks. 14. Chromosomes of nine species of Asian haemaphysalines. *Chromosoma* 45:445–56
63. Oliver, J. H. Jr., Wilkinson, P. R., Kohls, G. M. 1972. Observations on hybridization of three species of North American *Dermacentor* ticks. *J. Parasitol.* 58:380–84
64. Opperman, E. 1935. Die Entstehung der Riesenspermien von *Argas columbarum* (*reflexus* F.). *Z. Mikrosk. Anat. Forsch.* 37:538–60
65. Patau, K. 1936. Cytologisch Untersuchungen an der Haploid—Parthenogenetischen Milbe *Pediculoides ventricosus* Newp. *Berl. Zool. Jahrb. Abt. Allgen. Zool. Physiol. Tiere* 56:277–322
66. Peacock, A. D., Weidmann, U. 1961. Recent work on the cytology of animal parthenogenesis. *Przeglad Zool.* 5:5–27 (In Polish, English summary)

67. Pijnacker, L. P., Ferwerda, M. A. 1972. Diffuse kinetochores in the chromosomes of the arrhenokous spider mite *Tetranychus urticae. Experientia* 28: 354
68. Pijnacker, L. P., Ferwerda, M. A. 1975. Maturation divisions with double the somatic chromosome number in the privet mite *Brevipalpus obovatus. Experientia* 31:421–22
68a. Prasse, J. 1968. Untersuchungen über Oögenese, Befruchtung, Eifurchtung und Spermatogenese bei *Caloglyphus berlesei,* und *Caloglyphus michaeli. Biol. Zentralbl.* 87:757–75
69. Regev, S. 1974. Cytological and radioassay evidence of haploid parthenogenesis in *Cheyletus malaccensis. Genetica* 45:125–32
70. Samson, K. 1909. Zur Spermichistrogenese der Zecken. *S. Ber. Ges. Naturforsch. Freunde* 8:486–99
71. Schrader, F. 1923. Haploidy bei einer Spinnenmilbe. *Arch. Midrosk. Anat.* 97:610–22
72. Skaliy, P., Hayes, W. J. 1949. The biology of *Liponyssus bacoti. Am. J. Trop. Med.* 29:759–72
73. Sokolov, I. I. 1934. Untersuchungen uber die Spermatogenese bei den Arachniden. V. Uber die Spermatogenese der Parasitidae. *Z. Zellforsch.* 21:42–109
74. Sokolov, I. I. 1945. Karyological study of some Acari and the problem of sex determination in the group. *Izv. Akad. Nauk USSR* 6:654–63
75. Sokolov, I. I. 1954. The chromosome complex of mites and its importance for systematics and phylogeny. *Trud. Leningrad. Obshchest. Estestvoispyt.* 72: 124–59
76. Sokolov, I. I. 1958. Cytological studies of the development of the male germ cells in *Ornithodoros papillipes. Rev. Entomol. USSR* 37:260–81
77. Stella, E. 1938. Ovogenesi e spermatogenesi di *Rhipicephalus sanguineus. Arch. Zool. Ital.* 27:11–29
78. Taberly, G. 1958. La cytologie de la parthenogenese chez *Platynothrus peltifer. C. R. Acad. Sci.* 247:1655–57
79. Taberly, G. 1960. La regulation chromosomique chez *Trhypochthonius tectorum* espece parthenogenetique d'Oribate: un nouvel exemple de mixocinese *C. R. Acad. Sci.* 250:4200–1
80. Thor, S. 1904. Recherches sur l'anatomie comparee de Acariens prostigmatiques. *Ann. Sci. Nat.* 8:19
81. Treat, A. E. 1965. Sex-distinctive chromatin and the frequency of males in the moth ear mite. *J. N. Y. Entomol. Soc.* 73:12–18
82. Treat, A. E. 1966. A new *Blattisocius* from noctuid moths. *J. N. Y. Entomol. Soc.* 74:143–59
83. Tuzet, O., Millot, J. 1937. Recherches sur la Spermiogenese des *Ixodes. Bull. Biol. Fr. Belg.* 71:190–205
84. Wagner-Jevseenko, O. 1958. Fortpflanzung bei *Ornithodorus moubata* und genitale Ubertragung von *Borrelia duttoni. Acta Trop.* 15:119–68
85. Warren, E. 1931. The spermatogenesis of ticks. *Nature* 128:454–55
86. Warren, E. 1933. On atypical modes of sperm development in certain arachnids. *Ann. Natal Mus.* 7:151–94
87. Warren, E. 1940. On the genital system of *Dermanyssus gallinae* and several other Gamasidae. *Ann. Natal Mus.* 9:409–59
88. Wong, C. L. 1967. A study of the biology of two species of Podocinidae. *Univ. Kansas Sci. Bull.* 47:575–600
89. Wysoki, M. 1973. Further studies on karyotypes and sex determination of phytoseiid mites. *Genetica* 44:139–45
90. Wysoki, M., Swirski, E. 1968. Karyotypes and sex determination of ten species of phytoseiid mites. *Genetica* 39:220–28

Ann. Rev. Entomol. 1977. 22:431–50

INFLUENCE OF POPULATION DENSITY ON SIZE, FECUNDITY, AND DEVELOPMENTAL RATE OF INSECTS IN CULTURE

❖6137

T. Michael Peters and Pedro Barbosa
Department of Entomology, University of Massachusetts, Amherst, Massachusetts 01002

The effect of population density on an insect population is complex. Populations in nature interact with such a variable environment that the influence of density, per se, is subject to analysis only through highly sophisticated techniques that separate the direct effects of density from the indirect effects that are concurrent with increased numbers per unit area. The influence of population density under laboratory conditions has been investigated repeatedly (1, 3, 12, 32, 33, 76, 79, 105). However, the specific influence of density in self-perpetuating laboratory colonies, while important, has seldom been the subject of research (66, 115, 116), although it has been included tangentially in a recent book on mass rearing technology (121) and in a review of problems in producing biocontrol agents (78). Therefore this review must rely heavily upon the literature of population density effects in laboratory experiments to elucidate the effects and their mechanisms, where they are known. Note also that this review cites only a part of the voluminous literature on density, and since many papers simply reiterate various conclusions in different species, only representative research is included (see also 9).

The influence of population density on the biological parameters we are considering may be either a one-tailed response (*Drosophila* type of Watt; 136), with the intensity of effects increasing as numbers are increased, or a two-tailed response (Allee type of Watt; 136), with a low response at some optimal density range but with greater effects evident at densities both above and below the optimal. The two-tailed response indicates a biological parameter often referred to as undercrowding (90, 105, 125). There are also indications of a threshold response (73, 90) for specific parameters affected by density (i.e. mean total fecundity). The difficulty lies in predicting the type of response since it may require a greater understanding of the genetic controls than is currently available.

The characteristics of size, developmental rate, and fecundity at first glance seem to be inextricably locked together as the mutual expressions of the effects that rearing stresses have on insects. There is, however, considerable evidence that these parameters differ not only in their sensitivity to density (8, 127), but may be entirely independent of one another (112). Although there is considerable overlap of parameter expression (12, 110), each is considered in a separate section. A schematic model analyzing density effects on size, fecundity, and development is presented in Figure 1. Two other aspects warrant discussion and are presented in separate sections: behavioral modifications and the mechanisms that induce changes in various parameters affected by shifts in numbers.

SIZE

The general patterns of population density influence on size or weight are either an immediate effect in which changes in population density are directly reflected in the size or weight of resultant individuals (12, 68, 77, 84, 115), or a lag between differences in population density and changes in size or weight (84).

An example of direct influence is given by Henneberry & Kishaba (35). They demonstrated the decreased pupal weights of *Trichoplusia ni* with increased larval densities in small (waxed-cup) rearing containers. For house flies, Sullivan & Sokal (125) found that mean dry weights of pupae from increasingly crowded larval cultures yielded differences among Orlando strain adults from 4.15 ± 0.071 mg

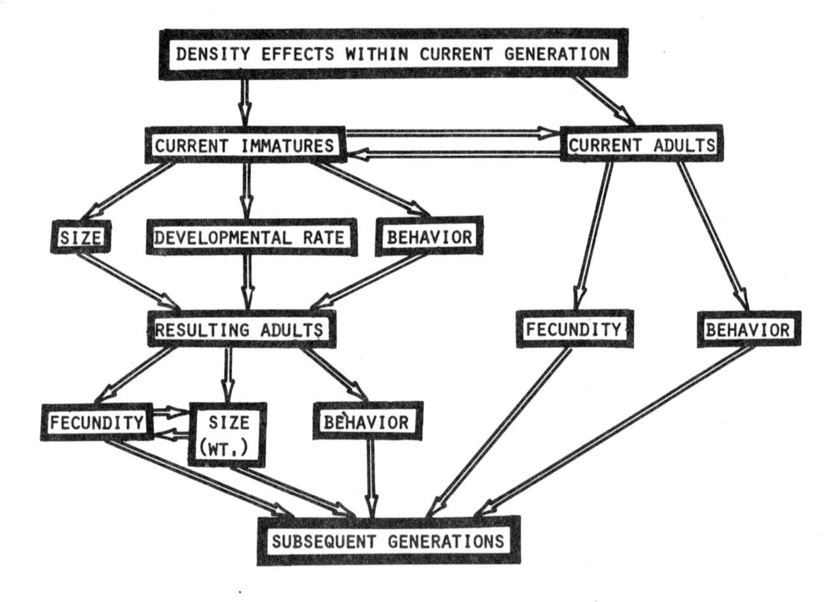

Figure 1 Schematic model for the influence of population density on developmental rate, size, fecundity, and behavior of insects.

(mean per fly) at 40 larvae per unit medium to 0.69 ± 0.021 mg (2560 larvae per unit); this is a 5/6ths weight reduction.

The relationship between pupal weights and increasingly stressed larval *Aedes aegypti* has also been demonstrated by Barbosa et al (12). However, they linked not only weight but also developmental rate and mortality to stress. The optimal density, 0.5 larvae per ml of medium (109), yielded females with mean weights of 0.75 mg (males averaged 0.42 mg). Weights dropped to 0.23 for females at 6.5 larvae per ml (0.16 mg for males). Regardless of density or sex the duration of development was directly linked to weight, and when the pupation period was divided into quartiles, the latest quartiles of the pupal population included the heaviest individuals. Interestingly, although rate of development was fastest and size was greatest for 0.5 larvae per ml of medium, preimaginal mortality was lowest under conditions of moderate stress (2 larvae per ml).

Leonard (68) demonstrated the link between the effect of stress on body weight and on reproductive capabilities. He showed that pupal weights diminished for increasingly crowded *Lymantria dispar,* the significance lying in the direct correlation between pupal size of female and the fecundity of the adult. Leonard reported a 20% reduction in male pupal weight and a 35% reduction for females.

Additional aspects of the relationship between size decrease and population density are the rebound effect and also the differences in the type of effect on the two sexes, especially as it applies to migratory locusts (see below).

In a comparison, Miller (84) showed a continued decrease in adult body weight of *Drosophila simulans* with increasing density (5 larvae per vial to 300 larvae per vial) but a rebound effect for *Drosophila melanogaster* weights at high densities, so that adults from 300 larvae per vial had greater individual weights than adults from the 200 larvae per vial. The rebound, usually attributed to necrophagy, has been reported for other species of Diptera, reviewed by Miller (84).

As for differences in effect between the sexes, crowding was reported to produce smaller adults in both *Plusia gamma* and *Pieris brassicae,* with a greater effect on females (77). However, reductions in weight and in wing length could not be correlated to density changes. Faure (28) also found that greater changes occurred in females than in males when density was altered. Gregaria phases of two migratory locust species, however, were larger than their solitaria-phase counterparts. Albrecht et al (1) showed that larger but fewer nymphs are produced by crowded female *Locusta migratoria migratorioides* as compared to isolated females. Thus, the total nymphal mass per ootheca is essentially the same but is distributed differently for crowded versus solitary locust parents.

Several authors have used differences in size rather than weight as a parameter to judge density effects. For example, Sang (115) reported that numbers of emerging *D. melanogaster* are maintained at the expense of size when preimaginal stages are crowded. Subsequently, Lints & Lints (73) emphasized the relationship that usually exists between size and weight. They showed that the mean size of emerging adult *D. melanogaster* hybrids diminishes with increasing preimaginal population densities whereas variance and range of size increases. Also, a decrease of 20% (or 30% for the hybrids from the reciprocal cross) was shown for adults from the highest density (480 per vial) as compared to the lowest density (3 per vial). In their discussion, they showed how great this effect really is because, if the thorax and abdomen are considered roughly spherical, a 20% reduction in the sphere radius corresponds to a 50% reduction in sphere volume. They also state that the decrease in mean thoracic size is a linear function of the logarithm of the preimaginal population density.

Size and weight changes are not always linked however. Continuous crowding of *Acyrthosiphon pisum* significantly reduced the weight and size of appendages and embryos. The weights of crowded individuals decreased significantly with three successive generations. Apterae reared at 64 per leaf cage were smaller and lighter than those reared at 16 per leaf cage. However, those at 16 per leaf cage, although significantly lighter than those at 4 per leaf cage, were not smaller.[1] Murdie (90) proposed that the difference in weight could be attributed to stress late in development when exoskeletal development might be accomplished with the available food; however, there would not be sufficient food for buildup of internal reserves or normal development of embryos.

That reduced size as a result of different stresses could be expressed differently in future generations was also demonstrated by Murdie (91). He showed that small aphids, produced from a crowded culture, were more fecund than similarly sized aphids that were small as a result of rearing at high temperature. The undersized first offspring from crowd-induced small aphids had an increased growth rate over normal-sized aphids. Successive offspring were progressively heavier at birth. Compensatory development in temperature-stressed small aphids required two generations to return to normal. Lints & Lints (72) showed with Diptera that there was no relationship between size and respiratory rate for similarly sized flies from different preimaginal population densities. Even though respiratory rate for flies from noncrowded preimaginal densities decreased with age, this trend was not evident when flies from crowded preimaginal densities were tested.

FECUNDITY

The total egg production (fecundity) for a species is the result of a complex of factors that affect the rate and extent of egg production, fertility, and hatchability, as well as various characteristics of the progeny. Each of these major factors is considered in the following section. The reader is referred to Karandinos & Axtell (61) for a schematic model analyzing adult density effects on birth rate, and to Watt (136) for a mathematical treatment of population effects on fecundity.

Total Fecundity Versus Mean Daily Fecundity

High-density, mixed-sex populations of *Hypera postica* adults have lowered total fecundity and fecundity per unit of time. This has been shown by alternately exposing test animals to limited and unlimited oviposition sites in both crowded (19.6 ml per weevil) and low-density (146.2 ml per weevil) conditions. Total fecundity of

[1]Size based on length of third antennal segment.

noncrowded females was significantly higher for those that had a male introduced 1 day per week than for those that mated only once in life. It was also higher than those for which equal numbers of males were present or for females that were multiply mated. Percentage fertility of all but the once-mated females was not significantly different (66). Adult density of house flies from 2.56–4.77 flies per $inch^3$ had no effect on total fecundity, but after oviposition sites were saturated with eggs they were deposited on the water containers (104). Daily fecundity was not reported.

Grasshoppers reared and maintained under different densities had lower mean daily fecundity (0.42 pods per female per day) and total fecundity (359 eggs per female) at 20 pairs of solitary grasshoppers per cage than for 10, 3, or 1 pair per cage. At 1 pair per cage, 605.8 eggs per female were produced with 0.55 pods per female per day (122). However, with *Locusta migratoria migratorioides,* Albrecht et al (1) found that crowding the laying females reduced the number of ovarioles that function during each reproduction cycle, as well as reducing the number of eggs (to about 300) as compared to egg production by isolated females (approximately 1000 per female). However, crowding females during their nymphal development had the reverse effect. Isolated females that were crowded as nymphs produced 1500 eggs per female whereas the reversed regime yielded 150 eggs per female.

Lints & Lints (72), in comparing reciprocal cross *D. melanogaster* hybrids at low (30 per vial) and high (240 per vial) preimaginal densities, showed that mean total fecundity is reduced for one crowded hybrid as compared to the ones from uncrowded conditions, but not for its reciprocal cross. The greatest difference ($P < 0.001$) was in mean daily egg production, with females from low densities laying 50.4 ± 3.2 or 56.7 ± 3.9 eggs and the same hybrids from high densities laying 37.2 ± 2.5 or 38.6 ± 2.2 eggs per day. Thus, they were unable to show a relationship between life span and fecundity, or between life span and fertility.

With a greater spread of absolute densities and shorter intervals between treatments, Lints & Lints (73) showed that mean total egg production remained constant for 3, 7, 15, 30, 60, and 120 per vial (or 240 for the second hybrid tested, 80 times the lowest density used) but changed, possibly in a curvilinear manner, at higher densities. On the other hand, mean daily egg production diminished with increasing density.

Thus, the effect of density on total and daily fecundity may vary even between closely related species, as when Nakamura (93) reported different responses in total fecundity of two species of *Callosobruchus* to density. With *C. chinensis,* a *Drosophila* type response in total fecundity was obtained. The same treatment yielded an Allee type response for *C. rhodesianus.*

Adult Longevity

High population densities not only tend to prolong the preimaginal developmental period of crowded immatures but have been shown to increase longevity of crowded adults. Ishida (50) reported that adult density affects the longevity of the sexes differently. As density increases male longevity decreases. Conversely, female longevity is increased with density, but only if oviposition sites are restricted and the females are forced to retain their eggs. LeCato & Pienkowski (66) showed that crowding *Hypera postica* reduced longevity because females were unable to oviposit due to continual mating pressure from males.

Crowding solitary grasshoppers throughout their lives not only shortened the total life span (47.0 ± 0.0 days for 1 pair per cage versus 38.4 ± 12.2 days for 20 pairs per cage) but modified fecundity (122). Of greater interest, however, is the effect on longevity of generations that are produced after the density stress has been applied to one generation (123).

Changes in longevity also may be correlated with preimaginal crowding as well as adult crowding. For example, Sang (115) reported increasing adult longevity due to preimaginal crowding. A more extensive study of this phenomenon by Miller & Thomas (85), comparing smaller flies produced under crowded preimaginal conditions with larger flies from less dense cultures, showed greater longevity for the smaller flies when they had access to food during the imaginal period, but the phenomenon was reversed by imaginal starvation.

The question of longevity has been investigated by Lints and co-workers as a part of their long series of thought-provoking papers on the effects of *Drosophila melanogaster* hybrids' preimaginal environment. When the adults from low-density (30 per vial) and high-density (240 per vial) preimaginal environments are compared, the adults from 240 per vial have an imaginal life span 26% longer (35% longer for the reciprocal cross hybrid) (72). In larval density experiments ranging from 3 per treatment to 480 per treatment the mean imaginal life span is increased with increasing larval densities. As compared with the shortest mean imaginal longevity (39.5 ± 5.2 days) for 7 flies per treatment, the mean at 480 per treatment is 66.3 ± 4.1 days (73).

Thus, although a generally linear relationship exists between increasing density and longevity, other factors modify the basic relationship. Food availability, sex differences, and genetic makeup, as well as the developmental stage subjected to crowding, can influence the direction and magnitude of a species' response.

Effect on Subsequent Generations

Dixon & Wratten (24) demonstrated that high density affected the fecundity of subsequent generations of aphids because the size of parents and their progeny at birth are directly correlated, with weight of the adult and number of progeny yielding straight line regressions somewhat higher for apterae than alatae. Since the degree of crowding has been shown by Lees (67) to increase production of alatae over apterae in *Megoura viciae,* the effect of parental crowding not only may be expressed as a decrease in size of the aphid and its progeny but of lowering fecundity by causing the production of alatae, which are less fecund than comparably sized apterae. The fecundity of alatae is as high as that of apterae early in adult life but drops lower than that of apterae later in the reproductive period. Recovery of fecundity, through increase in size, may take four generations.

As might be expected, the effect of continuous crowding throughout the life of solitary grasshoppers yielded shorter-lived, less fecund females (122) in the stressed generation (fewer egg pods per female and fewer eggs per pod). However, the next generation produced more pods per female (19.7 ± 7.6) if not crowded than produced by the progeny of uncrowded parents (17.2 ± 8.2 pods per female). The progeny of crowded parents also lived longer, 56.6 versus 41.1 ± 16.7 days for progeny from uncrowded parents. The greater efficiency of egg production by progeny of crowded parents was also expressed in the third generation. By using migratory locusts, Albrecht et al (1) demonstrated a relationship between the number of ovarioles and parental or grandparental crowding.

The effect of density on fecundity of subsequent generations, therefore, may be mediated through changes in size and consequent restricted reproductive capacity as well as through behavioral and physiological modifications.

Egg Hatchability

When Schmidt el al (116) varied cage size and density per cage of adult *Haematobia irritans,* they showed that hatchability of the eggs from more densely populated cages was always higher. Because adult survival was better at lower densities, without reduction in numbers of eggs produced, 0.6–9.5 cm^3 of space per fly was determined as optimal.

A probable explanation for the increased hatchability of fly eggs from denser adult populations was offered by Bryant (17). He showed that 640 eggs yielded better percentage hatch than 80 per treatment, not because of chemical interactions among eggs or newly hatched larvae but because high-density cultures were warmer and more moist around the egg than low-density egg cultures. Increased survival of late-hatching larvae was stimulated by larval interaction with yeast growth in the medium. On the other hand, Albrecht et al (1) showed that 93% of *Locusta migratoria migratorioides* eggs hatch for isolated pairs, but only 80% hatch at an oviposition density of 142 insects per cage.

Cannibalism

Although cultures of biological control agents are the most obvious exceptions, most insect cultures do not have cannibalism. With some species, however, population size is effectively controlled by cannibalism, as in *Tribolium* beetles (141). Predation by adults on pupae was a more effective regulating mechanism on abundance of adults than was larval feeding on eggs. In the American cockroach (137) the degree of cannibalism in cultures was linked to density but was not seen as beneficial to the colonies when the cannibalism was predatory, i.e. caused by the agitation and resultant injury induced by crowding. Such cannibalism interferes with appetite and growth through an increase in alertness and anxiety. Sanitary cannibalism, primarily necrophagy, was seen as beneficial in reducing culture contaminants (137).

Mating Frequency

Nishigaki (102) showed that copulation frequency increased exponentially but fecundity increased only proportionally with increasing density in *Callosobruchus chinensis.* However, Leppla (69) varied *Pectinophora gossypiella* adult populations from 5 to 20 pairs per pint container without statistically significant differences in the percentage of mated females or spermatophores transferred per female.

DEVELOPMENTAL RATE

The most direct and detrimental effect of density, usually occurring at higher numerical levels, is the death of individuals, i.e. mortality of the crowded stage or subsequent stages. The mortality response of most species is generally exhibited as density reaches a certain species-specific, environmentally modulated level. For example, in the Mediterranean fruit fly, populations ranging in density from 1–15 eggs per cm^2 exhibit increasing mortality with increasing density (80). Fifty percent of the population died on an average of 22.5 days at lower densities and only 17.0 days for higher densities. Similarly, it took an average of 60.5 days before the last insect of the population died at lower densities and 42.5 days at higher densities (80). Most species in which crowding leads to mortality respond similarly. The severity of mortality or the age-specific pattern of mortality varies from species to species and is in part a function of the rearing condition used and the species strain (4, 18, 40, 64, 117, 132, 138, 140).

Group Effect

In most species, in addition to density-induced mortality, growth retardation is often manifested at specific numerical levels. These changes in growth rate, developmental period, and associated physiological alterations have been reported in groups of insects as small as two individuals as well as in aggregations of several thousands. Many of these developmental shifts have been suggested to be a result of a group effect. Chauvin (21) defines this phenomenon as "the effects produced by proximity upon the individuals in populations of low density." Similarly, Iwao (56) has stated that "the grouping of individuals, even with small numbers, may exert notable influences upon survival, development and morpho-functional characters of animals." In comparing various insect species in culture, reared under set conditions, it is difficult to determine for any given species where group effect (low density) ends and where overcrowding (high density) begins; nevertheless, dramatic changes in development have been demonstrated for insects in culture at both numerical extremes.

Although many species in various orders respond to group effects, lepidopterous species exhibit some particularly fascinating developmental responses. Larvae of the noctuid *Leucania separata* reared in isolation (lower-density type) differ from those reared in groups (higher-density type). The higher-density type is darker and more active, has a high tolerance of unpalatable foods, and exhibits a fast and synchronized larval and pupal developmental rate (56). It must be emphasized that this only represents the *L. separata* response to grouping. Responses to grouping may not result in speeding up development but may retard development rate and prolong the developmental period. Thus, an evaluation of grouping or density in insects in culture may result in a series of apparently contradictory conclusions. A critical evaluation of grouping in Lepidoptera has led Iwao (56) to suggest a probable relationship between the changes that a species may undergo in culture and its mode of life in nature. Thus, he has suggested that the nature of group effect may be a function of the characteristic life style of the species. Thus, in those species that lay eggs singly, induced grouping may result in high mortality and extended larval development. Beyond the solitary mode, eggs may be laid in small groups, larvae may be gregarious, or gregariousness may differ in early instars compared to later instars. In species where eggs are laid in small groups moderate crowding may lead to accelerated development. In fully gregarious species positive effects dominate when individuals are grouped and may be critical to survival; these usually include rapid larval development. In all cases negative influences may be superimposed by food contamination or other fouling of rearing chambers (56). Within the wide range of solitary to fully gregarious species, many species may have a specific density level that allows for optimal development. These species develop optimally at a level usually above isolation but well below extremely high densities (31, 51, 137, 140). At present, group effect has been documented in crickets, cockroaches, aphids, locusts, and many other groups of insects (19, 20, 21, 58, 111, 137, 138).

Overcrowding

Although it is often difficult to determine what density can be considered low or high for any given insect group, certain numerical levels when compared to those commonly found in nature can be designated as overcrowded. Thus, in addition to the effects of lower density (group effect), we must examine the effects of high density on development. Among the more frequently reported outcomes of rearing at high densities is the retardation of growth and subsequent prolongation of larval developmental period in various species, including *Culex pipiens* (51), *Aedes polynesiensis* (49), *Aedes aegypti* (12, 87, 117, 134), *Culex quinquefasciatus* (45), *Trogoderma glabrum* (14), *Tipula oleracea* (65), *Endrosis sarcitrella* (4), and *Ptinus tectus* (31).

In cultures of Mediterranean fruit flies ranging from 1–15 eggs per cm^2, for example, the duration of larval development is extended from 24 to 59 days as density increases (80). However, the types of developmental shifts due to density are species specific and are not isolated from shifts caused by other rearing conditions. Thus, culture of *Drosophila melanogaster* in vials with 3 to 480 eggs per vial resulted in an 8-day increase in duration of development, which is a great deal less than that caused by a temperature drop of 15°C (74). In still other species, the developmental period is shortened by increasing density (22, 41, 68, 76).

For many species of insects in culture the effects of crowding are reflected in changes induced in available food. The interaction between the influence of food quality and quantity and the effects of overcrowding may be a key factor in developmental shifts. Each factor may also be difficult to differentiate or isolate as an independent variable. Thus, an increase in larvae of the hymenopterous parasitoid, *Pteromalus puparum,* within the host results in a reduction in developmental time (16). Perhaps the best illustration of a group for which food and habitat are synonomous is in the stored grain pests. Studies on various species of *Tribolium,* for example, have indicated that increasing density and conditioning of flour results not only in a rise in larval mortality but in a retardation of development (105–107).

Pupal and Adult Development

The effects of increasing density on insects in culture initially may be reflected in the delayed onset of pupation (12, 65, 107, 117). Thus, in underpopulated larval population of *Anopheles quadrimaculatus* pupation begins on day 8 of development and is completed on day 10. Overpopulated larvae begin pupating on day 10 and do not terminate until day 20 or 21. However, as with the responses of larvae, crowding may also be expressed in pupae as accelerated onset of pupation or the absence of any effect on onset of pupation (14, 95). Nayar and Sauerman (96, 98, 99) have also shown that larval aggregation formations in *Aedes taeniorhynchus* and other species affected synchronization of larval-pupal ecdysis. Temporary overcrowding in nine mosquito species may also enhance synchronization of pupation. Other species show similar density-induced synchronization (76).

The length of pupal period may be prolonged with increasing density (41, 65, 117). For example, the pupal period of *Culex nigripalpus* lasts 90 hr at a density of 75 larvae per rearing unit and 113 hr at a density of 200 larvae (95). The process of pupation may be inhibited by crowding (130, 139), or larvae may not successfully undergo larval-pupal ecdysis, and higher mortality may result from higher density.

Very few studies have been conducted on the influence of density on adult development. Most changes due to density represent the culmination of changes in pupal populations. Thus, reduction in adult emergence reflects changes in pupal development and survival.

BEHAVIOR

The kinds of activity reportedly exhibited by insects under different population density regimes are varied but essentially fall into a small number of categories: feeding, locomotory tendencies, and communication. Aggregative and colonial behavior also have been of major interest. Although the effects of population density on behavior were not specifically included in the originally accepted title for this review, the authors deem it an important inclusion to round out the topic by presenting Table 1. The line of demarcation between behavioral and physiological changes is tenuous at best. Many other changes that occur in insects with changing density, but not detailed in this review (see Table 2), have physiological bases and show clear implications for the behavior of insects.

Specific aspects of behavior in mass-rearing programs have been presented by Boller (15). The subject of behavior in laboratory colonies of insects would be an appropriate topic for a separate review.

CAUSE AND EFFECT

Although enumerating the effects of density on insects in culture could literally fill this volume, very little research has been conducted on the underlying mechanisms by which density exerts its influences. Among the frequently suggested causes for density-induced physiological shifts has been the mechanical stimulation of individuals in crowded cultures (90, 117, 125). Most references to this mode of action have been primarily speculative. However, Tschinkel & Willson (130) did test the hypothesis of overstimulation as a result of excessive contact by artificially subjecting populations of *Zophobas rigipes* (Coleoptera: Tenebrionidae) to vibration. These experiments demonstrated subsequent inhibition of pupation.

Table 1 Behavioral changes with population density

Change in:	Reference
Activity	7, 41, 76, 92
Exploratory activity	25
Positive phototaxis	36
Locomotory activity (larval)	52
Flight behavior	38, 97, 100
Ability to fly	131
Higher flight tendency (in swarm)	83
Migratory activity	101, 119
at emergence	99, 119, 120
Flight-reproductive relationship	120
Adult biting rate	128
Larval feeding behavior	57
Imbibition	37
Aggressive protection of food	6
Form of larval aggregation	95, 98, 99
Aggregation	26
Song length	94
Emission of sex pheromone	34
Learning ability	3
Behavior (social insects)	36, 63

A second hypothesis suggests that the essential factor in overcrowded populations is the abundance and quality of food. Bar Zeev (13), Jones (59), Wada (134), Nayar (95), Cohen (22), Judge & Schaefers (60), Lees (67), and Toba et al (129) have speculated that the lack of a sufficient quantity or quality of food (i.e. partial starvation) is the only direct cause of changes in development, size, fecundity, etc. Indeed, in many rearing situations lack of food is a significant parameter in overcrowded cultures and perhaps may interact with other causal agents. Thus, Moore & Whitacre (88) found that chemical factors elaborated by larval mosquitoes induce changes in mosquito development, but production of the chemical is a function of the amount of food available.

Perhaps the greatest amount of data on the induction of changes in crowded individuals demonstrates the significance of chemical mediation. The importance of biochemicals in low-density populations has been demonstrated in the cricket species, *Gryllodes sigillatus* and *Acheta domesticus* (81, 82). Fatty acids and their methyl esters were shown to affect growth of these species. Development in individuals exposed to various C12-C18 fatty acids or their methyl esters was different in isolated individuals compared to those in groups. The effects were generally a function of the species treated, the biochemical tested, and the length of exposure. The route of entry was cuticular and dietary introduction was ineffective (82).

The strongest evidence of chemical mediation of density-induced shifts in physiology has been demonstrated in overcrowded populations of mosquitoes. Although

Table 2 Changes manifested in crowded individuals

Stage affected	Change	Reference
Adult	Polymorphism development	29, 54, 60, 62, 67, 86, 118, 120, 126, 129, 133
	Lower wing loading (dry body weight: cube of wing length)	99
	Delayed emergence	89
	Decreased life span of females	123
Immature	Darker coloration	39, 41, 52, 54
	Hormonal control of coloration	103
	Higher proportion of extra-instar individuals	54, 68
	Fewer instars	76
	Reduced respiratory levels	10, 11
	Wider food tolerance	53
	Food ingestion increased in 1st and 2nd instars but decreased in 4th and 5th (total amount equal)	40
	Greater tolerance to starvation	55
	Increased incidence of disease	57, 100, 124
	Higher fat content and lower water content	76

Peters et al (108) suggested the action of biochemicals as the source of growth inhibition in mixed rearings of *Aedes aegypti* and *Culex pipiens,* Moore & Fisher (87) provided preliminary evidence for the activity of a toxic factor or growth retardant factors. Later, Moore & Whitacre (88) suggested that production of growth retardant factors was probably induced by food shortage. Evidence of chemical factors acting in dense populations was also demonstrated in *Culex pipiens quinquefasciatus* (45, 46). Overcrowded larval cultures were shown to produce a mixture of organic compounds. Subsequent analysis indicated that the mixture included heptadecane, octadecane, 7-methyloctadecane, and 8-methylnonadecane as well as 2-ethyl and 2-methyl substituted long-chain aliphatic carboxylic acids (47, 48).

The rate of development of overcrowded cultures is delayed by the activity of methyloctadecanes and methylnonadecanes (M. Mulla, Y. Hwang, J. Arias, G. Majori, manuscript in preparation). The action of other components include toxic effects inducing morphogenetic traits similar to those exhibited by larvae treated with juvenile hormones or other growth regulators. Ikeshoji (43) also demonstrated the production of toxic and growth-retarding factors in cultures of *Chironomus* sp., which delayed emergence and decreased emergence rates. The active component was identified as calcium nitrite (44). Finally, chemical secretions, primarily ethylquinones, have been reported to be produced by crowded flour beetles (2, 64, 75, 113). Other effects have also been reported in beetles in conditioned medium (114).

CONCLUDING REMARKS

At this point, the complex influences that population density has on insects in culture is obvious. This complexity appears to be a result of some striking, but frequently overlooked, fundamental biological differences among insect species used in experiments, compounded by differences in experimental design and technology. The biological differences exist among the kinds of responses genetically available within the experimental animals and the magnitude and combination of responses elicited. These are orchestrated by the degree of sensitivity by the animal for each response. For example, control of growth rate (ratio of size/developmental rate) in *Tribolium* and *Drosophila* is quite different. Developmental time is under a classical polygenic system in *Tribolium* (27), whereas selection for developmental time in *Drosophila* was not possible (71) and the system has not been identified. Although imaginal longevity and preimaginal growth rate have been directly related in both *Tribolium* and *Drosophila,* the responses are opposite (H. Soliman and F. A. Lints, manuscript in preparation). The lower the growth rate the shorter the life span of *Tribolium,* whereas lower growth rate lengthens the life span of *Drosophila.*

Andersen (5) also shows qualitative differences elicited in animals at increased population density in which those with chromosomal and random sex determination in a heterogametic male system demonstrated no effect of crowding on sex ratio, a positive feedback. Other systems, i.e. Lepidoptera with heterogametic females, had a negative feedback, as did parasitic Hymenoptera with chromosomal but nonrandom sex determination.

Not only may the kinds of responses to density vary, but totally different responses of the same insect to different environmental parameters may occur. For example, *Drosophila* exhibits both a decrease in developmental time and size with an increase in environmental temperature, whereas an increase in preimaginal population density, while also resulting in a decrease in size, exhibits an increase in developmental time (74). This is of particular interest since growth rate and longevity are so closely related (70).

The specificity of the characteristic responses to stress exhibited by different species compounds the already complex effects of density. With some groups, i.e. lepidopterous larvae, the responses may be categorized according to the species' degree of gregariousness (56, 76). For other groups, one species (i.e. *Drosophila simulans*) exhibits very different responses than another (i.e. *D. melanogaster*) to the same stresses (84).

The degree of sophistication with which scientists analyze populations for response to density is changing. Extreme phase changes in locust morphology are so striking that the differences can not be overlooked (28). However, polymorphism in aphids, once thought to be absolute and qualitative events, has only recently been shown to be gradual, both in morphology and in behavior (120).

The lags exhibited in a specific response (125) by some populations to different densities seem to be adequately explained as a difference in levels of sensitivity between the variety of compensatory responses of a population (12) (T. M. Peters and P. Barbosa, unpublished data).

Experimental and Management Problems

The primary difficulty in elucidating the effects of population density on parameters of insect biology is in isolating the density effects. Most high-density conditions also have concurrent effects from starvation and fouling of the environment with waste products as well as the potential for epizootics (124). Therefore, many studies on density effects have been confounded because problems such as pollution and access to food have not been adequately accounted for in the design.

Problems exist in comparing absolute population densities between species. Undoubtedly some of these problems are due to differences in sizes of the species under consideration. The importance of considering size of individuals is demonstrated by such results as reported by Hirata (40) and Cole (23) for earlier-instar larvae in which survival and/or growth rate is enhanced by dense populations. However, as body size of each individual increases, the number of contacts, absolute utilization of requisites, and pollution is increased and becomes a negative feedback to survival and/or growth in later life. Early crowding has also been beneficial with insects that are gregarious in early life but solitary in later instars, as in several Lepidoptera (56).

Everyone who has run density studies has faced the question of how to manage the changes in density and whether the changes are comparable when rearing vessels are changed for a given density as opposed to absolute density differences with a selected standard rearing vessel. Reports exist on comparable effects when smaller containers are used with proportionally smaller absolute populations (30). That shape of the rearing vessel is important in density studies is emphasized by Ikeshoji & Mulla (45). The control for their crowding experiments is based on the same volume of medium for a standard number of larval mosquitoes, but in a shallow pan rather than a beaker. This changes the surface per larva without changing volume per larva. Problems with food availability and distribution within the rearing container are similar.

Maintaining insect numbers in a culture must ultimately become a balance between the resources available for culture maintenance and the volume of animals required for experimental uses. Such a balance is affected by relatively greater vulnerabilities of small lab populations with an Allee-type fecundity response (136). Such vulnerability is increased if food of the insect cultured is live, as with parasitoids (135).

Thus, no overall generalization develops from a review of the work on the relationship between numbers and the changes induced in insects in culture. Perhaps this lack of a generalization brings into focus the complexity of each system and the need to understand the implications of culturing specific insect groups through further investigation.

Literature Cited

1. Albrecht, F. O., Verdier, M., Blackith, R. E. 1958. Détermination de la fertilité par l'effet de groupe chez le criquet migrateur (*Locusta migratoria migratorioides* R. & F.) *Bull. Biol. Fr. Beig.* 4:349–427
2. Alexander, P., Barton, P. H. R. 1943. The excretion of ethylquinone by the flour beetle. *Biochem. J.* 37:463–65
3. Allee, W. C. 1934. Recent studies in mass physiology. *Biol. Rev. Biol. Proc. Cambridge Phil. Soc.* 9:1–40
4. Andersen, F. S. 1956. Effects of crowding in *Endrosis sarcitrella. Oikos* 7:215–26
5. Andersen, F. S. 1961. Effect of density on animal sex ratio. *Oikos* 12:1–16
6. Ankersmit, G. W., van der Meer, F. Th. M. 1973. Studies on the sterile-male technique as a means of control of *Adoxophyes orana.* 1. Problems of mass rearing (crowding effects). *Neth. J. Plant Pathol.* 79:54–61
7. Atsuhiro, S. 1969. The locomotive activity of *Leucania separata* larvae in relation to rearing densities. *Jpn. J. Ecol.* 19:73–75
8. Bakker, K. 1969. Selection for rate of growth and its influence on competitive ability of larvae of *Drosophila melanogaster. Neth. J. Zool.* 19:541–95
9. Barbosa, P., Peters, T. M. 1970. The manifestations of overcrowding. *Bull. Entomol. Soc. Am.* 16:89–93
10. Barbosa, P., Peters, T. M. 1972. The effect of larval overcrowding on pupal respiration in *Aedes aegypti. Can. J. Zool.* 50:1179–81
11. Barbosa, P., Peters, T. M. 1973. Some effects of overcrowding on the respiration of larval *Aedes aegypti. Entomol. Exp. Appl.* 16:146–56
12. Barbosa, P., Peters, T. M., Greenough, N. C. 1972. Overcrowding of mosquito populations: response of larval *Aedes aegypti* to stress. *Environ. Entomol.* 1:89–93
13. Bar-Zeev, M. 1957. The effect of density on the larvae of a mosquito and its influence on fecundity. *Bull. Res. Counc. Isr.* 6(B):220–28
14. Beck, S. D. 1971. Growth and retrogression in larvae of *Trogoderma glabrum.* 2. Factors influencing pupation. *Ann. Entomol. Soc. Am.* 64:946–49
15. Boller, E. 1972. Behavioral aspects of mass-rearing insects. *Entomophaga* 17:9–25
16. Bouletreau, M., David, J. 1967. Influence de la densité de population larvaire sur le taille des adultes, le durée de développement et al fréquence de la diapause chez *Pteromalus puparum* L. *Entomophaga* 12:186–97
17. Bryant, E. H. 1970. The effect of egg density on hatchability in two strains of the housefly. *Physiol. Zool.* 43:288–95
18. Bryant, E. H., Sokal, R. R. 1967. The fate of immature housefly populations at low and high densities. *Res. Popul. Ecol. Kyoto* 9:19–44
19. Centre National de la Recherche Scientifique 1968. *L'effet de Groupe chez les Animaux. Colloq. Int. No. 173, Paris* 390 pp.
20. Chauvin, R. 1958. L'action du groupement sur la croissance des grillons (*Gryllulus domesticus*). *J. Insect Physiol.* 2:235–48
21. Chauvin, R. 1967. *The World of an Insect.* London: World Univ. Libr. 254 pp.
22. Cohen, B. 1968. The influence of larval crowding on the development time of populations of *Drosophila melanogaster* on chemically defined medium. *Can. J. Zool.* 46:493–97
23. Cole, W. C. 1973. Crowding effects among single-age larvae of the mountain pine beetle, *Dendroctonus ponderosae. Environ. Entomol.* 2:285–93
24. Dixon, A. F. G., Wratten, S. D. 1971. Laboratory studies on aggregation, size and fecundity in the black bean aphid, *Aphis fabae* Scop. *Bull. Entomol. Res.* 61:97–113
25. Ebeling, W., Reierson, D. A. 1970. Effect of population density on exploratory activity and mortality rate of German cockroaches in choice boxes. *J. Econ. Entomol.* 63:350–55
26. Ellis, P. E., Gillett, S. 1968. Social aggregation and an airborne gregarizing factor in locusts. See Ref. 19, pp. 173–82
27. Englert, D. C., Bell, A. E. 1970. Selection for time of pupation in *Tribolium castaneum. Genetics* 64:541–52
28. Faure, J. C. 1932. The phases of locusts in South Africa. *Bull. Entomol. Res.* 23:293–424
29. Forrest, J. M. S. 1974. The effects of crowding on morph determination of the aphid *Dysaphis devecta. J. Entomol.* (A.) 48:171–75
30. Greenough, N. C., Peters, T. M., Barbosa, P. 1971. Effects of crowding in

larval *Aedes aegypti,* using proportionally reduced experimental universes. *Ann. Entomol. Soc. Am.* 64:26–29
31. Gunn, D. L., Knight, R. H. 1945. The biology and behavior of *Ptinus tectus* Boie, a pest of stored products. VI. Culture conditions. *J. Exp. Biol.* 21:132–43
32. Hammond, E. C. 1938. Biological effects of population density in lower organisms. *Q. Rev. Biol.* 13:421–38
33. Hammond, E. C. 1939. Biological effects of population density in lower organisms. *Q. Rev. Biol.* 14:35–49
34. Happ, G. M., Wheeler, J. 1969. Bioassay, preliminary purification, and effect of age, crowding and mating on the release of sex pheromone by female *Tenebrio molitor. Ann. Entomol. Soc. Am.* 62:846–51
35. Henneberry, T. J., Kishaba, A. N. 1966. In *Insect Colonization and Mass Production,* ed. C. N. Smith, pp. 461–78. New York: Academic. 618 pp.
36. Hewitt, P. H., Nel, J. J. C. 1969. The influence of group size on the sarcosomal activity and the behavior of *Hodotermes mossambicus* alate termites. *J. Insect Physiol.* 15:2169–77
37. Hewitt, P. H., Nel, J. J. C., Schoeman, I. 1971. Influence of group size on water imbibition by *Hodotermes mossambicus* alate termites. *J. Insect Physiol.* 17:587–600
38. Hirata, S. 1956. Influence of larval density upon the variations observed in the adult stage on the phase variation of cabbage armyworm *Mamestra brassicae.* II. Influence of larval density on the variations observed in the adult. *Res. Popul. Ecol. Kyoto* 3:79–92
39. Hirata, S. 1957. On the phase variation of the cabbage armyworm *Barathra brassicae* L. IV. Some regulating mechanisms of development in the crowded population. *Jpn. J. Appl. Ent. Zool.* 1:204–8
40. Hirata, S. 1963. On the phase variation of the cabbage armyworm *Mamestra (Baratha) brassicae* (L.). VII. Effect of larval crowding on food consumption and body weight. *Jpn. J. Ecol.* 13: 125–27
41. Hodjat, S. H. 1969. The effects of crowding on the survival, rate of development, size, color and fecundity of *Dysdercus fasciatus* Sign. in the laboratory. *Bull. Entomol. Res.* 58:487–504
42. Hodjat, S. H. 1970. Effects of crowding on colour, size and larval activity of *Spodoptera littoralis. Entomol. Exp. Appl.* 13:97–106
43. Ikeshoji, T. 1973. Overcrowding factors of chironomid larvae. *Jpn. J. Sanit. Zool.* 24:149–53
44. Ikeshoji, T. 1974. Isolation and identification of the overcrowding factors of chironomid larvae. *Jpn. J. Sanit. Zool.* 24:201–6
45. Ikeshoji, T., Mulla, M. S. 1970. Overcrowding factors of mosquito larvae. *J. Econ. Entomol.* 63:90–96
46. Ikeshoji, T., Mulla, M. S. 1970. Overcrowding factors of mosquito larvae. 2. Growth retarding and bacteriostatic effects of the overcrowding factors of mosquito larvae. *J. Econ. Entomol.* 63:1737–43
47. Ikeshoji, T., Mulla, M. S. 1974. Overcrowding factors of mosquito larvae: isolation and chemical identification. *Environ. Entomol.* 3:482–86
48. Ikeshoji, T., Mulla, M. S. 1974. Overcrowding factors of mosquito larvae: activity of branched fatty acids against mosquito larvae. *Environ. Entomol.* 3:487–91
49. Ingram, R. L. 1954. A study of the bionomics of *Aedes polynesiensis* Marks under laboratory conditions. *Am. J. Hyg.* 60:169–85
50. Ishida, H. 1952. Studies on the density effect and the extent of available space in the experimental population of the azuki bean weevil. *Res. Popul. Ecol. Kyoto* 1:25–35
51. Ishii, T. 1963. The effect of population density on the larval development of *Culex pipiens pallens. Jpn. J. Ecol.* 13:128–32
52. Iwao, S. 1959. Comparisons of the manifestation of density-dependent variabilities in *Leucania unipuncta* Haworth, *L. loreyi* Dopunchel and *L. placida* Butler. *Jpn. J. Ecol.* 9:32–38
53. Iwao, S. 1959. Phase variation in the armyworm, *Leucania unipuncta* Haworth. IV. Phase difference in the range of food tolerance of the final instar larva. *Jpn. J. Appl. Entomol. Zool.* 3:164–71
54. Iwao, S. 1959. Some analyses on the effect of population density on larval coloration and growth in the armyworm *Leucania unipuncta* Haworth. *Physiol Ecol.* 8:107–16
55. Iwao, S. 1967. Resistance to starvation of pale and black larvae of the armyworm *Leucania separata* Walker. *Botyu-Kagaku* 32:44–46
56. Iwao, S. 1968. Some effects of grouping in lepidopterous insects. See Ref. 19, pp. 185–210

57. Jacques, R. P. 1962. Stress and nuclear polyhedrosis in crowded populations of *Trichoplusia ni. J. Insect Pathol.* 4:1–22
58. Johnson, W. S., McFarlane, J. E. 1973. Occurrence of the group effect in *Acheta domesticus* (L.) in relation to parental age. *Ann. Entomol. Soc. Quebec* 18: 61–65
59. Jones, W. L. 1960. *The effects of crowding on the larvae of Aedes aegypti* (L.). Ph.D. thesis. The Ohio State Univ. 65 pp.
60. Judge, F. D., Schaefers, G. A. 1971. Effects of crowding on alary polymorphism in the aphid *Chaetosiphon fragaefolii. J. Insect Physiol.* 17:143–48
61. Karandinos, M. G., Axtell, R. C. 1972. Population density effects on fecundity of *Hippelates pusio* Leow. *Oecologia Berlin* 9:341–48
62. Kawada, K. 1964. The development of winged forms in the cabbage aphid *Brevicoryne brassicae* L. *Berichte d. O'-hara Inst.* 12:189–95
63. King, E. G. Jr., Spink, W. T. 1974. Laboratory studies on the biology of the Formosan subterranean termite with primary emphasis on young colony development. *Ann. Entomol. Soc. Am.* 67:953–58
64. Kinkade, M. L., Erdman, H. E. 1973. Dietary and density dependent factors in the induction of population autocide of flour beetles *Tribolium confusum. Environ. Entomol.* 2:75–76
65. Laughlin, R. 1960. Biology of *Tipula oleracea* L.: growth of the larva. *Entomol. Exp. Appl.* 3:185–97
66. LeCato, G. L. III, Pienkowski, R. L. 1972. Fecundity, egg fertility, duration of oviposition, and longevity of alfalfa weevils from eight matings and storage conditions. *Ann. Entomol. Soc. Am.* 65:319–23
67. Lees, A. D. 1967. The production of the apterous and alate forms in the aphid *Megoura viciae* Buckton, with special reference to the role of crowding. *J. Insect Physiol.* 13:289–318
68. Leonard, D. E. 1968. Effects of density of larvae on the biology of the gypsy moth (*Porthetria dispar*). *Entomol. Exp. Appl.* 11:291–304
69. Leppla, N. C. 1974. Influence of density, age, and sex ratio on mating in pink bollworm moths. *J. Ga. Entomol. Soc.* 9:193–98
70. Lints, F. A. 1963. De l'influence de la Formule caryocytoplasmique et du Milieu sur les relations entre longevité et vitesse de croissance chez *Drosophila melanogaster. Bull. Biol. Fr. Belg.* 97:605–26
71. Lints, F. A., Gruwez, G. 1972. What determines the duration of development in *Drosophila melanogaster? Mech. Ageing Dev.* 1:285–97
72. Lints, F. A., Lints, C. V. 1969. Respiration in *Drosophila.* III. Influence of preimaginal environment on respiration and ageing in *Drosophila melanogaster* hybrids. *Exp. Gerontol.* 4:81–94
73. Lints, F. A., Lints, C. V. 1969. Influence of preimaginal environment on fecundity and ageing in *Drosophila melanogaster* hybrids. *Exp. Gerontol.* 4:231–44
74. Lints, F. A., Lints, C. V. 1971. Influence of preimaginal environment on fecundity and ageing in *Drosophila melanogaster* hybrids. III. Developmental speed and life-span. *Exp. Gerontol.* 6:427–45
75. Loconti, J. D., Roth, L. M. 1953. Composition of the odorous secretion of *Tribolium castaneum. Ann. Entomol. Soc. Am.* 46:281–89
76. Long, D. B. 1953. Effects of population density on larvae of Lepidoptera. *Trans. R. Entomol. Soc. London* 104:543–85
77. Long, D. B., Zaher, M. A. 1958. Effect of larval population density on the adult morphology of two species of Lepidoptera, *Plusia gamma* L. and *Pieris brassicae* L. *Entomol. Exp. Appl.* 1:161–73
78. Mackauer, M. 1976. Genetic problems in the production of biological control agents. *Ann. Rev. Entomol.* 21:369–85
79. MacLagan, D. S. 1932. The effect of population density upon rate of reproduction with special reference to insects. *Proc. R. Soc. London Ser. B* 111:437–54
80. Martinez-Beringola, M. L. 1966. Influencia de la densidad de población larvaria en la duración del desarrollo de *Ceratitis capitata* Wied. II. *Bol. R. Soc. Esp. Hist. Nat. Secc. Biol.* 64:351–60
81. McFarlane, J. E. 1966. Studies on group effect in crickets. I. Effect of methyl linolenate, methyl linoleate and vitamin E. *J. Insect Physiol.* 12:179–88
82. McFarlane, J. E. 1968. Chemical aspects of group effects in insects. See Ref. 19, pp. 105–11
83. Michel, R. 1971. Influence du groupement en essaim artificiel sur la tendence au vol du criquet Pélerin (*Schistocerca gregaria* Forsk). *Behavior* 39:58–72
84. Miller, R. S. 1964. Larval competition in *Drosophila melanogaster* and *D. simulans. Ecology* 45:132–47

85. Miller, R. S., Thomas, J. L. 1958. The effects of larval crowding and body size on the longevity of adult *Drosophila melanogaster*. *Ecology* 39:118–25
86. Mittler, T. E., Kunkel, H. 1971. Wing production by grouped and isolated apterae of the aphid *Myzus persicae* on artificial diet. *Entomol. Exp. Appl.* 14:83–92
87. Moore, C. G., Fisher, B. R. 1969. Competition in mosquitoes. Density and species ratio effects on growth, mortality, fecundity, and production of growth retardant. *Ann. Entomol. Soc. Am.* 62:1325–31
88. Moore, C. G., Whitacre, D. M. 1972. Competition in mosquitoes. 2. Production of *Aedes aegypti* larval growth retardant at various densities and nutrition levels. *Ann. Entomol. Soc. Am.* 65:915–18
89. Mori, S. 1954. Population effect on the daily periodic emergence of *Drosophila. Mem. Coll. Sci. Univ. Kyoto Ser. B* 21:49–54
90. Murdie, G. 1969. Some causes of size variation in the pea aphid, *Acyrthosiphon pisum* Harris. *Trans. R. Entomol. Soc. London* 121:423–42
91. Murdie, G. 1969. The biological consequences of decreased size caused by crowding or rearing temperatures in apterae of the pea aphid *Acyrthosiphon pisum* Harris. *Trans. R. Entomol. Soc. Lond.* 121:443–55
92. Nakamura, H. 1966. The activity-types observed in the adult of *Callosobruchus chinensis* L. *Jpn. J. Ecol.* 16:236–41
93. Nakamura, H. 1967. Comparative study of adaptability to the density in two species of *Callosobruchus. Jpn. J. Ecol.* 17:57–63
94. Nakao, S. 1958. On the diurnal variation of a song-length of *Meimuna opalifera* Walker and the effects of population density upon it. *Kontyu* 26:201–9
95. Nayar, J. K. 1968. Biology of *Culex nigripalpus* Theobald. Part I. Effects of rearing conditions on growth and the diurnal rhythm of pupation and emergence. *J. Med. Entomol.* 5:39–46
96. Nayar, J. K., Sauerman, D. M. Jr. 1968. Larval aggregation formation and population density interrelationships in *Aedes taeniorhynchus,* their effects on pupal ecdysis and adult characteristics at emergence. *Entomol. Exp. Appl.* 11:423–42
97. Nayar, J. K., Sauerman, D. M. Jr. 1969. Flight behavior and phase polymorphism in the mosquito *Aedes taeniorhynchus. Entomol. Exp. Appl.* 12: 365–75
98. Nayar, J. K., Sauerman, D. M. Jr. 1970. A comparative study of growth and development in Florida mosquitoes. Part 1. Effects of environmental factors on ontogenetic timings, endogenous diurnal rhythm and synchrony of pupation and emergence. *J. Med. Entomol.* 7:163–74
99. Nayar, J. K., Sauerman, D. M. Jr. 1970. A comparative study of growth and development in Florida mosquitoes. Part 3. Effects of *temporary* crowding on larval aggregation formation, pupal ecdysis and adult characteristics at emergence. *J. Med. Entomol.* 7:521–28
100. Nayar, J. K., Sauerman, D. M. Jr. 1973. A comparative study of growth and development in Florida mosquitoes. Part 4. Effects of *temporary* crowding during larval stages on female flight activity patterns. *J. Med. Entomol.* 10:37–42
101. Nayar, J. K., Sauerman, D. M. Jr. 1975. Flight and feeding behavior of autogenous and anautogenous strains of the mosquito *Aedes taeniorhynchus. Ann. Entomol. Soc. Am.* 68:791–96
102. Nishigaki, J. 1963. The effect of low population density on the mating chance and the fecundity of the azuki bean weevil, *Callosobruchus chinensis. Jpn. J. Ecol.* 13:178–84
103. Ogura, N. 1975. Hormonal control of larval coloration in the armyworm, *Leucania separata. J. Insect Physiol.* 21:559–76
104. Osborn, A. W., Shipp, E., Rodger, J. C. 1970. Housefly fecundity in relation to density. *J. Econ. Entomol.* 63:1020–21
105. Park, T. 1937. Experimental studies of insect populations. *Am. Nat.* 71:21–33
106. Park, T. 1938. Studies in population physiology. VIII. The effect of larval population density on the postembryonic development of the flour beetle *Tribolium confusum* Duval. *J. Exp. Zool.* 79:51–70
107. Park, T., Miller, E. V., Lutherman, C. Z. 1939. Studies in population physiology. IX. The effect of imago population density on the duration of the larval and pupal stages of *Tribolium confusum* Duval. *Ecology* 20:365–73
108. Peters, T. M., Chevone, B. I., Callahan, R. A. 1969. Interactions between larvae of *Aedes aegypti* (L.) and *Culex pipiens* L. in mixed experimental populations. *Mosq. News* 29:435–38
109. Peters, T. M., Chevone, B. I., Greenough, N. C., Callahan, R. A., Barbosa,

P. 1969. Intraspecific competition in *Aedes aegypti* (L.) larvae. I. Equipment, techniques, and methodology. *Mosq. News* 29:667–74

110. Peters, T. M., Greenough, N. C., Barbosa, P. 1969. Recent advances in mosquito research at the University of Massachusetts. *Proc. 56th Ann. Meet. N.J. Mosq. Exterm. Assoc.* pp. 95–101
111. Pettit, L. C. 1940. The effect of isolation on growth in the cockroach *Blattella germanica* (L). *Entomol. News* 51:293
112. Robertson, F. W. 1957. Studies in quantitative inheritance. XI. Genetic and environmental correlation between body size and egg production in *Drosophila melanogaster. J. Genet.* 55:428–43
113. Roth, L. M. 1943. Studies on the gaseous secretion of *Tribolium confusum* Duval. II. The odoriferous glands of *Tribolium confusum. Ann. Entomol. Soc. Am.* 36:397–424
114. Roth, L. M., Howland, R. B. 1941. Studies on the gaseous secretion of *Tribolium confusum* Duval. I. Abnormalities produced in *Tribolium* by exposure to a secretion given off by the adults. *Ann. Entomol. Soc. Am.* 34: 151–75
115. Sang, J. H. 1949. The ecological determinants of population growth in *Drosophila* culture. *Physiol. Zool.* 22:183–202
116. Schmidt, C. D., Ward, C. R., Eschle, J. L. 1973. Rearing and biology of the horn fly in the laboratory: effects of density on survival and fecundity of adults. *Environ. Entomol.* 2:223–24
117. Shannon, R. C., Putnam, P. 1934. The biology of *Stegomyia* under laboratory conditions. I. The analysis of factors which influence larval development. *Proc. Entomol. Soc. Wash.* 36:210–16
118. Shaw, M. J. P. 1970. Effects of population density on alienicolae of *Aphis fabae* Scop. I. The effect of crowding on the production of alatae in the laboratory. *Ann. Appl. Biol.* 65:191–96
119. Shaw, M. J. P. 1970. Effects of population density on alienicolae of *Aphis fabae* Scop. II. The effects of crowding on the expression of migratory urge among alatae in the laboratory. *Ann. Appl. Biol.* 65:197–203
120. Shaw, M. J. P. 1970. Effects of population density on alienicolae of *Aphis fabae* Scop. III. The effect of isolation on the development of form and behavior of alatae in a laboratory clone. *Ann. Appl. Biol.* 65:205–12
121. Smith, C. N., ed. 1966. *Insect Colonization and Mass Production.* New York: Academic. 618 pp.
122. Smith, D. S. 1970. Crowding in grasshoppers. I. Effect of crowding within one generation on *Melanoplus sanguinipes. Ann. Entomol. Soc. Am.* 63:1775–76
123. Smith, D. S. 1972. Crowding in grasshoppers. II. Continuing effects of crowding on subsequent generations of *Melanoplus sanguinipes. Environ. Entomol.* 1:314–17
124. Steinhaus, E. A. 1958. Crowding as a possible stress factor in insect disease. *Ecology* 39:503–14
125. Sullivan, R. L., Sokal, R. R. 1963. The effects of larval density on several strains of the house fly. *Ecology* 44:120–30
126. Sutherland, O. R. W. 1969. The role of crowding in the production of winged forms by two strains of the pea aphid, *Acyrthosiphon pisum. J. Insect Physiol.* 15:1385–410
127. Tantawy, A. D., Vetukhiv, M. O. 1959. Effects of size on fecundity, longevity and viability in populations of *Drosophila pseudoobscura. Am. Nat.* 94:395–403
128. Terzian, L. A., Stahler, N. 1955. The effects of larval population density on some laboratory characteristics of *Anopheles quadrimaculatus* Say. *J. Parasitol.* 35:487–95
129. Toba, H. H., Paschke, J. D., Friedman, S. 1967. Crowding as the primary factor in the production of the agamic alate form of *Therioaphis maculata. J. Insect Physiol.* 13:381–96
130. Tschinkel, W. R., Willson, C. D. 1971. Inhibition of pupation due to crowding in some tenebrionid beetles. *J. Exp. Zool.* 176:137–46
131. Utida, S. 1965. "Phase" dimorphism in the laboratory population of the cowpea weevil, *Callosobruchus maculatus.* IV. The mechanism of induction of the flight form. *Jpn. J. Ecol.* 15:193–99
132. Utida, S. 1967. Collective oviposition and larval aggregation in *Zabrotes subfasciatus* (Boh.). *J. Stored Prod. Res.* 2:315–22
133. Utida, S. 1972. Density dependent polymorphism in the adult of *Callosobruchus maculatus. J. Stored Prod. Res.* 8:111–26
134. Wada, Y. 1965. Effect of larval density on the development of *Aedes aegypti* (L.) and the size of the adult. *Quaest. Entomol.* 1:223–49

135. Walker, I. 1967. Effect of population density on the viability and fecundity in *Nasonia vitripennis*. *Ecology* 48:294–301
136. Watt, K. E. F. 1960. The effect of population density on fecundity in insects. *Can. Entomol.* 92:674–95
137. Wharton, D. R. A., Lola, J. E., Wharton, M. L. 1967. Population density, survival, growth, and development of the American cockroach. *J. Insect Physiol.* 13:699–716
138. Wharton, D. R. A., Lola, J. E., Wharton, M. L. 1968. Growth factors and population density in the American cockroach, *Periplaneta americana*. *J. Insect Physiol.* 14:637–53
139. Woolever, P., Pipa, R. 1970. Spatial and feeding requirements for pupation of last instar larval *Galleria mellonella*. *J. Insect Physiol.* 16:251–62
140. Youdeowei, A. 1967. Observations on some effects of population density on *Dysdercus intermedius* Distant. *Bull. Entomol. Soc. Nigeria* 1:18–26
141. Young, A. M. 1970. Predation and abundance in populations of flour beetles. *Ecology* 51:602–19

Ann. Rev. Entomol. 1977. 22:451–81

COTTON INSECT PEST MANAGEMENT

❖6138

Dale G. Bottrell[1] *and Perry L. Adkisson*
Department of Entomology, Texas A&M University, College Station, Texas 77843

Cotton, the world's most important fiber crop, is being cultivated on more than 30 million hectares of land in some 80 countries (17, 58). Although cotton is grown for its fiber, the seeds also are valued for their oil and for use in human and livestock feeds. The USSR, United States, China, India, Brazil, Pakistan, Mexico, Turkey, Egypt, and the Sudan presently are the largest cotton-producing nations; it is important to the economy of these countries and contributes significantly to a productive and balanced world agriculture. The crop also is of particular importance to many developing countries whose export earnings may be derived largely from the sale of cotton or cotton goods. Excellent articles on the production and utilization of cotton have been published by Brown & Ware (28) and Elliot et al (50). These aspects receive only cursory attention in this review.

Cotton lint is produced in a seed-bearing capsule, or boll, and the principal pest insects of the crop are those that attack the bolls or the flower buds (squares) that precede them. Insects of secondary importance are those that attack the leaves, stems, and planted seeds. The volume of written information on cotton insects and their control is quite large. Space will not permit a comprehensive review of even that information with the closest relevance to the present review, therefore as a supplement to this review, we refer the reader to other references (40, 62, 93, 97, 122, 127, 129, 135) for descriptions of the major pests of cotton and other relevant information. Discussions of recent developments in integrated control and management of cotton insect pests also have been provided by others (2, 5, 6, 43, 55, 58, 95, 113, 118, 131, 154–156, 171, 178).

HISTORICAL PATTERNS OF ARTHROPOD PEST CONTROL IN COTTON

An historical analysis of cotton production and associated pest control practices throughout the world shows a recurring pattern (45, 58, 148, 151), which is characterized by a series of successional phases, beginning with a subsistence phase typical

[1]Resigned from Texas A&M University on February 28, 1975.

of primitive, low-yielding cotton production and progressing toward a mechanized phase that initially produces much higher yields. In this review we trace the developments in cotton insect control that have coincided with each phase of production. We also discuss how these developments have led to the shaping of contemporary principles and tactics of insect pest management strategies for cotton.

Subsistence Phase

In this phase, cotton is usually grown under nonirrigated conditions in small plots as part of the subsistence agriculture. Yields are very low as there is no organized method of pest control and protection of the crop depends on natural control factors, the pest resistance inherent in native varieties, hand-picking the pests, cultural practices, and rare pesticidal treatments (45).

Exploitation Phase

A highly significant phase of cotton insect pest control in the United States and many other cotton-growing countries occurred during the late 1940s and early 1950s. For the first time, and perhaps the last, highly effective insecticides were available that controlled all of the serious pest insects of cotton; these were the new organic insecticides developed shortly after World War II. Although the use of inorganic insecticides for control of cotton insects preceded the new organics by some 40 years (129, 131), the latter materials produced more effective insect control and higher cotton yields by preventing damage from a broader spectrum of insect pests and thus allowing greater benefits to be produced from the use of fertilizers and irrigation than the narrower spectrum inorganic materials (62, 118, 129, 131).

Doutt & Smith (45) described the period when new organic insecticides and other crop production inputs were being used extensively to produce extremely high yields as the exploitation phase of cotton production. This phase began about 1946, concomitant with the introduction of the organochlorine insecticides. The following few years characterized a period when concerted efforts were made to maximize yields from land cultivated for cotton.

EMERGENCE OF CROP YIELD MAXIMIZATION STRATEGIES The new post-war organic insecticides produced an eventual impact perhaps unlike any other modern agricultural technological advancement. It is beyond the scope of this article to discuss the total impact of these chemicals on man and his environment; such discussions have been made many times in recent years by many authors (37, 47, 59, 107, 108, 111, 112, 115, 150). This review centers on the impact that post-war organic insecticides had in shaping a concept we refer to as crop yield maximization strategies. These strategies, which evolved in parallel with the introduction of the new organic insecticides, contributed greatly to the designing of the insecticidally dependent insect pest control practices for cotton and many other crops during the late 1940s and early 1950s. These methods were used successfully for several years afterwards.

More recently, however, problems of insecticide-resistant pest strains, secondary pest outbreaks, and environmental quality have greatly reduced, if not negated, the success of insecticidally dependent crop yield maximization strategies; these strategies are no longer dependable for most cotton producers. Nevertheless, many present-day cotton producers, who profited substantially from these practices during times when insecticides were effective and economical to use, have not yet discarded them in favor of currently available integrated pest control techniques. This attitude is a major hinderance to the implementation of new, integrated control technology.

The effectiveness of pest control produced by the new organic insecticides provided incentives for agriculturalists the world over to develop cotton varieties with the highest yield potential and the production technology to use in maximizing this potential. The greatest effort in this direction was made in countries that could afford to use the large quantities of insecticides, fertilizers, and cheap energy inputs required to squeeze maximum yield from the new varieties.

Protective insecticide treatments to the plant nurseries provided cotton breeders with a safeguard against the selection of anything other than the highest-yielding germ plasm. Insect-free plant nurseries thus became a salient feature of most postwar cotton-breeding programs. This umbrella of insecticide permitted the development of only the highest-yielding cotton varieties, those that would fruit as long as the season would permit and that produced great yield responses to fertilizers and irrigation; however, the new high-yielding cotton varieties released from the insecticidally shrouded nurseries often were highly vulnerable to insect pest attack and required frequent insecticide treatments to insure high yields when insect attacks occurred. Producers commonly applied these treatments even when pest attacks were below damaging levels because they provided good insurance for protection of investment in land, machinery, fertilizers, and other inputs required to produce the crop.

THE DIVIDENDS Most cotton farmers of the late 1940s and early 1950s became accustomed to using insecticides on a season-long, fixed or calendar preventive treatment schedule irrespective of pest numbers or crop damage (116, 178). Although these treatments were not always required to prevent yield losses, they did eliminate the necessity for scouting fields to determine the actual need for control; they also eliminated the risks of loss if no treatments were made. The costs for these treatments were relatively small compared with some of the other costs of cotton production (152, 159), hence extensive treatment of the crop with insecticides provided an inexpensive, reliable, and high-return form of crop insurance. This was especially appealing to farmers who were attempting to produce the highest yields possible from their farming operations.

Many banks and other lending agencies viewed these treatments as good insurance for high-risk investments and often required, as one condition to a crop loan, that a full-season, preventive insect control program be used. These treatments paid high dividends for many years.

This is apparent from examination of the trends of cotton yields before and after the introduction of the organic insecticides into the United States. During the 10-year period from 1936 to 1945, which preceded the first wide-scale use of hydrocarbon or organochlorine insecticides in 1946, cotton yields averaged about 251 lbs of lint cotton per acre. This compares to average yields of about 300 lbs of lint cotton per acre, or 16% more, which resulted during the first 10 years, 1946–1955, that

DDT and other organic insecticides were used extensively on cotton in the United States (calculated from data supplied by J. G. Thomas).

These yield data do not reflect all the gains to producers that the new insecticides made possible. Before the availability of the new organic insecticides, many cotton growers throughout the area of the United States infested with the boll weevil (*Anthonomus grandis*) relied almost exclusively on cultural control practices to combat this pest (174). Even though calcium arsenate was quite effective against the weevil (63, 118, 131), many cotton farmers of that period did not depend on it for insect control; instead, they grew earlier-maturing or shorter-season cotton varieties that were less damaged by late-season weevil attacks than the more indeterminate, longer-season cottons (87). When combined with the early fall destruction of the harvested cotton plants to reduce numbers of potential overwintering boll weevils, this system allowed a means for growing these cottons profitably with only minimal need for insecticides (86–88, 103); however the staple length of these shorter-season cottons often was inferior (shorter) to that of the longer-season cottons. This meant that the farmers often received lower prices for their lint cotton when they grew the early maturing varieties, and therefore the new organic insecticides, plus yield-boosting fertilizers and other post-war production advancements, made it more profitable for cotton producers to grow the longer-season cotton varieties. The new insecticides protected the plants from weevil damage throughout their extended fruiting period, thereby permitting greater yields of higher-quality, longer-staple lint, which brought a better price in the market (174).

The new organic insecticides also had a tremendous impact on cotton production in other countries of the developed world and, moreover, were a chief driving force behind the expansion of cotton into many countries of the under-developed world, whose economies eventually became dependent on the export earnings of this crop (17). Thus, the organic pesticides had a major role in increasing cotton yields the world over; they also were instrumental in reshaping the world's geographic cropping patterns.

The widescale use of post-war organic insecticides in cotton and their simultaneous introduction into the culture of other major agricultural crops led to an unprecedented reshaping of the science of pest control. This process of reshaping has, perhaps, been more visible in crop protection developments in cotton than in any other crop, which is apparent by comparing the status of cotton insect pest control today with that of two decades ago, as described by Gaines (62). A majority of the publications covered in Gaines' 1957 review focused almost entirely on perfecting insecticidal control techniques that provided the greatest level of pest suppression. Today nearly all of these materials have been exploited to their maximum. Many have even disappeared from the cotton scene because the target pests have become resistant to them or because they have been banned in the United States by the Environmental Protection Agency.

COLLAPSE OF THE INSECTICIDE-DRIVEN MAXIMIZATION STRATEGIES (THE DEVELOPMENT OF INSECTICIDE-RESISTANT PEST STRAINS) The earliest of the post-World War II cotton entomologists apparently did not seriously believe that the frequent use of organic insecticides would lead to the selection of resistant pest insect strains. For instance, none of the four articles (41, 53, 63, 129) appearing in the section on cotton insects of the 1952 United States Department of Agriculture Yearbook of Agriculture (devoted entirely to insects) placed importance on the potential threat of insecticide-resistant pest strains, although authors of another section in that book (90, 126) clearly expressed their growing concern over this matter. Perhaps ignoring the threat of resistance in cotton insect pests merely reflected the prevailing enthusiasm shared by entomologists working with the new insecticides; after all, these people had just taken command of the most powerful weapons ever used in the fight against insect pests. Furthermore, they had been given tremendous responsibilities to develop methodology for using these weapons, especially since initial experiments had suggested the new organic insecticides might offer the ultimate solution to certain pest problems.

DDT was the first of the new insecticides to be tested extensively in cotton. It was rapidly followed by the introduction of benzene hexachloride (BHC), toxaphene, and chlordane (129). All these chemicals were found to be highly effective against certain cotton pests and were used in the control recommendations of many states, but not one of these compounds alone controlled all the major arthropod pests of cotton. Therefore, many combinations were developed, such as a mixture containing 3% benezene hexachloride, 5% DDT, and 40% sulfur, or toxaphene combined with DDT, in attempt to find a good all-purpose treatment that would rid the fields of all the arthropod pests.

The synthesis of the new organics was greatly accelerated and thousands were tested for insecticidal efficacy during the late 1940s and early 1950s (129). None of them, however, satisfied the requirements for an all-purpose insecticide that would control all the arthropod pests of cotton. This obviously was considered a worthy goal to pursue for even minor pests as is evident by the following statement, which appeared in the 1952 USDA Yearbook (129). "Killing off many insects of minor importance helps the plants to grow faster, so that the crop is usually matured earlier and the farmer has a better chance to get it properly harvested." We include this quote to illustrate the prevailing philosophy of insect control of most, but not all, of the principal cotton entomologists of the late 1940s and early 1950s. A few, such as Dwight Isely of Arkansas, were cautioning that insecticides should be used only as a last resort after other methods of suppression had failed (95).

Whitcomb (178) and Newsom (116) described the approach most cotton entomologists advocated for the new organic materials. In general, these entomologists recommended that the farmers spray or dust regularly, usually once a week, from the time the cotton started squaring until all of its green bolls had hardened and were no longer susceptible to attack by fruit-feeding insects such as the boll weevil or bollworm, *Heliothis zea.* Some recommendations called for additional, early season treatments to protect the presquaring cotton plants. As a general rule, control recommendations released by the state extension services included information, as skimpy as it may have been, on economic thresholds. In most cases, early season treatments for thrips (*Frankliniella* spp.), aphids (*aphis gossypii*), the cotton fleahopper (*Pseudatomoscelis seriatus*), and overwintering boll weevils were applied as automatic preventive dusts or sprays based on the growth stage of the cotton plant, whereas late season treatments were based on pest numbers or percentages of fruit damaged. In other cases, producers followed automatic treatments from seedling emergence to harvest irrespective of pest numbers or crop damage.

Regardless of which method was used for determining insecticidal treatment need, most applications were applied before yield-damaging pest levels were reached. The economic thresholds used were very conservative and encouraged the early use of insecticides to prevent losses by key pests. Little thought was given to the problems of the unleashing of potential or secondary pests by these treatments; in fact, the terms *key, potential,* and *secondary* pests were not included in the jargon of the cotton entomologists of this period.

This more or less indiscriminate preventive strategy of insecticidal control was used for several years before insecticide-resistant strains became dominant in major pest groups. Resistant pest strains eventually emerged, incipiently at first in minor pest groups and then in major pest groups that were at first confined to localized geographic regions. Twenty-four species of insect and spider mite cotton pests in the United States are currently known to be resistant to one or more chemical insecticides or acaracides, and several more are suspected of resistance (113). Resistant strains of the arthropod cotton pests now are widespread in the United States and other cotton-producing countries of the world (58, 59, 155).

Crisis Phase

The spread of insecticide-resistant pest strains and the parallel emergence of companion problems of insect control mark the onset of what has been termed the crisis phase of cotton crop protection (45). The first major problem of resistance in cotton pests erupted in the mid-1950s when a strain of organochlorine-resistant boll weevils developed in the mid-South (136). This strain was first detected in Louisiana and Mississippi but soon was discovered in other southern and southwestern states and by 1960 was scattered widely in small pockets throughout its range of distribution in the United States (25, 118). This was quickly followed by the development of DDT- and endrin-resistant strains of the tobacco budworm (*Heliothis virescens*) and bollworm (1, 8, 26, 27, 68–70, 183).

Insecticide-resistant pest strains were no longer a figment of the imagination but a grim reality. Still, many entomologists of that time doubted the seriousness of the problems these pest strains would eventually create; these people were optimistic that the insecticide industry would continue to produce new effective insecticides that would counteract any problems created by the organochlorine-resistant pests.

The insecticide industry, whose future was largely governed by its ability to provide effective insecticides for cotton pests, responded favorably to this call. As Reynolds et al (131) discussed in detail, the trade-off was a switch to certain organophosphorus (OP) and carbamate insecticides that provided highly effective control of the boll weevil but were not so effective against *Heliothis.* The OP compounds, principally methyl parathion but to a lesser degree azinphosmethyl, EPN, malathion, and others, became most widely used to control organochlorine-resistant boll weevils. These materials were highly effective, even when used at low dosages; however they dealt severe destruction to the arthropod parasites and predators that reside in cotton fields, unleashing tremendous outbreaks of *Heliothis*. The organochlorines also had caused the death of insect natural enemies with the consequent unleashing of secondary and potential pests (22, 64, 117, 119, 134), although not to the extent caused by the OP compounds. The latter were much more ecologically disruptive than the chlorinated hydrocarbon insecticides for two reasons. First, the major arthropod natural enemies most commonly found in cotton were, in general, more susceptible to the OP than to the organochlorine insecticides (39, 134). The second reason was that OP materials were much less persistent than the organochlorines and thus had to be applied at more frequent intervals.

For all these reasons (the higher toxicity of the OP compounds to the boll weevil, the lower toxicity to *Heliothis*, and the more complete destruction of natural enemies) the boll weevil faded in importance as a yield-damaging cotton pest and the bollworm and tobacco budworm became the pests of primary concern to most cotton producers and entomologists.

The low dosages of OP insecticides used against the boll weevil ordinarily did not control the *Heliothis* outbreaks their use induced and so it was necessary to mix organochlorines or carbamates with the OP materials to control the pests (2, 5, 6, 9, 10, 131). In fact, the insecticide industry began marketing insecticidal mixtures, such as toxaphene-DDT-methyl parathion, endrin-methyl parathion, and monocrotophos-methyl parathion-chlordimeform, which provided control of boll weevils and the secondary pests their use engendered.

These insecticide mixtures were highly effective at first and rendered the cotton fields almost free of all insects (131), but an endless treadmill of treatments, involving greater quantities of insecticide applied at increasingly frequent intervals became necessary to counteract the outbreaks of the resistant *Heliothis*. This greatly increased the costs of insect pest control in cotton-growing regions infested with the boll weevil; nevertheless, although expensive, the high dosages and frequent treatment still provided effective control of the resistant pest complex and therefore allowed continued efforts to squeeze maximum yields from each area of land cultivated for cotton.

Maximizing yields in the United States also was being encouraged by governmental subsidies, with the amount of the payment being based on the historical yields of cotton produced per acre on each farm unit. The incentive was to produce the highest possible yield per acre of cotton. In spite of this incentive, it became less realistic for the cotton producer to strive for maximum yields since the increased costs of pest control greatly lessened the margin of profit (113). Even though the higher costs incurred for control of the resistant *Heliothis* accounted for only a small percentage of the total costs required for the farmer to produce a cotton crop, they were sufficient to place the producer in a tighter cost-price squeeze. In addition, the increased cost of pest control occurred at a time in the 1960s when other farm production expenses also were increasing and absorbing a higher percentage of the realized gross farm income (78), while at the same time the prices farmers received per pound of harvested cotton were decreasing (184) due to increasing competition from synthetic fibers and other man-made materials (17). Although the average cotton yield per acre increased slightly during this period, this was offset by the ever increasing costs of production, including the higher costs for *Heliothis* control. Cotton farming became much less profitable since the unit value of the marketable cotton did not increase proportionally with costs. Ironically, although the postwar organic insecticides had made it possible for cotton farmers to implement crop yield maximization strategies, the induction of insecticide-resistant pest strains now began to negate much of the benefit gained from their use.

Many cotton producers continued to strive for maximum yields despite the fact that they were no longer a realistic goal in areas where the *Heliothis* complex, especially the tobacco budworm, had developed a high level of resistance to the organochlorine, carbamate, and OP insecticides. Not surprisingly, however, many of these farmers eventually fell victim to disadvantaged odds. Some went bankrupt and others ceased to grow cotton; regardless, the price they paid was high.

Disaster Phase

The disaster phase of insect pest control in cotton is exemplified by happenings in the Lower Rio Grande Valley of Texas and the bordering area of Northeastern Mexico in the late 1960s when the tobacco budworm became resistant to the OP insecticides. Many cotton farmers in this region treated fields from 15 to 20 or more times with high dosages of highly toxic materials, such as methyl parathion, but still they suffered great losses in yield and in some cases fields were shredded without harvest (2, 3, 5, 6, 100). Others farmers produced relatively high yields but made small profits because of the large costs incurred for the insecticidal treatments. Without alternative insecticides to control outbreaks of tobacco budworms that now were resistant to the organochlorines, carbamates, and OP compounds, many farmers saw their cotton yields destroyed. Although the entire cotton industry of this part of Texas and Mexico suffered, the situation was most disasterous in Mexico. Cotton farms were abandoned; gins, compresses, and cottonseed oil mills were closed; farm workers were forced to migrate from the area; the entire economic and sociological structure of the area's small villages and rural communities was affected (2, 3, 5, 6, 100). In addition, the high-dosage treatments of the toxic OP insecticides made during the final collapse of the control program frequently poisoned spray applicators and agricultural workers (181).

Although resistance now occurs in tobacco budworms scattered throughout much of the United States boll weevil belt, major resistance problems are confined largely to populations residing in South Texas and the mid-South (30, 70, 75, 183). Consequently, the pest has yet to cause a complete restructuring of insect control practices in the boll weevil belt except in Texas. However, the budworm is beyond economical chemical control in localized areas in several states east of Texas; it caused severe yield losses in Louisiana in 1974 and in Mississippi in 1975.

Cotton producers in these and other weevil-infested states might expect chaos similar to that which occurred in Texas and Mexico unless new insecticides became available to control the OP-resistant budworm or unless improved non-insecticidal tactics are developed. These new controls are especially needed now that the Environmental Protection Agency has banned the use on cotton of DDT (131), which is still effective against *Heliothis* in many areas of the southern United States.

Presently available insecticides will not solve the problem because they will have to be applied at such high dosages and close intervals as to be economically unfeasible. The increased level of treatment will only increase the problems of *Heliothis* since resurgence of this pest complex after initial treatments, unless carefully made, may be especially severe. The more the fields are treated with these insecticides the more they will need treating. Thus many producers now are on a chemical treadmill leading to bankruptcy unless nonchemical alternative insect control strategies are developed or unless they cease to grow cotton.

The above example is from an area where the boll weevil is the primary target pest of concern. The control procedures set in motion for this pest led to the insecticidal treadmill that eventually produced disasterous consequences through the development of an OP resistant strain of the tobacco budworm.

In California, where the boll weevil does not exist, a similar sequence of events has occurred. In this case lygus bugs, particularly *Lygus hesperus* and the pink bollworm, *Pectinophora gossypiella,* are the primary key pests against which initial insecticidal control measures are used. Eventually, however, severe insecticidally induced outbreaks of nontarget, secondary pests such as cabbage loopers (*Trichoplusia ni*), beet armyworms (*Spodoptera exigua*), and bollworms exceed the key pests in importance and become the targets for most of the late-season treatments (48, 49, 52, 58, 155, 171).

The list of similar examples from other cotton-growing areas of the world is now extensive, including case histories reported from El Salvador, Peru, Nicaragua, Guatemala, Egypt, Australia, and other countries. Not all of these examples have described the evolution of insecticidal control to the point of total collapse of the cotton industry as experienced in Northeastern Mexico, but consequences as severe, or nearly as severe, have occurred in Peru, El Salvador, Australia, and Nicaragua (15, 55, 148). Each collapse has been accompanied by considerable impact on the social and political institutions of the affected areas with consequent restructuring of many farm and agribusiness enterprises.

The genesis of most of these collapses has come from the exclusive and indiscriminate use of insecticides to control highly damaging, perennially occurring key pests such as the boll weevil or pink bollworm. Insecticides often offered the most effective and convenient way for controlling these pests. They were especially required to minimize risk and to insure yields for those farming operations that strived for maximum productivity.

The above has not always been the case, however; some disaster-destined control programs have evolved in cotton areas where there are no key pests. A classic example occurred in the irrigated cotton-growing area around Pecos, Texas (121). There insecticides have been used extensively in cotton at a great expense to the producers but without a corresponding increase in yield. In fact, the insecticide treatments have frequently led to large crop losses because they have triggered outbreaks of *Heliothis* (5).

Modern-day pest management specialists frequently refer to the heavy use of insecticides without real need as the "pesticide syndrome" (45). Insecticide salesmen have been singled out as the instigators behind this problem. The fact is, however, that a farmer, like any manager, is apt to buy insurance that minimizes risk (even if the cost is high) if he believes the end result is protection of profit. It is no wonder that the pesticide syndrome has been commonplace in high-cost, high-yielding cotton production operations. Protection of capital investment, simplicity of application, and the assurance of good yields all argue, over the short term, in favor of insecticides for cotton insect control.

Integrated control systems, though more sound ecologically and possessing greater long-term benefits, are more complex to manage and therefore are harder to sell; also, there generally are more insecticide salesmen than extension personnel or qualified private consultants calling on producers. Thus, producers often may be influenced to rely entirely on insecticides rather than integrated control measures for pest suppression. Premature and exclusive use of insecticides has been the major inducer of the pest insect problems in Pecos, Texas, and for many other areas of the world. This is a major hinderance that must be overcome if producers are to implement the integrated control procedures needed to avert further pesticide-induced disasters (5). This is an especially acute problem that confronts developing countries where there is a shortage of the research and extension personnel needed to develop and demonstrate the integrated approach (154).

Future prospects for insect pest control in all cotton production areas are not as grim as the foregoing account might infer. Certain of the high-yielding areas have yet to experience great insecticide use; for example, the vast, predominately irrigated cotton-growing area of the Texas High Plains is one of the areas in the United States with the lowest use of insecticides (125). The cotton of this area, like that of the Pecos area, is not perennially attacked by serious key pests (141). However, High Plains cotton producers, unlike those in Pecos, have resisted the wholesale use of insecticides on their crop. They depend on natural mortality factors to suppress *Heliothis* and apply insecticides only when absolutely needed. This is also true for large areas of cotton grown in the northern Mississippi River Delta in the United States.

The Recovery Phase

Doutt & Smith (45) used the term *integrated control* or *recovery phase* to describe developments in cotton crop protection that utilize more efficient methods of insect pest control and place emphasis on a combination of suppression techniques in addition to pesticides. This phase has been characterized by expanded efforts to use natural and cultural methods of pest control, such as those provided by natural enemies, pest-resistant cotton varieties, phytosanitation measures, and the discrete use of insecticides. Although a varied array of promising pest control tactics has emerged from this, chemical insecticides still play an important role in the control of cotton insects. However, in the integrated control strategy, insecticides are used only to suppress pest population outbreaks that have attained damaging levels (163). Selection of the chemical used, the dosage applied, and the treatment time is carefully coordinated to minimize the hazards of key pest resurgence or the unleashing of secondary pest outbreaks. This is the same approach that Dwight Isley advocated much earlier but failed to articulate in such eloquent terms as used by more recent authors (95).

An example of an area in which recovery has occurred is the Lower Rio Grande Valley of Texas, where a successful integrated program has been implemented by many growers and cotton production has rebounded to nearly normal levels. This program is based on early crop maturity, insecticidal control of prehibernating boll weevils, and early mandatory stalk destruction on an area-wide basis. Economic threshold levels for initiating insecticide treatments for control of the cotton fleahopper and *Heliothis* have been increased. Low dosages of insecticides are used against the fleahopper and early season treatments for *Heliothis* are discouraged. The aim is to suppress the boll weevil during the harvest season so that insecticide treatments (other than spot treatments of heavy localized infestations) are avoided in the subsequent growing season. The objective is to conserve natural enemies and avoid induction of *Heliothis* attacks. This program has worked very well and is being greatly improved by introducing into the system new short-season cotton varieties and production techniques. Cotton varieties resistant to the fleahopper, boll weevil, and *Heliothis* are nearing release and should be introduced into the system within 3 to 5 years. This will provide even greater improvement (6).

Similar recoveries have been made in Peru and Nicaragua and, hopefully, one is in progress in Australia (58).

ORIGINS OF INTEGRATED PEST MANAGEMENT

Although most discussions of integrated control and pest management center on the developments of the past decade, the genesis of ecologically oriented pest control actually traces back to the late nineteenth and early twentieth centuries. Particularly outstanding contributions of this period included the pioneering work on control of cotton insect pests in the United States as discussed by Whitcomb (178) and Smith et al (153). Truly great historical figures of American entomology, such as R. V. Riley, J. H. Comstock, L. O. Howard, and W. D. Hunter, were among those who molded the foundations of modern integrated insect pest management as it is recognized in cotton and other agricultural crops today. The recent trend toward more sophisticated insect pest management, which embraces as a primary component the utilization of natural manipulatable factors, is a mere revival of the application of concepts that were well advanced by the early part of this century.

Exceptionally notable contributions of this classic period were by entomologists who worked on the boll weevil during the first few years after its invasion into this country from Mexico in about 1890. Without effective insecticides to combat this very serious pest, entomologists were forced to seek a strategy for optimally integrating all suppressive factors, such as resistant varieties, phytosanitation practices, and biological control, in a way that would prevent intolerable yield losses. These efforts evolved, beyond dispute, into one of the most sophisticated management systems ever developed for an insect pest, today's finest pest management systems notwithstanding.

The general chronological developments of the first boll weevil management system, fully developed and tested by 1920, are outlined in Table 1. It is difficult to determine just how effective this system was, but it was obviously adequate for

Table 1 Evolution of an early pest management system for the boll weevil in North America

Strategy concept or component[a]	Date of accomplishment (approx.)	Reference
Developing a conceptual model of the pest's life system	1900	(88)
Recognizing some of the ecological and economic consequences of the primary control tactics	1901–1903	(86, 103)
Recognizing need for community-wide pest suppression in preference to individual field or farm control measures	late 1800s	(103, 169)
Determining economic thresholds (essentially same as used today)	1920	(34)
Manipulating cotton environment to maximize benefits of natural control contraints; recognizing major natural enemies	1900	(88)
Outlining total management system based on long-term required needs in an economic perspective	1904	(88)
Advocating concepts of management as being more realistic than pest eradication attempts	1900	(88) and several others above

[a]Outline of major factors of emerging pest management strategies as stressed by Rabb (128), with slight modification.

many cotton farming operations, as shown by the fact that many farmers clung to the system even after calcium arsenate was introduced in the early 1920s. Nevertheless, calcium arsenate and the attendant aviation engineering improvements, which permitted wholesale aerial application of the material, did contribute significantly to the eventual erosion of ecologically oriented pest control in cotton that has only recently been revived (131, 153, 178).

RECENT DEVELOPMENTS IN INTEGRATED PEST MANAGEMENT

Smith and van den Bosch (156), Smith and Reynolds (155), Falcon & Smith (58), Adkisson (5, 6), Smith et al (154), and van den Bosch et al (171) have discussed special ecological features of the cotton agroecosystems and how manipulations in these systems can aggravate pest problems on one hand and be used for more effective management of pest populations on the other. These authors have clearly shown that manipulations in cotton production practices may result in a rather drastic attendant shift in the status of many pest species that inhabit the crop. Manipulations may include a switch to a new variety, rotation to another crop, alteration of fertilizer, row-spacing or irrigation schemes, shift from a cotton monoculture to a multi-crop mosaic, and changes in insecticide use patterns. The manipulations may be productive and have a damping effect on a damaging pest, but they also may be counterproductive and allow the establishment of new damaging pest hierarchies that result from small, subtle, man-controlled manipulations of the agroecosystem.

The recent revival of ecologically oriented pest control offers special opportunities for modern cotton protection specialists. The science of ecology has greatly advanced during the past 75 years; hence, modern specialists may draw on many skills that were not available to our early twentieth century predecessors (84).

Establishing the Need to Take Action

Establishing the need for remedial measures should be the first principle of insect control on any crop (33, 162). The truth is, however, that this need has seldom been properly established (84, 149). Articles dealing with yield losses due to the arthropods of cotton (118, 122, 131) have identified some of the major problems involved in the establishment of economic injury levels and the companion economic thresholds. It is certain that many cotton insect pests can cause devastating losses if allowed to go unchecked; however at what point remedial action is called for, or whether it might be delayed or entirely omitted, is not often accurately established. The ability of cotton to compensate for insect damage is great but is seldom accurately determined for a broad range of insect densities, conditions of weather, natural pest mortality, or agronomic practices, let alone put in the proper economic perspective (78, 118, 122, 149). The average economic thresholds that usually emerge in control recommendations, consequently, have only minimal utility in pest control decision making (162). For example, large populations of thrips, mainly *Frankliniella* spp., often produce a tattered effect on the leaves of seedling cotton plants that may appear to be very damaging; however the plants are quite efficient in overcoming this early damage and usually go on to produce normal yields in spite of the thrips. Newsom & Brazzel (118) reported that yields were increased by insecticidally controlling thrips in only 19 out of more than 150 field experiments conducted at several geographic locations and under a wide range of conditions that included varied thrips density, different crop yield situations, etc. Nevertheless, many extension control guides include only an average economic threshold value for initiating thrips control.

A common error often introduced by the experimental design of studies conducted to determine economic injury levels is the use of small experimental plots, situated side by side, each receiving a different level of insecticide treatment, usually producing from 0–100% control. This is an especially poor design when systemic insecticides, such as aldicarb, are used (130). The systemic insecticides often produce increased yields independently of the insect infestations; therefore the control or untreated plots have a yield disadvantage regardless of the pest density developing within them.

Another error in experimental design is the use of small plots situated side by side without sufficient space between any two plots to buffer insecticidal spray drift from one to the other. A good example of the problems encountered with this type of design was reported by Harp (74) from experiments designed to control *Heliothis*. He recorded the lowest yields from the untreated control (75% less than in the best insecticide treatment) in the small plot arrangement. However, an outside control (i.e. an experimental unit buffered from the insecticidal spray applied to the small plots where natural biological control was maintained) yielded more than that best insecticide treatment. It was obvious that in the small plot check the insecticidal drift killed the insect natural enemies residing there, unleashing a *Heliothis* outbreak that almost totally destroyed the yield in this treatment.

As discussed below and by Reynolds et al (131), computer models of cotton plant growth and crop yield (71) offer great potential in sharpening the approach to the establishment of economic injury levels and overall damage assessment needs for pest control in cotton.

Sampling and Measuring

Sampling and measuring are essential features of all aspects of integrated control from the elementary research stages to implementation. Without accurate estimates of pest and natural enemy population densities and their interactions with the crop plants, integrated control has no focal point from which to proceed (139).

Experiments designed to improve the sampling of pest populations and measuring of crop losses as related to integrated control have been extensive [recent articles of importance have been reported by Ruesink & Kogan (139), Chiarappa (33), and Stern (162)]. A number of authors have presented sampling plans, techniques, and concepts in sampling as related to specific groups of cotton pest insects (12, 67, 94, 124, 145).

Reynolds et al (131) discussed the broader problems of sampling in integrated control for cotton, which included plant measurement techniques, field checking, use of traps, etc. A rather novel development, emerging from pheromone trap studies, has been a technique for improved timing of insecticide treatments for control of the pink bollworm (168). Treatment time is determined by trends of moth catches in the traps. Other significant developments are comprehensively reviewed in the publications cited above.

It should be emphasized that pest suppression produced by manipulation of natural enemies still needs precise evaluation. This evaluation poses a difficult task, particularly with predators, which unlike parasites and pathogens do not leave clues with their hosts. The number of pest deaths produced by predators may not be assessed directly by examination of their hosts except in certain cases (see 20).

NATURAL MANIPULATABLE MECHANISMS

The new tactics of integrated control for cotton embrace all ecologically compatible measures and are centered around three manipulatable mechanisms for pest insect suppression: (*a*) varietal manipulation; (*b*) biological control by parasites, predators, and pathogens; and (*c*) environmental manipulations that work to the disadvantage of the pest or to the advantage of an entomophagous species.

Varietal Manipulation

The search for pest resistance in cotton germ plasm and transfer of resistance into commercial cultivars has attained great momentum in the past few years. Cotton stocks resistant to most of the major arthropod pests have been identified and the first truly insect-resistant varieties are in advanced stages of breeding. The overall status of cotton resistance to insect and spider mite pests and the use of resistant varieties in integrated control were recently reviewed by Maxwell et al (106), Leigh (92), and Maxwell & Adkisson (104); thus, only a brief summary of the utilization of host plant resistance in cotton pest management programs is given in this article.

Early maturing cottons (developed during the pre-calcium arsenate era) were used as a primary means of escaping late-season boll weevil attacks until the new organic insecticides were introduced in the late 1940s (174). Because these varieties often produced less yield and a shorter staple length than the indeterminate-fruiting types, they were abandoned when organic insecticides became available that produced highly effective late-season control of the weevil and *Heliothis.* This concept now has reemerged because of the high costs and poor control produced by presently available cotton insecticides and because of the need of cotton producers to minimize other production inputs and the effects of weather hazards.

One primary advantage of the short-season cottons is the ability to escape much of the late-season buildup of boll weevils, bollworms, and other insect pests that often decimate the crop. A second advantage is that these varieties may be harvested in many areas before the occurrence of fall rains that damage quality, reduce yield, and prevent the early shredding and plowing under of stalks for control of the boll weevil and pink bollworm. A number of experimental lines of semi-dwarf cotton are presently under test for bollworm and boll weevil escape and for adaptation to certain areas of Texas (13, 14). Bird et al (18) also are developing a number of multi-disease-resistant lines of okra-leaf, frego bract cottons that are early in maturity and highly resistant to the boll weevil.

Tamcot SP 37, a short-season cotton, when planted in a high-density pattern, has been grown in South Texas with greatly reduced amounts of nitrogen fertilizer (25 lbs/acre), irrigation, and insecticides, without sacrificing yield. The short-season variety produces yields (800 lbs of lint per acre) as great as those produced by standard commercial varieties grown in a conventional system and treated with four to eight times more nitrogen (100-200 lbs per acre), twice the irrigation, and three to five times (six to ten treatments per season) more insecticide. This type of system has great potential for conserving the use of the energy and fossil fuel-derived products required to produce cotton. It also has produced profits of over $100 per acre more than the conventional cotton production system of the area (157).

In addition, the short-season cotton may be harvested within 135 days of planting and approximately one month ahead of normal harvesting (157). This has great advantages because in many areas the food and breeding sites of the pink bollworm and boll weevil in the cotton may be removed by early harvest and stalk destruction before environmental conditions (i.e. short days, cool temperatures) are sufficient to force these pests into diapause. This means that if within a region plantings of cotton were restricted to these varieties, tremendous reductions in the average size of the overwintering populations of these pests could be achieved by normal production practices and without the added expense of insecticide treatments.

Also, high-gossypol, glabrous, nectariless varieties of cotton highly resistant to *Heliothis* are in advanced stages of breeding and should soon be available for commercial production (101, 102).

Biological Control

There is substantial evidence that, except in special situations, long-term suppression of a complex of pest species in an agroecosystem is unlikely to be achieved unless biological control by predators, parasites, or pathogens is included as a primary agent (83, 84). Although biological control is still grossly underemphasized as a significant factor in integrated control, in the past decade there has been an increasing appreciation of the importance of natural enemies in suppressing injurious or potentially injurious insects. Publications of Huffaker & Stinner (85), van den Bosch (170), DeBach (42), Huffaker (82), Glass (66), and Maxwell & Harris (105) treat biological control in the broadest sense and include many recent examples of applying the method to cotton insect control programs.

In integrated control, the use of natural enemies has been pursued by (*a*) importing exotic enemies (commonly from the native region of the target pest) and by (*b*) manipulating the pest host, the environment, and/or the enemies themselves to make the resident natural enemies more effective (84). The first approach (i.e. classic biological control) ranks as one of the most effective of all pest control tactics (42, 66); however, there has been only minimal effort in recent years to seek out exotic natural enemies for controlling cotton insect pests, and only a few entomologists (49, 118, 131) have stressed the importance of classic biological control in future integrated control programs for cotton.

Some of the most important cotton insect pests are of foreign origin and they often lack the suppressive force of natural enemies formerly associated with them in their native regions. Classic biological control has great possibilities in cotton pest management programs for suppressing these pests.

Augmenting and conserving resident natural enemies also offers particular promise (72, 110, 164). Most cotton agroecosystems have a rich complex of naturally occurring entomophagous arthropods and entomogenous microorganisms. Faunal check lists have shown as many as 600 species of arthropod predators associated with cotton in the higher-rainfall and more ecologically diverse regions of the United States (179); even the lower-rainfall, less diverse regions may have as many as 300 species of predatory and parasitic arthropods in cotton (172).

Some individual insect and spider mite cotton pests are attacked by a diverse group of natural enemies. Forty-two insect and mite species are known to parasitize the boll weevil (40). Ten to fifteen families of predators are important in the natural control of the bollworm-tobacco budworm complex in the United States, and over 40 species of arthropod parasites attack these pests in this country (132).

A large variety of naturally occurring entomogenous viruses, bacteria, protozoa, fungi, rickettsiae, and nematodes also attack cotton insect and spider mite pests (131).

Although seldom included in the analysis of naturally occurring mortality factors and usually ignored except under conditions of epizootics, pathogens are extremely important regulatory agents of some cotton pests. A nuclear polyhedrosis virus of the cabbage looper, for example, may be a major regulator of the pest late in the season (49). Besides causing out-right death, insect pathogens may interfere with the development and reproduction of certain cotton insect pests or increase their susceptibility to other pathogens, predators, parasites, or chemical insecticides (46, 54).

Many species of birds, mammals, reptiles, and amphibians prey on cotton pests; however, they rarely have been considered in recent studies of the analysis of naturally occurring mortality factors. Some very excellent work was done to quantify their importance in the early part of this century. Howell (80), for example, reported 43 species of avian predators of the boll weevil and proposed legislation to protect the more important species.

Although nearly all of the recent articles dealing with cotton pest management have stressed the importance of naturally occurring biological control, there actually is only limited information that precisely quantifies the effectiveness of natural enemies in holding cotton pests species below economic densities. Articles by Ridgway et al (133), Lingren et al (96), Whitcomb & Bell (179), van den Bosch et al (173), and Ehler & van den Bosch (49) are among the relatively few that provide quantitative insight into the absolute efficiency of insect natural enemies. Most evidence of the powerful suppressive or restraining action of insect natural enemies has come from studies that employed insecticides to disrupt this hidden natural control (52, 83); hence it is difficult to make definitive conclusions with regard to the actual or potential absolute value of natural enemies in regulating many pest species of cotton, but the long history of the resurgence and unleashing of primary and secondary pest outbreaks after insecticide treatment of cotton provides proof of their importance.

There are many examples of recently attempted efforts to augment and conserve indigenous natural enemies in cotton. Some of the more promising methods have been: (*a*) applying supplementary predator foods to retain, arrest, attract, and sustain natural enemies when natural prey populations are small or when non-prey food, such as honeydew or cotton pollen, is inadequate for the enemies (72); and (*b*) the selective use of chemical insecticides that are inherently broadly toxic to most natural enemies when applied in a conventional manner. Approaches to seek ecological selectivity in these materials have included using the minimal application dosages essential to produce satisfactory control of the target pests (160), insecticidal treatment restricted to pheromone-baited strip sections of the cotton fields (99), insecticidal control of migrant pests in alternate, non-cotton plant communities (110), and merely the timing of the insecticidal treatments in cotton to minimize damage to the natural enemy populations (5).

Hundreds of examples exist for the last approach. Perhaps the most successful, widely adopted cotton insect control program designed around this principle is the "diapause boll weevil control" approach developed by Brazzel (24). This approach, as opposed to the conventional in-season insecticidal boll weevil control method, utilizes insecticides near the cotton harvest period to control pre-hibernating weevils. If the numbers of diapausing weevils are sufficiently reduced (90–99%), the development of damaging infestations does not occur until late during the subsequent season and after the danger of unleashed *Heliothis* outbreaks has largely passed (6, 11).

The typical diapause boll weevil control described by Brazzel (24), and modifications thereof (11, 141, 142), have been used successfully in several cotton-growing regions of the United States (40). These have great promise on an even wider scale as a way to avert insecticidally induced flareups of crop-damaging populations of *Heliothis* and other potential and secondary pests, which almost invariably occur when insecticides are used against the weevil during the mid-cotton-growing season (143, 167). Nevertheless, diapause boll weevil control does not offer the ultimate panacea for this pest problem and caution must particularly be exercised in timing the late-season insecticidal treatments. These treatments commonly induce transient *Heliothis* outbreaks that can be damaging if the plants have many immature bolls (19, 142).

Treating cotton with commercial formulations of microbial insecticides also offers a form of biological control for many lepidopteran pests (29, 46, 54, 56, 57). However, many problems confront the commercial registration of pathogens. There are other obstacles that also must be overcome if microbial control is to reach its full potential in integrated control programs (131). This method does offer, however, great potential for certain cotton pests.

Although good progress has been made recently in efforts to increase the benefits of naturally occurring biological control agents of the arthropod pests of cotton, the most classic examples of such efforts still come from work conducted during the first part of this century. The monumental paper by Pierce et al (123) on natural enemies of the boll weevil is offered as an example of the sophistication and elegance of this early work.

Cultural Control

Cultural control practices have long been used for suppressing many cotton insect pests and provide the foundation for the successful integrated control of two of the most important pests, the pink bollworm and the boll weevil (4, 7, 109). Early cultural practices included hand-picking eggs, larvae, and infested plant parts, cutting and burning stalks, and destroying the overwintering habitat of such pests as the boll weevil. Hand-picking egg masses plays an important role in controlling the Egyptian cotton leafworm, *Spodoptera littoralis,* in Egypt today, and hand collecting and burning infested bolls and stalks still is employed as one phase of controlling the pink bollworm in China. Among the many cultural practices useful in reducing insect pest problems in cotton are area-wide uniform planting dates, a cotton-free period, early crop maturity and harvest, destruction of infested bolls and alternate hosts, early and uniform stalk destruction, and use of trap crops (4, 131).

In most cases cultural control practices are designed to severely disrupt a pest population with only minor effects on parasites and predators. In the successful program, considerable suppression of the pest species may be achieved without primary pest resurgence or the unleashing of a secondary pest outbreak (6).

The present cultural program for controlling the pink bollworm in Texas provides an example of a highly successful cultural control program (4, 7). The objective of this program is to reduce the overwintering population of pink bollworms to such an extent that infestations of damaging proportions do not develop during the subsequent growing season. This is accomplished by early maturity of the crop, using defoliants or desiccants to cause all bolls to open at nearly the same time to facilitate machine harvesting, early harvesting, early shredding of stalks and plowing under of crop debris on an area-wide basis by all cotton producers, winter and early-spring irrigations in desert areas, and uniform planting of cotton during a designated period timed to allow moths to emerge and die before cotton fruit is available for oviposition. The use of proper ginning techniques and sanitation at the cotton gin site also are important aspects to insure that no overwintering larvae are left alive in stored or waste cottonseed. These practices have reduced the pink bollworm in Texas to the status of a minor pest and insecticides are seldom required for its control (4, 7).

Early stalk destruction provides a powerful tool for controlling the boll weevil by removing food and breeding sites before environmental conditions force the pest into diapause (86–88, 104).

Another type of cultural control may be achieved by manipulating alternate crop hosts of certain key cotton pests, such as *Lygus* spp. in California. The *Lygus* infest both cotton and alfalfa, two of the major crops of the California desert, but prefer alfalfa where they do little damage. When the alfalfa is cut, the *Lygus* bugs then may infest cotton in damaging numbers. If the *Lygus* are not treated with insecticides they may reduce the yields of cotton. If the bugs are treated, insect natural enemies are killed unleashing secondary outbreaks of the bollworm and certain leaf-feeding caterpillars, which then may require season-long insecticide treatments. Since the bugs prefer alfalfa and do little damage to it, they should be kept there if possible. This can be achieved by cutting the alfalfa in alternate strips rather than by cutting the entire field. The adult *Lygus* then move to the uncut alfalfa rather than to the cotton. If the program is properly managed the *Lygus* may be alternatively moved from the cut to the uncut alfalfa throughout the season without becoming a pest in nearby cotton (161).

A second approach to *Lygus* control in California to prevent the unleashing of secondary pest outbreaks in cotton has been the insecticide treatment of an alternate crop host, safflower, to prevent migration of the bugs to cotton (110). In the San Joaquin Valley great numbers of *Lygus* may develop in safflower where they apparently cause little damage. When the safflower nears maturity the *Lygus* may migrate in tremendous numbers to cotton. It is more economical and less damaging to insect natural enemies if the *Lygus* are killed with insecticides applied to the safflower than to cotton. Thus, one or two carefully timed treatments made to the safflower are sufficient to control the *Lygus* and extensive insecticide treatment of cotton may be avoided.

Supportative Tactics

PHEROMONES During the past 10 years a burgeoning literature has been developing on insect pheromones, largely stimulated by the perfection of sophisticated methods for microchemical analysis and by the recognition that the materials may have practical uses in pest control (146). Nevertheless, except in isolated cases, the practicability of pheromones as a suppression tactic remains to be demonstrated. Their tremendous value in pest survey and detection, however, is undisputed. As already noted, pheromone trap records have been used to determine the timing of insecticide treatments for control of the pink bollworm (168). Also, pheromone traps have been extremely valuable in complementing or eliminating conventional cotton pest population survey and detection procedures (16, 40, 79).

There are examples, some rather novel, that demonstrate the potential value of pheromone tactics in suppressing pest populations. An elegant example is the use of gossyplure as a male confusant to suppress mating in the pink bollworm (147). Although the true value of this procedure as a practical control tactic has not been proven, it has shown considerable promise in preliminary field tests.

Pheromone control tactics have been tested more extensively against the boll weevil than with any other cotton insect pest (21, 40). Trap crop systems, employing early planted strips of cotton baited with grandlure to attract boll weevils early in the growing season, combined with insecticidal treatments to destroy the weevils, have shown promise in some regions of the United States (65, 99, 144). These results indicate that these tactics may be used effectively in regions where cotton plants in the trap crops can be managed so as to fruit earlier than the commercial cotton. Early fruiting seems to be necessary for this technique to offer much utility since this procedure has shown very little promise for weevil suppression in Texas where it is difficult to attain desirable earliness in the trap crops.

Some entomologists (23, 73, 98) claim that pheromone traps can be used to remove a high percentage (up to 100%) of a natural population of boll weevils. Others (21, 140) report that these claims lack valid quantitative evidence that pheromone traps have any utility in suppressing the boll weevil. This type of information is essential before pheromone trapping tactics may be recommended with confidence for adoption in integrated programs for control of the boll weevil or any other insect pest.

MISCELLANEOUS TACTICS Numerous miscellaneous control tactics, some showing great potential against certain cotton insect pests, have been explored in recent years. For example, hundreds of growth regulators (158) have been examined for use in insect control; however, none presently is used commercially against cotton insects. Growth regulators, although not currently developed to the point of practical use in cotton insect control programs, have many desirable attributes needed for the future. Other tactics such as light traps (76) or artificial lighting systems (114), although potentially effective against some pests, are not very practical on the whole and offer only minimal utility.

ERADICATION AS AN ALTERNATIVE

A discussion of pest eradication or elimination is beyond the scope of this article, but we should note that some entomologists believe this is a realistic alternative for certain key cotton insect pests. They argue logically that elimination of serious key pests, such as the boll weevil, would offer the surest means for permitting adoption of integrated control for secondary pests and for minimizing insecticidally induced outbreaks of potential pests. As technological skills advance in pest control science, eradication may emerge as the preferred strategy for dealing with some of the most serious cotton insect pests. However, there presently are many countermanding technological and operational problems that must be overcome before eradication of any of these pests becomes a realistic goal (112, 113).

Several recent articles have argued the pros and cons of eradication of the boll weevil from the United States (31, 40, 51, 60). A committee from the National Academy of Science seriously questions whether current technology is sufficient to achieve this goal (113).

IMPLEMENTATION OF PEST MANAGEMENT SYSTEMS

The past decade has produced many examples of the numerous advantages of controlling agricultural crop pests via integrated management strategies based on sound ecological principles. These systems promise to allow farmers to produce crops more profitably while lessening the problems of environmental pollution and human health hazards inherent to the insecticidal control programs that previously have prevailed. Moreover, more important today than a few years ago, these new pest control programs promise to greatly reduce the fossil fuel energy required for cotton production. This results from less dependence on petroleum-derived insecticides and a reduction in fuel required by tractor-driven sprayers, aircraft, and other equipment used in applying the chemicals. Examples of the potential value of some recent integrated programs for cotton in maximizing farmer profits, reducing the insecticide load in the environment, and conserving fossil-fuel energy have been reported by Larson et al (91), Casey et al (32), and Sprott et al (157). However, despite the many obvious potential benefits of adopting the new integrated systems described by these authors, integrated control has not really been successfully implemented in much of the world's cotton-growing regions. It is interesting that for the most part integrated control has been adopted only in areas that have witnessed a disaster, i.e. Peru and Texas, where conventional insecticide control programs have failed, or in countries such as China where it has been implemented by governmental decree.

Major efforts have been made in recent years to devise plans to make integrated control more attractive to cotton growers. The agricultural extension services in the United States recently initiated a large-scale pilot demonstration program intended to transfer the most current integrated control technology to the farmer-use level (185). This program has succeeded in greatly reducing the amount of insecticide used in cotton within the demonstration area, although actual farmer use of integrated programs is currently only minimal. The reluctance of farmers to implement such programs poses a major problem, and the final remedy may require a major overhaul of the present system of agricultural extension, legislation, and commercial pest control.

Glass (66) recently reviewed the problem of future needs of implementation. Townsend (169) outlined extensional, legislative, and farmer organizational needs that he believed necessary to maintain a long-term boll weevil suppression program. Townsend's article appeared in 1894; Glass's appeared in 1975. The basic underlying philosophy presented is identical in both articles.

MODELING AND AGROECOSYSTEM ANALYSIS

Though the terms modeling and systems analysis have only recently found their way into pest management literature, the underlying concepts have always been a part of pest control (36). Traditionally, entomologists have attempted to abstract and collate information from literature and from their own observations to build a model, whether mental or physical, of the whole system surrounding the target pest or groups of pests (35). Early workers with the boll weevil were very successful in developing a conceptual model of this pest's complex life system and the interacting variables affecting its abundance. Articles by Hunter & Hinds (88) and Pierce et al (123) offer proof of the sophisticated view these workers had of the boll weevil's life system and the environmental factors influencing generation-to-generation and year-to-year changes in population density.

Although not equipped with computers, these early entomologists employed a form of systems analysis. Because there were no effective insecticides, they were forced to study collectively the action, reactions, and interactions of the boll weevil's life system as a whole. This was the only way they could make maximum use of the natural suppressive forces that were regulating the density of the pest. System analysis, as defined by Watt (176), is merely a body of techniques and theories for analyzing complex problems, viewed as systems of interlocking cause-effect pathways. These are the same kind of pathways that early entomologists sought in their fight against the boll weevil.

We strongly suspect that the modern thrust toward modeling and systems analysis, as related to cotton insect research and control, might have come of age 50 years earlier if electronic computers had surfaced then rather than calcium arsenate. The emergence of calcium aresenate, however, largely eliminated the need for the computer and a continuation of the systems approach to boll weevil control. As Fye (61) pointed out, the "sledge-hammer" approach of insecticidal control naturally led to less sophistication, less complexity and, therefore, less need for computers or systems analysis.

The recent trend back to more ecologically oriented pest control in cotton has created a need for the sophisticated analytical and synthesis techniques provided by mathematical models, systems analysis, and their companion methodologies. Experimental design, data collection, and decision-making processes have become increasingly complex and have created the need for more than empirical methodologies, therefore methods of systems analysis are being widely adopted at many different levels in recent research programs that have taken a holistic approach to cotton pest problem solving.

Computer population models have been developed in conjunction with studies to determine the influence of cotton plant phenology on pest population increase under different environmental conditions (43, 44, 180) and to quantify pest migration (165) or pheromone trap response (89). The value of these computer models for pest management systems still must be validated; however, numerous benefits have emerged from models in terms of unifying and guiding research and gaining a clearer understanding of the various interactions in the system.

Articles by Conway (35), Conway & Murdie (36), Ruesink (137, 138) and Witz (182) provide good discussions of the different types of models, theories and techniques of systems analysis and recent developments in the general field of pest management modeling. Other articles deal specifically with developments in the integrated control of the insect pests of cotton (43, 61, 120, 154).

The long-term goal of several of the efforts underway for cotton agroecosystem modeling is the development of a computer model that realistically simulates and represents all occurrences that take place in a particular cotton field, or that field and the associated agricultural and non-agricultural surroundings at any given time from planting through harvest. The apparent utility of such a model (120) is that it could be used in simulation studies to determine how the cotton ecosystem might be manipulated, i.e. use of a particular combination of cotton plant varieties, fertilizer, insect control practices, etc, to achieve optimal yield. Models must be built that yield the established probability of a whole spectrum of possible outcomes and estimate the accuracy (realism) of the answers that the models supply (43). Although the value of a realistic simulation model for the cotton agroecosystem is undisputed, such a model does not currently exist. However, progress has been made recently toward developing cotton growth (177) and crop yield (71) models, models for improving the timing and efficiency of insecticide treatments (81, 166), and several other models of the interlocking natural and man-controlled cotton agroecosystem components (77, 89, 175, 180).

Research and extension specialists in several cotton regions of the United States are exploring the use of on-line computer delivery systems, such as those used in apple pest management programs in Michigan (38). Such systems have great immediate potential in cotton pest management programs in spite of the present lack of realistic crop-insect management models in cotton. Algorithms have been developed for these systems that provide instant information of present and predicted pest abundance, weather patterns, crop variety performance, etc, upon command by the farmer or his extension specialist. These systems should greatly strengthen the farmer's confidence in the pest control decisions he must make. They should also facilitate the adoption of more advanced integrated strategies for cotton production and pest control that will optimize yields and maximize producer profits.

Acknowledgments

This work was supported in part by the National Science Foundation and the Environmental Protection Agency, through a grant (NSF GB-34718) to the University of California. The findings, opinions, and recommendations expressed herein are those of the authors and not necessarily those of the University of California, the National Science Foundation, or the Environmental Protection Agency.

We extend our sincere thanks to the many colleagues who provided reprints, copies of manuscripts, and other material pertaining to the review subject.

Literature Cited

1. Adkisson, P. L. 1965. Present status of insecticide resistance in certain geographical populations of bollworms in Texas. *Tex. Agric. Exp. Stn. PR 2358* 5 pp.
2. Adkisson, P. L. 1969. How insects damage crops. In *How Crops Grow—A Century Later. Conn. Agric. Exp. Stn. Bull.* 708:155–64
3. Adkisson, P. L. 1971. Objective uses of insecticides in agriculture. *Proc. Symp. Agric. Chem.-Harmony Discord Food People Environ.* pp. 43–51
4. Adkisson, P. L. 1972. Use of cultural practices in insect pest management. In *Implementing Practical Pest Management Strategies. Proc. Natl. Ext. Insect-Pest Manage. Workshop,* pp. 37–50
5. Adkisson, P. L. 1973. The principles, strategies, and tactics of pest control in cotton. In *Insects: Studies in Population Management,* ed. P. W. Geier, L. R. Clark, D. J. Anderson, H. A. Nix, pp. 274–83
6. Adkisson, P. L. 1973. The integrated control of the insect pests of cotton. *Proc. Tall Timbers Conf. Ecol. Anim. Control Habitat Manage.* 4:175–88
7. Adkisson, P. L., Gaines, J. C. 1960. Pink bollworm control as related to the total cotton insect control program of central Texas. *Texas Agric. Exp. Stn. Misc. Publ. 444* 7 pp.
8. Adkisson, P. L., Nemec, S. J. 1965. Efficiency of certain insecticides for killing bollworms and tobacco budworms. *Tex. Agric. Exp. Stn. PR 2357* 11 pp.
9. Adkisson, P. L., Nemec, S. J. 1966. Comparative effectiveness of certain insecticides for killing bollworms and tobacco budworms. *Tex. Agric. Exp. Stn. Bull. 1048* 4 pp.
10. Adkisson, P. L., Nemec, S. J. 1967. Insecticides for controlling the bollworm, tobacco budworm and boll weevil. *Tex. Agric. Exp. Stn. Misc. Publ. 837* 7 pp.
11. Adkisson, P. L., Rummel, D. R., Sterling, W. L., Owen, W. L. Jr. 1966. Diapause boll weevil control: a comparison of two methods. *Texas Agric. Exp. Stn. Misc. Publ. 1054* 11 pp.
12. Allen, J., Gonzalez, D., Gokhale, D. V. 1972. Sequential sampling plans for the bollworm, *Heliothis zea. Environ. Entomol.* 1:771–80
13. Baldwin, J. L., Walker, J. K., Gannaway, J. R., Niles, G. A. 1974. Bollworm attack on experimental semi-dwarf cottons. *Texas Agric. Exp. Stn. Bull. 1144* 12 pp.
14. Baldwin, J. L., Walker, J. K., Gannaway, J. R., Niles, G. A. 1974. Semi-dwarf cottons and bollworm attack. *J. Econ. Entomol.* 67:779–82
15. Barducci, T. B. 1971. Ecological consequences of pesticides used for the control of cotton insects in Cañete Valley, Peru. In *The Careless Technology—Ecology and International Development,* ed. M. T. Farran, J. P. Milton, pp. 423–38. New York: Natl. Hist.
16. Bariola, L. A., Keller, J. C., Turley, D. C., Farris, J. R. 1973. Migration and population studies of the pink bollworm

in the arid west. *Environ. Entomol.* 2:205–07
17. Berger, J. 1969. *The World's Major Fiber Crops: Their Cultivation and Manuring.* Zurich: Conzett and Huber. 294 pp.
18. Bird, L. S., Bourland, F. M., Hood, J. E., Bush, D. L. 1974. Multi-disease adversity resistant okra leaf and frego bract cottons. *Proc. 1974 Beltwide Cotton Prod. Res. Conf.* 27:95–97
19. Bottrell, D. G., Almand, L. K. 1969. The effects of reproductive-diapause boll weevil control programs on populations of the bollworm and the tobacco budworm in cotton, 1968. *Tex. Agric. Exp. Stn. PR 2702* 6 pp.
20. Bottrell, D. G., Huffaker, C. B., ed. 1974. Evaluation of the role of predators in crop ecosystems. *Integrated Pest Manage. Workshop Rep. New Orleans, Feb.* 32 pp.
21. Bottrell, D. G., Rummel, D. R. 1976. Suppression of boll weevil populations with pheromone traps. In *Detection and Management of the Boll Weevil with Pheromone. Tex. Agric. Exp. Stn. Res. Monogr. No. 8.* pp. 37–44
22. Boyer, W. P., Bell, R. 1961. The relationship of spider mite infestations in cotton to early season use of insecticides. *J. Kans. Entomol. Soc.* 34:132–34
23. Boyd, F. J. Jr., Brazzel, J. R., Helms, W. F., Moritz, R. J., Edwards, R. R. 1973. Spring destruction of overwintered boll weevils in West Texas with wing traps. *J. Econ. Entomol.* 66: 507–10
24. Brazzel, J. R. 1959. The effect of late-season applications of insecticides on diapausing boll weevils. *J. Econ. Entomol.* 52:1042–45
25. Brazzel, J. R. 1961. Boll weevil resistance to insecticides in Texas in 1960. *Tex. Agric. Exp. Stn. PR 2171* 4 pp.
26. Brazzel, J. R. 1963. Resistance to DDT in *Heliothis virescens. J. Econ. Entomol.* 56:571–74
27. Brazzel, J. R. 1964. DDT resistance in *Heliothis zea. J. Econ. Entomol.* 57:455–57
28. Brown, H. B., Ware, J. O. 1958. *Cotton.* New York: McGraw-Hill. 566 pp.
29. Burgess, H. D., Hussey, N. W. (ed.) 1971. *Microbial Control of Insects and Mites.* New York: Academic. 861 pp.
30. Canerday, T. D. 1974. Response of bollworm and tobacco budworm in Georgia to methyl parathion. *J. Econ. Entomol.* 67:299
31. Carter, L. J. 1974. Eradicating the boll weevil: would it be a no-win battle? *Science* 183:494–99
32. Casey, J. E., Lacewell, R. D., Sterling, W. L. 1974. Economic and environmental implications of cotton production under a new cotton pest management system. *Tex. Agric. Exp. Stn. Misc. Publ. 1152.* 19 pp.
33. Chiarappa, L., ed. 1971. Crop loss assessment methods: FAO manual on the evaluation and prevention of losses by pests, disease and weeds. Farmham Royal, Slough, England: CAB 162 pp.
34. Coad, B. R., Cassidy, T. P. 1920. Cotton boll weevil control by the use of poison. *US Dep. Agric. Bull. 875* 31 pp.
35. Conway, G. R. 1973. Experience in insect pest modeling: a review of models, uses and future directions. In *Insects: Studies in Population Management,* ed. P. W. Geier, L. R. Clark, D. J. Anderson, H. A. Nix, pp. 103–30
36. Conway, G. R., Murdie, G. 1972. Population models as a basis for pest control. In *Mathematical Models in Ecology,* ed. J. N. R. Jeffers, pp. 195–213. Oxford, England: Blackwell
37. Cope, O. B. 1971. Interactions between pesticides and wildlife. *Ann. Rev. Entomol.* 16:325–64
38. Croft, B. A. 1975. Tree fruit pest management. In *Introduction to Pest Management,* ed. R. L. Metcalf, W. Luckmann, pp. 471–507. New York: Wiley. 587 pp.
39. Croft, B. A., Brown, A. W. A. 1975. Response of arthropod natural enemies to insecticides. *Ann. Rev. Entomol.* 20:285–335
40. Cross, W. H. 1973. Biology, control, and eradication of the boll weevil. *Ann. Rev. Entomol.* 18:17–46
41. Curl, L. F., White, R. W. 1952. The pink bollworm. *US Dep. Agric. Yearb. Agric.* pp. 505–11
42. DeBach, P. 1974. *Biological Control by Natural Enemies.* Cambridge: Cambridge Univ. 323 pp.
43. De Michele, D. W., Bottrell, D. G. 1976. Systems approach to cotton insect pest management. In *Integrated Pest Management,* ed. J. L. Apple, R. F. Smith, pp. 107–32. New York: Plenum
44. De Michele, D. W., Curry, G. L., Sharpe, P. H. G., Barfield, C. S. 1976. Cotton square drying. 1. A theoretical model. *Environ. Entomol.* In press
45. Doutt, R. L., Smith, R. F. 1971. The pesticide syndrome. In *Biological Con-*

trol, ed. C. B. Huffaker, pp. 3–15. New York: Plenum. 511 pp.
46. Dulmage, H. T. 1973. Assay and standardization of microbial insecticides. *Ann. NY Acad. Sci.* 217:187–99
47. Durham, W. F., Williams, C. H. 1972. Mutagenic, teratogenic, and carcinogenic properties of pesticides. *Ann. Rev. Entomol.* 17:123–48
48. Ehler, L. E., Eveleens, K. G., van den Bosch, R. 1973. An evaluation of some natural enemies of cabbage looper on cotton in California. *Environ. Entomol.* 2:1009–15
49. Ehler, L. E., van den Bosch, R. 1974. An analysis of the natural biological control of *Trichoplusia ni* on cotton in California. *Can. Entomol.* 106:1067–73
50. Elliot, F. C., Hoover, M., Porter, W. K. Jr. 1968. *Advances in Production and Utilization of Quality Cotton: Principles and Practices.* Ames: Iowa State Univ. 532 pp.
51. Entomological Society of America 1973. The pilot boll weevil eradication experiment. *Bull. Entomol. Soc. Am.* 19:218–21
52. Eveleens, K. G., van den Bosch, R., Ehler, L. E. 1973. Secondary outbreak induction of beet armyworm by experimental insecticide applications in California. *Environ. Entomol.* 2:497–503
53. Ewing, K. P. 1952. The bollworm. *US Dep. Agric. Yearb. Agric.* pp. 511–14
54. Falcon, L. A. 1971. Microbial control as a tool in integrated control programs. In *Biological Control,* ed. C. B. Huffaker, pp. 346–64. New York: Plenum. 511 pp.
55. Falcon, L. A. 1971. Progreso del control integrado en el algodón de Nicaragua. *Rev. Peru. Entomol. Agric.* 14: 376–78
56. Falcon, L. A. 1973. Biological factors that affect the success of microbial insecticides: development of integrated control. *Ann. NY Acad. Sci.* 217:173–86
57. Falcon, L. A. 1974. Insect pathogens: integration into a pest management system. *Proc. Summer Inst. Biol. Control Plant Insects Dis.* Jackson: Univ. Miss. pp. 618–27
58. Falcon, L. A., Smith, R. F. 1973. *Guidelines for Integrated Control of Cotton Insect Pests.* FAO AGPP: Misc. 8. 92 pp.
59. FAO. 1970. Pest resistance to pesticides in agriculture. Importance, recognition, and countermeasures. AGP/CP/26. Rome: UN/FAO. 32 pp.
60. Feltner, R. L. 1975. Boll weevil eradication programs and plans. *Proc. Beltwide Cotton Prod.-Mech. Conf. Natl. Cotton Counc.* pp. 14–15
61. Fye, R. E. 1974. Modeling of cotton insect populations. *US Dep. Agric. West. Reg. ARS W-14* 6 pp.
62. Gaines, J. C. 1957. Cotton insects and their control in the United States. *Ann. Rev. Entomol.* 2:319–38
63. Gaines, R. C. 1952. The boll weevil. *US Dep. Agric. Yearb. Agric.* pp. 501–04
64. Gaines, R. C. 1954. Effect on beneficial insects of several insecticides applied for cotton insect control. *J. Econ. Entomol.* 47:543–44
65. Gilliland, F. R. Jr., Lambert, W. R., Weeks, J. R., Davis, R. L. 1973. A pest management system for cotton insect pest suppression. *Ala. Agric. Exp. Stn. PR Ser. 105* 6 pp.
66. Glass, E. H. 1975. Integrated pest management: rationale, potential, needs and implementation. *Entomol. Soc. Am. Spec. Publ. 75–72* 141 pp.
67. Gonzalez, D. 1970. Sampling as a basis for pest management strategies. *Proc. Tall Timbers Conf. Ecol. Anim. Control Habitat Manage.* 2:83–101
68. Graves, J. B., Roussel, J. S., Phillips, J. R. 1963. Resistance to some chlorinated hydrocarbon insecticides in the bollworm, *Heliothis zea. J. Econ. Entomol.* 56:442–44
69. Graves, J. B., Clower, D. F., Bradley, J. R. 1967. Resistance of the tobacco budworm to several insecticides in Louisiana. *J. Econ. Entomol.* 60:887–88
70. Graves, J. B., Clower, D. F. 1975. Status of resistance in *Heliothis* species. *Proc. Beltwide Cotton Prod. Res. Conf. Natl. Cotton Counc.* pp. 142–5
71. Gutierrez, A. P., Falcon, L. A., Loew, W., Leipzig, P. A., van den Bosch, R. 1975. An analysis of cotton production in California: a model for acala cotton and the effects of defoliators on its yields. *Environ. Entomol.* 4:125–36
72. Hagen, K. S., Hale, R. 1974. Increasing natural enemies through use of supplementary feeding and non-target prey. *Proc. Summer Inst. Biol. Control Plant Insects Dis.* Jackson: Univ. Miss. pp. 170–181.
73. Hardee, D. D., Lindig, O. H., Davich, T. B. 1971. Suppression of populations of boll weevils over a large area in West Texas with pheromone traps in 1969. *J. Econ. Entomol.* 64:928–33
74. Harp, S. J. 1974. Field tests for control of *Heliothis* on cotton, Hillsboro,

Texas. *Tex. Agric. Exp. Stn. Dep. Entomol. Tech. Rep. No. 22.* pp. 13–15
75. Harris, F. A., Graves, J. B., Nemec, S. J., Vinson, S. B., Wolfenbarger, D. A. 1972. Insecticide resistance. In *Distribution, Abundance and Control of Heliothis Species in Cotton and Other Host Plants. Southern Coop. Ser. Bull. 196* pp. 17–27
76. Hartstack, A. W., Hollingsworth, J. P., Ridgway, R. L., Hunt, H. H. 1971. Determination of trap spacings required to control an insect population. *J. Econ. Entomol.* 64:1090–100
77. Hartstack, A. W., Witz, J. A., Ridgway, R. L. 1975. Suggested applications of a dynamic *Heliothis* model (MOTHZV-1) in pest management decision making. *Proc. Beltwide Cotton Prod. Res. Conf. Natl. Cotton Counc.* pp. 118–22
78. Headley, J. C. 1972. Defining the economic threshold. In *Pest Control Strategies for the Future.* Washington DC: NAS. pp. 100–08
79. Hendricks, D. E., Graham, H. M., Bureea, R. J., Perez, C. T. 1973. Comparison of the numbers of tobacco budworms and bollworms caught in sex pheromone traps versus black-light traps in Lower Rio Grande Valley of Texas. *Environ. Entomol.* 2:911–14
80. Howell, A. H. 1907. The relation of birds to the cotton boll weevil. *US Dep. Agric. Bur. Biol. Survey Bull. 29* 30 pp.
81. Hueth, D., Regev, U. 1974. Optimal agricultural pest management with increasing pest resistance. *Am. J. Agric. Econ. Aug.* pp. 764–73
82. Huffaker, C. B., ed. 1971. *Biological Control.* New York: Plenum. 511 pp.
83. Huffaker, C. B. 1971. The ecology of pesticide interference with insect populations. *Proc. Symp. Agric. Chem.—Harmony Discord Food, People Environ.* pp. 92–104
84. Huffaker, C. B. 1972. Ecological management of pest systems. In *Challenging Biological Problems: Directions Toward Their Solution,* ed. J. A. Behnke. New York: Oxford Univ. pp. 313–42
85. Huffaker, C. B., Stinner, R. E. 1971. The role of natural enemies in pest control programs. In *Entomological Essays to Commemorate the Retirement of Professor K. Yasumatsu.* Tokyo: Hokuryukan. pp. 333–50
86. Hunter, W. D. 1904. The most important step in the cultural system of controlling the boll weevil. *US Dep. Agric. Bur. Entomol. Circ. 56.* 7 pp.
87. Hunter, W. D. 1911. The boll weevil problem with special reference to means of reducing damage. *US Dep. Agric. Farmers Bull.* 344:25–26
88. Hunter, W. D., Hinds, W. E. 1904. The Mexican cotton boll weevil. *US Dep. Agric. Div. Entomol. Bull. 45.* 116 pp.
89. Jones, J. W., Thompson, A. C., McKinion, J. M. 1975. Developing a computer model with various control methods for eradication of boll weevils. *Proc. Beltwide Cotton Prod. Res. Conf. Natl. Cotton Counc.* 118 pp.
90. King, W. V. 1952. Mosquitoes and DDT. *US Dep. Agric. Yearb. Agric.* pp. 327–30
91. Larson, J. L., Lacewell, R. D., Casey, J. E., Heilman, M. D., Namken, L. N., Parker, R. D. 1975. Economic, environmental and energy use implications of short-season cotton production: Texas Lower Rio Grande Valley. *So. J. Agric. Econ.* pp. 171–77
92. Leigh, T. F. 1975. Insect resistance in cotton—what for the future. *Proc. Beltwide Cotton Prod. Res. Conf. Natl. Cotton Counc.* pp. 140–41
93. Lincoln, C., Phillips, J. R., Whitcomb, W. H., Dowell, G. C., Boyer, W. P., Bell, K. O. Jr., Dean, G. L., Matthews, E. J., Graves, J. B., Newsom, L. D., Clower, D. F., Bradley, J. R. Jr., Bagent, J. L. 1967. The bollworm-tobacco budworm problem in Arkansas and Louisiana. *Ark. Agric. Exp. Stn. Bull. 720.* 66 pp.
94. Lincoln, C., Boyer, W. P., Dowell, G. C., Barnes, G., Dean, G. 1970. Six years experience with point-sample cotton insect scouting. *Ark. Agric. Exp. Stn. Bull. 754.* 40 pp.
95. Lincoln, C., Boyer, W. P., Miner, F. D. 1975. The evolution of insect pest management in cotton and soybeans: past experience, present status, and future outlook in Arkansas. *Environ. Entomol.* 4:1–7
96. Lingren, P. D., Ridgway, R. L., Cowan, C. B. Jr., Davis, J. W., Watkins, W. C. 1968. Biological control of the bollworm and the tobacco budworm by arthropod predators affected by insecticides. *J. Econ. Entomol.* 61:1521–25
97. Little, V. A., Martin, D. F. 1942. *Cotton Insects of the United States.* Minneapolis: Burgess. 130 pp.
98. Lloyd, E. P., Merkl, M. E., Tingle, F. C., Scott, W. P., Hardee, D. D., Davich, T. B. 1972. Evaluation of male-baited traps for control of boll weevils following a reproduction-diapause program in

Monroe County, Mississippi. *J. Econ. Entomol.* 65:552–55
99. Lloyd, E. P., Scott, W. P., Shaunak, K. K., Tingle, F. C., Davich, T. B. 1972. A modified trapping system for suppressing low-density populations of overwintered boll weevils. *J. Econ. Entomol.* 65:1144–47
100. Lukefahr, M. J. 1970. The tobacco budworm situation in the Lower Rio Grande Valley and northern Mexico. *Proc. Second Ann. Tex. Conf. Insects Plant Dis. Weed Brush Control* College Station: Texas A&M Univ. pp. 140–45
101. Lukefahr, M. J., Houghtaling, J. E. 1969. Resistance of cotton strains with high gossypol content to *Heliothis* spp. *J. Econ. Entomol.* 62:588–91
102. Lukefahr, M. J., Houghtaling, J. E., Graham, H. M. 1971. Suppression of *Heliothis* populations with glabrous cotton strains. *J. Econ. Entomol.* 64:486–88
103. Malley, F. W. 1901. The Mexican cotton-boll weevil. *US Dep. Agric. Farmers Bull. 130.* 29 pp.
104. Maxwell, F. G., Adkisson, P. L. 1976. Host plant resistance to cotton insects. *Proc. US/USSR Symp. Integrated Control Arthropod Dis. Weed Pests Cotton Sorghum Decidious Fruits Tex. Agric. Exp. Stn. Misc. Publ. 1276.* pp. 143–51
105. Maxwell, F. G., Harris, F. A. 1974. *Proceedings of the Summer Institute on Biological Control of Plant Insects and Diseases.* Jackson: Univ. Miss. 647 pp.
106. Maxwell, F. G., Jenkins, J. N., Parrott, W. L. 1972. Resistance of plants to insects. *Adv. Agron.* 24:187–265
107. Menzie, C. M. 1972. Fate of pesticides in the environment. *Ann. Rev. Entomol.* 17:199–222
108. Metcalf, R. L. 1975. Insecticides in pest management. In *Introduction to Insect Pest Management,* ed. R. L. Metcalf and W. Luckmann, pp. 235–73. New York: Wiley
109. Mueller, A. J., Sharma, R. K., Reynolds, H. T., Toscano, N. C. 1974. Effect of crop rotations on emergence of overwintered pink bollworm populations in the Imperial Valley, California. *J. Econ. Entomol.* 67:227–28
110. Mueller, A. J., Stern, V. M. 1974. Timing of pesticide treatments on safflower to prevent *Lygus* from dispersing to cotton. *J. Econ. Entomol.* 67:77–80
111. NAS. 1969. *Principles of plant and animal pest control: Insect-Pest Management and Control. Natl. Acad. Sci. Publ. 1695(3)* 508 pp.
112. NAS. 1976. *Pest Control: An Assessment of Present and Alternative Technologies.* Vol. 1. *Contemporary Pest Control Practices and Prospects.* Washington DC: Natl. Acad. Sci. 506 pp.
113. NAS. 1976. *Pest Control: An Assessment of Present and Alternative Technologies.* Vol. 3. *Cotton Pest Control.* Washington DC: Natl. Acad. Sci. 139 pp.
114. Nemec, S. J. 1970. Artificial lighting for bollworm control. *Cotton Gin Oil Mill Press.* 71:10
115. Newsom, L. D. 1967. Consequences of insecticide use on nontarget organisms. *Ann. Rev. Entomol.* 12:257–86
116. Newsom, L. D. 1970. The end of an era and future prospects for insect control. *Proc. Tall Timbers Conf. Ecol. Anim. Control Habitat Manage.* 2: 117–36
117. Newsom, L. D. 1974. Predator-insecticide relationships. *Entomophaga Mem. Ser. 7* 88 pp.
118. Newsom, L. D., Brazzel, J. R. 1968. Pests and their control. In *Advances in production and Utilization of Quality Cotton: Principles and Practices,* ed. F. C. Elliot, M. Hoover, W. K. Porter, Jr. Ames: Iowa State Univ. 532 pp.
119. Newsom, L. D., Smith, C. E. 1949. Destruction of certain insect predators by applications of insecticides to control cotton pests. *J. Econ. Entomol.* 42: 904–8
120. Parvin, D. W. Jr., Tyner, F. H. 1974. The systems approach—research or research management. *So. J. Agric. Econ. July* pp. 258–66
121. Pate, T. L., Hefner, J. J., Neeb, C. W. 1972. A cotton management program to reduce cost of insect control in the Pecos area. *Tex. Agric. Exp. Stn. Misc. Publ. 1023* 7 pp.
122. Pearson, E. O., Maxwell-Darling, R. C. 1958. *The Insect Pests of Cotton in Tropical Africa.* London: Eastern. 355 pp.
123. Pierce, W. D., Cushman, R. A., Hood, C. E. 1912. The insect enemies of the cotton boll weevil. *US Dep. Agric. Bur. Entomol. Bull. 100.* 99 pp.
124. Pieters, E. P., Sterling, W. L. 1974. A sequential sampling plan for the cotton fleahopper, *Pseudatomoscelis seriatus. Environ. Entomol.* 3:101–6
125. Pimentel, D. 1973. Extent of pesticide use, food supply, and pollution. *J. NY Entomol. Soc.* 81:13–33
126. Porter, B. A. 1952. Insects are harder to kill. *US Dep. Agric. Yearb. Agric.* pp. 317–20

127. Quaintance, A. L., Brues, C. T. 1905. The cotton bollworm. *US Dep. Agric. Bull. 50* 155 pp.
128. Rabb, R. L. 1970. Introduction to the conference. In *Concepts of Pest Management,* ed. R. L. Rabb, F. E. Guthrie. Raleigh: NC State Univ. pp. 1–5
129. Rainwater, C. F. 1952. Progress in research on cotton insects. *US Dep. Agric. Yearb. Agric.* pp. 497–500
130. Reynolds, H. T. 1971. Recent developments with systemic insecticides for insect control on cotton. *Proc. West. Cotton Prod. Conf. Natl. Cotton Counc.* pp. 18–20
131. Reynolds, H. T., Adkisson, P. L., Smith, R. F. 1975. Cotton insect pest management. In *Introduction to Pest Management,* ed. R. L. Metclaf, W. Luckmann. New York: Wiley. pp. 397–443
132. Ridgway, R. L., Lingren, P. D. 1972. Predaceous and parasitic arthropods as regulators of *Heliothis* populations. In *Distribution, Abundance and Control of Heliothis Species in Cotton and Other Host Plants. South. Coop. Ser. Bull. 169* pp. 48–56
133. Ridgway, R. L., Lingren, P. D., Cowan, C. B. Jr., Davis, J. W. 1967. Populations of arthropod predators and *Heliothis* spp. after applications of systemic insecticides to cotton. *J. Econ. Entomol.* 60:1012–16
134. Ripper, W. E. 1956. Effect of pesticides on balance of arthropod populations. *Ann. Rev. Entomol.* 1:403–38
135. Ripper, W. E., George, L. 1965. *Cotton Pests of the Sudan: Their Habitats and Control.* Oxford: Blackwell. 345 pp.
136. Roussel, J. S., Clower, D. F. 1955. Resistance to the chlorinated hydrocarbon insecticides in the boll weevil (*Anthonomus grandis Boh.*). *Louisiana Agric. Exp. Stn. Circ. 41.* 9 pp.
137. Ruesink, W. G. 1975. Analysis and modeling in pest management. In *Introduction to Pest Management,* ed. R. L. Metcalf, W. H. Luckmann. New York: Wiley. pp. 353–76
138. Ruesink, W. G. 1976. Status of the systems approach to pest management. *Ann. Rev. Entomol.* 21:27–44
139. Ruesink, W. G., Kogan, M. 1975. The quantitative basis of pest management: sampling and measuring. In *Introduction to Insect Pest Management,* ed. R. L. Metcalf and W. Luckmann. New York: Wiley. pp. 309–49
140. Rummel, D. R., Bottrell, D. G. 1976. Seasonally related decline in response of boll weevils to pheromone traps during mid-season. *Environ. Entomol.* 5: 783–87
141. Rummel, D. R., Bottrell, D. G., Adkisson, P. L., McIntyre, R. C. 1975. An appraisal of a 10-year effort to prevent the westward spread of the boll weevil. *Bull. Entomol. Soc. Am.* 21:6–11
142. Rummel, D. R., Jordan, L. B. 1973. A rescheduled reproductive-diapause control program for maximum suppression of potential overwintering boll weevils in the Rolling Plains of Texas. *Tex. Agric. Exp. Stn. PR.-3210* 11 pp.
143. Rummel, D. R., Reeves, R. E. 1971. Response of bollworm and predaceous arthropod populations to aldicarb treatments in cotton. *J. Econ. Entomol.* 64:907–11
144. Scott, W. P., Lloyd, E. P., Bryson, J. O., Davich, T. B. 1974. Trap plots for suppression of low density overwintered populations of boll weevils. *J. Econ. Entomol.* 67:281–83
145. Sevacherian, V., Stern, V. M. 1972. Sequential sampling plans for lygus bugs in California cotton fields. *Environ. Entomol.* 1:704–10
146. Shorey, H. H. 1973. Behavioral response to insect pheromones. *Ann. Rev. Entomol.* 18:349–80
147. Shorey, H. H., Kaae, R. S., Gaston, L. K. 1974. Sex pheromones of Lepidoptera. Development of a method for pheromonal control of *Pectinophora gossypiella* in cotton. *J. Econ. Entomol.* 67:347–50
148. Smith, R. F. 1969. Patterns of crop protection in cotton ecosystems. *Proc. Cotton Symp. Insect and Mite Control Problems Calif.* Berkeley: Univ. Calif. pp. 5–11
149. Smith, R. F. 1969. The importance of economic injury levels in the development of integrated pest control programs. *Qual. Plant. Mater. Veg. XVII:* 81–92
150. Smith, R. F. 1970. Pesticides: their use and limitations in pest management. In *Concepts of Pest Management,* ed. R. L. Rabb, F. G. Guthrie. Raleigh: NC State Univ. pp. 103–18
151. Smith, R. F. 1971. Fases en el desarrollo del control integrado. *Bol. Soc. Entomol. Peru* 6:54–56
152. Smith, R. F. 1971. Economic aspects of pest control. *Proc. Tall Timbers Conf. Ecol. Anim. Control. Habitat Manage.* 3:53–83
153. Smith, R. F., Apple, J. L., Bottrell, D. G. 1976. The origins of integrated pest management concepts for agricultural

crops. In *Integrated Pest Management,* ed. J. L. Apple, R. F. Smith, pp. 1–16. New York: Plenum

154. Smith, R. F., Huffaker, C. B., Adkisson, P. L., Newsom, L. D. 1974. Progress achieved in the implementation of integrated control projects in the USA and tropical countries. *EPPO Bull.* 4:221–39
155. Smith, R. F., Reynolds, H. T. 1972. Effects of manipulations of cotton agroecosystems on insect pest populations. In *The Careless Technology-Ecology and International Development,* ed. M. T. Farvar, J. P. Milton. New York: Natural History. pp. 373–406
156. Smith, R. F., van den Bosch, R. 1967. Integrated control. In *Pest Control: Biological, Physical, and Selected Chemical Methods,* ed. W. W. Kilgore, R. L. Doutt. New York: Academic. pp. 295–340
157. Sprott, J. M., Lacewell, R. D., Niles, G. A., Walker, J. K., Gannaway, J. R. 1976. Agronomic, economic, energy use and environmental implications of short-season, narrow-row cotton production: Frio County, Texas. *Tex. Agric. Exp. Stn. Misc. Publ. 1250* 23 pp.
158. Staal, G. B. 1975. Insect growth regulators with juvenile hormone activity. *Ann. Rev. Entomol.* 20:417–60
159. Starbird, I. R., French, B. L. 1972. Costs of producing upland cotton in the United States, 1969. *US Dep. Agric. Econ. Res. Serv. Agric. Econ. Rep. 229* 47 pp.
160. Sterling, W., Haney, R. L. 1973. Cotton yields climb, costs drop through pest management systems. *Tex. Agric. Prog.* 19:4–7
161. Stern, V. M. 1969. Interplanting alfalfa in cotton to control lygus bugs and other insect pests. *Proc. Tall Timbers Conf. Ecol. Anim. Control Habitat Manage.* 1:55–69
162. Stern, V. M. 1973. Economic thresholds. *Ann. Rev. Entomol.* 18:259–80
163. Stern, V. M., Smith, R. F., van den Bosch, R., Hagen, K. S. 1959. The integrated control concept. *Hilgardia* 29:81–101
164. Stinner, R. E. 1977. Efficacy of inundative releases. *Ann. Rev. Entomol.* 22: 515–31
165. Stinner, R. E., Rabb, R. L., Bradley, J. R. 1974. Population dynamics of *Heliothis zea* and *H. virescens* in North Carolina: a simulation model. *Environ. Entomol.* 3:163–68
166. Talpaz, H., Borosh, I. 1974. Strategy for pesticide use: frequency and applications. *Am. J. Agric. Econ. Nov.* pp. 738–44
167. Timmons, F. D., Brook, T. S., Harris, F. A. 1973. Effects of aldicarb applied side-dress to cotton on some arthropods in the Monroe County, Mississippi, boll weevil diapause-control area in 1969. *J. Econ. Entomol.* 66(1):151–53
168. Toscano, N. C., Mueller, A. J., Sevacherian, V., Sharma, R. K., Niilus, T., Reynolds, H. T. 1974. Insecticide applications based on hexalure® trap catches versus automatic schedule treatments for pink bollworm moth control. *J. Econ. Entomol.* 67:522–24
169. Townsend, C. H. T. 1894. Report on the Mexican cotton-boll weevil in Texas. *US Dep. Agric. Div. Entomol. Insect Life* 7:295–309
170. van den Bosch, R. 1971. Biological control of insects. *Ann. Rev. Ecol. Syst.* 2:45–66
171. van den Bosch, R., Leigh, T. F., Falcon, L. A., Stern, V. M., Gonzalez, D., Hagen, K. S. 1971. The developing program of integrated control of cotton pests in California. In *Biological Control,* ed. C. B. Huffaker. New York: Plenum. pp. 377–94
172. van den Bosch, R., Hagen, K. S. 1966. Predaceous and parasitic arthropods in California cotton fields. *Calif. Agric. Exp. Stn. Bull. 820* 32 pp.
173. van den Bosch, R., Leigh, T. F., Gonzalez, D., Stinner, R. E. 1969. Cage studies on predators of the bollworm in cotton. *J. Econ. Entomol.* 62:1486–89
174. Walker, J. K., Jr., Niles, G. A. 1971. Population dynamics of the boll weevil on modified cotton types: implications for pest management. *Tex. Agric. Exp. Stn. Bull.* 1109. 14 pp.
175. Wanjura, D. F., Colwick, R. F., Jones, J. W. 1975. Status of cotton-production-system modeling in regional research project S-69. *Proc. Beltwide Cotton Prod. Res. Conf. Natl. Cotton Counc.* pp. 159–61
176. Watt, K. E. F. 1966. The nature of systems analysis. In *Systems Analysis in Ecology,* ed. K. E. F. Watt. New York: Academic. pp. 1–14
177. Weaver, R. E. C., Law, V. J., Bailey, R. V. 1975. Bases for the use of single plant models in field studies. *Proc. Beltwide Cotton Prod. Res. Conf. Natl. Cotton Counc.* pp. 168–70
178. Whitcomb, W. H. 1970. History of integrated control as practiced in the cotton

fields of the south central United States. *Proc. Tall Timbers Conf. Ecol. Anim. Control Habitat Manage.* 2:147–55

179. Whitcomb, W. H., Bell, K. 1964. Predaceous insects, spiders, and mites of Arkansas cotton fields. *Ark. Agric. Exp. Stn. Bull. 690* 84 pp.
180. Wilson, A. G. L., Hughes, R. D., Gilbert, N. 1972. The response of cotton to pest attack. *Bull. Entomol. Res.* 61: 405–14
181. Wiseman, J. S., Smith, D. A. 1971. Epidemiology of pesticide poisoning in the Lower Rio Grande Valley in 1969. *Tex. Med.* 67:56–59
182. Witz, J. A. 1973. Integration of systems science methodology and scientific research. *Agric. Sci. Rev.* 11:37–48
183. Wolfenbarger, D. A., Lukefahr, M. J., Graham, H. M. 1973. LD^{50} values of methyl parathion and endrin to tobacco budworms and bollworms collected in the Americas and hypothesis on the spread of resistance in these lepidopterans to these insecticides. *J. Econ. Entomol.* 66:211–16
184. *The World Almanac and Book of Facts.* 1976. Newspaper Enterprise Association, Inc. 984 pp.
185. Young, D. F. Jr. 1975. An assessment of the three-year cooperative cotton pest management program. *Proc. Beltwide Cotton Prod. Res. Conf. Natl. Cotton Counc.* pp. 124–27

Ann. Rev. Entomol. 1977. 22:483–513

DEGRADATION OF ORGANOPHOSPHORUS AND CARBAMATE INSECTICIDES IN THE SOIL AND BY SOIL MICROORGANISMS[1]

❖6139

James Laveglia and Paul A. Dahm
Department of Entomology, Iowa State University, Ames, Iowa 50011

INTRODUCTION

There are many reviews dealing with pesticide (including insecticide) degradation in soil and metabolism by microorganisms (3, 10, 33, 37, 55, 58, 62, 67, 83–85, 88, 100, 102, 109, 130). Of all the insecticides included in these reviews, organochlorine insecticides have received the most attention (3, 10, 37, 55, 56, 58, 62, 83–85, 88, 100, 102, 109) because they have been used longer and more extensively than organophosphorus and carbamate insecticides. The most comprehensive review on the behavior of pesticides in soils has been prepared by Helling, Kearney & Alexander (58); they discuss physicochemical and metabolic processes affecting pesticides, the effect of pesticides on the soil microbiota, and the implications of these processes on persistence, bioactivity, and plant uptake. Edwards (37) and Harris (56) describe numerous processes and factors that affect the persistence of insecticides in soil, Bollag (10) reviews the biochemical transformation of pesticides by soil fungi, and Matsumura (84, 85) reviews the degradation of pesticides with emphasis on environmental significance. Matsumura (83) and Matsumura & Boush (88) have reviewed examples of insecticide metabolism by microorganisms.

In this review we limited our coverage to the growing number of papers dealing with the degradation of organophosphorus and carbamate insecticides in the soil and by soil microorganisms. We include both field and laboratory experiments and

[1]Journal Paper No. J-8427 of the Iowa Agriculture and Home Economics Experiment Station, Ames, Iowa. Project No. 2121. The survey of literature for this review was completed in December 1975.

information on soil types, insecticide concentrations, and microorganisms. Commercial insecticide formulations are designated whenever they were used; otherwise technical or analytical grade insecticides were used.[2] Insecticides are referred to by their common names (68). Some of the papers reviewed give no recovery rates for insecticides added in the experiments described. However, we have given insecticide dosages in pounds of active ingredient per acre whenever this information was available.

ORGANOPHOSPHORUS INSECTICIDES

Parathion

DEGRADATION IN SOIL Lichtenstein & Schulz (79) found that 90 days after applying parathion to Carrington silt loam plots at 5 lb acre^{-1} 3.1% (0.1 ppm) of the dosage remained. Under laboratory conditions 30% of the parathion was lost after 12 days. Kasting & Woodward (64) found no parathion present after 16 days when applied at 2 lb acre^{-1} to a clay-loam soil. At 12 lb acre^{-1} it took 79 days for the parathion to disappear completely and when 100 lb acre^{-1} was applied, 1.2 ppm (2.4%) remained 165 days after treatment. Iwata, Westlake & Gunther (61) studied parathion degradation in six different California soil types. When applied at 20 ppm, the insecticide degraded rapidly in a loamy sand, and 30 days later 0.2 ppm remained. In a silt loam, a clay, and a sandy loam, the parathion level dropped to between 1 and 2 ppm in 30 days and gradually decreased to about 0.2 ppm in 130 days. In a loam soil, the residue after 8 months remained above 3 ppm.

When higher concentrations of parathion were applied to soil, persistence increased (23, 132). Carlo, Ashdown & Heller (23) applied 200 ppm of insecticide to Stephenville fine sandy loam, and at the end of 8 weeks about 33% remained in the soil. The most rapid degradation took place during the first 2 weeks when 45–55% of the parathion was lost. Wolfe et al (132), simulating accidental spillage, treated field plots with undiluted commercial grade 45.6% liquid emulsifiable concentrate (30,000 to 95,000 ppm). The levels of parathion were considerably lower by the end of the first year; during the second year the rate of disappearance was much less. Appreciable quantities of parathion remained in both the top 1 inch and the 1- to 3-inch levels at the end of year 5, with the lowest concentration at 13,000 ppm.

[2]Insecticide nomenclature: aldicarb – Ambush®, Temik®; azinphosmethyl – Gusathion®, Guthion®; carbaryl – Sevin®; carbofuran – Curaterr®, Furadan®; carbophenothion – Acarithion®, Garrathion®, Trithion®; chlorfenvinphos – Birlane®, Sapecron®, Supona®; diazinon – Diazinon®, Spectracide®; dichlorvos – Herkol®, Linden®, Nogos®, Vapona®; dicrotophos – Bidrin®, Carbicron®; dimethoate – Cygon®, De-Fend®, Perfekthion®, Rogor®; disulfoton – Di-Syston®, Frumin AL®, Solvirex®; fenitrothion – Accothion®, Folithion®, Sumithion®; fensulfothion – Dasanit®, Terracur-P®; fonofos – Dyfonate®; malathion – Cythion®; mecarbam – Afos®, Murfotox®, Pestan®; methidathion – Supracide®, Ultracide®; methyl parathion – Bladan M®, Folidol M®, Metacide®, Wofotox®; mevinphos – Phosdrin®; parathion – Alkron®, Bladan®, Folidol E605®, Niran®; phorate – Thimet®; phosmet – Imidan®, Prolate®; terbufos – Counter®; thionazin – Nemaphos®, Zinophos®.

Stewart, Chisholm & Ragab (119) studied the long-term persistence of parathion. Field plots received annual spring applications of 31.4 lb acre^{-1} for 5 years. Sixteen years after the last treatment, about 0.1% of the total parathion applied to the plots remained. It was suggested that the insecticide might be dissolved in lipids of the soil organic matter and thus be protected from bacterial degradation and hydrolysis.

Chopra & Khullar (28) found that the degradation of parathion increased with greater insecticide concentration, higher incubation temperature, longer exposure to ultraviolet light, higher relative humidity, longer incubation, and higher pH. Carlo, Ashdown & Heller (23) also reported that the recovery of parathion from basic soil was 20–30% less than from acid or neutral soil.

Soil sterilization also influenced parathion degradation. The insecticide persisted longer in sterilized soils than in unsterilized soils (47, 79, 108, 113). Lichtenstein & Schulz (79) incubated parathion in autoclaved (15 psi, 120°C for 7 hr daily for 5 days) and nonautoclaved soil. After 6 days, about 85% of the insecticide remained in the autoclaved soil, whereas only 54% was present in the nonautoclaved soil. Sethunathan & Yoshida (113) studied the persistence of parathion in autoclaved (121°C for 1 hr daily for 3 days) flooded soil and nonautoclaved flooded soil. Parathion disappeared more rapidly from the nonautoclaved soil samples. Half-life values for parathion in nonautoclaved samples of four soil types were 6.2, 12.3, 1.6, and 9.0 days; the corresponding half-life values in the same four soils autoclaved were 72.5, 68.0, 37.7, and 80.2 days. After a 2-week incubation period, Getzin & Rosefield (47) recovered twice as much parathion in autoclaved (1 hr at 15 psi three times at weekly intervals) soil as in nonautoclaved soil. Sethunathan (108) also investigated the effect of autoclaving (15 psi for 1 hr for 3 days) flooded acid sulfate soil on the persistence of parathion. After 28 days of incubation, 61.7% of the insecticide was recovered from the autoclaved soil, whereas only 9.5% remained when the soil was not autoclaved.

Chemical sterilization of soil by sodium azide also decreased the degradation of parathion (77). Adding the surfactants alkyl benzene sulfonate and linear alkyl benzene sulfonate to insecticide-treated soil increased the persistence of parathion. Two months after the insecticide and surfactant applications, 13 times more parathion was recovered from these soils than from soils that had only been treated with insecticide (76).

Metabolism of parathion in soil follows two pathways, hydrolysis to *p*-nitrophenol and diethyl thiophosphoric acid and reduction to aminoparathion (50, 79). Lichtenstein & Schulz (79) found *p*-nitrophenol but no aminoparathion when parathion was added to autoclaved soils. Iwata, Westlake & Gunther (61) detected no aminoparathion when a sandy loam was fortified with 200 ppm of parathion and sampled every 3 to 4 days for 30 days. When the same soil was flooded with water, aminoparathion was detected. Sethunathan & Yoshida (113) studied parathion degradation in submerged rice soils. Most of the parathion added to these soils rapidly decomposed, and 14 days after incubation most of the parent insecticide had disappeared. The only metabolite found under these anaerobic conditions was aminoparathion. In contrast, during the 2-week incubation period no appreciable degradation of the insecticide occurred under aerobic conditions.

Lichtenstein & Schulz (79) studied the persistence of some parathion metabolites (paraoxon, aminoparathion, *p*-aminophenol, and *p*-nitrophenol) when applied at 20 ppm to the soil. Paraoxon was metabolized rapidly; one-half the applied dosage disappeared in 5.5 hr. The major metabolite was *p*-nitrophenol. Most of the aminoparathion disappeared 1 day after treatment and none was detected after an additional 3 days of soil incubation. No residues of *p*-aminophenol were detected; *p*-nitrophenol was present in the soil 7 days after application. When the 16-day soil sample was analyzed, no *p*-nitrophenol was recovered. Griffiths & Walker (51) isolated a *Pseudomonas* from soil that was responsible for the decomposition of *p*-nitrophenol. When parathion was exposed to this organism, no metabolism occurred.

DEGRADATION BY MICROORGANISMS Sethunathan & Yoshida (113) tested the influence of a *Flavobacterium* sp. on parathion degradation. When this bacterium was added to submerged soils containing parathion, the insecticide was degraded faster than in uninoculated soil. The half-life for parathion in the inoculated soil was 0.76 day and in the uninoculated soil 3.5 days. The only metabolite found in the inoculated soil was *p*-nitrophenol; the major metabolite in the uninoculated soil was aminoparathion. These results suggest that at least one *Flavobacterium* sp. has a high capacity to hydrolyze parathion and may be useful in decontaminating parathion-polluted environments or containers. Lichtenstein & Schulz (79) added glucose (14 ml of a 0.1 M solution) to parathion-treated soil and 1.5 g of dried yeast to parathion-treated soil water and found that the insecticide disappeared faster than in soil not treated with glucose or soil water not treated with yeast. The major metabolite detected was aminoparathion.

An alga, *Chlorella pyrenoidosa,* metabolized parathion (82, 135). Zuckerman et al (135), after adding parathion to algal cultures and incubating for 7 days, found that the principal compound within the cells was parathion, whereas the major one in the medium was aminoparathion. It was suggested that aminoparathion is formed within the cells, and once formed it is rapidly released. Mackiewicz et al (82) grew bean plants under aseptic conditions and exposed the roots to either parathion or parathion and *C. pyrenoidosa.* After 7 days, extracting the plants exposed only to the insecticide showed parathion exclusively. The plants exposed to parathion and *C. pyrenoidosa* contained not only parathion but also aminoparathion and an unidentified sulfur-containing metabolite. It was concluded that parathion metabolites found in plants may be a product of microbial metabolism rather than metabolism by the plant.

A fungus, *Penicillium waksmani,* isolated from an acid sulfate soil under flooded conditions degraded parathion (103). After incubation for 14 days, the fungus reduced the insecticide from 2305–425 μg. Aminoparathion was formed as well as two unidentified polar metabolites. Matsumura & Boush (87) studied parathion degradation by a soil fungus, *Trichoderma viride.* After 24 hr of incubation, 22% of the applied insecticide (10^{-6} M) was converted to water-soluble metabolites. Solvent-soluble metabolites accounted for 3% of the original dose.

Munnecke & Hsieh (96) studied the degradation of parathion when exposed to a mixed microbial culture (*Pseudomonas, Xanthomonas, Azotomonas,* and *Brevibacterium*). The culture degraded 50 mg of parathion $liter^{-1}$ hr^{-1}. The major metabolite was *p*-nitrophenol. When a pseudomonad, isolated from the mixed culture, was exposed to *p*-nitrophenol, nitrite and hydroquinone were produced. Hsieh & Munnecke (60) hypothesized complete conversion of the carbon atoms in *p*-nitrophenol and diethyl thiophosphoric acid to CO_2 by microorganisms. The two ethyl groups may be oxidized and metabolized through the glyoxylic acid cycle, whereas the phenol may undergo oxidative cleavage of the aromatic ring, followed by breakdown through the β-ketoadipic acid cycle. This complete breakdown of parathion hypothesis was supported by degradation studies of ^{14}C-ring-labeled and ^{14}C-ethyl-labeled parathion. In both cases, $^{14}CO_2$ was produced. Siddaramappa, Rajaram & Sethunathan (115) isolated two bacteria, *Bacillus* sp. and *Pseudomonas* sp., from flooded soil. *Pseudomonas* sp. hydrolyzed parathion and then released nitrite from the *p*-nitrophenol, but *Bacillus* sp. was unable to hydrolyze parathion.

Boush & Matsumura (16) showed that parathion was degraded by *Pseudomonas melophthora,* the bacterial symbiote of the apple maggot, *Rhagoletis pomonella.* After a 24-hr incubation period, 79.6% of the original insecticide remained, whereas 15.5% was recovered as water-soluble metabolites and 4.8% as solvent-soluble metabolites. Hirakoso (59) studied the effects of eight bacteria on the insecticidal activity of parathion (presumably at a concentration of 20 ppm). There was complete loss of insecticidal activity caused by *Pseudomonas aeruginosa* and *Escherichia freundii.* Parathion activity was greatly reduced by five of the remaining six bacteria, four of which were in the genus *Pseudomonas.*

Yasuno et al (133) found *Bacillus subtilis* quite effective in reducing the insecticidal activity of parathion. When 1 ppm of the insecticide was incubated with the bacteria, complete inactivation took place in 2 days; at 20 ppm, 4 days were needed before complete inactivation occurred. The major metabolic route of parathion in the presence of *B. subtilis* was reduction to aminoparathion. Singh (116) isolated *Bacillus cereus* from a coreid bug and found that it degraded 820 μg of parathion in 6 hr. Mick & Dahm (92) found that 85% of the initial concentration of parathion was reduced to aminoparathion by two species of *Rhizobium*; 10% was hydrolyzed to diethyl thiophosphoric acid.

Rapid reduction of parathion occurred in bovine rumen fluid (2, 30). Cook (30) reported that aminoparathion was formed principally within the first 30 min of incubation. Ahmed, Casida & Nichols (2) found that the yield of aminoparathion in rumen fluid varied with the concentration of parathion. After 2 hr of incubation with 2 ppm of insecticide, 62% was recovered as aminoparathion. When 300 ppm were used, 25% was recovered in 2.7 hr, and with 500 ppm only 53% of aminoparathion was recovered after 24 hr. Ahmed, Casida & Nichols (2) also found that paraoxon was reduced to aminoparaoxon by bovine rumen fluid.

Methyl Parathion and Fenitrothion

DEGRADATION IN SOIL Lichtenstein & Schulz (79) treated a Carrington silt loam with 200 ppm of methyl parathion and found that the insecticide was degraded rapidly. After 3 days of incubation at 30°C, 26% of the applied methyl parathion was recovered. Only 6% of the insecticide remained 12 days after treatment. In field trials, methyl parathion also disappeared rapidly: 15 days after application at 5 lb acre^{-1} about 8% remained in the soil, and after 30 days only 3.1% of the applied dosage was recovered.

DEGRADATION BY MICROORGANISMS Miyamoto, Kitagawa & Sato (93) studied the degradation of fenitrothion by *B. subtilis.* The major metabolite, accounting for 65% of the added insecticide, was aminofenitrothion; other minor metabolites found were dimethyl thiophosphoric acid and desmethyl fenitrothion. The bacteria degraded aminofenitrothion slower than the parent compound, and desmethyl aminofenitrothion was identified as a metabolite. No reduction of desmethyl fenitrothion to desmethyl aminofenitrothion was detected, and dimethyl thiophosphoric acid was not formed from aminofenitrothion.

Methyl parathion was metabolized the same way fenitrothion was but twice as fast (93). After 36 hr, only 0.9% of the added (2×10^{-4} M) methyl parathion remained, whereas 0.8% of the added (2×10^{-4} M) fenitrothion was present 72 hr after incubation.

Hirakoso (59) studied the effect of eight bacteria on the insecticidal activity of fenitrothion against mosquito larvae. Two species in the genus *Pseudomonas* and one in the genus *Escherichia* greatly reduced the toxicity of the insecticide after 4 days of incubation. Complete loss of insecticidal activity was caused by *E. freundii.* Yasuno et al (133) found that *B. subtilis* inactivated the insecticidal effect of fenitrothion. When 20 ppm of insecticide were added to culture medium containing *B. subtilis,* 40% remained 1 day later; after 4 days less than 7% was present; and only 1% was detected after 8 days of incubation. When fenitrothion was added to *B. subtilis* in the presence of various antibiotics, full insecticidal toxicity was retained.

Malathion

DEGRADATION IN SOIL Degradation of malathion in soil is quite rapid. Lichtenstein & Schulz (79) found that after the application of malathion at 5 lb acre^{-1} only 15% of the applied dosage was recovered 3 days later; after 1 week, 95% of the applied malathion had disappeared. Chopra & Girdhar (27) studied the influence of time, temperature, and pH on the degradation of malathion. In Ludhiana, Palampur, and Kamma soils, malathion was degraded 100% in 4 days; higher temperatures produced more rapid degradation. At pH 3.9 and 12.1, 80 and 90%, respectively, of the malathion were degraded in 24 hr. At pH 7.2 only 50% was degraded.

There is some discrepancy between reports about the degradative process of malathion (72, 127). Konrad, Chesters & Armstrong (72) reported that rates of malathion degradation in soils were related directly to the extent of malathion adsorption, suggesting that degradation occurred by a chemical mechanism that was catalyzed by adsorption. Soil sterilization by electron beam irradiation (5×10^6 rads) had little effect on the degradation rates, and it was concluded that malathion degradation was nonbiological. Information also was obtained on the pathways and products of malathion degradation. Hydrolysis did not occur in acid systems and was slow at pH 9 and rapid at pH 11. At pH 9, hydrolysis resulted in the formation of thiomalic acid and dimethyl phosphorothioate (Figure 1) as final products, with accumulation of diethyl thiomalate as an intermediate owing to unequal rates of hydrolysis of the ester linkages.

Walker & Stojanovic (127) also found that malathion was quite stable under neutral or acid pH conditions and that susceptibility to hydrolysis increased with increasing alkalinity. In contrast to the study by Konrad, Chesters & Armstrong (72), Walker & Stojanovic (127) demonstrated that malathion disappearance was much more rapid under nonsterile than under sterile conditions (autoclaving twice for 15 min at 15 psi at 121°C) and indicated that malathion disappearance is stimulated by the various microbiological systems in the soil. Under sterile conditions, the observed malathion remaining in Trinity loam, Okolona clay, and Free-

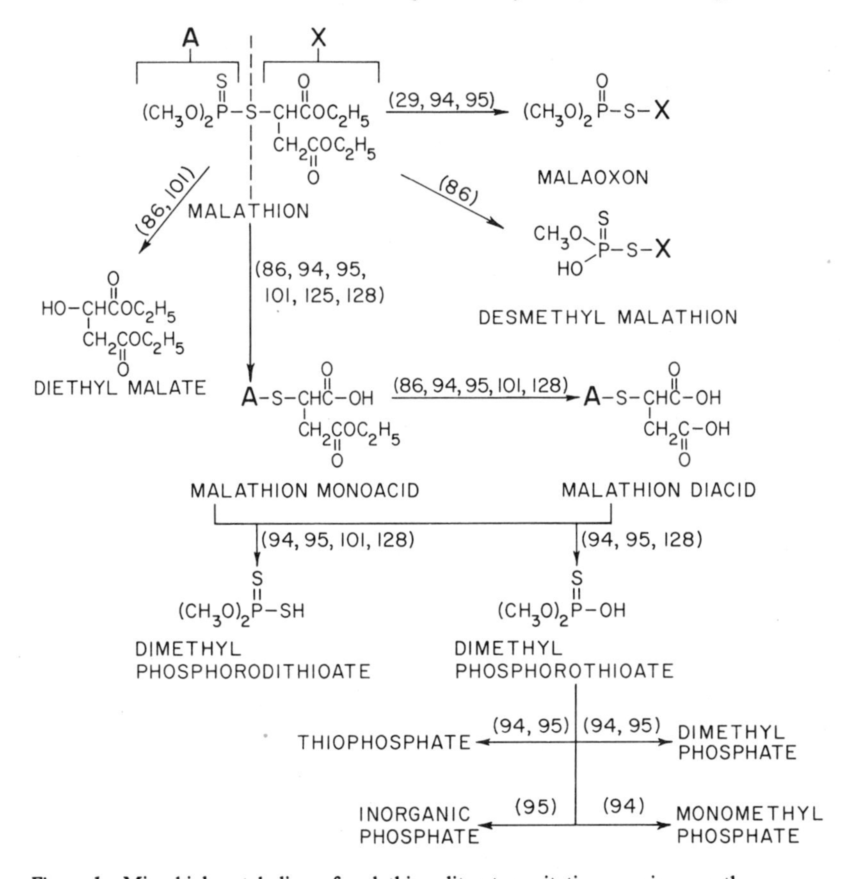

Figure 1 Microbial metabolism of malathion; literature citations are in parentheses.

stone sandy loam after 10 days of incubation was 91, 77, and 95%, respectively. In nonsterile soil, complete disappearance had occurred.

Although these two studies (72, 127) give conflicting data in regard to the ability of sterile and nonsterile soils to degrade malathion, it must be realized that two different methods were used to sterilize the soil. In one instance (72), electron beam irradiation (5 $\times$ 10^6 rads) was used; in the other (127), soil was autoclaved for 15 min at 15 psi at 121°C. Both methods sterilize the soil, but when irradiation is used the heat-labile enzymes are not destroyed. This was probably the reason Konrad, Chesters & Armstrong (72) found malathion degradation occurring in sterilized soil.

Getzin and Rosefield (47) studied the degradation of malathion in nonsterile, heat-sterilized (1 hr at 15 psi three times at weekly intervals), and gamma radiation-sterilized (4-mrad dose at a rate of 250,000 rads hr^{-1}) soil. After 1 day, 97% of the malathion was degraded in the nonsterile soil, 90% in the irradiated soil, and only 7% in the autoclaved soil. Because of the rapid degradation in nonsterile soil, it was suggested that microorganisms were partly responsible for degradation of the compounds. The difference between the amount of malathion decomposed in irradiated soil and autoclaved soil was attributed to nonviable, heat-labile substances. A stable, heat-labile enzyme, which catalyzed the hydrolysis of malathion to its monoacid, was partially purified by $MnCl_2$ treatment, $(NH_4)_2SO_4$ precipitation, dialysis, and ion-exchange chromatography (48). This esterase was extracted from irradiation-sterilized soil as well as from nonirradiated soil. When partly purified enzyme was applied to the soil, it possessed many of the characteristics necessary for prolonged activity in a cell-free state in the soil. It was less heat-labile than most enzymes, lost little or no activity upon prolonged storage, and survived desiccation in soils.

DEGRADATION BY MICROORGANISMS There have been many reports of microorganisms metabolizing malathion to various end products (29, 86, 94, 95, 101, 125, 128; Table 1). Carboxyesterase activity, which degrades malathion to its monoacid and diacid, is the predominant degradative pathway (86, 94, 95, 101, 125, 128). Phosphatase activity also has been reported (86, 94, 95, 101, 128); oxidative desulfuration and demethylation seem to be rather minor metabolic routes (29, 86, 94, 95). A scheme for microbial metabolism of malathion is shown in Figure 1.

Diazinon

DEGRADATION IN SOIL In the soil, diazinon has a half-life of between 2 and 8 weeks depending on the soil conditions (17, 46, 110, 121). Sethunathan & MacRae (110) found that 2–6% of the applied diazinon (50 ppm) remained 7–10 weeks after application, and Getzin & Rosefield (46) reported that less than 8% was present 20 weeks after applying 5 μg of diazinon cm^{-3} of soil. The major metabolic route for diazinon in the soil is hydrolysis to 2-isopropyl-4-methyl-6-hydroxypyrimidine (41, 71, 74, 112), followed by cleavage of the cyclic moiety with formation of CO_2 (41, 46, 74, 110, 111, 112). No detectable amounts of diazoxon (17, 41, 112), sulfotepp (41), or 2-isopropyl-4-methyl-6-mercaptopyrimidine (41) were formed. Laanio, Dupuis & Esser (74) also found a small quantity of 2-(1'-hydroxy-1'-methyl)ethyl-4-methyl-6-hydroxypyrimidine present after treating a rice-paddy soil system with

Table 1 Formation of malathion metabolites by various microorganisms

Metabolite	Microorganism	Ref.
Malaoxon	*Aspergillus niger*	94
	Chlorella pyrenoidosa	29
	Penicillium notatum	94
	Rhizobium trifolii	95
	Rhizoctonia solani	94
Inorganic phosphate	*Rhizobium leguminosarum*	95
	Rhizobium trifolii	95
Thiophosphate	*Aspergillus niger*	94
	Penicillium notatum	94
	Rhizobium leguminosarum	95
	Rhizobium trifolii	95
Monomethyl phosphate	*Aspergillus niger*	94
	Penicillium notatum	94
Dimethyl phosphate	*Aspergillus niger*	94
	Penicillium notatum	94
	Rhizobium leguminosarum	95
	Rhizobium trifolii	95
Dimethyl phosphorothioate	*Arthrobacter* sp.	128
	Aspergillus niger	94
	Penicillium notatum	94
	Rhizobium leguminosarum	95
	Rhizobium trifolii	95
Dimethyl phosphorodithioate	*Arthrobacter* sp.	128
	Aspergillus niger	94
	Penicillium notatum	94
	Rhizobium leguminosarum	95
	Rhizobium trifolii	95
Malathion diacid	*Arthrobacter* sp.	128
	Aspergillus niger	94
	Penicillium notatum	94
	Pseudomonas sp.	86
	Rhizobium leguminosarum	95
	Rhizobium trifolii	95
	Trichoderma viride	86
Malathion monoacid	*Arthrobacter* sp.	128
	Aspergillus niger	94
	Penicillium notatum	94
	Pseudomonas sp.	86, 125
	Rhizobium leguminosarum	95
	Rhizobium trifolii	95
	Trichoderma viride	86
Diethyl malate	*Pseudomonas* sp.	86
	Trichoderma viride	86
Desmethyl malathion	*Pseudomonas* sp.	86
	Trichoderma viride	86

^{14}C-ring-labeled diazinon. Tentative identification of hydroxydiazinon, in low concentrations, was also reported but not confirmed.

Konrad, Armstrong & Chesters (71) used both ^{14}C-ring-labeled and ^{14}C-ethoxy-labeled diazinon and found that hydrolysis occurred forming 2-isopropyl-4-methyl-6-hydroxypyrimidine and diethyl thiophosphoric acid. Because of the rapid rate of chemical hydrolysis and the seemingly slow adaptation of diazinon-degrading microorganisms, it was proposed that diazinon degradation in soils was a chemical process and hydrolysis a possible mechanism. To confirm the occurrence of diazinon degradation, the rate of liberation of diazinon degradation products into solution was determined. Benzene-extractable ^{14}C was attributed to intact diazinon. A measure of accumulation of degradation products was obtained by subtracting the benzene-extractable ^{14}C activity from the total ^{14}C activity. The total ^{14}C activity added to the soil decreased initially because of adsorption of diazinon, but it increased on prolonged incubation as the degradation products of diazinon were liberated into solution. This rapid adsorption, followed by degradation at the adsorption sites and release of the degradation products into solution, suggested that hydrolysis was controlled by adsorption and not by pH.

Bro-Rasmussen, Noddegaard & Voldum-Clausen (17) studied the disappearance of diazinon in a loam and a sandy loam, at 25 and 75% of water capacity, with two different diazinon concentrations (8.5 and 0.85 lb acre^{-1}), and in steam-treated (100°C for 1 hr) and untreated soils. The factor that had the most influence on diazinon degradation was the steam treatment. Soil type and moisture level also were significant factors contributing to diazinon disappearance.

In a preliminary study, Getzin & Rosefield (47) found that the amount of diazinon degraded was approximately the same in nonsterile, autoclaved (1 hr at 15 psi three times at weekly intervals), irradiated (4-mrad dose at a rate of 250,000 rads hr^{-1}), and irradiated and autoclaved soil. In a later study, Getzin (42) again found that degradation rates for diazinon were similar in autoclaved and nonautoclaved soil. Lowering the soil pH from 8.1 to 4.3 accelerated the nonbiological degradation of diazinon. Higher temperatures (from 15°–25°C) and soil moisture levels (from 2–30%) also increased insecticide decomposition. He concluded that diazinon was primarily degraded through nonbiological pathways.

Lichtenstein (76) studied the effect of two surfactants, alkyl benzene sulfonate and linear alkyl benzene sulfonate, on diazinon persistence. Without surfactants, only one-third of the applied insecticidal dosage (10 ppm) was recovered 15 days after soil treatment; the presence of either surfactant, however, almost doubled the persistence of the insecticide. This effect was thought to be caused by the polar (hydrophilic) portion of the surfactant molecule that induced a stronger binding to soil colloids of the slightly water-soluble insecticide. These results were confirmed in a later study (77), which also reported on the effects of sodium azide (0.05% of soil weight) and autoclaving (6 hr day^{-1} for 5 days) on diazinon in soil. Although autoclaving the soils reduced the numbers of microorganisms even more than sodium azide, the persistence of diazinon was not appreciably affected by soil autoclaving. Sodium azide caused the complete disappearance of diazinon. The effect of sodium azide on the stability of diazinon in soil seemingly is not directly related to microbiological activity.

Some reports have shown increased diazinon decomposition in unsterilized soil compared with sterilized soil (97, 110, 112). Sethunathan & MacRae (110) studied the persistence of diazinon in Maahas clay, Luisiana clay, and Pila clay loam. In two of these soils, the insecticide disappeared more rapidly from the unsterilized than from the sterilized samples. Luisiana clay is an acid soil, and diazinon decomposed more rapidly in the sterilized soil samples. These findings suggest that in some soils microorganisms play an important role in diazinon degradation, whereas in other soils, especially acidic soils, nonbiological factors are the most important.

When unsterilized soil samples were left in contact with aqueous diazinon, a brownish material a few millimeters thick was present (97, 112). This layer was not formed in sterilized soil. Nasim, Baig & Lord (97) cultured this material and found that it decomposed diazinon quite readily.

The breakdown of the pyrimidine ring of the diazinon molecule, or its hydrolysis products, to CO_2 is microbial in nature (41, 110, 112). Getzin (41) reported that after 20 weeks 35% of the ^{14}C from ring-labeled diazinon was released as $^{14}CO_2$. In soil fumigated with 0.1 ml of propylene oxide, larger amounts of 2-isopropyl-4-methyl-6-hydroxy-pyrmidine were recovered than in nonfumigated soil. Conversely, little $^{14}CO_2$ was produced in the fumigated soil, whereas large amounts were released from nonfumigated soil. When 0.25 mg of ^{14}C-ring-labeled diazinon was applied to flooded soil only small quantities of $^{14}CO_2$ were liberated (110). It was suggested that the anaerobic nature of the flooded soil was not conducive to oxidation. Sethunathan & Yoshida (112) found that nonsterilized soil released more $^{14}CO_2$ than sterilized soil when ^{14}C-ring-labeled diazinon was applied to submerged soil.

DEGRADATION BY MICROORGANISMS Williams et al (131) demonstrated that ^{14}C-ethoxy-labeled diazinon uptake occurred quite rapidly with both *Entodinium simplex* and *Isotricha intestinalis* obtained from the ruminal ingesta of calves.

Butler, Deason & O'Kelley (22) studied the loss of 1 ppm of diazinon from actively growing cultures of 21 isolates of planktonic algae. After a 2-week incubation period, insecticide recovery ranged from 62–96% depending on the isolate. Boush & Matsumura (16) found 71% of the added diazinon (10^{-6} M) remaining after incubation for 24 hr at 30°C with *Pseudomonas melophthora.* Of the metabolites recovered, 26.7% were water soluble whereas only 2.4% were solvent soluble. A soil fungus, *Trichoderma viride,* also was capable of degrading diazinon (87). In 24 hr at 30°C, 24.4% of the applied insecticide (10^{-6} M) was recovered as water-soluble metabolites and 0.1% as solvent-soluble metabolites.

Yasuno et al (133) added 20 ppm of diazinon to *B. subtilis* culture media and at various time intervals studied the insecticidal effects of the media against mosquito larvae. Diazinon retained its toxicity throughout the 16-day incubation period. Hirakoso (59) found that only two species of bacteria (*Escherichia coli* and *Pseudomonas fluorescens*) out of the eight tested reduced the insecticidal activity of 20 ppm of diazinon.

Sethunathan & Yoshida (114) incubated ^{14}C-ring-labeled diazinon with *Flavobacterium* sp. isolated from paddy water by enrichment culture. Within 24 hr after inoculation with the bacterium about 95% of the diazinon was decomposed and 66% was recovered as 2-isopropyl-4-methyl-6-hydroxypyrimidine; by 72 hr after inoculation, no 2-isopropyl-4-methyl-6-hydroxypyrimidine was detected. More than 30% of the added radioactivity was liberated as $^{14}CO_2$, indicating that the pyrimidine ring was degraded. The bacterium hydrolyzed diazinon under both aerobic and anaerobic conditions, but it did so more rapidly under aerobic conditions. The pyrimidine ring of 2-isopropyl-4-methyl-6-hydroxypyrimidine was degraded only under aerobic conditions.

Trela, Ralston & Gunner (126) found that microorganisms isolated from diazinon-treated soil converted the pyrimidinyl portion of diazinon to water-soluble moieties. One of the breakdown products of diazinon was tentatively identified as tetraethylpyrophosphate. Gunner & Zuckerman (52) studied the influence of *Arthrobacter* sp. and *Streptomyces* sp. on ^{14}C-ring-labeled diazinon degradation. When the two bacteria were incubated separately with diazinon, no degradation was detected. When the bacteria were incubated together with the insecticide rapid metabolism occurred. After 7 days, only 6% of the parent molecule remained, whereas 84% was converted to one unidentified metabolite (I) and 10% to a second unidentified metabolite (II). Diazinon was no longer detected at 21 days, whereas 86% of the radioactivity was in metabolite I and 14% was present in metabolite II. It was proposed that a synergistic relationship existed between the two bacteria in the degradation of diazinon.

Thionazin

DEGRADATION IN SOIL Kiigemagi & Terriere (70) reported that the half-life of thionazin under summer conditions was 13–17 days. Less than 10% of the insecticide remained in the soil 100 days after application. Getzin & Rosefield (46) found that the persistence of thionazin varied in four different soil types. The half-lives in organic soil, Puyallup sandy loam, Puget silt loam, and Chehalis clay loam were approximately 10, 6, 4, and 1.5 weeks, respectively. After 24 weeks, the amount of extractable thionazin ranged from 2–27%.

A primary degradative pathway for thionazin was determined by Getzin (41) by using ^{14}C-thionazin labeled in the 3 position of the pyrazinol ring. Hydrolysis at the heterocyclic phosphate bond was followed by cleavage of the cyclic moiety with subsequent formation of $^{14}CO_2$. An expected hydrolysis product, 2-pyrazinol, was not positively identified in nonfumigated soil, and only a small amount was detected in fumigated soil. Since larger amounts of 2-pyrazinol were recovered from fumigated soil, and larger amounts of $^{14}CO_2$ were released from nonfumigated soil, it was suggested that microorganisms degraded 2-pyrazinol to CO_2. It was hypothesized that the failure to recover significant amounts of 2-pyrazinol was in part due to its rapid decomposition by soil microorganisms. Attempts to identify the oxygen analog of thionazin in soil were unsuccessful (41, 70). Getzin (41) applied 1.5 mg of the oxygen analog of thionazin per 300 cm^3 of soil and found that a linear decay occurred with a half-life of 4 hr. The rapid decomposition of this compound makes detection rather difficult.

Getzin (42) studied the effects of autoclaving, temperature, moisture, and acidity on the persistence of thionazin. The time required for 50% degradation of 20 ppm of thionazin in the autoclaved and nonautoclaved soil was 9 and 2 weeks, respectively. After 16 weeks, 30% of the initial dose remained in the autoclaved soil; only 5% was present in the nonautoclaved soil. Degradation increased with increasing temperature. After 4 weeks, 40% of the applied insecticide (20 ppm) remained at 15°C, whereas at 35°C only 10% was present. Soil moisture also affected thionazin persistence. After 8 weeks at soil moisture contents of 2, 10, 20, and 30%, the amounts of added thionazin (20 ppm) that remained were 60, 30, 15, and 13%, respectively. Soil pH levels between 4.3 and 8.1 caused moderate changes in the persistence of the insecticide. Thionazin was most persistent at pH 5.5 and least persistent at pH 8.1. Intermediate persistence curves were obtained at pH 4.3 and 6.7.

Phorate, Terbufos, and Disulfoton

DEGRADATION IN SOIL Degradation of phorate in soil is primarily a two-step oxidation reaction (4, 45, 49, 78, 91, 106, 121, 129). The first step is a rapid oxidation of phorate to phorate sulfoxide, the second step is a slower oxidation of phorate sulfoxide to phorate sulfone (Figure 2). In field plots, Menzer, Fontanilla & Ditman (91) also found a large amount of the oxygen analog of phorate sulfoxide, as well as minute amounts of the oxygen analogs of phorate and phorate sulfone.

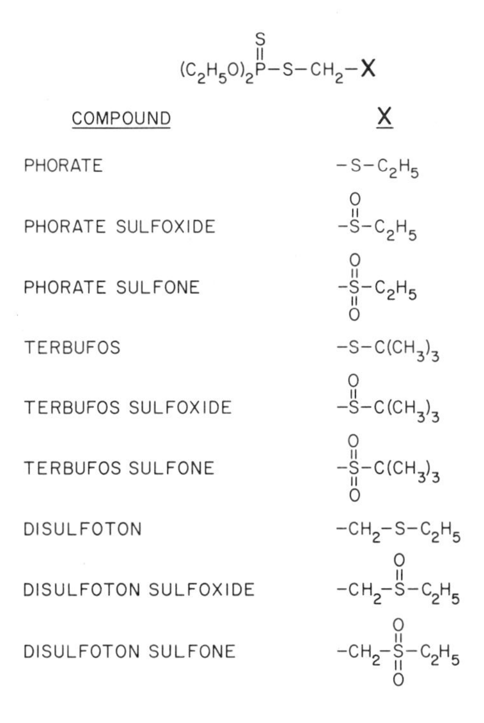

Figure 2 Structures of phorate, terbufos, disulfoton, and some related compounds.

The oxidation of phorate in soil can be a rather rapid reaction. Suett (121) found, under field conditions, that phorate was diminished by 50% in mineral and fen soils in 2 days, and to less than 1% of the applied amount (1.5 ppm in mineral soil and 3.0 ppm in fen soil) in 7 and 14 weeks in a mineral soil and a fen soil, respectively. Waller & Dahm (129) reported that after 2 days of incubation at 30°C in Clarion, Harps, and Webster soils about 50% of the recovered radioactivity from an added dosage of 0.5 ml of 10^{-4} M phorate was phorate sulfoxide, 29% was phorate sulfone, and 21% was phorate. Getzin & Chapman (45) detected rapid oxidation of phorate to phorate sulfoxide and phorate sulfone. As much as 13% of these oxidative products was present in zero-day samples after phorate had been applied to three soils.

Phorate sulfoxide and phorate sulfone remain in the soil for many weeks. After 34 weeks, Suett (121) observed that in a mineral soil and a fen soil 20–30% and 30–40%, respectively, of the applied phorate remained as phorate sulfoxide. Lichtenstein et al (78) found phorate, phorate sulfoxide, and phorate sulfone persistence was increased when the insecticide was mixed with the upper 4- to 5-inch soil layer. Thirty days were required before one-half of the applied dosage could no longer be detected as phorate or its metabolites, compared with 6 days when phorate was applied only to the soil surface. Phorate sulfoxide reached its highest concentration 6 days after phorate application. Subsequently, the sulfoxide level declined and the amount of phorate sulfone increased, reaching its peak at 1 month after soil treatment. Residues consisting entirely of phorate sulfone were found 2 months after applying phorate to the soil surface, but when the insecticide was mixed with the upper soil layer the sulfone was detected after 5 months.

Schulz et al (106) demonstrated that phorate metabolism in Plano silt loam was relatively slower when the insecticide was applied at higher rates. Of the total residues recovered from field-plot cores treated 4.5 months previously with phorate at 5 lb acre^{-1}, 38 and 54% were in the form of phorate sulfoxide and phorate sulfone, respectively. At 10 lb acre^{-1}, however, these figures amounted to 14.5 and 24.9% of the two metabolites, respectively, still leaving 61% of the totally recovered residue in the form of phorate.

Getzin & Shanks (49) treated Sultan silt loam (usually 400 cm^3) with 20 μg of phorate sulfoxide. Two to nine weeks after treatment, phorate sulfone reached its maximum level, 5.8 μg. This value is in contrast to the reported value (49) because an error was found in the published results. In converting the data from micrograms to micromoles, the values are incorrectly stated by a factor of 100 (this was confirmed in a letter from Dr. Getzin dated November 7, 1975). The reported peak for phorate sulfone was 2.0 μmoles; the actual value is 0.02 μmole. Both the sulfoxide and sulfone (1.4 and 4.4 μg, respectively) were present 16 weeks after phorate sulfoxide application. About 0.7 μg of phorate also was found 1 to 2 weeks after soil treatment. Because of this error in converting micrograms to micromoles, it is doubtful that significant quantities of phorate sulfoxide were reduced to phorate in the soil.

The major metabolites of disulfoton in the soil are disulfoton sulfoxide and disulfoton sulfone (117, 122, 123). Takase, Tsuda & Yoshimoto (123) compared disulfoton degradation in autoclaved (1.2 atmospheric pressure, 120°C, three times for 20 min) and nonautoclaved clay loam. Disulfoton had a half-life of about 2 days in unsterilized soil; 55% of the disulfoton remained in the sterilized soil after 28 days. It was concluded that soil microorganisms were involved with the oxidation process. Disulfoton also undergoes rapid oxidation in soil under flooded conditions (122, 123). The major metabolite was disulfoton sulfoxide. Disulfoton sulfoxide and disulfoton sulfone, which were hardly detectable in field soils, were found in appreciable amounts in both the water layer and the soil layer under flooded conditions (123). Menzer, Fontanilla & Ditman (91) reported large amounts of disulfoton sulfone and the oxygen analog of disulfoton sulfone when disulfoton was applied to the soil. Only minute amounts of disulfoton sulfoxide, as well as oxygen analogs of disulfoton and disulfoton sulfoxide, were detected.

Laveglia & Dahm (75) studied the oxidation of terbufos (1 ppm) in Clarion, Harps, and Webster soils. Terbufos, having a half-life of 4 to 5 days, was oxidized rapidly to its sulfoxide. The sulfoxide reached a peak (0.3 ppm) after 2 weeks of incubation. At the end of 1 week, terbufos sulfone appeared and increased steadily over the next 2 weeks.

DEGRADATION BY MICROORGANISMS Ahmed & Casida (1) studied the metabolism of phorate by various microorganisms. *Pseudomonas fluorescens* and *Thiobacillus thiooxidans,* when incubated with phorate for 8 days, hydrolyzed 58 and 75% of the insecticide, respectively, but no oxidation occurred. Although *T. thiooxidans* requires sulfur for growth, when phorate was added to the medium as the only sulfur source the bacterium failed to grow. They also found that with *Chlorella pyrenoidosa* phorate was oxidized to its sulfoxide, which was very stable to hydrolysis; the sulfoxide was converted slowly to its oxygen analog. *Torulopsis utilis* reacted nearly the same as *C. pyrenoidosa* but was less effective in oxidizing phorate sulfoxide to the oxygen analog of phorate sulfoxide.

Bhaskaran et al (8) demonstrated that most of the 20 fungal cultures and 10 isolates of *Streptomyces* spp. tested could use disulfoton as a phosphorus source. The isolates tested, however, showed poor utilization of disulfoton when supplied as the only source of carbon. Among the microorganisms tested, species of *Aspergillus, Helminthosporium,* and *Streptomyces* showed maximum utilization of disulfoton as carbon and phosphorus sources. Even though these species could use disulfoton as a source of carbon and phosphorus, there was a marked inhibition of their growth in the presence of the insecticide when compared with the control.

Chlorfenvinphos, Dichlorvos, and Mevinphos

DEGRADATION IN SOIL Chlorfenvinphos is a rather persistent insecticide when applied to the soil (5, 18, 121). Beynon, Davies & Elgar (5) found that when field plots were treated with chlorfenvinphos at 4 lb acre^{-1}, the insecticide was most persistent in peat and least persistent in sandy loam and brick earth. The half-life in peat was 16 weeks or more, 12 weeks or more in loam, and 2 weeks or more in sandy loam and brick earth. Bro-Rasmussen, Noddegaard & Voldum-Clausen (18) compared the persistence of chlorfenvinphos in field and laboratory experiments.

About 1.8 lb acre^{-1} of insecticide in the field and 8.5 lb acre^{-1} in the laboratory were applied to a loam soil. In the laboratory experiments, the half-life ranged from 84 to 137 days, whereas in the field experiment the average half-life was 135 days. Ten months after treatment of the field plots 33% of the insecticide remained. In a field experiment, Suett (121) applied chlorfenvinphos at about 2 lb acre^{-1} to mineral and fen soils. Half the insecticide had disappeared in 9 and 18 weeks in the mineral and fen soils, respectively. Seven months after the initial application, 20–30% of the applied amount remained in the mineral soil, whereas 40–50% remained in the fen soil.

Beynon & Wright (7) applied 15 ppm of ^{14}C-chlorfenvinphos (labeled at the two vinyl carbon atoms) to sandy loam, loam, clay, and peat soils. Four months after treatment the soils were analyzed for degradation products. Most of the radioactivity recovered (1–4.7 ppm) was identified as chlorfenvinphos. Besides the parent compound, three major degradation products were found. Dealkylation occurred, and 0.1–0.2 ppm of desethyl chlorfenvinphos was detected. The other two metabolites were 2,4-dichloroacetophenone (0.1–0.5 ppm) and 1-(2',4'-dichlorophenyl)-ethan-1-ol (0.06–1.0 ppm).

Beynon et al (6) also studied chlorfenvinphos breakdown in field plots. Residues of 1-(2',4'-dichlorophenyl)-ethan-1-ol and 2,4-dichloroacetophenone did not exceed 0.2 ppm in soils within 6 months of application of chlorfenvinphos at up to 8 lb acre^{-1}.

Nowak, Swarcewicz & Wybieralski (99) studied the effect of temperature and organic carbon content on the degradation of dichlorvos and found that the degradation rate increased with organic carbon content and with higher incubation temperatures.

Burns (21) studied the loss of mevinphos from five Sacramento Delta soils. When the insecticide was applied at 100 ppm, the half-life ranged from 2 to 12 hr, depending on the soil type. There was a rapid loss of mevinphos from the muck, clay, and peat soils, and a somewhat slower rate of disappearance from the peat/sand and sand soils. The difference in mevinphos loss between autoclaved (30 min at 15 psi) and nonautoclaved soils was small in peat/sand and sand soils. In muck, clay, and peat soils, up to 80% of the insecticide was lost in the nonautoclaved soil after 12 hr, whereas only 15–40% was lost in the autoclaved soil.

Getzin and Rosefield (47) studied the decomposition of dichlorvos and mevinphos in nonsterile, irradiated (4-mrad dose at a rate of 250,000 rad hr^{-1}), and autoclaved (1 hr at 15 psi three times at weekly intervals) Chehalis clay loam. Both insecticides were degraded more rapidly in nonsterile soil than in irradiated soil. After a 1-day incubation period, 17 and 88% of the added dichlorvos were degraded in autoclaved and irradiated soil, respectively, whereas for mevinphos 1 and 38% were degraded in autoclaved and irradiated soil, respectively. The difference between the amounts of each insecticide decomposed in autoclaved soil and irradiated soil is attributed to nonviable, heat-labile substances.

DEGRADATION BY MICROORGANISMS Boush & Matsumura (16) studied ^{14}C-dichlorvos degradation by *Pseudomonas melophthora.* After the insecticide

(10^{-6} M) was added to the medium and incubated for 24 hr at 30°C, the rate of metabolite formation was 83.3% water-soluble metabolites, 8.5% solvent-soluble metabolites, and 8.3% original insecticide. These same investigators (87) conducted the same experiment (16) with the soil fungus *Trichoderma viride.* Only 0.5% of the added dichlorvos was recovered as solvent-soluble metabolites, whereas 95.4% was identified as water-soluble metabolites.

Hirakoso (59) studied the effects of eight species of bacteria (*Escherichia, Protaminobacter,* and *Pseudomonas*) on the insecticidal activity of dichlorvos against mosquito larvae. Six of the eight bacteria significantly reduced the knock-down activity of the added insecticide (20 ppm) to mosquito larvae after 4 days. One of the bacteria, *Pseudomonas incognita,* completely inactivated dichlorvos. Yasuno et al (133) found no dichlorvos after 20 ppm of the insecticide were incubated with *B. subtilis* for 16 days.

Dimethoate and Mecarbam

DEGRADATION IN SOIL Bache & Lisk (4) added 100 ppm of dimethoate to 10-g soil samples and found that the half-life was about 13 days. After 30 days of incubation, 26% of the applied dosage remained in the soil. Bro-Rasmussen, Noddegaard & Voldum-Clausen (18) conducted field and laboratory studies on the persistence of dimethoate and mecarbam. In the laboratory after application of 8.5 lb $acre^{-1}$ to a loam soil, it was found that half-lives were 77 days for mecarbam and 33 days for dimethoate. After application of 1.8 lb $acre^{-1}$ to a loam soil in the field, degradation was much faster; half-lives were 13 days for mecarbam and 11 days for dimethoate. These marked differences were attributed to the changing climatic conditions such as fluctuating temperatures and variations in soil moisture during the field study.

Bohn (9) found that, with no rainfall, half of the dimethoate applied (1 lb $acre^{-1}$) to a sandy loam soil disappeared in 4 days. When the insecticide application was followed by a moderate rainfall, the half-life was reduced to 2.5 days. Novozhilov, Volkova & Rozova (98) reported that the rate of dimethoate degradation in the soil was a function of the water content. Total degradation took 21 days in soil with a water content of 29%, 43 days at 13% water content, and 75 days when water content was 5%.

Getzin & Rosefield (47) studied the degradation of dimethoate in autoclaved (1 hr at 15 psi three times at weekly intervals), irradiated (4-mrad dose at a rate of 250,000 rad hr^{-1}), and nonsterile soil. After a 2-week incubation period, 77% of the applied insecticide (about 2 ppm) was degraded in the unsterilized soil, whereas only 17 and 20% disappeared in the autoclaved and irradiated soil, respectively. This suggested that microorganisms were partly responsible for dimethoate degradation.

It has been reported that, in the soil, dimethoate is metabolized to dimethoxon (4, 35). Bache & Lisk (4) recovered 7% of the added dimethoate as dimethoxon after 30 days of incubation. Duff & Menzer (35) found that dimethoate was converted to dimethoxon and two unknown metabolites. In field experiments, conversion to dimethoxon was faster in soil that received more rainfall. Dimethoate carboxylic acid was the only hydrolytic metabolite identified.

Fonofos

DEGRADATION IN SOIL Fonofos was not degraded rapidly when applied to the soil (69, 70, 104, 105, 121). Kiigemagi & Terriere (69) found no significant loss of fonofos from Chehalis silt loam during the 24-hr period after application (3.87 and 4.78 lb $acre^{-1}$), regardless of whether the plot was disced or not disced after treatment. In a later study, Kiigemagi & Terriere (70) recovered 74% of the insecticide 1 week after application and 27% after 4 months. The results of Schulz & Lichtenstein (105) were much the same for Carrington silt loam. Four months after application at 10 lb $acre^{-1}$, 24% of the fonofos was recovered. Suett (121) studied the persistence of fonofos in a mineral and a fen soil; a half-life of 11 weeks was reported in the mineral soil and 22 weeks in the fen soil. Longer studies by Saha et al (104) showed that after 12 months about 37.2 and 30.3% of the applied insecticide (5 lb $acre^{-1}$) remained in Wood mountain light loam and Elstow loam, respectively. At 10 lb $acre^{-1}$, 54.8 and 67.0% remained in Wood mountain light loam and Elstow loam, respectively. Twenty-nine months after treatment, about 6% of the fonofos, when applied at either 5 or 10 lb $acre^{-1}$, was present in both loam soils. No metabolites of fonofos were detected in the soil by thin-layer or gas chromatography (104, 105).

DEGRADATION BY MICROORGANISMS Flashinski & Lichtenstein (38) tested the ability of nine fungal species to metabolize fonofos. *Mucor plumbeus* and *Rhizopus arrhizus* were the most active in degrading the insecticide. A variety of metabolites were found by using ^{14}C-ethoxy-labeled and ^{14}C-ring-labeled fonofos. Oxidative desulfuration to fonofoxon was followed by cleavage to *O*-ethyl ethoxyphosphonothioic acid (ETP) and *O*-ethyl ethoxyphosphonic acid (EOP). The thiophenyl moiety was oxidized to methylphenylsulfoxide (MPSO) and methylphenylsulfone ($MPSO_2$).

In a later study, Flashinski & Lichtenstein (39) added the fungus *R. arrhizus* to soil treated with 2 ppm of ^{14}C-ethoxy-labeled fonofos. The insecticide was degraded during a 21-day incubation period as indicated by recovery of only 11.9% of the applied insecticide. A significant amount of fonofoxon was found, as well as water-soluble radioactivity. In the soil samples to which only water had been added, 40.5% of the applied insecticide was recovered and only trace amounts of fonofoxon were detected.

Flashinski & Lichtenstein (40) tested the ability of three fungal species (*M. plumbeus, Penicillium notatum,* and *R. arrhizus*) to degrade fonofos under various environmental conditions. Different nutrient sources affected the fungal degradation of fonofos. Time-course studies on the metabolism of the insecticide indicated that fonofos was first absorbed by the fungal mycelium, where it was metabolized. This was followed by the release of the water-soluble metabolites ETP, EOP, MPSO, and $MPSO_2$. Optimum conditions for the degradation of fonofos by *R. arrhizus* were pH 6.0–7.0 and 15°–25°C.

Other Organophosphorus Insecticides

DEGRADATION IN SOIL Schulz et al (107) studied the persistence and metabolism of azinphosmethyl in Carrington silt loam under field and laboratory conditions. Under field conditions, azinphosmethyl (5 lb acre^{-1}) was least persistent after its application as an emulsion; when left on the soil surface there was a 50% loss within 12 days after soil treatment. The insecticide was most persistent after it was applied at the same rate in granular form and rototilled into the soil to a depth of 4–5 inches. Within 28 days 50% of the applied dosage disappeared. One year after the azinphosmethyl granules had been incorporated into the soil azinphosmethyl plus eight other compounds were detected. Four of the metabolites were identified as *N*-mercaptomethyl benzazimide, *N*-methyl benzazimide, *N*-methyl benzazimide sulfide or *N*-methyl benzazimide disulfide, and benzazimide. Under laboratory conditions, azinphosmethyl broke down in the loam soil with the appearance of benzazimide and an unknown compound of high polarity.

Staiff et al (118) applied azinphosmethyl as undiluted commercial grade 18.1% liquid-emulsifiable concentrate to field plots of sandy loam soil to simulate accidental spillage. One day after application 49,946 ppm of the insecticide were detected, 1 year later 25,011 ppm were present, and after 8 years 850 ppm still persisted in the soil.

Menn, Patchett & Batchelder (90) studied the persistence of carbophenothion when applied at 10 ppm to Santa Cruz loamy sand, Sorrento loam, and Yolo silty clay loam. Carbophenothion persisted longest in Sorrento loam with a half-life of 200 days, followed by Santa Cruz loamy sand (170 days) and Yolo silty clay loam (100 days). When Santa Cruz loamy sand was autoclaved, the time required for 50% degradation was increased to 525 days. Fumigation of Sorrento loam with metham-sodium (1 qt in 4 gal of water per 100 ft^2) before applying 10 ppm of carbophenothion increased insecticide persistence four times longer than in nonfumigated soil.

Corey (32) found that dicrotophos degraded faster when soil moisture was at a higher level. When a sandy loam soil was moist (50-100% moisture content) 18.5–6.8% of the applied insecticide (10 ppm) was recovered 7–8 days after treatment. Seven to eight days after application of the insecticide to air-dried soil, 80.5% was recovered.

Hall & Sun (54) studied the effect of soil sterilization on dicrotophos persistence. Eight days after application of 100 μl of 0.1% dicrotophos solution to autoclaved (15 psi for 30 min) Ripperdan sandy loam, 59.7% was recovered. When the same amount was applied to nonautoclaved soil only 15.2% was recovered.

Harris, Thompson & Tu (57) added 25 ppm of fensulfothion to Plainfield sand and found rapid oxidation to fensulfothion sulfone. After 20 weeks, the amount of fensulfothion present in the soil was approximately 3% of the initial application whereas the sulfone amounted to about 38%. Chisholm (26) also found fensulfothion sulfone to be the major metabolic product when fensulfothion (8.9 lb acre^{-1}) was added to Somerset sandy loam. About 98% of the applied insecticide disappeared 4 months after treatment.

Getzin (43) treated Puyallup sandy loam, Sultan silt loam, Chehalis clay loam, and organic soil with 4 mg of ^{14}C-alkyl-labeled methidathion. The insecticide degraded fastest in the Chehalis clay loam and organic soil; 71 and 39%, respectively, had disappeared 1 week after treatment. In all four soils, less than 10% of the initial concentration remained after 16 weeks. The expiration of $^{14}CO_2$ from the soils paralleled the degradation of the parent compound. After 16 weeks, 42–62% of the ^{14}C was recovered as $^{14}CO_2$. To test the hypothesis that microorganisms contributed to the degradation of methidathion, Sultan silt loam, Chehalis clay loam, and organic soil were fumigated with Vorlex® (80% dichloropropenes and 20% sodium isothiocyanate, with 0.4 ml applied to 400 cm^3 of soil) after which 290 μg of ^{14}C-ring-labeled methidathion were added. The insecticide degraded rapidly in the nonfumigated soils and slowly in the fumigated soils. Over 50% of the insecticide disappeared within a week from the three nonfumigated soils. After 16 weeks, 49–60% of the methidathion remained in the fumigated soils. Recovery of $^{14}CO_2$ again paralleled the degradation of the insecticide but only in the nonfumigated soil. After 16 weeks, only 2% of the radioactivity was expired as $^{14}CO_2$ in fumigated soil.

Dupuis, Muecke & Esser (36) also found that methidathion was rapidly metabolized to CO_2. When 292 μg of ^{14}C-labeled insecticide were added to 20 g of humus-rich and loamy soil, 9.6 and 27.9%, respectively, of the radioactivity applied was evolved as $^{14}CO_2$ after 8 days. The loamy soil previously autoclaved (30 min at 125°C three times at 24-hr intervals) gave rise to only 2% $^{14}CO_2$ after 26 days.

Menn et al (89) studied the degradation of phosmet (10 ppm) in Sorrento loam and Santa Cruz loamy sand. The time for 50% degradation of the insecticide in autoclaved (15 psi for two 1-hr periods at 3-day intervals) and nonautoclaved loam soil was 4.5 and 3 days, respectively. In autoclaved and nonautoclaved loamy sand, 50% degradation occurred in 12.2 and 8 days, respectively.

CARBAMATE INSECTICIDES

Carbaryl

DEGRADATION IN SOIL In a field study, Johnson & Stansbury (63) treated Norfolk sandy loam with carbaryl (Figure 3) at concentrations of 1.5, 4.5, and 13.5 ppm. The insecticide was tilled into the soil and the soil was sampled periodically. The half-life of carbaryl was approximately 8 days at all concentrations, and the insecticide was completely degraded within 40 days.

Kazano, Kearney & Kaufman (65) applied ^{14}C-carbonyl-labeled carbaryl at 2 and 200 ppm to five different soil types (Konosu clay, Utsunomiya loam, Fukuyama sandy loam, Chikugo clay loam, and Chiba loamy sand). After 32 days, the greatest degradation of the insecticide occurred in Chikugo clay loam; only 25% of the carbaryl remained in the soil at both concentrations. The least degradation occurred in Chiba loamy sand; about 85% of the added carbaryl was present at both concentrations. $^{14}CO_2$ evolution also varied from about 2% of the applied radioactivity in

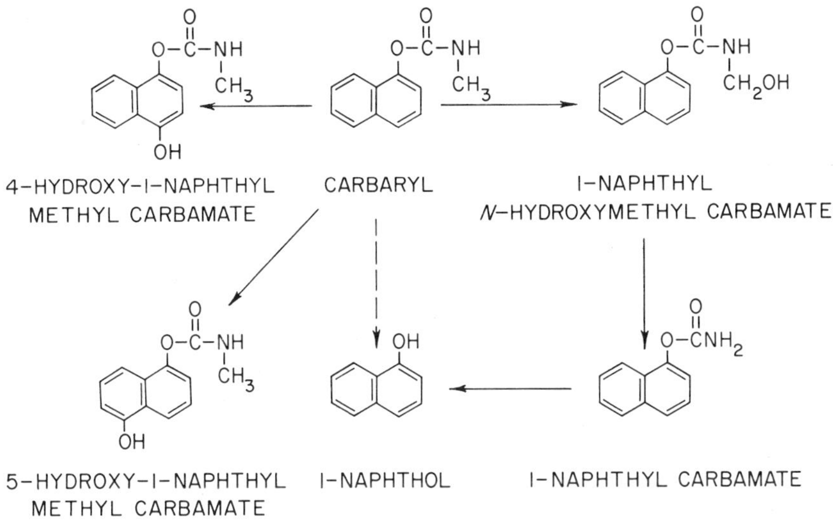

Figure 3 Microbial metabolism of carbaryl according to Liu & Bollag (81).

Chiba loamy sand to about 37% in Chikugo clay loam. Hydrolysis was the main pathway of degradation because less than 1% of the initial radioactivity corresponded to ^{14}C-carbonyl metabolites. Similar results were obtained when 2 ppm of ^{14}C-ring-labeled 1-naphthol were added to Chikugo clay loam and Fukuyama sandy loam. After 60 days of incubation in Chikugo clay loam, about 85% of the added ^{14}C remained in the soil and about 8% was collected as $^{14}CO_2$; in Fukuyama sandy loam, about 89% remained in the soil and about 6% was collected as $^{14}CO_2$.

In a field study, Caro, Freeman & Turner (25) found no significant loss of carbaryl during the first 40 days after application of 4.5 lb of carbaryl per acre to Coshocton silt loam. About 135 days were required for 95% of the insecticide to disappear. It was concluded that the initial 40 days, when no significant carbaryl degradation took place, was a lag phase during which active soil microorganisms grew and adapted to the insecticide.

After foliar application of carbaryl to an apple orchard at 3 lb acre^{-1} in June, July, and August, Kuhr, Davis & Bourke (73) found 3 ppm of the insecticide in soil samples 1 day after treatment, except in July when the average was 13.8 ppm. Two weeks after treatment, even in July, the amount of residue fell below detectable limits of 1 ppm.

DEGRADATION BY MICROORGANISMS Boush & Matsumura (16) studied the degradation of 10^{-6} M ^{3}H-ring-labeled carbaryl incubated at 30°C for 24 hr with *Pseudomonas melophthora.* Of the total insecticide applied, 6.4% was recovered as water-soluble metabolites and 45.5% was identified as 1-naphthol, a solvent-soluble metabolite. Matsumura & Boush (87) conducted the same experiment (16) but used a soil fungus, *Trichoderma viride.* This fungus showed little degradative activity toward carbaryl because only 2.1% of the applied insecticide was recovered as water-soluble metabolites and 9.2% as solvent-soluble metabolites.

Tewfik & Hamdi (124) added an unidentified soil bacterium to a mineral salts medium containing 350 ppm of carbaryl. After 5 days of incubation, only 8 ppm of the insecticide remained in the medium. Zuberi & Zubairi (134) incubated (37°C) a 100-ml suspension of *Pseudomonas phaseolicola* in peptone broth with 100 mg of carbaryl. After 48 hr, 0.31 mg of 1-naphthol was produced.

Sud, Sud & Gupta (120) reported that *Achromobacter* sp. utilized carbaryl (0.5 g 1000 ml^{-1} of salt medium) as a carbon source. Four degradation products were tentatively identified as 1-naphthol, hydroquinone, catechol, and pyruvate.

Bollag & Liu (13) inoculated a culture medium containing 4.5×10^{-2} μM of ^{14}C-methyl-labeled carbaryl with three separate microorganisms: a fungus, *Fusarium solani*; a gram-negative coccus; and a gram-positive rod. *F. solani* was most effective in decomposing the insecticide. Five days after inoculation, the radioactivity decreased by 24%, and after 12 days by 82%. The gram-negative coccus degraded approximately 51% of the insecticide after 7 days; the gram-positive rod was the least effective in degrading carbaryl. Mixed cultures of any two or all three of the microorganisms usually resulted in a greater loss of radioactivity than single cultures possibly because of synergistic effects. The only identified metabolite of carbaryl was 1-naphthol.

Liu & Bollag (80) studied the metabolism of ^{14}C-methyl-labeled carbaryl (3.8×10^{-5} μmoles) in 1.5 liters of medium inoculated with a soil fungus, *Gliocladium roseum,* and incubated for 7 days at 28°C. Three hydroxylation products were identified, 1-naphthyl *N*-hydroxymethylcarbamate, 4-hydroxy-1-naphthyl methylcarbamate, and 5-hydroxy-1-naphthyl methylcarbamate (Figure 3). No hydrolytic metabolites were found.

Liu & Bollag (81) incubated a soil fungus, *Aspergillus terreus,* with ^{14}C-methyl-labeled and ^{14}C-ring-labeled carbaryl (0.02 μCi ml^{-1}). After 6 days, 20% of the added radioactivity was recovered as 1-naphthyl *N*-hydroxymethylcarbamate, 18% was detected as 1-naphthyl carbamate, and 50% remained unchanged. Only minute amounts of the 4- and 5-hydroxy-1-naphthyl methylcarbamates were identified. These workers showed that 1-naphthyl carbamate was the intermediary metabolite between 1-naphthyl *N*-hydroxymethylcarbamate and 1-naphthol (Figure 3). When sesamex was added to the medium containing *A. terreus* and carbaryl, metabolism of the insecticide was partly inhibited at 50 to 100 ppm and completely inhibited at 200 to 500 ppm of sesamex (12).

Bollag & Liu (14) found many species of fungi (*Aspergillus, Fusarium, Geotrichum, Gliocladium, Helminthosporium, Mucor, Penicillium, Rhizopus,* and *Trichoderma*) capable of hydroxylating carbaryl. The metabolic products varied qualitatively as well as quantitatively with the various fungi. The major metabolite of all the *Aspergillus* species, except *A. fumigatus,* was 1-naphthyl *N*-hydroxymethylcarbamate. In contrast, isolates of *Penicillium* sp., *Mucor* sp., and *Rhizopus* sp. had a stronger tendency to hydroxylate carbaryl in the ring position.

Degradation of 1-naphthol by soil microorganisms also has been reported (11, 13, 15, 65). Bollag & Liu (13) incubated 3.06×10^{-5}mM ^{14}C-ring-labeled 1-naphthol with three microbial isolates. A gram-negative coccus decomposed about half of the applied ^{14}C-1-naphthol within 12 days. *Fusarium solani* and a gram-positive rod degraded less than 20% of the radioactive compound in 12 days. In a later study, also using *F. solani,* Bollag & Liu (15) added ^{14}C-ring-labeled 1-naphthol to a liquid culture medium at 20 μg per ml. After 12 days of incubation, loss of radioactivity from the growth medium was less than 20%. When a mycelial cell extract of *F. solani* was incubated with ^{14}C-1-naphthol, degradation was rapid. Only 2.5% of the radioactivity was recovered from the reaction mixture after 1 hr; 88% was trapped as $^{14}CO_2$ in NaOH solution.

Kazano, Kearney & Kaufman (65) incubated a *Pseudomonas* sp. with 100 ppm of ^{14}C-ring-labeled 1-naphthol and recovered only 7.4% of the initial radioactivity as $^{14}CO_2$ after 7 hr. After ether extraction of the inoculum, four metabolities plus 1-naphthol were found. One metabolite was identified as coumarin; concentrations of the other three were too small to be identified.

Bollag, Czaplicki & Minard (11) incubated an unidentified bacterium (gram-negative, aerobic rod) isolated from river water with 20 ppm of ^{14}C-ring-labeled naphthol. After 60 hr, 44% of the radioactivity was trapped as $^{14}CO_2$, 22% was recovered in the bacterial cells, and 17% remained in the growth medium. After diethyl ether extraction of the growth medium, the main product was 4-hydroxy-1-tetralone.

Carbofuran

DEGRADATION IN SOIL In a field study, Caro et al (24) applied carbofuran to two plots (one was made up of Keene and Rayne silt loam and the other was Coshocton silt loam). The first received a broadcast treatment (4.83 lb acre^{-1}) and the second an in-furrow treatment (3.17 lb acre^{-1}). The half-lives of carbofuran in the two treatments were 46 and 117 days, respectively. Other factors accelerating carbofuran degradation were high pH and heavy soil texture. An oxidation product, 3-ketocarbofuran, reached a peak of 4.6% 29 days after the broadcast treatment, whereas the peak for the in-furrow application was 7.8% 60–80 days after treatment. Only sporadic traces of a hydrolysis product, 3-hydroxycarbofuran, were found in the soil samples.

Gupta & Dewan (53) applied carbofuran at the rate of 0.29 lb acre^{-1} (in-furrow treatment) and sampled for residues over a 44-day period. After 24 days, about 0.25 ppm remained; the final residue value was about 0.12 ppm. At approximately 30 days after application of the insecticide 0.7 ppm of 3-ketocarbofuran was noted.

Getzin (44) studied the degradation of carbofuran (4 mg of carbofuran containing 7.4×10^{-4} mM of ^{14}C-carbonyl-labeled insecticide 400 cm^{-3} of soil) in Ritzville silt loam, Chehalis clay loam, Sultan silt loam, and organic muck. Approximate times required for 50% loss of carbofuran in the four soils were 4, 8, 54, and 54 weeks, respectively. Sterilization (5 mrads of gamma radiation) had no effect upon the degradation rate of carbofuran in the Ritzville silt loam and only a slight effect in the Sultan silt loam and organic muck. The insecticide degraded significantly faster in nonirradiated Chehalis clay loam (28% remained after 16 weeks) than in the irradiated soil (78% remained after 16 weeks). After 54 weeks of incubation most of the radioactivity (as high as 79%) from the ^{14}C-carbofuran was expired as $^{14}CO_2$. Nonextractable soil-bound radioactivity was not a significant factor except in irradiated Ritzville silt loam where a maximum of 20% of the radioactivity was bound to the soil. Neither 3-hydroxy nor 3-ketocarbofuran was detected.

^{14}C-ring-labeled carbofuran (5 $\mu g\ cm^{-3}$ of soil) was applied to Ritzville silt loam and Chehalis clay loam (44). The breakdown of ring-labeled carbofuran was associated with an accumulation of nonextractable, soil-bound radioactivity (as high as 53% of the applied radioactivity in Ritzville silt loam and 46% in Chehalis clay loam) and a gradual evolution of $^{14}CO_2$.

Because carbofuran phenol was the expected degradation product of carbofuran, 5 μg of ^{14}C-carbofuran phenol cm^{-3} of soil were added to Ritzville silt loam and Chehalis clay loam (44). The compound was bound rapidly to the soils. Immediately after treatment, nonextractable radioactivity amounted to 21 and 24%, respectively, in the two loam soils. Soil-bound residues reached a maximum of 70–80% 2 weeks after treatment. After 32 weeks incubation, 25% of the added ^{14}C-carbofuran phenol was expired as $^{14}CO_2$.

Getzin (44) has suggested that the main pathway of carbofuran degradation in soil is hydrolysis at the carbamate linkage. The carbamate moiety is degraded to CO_2, and the carbofuran phenol is rapidly bound to the soil. A gradual degradation of the carbofuran phenol follows with the release of CO_2.

Aldicarb

DEGRADATION IN THE SOIL Coppedge et al (31) applied 20 ppm of ^{35}S-aldicarb to Houston clay, Norwood silty clay loam, and Lakeland fine sand. The approximate half-lives of this insecticide in these three soils were 9, 7, and 12 days, respectively. Four weeks after treatment, the three soils contained 6.1, 0.3, and 27.2%, respectively, of the applied dose of aldicarb. The major metabolic product recovered from the soil was aldicarb sulfoxide. One week after treatment, 31.5, 41.5, and 27.5%, respectively, of the applied aldicarb was oxidized to its sulfoxide. Further oxidation to aldicarb sulfone was slow and, except in Lakeland fine sand at 6 and 12 weeks after treatment, all values were under 10% of the applied dose. The oxime sulfoxide was also recovered at low concentrations (less than 5% of the applied dose).

Bull (19) applied ^{35}S-aldicarb (2 mg per 100 g of soil) to Lufkin fine sandy loam and placed the containers in the field. Rapid loss of aldicarb was observed. After 1 week, only 1.6% of the applied aldicarb was recovered, although 36.8% was recovered as aldicarb sulfoxide. Aldicarb sulfone reached a peak of 4.6% 1 week after treatment.

Kearby, Ercegovich & Bliss (66) found that the half-life of aldicarb in Gilpin fine loam was about 15 days when applied to the soil at 0.5, 1.0, and 2.0 lb per 12 trees in a stand of 10-year-old Scotch pines. When applied at 0.5 and 1.0 lb, no residue was found at 36 and 63 days, respectively, and only a trace amount, 0.07 ppm, was detected at 63 days after an application of 2.0 lb.

Bull et al (20) studied the fate of ^{14}C-aldicarb (20 ppm) in four different soil types (Lufkin fine sandy loam, Woodward fine sandy loam, Houston clay, and muck), at three moisture levels (0, 50, and 100% of field capacity), and at three pH values (6, 7, and 8). No significant differences were attributed to pH within the range tested. Aldicarb and its toxic derivatives (aldicarb sulfoxide and aldicarb sulfone) were relatively stable in all the dry soils, in sand at all moisture levels, and in loam at 50% moisture. Very rapid volatilization of the applied radioactivity occurred from the dry, pH 8 sand sample. The 50% moisture level seemingly was optimum for the oxidation of aldicarb to its sulfoxide and sulfone in loam, clay, and muck. The results obtained at a moisture level of 100% were difficult to interpret because of the poor recoveries of the applied radioactivity, although a faster rate of decomposition to nontoxic products (oxime sulfoxide, oxime sulfone, nitrile sulfoxide, and nitrile sulfone) was noted.

CONCLUSIONS

Most insecticides end up on or in the soil. A combination of characteristics of the soil, climate, and insecticide structure determines the persistence of insecticides in the soil. The soil is a complex and variable environment that is difficult to normalize in relation to insecticide degradation. The most prominent feature of the fate of organophosphorus and carbamate insecticides in soil is their lesser persistence as toxic molecules as compared with chlorinated hydrocarbon insecticides. The effective life of organophosphorus and carbamate insecticides in soil usually is weeks and months, compared with months and years for chlorinated hydrocarbon insecticides. Most organophosphorus and carbamate insecticides are very susceptible to chemical alteration. The soil is an excellent medium for both biological and nonbiological modifications of organophosphorus and carbamate insecticides. Microbial metabolism of these two classes of insecticides seems to be the most important factor in accounting for their degradation in soil. It is difficult, however, to quantify the relative importance of biological and nonbiological degradation of these insecticides from the data in some of the studies.

The principal chemical reactions of these insecticides in soil are hydrolysis, reduction, oxidation, and dealkylation, although not all reactions occur with any one insecticide. Both phosphoryl and carbonyl groups are susceptible to hydrolysis. The reduction of nitro groups is well established. Oxidation of carbonyl groups and of benzyl, naphthyl, and heterocyclic rings proceeds to the formation of carbon dioxide. The sulfur atom associated with thioalkyl and thiophenyl groups is progressively oxidized to a sulfoxide and a sulfone; the formation and persistence of compounds with these two groups, however, seem to vary according to the structure of the parent compound. The sulfoxide and sulfone analogues of these insecticides are, like the parent compounds, quite toxic to insects and increase overall effectiveness. Both *O*-dealkylation and *N*-dealkylation are important reactions with organophosphorus and carbamate insecticides. Hydroxylation of the nitrogen methyl group of carbamates and the naphthyl group of carbaryl and formation of 3-ketocarbofuran from carbofuran have also been reported. Several instances of oxidative desulfuration of thiophosphates and dithioates have been reported, but so far, this reaction seems to be of secondary importance. It is known that microorganisms differ widely in their reactions to foreign chemicals (10, 34). We have given numerous examples in this review of differences in metabolic attack by microorganisms.

Other factors affecting degradation of organophosphorus and carbamate insecticides in soil include pH, time, temperature, adsorption, moisture, and soil type. A strongly alkaline soil usually accelerates hydrolysis, but there are instances in which a strongly acid soil also will hydrolyze insecticides. The initial degradation process usually is quite rapid, even at dosages far greater than normal. An asymptotic persistence of small amounts of some insecticides may be owing to adsorption and soil type. In at least one instance (diazinon), hydrolysis of an insecticide seems to be controlled by adsorption onto soil more than by a pH effect. An increase in temperature accelerates degradation with the most frequent range for temperature studies from about 15°–35°C. Numerous studies have shown that degradation of organophosphorus and carbamate insecticides is strongly correlated with increasing soil moisture. Many soil types have been used for insecticide degradation studies; it is unfortunate that a few predominant soils have not been included in more of these studies to provide baseline data for comparison purposes. The reviews by Edwards (37) and Harris (56) provide especially useful summaries of the influence of soil characteristics on persistence of insecticides in soil; therefore, the topic has not been extensively developed in this review.

Future dependence on organophosphorus and carbamate insecticides seems quite likely. Their fate in soil is rather well understood, but some additional investigations are needed. More precise data would help differentiate between biological and nonbiological degradation of insecticides in soil. Further studies of their sorption and degradation products with soil also are needed. A few reports of synergistic effects of two or more species of microorganisms on the metabolism of insecticides suggest a need for research to resolve this question. Soil-sterilization studies have revealed heat-labile substances, not destroyed by irradiation, that can degrade insecticides. The nature and action of these substances should be determined.

Acknowledgments

This review was supported by National Institutes of Health research grant number ES-00205 from the National Institute of Environmental Health Sciences and North Central Regional Project NC-96.

We thank Diane Lohr, Yvonne Dubberke, Betty Dahm, and Jan Millard for their assistance in preparing this manuscript.

Literature Cited

1. Ahmed, M. K., Casida, J. E. 1958. Metabolism of some organophosphorus insecticides by microorganisms. *J. Econ. Entomol.* 51:59–63
2. Ahmed, M. K., Casida, J. E., Nichols, R. E. 1958. Bovine metabolism of organophosphorus insecticides: significance of rumen fluid with particular reference to parathion. *J. Agric. Food Chem.* 6:740–46
3. Alexander, M. 1969. In *Soil Biology*, pp. 209–40. Paris: UNESCO. 240 pp.
4. Bache, C. A., Lisk, D. J. 1966. Determination of oxidative metabolites of dimethoate and Thimet in soil by emission spectroscopic gas chromatography. *J. Ass. Off. Agric. Chem.* 49:647–50
5. Beynon, K. I., Davies, L., Elgar, K. 1966. Analysis of crops and soils for residues of diethyl 1-(2,4-dichlorophenyl)-2-chlorovinyl phosphate. II. Results. *J. Sci. Food Agric.* 17:167–74
6. Beynon, K. I., Edwards, M. J., Elgar, K., Wright, A. N. 1968. Analysis of crops and soils for residues of chlorfenvinphos insecticide and its breakdown products. *J. Sci. Food Agric.* 19:302–7
7. Beynon, K. I., Wright, A. N. 1967. The breakdown of ^{14}C-chlorfenvinphos in soils and in crops grown in the soils. *J. Sci. Food Agric.* 18:143–50
8. Bhaskaran, R., Kandasamy, D., Oblisami, G., Subramaniam, T. R. 1973. Utilization of Disyston as carbon and phosphorus sources by soil microflora. *Curr. Sci.* 42:835–36
9. Bohn, W. R. 1964. The disappearance of dimethoate from soil. *J. Econ. Entomol.* 57:798–99
10. Bollag, J.-M. 1972. Biochemical transformation of pesticides by soil fungi. *Crit. Rev. Microbiol.* 2:35–58
11. Bollag, J.-M., Czaplicki, E. J., Minard, R. D. 1975. Bacterial metabolism of 1-naphthol. *J. Agric. Food Chem.* 23:85–90
12. Bollag, J.-M., Liu, K.-C. 1974. Effect of methylenedioxyphenyl synergists on metabolism of carbaryl by *Aspergillus terreus. Experientia* 30:1374–75
13. Bollag, J.-M., Liu, S.-Y. 1971. Degradation of Sevin by soil microorganisms. *Soil Biol. Biochem.* 3:337–45
14. Bollag, J.-M., Liu, S.-Y. 1972. Hydroxylations of carbaryl by soil fungi. *Nature* 236:177–78
15. Bollag, J.-M., Liu, S.-Y. 1972. Fungal degradation of 1-naphthol. *Can. J. Microbiol.* 18:1113–17
16. Boush, G. M., Matsumura, F. 1967. Insecticidal degradation by *Pseudomonas melophthora,* the bacterial symbiote of the apple maggot. *J. Econ. Entomol.* 60:918–20
17. Bro-Rasmussen, F., Noddegaard, E., Voldum-Clausen, K. 1968. Degradation of diazinon in soil. *J. Sci. Food Agric.* 19:278–81
18. Bro-Rasmussen, F., Noddegaard, E., Voldum-Clausen, K. 1970. Comparison of the disappearance of eight organophosphorus insecticides from soil in laboratory and in outdoor experiments. *Pestic. Sci.* 1:179–82
19. Bull, D. L. 1968. Metabolism of UC-21149 (2-methyl-2-(methylthio)-propionaldehyde *O*-(methylcarbamoyl)oxime) in cotton plants and soil in the field. *J. Econ. Entomol.* 61:1598–1602
20. Bull, D. L., Stokes, R. A., Coppedge, J. R., Ridgway, R. L. 1970. Further studies of the fate of aldicarb in soil. *J. Econ. Entomol.* 63:1283–89
21. Burns, R. G. 1971. The loss of Phosdrin and phorate insecticides from a range of soil types. *Bull. Environ. Contam. Toxicol.* 6:316–21
22. Butler, G. L., Deason, T. R., O'Kelley, J. C. 1975. Loss of five pesticides from cultures of twenty-one planktonic algae. *Bull. Environ. Contam. Toxicol.* 13:149–52
23. Carlo, C. P., Ashdown, D., Heller, V. G. 1952. The persistence of parathion, toxaphene, and methoxychlor in soil. *Okla. Agric. Exp. Stn. Tech. Bull.* 42:3–11
24. Caro, J. H., Freeman, H. P., Glotfelty, D. E., Turner, B. C., Edwards, W. M. 1973. Dissipation of soil-incorporated carbofuran in the field. *J. Agric. Food Chem.* 21:1010–15
25. Caro, J. H., Freeman, H. P., Turner, B. C. 1974. Persistence in soil and losses in runoff of soil-incorporated carbaryl in a small watershed. *J. Agric. Food Chem.* 22:860–63
26. Chisholm, D. 1974. Persistence of fensulfothion in soil and uptake by rutabagas and carrots. *Can. J. Plant Sci.* 54:667–71
27. Chopra, S. L., Girdhar, K. C. 1971. Persistence of malathion *S*-1,2-bis (ethoxy carboxyl) ethyl *O,O*-dimethyl phosphorodithioate in Punjab soils. *Indian J. Appl. Chem.* 34:201–7
28. Chopra, S. L., Khullar, F. C. 1971.

Degradation of parathion in soils. *J. Indian Soc. Soil Sci.* 19:79–85
29. Christie, A. E. 1969. Effects of insecticides on algae. *Water Sewage Works* 116:172–76
30. Cook, J. W. 1957. In vitro destruction of some organophosphate pesticides by bovine rumen fluid. *J. Agric. Food Chem.* 5:859–63
31. Coppedge, J. R., Lindquist, D. A., Bull, D. L., Dorough, H. W. 1967. Fate of 2-methyl-2-(methylthio)propionaldehyde *O*-(methylcarbamoyl) oxime (Temik) in cotton plants and soil. *J. Agric. Food Chem.* 15:902–10
32. Corey, R. A. 1965. Laboratory tests with Bidrin insecticide. *J. Econ. Entomol.* 58:112–14
33. Crosby, D. G. 1973. The fate of pesticides in the environment. *Ann. Rev. Plant Physiol.* 24:467–92
34. Dagley, S. 1975. Microbial degradation of organic compounds in the biosphere. *Am. Scientist* 63:681–89
35. Duff, W. G., Menzer, R. E. 1973. Persistence, mobility, and degradation of ^{14}C-dimethoate in soils. *Environ. Entomol.* 2:309–18
36. Dupuis, G., Muecke, W., Esser, H. O. 1971. The metabolic behavior of the insecticidal phosphorus ester GS-13005. *J. Econ. Entomol.* 64:588–97
37. Edwards, C. A. 1972. In *Organic Chemicals in the Soil Environment,* eds. C. A. I. Goring, J. W. Hamaker, Vol. 2, pp. 515–68. New York: Marcel Dekker. 468 pp.
38. Flashinski, S. J., Lichtenstein, E. P. 1974. Metabolism of Dyfonate by soil fungi. *Can. J. Microbiol.* 20:399–411
39. Flashinski, S. J., Lichtenstein, E. P. 1974. Degradation of Dyfonate in soil inoculated with *Rhizopus arrhizus. Can. J. Microbiol.* 20:871–75
40. Flashinski, S. J., Lichtenstein, E. P. 1975. Environmental factors affecting the degradation of Dyfonate by soil fungi. *Can. J. Microbiol.* 21:17–25
41. Getzin, L. W. 1967. Metabolism of diazinon and Zinophos in soils. *J. Econ. Entomol.* 60:505–8
42. Getzin, L. W. 1968. Persistence of diazinon and Zinophos in soil: effects of autoclaving, temperature, moisture, and acidity. *J. Econ. Entomol.* 61: 1560–65
43. Getzin, L. W. 1970. Persistence of methidathion in soils. *Bull. Environ. Contam. Toxicol.* 5:104–10
44. Getzin, L. W. 1973. Persistence and degradation of carbofuran in soil. *Environ. Entomol.* 2:461–67
45. Getzin, L. W., Chapman, R. K. 1960. The fate of phorate in soils. *J. Econ. Entomol.* 53:47–51
46. Getzin, L. W., Rosefield, I. 1966. Persistence of diazinon and Zinophos in soils. *J. Econ. Entomol.* 59:512–16
47. Getzin, L. W., Rosefield, I. 1968. Organophosphorus insecticide degradation by heat-labile substances in soil. *J. Agric. Food Chem.* 16:598–601
48. Getzin, L. W., Rosefield, I. 1971. Partial purification and properties of a soil enzyme that degrades the insecticide malathion. *Biochim. Biophys. Acta* 235:442–53
49. Getzin, L. W., Shanks, C. H. Jr. 1970. Persistence, degradation, and bioactivity of phorate and its oxidative analogues in soil. *J. Econ. Entomol.* 63:52–58
50. Graetz, D. A., Chesters, G., Daniel, T. C., Newland, L. W., Lee, G. B. 1970. Parathion degradation in lake sediments. *J. Water Pollut. Control Fed.* 42:76–94
51. Griffiths, D. C., Walker, N. 1970. Microbiological degradation of parathion. *Meded. Fac. Landbouwwet. Rijksuniv. Gent.* 35:805–10
52. Gunner, H. B., Zuckerman, B. M. 1968. Degradation of 'diazinon' by synergistic microbial action. *Nature* 217:1183–84
53. Gupta, R. C., Dewan, R. S. 1974. Residues and metabolism of carbofuran in soil. *Pesticides* 8:36–39
54. Hall, W. E., Sun, Y. P. 1965. Mechanism of detoxication and synergism of Bidrin insecticide in house flies and soil. *J. Econ. Entomol.* 58:845–49
55. Harris, C. R. 1970. In *International Symposium on Pesticides in the Soil,* pp. 58–64. East Lansing: Mich. State Univ. 144 pp.
56. Harris, C. R. 1972. Factors influencing the effectiveness of soil insecticides. *Ann. Rev. Entomol.* 17:177–98
57. Harris, C. R., Thompson, A. R., Tu, C. M. 1971. Insecticides and the soil environment. *Proc. Entomol. Soc. Ont.* 102:156–68
58. Helling, C. S., Kearney, P. C., Alexander, M. 1971. Behavior of pesticides in soils. *Adv. Agron.* 23:147–240
59. Hirakoso, S. 1969. Inactivating effects of micro-organisms on insecticidal activity of Dursban. *Jpn. J. Exp. Med.* 39:17–20
60. Hsieh, D. P. H., Munnecke, D. M. 1972. Accelerated microbial degrada-

tion of concentrated parathion. *Proc. Int. Ferment. Symp. 4th Technol. Today* pp. 551–54

61. Iwata, Y., Westlake, W. E., Gunther, F. A. 1973. Persistence of parathion in six California soils under laboratory conditions. *Arch. Environ. Contam. Toxicol.* 1:84–96
62. Johnsen, R. E. 1976. DDT metabolism in microbial systems. *Residue Rev.* 61:1–28
63. Johnson, D. P., Stansbury, H. A. 1965. Adaptation of Sevin insecticide (carbaryl) residue method to various crops. *J. Agric. Food Chem.* 13:235–38
64. Kasting, R., Woodward, J. C. 1951. Persistence and toxicity of parathion when added to the soil. *Sci. Agric.* 31:133–38
65. Kazano, H., Kearney, P. C., Kaufman, D. D. 1972. Metabolism of methylcarbamate insecticides in soils. *J. Agric. Food Chem.* 20:975–79
66. Kearby, W. H., Ercegovich, C. D., Bliss, M. Jr. 1970. Residue studies on aldicarb in soil and Scotch pine. *J. Econ. Entomol.* 63:1317–18
67. Kearney, P. C., Helling, C. S. 1969. Reactions of pesticides in soils. *Residue Rev.* 25:25–44
68. Kenaga, E. E., End, C. S. 1974. Commercial and experimental organic insecticides. *Entomological Society of America Special Publication 74-1* (1974 Revision). 77 pp.
69. Kiigemagi, U., Terriere, L. C. 1971. Losses of organophosphorus insecticides during application to the soil. *Bull. Environ. Contam. Toxicol.* 6: 336–42
70. Kiigemagi, U., Terriere, L. C. 1971. The persistence of Zinophos and Dyfonate in soil. *Bull. Environ. Contam. Toxicol.* 6:355–61
71. Konrad, J. G., Armstrong, D. E., Chesters, G. 1967. Soil degradation of diazinon, a phosphorothioate insecticide. *Agron. J.* 59:591–94
72. Konrad, J. G., Chesters, G., Armstrong, D. E. 1969. Soil degradation of malathion, a phosphorodithioate insecticide. *Soil Sci. Soc. Am. Proc.* 33: 259–62
73. Kuhr, R. J., Davis, A. C., Bourke, J. B. 1974. Dissipation of Guthion, Sevin, Polyram, Phygon and Systox from apple orchard soil. *Bull. Environ. Contam. Toxicol.* 11:224–30
74. Laanio, T. L., Dupuis, G., Esser, H. O. 1972. Fate of ^{14}C-labeled diazinon in rice, paddy soil, and pea plants. *J. Agric. Food Chem.* 20:1213–19
75. Laveglia, J., Dahm, P. A. 1975. Oxidation of terbufos (Counter®) in three Iowa surface soils. *Environ. Entomol.* 4:715–18
76. Lichtenstein, E. P. 1966. Increase of persistence and toxicity of parathion and diazinon in soils with detergents. *J. Econ. Entomol.* 59:985–93
77. Lichtenstein, E. P., Fuhremann, T. W., Schulz, K. R. 1968. Effect of sterilizing agents on persistence of parathion and diazinon in soils and water. *J. Agric. Food Chem.* 16:870–73
78. Lichtenstein, E. P., Fuhremann, T. W., Schulz, K. R., Liang, T. T. 1973. Effects of field application methods on the persistence and metabolism of phorate in soils and its translocation into crops. *J. Econ. Entomol.* 66:863–66
79. Lichtenstein, E. P., Schulz, K. R. 1964. The effects of moisture and microorganisms on the persistence and metabolism of some organophosphorous insecticides in soils, with special emphasis on parathion. *J. Econ. Entomol.* 57:618–27
80. Liu, S.-Y., Bollag, J.-M. 1971. Metabolism of carbaryl by a soil fungus. *J. Agric. Food Chem.* 19:487–90
81. Liu, S.-Y., Bollag, J.-M. 1971. Carbaryl decomposition to 1-naphthyl carbamate by *Aspergillus terreus. Pestic. Biochem. Physiol.* 1:366–72
82. Mackiewicz, M., Deubert, K. H., Gunner, H. B., Zuckerman, B. M. 1969. Study of parathion biodegradation using gnotobiotic techniques. *J. Agric. Food Chem.* 17:129–30
83. Matsumura, F. 1972. In *Environmental Quality and Safety,* eds. F. Coulston, F. Korte, Vol. 1, pp. 96–106. New York: Academic. 267 pp.
84. Matsumura, F. 1973. In *Environmental Pollution by Pesticides,* ed. C. A. Edwards, pp. 494–513. New York: Plenum. 542 pp.
85. Matsumura, F. 1974. In *Survival in Toxic Environments,* eds. M. A. Q. Khan, J. P. Bederka, Jr., pp. 129–54. New York: Academic. 553 pp.
86. Matsumura, F., Boush, G. M. 1966. Malathion degradation by *Trichoderma viride* and a *Pseudomonas* species. *Science* 153:1278–80
87. Matsumura, F., Boush, G. M. 1968. Degradation of insecticides by a soil fungus, *Trichoderma viride. J. Econ. Entomol.* 61:610–12
88. Matsumura, F., Boush, G. M. 1971. In *Soil Biochemistry,* eds. A. D. McLaren,

J. Skujins, Vol. 2, pp. 320–36. New York: Marcel Dekker. 527 pp.

89. Menn, J. J., McBain, J. B., Adelson, B. J., Patchett, G. G. 1965. Degradation of *N*-(mercaptomethyl)phthalimide-*S*-(*O,O*-dimethylphosphorodithioate) (Imidan) in soils. *J. Econ. Entomol.* 58:875–78
90. Menn, J. J., Patchett, G. G., Batchelder, G. H. 1960. The persistence of Trithion, an organophosphorus insecticide, in soil. *J. Econ. Entomol.* 53: 1080–82
91. Menzer, R. E., Fontanilla, E. L., Ditman, L. P. 1970. Degradation of disulfoton and phorate in soil influenced by environmental factors and soil type. *Bull. Environ. Contam. Toxicol.* 5:1–5
92. Mick, D. L., Dahm, P. A. 1970. Metabolism of parathion by two species of *Rhizobium*. *J. Econ. Entomol.* 63: 1155–59
93. Miyamoto, J., Kitagawa, K., Sato, Y. 1966. Metabolism of organophosphorus insecticides by *Bacillus subtilis,* with special emphasis on Sumithion. *Jpn. J. Exp. Med.* 36:211–25
94. Mostafa, I. Y., Bahig, M. R. E., Fakhr, I. M. I., Adam, Y. 1972. Metabolism of organophosphorus insecticides. XIV. Malathion breakdown by soil fungi. *Z. Naturforsch.* 27(b):1115–16
95. Mostafa, I. Y., Fakhr, I. M. I., Bahig, M. R. E., El-Zawahry, Y. A. 1972. Metabolism of organophosphorus insecticides. XIII. Degradation of malathion by *Rhizobium* spp. *Arch. Mikrobiol.* 86:221–24
96. Munnecke, D. M., Hsieh, D. P. H. 1974. Microbial decontamination of parathion and *p*-nitrophenol in aqueous media. *Appl. Microbiol.* 28:212–17
97. Nasim, A. I., Baig, M. M. H., Lord, K. A. 1972. Loss of diazinon from Dacca paddy field soils. *Pak. J. Sci. Ind. Res.* 15:330–32
98. Novozhilov, K. V., Volkova, V. A., Rozova, V. N. 1974. Dynamics of the degradation of dimethoate in plants and soil. *Khim. Sel'sk. Khoz.* 12:199–201 [*Pesticides Abstracts* 7:773 (abstract 74-2803). 1974]
99. Nowak, A., Swarcewicz, M., Wybieralski, J. 1972. The effect of soil organic carbon content and temperature on the speed of decomposition of Nogos 50 EC. *Zesz. Nauk. Akad. Roln. Szczecinie* 38:293–98 [*Pesticides Abstracts* 7:780 (abstract 74-2834). 1974]
100. Paris, D. F., Lewis, D. L. 1973. Chemical and microbial degradation of ten selected pesticides in aquatic systems. *Residue Rev.* 45:95–124
101. Paris, D. F., Lewis, D. L., Wolfe, N. L. 1975. Rates of degradation of malathion by bacteria isolated from aquatic system. *Environ. Sci. Technol.* 9:135–38
102. Pfister, R. M. 1972. Interactions of halogenated pesticides and microorganisms: a review. *Crit. Rev. Microbiol.* 2:1–33
103. Rao, A. V., Sethunathan, N. 1974. Degradation of parathion by *Penicillium waksmani* Zaleski isolated from flooded acid sulphate soil. *Arch. Microbiol.* 97:203–8
104. Saha, J. G., Burrage, R. H., Lee, Y. W., Saha, M., Sumner, A. K. 1974. Insecticide residue in soil, potatoes, carrots, beets, rutabagas, wheat plants and grain following treatment of the soil with Dyfonate. *Can. J. Plant Sci.* 54:717–23
105. Schulz, K. R., Lichtenstein, E. P. 1971. Field studies on the persistence and movement of Dyfonate in soil. *J. Econ. Entomol.* 64:283–87
106. Schulz, K. R., Lichtenstein, E. P., Fuhremann, T. W., Liang, T. T. 1973. Movement and metabolism of phorate under field conditions after granular band applications. *J. Econ. Entomol.* 66:873–75
107. Schulz, K. R., Lichtenstein, E. P., Liang, T. T., Fuhremann, T. W. 1970. Persistence and degradation of azinphosmethyl in soils, as affected by formulation and mode of application. *J. Econ. Entomol.* 63:432–38
108. Sethunathan, N. 1973. Degradation of parathion in flooded acid soils. *J. Agric. Food Chem.* 21:602–4
109. Sethunathan, N. 1973. Microbial degradation of insecticides in flooded soil and in anaerobic cultures. *Residue Rev.* 47:143–65
110. Sethunathan, N., MacRae, I. C. 1969. Persistence and biodegradation of diazinon in submerged soils. *J. Agric. Food Chem.* 17:221–25
111. Sethunathan, N., Pathak, M. D. 1972. Increased biological hydrolysis of diazinon after repeated application in rice paddies. *J. Agric. Food Chem.* 20: 586–89
112. Sethunathan, N., Yoshida, T. 1969. Fate of diazinon in submerged soil. *J. Agric. Food Chem.* 17:1192–95
113. Sethunathan, N., Yoshida, T. 1973. Parathion degradation in submerged rice soils in the Philippines. *J. Agric. Food Chem.* 21:504–6

114. Sethunathan, N., Yoshida, T. 1973. A *Flavobacterium* sp. that degrades diazinon and parathion. *Can. J. Microbiol.* 19:873–75
115. Siddaramappa, R., Rajaram, K. P., Sethunathan, N. 1973. Degradation of parathion by bacteria isolated from flooded soil. *Appl. Microbiol.* 26:846–49
116. Singh, G. 1974. Endosymbiotic microorganisms in *Cletus signatus* Walker. *Experientia* 30:1406–7
117. Singh, K., Gulati, K. C., Dewan, R. S. 1972. Persistence of Disyston residues in soil and plant. *Indian J. Agric. Sci.* 42:1135–38
118. Staiff, D. C., Comer, S. W., Armstrong, J. F., Wolfe, H. R. 1975. Persistence of azinphosmethyl in soil. *Bull. Environ. Contam. Toxicol.* 13:362–68
119. Stewart, D. K. R., Chisholm, D., Ragab, M. T. H. 1971. Long term persistence of parathion in soil. *Nature* 229:47
120. Sud, R. K., Sud, A. K., Gupta, K. G. 1972. Degradation of Sevin (1-naphthyl-*N*-methyl carbamate) by *Achromobacter* sp. *Arch. Mikrobiol.* 87: 353–58
121. Suett, D. L. 1971. Persistence and degradation of chlorfenvinphos, diazinon, fonophos and phorate in soils and their uptake by carrots. *Pestic. Sci.* 2:105–12
122. Takase, I., Nakamura, H., Kobayashi, M., Tsuboi, A., Wakabayashi, S. 1973. The fate of disulfoton in paddy field soil. *Noyaku Kenkyu* 19:58–64 [*Pesticides Abstracts* 7:223–24 (abstract 74-0799). 1974]
123. Takase, I., Tsuda, H., Yoshimoto, Y. 1972. The fate of ®Disyston active ingredient in soil. *Pflanzenschutz-Nachr.* 25:43–63
124. Tewfik, M. S., Hamdi, Y. A. 1970. Decomposition of Sevin by a soil bacterium. *Acta Microbiol. Pol. Ser. B* 2:133–35
125. Tiedje, J. M., Alexander, M. 1967. Microbial degradation of organophosphorus insecticides and alkyl phosphates. Presented at Ann. Meet. Am. Soc. Agron., Washington, D.C.
126. Trela, J. M., Ralston, W. J., Gunner, H. B. 1968. Metabolism of diazinon by soil microflora. Presented at Ann. Meet. Am. Soc. Microbiol., Detroit, Mich.
127. Walker, W. W., Stojanovic, B. J. 1973. Microbial versus chemical degradation of malathion in soil. *J. Environ. Qual.* 2:229–32
128. Walker, W. W., Stojanovic, B. J. 1974. Malathion degradation by an *Arthrobacter* species. *J. Environ. Qual.* 3:4–10
129. Waller, J. B., Dahm, P. A. 1973. Phorate loss from Iowa soils as affected by time, temperature and soil sterilization. *Proc. North Cent. Branch Entomol. Soc. Am.* 28:171
130. Ware, G. W., Roan, C. C. 1970. Interactions of pesticides with aquatic microorganisms and plankton. *Residue Rev.* 33:15–45
131. Williams, P. P., Robbins, J. D., Gutierrez, J., Davis, R. E. 1963. Rumen bacterial and protozoal responses to insecticide substrates. *Appl. Microbiol.* 11:517–22
132. Wolfe, H. R., Staiff, D. C., Armstrong, J. F., Comer, S. W. 1973. Persistence of parathion in soil. *Bull. Environ. Contam. Toxicol.* 10:1–9
133. Yasuno, M., Hirakoso, S., Sasa, M., Uchida, M. 1965. Inactivation of some organophosphorous insecticides by bacteria in polluted water. *Jpn. J. Exp. Med.* 35:545–63
134. Zuberi, R., Zubairi, M. Y. 1971. Carbaryl degradation by *Pseudomonas phaseolicola* and *Aspergillus niger*. *Pak. J. Sci. Ind. Res.* 14:383–84
135. Zuckerman, B. M., Deubert, K., Mackiewicz, M., Gunner, H. 1970. Studies on the biodegradation of parathion. *Plant Soil* 33:273–81

Ann. Rev. Entomol. 1977. 22:515–31

EFFICACY OF INUNDATIVE RELEASES[1]

❖6140

R. E. Stinner
Department of Entomology, North Carolina State University, Raleigh, North Carolina 27607

INTRODUCTION

The value of entomophaga for regulating and controlling agricultural pests has long been recognized (23), but conscious manipulation and utilization of arthropods was all but abandoned by most entomologists with the advent of modern pesticides in the 1940s (89). Subsequent concern about resistance, pesticide misuse, and contamination (17) has reestablished the use of arthropods for pest control. The work up to the early 1960s has been summarized (26); therefore, this article is limited to research reported since that time.

Two distinct categories for augmenting the use of beneficial arthropods have been recognized (26): inoculative and inundative releases. The former involves releasing relatively small numbers of beneficial arthropods as colonizing populations, with the purpose of providing relatively long-term pest regulation through in-field reproduction of the released species. Inundative releases, on the other hand, release large numbers to cause an immediate and direct mortality in the pest population, with no expectation of long-term regulation. It is this latter usage of entomophaga as biological insecticides to which this article is addressed. In practice, however, it must be realized that the distinction between these categories is often vague. Inoculative releases may be made over several weeks and involve relatively large numbers of individuals released, whereas the inundative approach may only require, at certain times, a single release of relatively low numbers to achieve a ratio of predators or parasitoids to prey sufficient for control.

There are numerous reviews available on inundative releases in the broader context of biological control and integrated pest management (6, 25, 27, 45, 78, 90, 94, 98, 101) and for specific crops or pest complexes (21, 29, 38, 61, 62, 92, 104, 117, 131, 135), therefore the purpose of this paper is not to give an exhaustive review

[1]Paper number 4940 of the Journal Series of the North Carolina Agricultural Experiment Station, Raleigh, North Carolina 27607.

but to briefly summarize a spectrum of efforts to use inundative releases in diverse systems and to discuss the research.

The major problem encountered in utilizing these biological insecticides centers on their cost/benefit ratio as compared to alternatives, such as pesticides. To evaluate the potential of inundative releases properly, it is necessary to examine the factors that affect this cost/benefit, and as research in this area is critical to realizing that potential, review of the research base necessary for decreasing the cost/benefit ratio represents a major part of this article.

RELEASE EFFORTS

Mass releases of entomophaga have given mixed results, often yielding data that are difficult to evaluate accurately. In many of these trials, mortality was too variable for proper interpretation. A measure of mortality alone, however, has little value in determining the monetary benefit from a control tactic since pest density and resultant damage must be taken into account. Therefore, an effort is made in this review to provide information on both costs and economic benefits.

Experimental Releases

Most of the insect predators utilized belong to two families, Chrysopidae and Coccinellidae. In field-cage releases on cotton (95), *Chrysopa carnea* reduced populations of *Heliothis zea* and *H. virescens* by 74–99+%. In two field releases on 0.02-ha plots, *C. carnea* larvae (totaling 730,000 per ha) reduced these pest species by as much as 96%, depending on release rate, and increased yield threefold (93, 96). In the same study, releases of 125,000–500,000 *C. carnea* eggs per ha were also shown to be effective. In experimental releases on up to 12 ha of cotton, weekly releases of 250,000 2- to 3-day-old larvae per ha maintained *Heliothis* spp. below economic thresholds (100). *C. carnea* larvae have also been used to control the green peach aphid, *Myzus persicae,* on greenhouse chrysanthemums in which release ratios (predator:aphid) of 1:150 gave adequate control, but a 1:50 ratio was more satisfactory (106).

Releases of a *Chrysopa* sp. (probably *carnea*) for control of aphids on potatoes were only partially successful (109). Season-long reductions of up to 96% for the buckthorn aphid, *Aphis nasturii,* and 83% for *Myzus persicae* were achieved with approximately 84,000 larvae per ha; however, these releases gave virtually no control of the potato aphid *Macrosiphum euphorbiae,* the most abundant species in the experimental plots. Yield comparisons were not provided.

The species of Coccinellidae utilized in experimental mass release programs include *Stethorus picipes, Cycloneda sanguinea,* and *Coccinella septempunctata.* Releases of 500–1000 *S. picipes* per tree against the avocado brown mite, *Oligonychus punicae,* in selected California avocado orchards decreased the peak mite populations (72). Large scale experimentation involving three to five releases, totaling 400–500 predators per tree, yielded consistently lower mite populations and damage ratings (based on leaf bronzing) in release plots than check plots. Damage in release plots was tolerable whereas that in check plots caused appreciable defoliation (73).

Cycloneda sanguinea has been used successfully in controlling the cotton aphid, *Aphis gossypii,* on greenhouse cucumbers. Releases of third-instar larvae at a 1:20 ratio (predator:aphid) gave almost complete control on both young and mature plants within 18 days. Releases of second-instar *C. sanguinea* also gave control but required a somewhat longer time (47). *Coccinella septempunctata* has been released on potatoes for the control of aphids. Four releases of second- and third-instar larvae, at a rate totaling 76,250 per ha, in small fields yielded up to 57% reduction of *Aphis nasturii* but no reduction in *Myzus persicae* or *Macrosiphum euphorbiae.* However, combined releases of both *Coccinella* and *Chrysopa* produced reductions of 33–97% in the three aphid species with an average of 58% for all species (109). A number of experimental and practical releases of *Hippodamia convergens* also have been made using adults from overwintering aggregations, but these have been ineffective due to rapid dispersal of the beetles (38).

The predaceous mirid *Cyrtorhinus mundulus* has been used for controlling the sugarcane leafhopper, *Perkinsiella saccharicida,* in Taiwan. A single release of 850 nymphs per ha caused an egg mortality in excess of 80% within 22 days after release (20).

Utilizing acarine predators, notably *Phytoseiulus* spp., for control of phytophagous mites, such as *Tetranychus urticae,* on numerous hosts (12, 13, 48, 81, 82) has received much attention, although many of these releases must be considered inoculative rather than inundative. In an excellent review of biological control in greenhouses, Hussey & Bravenboer (47) recommended monthly releases under certain conditions. In field strawberries, releases of 8,000,000 mites per ha per week for 8 weeks did not significantly increase yield (81). The use of *Phytoseiulus* is more fully discussed in the next section.

Mass releases of the green hydra, *Chlorohydra viridissima* (139), for the control of mosquitoes in California is one of the most unique uses of a non-arthropod predator in biological control trials. Releases of 1000 hydra per m^2 of water surface produced population reductions of up to 67% for *Aedes nigromaculis* and from 34–79% for *Culex tarsalis,* depending on habitat. Although these experimental releases were inoculative, periodic releases probably would be required for adequate control since survival of the hydra was low (5%) in temporary aquatic habitats.

Parasitic organisms have also been used in experimental releases, with nematodes and parasitoids in the insect families Trichogrammatidae and Braconidae the most widely used. Several recent reviews on the use of nematodes are available (7, 32, 87, 121, 132), as is a review of *Trichogramma* release programs through the early 1960s (54).

Many species of egg parasitoids in the genus *Trichogramma* have been released for the control of lepidopterous pests. *T. semifumatum* released in up to 19.4 ha of cotton in Texas (three releases of 194,000–289,000 per ha per release) caused 21–85% (61% average) parasitization of *Heliothis* spp. eggs and a reduction (67%) in the resultant larval population to a level below the economic threshold. Higher release rates yielded 95% egg parasitization and an 80% reduction in the larval population (120). However, similar releases in field corn, even with 40–60% natural parasitism, did not reduce damage or resultant populations of large larvae (R. E. Stinner, unpublished data). In India, weekly releases of *T. chilotraeae* against *Helio-*

this armigera on tomato and potato (250,000 per ha per release) gave parasitization rates of up to 92 and 94%, respectively, with a reduction in damaged tomato fruit of 50–75% (R. E. Patel, personal communication).

Releases of 100,500 *T. pretiosum* per ha per week in California processing tomatoes gave parasitization rates averaging 81, 59, and 77% for *Heliothis zea, Trichoplusia ni,* and *Manduca sexta,* respectively. Damage to fruit was 2.2–2.6% in release plots versus 3.9–8.5% in control plots (80). Similar releases of *Trichogramma* sp. (50,000–100,000 per ha) against *Trichoplusia ni* on cabbage in the Moldavian SSR yielded 62–83% parasitization and a reduction of 73–92% in head injury on five separate collective farms (140). A total of almost 30 releases of *T. evanescens* against *Pieris rapae* on mustard and rape in Missouri produced egg parasitization rates of 20–75% per host generation (84).

Trichogramma sp. released at levels of 40,000–100,000 per ha against the European corn borer *Ostrinia nubilalis* on corn in the Moldavian SSR decreased injury of ears by 39–67% (140). Preliminary trials in Delaware, using *T. nubilalum,* indicated potential for control of *O. nubilalis.* The release of 20–40 parasitized egg masses, four to five times per week at a single site, yielded parasitization over 50% on artificially placed egg masses surrounding the release site (P. P. Burbutis and G. Curl, personal communication).

In a summary of biological control attempts against the surgarcane borer complex in Taiwan, Chen (19) reports that the use of 40,000 *T. australicum* per ha (2500–5000 per week) caused 62% fewer bored stalks and 80% fewer bored joints.

A review of the utilization of *Trichogramma* against the codling moth *Laspeyresia pomonella* prior to 1960 is available (117). In more recent work (28), continuous releases of *Trichogramma minutum* and *T. cacoeciae* against *L. pomonella* and the red-banded leafroller *Argyrotaenia velutiana* in apple orchards caused egg mass parasitization of up to 80 and 66%, respectively. The percentage of damaged fruit, however, was not significantly decreased.

A number of braconid parasitoids have also been used in release programs with varied results. *Chelonus eleaphilus* releases (500 parasitoids per tree every 15 days) on olive trees in the Antibes produced 92% mortality (3). *C. blackburni* released in Arizona cotton for control of the pink bollworm *Pectinophora gossypiella* had little effect on the pest populations (14, 15), but relatively few *C. blackburni* (totaling less than 6000 per ha for 13 releases) were utilized. In the same studies, releases of *Bracon kirkpatricki* so reduced the pink bollworm populations that only one insecticide treatment was required as compared to four applications in a nonrelease area.

Apanteles melanoscelus was released in Connecticut woodlands against the gypsy moth *Porthetria dispar,* with parasitism of larvae ranging up to 44% in specific plots having low pest populations. In plots with high host densities parasitization was relatively low, suggesting that the release ratio of parasitoid/host was insufficient. Mean parasitization ranged from 6–25% (133). Releases of *Lysiphlebus testaceipes* against the green bug *Schizaphis graminum* in Oklahoma sorghum also produced mixed results, possibly due to variability caused by hyperparasitoids (K. J. Starks, personal communication). Preliminary releases of *Rhizarcha* spp. on greenhouse chrysanthemums, in combination with parasitoids of the genus *Chrysocharis,* controlled the leafminer *Phytomyza atricornis,* but economic production of these parasitoids is not possible at this time (47).

Aphidiids have been successfully used to control aphids in several diverse situations. Releases of *Aphidius matricariae* have controlled *M. persicae* on greenhouse chrysanthemums (47), but release timing is critical.

Aphidius smithi was released (rate unknown) in alfalfa in Washington (state) against the pea aphid, *Acyrthosiphon pisum,* to prevent its buildup and migration from alfalfa to peas (39). Releases were made from portable field cages equipped with temperature-controlled release vents. The release field, as well as insecticide-treated and untreated controls, were monitored for alate populations from March through August. In 1967 the releases reduced alate populations by 47 and 77% when compared to the untreated and insecticide-treated fields, respectively. In 1969, a similar release program yielded a reduction in the mean weekly aphid population of over 99% when compared with untreated check fields. Unfortunately, insecticide treatments for pests other than aphids are necessary, making the utilization of this parasitoid impractical at this time.

Several eulophids have been mass released against homopterous pests in greenhouses. *Aphelinus flavipes* is reported capable of controlling *Aphis gossypii* (47). Under low temperatures (19°C), which slow aphid reproduction, a single release is sufficient but multiple releases are required at higher temperatures (47). The use of *Encarsia formosa* for control of the greenhouse whitefly *Trialeurodes vaporiorum* on various greenhouse vegetables and flowers has been the subject of much investigation (5, 47, 105) and is discussed in the next section. The encyrtid *Microterys flavus* has been used to control brown soft scale, *Coccus hesperidum,* for prolonged periods in south Texas citrus groves (40). *M. flavus* occurs naturally in this region, but insecticide treatments of cotton adjacent to citrus apparently exterminate it locally. Its maintainance as a control agent in citrus requires periodic releases (40), which have characteristics of both inoculative and inundative releases. Another encyrtid, *Tachinaephagus zealandicus,* and three pteromalids, *Muscidifurax raptor, M. zaraptor,* and *Spalangia endius,* have been used in the control of filth-breeding flies in California (*Fannia* spp., *Musca domestica, Muscina stabulans,* and others) through an integrated approach, which utilizes periodic releases of the above parasitoids and management practices favoring the increase of naturally occurring entomophaga (57–59, 63, 64).

Practical Releases

In the practical application of beneficials as biological insecticides, trichogrammatids lead the list. The use of these egg parasites against numerous lepidopterous pests is common in the Soviet Union (29, 74), South America (18), Taiwan (21), the People's Republic of China (C. B. Huffaker, unpublished report), and to some extent the United States (94).

In the Soviet Union, *Trichogramma* spp. are maintained at seven regional laboratories and are provided to many smaller production units (29). A large production facility is presently producing 3,000,000 adults per day (R. E. Stinner, unpublished report). For control of *Mamestra brassicae,* a noctuid pest of cabbage, three releases per generation at levels of 40,000–60,000 per ha are used. Similarly, *Ostrinia nubilalis* is controlled with three releases per generation of 60,000–100,000 per ha. The codling moth, *L. pomonella,* is controlled in northern regions where it has one generation per year, but control breaks down in southern areas of the USSR where two to three generations per year occur (R. E. Stinner, unpublished report). Additional releases are routinely made against *Agrotis segetum, Loxostege sticticalis,* and *Spodoptera exigua.* Costs for treatment are approximately $0.23 per ha per release and trichogrammatids are presently used on over 6,000,000 ha (29). Extension circulars on the proper use of these wasps are available to state and collective farms (74).

In South America, releases of *Trichogramma* against *Heliothis* spp. in cotton are common, but there is a great deal of controversy as to their effectiveness in controlling the pests (18). Some observers justify these releases because of their psychological effect in preventing unnecessary and even detrimental pesticide treatments. Similarly, the effectiveness of releases of *T. australicum,* routinely made against the sugarcane borer complex in Taiwan, has not been well documented (21).

In contrast, the use of various species of *Trichogramma* in the People's Republic of China is considered quite effective for controlling the rice leaf roller, *Cnapholocrosis medinalis* (five releases totaling 750,000 per ha), the European corn borer *Ostrinia nubilalis* (three releases totaling 225,000–325,000 per ha, total cost $3 per ha), and *Samia cynthea* in forests (10,000 per ha), as well as numerous other lepidopterous pests. Other species used routinely for pest control include a scelionid, *Anastatus* sp., against the lichee stink bug *Tessarotoma papillosa, Bracon greeni* against *Eublemma amabilis,* and *Coccinella septempunctata* against various aphid species in cotton (C. B. Huffaker, unpublished report).

In the United States, more than 3.5 billion *Trichogramma* are produced annually, with more than 2 billion of these reared for commercial release. Numerous other species, including *Metaphycus* sp. (5 million), *Aphytis* sp. (50 million), *Leptomastix* sp. (56 million), *Cryptolaemus* sp. (71 million), *Chrysopa* sp. (50 million), and *Hippodamia* sp. (700 million) (98), are also in commercial production, primarily in California.

SUPPORTING RESEARCH

The major obstacle in utilizing beneficial arthropods as biological insecticides appears to be economic rather than ecological. Much research is presently devoted to methods of reducing costs and increasing benefits. The former generally involves technological advances and the latter a fuller knowledge and utilization of the behavior and genetics of entomophaga. A general review of the concepts involved in these supportive efforts is provided by Rabb et al (90).

Decreasing Costs

Cost minimization involves four major areas of research: artificial diets for entomophagous species, mass rearing techniques, storage techniques, and release procedures. Since the provision of hosts for rearing entomophaga is expensive (in terms of both host-rearing costs and synchronizing suitable host numbers and stages), the development of artificial ovipositional substrates and diets is critical to cost reduction. The basic nutritional requirements, particularly for parasitoids and host-specific predators, are often quite specific, although they may lack the direct metabolic dependence on hosts, which is characteristic of the true parasite (S. N. Thompson, personal communication). At present these nutritional requirements are being explored (123, 125–127) and an artificial diet has been developed for *Chrysopa carnea* (34–36, 88, 129). *C. carnea* adults fed a mixture of yeast protein hydrolysate, a carbonate, choline chloride, and distilled water produced 700 eggs per female over a 36-day period (35, 36), not significantly different from the fecundity of adults fed on mealybug honeydew and honey (37). Liquid diets for *Chrysopa* larvae have also been developed using various enzymatic hydrolysates, choline, ascorbic acid, fructose, and other compounds (34, 129). Since satisfactory presentation of the diet to larvae has been a difficult problem, yields of adults have varied from 56–65%. In addition, the larval period has been more than twice that of larvae fed on *Sitotroga cerealella* eggs (129).

Defined or semidefined diets have been developed for several parasitoids, including the pupal endoparasitoid *Pteromalus puparum* (11, 43), the larval ectoparasitoids *Exeristes roborator* (124) and *Itoplectis conquisitor* (137, 138), and the egg parasitoid *Trichogramma pretiosum* (42).

Pteromalus puparum has been reared in extracted hemolymphs of both its normal host (*Pieris* spp.) and *Heliothis zea.* On *Pieris* hemolymph egg to adult survival was 67% (11); on *H. zea* hemolymph plasma survival ranged from 39–47%, with hemolymph-water dilutions having greater than 50% hemolymph (43).

E. roborator, a parasitoid of the pink bollworm *Pectinophora gossypiella,* has been reared from egg to adult on meridic and holidic semisolid diets (124). Development rates were lower than on the normal host, but survival was higher and a much higher female:male sex ratio (3:1 as compared to 1:2 or 1:3) was obtained.

For *Itoplectis conquisitor,* both a chemically defined synthetic medium that induces oviposition (4) and a diet (137, 138) have been developed. The ovipositional medium contained three amino acids and magnesium chloride in distilled water (pH adjusted to 7) enclosed in parafilm tubes. The artificial medium was more effective than host hemolymph in inducing oviposition, as determined by total eggs laid. Egg to adult survival on the artificial diet varied from 14 to 86%, depending on the amino acid:glucose ratio. Mean developmental times ranged from 24 to 27 days. For both *E. roborator* and *I. conquisitor* the physical properties of the synthetic larval diets were found to play a critical role in survival (124, 138).

Artificial substrates and diets have also been developed for *Trichogramma* spp. Neisenheimer's saline solution encapsulated in parafilm droplets exposed to approximately 50 *T. californicum* females for 24 hr at 25°C received almost 2700 eggs (as opposed to none for distilled water) (91). Another species, *T. pretiosum,* has been reared from egg to adult in vitro in processed hemolymph plasma of *Heliothis zea* (42). No significant difference in survival between diet-reared and *Trichoplusia ni*-reared parasitoids was observed when excess medium was withdrawn after 24 hr

(85% survival for both). Development, however, was approximately 13% slower in the artificial medium.

Mechanization in mass rearing is one of the most obvious avenues to lowering costs of release programs. Most of the research on the mechanization of rearing procedures has been with *Chrysopa* and *Trichogramma*. With *Chrysopa*, adult handling is accomplished with a vacuum device. Eggs are removed with a sodium hypochlorate solution or mechanically (99, 75) and placed in plastic grid panels (99, 102) or ornamental Masonite® (75, 76) for larval rearing.

In the United States, mechanization of *Trichogramma* rearing has begun (R. K. Morrison, R. E. Stinner, and R. L. Ridgway, unpublished data), but it has not reached the level of sophistication found in the USSR. In a recent visit, the author observed a *Trichogramma* factory in operation. With the exception of host-egg collection, the entire rearing process of both parasitoid and host was controlled by a mini-computer. A key element in consistent, high, and uniform parasitism observed in their system appears to be an automatic light-switching mechanism. Since the adults are positively phototactic, alternating a dim light source from side to side at 2-hr intervals causes the parasitoids to move slowly back and forth across the host eggs, avoiding uneven parasitism, which has been a source of concern in the United States' efforts (118). This light alternation technique is presently being used experimentally in the United States with very encouraging results (R. K. Morrison, personal communication). An area of research in mass rearing that has received little attention is accurate production estimation, although it is necessary for determining quality and maintaining consistency in production. Rabb et al (90) point out that it is "of more importance to have consistent production than to maximize production at the risk of wide variation in quantity and quality of the insects, particularly in periodic mass releases where a short time-lag exists between the recognition of the infestation and the need to treat." Unfortunately, with few exceptions (75, 118), this aspect of mass rearing has been largely ignored in the literature.

One method for counteracting low-level or variable production is the development of storage techniques, generally involving low temperature, that can be used in maintaining readily available stock from which material can be drawn as needed. Such techniques have been developed for a wide variety of entomophaga or their hosts. *Chrysopa carnea* can be stored for 150 days as adults or 20–35 days as eggs at 5°C (60–80% RH) with no apparent detrimental effects (2). Likewise, mummies of *Lysiphlebus fabarum* can also be kept at 5°C for relatively short periods (1). Storage of another aphid parasite, *Aphidius smithi*, as mummies at 1° to 4°C for over 2 months has been reported (116), but no data were provided on mortality or effects on longevity and fecundity of the resultant adults. A storage technique for *Trichogramma pretiosum* pupae has also been developed utilizing 16.7°C for the first 5 days of storage and then 15°C for up to 7 additional days of storage (118, 119). Adults from previously stored pupae demonstrated no detrimental effects of storage on longevity and fecundity under nonstress environmental conditions. Under conditions of temperature, moisture, and food stress, both longevity and fecundity were significantly lower for individuals previously stored than for nonstored individuals. These results point out the importance of testing for detrimental effects due to storage under conditions that to some degree simulate field stresses, rather than under more optimum laboratory conditions.

Another approach for the egg parasitoid, *Trichogramma dendrolimi,* involves holding pupae of the host *Samia cynthea* in cold storage and then crushing them to obtain eggs for parasitization (C. B. Huffaker, unpublished report).

Increasing Benefits

Optimizing benefits from inundative releases generally involves acquiring, maintaining, and if possible improving those species and biotypes that provide the greatest degree of control per unit cost. Research in this area includes selecting entomophagous and host species; measuring and controlling qualitative changes; increasing infield efficiency through prerelease conditioning, timing of release, and provision of hosts, kairomones, and artificial food; and integrating releases with other management practices.

Criteria for selecting entomophagous species for use in biological control have been enumerated (46): "(1) the general adaptive features of the enemy, including its synchronies and adaptations relative to the environment and its host or prey, (2) its searching capacity, (3) its power of increase relative to that of its host or prey and relative to its power of host or prey consumption, and (4) its general mobility (interrelated with searching capacity)." It is practical, however, to relax some of these criteria when using an entomophage inundatively (87). Characteristics relative to survival for more than a few days are much less important than those characteristics required for finding and attacking the host (prey) quickly. Such characteristics vary widely among demes or strains of entomophaga. For example, although species of *Trichogramma* are relatively non-host-specific, large differences exist in the searching and parasitization rates among species and among different strains (8, 9). The choice of hosts is likewise important, as the use of certain factitious hosts has been shown to cause measurable differences in efficiency (70, 111).

A second major area of concern in increasing efficiency lies in detecting and preventing the loss of entomophagous efficiency through genetic changes induced by the mass rearing programs. This problem can be minimized by acquiring a broad genetic base and providing adequate maintenance and renewal of this variability (90). In mass production, genetic changes in the culture can be a major source of decreased vigor in general, as well as specific physiological and behavioral modifications. These changes may be subtle, such as decreased searching efficiency, reported for both *Chrysopa* and *Trichogramma* (94), or they may be blatantly noticeable, as in the loss of diapause in the fly *Pseudosarcophaga affinis* (44). Davis & Burbutis (22) have demonstrated that age-selective rearing had no measurable effect on the biological quality of *Trichogramma nubilalum* but that the quality of care in handling was an important factor. This area has as yet received only minimum attention in practice, although the problem has been discussed in several reviews (10, 24, 68, 69).

Another potential area for increasing efficiency lies in the genetic selection of more efficient biotypes in the laboratory, as reviewed in (31, 136). Certain predators have already demonstrated reasonably high levels of pesticide resistance (44, 67, 130).

Although possible sublethal effects such as reduced fecundity will have to be carefully considered (55), selection for increasing this resistance should help to increase entomophage suitability in integrated programs. The feasibility of artificial selection for adaptation to temperature extremes has been shown experimentally with *Aphytis lingnanensis* (134); similar research is being conducted in the USSR on *Perillus bioculatus,* a predator of the Colorado potato beetle, *Leptinotarsa decemlineata* (A. Gusev, personal communication). However, the practicality of using such strains has not been demonstrated.

General vigor, as measured by longevity and fecundity, has been increased through selection methodology for *T. minutum* (128) and through hybrid crosses for the synanthropic fly parasitoids *Muscidifurax raptor, Spalangia enduis,* and *S. cameroni* (56).

Prerelease conditioning is an important aspect of mass release programs (90), with short-term acclimatization greatly affecting survival and efficiency (31). Unfortunately there appears to be little quantitative research in this area, which should be explored more vigorously as a means to increasing benefits from mass releases.

Determining optimum timing, distribution, and numbers has been one of the major areas of research in the United States. The optimization of these three variables can greatly reduce costs as well as increase efficiency (90), particularly against highly vagile and fecund host (prey) species, or where insecticidal treatments in adjacent fields may cause severe losses of the released entomophage (120). Wind also may greatly affect dispersal, as has been shown for *Trichogramma semifumatum* (41). With few exceptions (16, 47, 107, 108, 120), the detailed research needed to optimize timing, distribution, and rates of release under variable conditions of insecticide use, weather conditions, and other extraneous factors has not been conducted; however, the effects of these extraneous factors must be considered in developing release procedures. Likewise, research on mechanizing releases seems limited. Eggs of both *Chrysopa* sp. and *Coccinella septempunctata* suspended in a 0.125% agar solution were applied to potato foliage with a compressed air sprayer (110). Egg viability was high but adherence to the foliage was low. Similar results were reported for applications of *Chrysopa* eggs on cotton using several different suspension materials (97). Mechanized techniques developed for applying lepidopterous eggs to host plants (79, 122) could also serve in applying parasitized eggs in an inundative release.

The supplemental provision of food for entomophagous species in fields to be protected may increase the fecundity of the entomophage and slow its dispersal rate. Rates of dispersal from release sites also may be reduced through the artificial provision of hosts or kairomones. The rapid dispersal of released entomophaga was reported as the main factor in the failure of several releases (30, 115). Field trials with sprays of Wheast® yielded increased fecundity and longevity for field populations of *Chrysopa,* although at least part of this increase can be attributed to attraction of the natural population to concentrations of the artificial diet (33, 37). The release of supplemental hosts to maintain released entomophage populations during low host periods has been shown to be effective for control of *Pieris rapae* on cabbage with *Trichogramma evanescens* and *Apanteles rubecula* (84–86). Utilizing kairomones also has considerable potential; kairomones from *Heliothis zea* were found to stimulate *Microplitis croceipes* (50, 65) and *T. evanescens* (49). A hexane extract of *H. zea* moth scales sprayed on cotton in both a greenhouse and field experiment increased host finding, parasitization, and reproduction by *T. evanescens* (66).

FUTURE PROSPECTS

The future use of inundative releases will likely depend upon the successful integration of releases with other control strategies. The feasibility of this integration has been demonstrated quite clearly in several instances. The use of *Encarsia* spp. in combination with limited insecticidal sprays and fumigation gave excellent control of *Trialeuroides* on greenhouse tomatoes and cucumbers. The chemicals were selected and used in a way that had little effect on the immature stages of the parasitoid (5, 71).

Releases of *Phytoseiulus* spp. have been integrated in a system involving chemical applications, cultural manipulations, and natural predation for control of *T. urticae* on California strawberries (78). Initial results indicated yield increases 18% over the check.

In the control of animal waste flies, parasitoid manipulation alone has not given adequate control (57, 61, 77), but the integration of pesticides, waste management procedures, and parasitoid releases can be effective in controlling these populations (59, 60, 64).

Utilizing released and naturally occurring entomophaga in conjunction with host plant resistance also holds promise, as has been demonstrated for control of the green bug *S. graminum* through resistant sorghum and the combined action of both naturally occurring and released *Lysiphlebus testaceipes* (112–114).

There seems to have been no concerted effort to determine under what conditions either inundative or inoculative releases would have the greatest or least chance of providing economically feasible control. It seems intuitively obvious that situations involving pest complexes and heavy insecticidal use would not be conducive to releases against a single pest in the complex. On the other hand, where single key pest species on a single crop are involved or where a pest species exhibits a rapid invasion of a crop, inundative or inoculative releases could provide control for no more cost than presently used pesticide treatments, if rearing and release costs are competitive. The complex interactions involved in examining this problem from a theoretical base requires a systems approach. Applying systems analysis as a tool in pest management has been widely accepted, yet with the exception of several simple models (51–53) there seem to be no modeling efforts with direct relevence to inundative release programs. A number of theoretically based models of host-parasitoid systems have been developed, but only one even considers a fluctuating environment (83). In a recent review of the systems effort in pest management, Ruesink (103) could find only two reasonably accurate models of parasitoid population dynamics. The utilization of this approach would seem to be an efficient mechanism for developing and assessing practical inundative release programs.

Literature Cited

1. Adashkevich, B. P. 1973. Concerning the biology of *Lysiphlebus fabarum* Marsh. *J. Biol. Plant Prot. Probl. Kishinev* 2:27–33 (In Russian)
2. Adashkevich, B. P., Kuzina, N. P., Shijko, E. S. 1972. Rearing, storage and usage of *Chrysopa carnea* Steph. *J. Biol. Plant Prot. Probl. Kishinev* 1:8–13 (In Russian)
3. Arambourg, Y. 1970. Technique d'elevage et essais experimentaux de lachers de *Chelonus eleaphilus* Silv. Parasite de *Prays oleae* Bern. *Inst. Nac. Rech. Agron. Publ. 70–73* pp. 57–61
4. Arthur, A. P., Hegdekar, B. M., Batsch, W. W. 1972. A chemically defined synthetic medium that induces oviposition in the parasite *Itoplectis conquisitor. Can. Entomol.* 104:1251–58
5. Balevski, A. 1965. *Encarsia,* a parasite of *Trialeurodes vaporariorum* Westw. *Rev. Appl. Entomol.* 53(A):278
6. Beglyarov, G. A., Uschekov, A. T., Ponomareva, I. A. 1970. Biological control attempts against greenhouse aphids. *Proc. VII Int. Congr. Plant Prot. Paris* X:489–99
7. Benham, G. S. Jr., Poinar, G. O. Jr. 1973. Tabulation and evaluation of recent field experiments using the DD-136 strain of *Neoaplectana carpocapsae* Weiser: a review. *Exp. Parasitol.* 33:248–52
8. Biever, K. D. 1972. Effect of temperatures on the rate of search by *Trichogramma* and its potential application in field releases. *Environ. Entomol.* 1: 194–97
9. Boldt, P. E., Marston, N., Dickerson, W. A. 1973. Differential parasitism of several species of lepidopteran eggs by two species of *Trichogramma. Environ. Entomol.* 2:1121–22
10. Boller, E. 1972. Behavioral aspects of mass rearing of insects. *Entomophaga* 17:9–25
11. Bouletreau, M. 1972. Development et croissance larvories en conditions semi-artificielles et artificielles chez un hymenoptere entomophage: *Pteromalus puparum* L. (Chalc.). *Entomophaga* 17: 265–73
12. Bravenboer, L. 1963. Experiments with the predator *Phytoseiulus riegeli* Posse on glasshouse cucumbers. *Mitt. Schweiz. Entomol. Ges.* 36:53
13. Bravenboer, L., Dosse, G. 1962. *Phytoseuilus riegeli* Dosse als Predator einiger Schadmilben aus der *Tetranychus urticae* Gruppe. *Entomol. Exp. Appl.* 5:291–304
14. Bryan, D. E., Fye, R. E., Jackson, C. G., Patana, R. 1973. Releases of *Bracon kirkpatricki* (Wilkinson) and *Chelonus blackburni* Cameron for pink bollworm control in Arizona. *U.S. Dept. Agric. Res. Serv. Prod. Res. Rep.* 150:1–22
15. Bryan, D. E., Fye, R. E., Jackson, C. G., Patana, R. 1973. Releases of parasites for suppression of pink bollworm in Arizona. *U.S. Dept. Agric. Res. Serv.* W-7:1–8
16. Butler, G. D. Jr., Hungerford, C. M. 1971. Timing field releases of eggs and larvae of *Chrysopa carnea* to insure survival. *J. Econ. Entomol.* 64:311–12
17. Carson, R. 1962. *Silent Spring.* New York: Houghton Mifflin. 304 pp.
18. Castilla Chacon, R. 1973. Use of the parasite *Trichogramma* in the reduction of oviposition by *Heliothis* and other Lepidoptera, techniques for reproduction in the insectary, methods of field release, and sampling of host eggs for evaluation of results. *Algodon Mex.* 75:34–41 (In Spanish)
19. Chen, C. B. 1967. The biological control of sugarcane borers in Taiwan 1948–1966. *Mushi* 39:103–8
20. Chen, C. B. 1973. Research result and utilization of beneficial insects for the control of sugarcane pests in Taiwan. *Sci. Dev.* 1:10–15 (In Chinese)
21. Cheng, W. Y. 1975. *Biological control of sugarcane pests in the Republic of China.* Presented at Sem. Biol. Tactics Integrated Pest Manage. Syst. N. C. State Univ., Raleigh
22. Davis, C. P., Burbutis, P. P. 1974. The effect of age-selective rearing on the biological quality of females of *Trichogramma nubilalum. Ann. Entomol. Soc. Am.* 67:765–66
23. DeBach, P. 1964. The scope of biological control. In *Biological Control of Insect Pests and Weeds.* New York: Reinhold. 844 pp.
24. DeBach, P. 1965. Some biological and ecological phenomena associated with colonizing entomophagous insects. In *The Genetics of Colonizing Species,* ed. H. G. Baker, G. L. Stebbins, pp. 287–303. New York: Academic. 588 pp.
25. DeBach, P. 1974. Augmentation of natural enemies. In *Biological Control by Natural Enemies,* pp. 220–27. London: Cambridge Univ. Press. 323 pp.

26. DeBach, P., Hagen, K. S. 1964. Manipulation of entomophagous species. In *Biological Control of Insect Pests and Weeds,* ed. P. DeBach, pp. 429–58. London: Chapman Hall. 844 pp.
27. DeLucchi, V. 1975. Die konventionelle biologische Bekampfung–ein Stiefkind des Pflanzenschutzes. *A. Angew. Entomol.* 77:367–77
28. Dolphin, R. E., Cleveland, M. L., Mouzin, T. E., Morrison, R. K. 1972. Releases of *Trichogramma minutum* and *T. cacoeciae* in an apple orchard and the effects on populations of codling moths. *Environ. Entomol.* 1:481–84
29. Dysart, R. J. 1973. The use of *Trichogramma* in the USSR. *Proc. Tall Timbers Conf. Ecol. Anim. Control Habitat Manage.* 4:165–73
30. Elsey, K. D. 1975. *Jalysus spinosus:* increased numbers produced on tobacco by early-season releases. *Tob. Sci.* 19:13–15
31. Force, D. C. 1967. Genetics in the colonization of natural enemies for biological control. *Ann. Entomol. Soc. Am.* 60:772–29
32. Gordon, R., Ebsary, B. A., Bennett, G. F. 1973. Potentialities of mermithid nematodes for the biocontrol of blackflies. A review. *Exp. Parasitol.* 33: 226–38
33. Hagen, K. S., Hale, R. 1974. Increasing natural enemies through use of supplementary feeding and non-target prey. In *Proc. Summer Inst. Biol. Control Plant Insects Dis.,* ed. F. G. Maxwell, F. A. Harris, pp. 170–81. Jackson: Univ. Press Miss. 647 pp.
34. Hagen, K. S., Tassen, R. L. 1965. A method of providing artificial diets to *Chrysopa* larvae. *J. Econ. Entomol.* 58:999–1000
35. Hagen, K. S., Tassen, R. L. 1966. Artificial diet for *Chrysopa carnea* Stephens. In *Ecology of Aphidophagous Insects,* pp. 83–87. Prague: Academia Czech. Acad. Sci.
36. Hagen, K. S., Tassen, R. L. 1966. The influence of protein hydrolysates of yeasts and chemically defined diets upon the fecundity of *Chrysopa carnea* Stephens. *Acta Soc. Zool. Bohemoslov.* 30:219–27
37. Hagen, K. S., Sawall, E. F. Jr., Tassen, R. L. 1970. The use of food sprays to increase effectiveness of entomophagous insects. *Proc. Tall Timbers Conf. Ecol. Anim. Control Habitat Manage.* 2:59–81
38. Hagen, K. S., Viktorov, G. A., Yasumatsu, K., Schuster, M. F. 1976. Range, forage and grain crops. In *Theory and Practice of Biological Control,* ed. C. B. Huffaker, P. S. Messenger. New York: Academic. In press
39. Halfhill, J. E., Featherston, P. E. 1973. Inundative releases of *Aphidius smithi* against *Acyrthosiphon pisum. Environ. Entomol.* 2:469–72
40. Hart, W. G. 1972. Compensatory releases of *Microterys flavus* as a biological control agent against brown soft scale. *Environ. Entomol.* 1:414–19
41. Hendricks, D. E. 1967. Effect of wind on dispersal of *Trichogramma semifumatum. J. Econ. Entomol.* 60:1367–73
42. Hoffman, J. D., Ignoffo, C. M., Dickerson, W. A. 1975. *In vitro* rearing of the endoparasitic wasp, *Trichogramma pretiosum. Ann. Entomol. Soc. Am.* 68:335–36
43. Hoffman, J. D., Ignoffo, C. M., Long, S. H. 1973. *In vitro* cultivation of an endoparasitic wasp, *Pteromalus puparum. Ann. Entomol. Soc. Am.* 66:633–34
44. House, H. L. 1967. The decreasing occurrence of diapause in the fly *Pseudosarcophaga affinis* through laboratory-reared generations. *Can. J. Zool.* 45:149–53
45. Hoyt, S. C., Caltagirone, L. 1971. The developing programs of integrated control of pests of apples in Washington and peaches in California. In *Biological Control,* ed. C. B. Huffaker, pp. 395–421. New York: Plenum. 511 pp.
46. Huffaker, C. B., Kennett, C. E. 1969. Some aspects of assessing efficiency of natural enemies. *Can. Entomol.* 101: 425–47
47. Hussey, N. W., Bravenboer, L. 1971. Control of pests in glasshouse culture by the introduction of natural enemies. See Ref. 45, pp. 195–216
48. Hussey, N. W., Parr, W. J., Gould, H. J. 1965. Observations on the control of *Tetranychus urticae* Koch on cucumbers by the predatory mite *Phytoseiulus riegeli* Dosse. *Entomol. Exp. Appl.* 8:271–81
49. Jones, R. L., Lewis, W. J., Beroza, M., Bierl, B. A., Sparks, A. N. 1973. Host-seeking stimulants (kairomones) for the egg parasite, *Trichogramma evanescens. Environ. Entomol.* 2:593–96
50. Jones, R. L., Lewis, W. J., Bowman, M. C., Beroza, M., Bierl, B. A. 1971. Host-seeking stimulant for parasite of corn

earworm: isolation, identification, and synthesis. *Science* 173:842–43

51. Knipling, E. F. 1970. Influence of host density on the ability of selective parasites to manage insect populations. *Proc. Tall Timbers Conf. Ecol. Anim. Control Habitat Manage.* 2:3–21
52. Knipling, E. F. 1972. Simulated population models to appraise the potential for suppressing sugarcane borer populations by strategic releases of the parasite, *Lixophaga diatraeae. Environ. Entomol.* 1:1–6
53. Knipling, E. F., McGuire, J. U. Jr. 1968. Population models to appraise the limitations and potentialities of *Trichogramma* in managing host insect populations. *U.S. Dept. Agric. Tech. Bull.* 1387. 44 pp.
54. Kot, J. 1964. Experiments in the biology and ecology of species of the genus *Trichogramma* Westwood and their use in plant protection. *Ekol. Pol.* 12(A):243–303
55. Lawrence, P. O., Kerr, S. H., Whitcomb, W. H. 1973. *Chrysopa rufilabris:* effect of selected pesticides on duration of third larval stadium, pupal stage, and adult survival. *Environ. Entomol.* 2: 477–80
56. Legner, E. F. 1972. Observations on hybridization and heterosis in parasitoids of synanthropic flies. *Ann. Entomol. Soc. Am.* 65:254–63
57. Legner, E. F., Brydon, H. W. 1966. Suppression of dung-inhabiting fly populations by pupal parasites. *Ann. Entomol. Soc. Am.* 59:638–51
58. Legner, E. F., Dietrich, E. I. 1972. Inundation with parasitic insects to control filth breeding flies in California. *Proc. PAP Ann. Conf. Calif. Mosq. Control Assoc.* 40:129–30
59. Legner, E. F., Dietrich, E. I. 1974. Effectiveness of supervised control practices in lowering population densities of synanthropic flies on poultry ranches. *Entomophaga* 19:467–78
60. Legner, E. F., Olton, G. S. 1968. The biological method and integrated control of house and stable flies in California. *Calif. Agric.* 22(6):2–4
61. Legner, E. F., Poorbaugh, J. H. Jr. 1972. Biological control of vector and noxious synanthropic flies: a review. *Calif. Vector Views* 19(11):81–100
62. Legner, E. F., Sjogren, R. D., Hall, I. M. 1974. The biological control of medically important arthropods. *CRC Rev. Environ. Control* 4(1):85–113
63. Legner, E. F., Bay, E. C., Brydon, H. W., McCoy, C. W. 1966. Research with parasites for biological control of house flies in southern California. *Calif. Agric.* 20(4):10–12
64. Legner, E. F., Bowen, W. R., Rooney, W. F., McKeen, W. D., Johnston, G. W. 1975. Integrated fly control in poultry ranches. *Calif. Agric.* 29(5):8–10
65. Lewis, W. J., Jones, R. L. 1971. Substance that stimulates host-seeking by *Microplitis croceipes,* a parasite of *Heliothis* species. *Ann. Entomol. Soc. Am.* 64:471–73
66. Lewis, W. J., Jones, R. L., Sparks, A. N. 1972. A host-seeking stimulant for the egg parasite *Trichogramma evanescens:* its source and a demonstration of its laboratory and field activity. *Ann. Entomol. Soc. Am.* 65:1087–89
67. Lingren, P. H., Ridgway, R. L. 1967. Toxicity of five insecticides to several insect predators. *J. Econ. Entomol.* 60:1639–41
68. Mackauer, M. 1972. Genetic aspects of insect production. *Entomophaga* 17: 27–48
69. Mackauer, M. 1976. Genetics problems in the production of biological control agents. *Ann. Rev. Entomol.* 21:369–85
70. Marston, N., Ertle, L. R. 1973. Host influence on the bionomics of *Trichogramma minutum. Ann. Entomol. Soc. Am.* 66:1155–62
71. McClanahan, R. J. 1970. Integrated control of the greenhouse whitefly on cucumbers. *J. Econ. Entomol.* 63:599–601
72. McMurtry, J. A., Johnson, H. G. 1967. Preliminary studies on releasing *Stethorus* beetles for control of the avocado brown mite. *Calif. Avocado Soc. Yearb.* 51:173–76
73. McMurtry, J. A., Johnson, H. G., Scriven, G. T. 1969. Experiments to determine effects of mass releases of *Stethorus picipes* on the level of infestation of the avocado brown mite. *J. Econ. Entomol.* 6:1216–21
74. Moldavian SSR. 1972. Recommendations for the use of trichogrammatids in pest control in field and vegetable crops. *Bull. Minis. Agric. Moldavian SSR* 3661:1–20
75. Morrison, R. K., Ridgway, R. L. 1976. Improvements in production techniques and equipment of a common green lacewing, *Chrysopa carnea. U.S. Dept. Agric. Res. Serv. SR Ser.* In press
76. Morrison, R. K., House, V. S., Ridgway, R. L. 1975. Improved rearing unit

for larvae of a common green lacewing. *J. Econ. Entomol.* 68:821–22

77. Mourier, H. 1972. Release of native pupal parasitoids. Vidensk. *Medd. Dan. Naturhist. Foren. Khobenhavn* 135: 129–37
78. National Academy of Science. 1969. Control by parasites, predators and competitors. In *Insect Pest Management and Control,* NAS Publ, Vol. 3. 508 pp.
79. Nordlund, D. A., Lewis, W. J., Gross, H. R. Jr., Harrell, E. A. 1974. Description and evaluation of a method for field application of *Heliothis zea* eggs and kairomones for *Trichogramma. Environ. Entomol.* 3:981–84
80. Oatman, E. R., Platner, G. R. 1971. Biological control of the tomato fruitworm, cabbage looper, and hornworms on processing tomatoes in southern California, using mass releases of *Trichogramma pretiosum. J. Econ. Entomol.* 64:501–6
81. Oatman, E. R., McMurtry, J. A., Voth, V. 1968. Suppression of the two-spotted spider mite on strawberry with mass releases of *Phytoseiulus persimilis. J. Econ. Entomol.* 61:1517–21
82. Oatman, E. R., McMurtry, J. A., Shorey, H. H., Voth, V. 1967. Studies on integrating *Phytoseiulus persimilis,* releases, chemical applications, cultural manipulations, and natural predation for control of the two-spotted spider mite on strawberry in southern California. *J. Econ. Entomol.* 60:1344–51
83. Oster, G., Takahashi, Y. 1974. Models for age-specific interactions in a periodic environment. *Ecol. Monogr.* 44: 483–501
84. Parker, F. D. 1970. Seasonal mortality and survival of *Pieris rapae* in Missouri and the effect of introducing an egg parasite, *Trichogramma evanescens. Ann. Entomol. Soc. Am.* 63:985–94
85. Parker, F. D., Pinnell, R. E. 1972. Further studies of the biological control of *Pieris rapae* using supplemental host and parasite releases. *Environ. Entomol.* 1:150–57
86. Parker, F. D., Lawson, F. R., Pinnell, R. E. 1971. Suppression of *Pieris rapae* using a new control system: mass releases of both the pest and its parasite. *J. Econ. Entomol.* 64:721–35
87. Peterson, J. 1973. Role of mermithid nematodes in biological control of mosquitoes. *Exp. Parasitol.* 33:239–47
88. Ponomareva, I. A., Beglyarov, G. A. 1973. Investigation of artificial nutrient media for cultivation of the common chrysop (*Chrysopa carnea* Steph.). *Probl. Plant Prot. Kishinev* 2:67–77
89. Rabb, R. L. 1970. Introduction to the conference. In *Concepts of Pest Management,* ed. R. L. Rabb, F. E. Guthrie, pp. 1–5. Proc. Conf. held at N. C. State Univ. Raleigh: N. C. State Univ. 242 pp.
90. Rabb, R. L., Stinner, R. E., van den Bosch, R. 1976. Conservation and augmentation of natural enemies. See Ref. 38. In press
91. Rajendram, G. F., Hagen, K. S. 1974. *Trichogramma* oviposition into artificial substrates. *Environ. Entomol.* 3: 399–401
92. Ramade, F. 1972. Evolution et avenir de la lutte biologique. *Bull. Soc. Zool. Fr.* 97:629–51 (English summary)
93. Ridgway, R. L. 1969. Control of the bollworm and tobacco budworm through conservation and augmentation of predaceous insects. *Proc. Tall Timbers Conf. Ecol. Anim. Control Habitat Manage.* 1:127–44
94. Ridgway, R. L. 1972. Use of parasites, predators, and microbial agents in management of insect pests of crops. *Proc. Natl. Ext. Insect-Pest Manage. Worksh.* pp. 50–62
95. Ridgway, R. L., Jones, S. L. 1968. Field-cage releases of *Chrysopa carnea* for suppression of populations of the bollworm and the tobacco budworm on cotton. *J. Econ. Entomol.* 61:892–98
96. Ridgway, R. L., Jones, S. L. 1969. Inundative releases of *Chrysopa carnea* for control of *Heliothis* on cotton. *J. Econ. Entomol.* 62:177–80
97. Ridgway, R. L., Jones, S. L. 1976. Development of methods for field distribution of eggs of an insect predator, *Chrysopa carnea* Stephens. *U.S. Dept. Agric. Res. Serv. SR Ser.* In press
98. Ridgway, R. L., Kinzer, R. E., Morrison, R. K. 1974. Production and supplemental releases of parasites and predators for control of insect and spider mite pests of crops. See Ref. 33, pp. 110–16
99. Ridgway, R. L., Morrison, R. K., Badgley, M. 1970. Mass rearing a green lacewing. *J. Econ. Entomol.* 63:834–36
100. Ridgway, R. L., Morrison, R. K., Kinzer, R. E., Stinner, R. E., Reeves, B. G. 1973. Programmed releases of parasites and predators for control of *Heliothis* spp. on cotton. *Proc. 1973 Beltwide Cotton Prod. Res. Conf.,* pp. 92, 94

101. Rivnay, E. 1968. Biological control of pests in Israel. *J. Entomol. Soc. Isr.* 3:1–156
102. Ru, N., Whitcomb, W. H., Murphy, M. 1976. Culturing of *Chrysopa rufilabris. Fla. Entomol.* 59:21–26
103. Ruesink, W. G. 1976. Status of the systems approach to pest management. *Ann. Rev. Entomol.* 21:27–44
104. Scepetilnikova, V. A. 1970. Perspektiven der Kenntnis und Anwendung von Eiparasiten der Gattung *Trichogramma* zur Bekämpfung land-und fortwirtschaftlicher Schädlinge. *Tagungsber. Dtsch. Akad. Landwirtschaftwiss. Berlin* 110:117–36
105. Scopes, N. E. A. 1969. The economics of mass rearing *Encarsia formosa,* a parasite of the whitefly, *Trialeurodes vaporariorum,* for use in commercial horticulture. *Plant Pathol.* 18:130–32
106. Scopes, N. E. A. 1969. The potential of *Chrysopa carnea* as a biological control agent of *Myzus persicae* on glasshouse chrysanthemums. *Ann. Appl. Biol.* 64: 433–39
107. Shands, W. A., Simpson, G. W., Gordon, C. C. 1972. Insect predators for controlling aphids on potatoes. 5. Numbers of eggs and schedules for introducing them in large field cages. *J. Econ. Entomol.* 65:810–17
108. Shands, W. A., Simpson, G. W. 1972. Insect predators for controlling aphids on potatoes. 4. Spatial distribution of introduced eggs of two species of predators in small fields. *J. Econ. Entomol.* 65:514–18
109. Shands, W. A., Simpson, G. W., Storch, R. H. 1972. Insect predators for controlling aphids on potatoes. 3. In small plots separated by aluminum flashing strip-coated with a chemical barrier and in small fields. *J. Econ. Entomol.* 65:799–805
110. Shands, W. A., Gordon, C. C., Simpson, G. W. 1972. Insect predators for controlling aphids on potatoes. 6. Development of a spray technique for applying eggs in the field. *J. Econ. Entomol.* 65:1099–1103
111. Simmonds, F. J. 1966. Insect parasites and predators. In *Insect Colonization and Mass Production,* ed. C. N. Smith, pp. 489–99. New York: Academic
112. Starks, K. J., Muniappan, R., Eikenbary, R. D. 1972. Interaction between plant resistance and parasitism against the greenbug on barley and sorghum. *Ann. Entomol. Soc. Am.* 65:650–55
113. Starks, K. J., Wood, E. A. Jr., Burton, R. L. 1974. Relationships of plant resistance and *Lysiphlebus testaceipes* to population levels of the greenbug on grain sorghum. *Environ. Entomol.* 3: 950–52
114. Starks, K. J., Burton, R. L., Teetes, G. L., Wood, E. A. Jr. 1976. Release of parasitoids to control greenbugs on sorghum. *U.S. Dept. Agric. Res. Serv. S Ser.* In press
115. Starks, K. J., Wood, E. A. Jr., Burton, R. L., Somsen, H. W. 1975. Behavior of convergent lady beetles in relation to greenbug control in sorghum. *U.S. Dept. Agric. Res. Serv.* 53:1–10
116. Stary, P. 1970. Methods of mass-rearing, collection and release of *Aphidius smithi* in Czechoslovakia. *Acta Entomol. Bohemoslov.* 5:339–46
117. Stein, W. 1960. Versuche zur biologischen Bekämpfung des Apfelwicklers [*Carpocapsa pomonella* (L.)] durch Eiparasiten der Gattung *Trichogramma. Entomophaga* 5:237–59
118. Stinner, R. E., Ridgway, R. L., Kinzer, R. E. 1974. Storage manipulation of emergence, and estimation of numbers of *Trichogramma pretiosum. Environ. Entomol.* 3:505–7
119. Stinner, R. E., Ridgway, R. L., Morrison, R. K. 1974. Longevity, fecundity, and searching ability of *Trichogramma pretiosum* by three methods. *Environ. Entomol.* 3:558–60
120. Stinner, R. E., Ridgway, R. L., Coppedge, J. R., Morrison, R. K., Dickerson, W. A. Jr. 1974. Parasitism of *Heliothis* eggs after field releases of *Trichogramma pretiosum* in cotton. *Environ. Entomol.* 3:497–500
121. Stoll, N. R. 1973. Rudolf William Glaser, and Neoaplectana. *Exp. Parasitol.* 33:226–38
122. Thewka, S. E., Puttler, B. 1970. Aerosol application of lepidopterous eggs and their susceptibility to parasitism by *Trichogramma. J. Econ. Entomol.* 63: 1033–34
123. Thompson, S. N. 1974. Aspects of amino acid metabolism and nutrition in the ectoparasitoid wasp, *Exeristes roborator. J. Insect Physiol.* 20:1515–28
124. Thompson, S. N. 1975. Defined meridic and holidic diets and asceptic feeding procedures for artificially rearing the ectoparasitoid *Exeristes roborator* (Fabricius). *Ann. Entomol. Soc. Am.* 68: 220–26
125. Thompson, S. N., Barlow, J. S. 1972. Synthesis of fatty acids by the parasite

Exeristes comstockii and two hosts, *Galleria mellonella* and *Lucilia sericata*. *Can. J. Zool.* 50:1105–10
126. Thompson, S. N., Barlow, J. S. 1973. The inconsistent phospholipid fatty acid composition in an insect parasitoid, *Itoplectis conquisitor* (Say). *Comp. Biochem. Physiol.* 44(B):59–64
127. Thompson, S. N., Barlow, J. S. 1974. The fatty acid composition of parasitic hymenoptera and its possible biological significance. *Ann. Entomol. Soc. Am.* 67:627–32
128. Urquijo, L. P. 1951. Application de la Genetica al aumento de la eficacia del *Trichogramma minutum* en la lucha biologica. *Bol. Patol. Veg. Entomol. Agric.* 18:1–12
129. Vanderzant, E. S. 1969. An artificial diet for larvae and adults of *Chrysopa carnea*, an insect predator of crop pests. *J. Econ. Entomol.* 62:256–57
130. van den Bosch, R., Hagen, K. S. 1966. Predaceous and parasitic arthropods in California cotton fields. *Univ. Calif. Agric. Exp. Sta. Bull. 820.* 32 pp.
131. van den Bosch, R., Beingolea, O., Hafez, M., Falcon, L. A. 1976. Biological control of insect pests of row crops. See Ref. 38. In press
132. Webster, J. M. 1973. Manipulation of environment to facilitate use of nematodes in biocontrol of insects. *Exp. Parasitol.* 33:197–206
133. Weseloh, R. M., Anderson, J. F. 1975. Inundative release of *Apanteles melanoscelus* against the gypsy moth. *Environ. Entomol.* 4:33–36
134. White, E. B., DeBach, P., Garber, M. J. 1970. Artificial selection for genetic adaptation to temperature extremes in *Aphytis lingnanensis* Compere. *Hilgardia* 40:161–92
135. Williams, J. R., Greathead, D. J. 1973. The sugarcane scale insect *Aulacaspis tegalensis* (Zhnt.) and its biological control in Mauritius and East Africa. *PANS* 19:353–67
136. Wilson, F. 1965. Biological control and the genetics of colonizing species. See Ref. 111, pp. 307–29
137. Yazgan, S. 1972. A chemically defined synthetic diet and larval nutritional requirement of the endoparasitoid *Itoplectis conquisitor. J. Insect Physiol.* 18: 2123–41
138. Yazgan, S., House, H. L. 1969. A hymenopterous insect, the parasitoid *Itoplectis conquisitor,* reared axenically on a chemically-defined symthetic diet. *Can. Entomol.* 102:1304–6
139. Yu, H. S., Legner, E. F., Sjogren, R. D. 1974. Mass release effects of *Chlorohydra viridissima* (Coelenterata) on field populations of *Aedes nigromaculis* and *Culex tarsalis* in Kern County, California. *Entomophaga* 19:409–20
140. Zilberg, L. P. 1972. Efficiency of *Trichogramma* in Northern Zone of Moldavia. *J. Biol. Plant Prot. Probl. Kishinev* 1:47–53 (In Russian)

AUTHOR INDEX

M

N

O

P

SUBJECT INDEX

P

CUMULATIVE INDEXES

CONTRIBUTING AUTHORS VOLUMES 13-22

CHAPTER TITLES VOLUMES 13-22